MIMO Power Line Communications

Narrow and Broadband Standards, EMC, and Advanced Processing

Devices, Circuits, and Systems

Series Editor
Krzysztof Iniewski
CMOS Emerging Technologies Research Inc.,
Vancouver, British Columbia, Canada

FORTHCOMING TITLES:

Gallium Nitride (GaN): Physics, Devices, and Technology
Farid Medjdoub and Krzysztof Iniewski

High Frequency Communication and Sensing: Traveling-Wave Techniques
Ahmet Tekin and Ahmed Emira

High-Speed Devices and Circuits with THz Applications
Jung Han Choi and Krzysztof Iniewski

Labs-on-Chip: Physics, Design and Technology
Eugenio Iannone

Laser-Based Optical Detection of Explosives
Paul M. Pellegrino, Ellen L. Holthoff, and Mikella E. Farrell

Metallic Spintronic Devices
Xiaobin Wang

Microfluidics and Nanotechnology: Biosensing to the Single Molecule Limit
Eric Lagally and Krzysztof Iniewski

Mobile Point-of-Care Monitors and Diagnostic Device Design
Walter Karlen and Krzysztof Iniewski

Nanoelectronics: Devices, Circuits, and Systems
Nikos Konofaos

Nanomaterials: A Guide to Fabrication and Applications
Gordon Harling and Krzysztof Iniewski

Nanopatterning and Nanoscale Devices for Biological Applications
Krzysztof Iniewski and Seila Selimovic

Optical Fiber Sensors and Applications
Ginu Rajan and Krzysztof Iniewski

Organic Solar Cells: Materials, Devices, Interfaces, and Modeling
Qiquan Qiao and Krzysztof Iniewski

Power Management Integrated Circuits and Technologies
Mona M. Hella and Patrick Mercier

Radio Frequency Integrated Circuit Design
Sebastian Magierowski

Semiconductor Device Technology: Silicon and Materials
Tomasz Brozek and Krzysztof Iniewski

Smart Grids: Clouds, Communications, Open Source, and Automation
David Bakken and Krzysztof Iniewski

Soft Errors: From Particles to Circuits
Jean-Luc Autran and Daniela Munteanu

FORTHCOMING TITLES:

VLSI: Circuits for Emerging Applications
Tomasz Wojcicki and Krzysztof Iniewski

Wireless Transceiver Circuits: System Perspectives and Design Aspects
Woogeun Rhee and Krzysztof Iniewski

MIMO Power Line Communications

Narrow and Broadband Standards, EMC, and Advanced Processing

Edited by

Lars T. Berger ● Andreas Schwager
Pascal Pagani ● Daniel M. Schneider

CRC Press
Taylor & Francis Group
Boca Raton London New York

CRC Press is an imprint of the
Taylor & Francis Group, an **informa** business

CRC Press
Taylor & Francis Group
6000 Broken Sound Parkway NW, Suite 300
Boca Raton, FL 33487-2742

First issued in paperback 2017

© 2014 by Taylor & Francis Group, LLC
CRC Press is an imprint of Taylor & Francis Group, an Informa business

No claim to original U.S. Government works

Version Date: 20131227

ISBN 13: 978-1-138-07243-5 (pbk)
ISBN 13: 978-1-4665-5752-9 (hbk)

Library of Congress Cataloging-in-Publication Data

MIMO power line communications : narrow and broadband standards, EMC, and advanced
  processing / editors, Lars Torsten Berger, Andreas Schwager, Pascal Pagani, Daniel Schneider.
    pages cm. -- (Devices, circuits, and systems)
  Includes bibliographical references and index.
  ISBN 978-1-4665-5752-9 (hardback)
  1. Electric lines--Carrier transmission. 2. MIMO systems. I. Berger, Lars Torsten, editor of
compilation.

TK5103.15.M56 2014
621.382'16--dc23   2013048162

Visit the Taylor & Francis Web site at
http://www.taylorandfrancis.com

and the CRC Press Web site at
http://www.crcpress.com

Contents

Part I Power Line Channel and Noise: Characteristics and Modelling

Part II Regulations, Electromagnetic Compatibility and MIMO Capacity

Part III   Current PLC Systems and Their Evolution

Part IV   Advanced PHY and MAC Layer Processing

Part V Implementations, Case Studies and Field Trials

Preface

Multiple-input multiple-output (MIMO) systems have been heavily investigated since the mid 1990s, targeting wireless communications. Nowadays, different MIMO processing options, with the aim of increasing data rates and communication reliability, are in operation in major wireless cellular systems such as *WCDMA, LTE* and *WiMAX*, as well as *wireless local area networks* (WLANs) based on *IEEE 802.11n*. Also, in the wireline world, *digital subscriber line* (DSL) systems have to deal with *near-end* and *far-end crosstalk* between individual modems. Recent developments treat DSL cable binders as MIMO communication channels, with the aim of applying multi-user coordination and interference mitigation techniques.

For a long time, the power line channel has been regarded as a *single-input single-output* (SISO) channel based on two conductors. In reality, many *in-home* installations make use of three wires, and *medium- and high-voltage* installations often have four or more conductors. Although the theoretical foundation of multi-conductor transmission line theory was extensively laid out in the last century, large-scale measurement results on MIMO power line channel and noise characteristics became available only in recent years, and only in 2011 did the *International Telecommunications Union* (ITU) publish a MIMO transceiver extension (G.9963) to their G.hn standard family. Simultaneously, the industry alliance, HomePlug, introduced MIMO signal processing within the HomePlug AV2 specification.

Demand for higher data rates and multi-user support is driven by increasing home entertainment and communication needs. For example, one can imagine a household whose inhabitants consume several high-definition television streams, while simultaneously browsing the Internet and making *voice over IP* (VoIP) calls. In this scenario, the focus is on high data rate applications with advanced quality of service differentiation. On the other hand, demand is also driven by the increasing number of intelligent devices that form part of the emerging *Smart Grid*. In such cases, coverage, reliability, and scalability are important attributes. Last but not least, the ever growing *Internet of things* (IoT) requires a huge amount of low data rate devices to communicate simultaneously, with stringent demands on power consumption and re-configurability.

All these aspects are reflected in this book, which gives a profound introduction to present day *power line communications* (PLC), as follows:

Part I: Power Line Channel and Noise: Characteristics and Modelling

Part II: Regulations, Electromagnetic Compatibility and MIMO Capacity

Part III: Current PLC Systems and Their Evolution

Part IV: Advanced PHY and MAC Layer Processing

Part V: Implementations, Case Studies and Field Trials

The individual chapters of Part I focus on narrow- and broadband channel characterisation, presenting measurement and modelling results from the United States, Europe and China. Building on this foundation, and taking into account current *regulations* and aspects of *electromagnetic compatibility* (EMC), Part II describes *MIMO signal processing* strategies and related *MIMO capacity* and *throughput* estimates. Current narrow- and broadband PLC standards and specifications are described in detail in the various chapters of Part III. Especially, the narrowband power line standards ITU-T G.9902, G.9903,

G.9904 and IEEE 1901.2 are discussed, followed by individual broadband PLC chapters on ITU G.hn, IEEE 1901 and HomePlug AV2. Part III is rounded off by an introduction to hybrid systems based on *IEEE 1905.1*, combining PLC with other wireless and wireline technologies. Advanced PLC processing options are treated in Part IV, drawing from a wide variety of research areas such as *beamforming/precoding*, *time reversal*, *multi-user processing* and *relaying*. The book is concluded by Part V, which contains *case studies* and *field trials* where advanced technologies are explained with practical examples. Topics are as diverse as *channel and noise emulation*, *cognitive notching*, *interference mitigation* and *MIMO PLC hardware feasibility*.

All chapters are written as much as possible in a self-contained manner. Additionally, the chapters have been extensively cross-referenced, allowing each reader to pursue a personal reading path. We hope you enjoy the diverse contents contributed by experts from industry and academia, making this book one of the first of its kind in the exciting world of *MIMO PLC*.

Lars T. Berger
Andreas Schwager
Pascal Pagani
Daniel M. Schneider

Significant portions of this book, particularly Chapters 13 and 15, were excerpted from IEEE 1901-2010, IEEE Standard for Broadband over Power Line Networks: Medium Access Control and Physical Layer Specifications, and from IEEE 1905.1-2013, IEEE Standard for a Convergent Digital Home Network for Heterogeneous Technologies, copyright IEEE, and reproduced with permission by limited license from IEEE. Permission for further use of this material must be obtained from IEEE. Requests may be sent to stds-ipr@ieee.r/sg. For a full copy of the standards, please visit IEEE's online store at standards.ieee.org/store.

MATLAB® is a registered trademark of The MathWorks, Inc. For product information, please contact:

The MathWorks, Inc.
3 Apple Hill Drive
Natick, MA 01760-2098 USA
Tel: 508-647-7000
Fax: 508-647-7001
E-mail: info@mathworks.com
Web: www.mathworks.com

Editors

Dr. Lars Torsten Berger is director of the R&D department of Kenus Informática, Paterna, Spain, and founder of BreezeSolve, a Valencia-based company offering engineering and consultancy services in telecommunications, signal processing and Smart Grid.

Dr. Berger received his Dipl-Ing in electrical engineering, MSc in communication systems and signal processing and PhD in wireless communications from the Ravensburg University of Cooperative Education, Germany; the University of Bristol, United Kingdom; and Aalborg University, Denmark, in 1999, 2001 and 2005, respectively.

While working for Nortel Networks, United Kingdom, and during his four years stay at Aalborg University financed by Nokia Networks, he focused extensively on *MIMO channel modelling*, *MIMO signal processing* and *MIMO scheduling algorithms* for 3G cellular systems. During a visiting professor term at the University Carlos III of Madrid, Spain, he extended his work to 4G cellular, as well as wireless LAN systems and sensor networks. From 2006 to 2010, he became senior engineer at *Design of Systems on Silicon* (DS2, Spain, in 2010 acquired by Marvell Semiconductors), forming part of the key system architecture team responsible for MIMO and multi-user-enabled power line silicon solutions. More recently, he has extended his area of interest to Smart Grid technologies.

Dr. Berger has published numerous MIMO, scheduling and PLC-related articles in conference proceedings and in international journals. He is the inventor of an international patent related to MIMO PLC signal transmission, has several patent applications pending and is editor of the book entitled *Smart Grid Applications, Communications and Security*.

Dr. Andreas Schwager has had interest in PLC for over 15 years. Currently, he works as a principal engineer in Sony's European Technology Center in Stuttgart, Germany. Sony's R&D in Stuttgart includes the areas of digital radio and video broadcasting, as well as home networking.

Dr. Schwager earned a diploma in telecommunication engineering from the University of Cooperative Education, Stuttgart, in 1993. In May 2010, the University of Duisburg-Essen awarded him a PhD (doctor of science [engineering]) following the publication of his thesis "Powerline Communications: Significant Technologies to Become Ready for Integration," which discussed the utilisation of MIMO PLC and the concept of dynamic notching to solve the vast EMC discussion on PLC.

Dr. Schwager's career began at the advanced development labs at ANT Bosch Telecom, Backnang, Germany, and Grundig, Nürnberg, Germany, where he developed narrowband PLC functions for satellite headends. He joined Sony in 1997, where the development of both software and hardware components for home-networking applications ignited his enthusiasm for PLC. Today, Dr. Schwager represents Sony at various standardisation committees at IEEE, ITU, CISPR, CENELEC, Homeplug and ETSI. He is the rapporteur of more than ten work items where technical standards and reports are published, the most recent being the three-part ETSI TR specifying MIMO PLC field measurements and presenting results of MIMO PLC properties. He also led several task forces at international standardisation bodies.

Dr. Schwager is the author of numerous papers presented at research conferences and the inventor of tens of granted IPR in the area of PLC and communications.

Dr. Pascal Pagani is an associate professor at the graduate engineering school Telecom Bretagne in Brest, France, and is a member of the Lab-STICC laboratory (UMR CNRS 6285) dedicated to information and communication science.

He received his MSc in communication systems and signal processing from the University of Bristol, United Kingdom, in 2002, and his PhD in electronics from INSA Rennes, France, in 2005.

Prior to joining Telecom Bretagne in 2012, he worked with France Telecom Orange Labs, where he was active in various projects, including UWB propagation channel characterisation and modelling, short range wireless system design and development of in-home wireline communications. He led France Telecom standardisation activities in the IEEE 802.15.3c, ITU SG15 G.hn and HomePlug TWG groups and participated in the ETSI Specialist Task Force 410 for the experimental assessment of the PLC MIMO transmission channel.

Dr. Pagani is a member of the Technical Committee on PLC within the IEEE Communications Society. He is the author of more than 50 publications, including books, book chapters and journal and conference papers in the fields of wireless and wired communication and has filed several pending patents. His current research interests are in the field of radio and wireline transmission, particularly long-haul radio wave propagation and advanced PLC.

Dr. Daniel Schneider currently works as a senior engineer in the European Technology Center of Sony in Stuttgart, Germany, where his work is concerned with communications systems and PLC in particular. He contributed to the work of the PLC HomePlug AV2 specification and was involved in the ETSI MIMO PLC field measurement campaign.

He received his Dipl-Ing in electrical engineering, with a focus on signal processing and communications, and his Dr-Ing for his thesis "In-Home Power Line Communications Using Multiple Input Multiple Output Principles" from the University of Stuttgart in 2006 and 2012, respectively.

Dr. Schneider has published multiple papers related to PLC and MIMO and is inventor of several international patents.

Contributors

Jose Abad
Broadcom Corporation
Malaga, Spain

Kaywan Afkhamie
Systems Group
Qualcomm Atheros
Ocala, Florida

Jean-Yves Baudais
Centre National de la Recherche
    Scientifique
Institut d'Electronique et de
    Télécommunications de Rennes
Rennes, France

Erez Ben-Tovim
Sigma Designs Israel S.D.I Ltd.
Tel-Aviv, Israel

Lars T. Berger
Kenus Informatica
Paterna, Spain

and

BreezeSolve
Valencia, Spain

Paola Bisaglia
DORA S.p.A., STMicroelectronics Group
Aosta, Italy

Thierry Chonavel
Telecom Bretagne
Brest, France

Guangbin Chu
Department of Information and
    Communication
China Electrical Power Research Institute
Fangshan, Beijing, People's Republic
    of China

Etan G. Cohen
Qualcomm Atheros
San Jose, California

Matthieu Crussière
Institut National des Sciences Appliquées
Institut d'Electronique et de
    Télécommunications de Rennes
Rennes, France

Salvatore D'Alessandro
WiTiKee S.r.l.
Udine, Italy

Klaus Dostert
Karlsruhe Institute of Technology
Institute of Industrial Information
    Technology
Karlsruhe, Germany

Stefano Galli
ASSIA, Inc.
CTO Office
New York City, New York

Lorenzo Guerrieri
DORA S.p.A., STMicroelectronics Group
Aosta, Italy

Rehan Hashmat
Eurecom
Valbonne, France

Duncan Ho
Qualcomm Inc.
San Diego, California

Srinivas Katar
Systems Group
Qualcomm Atheros
Ocala, Florida

Manjunath Krishnam
Qualcomm Atheros
Ocala, Florida

James Le Clare
Maxim Integrated Products
San Jose, California

Jianqi Li
Department of Information and
    Communication
China Electrical Power Research Institute
Fangshan, Beijing, People' Republic
    of China

Hidayat Lioe
Marvell Semiconductor Inc.
Santa Clara, California

Weilin Liu
Department of Information and
    Communication
China Electrical Power Research Institute
Fangshan, Beijing, People's Republic
    of China

Wenqing Liu
Karlsruhe Institute of Technology
Institute of Industrial Information
    Technology
Karlsruhe, Germany

Yang Lu
Department of Information and
    Communication
China Electric Power Research Institute
Fangshan, Beijing, People's Republic
    of China

Amilcar Mescco
Telecom Bretagne
Brest, France

Bibhu P. Mohanty
Qualcomm Inc.
San Diego, California

Arun Nayagam
Qualcomm Atheros
Ocala, Florida

Michel Ney
Telecom Bretagne
Brest, France

Fabio Osnato
STMicroelectronics Srl
Advanced System Technology
Agrate Brianza, Italy

Pascal Pagani
Telecom Bretagne
Brest, France

Purva R. Rajkotia
Qualcomm Atheros
Ocala, Florida

Deniz Rende
Systems Group
Qualcomm Atheros
Ocala, Florida

Markus Rindchen
Power Plus Communications AG
Mannheim, Germany

Raffaele Riva
STMicroelectronics Srl
Advanced System Technology
Agrate Brianza, Italy

Yago Sánchez Quintas
Sony Deutschland GmbH
Stuttgart, Germany

Daniel M. Schneider
Sony Deutschland GmbH
Stuttgart, Germany

Andreas Schwager
Sony Deutschland GmbH
Stuttgart, Germany

Martin Sigle
Karlsruhe Institute of Technology
Institute of Industrial Information
   Technology
Karlsruhe, Germany

Matthias Stephan
Power Plus Communications AG
Mannheim, Germany

Andrea M. Tonello
WiPLi Lab
Dipartimento di Ingegneria Elettrica,
   Gestionale e Meccanica
Università degli Studi di Udine
Udine, Italy

Piet Janse van Rensburg
Walter Sisulu University
East London, South Africa

Larry Yonge
Systems Group
Qualcomm Atheros
Ocala, Florida

Ahmed Zeddam
France Telecom Orange
Lannion, France

Part I

Power Line Channel and Noise: Characteristics and Modelling

1

Introduction to Power Line Communication Channel and Noise Characterisation

Lars T. Berger, Pascal Pagani, Andreas Schwager and Piet Janse van Rensburg

CONTENTS

1.1 Introduction*

Since the late 1990s, an increased effort has been put into the characterisation of *power line communication* (PLC) channels with the aim of designing communication systems that use the electrical power distribution grid as data transmission medium.

Reliable PLC systems, for home networking, *Internet protocol television* (IPTV), Smart Grid and smart building applications are now a reality. However, power lines have not been designed for communication purposes and constitute a difficult environment to convey information via early analogue signalling or nowadays widespread advanced digital PLC systems. The PLC channel exhibits *frequency-selective multi-path fading*, a *low-pass behaviour*,

---

* Chapter in parts based on material from [1–4].

cyclic short-term variations and *abrupt long-term variations* that are introduced in Section 1.5. Further, power line noise can be grouped based on temporal as well as spectral characteristics. Following, for example [5,6], one can distinguish *coloured background noise, narrowband* (NB) *noise, periodic impulsive noise* (asynchronous or synchronous to the *alternating current* [AC] frequency), as well as *aperiodic impulsive noise* (see Section 1.6). These impairments are leading some researchers to speak of a 'horrible channel' [7].

Apart from these, the very principle of PLC implies that small-signal, high-frequency technologies are being deployed over power-carrying cables and networks that were designed for electricity transmission at low frequencies. In terms of voltage, the equipments' communication ports would fail if they were connected directly to the power grid. This is similarly true when looking at PLC testing and measurement equipment, such as a spectrum analyser, which is why PLC couplers are needed to couple the communication signal into and out of the power line while at the same time protecting the communication equipment. Couplers may be of either *inductive* or *capacitive* nature with detailed coupling schemes introduced in Section 1.3. Before that, however, this chapter looks at PLC frequency bands and common topologies in Section 1.2 as these are possibly the most profound stage setters when characterising PLC channel and noise scenarios. In the sequel, the aim of Section 1.4 is to provide information on measurement equipment and procedures that have been used to generate a plurality of results for various chapters throughout this book. Further, Sections 1.5 and 1.6 introduce the underlying concepts of PLC channel and noise modelling, respectively, and guide the reader to the more detailed chapters on each topic. This chapter is rounded off by an appendix that explains the basics of dual conductor transmission line theory, considered interesting background reading for those new to PLC signal propagation.

1.2 PLC Frequency Bands and Topologies

The frequency bands – as agreed upon by the *International Telecommunications Union* [8] – are shown in Figure 1.1. The band name abbreviations stand for *super low, ultra low, very low, low, medium, high, ultra high, very high, super high, extremely high* and *tremendously high frequency*, respectively. As indicated in Figure 1.1, currently only the VLF up to the UHF bands are interesting for PLC systems. These systems are usually subdivided into *narrowband* (NB) and *broadband* (BB) PLC; the former operating below 1.8 MHz, the latter operating above [9]. Details on the regulations corresponding to these frequency bands can be found in Chapter 6. An overview on systems that belong to either the class of NB-PLC or the class of BB-PLC can be found in Chapter 10.

Besides the distinction into NB-PLC and BB-PLC, it has been common practice to distinguish power line topologies according to operation voltages of the power lines [2,9,10].

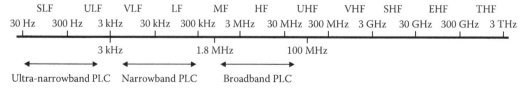

FIGURE 1.1
ITU frequency bands and their usage in power line communications.

High-voltage (HV) lines, with voltages in the range from 110 to 380 kV, are used for nation-wide or even international power transfer and consist of long overhead lines with little or no branches. This makes them acceptable wave guides with less attenuation per line length as for their *medium-voltage* (MV) and *low-voltage* (LV) counterparts. However, their potential for BB communication services has up to the present day been limited. Time-varying HV arcing and corona noise with noise power fluctuations in the order of several tens of dBs as well as the practicalities and costs of coupling communication signals in and out of these lines have been an issue. Further, there is a fierce competition of fibre optical links. In some cases, these links might even be spliced together with the ground conductor of the HV system [11,12]. Nevertheless, several successful trials using HV lines have been reported in [13–16].

MV lines, with voltages in the range from 10 to 30 kV, are connected to the HV lines via primary transformer substations. The MV lines are used for power distribution between cities, towns and larger industrial customers. They can be realised as overhead or underground lines. Further, they exhibit a low level of branches and directly connect to *intelligent electronic devices* (IED) such as reclosers, sectionalisers, capacitor banks and phasor measurement units. IED monitoring and control requires only relatively low data rates and NB-PLC can provide economically competitive communication solutions for these tasks. MV-related studies and trials can be found in [17–20].

LV lines, with voltages in the range from 110 to 400 V, are connected to the MV lines via secondary transformer substations. A communication signal on an MV line can pass through the secondary transformer onto the LV line, however, with a heavy attenuation in the order of 55–75 dB [21]. Hence, a special coupling device (inductive, capacitive) or a PLC repeater is frequently required if one wants to establish a high data rate communications path. The LV lines lead directly or over street cabinets to the end customers' premises. Note that considerable regional topology difference exits. For example, in the United States, a smaller secondary transformer on a utility pole might service a single house or a small number of houses. In Europe, however, it is more common that up to 100 households get served from a single secondary transformer substation. Further, as pointed out in [22], significant differences exist between building types. They may be categorised as *multi-flat buildings with riser, multi-flat buildings with common meter room, single family houses* and *high-rise buildings*. Their different electrical wiring topologies influence signal attenuation as well as interference between neighbouring PLC networks [23]. In most cases, the electrical grid enters the customer's premises over a *house access point* (HAP) followed by an *electricity meter* (M) and a distribution board (fuse box). From there, the LV lines run in a tree or star topology up to the different power sockets in every room. One frequently refers to PLC systems operating from outside to inside a customer's premises as *access* systems while systems operating within the premises are referred to as *in-home*. It can be summarised that the access scenario establishes data connections to a group of customers through the overhead and/or underground electrical power distribution grid [7,24–26]. The in-home scenario enables the communication of different devices within a user's premises [7,27–32].

Besides these operation voltage-oriented distinctions, one may distinguish *in-vehicle PLC* and *MIMO PLC*.

In-vehicle PLC has been considered to provide data access to moving vehicles like trains [33], as well as within the vehicles themselves, for example, in cars [34], aerospace and outer space applications [35,36], ships [37] and submarines [38]. Among others, advantages of PLC over other wireline communication techniques are weight reduction, reduced pin numbers for device internal connections and, in general, wiring complexity reductions. Especially, communicating with parked plug-in electrical vehicles has been at the focus of attention with respect to integrating electrical vehicle fleets into the *Smart Grid* that is currently built up all around the world [39]. Besides, *in-cars* PLC is starting to revitalise the after-sales market business for consumer electronics.

MIMO PLC – Building on the success of *multiple-input multiple-output* (MIMO) signal processing within wireless communications [40,41], it is worth noting that MV and HV installations often make use of four or more conductors. In this respect, a theoretical framework of *multi-conductor transmission line* (MTL) theory is extensively treated in [42]. Further, in many in-home installations three wires, namely *live* (L) (also called *phase*), *neutral* (N) and *protective earth* (PE), are common. Exactly how common on a worldwide scale was investigated by the *European Telecommunications Standards Institute* (ETSI).* The investigation of *Specialist Task Force* (STF) 410 [43] was based on

- A study of individual grounding systems and investigations into which grounding systems are used in which countries; such information could be derived from, for example, education material for electricians.

- The creation of a list of AC wall socket types and their respective usage area. For example, universal travel adapters indicate how many different power outlets exist in the world. Plugs and sockets could easily be checked for the presence of a PE pin. However, the existence of the pin does not guarantee the existence of a PE wire leading up to the socket. When renewing older buildings frequently, the protective earth of the socket is shortcut with the neutral wire behind the outlet or simply not connected at all.

- Research on the dates when the PE installation became mandatory in a country and an estimate of how many electrical installations have taken place since then.

- A survey of sales information, for example, worldwide sales numbers of *residual-current devices* (RCDs) or power cables including ground which in the sequel allows estimates on the PE availability level in a country.

- A worldwide survey of data from electrical standardisation committees and engineering clubs for each country.

Reference [43] lists detailed information and statistical evaluations on each of these points. It is almost impossible to summarise all this information into a single sentence, but when trying to do so, the result would be something like

> The third wire is present at all outlets in China and the Commonwealth of Nations, at most outlets in the western countries and only at very few outlets in JP and Russia.

1.3 Coupling Methods

When turning to coupling methods, one may generally distinguish between *inductive* and *capacitive* couplers. It should be noted that inductive couplers guarantee a balance between the lines whereas capacitive couplers often introduce asymmetries due to component

---

* ETSI is an independent, non-profit, standardisation organisation formed by equipment makers, network operators and other stakeholders from telecommunications industry.

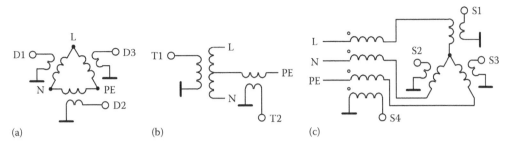

FIGURE 1.2
Inductive MIMO PLC couplers. (a) *Delta-style* (D), (b) *T-style* (T) and (c) *star-style* (S).

manufacturing tolerances. Besides symmetry, signal bandwidth and the dimensioning to protect the communications equipment, for example, against lightning strikes or other HV spikes on the grid side are decisive coupler properties. Moreover, the observed channel characteristics are not independent from the coupling devices used to inject and receive the power line signal. Inductive as well as capacitive couplers especially tailored to MV, HV and even up to extra HV lines can be found in ([44, Section 5.5.1]). Further, details on LV inductive *single-input single-output* (SISO) couplers may, for example, be found in [45,46]. The following will focus on LV inductive MIMO coupling options that play an important role throughout various chapters of this book.

Figure 1.2 presents three inductive MIMO coupler options, that is, a *delta-style* (D) coupler [47], a *T-style* (T) coupler [48] and a *star-style* (S) coupler [47]. Coupler designs are tightly related to radiated emission treated in more detail in Chapters 6 and 7. According to the *Biot–Savart law*, the main source of radiated emission is the *common-mode* (CM) current denoted I_{CM}. To avoid radiated emission, traditionally PLC modem manufacturers aim at injecting the signal as symmetrically as possible. This way, 180° out of phase electric fields are generated that neutralise each other resulting in little radiated emission. This desired symmetrical way of propagation is also known as *differential-mode* (DM) with its associated signal voltage U_{DM}. In a symmetrical network, the differential current I_{DM} flows from its feeding point via the network back to its source as indicated in Figure 1.3. In case of asymmetries, for example, caused by parasitic capacitances (inside the refrigerator in Figure 1.3), a small part of the differentially injected *radio frequency* (RF) current I_{DM} turns into CM current I_{CM}. It flows to ground or to any other consumer device and returns to its source via a series of asymmetries in the network. Normally, there are many asymmetries inside a PLC topology. For example, an open light switch causes an asymmetric circuit and, hence, even if only DM is injected by a PLC coupler, DM to CM conversion occurs [49].

Specifically, to avoid additional CM currents, feeding MIMO PLC signals can be done using the delta or T-style couplers from Figure 1.2 while it is not recommended using the star-style coupler – known also as *longitudinal coupler* – due to the risk of CM signal injection. As shown in Figure 1.2, the delta-style coupler, also called *transversal probe*, consists of three baluns arranged in a triangle between L, N and PE. The sum of the three voltages injected has to be zero (following Kirchhoff's law). Hence, only two of the three signals are independent. Turning to the T-style coupler, it feeds a DM signal between L and N plus a second signal between the middle point of L-N to PE. Further details on the pros and cons of each coupler type are discussed in [43].

All three coupler types are well suited for reception. However, especially the star-style coupler, where three wires are connected in a star topology to the centre point, is

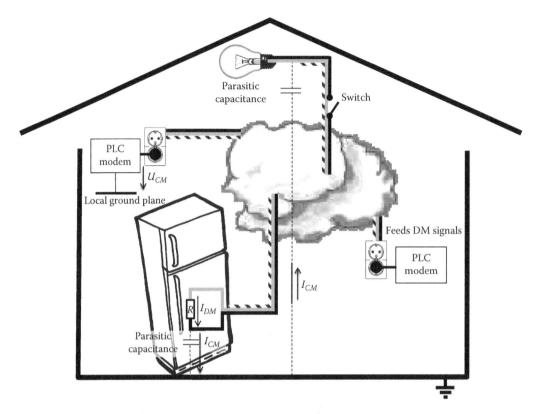

FIGURE 1.3
Generation of CM signals in a building.

interesting. Kirchhoff's law forces the sum of all currents arriving at the centre point to be zero. Thus, only two of the three *receive* (Rx) signals are independent. Nevertheless, due to parasitic components the signals at the third port may additionally improve the capacity of MIMO PLC system (see Chapter 5 for details). However, a more significant benefit is the possibility to receive CM signals, that is, a fourth reception path. The CM transformer is magnetically coupled (Faraday type). On average, CM signals are less attenuated than DM signals which makes their reception interesting, especially for highly attenuated channels [47] (see also Chapter 5).

In general, looking at the experienced input impedance of a mains network, there might be an impedance unbalanced between L-N and L-PE. This is especially true for lower frequencies where 50 Hz loads play an important role [27]. Further, the effective CM impedance is also very different from DM impedances (see Chapter 5 for details). Table 1.1 shows a summary of approximate expected power line input impedances as obtained from the open literature and other chapters in this book. Table 1.1 shows that, for example, impedances in Europe and the United States are very similar. Comparing L-N with L-PE impedances, a marked difference can be observed. For example, for frequencies below 500 kHz this difference can be more than three times and for frequencies between 1 and 30 MHz the L-PE impedance may be around two times larger than in the L-N impedance. However, when looking at the average statistics up to 100 MHz, impedance levels converge, yielding a more balanced MIMO system, the reason possibly also

TABLE 1.1

Power Line Impedances

Frequencies	Country	L-N in Ω	L-PE in Ω	N-PE in Ω	Source
50–500 kHz	JP	0.5–20 (6.5)	na	na	[50,51]
	Germany	1–60 (10)	na	na	Chapter 2
	Europe	na	1–200 (30)	na	[52]
	China	1–9 (5)	na	na	Chapter 3
	United States	na	1–150 (18)	na	[52,53]
1–30 MHz	JP	3–1 k (83)	na	na	[54]
	Germany	10–300 (30)	20–400 (60)	na	[27]
	Europe	(102)	9–400 (90)	na	[47,52]
	United States	na	6–400 (95)	na	[52,53]
1–100 MHz	Europe	10–190 (86)	10–190 (89)	10–190 (87)	Chapter 5

Note: Value in brackets indicates median. If there is more than one reference, the range was taken over the minima/maxima from the references and the arithmetic mean was calculated to obtain a single median value.

being that MIMO couplers used in the related measurement campaign terminated all three ports at signal injection. More details are found in Chapter 5.

Statistical knowledge of the input impedances may be taken into account in the MIMO coupler design. That is, isolation transformers/baluns are required for most MIMO coupling strategies to allow the multiple signals to float independently. In this respect, to obtain BB channel and noise characterisation and EMC results in Chapters 5 and 7, a single coupling circuit – capable of implementing the T-style, delta-style and star-style coupler – as introduced in Figure 1.2, has been designed with its schematic shown in Figure 1.4.

The physical connection to the power grid via a Schuko-type plug is shown on the left. Selection between the coupling types is performed via the switches labelled Sw1 and Sw2, shown in the centre. On the right-hand side, there are the terminals T1 to T2, D1 to D3 and S1 to S4 for connecting the coupler to measurement equipment such as a *network analyser* (NWA) or a *digital sampling oscilloscope* (DSO). The delta-style and T-style terminals are connected through current baluns to facilitate floating signals, and further minimise CM injection as well as any subsequent increase in radiation. The baluns in the delta-style coupler are 1:4 Guanella transformers with very low loss. They perform a 50–200 Ω impedance conversion. Considering a DM impedance of any pair of the three wires of 102 Ω where in the MIMO case this impedance exists twice in series plus once in parallel (resulting in 68 Ω), the 200 Ω output of the Guanella transformer appears to be the optimal matching compromise. Sw3 allows toggling between MIMO and SISO terminations of the feeding ports. The star-style terminals use coupled current transformers (CM chokes). Their function is to measure the CM current flowing through the three wires (L || PE || N). The inductance of each winding is selected small enough not to filter the PLC frequencies of interest (1–100 MHz). An additional switch (not shown in Figure 1.4) might short circuit the secondary (and thus magnetising inductance) of these CM chokes, thus making them transparent to the overall circuit when not in use.

Looking at the other electronic components in Figure 1.4, one can see that in series with each line, there is a 4.7 nF coupling capacitor, fulfilling the main coupling function – protection by means of filtering, while the capacitor to the Earth wire is implemented to maintain symmetry. Furthermore, it lowers the leakage current that could otherwise cause

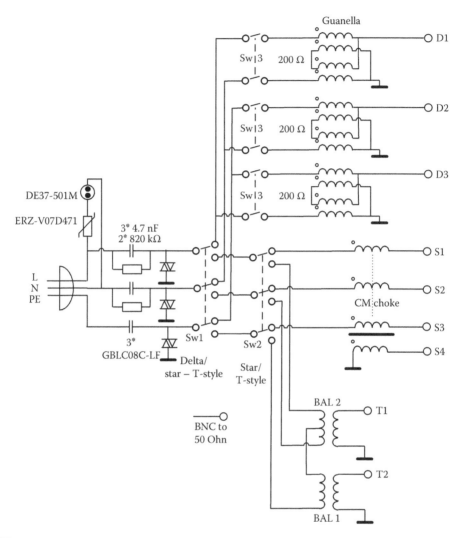

FIGURE 1.4
A 2-Tx 4-Rx MIMO PLC coupler with added protective circuitry. (Based on European Telecommunication Standards Institute (ETSI), Powerline Telecommunications (PLT); MIMO PLT Universal Coupler, Operating Instructions – Description, May 2011. Available online at: http://www.etsi.org/deliver/etsi_tr/101500_101599/ 101562/01.01.01_60/tr_101562v010101p.pdf.)

a *residual current device* (RCD) to fail. The breakdown voltage of these capacitors are rated roughly twice the expected *root mean square* (rms) value, that is, 1.5 times the expected peak value of the power waveform to be blocked. The mains connection can be unplugged at any time. Hence, the capacitors may be charged to dangerous voltages, and they are in this case discharged by a parallel resistor, large enough not to short circuit significant current past the blocking capacitor – thereby impacting on the filtering action of this capacitor. The RC time constant is adjusted so that the blocking capacitor discharge within 15 ms to protect against accidentally touching the prongs of the attached Schuko when unplugging it. Apart from the protection offered by the coupling capacitor, supplementary protective devices are deployed to absorb voltage transients. Surge protection diodes can be seen in parallel with the three lines. The input impedance of these and other protective devices are sufficiently high not to drain or filter the PLC signal. On the mains side of the coupling capacitors (across L-N), gas-discharge devices and *metal-oxide varistors* (MOVs) are used. They serve as over-voltage protection and are deployed in parallel as they possess different speed versus power characteristics and are, hence, complementary. Supplementary information on surge protection in general may be found in [55–58]. The specific safety components of the coupler as shown in Figure 1.4 are given in [45] alongside the coupler's calibration data.

1.4 Channel and Noise Measurement Set-Up

This section describes the measurement set-up and the equipment used to record channel, reflection and noise properties of the MIMO PLC channel on a LV grid. It therewith forms the basis for the results in Chapters 5, 7 through 9, 16, 17 and 22. The following measurement campaign objectives are defined:

- Frequency range from 1 to 100 MHz.
- *Channel transfer function* (CTF) measurements.
- Noise-level measurements.
- Input impedance measurements.
- Coupling factor measurements (i.e. the coupling factor or *k*-factor is the ratio between the electric field caused by signal radiation and the signal power fed into the main grid [59] [see also Chapter 7]).
- All measurements are to be performed with the same coupler design to allow for straightforward comparability of results.

1.4.1 Transfer Function Measurements

The CTF can be measured in the frequency domain by recording the scattering parameter S_{21} using a conventional NWA (for an introduction to scattering parameters see, for example, [60]). Figure 1.5 depicts the set-up for channel measurements in a private home.

A NWA is connected to two PLC couplers via its *transmit* (Tx) and *receive* (Rx) ports. The couplers follow the schema as introduced in Figure 1.4. All terminals T1 to T2, D1 to D3 and S1 to S4 are terminated by 50 Ω each. Terminating the ports avoids additional signal reflections caused by connecting measurement equipment. When terminating the three-wire system, the impedance of each wire pair is present three times in parallel. A final PLC modem implementation might terminate two or three of the three wires with, for example, a low impedance when transmitting and a high impedance when receiving.

Further, a ground plane as shown in Figure 1.6 is connected tight to the receiving coupler. To achieve a low impedance or high capacity connection to ground, a huge ground plane is necessary, especially also for reproducibility of the received CM signal. The size of the ground plane is sufficiently large when human contact no longer influences measurement results. From a physical point of view, human contact results in a capacitance increase towards ground, which will not influence the result when the ground plane's stand-alone capacity is already sufficiently large. In this respect also, the reception of lower

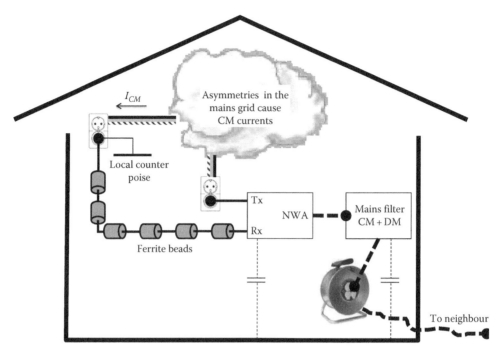

FIGURE 1.5
Private home with channel measurement equipment.

FIGURE 1.6
Ground plane with connected PLC coupler.

frequencies requires a larger ground plane than the reception of higher frequencies. In a practical setting, the backplane, for example, of a *high-definition television* (HDTV) could serve as a ground plane for an incorporated PLC modem. Coming back to the NWA, it provides a dynamic range of 120 dB for 'through', that is, S_{21}, measurements. Considering that the dynamic range of many commercial PLC modems lies in the order of 90 dB, 120 dB seems sufficient. However, to obtain meaningful results, the coaxial cables – especially the one connecting the Rx coupler with the Rx port of the NWA – have to support such a dynamic range, which means double-shielded cables are required. Moreover, due to long distances inside the buildings, low attenuation cables are preferred. RG214 or Ecoflex 10 cables fulfil these requirements. Further, to avoid signal ingress to the cable going back from the Rx coupler to the NWA, the cable is surrounded by suppression axial ferrite bead with a 15 cm inter-bead spacing as indicated in Figure 1.5.

If the test instruments (namely the NWA) are connected to the mains section for *electromagnetic interference* (EMI), impedance and CTF measurements, the instruments represent an additional load and may cause measurement errors. Hence, whenever possible the power for the measurement equipment is drawn from a neighbouring flat via an extension cable. Further, on the right hand side of Figure 1.5, one can see a 'mains filter CM+CD'. This filter is used to isolate the PE wire and consists of an isolation transformer and a *line impedance stabilisation network* (LISN). Additionally, a MIMO mains filter is used on each of the three wires eliminating DM signals, plus a CM choke to get rid of potential longitudinal signals. Such MIMO mains filter is not commercially available and was specifically manufactured for this measurement campaign. S_{21} measurement results using this set-up may be found in Chapter 5.

1.4.2 Reflection Measurements and Input Impedance Calculation

The *LV distribution network* (LVDN) is a network with undefined complex characteristic impedance. The often measured absolute value of the input impedance has little practical significance. Adding a short piece of mains cable may change the results considerably. Thus, ETSI STF 410 measured the reflection loss expressed through the scattering parameter S_{11} [60] at the 'delta' terminals of the couplers instead.

Generally, reflection measurement signals are fed and received at one and the same NWA port and require only the Tx coupler. The set-up is as shown in Figure 1.5 but without connecting the Rx coupler. The NWA is calibrated with a 'short', 'open' and a '50 Ω termination' connected at the end of the coaxial cable. The MIMO PLC coupler is considered to be part of the PLC channel.

S_{11} is a complex value which is a function of the load impedance and of the characteristic impedance of the measurement system (see Equation 1.14 for a general definition). Here, the measurement system consists of the NWA, which has a characteristic impedance of 50 Ω and the balun inside the MIMO coupler, which transforms the 50–200 Ω, that is, $Z_{coup} = 200$ Ω. Theoretically, S_{11} on the 50 Ω side is identical to S_{11} if measured on the 200 Ω side, except for a phase shift due to the length of the transmission lines inside the balun. The real and the imaginary parts of S_{11} are recorded. For engineering purposes, the absolute value $|S_{11}|$ is often sufficient. It allows calculating the maximum line input impedance $Z_{DM\,max}$, that is,

$$Z_{DM\,max} = Z_{coup} \cdot \frac{\left(1 + |S_{11}|\right)}{\left(1 - |S_{11}|\right)}. \tag{1.1}$$

However, the phase is also required to calculate Z_{DM} as a function of frequency and line length of the balun ($\approx$0.3 m^{-1}) denoted x, that is,

$$Z_{DM} = Z_{coup} \cdot \frac{\left(1 + S_{11} \cdot e^{j\beta x}\right)}{\left(1 - S_{11} \cdot e^{j\beta x}\right)},\tag{1.2}$$

where the phase constant β may be obtained as outlined in Appendix 1.A, Equation 1.12, assuming a wave speed v in the balun to be $\approx$200,000,000 m/s.

1.4.3 Noise Measurements

The noise measurement set-up is depicted in Figure 1.7. The MIMO coupler from Figure 1.4 is used in the star configuration (ports S1, S2, S3 and S4). Thus, the noise voltage present at the L, N and PE wires as well as the CM voltage can be directly sampled in the time domain by connecting a DSO (with *digital signal processing* [DSP] probe P1 to P4). The sampling rate is 500 Msamples/s. Further, care is taken that the DSP memory can store the four signals over 20 ms, corresponding to a single period of the AC line cycle at 50 Hz.

One drawback of the time domain measurement is that out-of-band noise can easily influence the result. Hence, DSO probes are using band-pass filters in order to reduce out-of-band noise. In each configuration, four different bands were tested, with

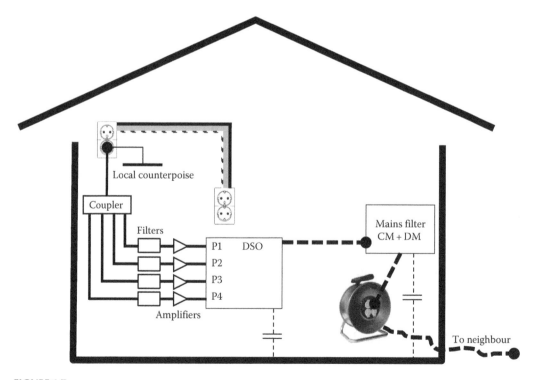

FIGURE 1.7
Private home with noise measurement equipment.

the respective frequency ranges 2–100 MHz, 2–88 MHz, 30–100 MHz and 30–88 MHz. Further, in environments with low noise levels, low noise amplifiers are used to boost the input signal before recording. Each amplifier presented a flat frequency response up to 100 MHz and a gain of 28 dB.

1.5 Channel Characterisation and Modelling Approaches

The power line channel and noise situations heavily depend on the scenario as outlined in Section 1.2 and, hence, span a very large range. Generally, frequency-selective multi-path fading, a low-pass behaviour, AC-related cyclic short-term variations and abrupt long-term variations can be observed (see, for example, Chapters 2 through 5).

1.5.1 Channel Characteristics

Multi-path fading is caused by inhomogeneities of the power line segments where cabling and connected loads with different impedances give rise to signal reflections and in the sequel in-phase and anti-phase combinations of the arriving signal components. The corresponding transfer function can readily be derived in close form as *infinite impulse response* (IIR) filter [1] and underlying basics are outlined in Appendix 1.A. One important parameter capturing the frequency-selective characteristics is the rms *delay spread* (DS). For example, designing *orthogonal frequency-division multiplexing* (OFDM) systems, the guard interval might be chosen as 2–3 times the rms DS to deliver good system performance [61]. To provide an orientation, the mean of the observed rms DS for a band from 1 MHz up to 30 MHz in the MV, LV-access and LV-in-home situations in [21,61] was reported to be 1.9, 1.2 and 0.73 µs, respectively. Similarly, in Chapter 4, rms DS in the range 0.2–2.5 µs are reported from LV-in-home measurements performed within the 1.8–88 MHz frequency band.

Besides multi-path fading, the PLC channel exhibits time variation due to loads and/or line segments being connected or disconnected [62]. Further, through synchronising channel measurements with the electrical grid AC mains cycle, Cañete et al. were able to show that the in-home channel changes in a cyclostationary manner [47,63] (see also especially Chapters 2 and 21).

Until now, the low-pass behaviour of PLC channels has not been considered. It results from dielectric losses in the insulation between the conductors and is more pronounced in long cable segments such as outdoor underground cabling. Transfer function measurements on different cable types and for different length can be found in [6,25].

1.5.2 Channel Modelling Overview

Channel characterisation and modelling are tightly intervened. Channel characterisation in terms of channel measurements is indispensable to derive, validate and fine-tune channel models, while the channel models themselves often provide valuable understanding and insight that might stimulate more advanced channel characterisation.

In general, PLC channel models can be grouped into *physical* and *parametric* models (or into bottom-up and top-down models as in [7]). While physical models describe the

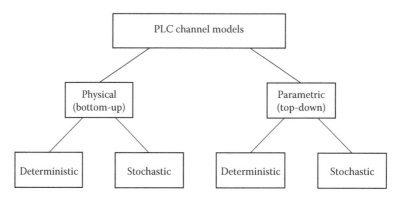

FIGURE 1.8
PLC channel modelling options.

electrical properties of a transmission line, for example, through the specification of cable type (line parameters), the cable length and the position of branches [48,64–67] (see also Appendix 1.A), parametric models use a much higher level of abstraction from the physical reality and describe the channel, for example, through its impulse response or transfer function [25,68,69] (see also especially Chapters 4 and 5).

Further, within each group, it can be distinguished between *deterministic* and *stochastic* models. While deterministic models aim at the description of one or a small set of specific reproducible PLC channels realisations, stochastic models aim at reflecting a wide range of channel realisations according to their probability of occurrence. This categorisation of channel modelling options is reflected in Figure 1.8 with a short description of each in the following.

Physical-deterministic modelling describes the electrical properties of a transmission line, for example, through the specification of cable parameters, cable length, the position of branches, etc. [64–67,70]. Most physical models are based on representing power line elements and connected loads in the form of their *ABCD* or *S-parameters* [60], which are subsequently interconnected to produce the channel's frequency response [64–67,70–73]. Alternatively, [1,74] introduced power line elements as well as connected loads as *IIR* filters, which is a novel and still intuitive approach if one considers that a communication signal travels in the form of an electromagnetic wave over the PLC channel and may bounce an infinite amount of times between neighbouring line discontinuities. Physical-deterministic models are especially well suited to represent and test deterministic power line situations. An introduction on to the underlying transmission line theory is provided in more detail in Appendix 1.A. *MTL* theory [42] is particularly well suited in the case of PLC propagation, since it allows describing the signal transmission along any arbitrary topology of interconnected wires and any set of connected loads. Physical-deterministic techniques are sometimes referred to as 'bottom up', as they start from a precise description of the electrical network under consideration in order to derive a global behaviour of the propagation channel. For a given electrical network, the physical-deterministic method can provide a *CTF* model very close to actual measurement. The drawback is that it requires a significant amount of input data and computational resources, especially if one wishes to derive channel statistics from a large number of different network topologies.

Physical-stochastic models combine the aforementioned deterministic approach with stochastic elements. In [75,76] a so-called 'statistical bottom-up' approach is presented, where the CTF is computed from the exact network topology using a deterministic algorithm. The stochastic nature of the model arises from the random generation of realistic electrical network topologies, based on a number of rules derived from observed cabling practices, an approach that has also been proposed in [1]. Physical stochastic models include the advantages of the deterministic approach in terms of accuracy with respect to physical transmission phenomena, while allowing for the random generation of statistically representative channel realisations. System engineers generally run digital simulations of the full system, which allow evaluating the behaviour and efficiency of different signal processing algorithms. A stochastic channel model is thus expected to reproduce the main effects of the propagation channel by generating a large number of random channel realisations, which are statistically representative of real-world observations.

Parametric-deterministic models are possibly one of the categories most used but not usually labelled as 'parametric-deterministic'. Here, this label is referring to a database of measured parameters, such as the CTF, where simple playback of the measurement results can be used in PLC system simulations and performance studies. The advantage is that the exact parameters as observed in real situations are used without the risk to generate unrealistic channels due to modelling inaccuracies. On the other hand, a large and diverse database is needed to obtain meaningful results on a more general level. An example of such a huge database sourced from six European countries is documented in Chapter 5.

Parametric-stochastic models use a high level of abstraction and describe the channel, for example, through its impulse response characteristics [6,68,77]. The analysis of collected measurement data allows defining a model in the form of a mathematical expression. The mathematical form of the model is not necessarily linked to the physical phenomena taking place in the transmission of electromagnetic signals, but is designed to faithfully reproduce the main characteristics of the channel under study. The model parameters are defined in a statistical way, which allows generating different random realisations of the CTF (or the channel's impulse response) exhibiting similar statistics as the experimental data. This modelling strategy is sometimes referred to as a 'top-down' approach, in the sense that it first considers global statistics of the propagation channel in order to define deeper details of the channel structure. This approach generally provides realistic results, with the drawback of requiring a large amount of experimental data to produce the model. The model in [25] is an early example of such statistical channel model, where a general CTF model has been defined based on physical considerations of the signal propagation through simple electrical network topologies. The model parameters were then obtained by fitting the mathematical model to a number of experimental measurements taken in the 0–20 MHz range. A more recent example of this modelling strategy is given in [78,79], where a channel model is defined on the basis of 144 measurements taken in different dwelling units in the 0–100 MHz range. The measurements are subdivided into nine different classes according to the observed channel capacity, and a statistical channel model is provided for each class.

Table 1.2 provides a comparison between the different PLC channel modelling options. Each of the four exists in its own right and bears advantages and disadvantages when it comes to specific applications. Thus, before deciding the type of channel model, the question

TABLE 1.2

Comparison of PLC Channel Modelling Options

Feature	Physical Deterministic	Physical Stochastic	Parametric Deterministic	Parametric Stochastic
Modelling principle	Electromagnetic transmission theory	Electromagnetic theory and topology generator	Playback of experimental measurement parameters	Statistical fit to experimental measurement parameters
Measurement requirements	None	None	Large data base	Large data base
Topology knowledge	Detailed	Detailed stochastic models	None	None
Complexity of model design	Medium	High	Low	Medium
Complexity of channel generation	High	High	Very low	Low
Correlated multi-user studies	Straightforward	Straightforward	Straightforward	Difficult
Closeness to experimental data	Accurate for considered topology	On a statistical basis	Exact	On a statistical basis
Ability to extrapolate	Yes	Yes	No	No

What is the channel model supposed to do? is key. Some desirable properties of a channel model could, for example, be to

1. Describe the influence of the time-variant channel on the received signal quality in link and system simulations and algorithm testing, for example, on impulsive noise avoidance, signal-to-noise ratio estimation, channel filter tracking, MIMO schemes.
2. Model the correlation between temporal and spatial channel and noise variations.
3. Support the investigation of multi-point (multi-user) PLC systems.
4. Allow extension to various propagation scenarios based only on a small set of additional scenario measurements.
5. Describe modal coupling to be used in the design of MIMO couplers.
6. Assist in the development of the modem's analogue front-end.

The objectives (1) to (3) and (5) can with more or less effort be realised with either a physical (bottom-up) or a parametric (top-down) model implementation. However, the objectives (4) and (6) are hard to realise with a parametric model. In general, parametric models require a larger range of measurement results to adjust the model parameters. On the contrary, physical models allow deploying knowledge, for example, on the physical dimensions of a new scenario, to adjust physical model parameters. Afterwards, only a smaller set of measurements is needed for rough verification purposes. Looking at signal processing–related issues with respect to MIMO PLC systems, a parametric model has certain advantages. It might be more easily deployed and parameters such as spatial correlation are well understood due to related studies in the wireless world [80]. However, looking at the practical implementation of, for example, MIMO couplers or the adjusting of analogue front-ends, a physical model, being significantly closer to the reality of electronic components, might be more useful. After these examples, it becomes clear that channel model selection has to be carried out on a case-by-case basis.

1.5.3 MIMO Channel Models

Turning specifically to MIMO channels, besides in the early patents [81–84], one of the first public investigations of the MIMO access case appears in [85] with perfectly isolated multiple phase wires. Sparked also through the success of MIMO signal processing in the wireless world [41], larger public *parametric-deterministic* investigations of MIMO signal processing for BB in-home PLC appears in [86]. Similar evaluations, based on a limited set of measurements, are conducted in [87,88]. Following this trend, experimental channel and noise characterisation for MIMO PLC systems have been conducted in [47,89–92]. For example, [87,88] conclude that the application of 2 × 2 MIMO signal processing to in-home PLC provides a capacity gain in the order of 1.9. Further, [86] showed that this capacity gain even increased with the number of Rx ports. In a 2 × 3 MIMO configuration, the average capacity gain ranged between 1.8 and 2.2 depending on the Tx power level. When adding CM reception, average gains between 2.1 and 2.6 were observed. Along these lines, additional MIMO capacity and throughput results can be found in Chapter 9.

Only a few proposals for *physical-deterministic* MIMO channel modelling have been made so far. The most straightforward bottom-up approach of modelling a channel composed of several wires is to apply MTL theory [42,60]. As shown in Figure 1.9, MTL theory can be applied to compute the currents $i_1(x,t)$, $i_2(x,t)$ and $i_3(x,t)$ flowing in a three-wire transmission line as well as the corresponding differential voltages $v_1(x,t)$, $v_2(x,t)$ and $v_3(x,t)$ for a given line position x and a given time t. To do so, a long list of per unit length line parameters needs to be known, that is, the inductances L_{11}, L_{22} and L_{33} and the resistances R_{11}, R_{22} and R_{33} of wires 1, 2 and 3, respectively, the mutual inductance between any pair of wires

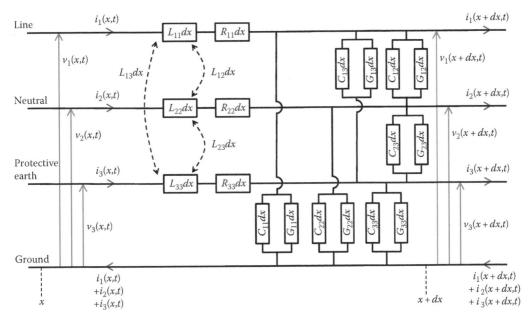

FIGURE 1.9
MTL theory: equivalent circuit of a per unit length section of a three-wire transmission line.

(L_{12}, L_{13} and L_{23}), the capacitances (C_{11}, C_{22} and C_{33}) and conductances (G_{11}, G_{22} and G_{33}) of each wire with respect to the ground, the mutual capacitances (C_{12}, C_{13} and C_{23}) and finally the conductances (G_{12}, G_{13} and G_{23}) between any pair of wires. Note that some authors consider a simplified model with three conductors only, where the PE wire is assumed to be equivalent to the ground [93]. At high frequencies this assumption is not valid, especially when the reception of CM signals is expected as introduced in Figure 1.3. In such cases, a more complete model such as the one presented in Figure 1.9 is necessary to provide accurate results. The MTL modelling approach has been used in the work of Banwell and Galli for in-home LV electrical networks [48,66,94], and by Anatory et al. [73] for overhead MV or HV networks. However, these studies do not consider the use of three electrical wires for the purpose of MIMO communication.

The first use of MTL theory to explicitly model a MIMO PLC channel in a *physical-stochastic* approach appears in [93,95]. The work therein extends the physical-stochastic SISO channel model presented in [76] by recomputing the MTL equations in the case of three conductors. Using the stochastic topology generator from [76], it is then possible to produce the CTF matrix for a large set of random electrical networks.

On the other hand, a *parametric-stochastic* approach has been applied by several research teams to devise models of the MIMO PLC channel. The first attempt is described in [96]. This study considers a 2 × 4 MIMO channel, where two differential input ports can be addressed simultaneously, and up to four Rx ports are considered, including the CM path. The model first considers a SISO PLC *channel impulse response* (CIR) composed of 5–20 taps according to the model defined within the European project OPERA [21]. It then builds the 2 × 4 MIMO channel matrix by producing eight variants of this CIR. Each of the variants has the same tap structure, but the amplitudes of some of these taps are multiplied using different random phases uniformly drawn from the interval [0,2π]. The more taps are modified using a random phase shift, the more uncorrelated the channel becomes. The model produces MIMO channels that exhibit similar correlation values as observed in the measurements of [87]. The same approach is further developed in [97], where a 3 × 3 MIMO channel model has been designed to fit observations from a measurement campaign in France. In total, 42 3 × 3 MIMO channels were measured in five different houses using a vector NWA. The proposed MIMO channel model builds on the SISO channel model first defined by Zimmermann and Dostert [25], and later extended by Tonello by providing complementary channel statistics [98]. In the following, the notation adopted during the OMEGA project [99] will be used, where the CTF $H(f)$ is given as a function of the frequency f by

$$H(f) = A \sum_{p=1}^{N_p} g_p e^{-j(2\pi d_p/v)f} e^{-(a_0 + a_1 f^K)d_p}, \qquad (1.3)$$

where

  v represents the speed of the electromagnetic wave in the copper wire (which may be approximated as two-thirds the speed of light*), that is, 200,000,000 m/s
  d_p and g_p represent the length and gain of the propagation path
  N_p represents the number of propagation paths
  Parameters a_0, a_1, K and A are attenuation factors

---

* Speed of light in vacuum 299,792,458 m/s.

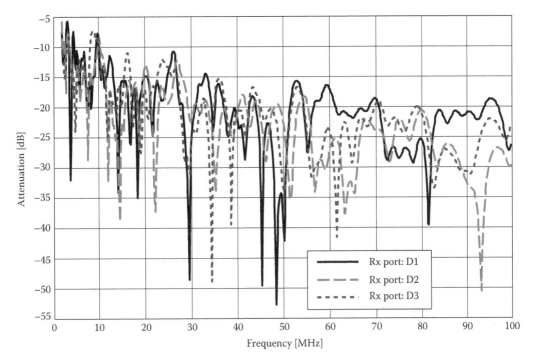

FIGURE 1.10
Example of CTF simulated using the MIMO PLC channel model of Hashmat et al. [97]. Tx port D1 only.

An example of channel realisation is given in Figure 1.10 and the approach will be revisited in Chapter 5 to devise a novel channel model on the basis of European field measurements.

An alternative *parametric-stochastic* approach is presented in [100]. This study characterised the MIMO channel covariance matrix R_h, by analysing 96 MIMO channel measurements recorded in five houses in North America. Following a similar approach as in [80], the MIMO channel matrix $H(f)$ is then modelled for each frequency f as

$$H(f) = K(f) \cdot R_r^{1/2}(f) \cdot H'(f) \cdot R_t^{1/2}(f),$$ (1.4)

where
   $K(f)$ is a normalising constant
   $H'(f)$ is a channel matrix composed of independent and identically distributed complex
      Gaussian variables
   $R_r(f)$ and $R_t(f)$ represent the Rx and Tx correlation matrices, respectively

Each channel correlation matrix is modelled by its decomposition in eigenvectors and eigenvalues. Details on this alternative model that allows very straightforward reproduction of the MIMO channel's correlation properties can be found in Chapter 4.

1.6 Noise Characterisation and Modelling Approaches

Turning from the channel to the noise situation, one should note that in contrast to many other communication channels the noise in a power line channel cannot be described as *additive white Gaussian noise* (AWGN).

1.6.1 Noise Characteristics

The noise observed on indoor power line networks has been traditionally categorised into several classes, depending on its origin, its level and its time domain signature [101]. Power line noise can be grouped based on temporal as well as on spectral characteristics. Following, for example [5,6], one can distinguish *coloured background noise, narrowband (NB) noise, periodic impulsive noise asynchronous to the mains frequency, periodic impulsive noise synchronous to the mains frequency* and *aperiodic impulsive noise* as indicated in Figure 1.11.

A first class consists of the *impulsive noise* generated by electronic devices connected to the mains grid, such as switched mode power supplies, light dimmers or compact fluorescent lamps. This type of noise is of short duration (a few μs) but of relatively high level in the order of tens of mV. Due to the periodic nature of the mains, noisy devices can generate impulses in a synchronous way with the mains period. In this case, the impulsive noise is said to be *periodic and synchronous to the mains frequency* and presents a repetition rate at multiples of 50 or 60 Hz dependent on the mains frequency. Other noise sources generate impulses at a higher periodical rate up to several kHz, which are classified as *periodic and asynchronous to the mains frequency*. Finally, strong impulses can also be observed more sporadically, without any periodicity with the mains or with itself. This type of noise is sometimes referred to as *aperiodic impulsive noise*. Examples of such noise types recorded during field measurements are given in Chapter 5. The different characteristics of the impulsive noise have been statistically analysed through the observation of experimental data in [102]. A comprehensive model of the PLC impulsive noise has been proposed in [101]. The pulses are first statistically characterised in terms of amplitude, duration and repetition rate, and the global noise scenario is then modelled in the form of a Markov chain of noise states.

A second class of noise consists of *narrowband (NB) noise*. This type generally corresponds to ingress noise from broadcasting radio sources, in particular from the *short-wave* (SW)

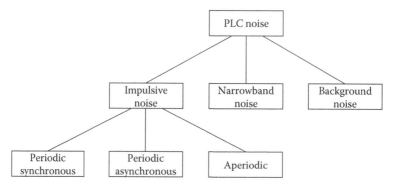

FIGURE 1.11
Classification of PLC noise.

and *frequency-modulation* (FM) bands. Other ingress noise corresponds to leakages from nearby electrical or industrial equipment. This type of noise usually generates strong interference over long durations in a narrow frequency bandwidth in the order of tens of kHz.

Finally, the remaining noise sources, presenting a lower level of interference, form a third class of noise called *background noise*. The background noise is generally coloured, in the sense that its *power spectral density* (PSD) is usually stronger at lower frequencies. In [30], the background noise PSD is modelled with decreasing power as a function of frequency. A similar approach has been adopted in the OMEGA Project [99], where the decaying function of frequency is complemented with a NB model representing the broadcast interference. Specifically, a statistical approach to average coloured background noise modelling is presented in [21] based on a large amount of noise measurements in MV as well as LV-access and LV-in-home situations. One general finding is that the mean noise power falls off exponentially with frequency. Alternatively, an interesting, quite different approach to model SISO PLC background noise is taken in [103], where a neural network technique is deployed for model generation.

An important feature relevant to all types of noise is their time dependency. Due to the uncontrolled nature of the noise source, the characteristics of the noise perceived at a given outlet can drastically change over time. For instance, the increase of human activity in the premises, for example, after working hours, leads to a stronger contribution of domestic appliances to the noise. It is less obvious that noise presents a cyclostationary structure, with a period related to the mains signal. This effect is mainly due to the periodic change of impedance at the network termination loads depending on the mains cycle. A thorough study of the time variation of the SISO PLC noise is presented in [104].

Further, Chapter 2 presents NB SISO noise measurement results for a small LV access network supplied by a single transformer substation. Beside the well-investigated noise classes, *background* and *impulsive noise*, a detailed look on the noise scenario identified two steady appearing noise classes: *NB interferers* and *switched frequency noise* (see Chapter 2 for details).

On the other hand, Chapter 3 introduces NB SISO (but multi-phase) noise measurements for typical underground LV access network in China. It is found that the noise power decreases with the frequency significantly. Measurements show a reduction of average noise power from a range between 50 and 90 dB µV at 50 kHz to a range between 30 and 60 dB µV at 500 kHz. Also, a considerable different noise level is reported at different times of the day.

1.6.2 Noise Modelling for MIMO PLC

Turning specifically to the MIMO noise situation, only a few modelling proposals have been made so far. For example, [89,105–107] are developing models of background noise on the basis of experimental noise measurements performed in five houses in France. The measurements were undertaken in the time domain using a DSO (as outlined previously in Section 1.4). However, the models are mainly targeting a reproduction of the frequency domain noise characteristics. In [106], the measurements are compared against two parametric SISO background noise models, namely, the Emsailian model [30] and the OMEGA model [99]. The models are fitted to the noise received on each of the MIMO Rx ports, and statistics of the model parameters are derived separately for each Rx port. In [107], the MIMO noise is regarded as a *multi-variate time series* (MTS), which allows capturing both the intrinsic characteristics of the noise received on each port, but also their cross-correlation. The noise MTS is then modelled using an *autoregressive* (AR) filtering procedure. The modelled noise PSD presents a high degree of similarity with the experimental observations. However, the model leaves room for improvements, especially considering its ability to reproduce sporadic time domain events, such as impulsive noise. Along the same lines, Chapter 5 presents MIMO noise measurements and noise modelling results. The measurements are conducted in 31 different dwelling units in five European countries, including Belgium, France, Germany, Spain and the United Kingdom. It was observed that the CM signal is affected on average by 5 dB more noise than the DM signals received on any wire combination. This difference can be explained by the higher sensitivity of the CM signal to interference from external sources, such as radio broadcasting. Moreover, it was observed that the L, PE and N ports present similar noise statistics. However, when considering large noise records (5% percentile), one can observe that the PE port is more sensitive to noise by ~2 dB than the N or L ports. Similarly, for low noise levels (95% percentile), the L port is less sensitive to noise by ~1 dB than the N or PE ports.

On the other hand, Chapter 4 addresses MIMO noise based on experimental measurements collected in the United States. In particular, it has been shown that the noise is correlated on the L-N, L-PE and N-PE receiver ports. The strongest correlation is measured between the L-PE and N-PE receiver ports. Moreover, the correlation is stronger for lower frequencies when compared to higher frequencies. Effects of noise correlation on MIMO capacity of a system with two transmit ports and three receive ports are studied. It is observed that noise correlation indeed helps to increase the MIMO channel capacity.

Besides all these initial efforts to characterise and model PLC noise specifically with respect to MIMO systems, a number of features of MIMO PLC noise still need to be investigated and modelled. In particular, the occurrence of impulsive noise, its time domain variations and the correlation of noise pulses observed on different Rx ports require further analysis.

1.7 Conclusion

MIMO PLC channel and noise characterisation and modelling have been introduced. As channel and noise are not independent of the deployed coupling strategies, these strategies have been covered in detail, specifically for LV in-home (or in building) configurations where a MIMO topology is formed by the live, neutral and protective earth wire. Various MIMO PLC transmitter and receiver topologies are possible. A major difference between transmitter and receiver side though is that CM injection is normally prevented at the transmitter, as CM signals are a direct cause of radiated emissions, which are limited by government regulations for obvious reasons. However, the power line channel does produce CM signals as a side effect because of asymmetry as well as parasitic capacitances. Thus at the receiver, the CM signal may as well be measured and processed for maximum receiver diversity and data throughput if receiver complexity is permissible.

Looking at characterisation and modelling, the characteristics of the SISO channel were presented citing a large amount of relevant literature. Two main modelling approaches, the physical (bottom-up) and the parametric (top-down) approaches, may be distinguished. Each approach may be paired with either deterministic or stochastic elements. In general, parametric models are based on series of field measurements and provide a compact description capturing the observed experimental characteristics. Physical models describe the underlying propagation phenomena in more detail. They are useful to predict the transmission conditions in a known environment or to launch physical-stochastic studies with the help of random topology generators.

With respect to SISO noise phenomena, important references in the field were cited capturing the main types of impairments. Impulsive noise consists of bursts of short-term over-voltage transients, which are generally considered in the time domain. NB noise and background noise are analysed in the frequency domain and correspond to long-term perturbations.

Due to the recent emergence of MIMO signal processing techniques for PLC systems, a few proposals only have been made so far to model the PLC MIMO channel and MIMO noise characteristics. The main existing studies in this field were reviewed and commented. The chapter's intention was not to cover any characterisation or modelling approach in detail but rather to set the scene for the individual chapters on this topic. Specifically, Chapter 2 provides an approach for determining the link quality in a LV grid under the influence of the time-variant features of the NB transmission channel up to 500 kHz. Besides noise scenarios and CTF, there is a focus on the access impedances of the LV grid. Results obtained for a LV grid of a small university campus are presented.

Chapter 3 looks at the NB power line channel for LV access networks in China. Major channel characteristics such as access impedances, interferences and attenuations for frequencies between 30 and 500 kHz are evaluated based on measurement results in a typical urban underground network.

Chapter 4 deals with the characterisation of in-home MIMO power line channels and noise in the band from 1.8 to 88 MHz. The characterisation is based on channel measurements in multiple US homes. It focuses particularly on the spatial properties of the MIMO power line channel and introduces a parametric-stochastic MIMO channel model consistent with the measurements. Further, noise measurements are analysed for their spectral and correlation properties, which are shown to affect PLC MIMO capacity.

Chapter 5 takes a further step in the field of BB (1–100 MHz) LV in-home MIMO channel and noise characterisation and modelling. Novel parametric-stochastic models are derived based on a series of European measurements.

Appendix 1.A: Introduction to Transmission Line Theory

The following two subsections introduce a model for the dual conductor transmission line that is worth being familiar with before looking at MTL line theory as introduced in Section 1.5. Basic concepts from line parameters to propagation constant, characteristic impedance up to the physical (bottom-up) model of a dual conductor stub-line example network are described, showing how the concatenation of line segments leads to the multipath effects and in the sequel to frequency-selective fading.

1.A.1 Propagation Constant, Characteristic Impedance and Phase Velocity

The dual conductor transmission line can be described as a concatenation of lumped elements as displayed in Figure 1.12. R, L, G and C describe a shunt resistance per unit length (Ω/m), a series inductance per unit length (H/m), a shunt conductance per unit length (S/m) and a shunt capacitance per unit length (F/m), respectively. With the help of these lumped

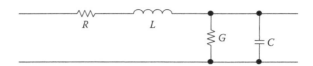

FIGURE 1.12
Lumped elements transmission line representation.

element parameters, also known as *primary line parameters*, the propagation constant γ and characteristic impedance Z_0 of the line can be calculated as shown later in this section.

The main function of the cables/conductors constituting the electrical grid is to deliver electricity. Hence, electrical cable manufactures usually do not specify line parameters in the frequency range of interest for PLC. Alternatively, the line parameters can be derived using electromagnetic theory as outlined in [29,108,109] and references therein. However, the math behind this direct approach is quite involved. Instead, different electromagnetic simulation tools like [110,111] may be used to obtain the line parameters. In [108], the primary line parameters are derived as a function of the cable geometry, the experienced permittivity, permeability and conductivity. All but R are approximated as frequency independent. R is approximated proportional to the square root of the frequency of interest, that is, $R \propto \sqrt{f}$. The admittance G is further approximated as 0. Similarly, [6] describes the primary line parameters based on a specific cable geometry, permittivity, permeability and a dissipation factor. In contrast to [108], G is, however, described as a linear function of frequency. An extensive set of measurement-based primary line parameter models are given in [112] tailored to overhead LV power distribution networks. Finally, line parameters might be determined by measurements of the open- and short-circuited line input impedance as outlined in Appendix C of [108]. As an example, the line parameter values from [108] for dual conductor in-home cabling are reproduced in Table 1.3.

TABLE 1.3

Example Line Parameter Sets

Parameters	Line 1	Line 2	Line 3	Line 4	Line 5
Name/type	H07V-U-1	H07V-U-2	H07V-R-1	H07V-R-2	H07V-R-3
Cross section, mm²	1.5	2.5	4	6	10
Radius, mm	0.691	0.892	1.128	1.382	1.784
Isolation thickness, mm	0.960	1.060	1.072	1.32	1.616
Equivalent relative permittivity	1.45	1.52	1.56	1.75	2
Cable geometry factor	2.7	2.4	2.17	1.96	1.69
C, pF/m	15	17.5	20	25	33
L, µH/m	1.08	0.96	0.87	0.78	0.68
R, $\sqrt{(f)} \cdot \Omega/m$	1.20e−04	9.34e−5	7.55e−5	6.25e−5	4.98e−5
G, S/m	0	0	0	0	0
Z_0, Ω	270	234	209	178	143

Source: Based on Cañete Corripio, F.J., Caracterizacion y modelado de redes electricas interiorescomo medio de transmision de banda ancha, Dissertation, Universidad de Malaga, Escuela Tecnica Superior de Ingenierya de Telecomunicacion, Malaga, Spain, 2006.

Starting with the lumped element representation of the transmission line, and using Krichhoff's laws, the complex propagation constant γ can be derived from the wave equations in the frequency domain as in [60], that is,

$$\gamma = \sqrt{(R + j\omega L)(G + j\omega C)}$$

$$= \alpha + j\beta, \tag{1.5}$$

where
  ω is the angular frequency, that is, $\omega = 2\pi \times f$, and f is the frequency of interest
  Re{γ} = α is the attenuation of the wave when travelling along the line, also known as *attenuation constant*, while Im{γ} = β represents the wave's phase rotation, also called *phase constant* or *wave number*
  γ, α and β have the unit 1/m
  As β corresponds to phase rotation, its unit is sometimes expressed as rad/m

Under the condition of a *low loss line*, that is, $G \ll \omega C \wedge R \ll \omega L$, γ may be approximated as [6,60]

$$\gamma \big|_{\text{low loss app}} = j\omega\sqrt{LC}\left(1 - \frac{j}{2}\left(\frac{R}{\omega L} + \frac{G}{\omega C}\right)\right)$$

$$= j\omega\sqrt{LC} + \frac{1}{2}\left(R\sqrt{\frac{C}{L}} + G\sqrt{\frac{L}{C}}\right) \tag{1.6}$$

and thus

$$\alpha \big|_{\text{low loss app}} = \frac{1}{2}\left(R\sqrt{\frac{C}{L}} + G\sqrt{\frac{L}{C}}\right), \tag{1.7}$$

$$\beta \big|_{\text{low loss app}} = \omega\sqrt{LC}. \tag{1.8}$$

The line's characteristic impedance Z_0 is of importance as it helps to specify the reflection and transmission coefficients that occur at line discontinuities, for example, when different line segments and loads are connected. On the basis of the wave equations, the characteristic line impedance as the ratio of voltages and currents on the line can be derived as in [60], that is,

$$Z_0 = \frac{V_0^+}{I_0^+}$$

$$= \frac{-V_0^-}{I_0^-}$$

$$= \frac{R + j\omega L}{\gamma}$$

$$= \sqrt{\frac{R + j\omega L}{G + j\omega C}}, \tag{1.9}$$

where V_0^+, I_0^+, V_0^-, I_0^- are the forward and backward travelling voltages and currents on the line, respectively. Also, Z_0 can under the low loss line approximation be simplified to [60]

$$Z_0|_{\text{low loss app}} = \sqrt{\frac{L}{C}}. \tag{1.10}$$

Looking at the voltage wave in the time domain additionally allows to determine the wavelength and the phase velocity as [60]

$$\lambda = \frac{2\pi}{\beta} \tag{1.11}$$

and

$$v = \frac{\omega}{\beta}$$
$$= \lambda f, \tag{1.12}$$

which under the low loss line approximation with the help of Equation 1.8 simplifies to

$$v\big|_{\text{low loss app}} = \frac{1}{\sqrt{LC}}. \tag{1.13}$$

1.A.2 Transmission Line Transfer Function

To understand the effects that lead to frequency-selective fading, consider the open stub line example in Figure 1.13 adapted from [77]. An impedance matched transmitter is placed at A. An impedance matched receiver is placed at C. Hence, in this simple example, there is no need to bother about impedance discontinuities at the input and the output of the network. D represents a 70 Ω parallel load. B marks the point of an electrical T-junction. l_x and Z_x represent the line lengths and characteristic impedances. t_{xy} indicates the transmission and r_{xy} the reflection coefficient encountered at impedance discontinuities whose dependencies on the characteristic impedances are derived in [60]. Generally, at an impedance discontinuity from Z_a to Z_b the reflection and transmission coefficients are given by

$$r_{ab} = \frac{Z_b - Z_a}{Z_b + Z_a} \tag{1.14}$$

and

$$t_{ab} = 1 + r_{ab}. \tag{1.15}$$

Specifically for the situation in Figure 1.13, r_{1B} is given by

$$r_{1B} = \frac{(Z_2 \| Z_3) - Z_1}{(Z_2 \| Z_3) + Z_1}, \tag{1.16}$$

where $(Z_2 \| Z_3)$ represents the impedance of Z_2 and Z_3 when connected in parallel.

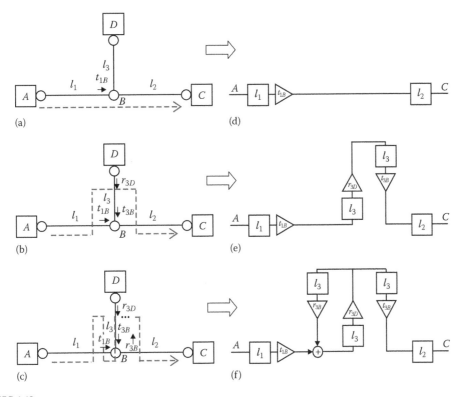

FIGURE 1.13
Relationship of wave propagation and filter elements in the stub line example. (a–c) Represent wave propagation paths and (d–f) are the respective IIR-filter elements. (From Berger, L.T. and Moreno-Rodríguez, G., *Acad. Publ. J. Commun.*, 4(1), 41. Copyright © 2009. Academy Publisher.)

A PLC signal travels in the form of a direct wave from A over B to C as displayed in Figure 1.13a. Another wave travels from A over B to D, bounces back to B and reaches C, as depicted in Figure 1.13b. All further waves travel from A to B, and undergo multiple bounces between B and D before they finally reach C (Figure 1.13c). The number of bounces between B and D is infinite, motivating the idea that an IIR filter may be used to represent the power line network. Considering the reflection and transmission coefficients as gains, and considering ideal transmission lines whose length relates only to a time delay, the simple stub line example may be transformed into IIR-filter elements as displayed in Figure 1.13d–f, where the boxes represent delays and the triangles represent filter coefficients.

The complete filter obtained for the stub line example is displayed in Figure 1.14 in its more conventional canonical form. A discrete time representation is considered. The smallest time step T_s relates to the system's sampling frequency via $f_s = 1/T_s$. Thus, every line length relates to a delay which, measured in samples, can be expressed as

$$N_{\text{delay},x} = \frac{\text{delay}_x}{T_s}$$

$$= \frac{l_x}{T_s \cdot v_x},$$ (1.17)

where l_x and v_x represent the length and the wave speed of line x, respectively ($\sim$200,000,000 m/s). The filter from Figure 1.14 may then be expressed through its z-transfer

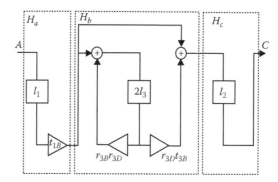

FIGURE 1.14
IIR-filter representation of stub line example. (From Berger, L.T. and Moreno-Rodríguez, G., *Acad. Publ. J. Commun.*, 4(1), 41. Copyright © 2009. Academy Publisher.)

function where a time delay is given as $z^{-\text{Round}\{l_x/T_s.v_x\}}$. Round$\{\cdot\}$ stands for rounding to the nearest integer. Instead of using this lengthy notation, the z-transform of the time delay caused by line length l_x is denoted by z^{-l_x}. The z-transfer functions of the sub-filter blocks marked in Figure 1.14 as H_a, H_b and H_c are

$$H_a = t_{1B} \cdot z^{-l_1},$$
(1.18)

$$H_b = 1 + \frac{t_{3B} \cdot r_{3D} \cdot z^{-2l_3}}{1 - r_{3B} \cdot r_{3D} \cdot z^{-2l_3}},$$
(1.19)

$$H_c = z^{-l_2}.$$
(1.20)

The overall z-transfer function of the stub line example is obtained through the concatenation of the three sub-filter blocks:

$$
\begin{aligned}
H &= H_a \cdot H_b \cdot H_c \\
&= t_{1B} \cdot z^{-l_1} \cdot \left(1 + \frac{t_{3B} \cdot r_{3D} \cdot z^{-2l_3}}{1 - r_{3B} \cdot r_{3D} \cdot z^{-2l_3}} \right) \cdot z^{-l_2} \\
&= \frac{t_{1B} \cdot z^{-(l_1+l_2)}}{1 - r_{3B} \cdot r_{3D} \cdot z^{-2l_3}} + \frac{(t_{3B} - r_{3B}) \cdot r_{3D} \cdot t_{1B} \cdot z^{-(l_1+l_2+2l_3)}}{1 - r_{3B} \cdot r_{3D} \cdot z^{-2l_3}}.
\end{aligned}
$$
(1.21)

The derived H is the *through-forward filter* of the stub line from input A to output B. Further detailed derivations of elements like a *star junction* or a general load can be found in [1]. It becomes clear that although accurate, this type of physical (bottom-up) modelling can become mathematically quite involved, especially when considering MTL lines. Nevertheless, it serves as an interesting example to understand the processes that lead to multi-path propagation and in the sequel frequency-selective fading of the PLC channel.

Acknowledgements

The work has been partially supported by the Spanish Ministry of Science and Innovation (MICINN) Program INNCORPORA-Torres Quevedo 2011. Much gratitude also goes to ETSI, the experts participating in STF 410 and all volunteers supporting the MIMO PLC field measurements.

References

1. L. T. Berger and G. Moreno-Rodríguez, Power line communication channel modelling through concatenated IIR-filter elements, *Academy Publisher Journal of Communications*, 4(1), 41–51, February 2009.
2. L. T. Berger, Broadband powerline communications, in *Convergence of Mobile and Stationary Next Generation Networks*, K. Iniewski, Ed. John Wiley & Sons, Hoboken, NJ, 2010, ch. 10, pp. 289–316.
3. L. T. Berger, Wireline communications in smart grids, in *Smart Grid – Applications, Communications and Security*, L. T. Berger and K. Iniewski, Eds. John Wiley & Sons, Hoboken, NJ, April 2012, pp. 191–230.
4. L. T. Berger, A. Schwager and J. J. Escudero-Garzás, Power line communications for smart grid applications, *Hindawi Publishing Corporation Journal of Electrical and Computer Engineering*, no. ID 712376, pp. 1–16, 2013, Received 3 August 2012; Accepted 29 December 2012, Academic Editor: Ahmed Zeddam. [Online] Available: http://www.hindawi.com/journals/jece/aip/712376/.
5. M. Zimmermann and K. Dostert, An analysis of the broadband noise scenario in power-line networks, in *International Symposium on Power Line Communications and Its Applications (ISPLC)*, Limerick, Ireland, April 2000, pp. 131–138.
6. M. Babic, M. Hagenau, K. Dostert and J. Bausch, Theoretical postulation of PLC channel model, The OPERA Consortium, IST Integrated Project Deliverable D4v2.0, March 2005.
7. E. Biglieri, Coding and modulation for a horrible channel, *IEEE Communications Magazine*, 41(5), 92–98, May 2003.
8. International Telecommunications Union (ITU), ITU radio regulations. [Online] Available: http://life.itu.int/radioclub/rr/art02.htm (accessed March 2013), 2008, Vol. 1, Article 2; Edition of 2008.
9. S. Galli, A. Scaglione and Z. Wang, For the grid and through the grid: The role of power line communications in the smart grid, *Proceedings of the IEEE*, 99(6), June 2011.
10. K. Dostert, *Powerline Communications*. Prentice Hall, Upper Saddle River, NJ, 2001.
11. G. Held, *Understanding Broadband over Power Line*. CRC Press, Boca Raton, FL, 2006.
12. P. Sobotka, R. Taylor and K. Iniewski, Broadband over power line communications: Home networking, broadband access, and smart power grids, in *Internet Networks: Wired, Wireless, and Optical Technologies, ser. Devices, Circuits, and Systems*, K. Iniewski, Ed., CRC Press, Boca Raton, FL, December 2009, ch. 8.
13. R. Pighi and R. Raheli, On multicarrier signal transmission for high voltage power lines, in *IEEE International Symposium on Power Line Communications and its Applications (ISPLC)*, Vancouver, British Columbia, Canada, April 2005.
14. D. Hyun and Y. Lee, A study on the compound communication network over the high voltage power line for distribution automation system, in *International Conference on Information Security and Assurance (ISA)*, Busan, Korea, April 2008, pp. 410–414.
15. R. Aquilu, I. G. J. Pijoan and G. Sanchez, High-voltage multicarrier spread-spectrum system field test, *IEEE Transactions on Power Delivery*, 24(3), 1112–1121, July 2009.

16. N. Strandberg and N. Sadan, HV-BPL phase 2 field test report, U.S. Department of Energy, Technical Report DOE/NETL-2009/1388, 2009, http://www.netl.doe.gov/smartgrid/referenceshelf/reports/HV-BPL_Final_Report.pdf (accessed December 2010).

17. P. Wouters, P. van der Wielen, J. Veen, P. Wagenaars and E. Steennis, Effect of cable load impedance on coupling schemes for MV power line communication, *IEEE Transactions on Power Delivery*, 20(2), 638–645, April 2005.

18. R. Benato and R. Caldon, Application of PLC for the control and the protection of future distribution networks, in *IEEE International Symposium on Power Line Communications and Its Applications (ISPLC)*, Pisa, Italy, March 2007.

19. A. Cataliotti, A. Daidone and G. Tiné, Power line communication in medium voltage systems: Characterization of MV cables, *IEEE Transactions on Power Delivery*, 23(4), 1896–1902, October 2008.

20. N. Pine and S. Choe, Modified multipath model for broadband MIMO power line communications, in *IEEE International Symposium on Power Line Communications and Its Applications*, Beijing, China, April 2012.

21. P. Meier, M. Bittner, H. Widmer, J.-L. Bermudez, A. Vukicevic, M. Rubinstein, F. Rachidi, M. Babic and J. Simon Miravalles, Pathloss as a function of frequency, distance and network topology for various LV and MV European powerline networks, The OPERA Consortium, Project Deliverable, EC/IST FP6 Project No. 507667 D5v0.9, April 2005.

22. A. Rubinstein, F. Rachidi, M. Rubinstein, A. Vukicevic, K. Sheshyekani, W. Bschelin and C. Rodríguez-Morcillo, EMC guidelines, The OPERA Consortium, IST Integrated Project Deliverable D9v1.1, October 2008, IST Integrated Project No. 026920.

23. A. Vukicevic, Electromagnetic compatibility of power line communication systems, Dissertation, École Polytechnique Fédérale de Lausanne, Lausanne, Switzerland, June 2008, no. 4094.

24. N. Gonzáez-Prelcic, C. Mosquera, N. Degara and A. Currais, A channel model for the Galician low voltage mains network, in *International Symposium on Power Line Communications (ISPLC)*, Malmö, Sweden, March 2001, pp. 365–370.

25. M. Zimmermann and K. Dostert, A multipath model for the powerline channel, *IEEE Transactions on Communications*, 50(4), 553–559, April 2002.

26. H. Liu, J. Song, B. Zhao and X. Li, Channel study for medium-voltage power networks, in *IEEE International Symposium on Power Line Communications (ISPLC)*, Orlando, FL, March 2006, pp. 245–250.

27. H. Philipps, Performance measurements of powerline channels at high frequencies, in *International Symposium in Power Line Communications*, Tokyo, Japan, March 1998, pp. 229–237.

28. D. Liu, E. Flint, B. Gaucher and Y. Kwark, Wide band AC power line characterization, *IEEE Transactions on Consumer Electronics*, 45(4), 1087–1097, 1999.

29. H. Philipps, Hausinterne Stromversorgungsnetze als übertragungswege für hochratige digitale Signale, Dissertation, Technical University Carolo-Wilhelmina zu Braunschweig, Braunschweig, Germany, 2002.

30. T. Esmailian, F. R. Kschischang and P. G. Gulak, In-building power lines as high-speed communication channels: Channel characterization and a test channel ensemble, *International Journal of Communication Systems*, 16, 381–400, 2003.

31. ETSI Technical Committee PowerLine Telecommunication (PLT), PowerLine Telecommunications (PLT); Hidden Node review and statistical analysis, Technical Report TR 102 269 V1.1.1, December 2003.

32. A. Schwager, L. Stadelmeier and M. Zumkeller, Potential of broadband power line home networking, in *Second IEEE Consumer Communications and Networking Conference*, January 2005, pp. 359–363.

33. P. Karols, K. Dostert, G. Griepentrog and S. Huettinger, Mass transit power traction networks as communication channels, *IEEE Journal on Selected Areas in Communications*, 24(7), 1339–1350, July 2006.

34. T. Huck, J. Schirmer, T. Hogenmuller and K. Dostert, Tutorial about the implementation of a vehicular high speed communication system, in *International Symposium on Power Line Communications and Its Applications (ISPLC)*, Vancouver, British Columbia, Canada, April 2005, pp. 162–166.

35. S. Galli, T. Banwell and D. Waring, Power line based LAN on board the NASA space shuttle, in *IEEE 59th Vehicular Technology Conference*, Milan, Italy, Vol. 2, May 2004, pp. 970–974.

36. J. Wolf, Power line communication (PLC) in space – Current status and outlook, in *ESA Workshop on Aerospace EMC*, Venice, Italy, 2012, pp. 1–6.

37. S. Tsuzuki, M. Yoshida and Y. Yamada, Characteristics of power-line channels in cargo ships, in *International Symposium on Power Line Communications and Its Applications (ISPLC)*, Pisa, Italy, March 2007, pp. 324–329.

38. J. Yazdani, K. Glanville and P. Clarke, Modelling, developing and implementing sub-sea power-line communications networks, in *International Symposium on Power Line Communications and Its Applications*, Vancouver, British Columbia, Canada, 2005, pp. 310–316.

39. L. T. Berger and K. Iniewski, *Smart Grid – Applications, Communications and Security*. John Wiley & Sons, New York, April 2012.

40. G. J. Foschini and M. J. Gans, On limits of wireless communications in a fading environment when using multiple antennas, *Wireless Personal Communications*, (6), 311–335, 1998.

41. L. Schumacher, L. T. Berger and J. Ramiro Moreno, Recent advances in propagation characterisation and multiple antenna processing in the 3GPP framework, in *XXVIth URSI General Assembly*, Maastricht, the Netherlands, August 2002, session C2.

42. C. R. Paul, *Analysis of Multiconductor Transmission Lines*. John Wiley & Sons, New York, 1994.

43. European Telecommunication Standards Institute (ETSI), Powerline Telecommunications (PLT); MIMO PLT; Part 1: Measurement Methods of MIMO PLT, February 2012. [Online] Available: http://www.etsi.org/deliver/etsi_tr/101500_101599/10156201/01.03.01_60/tr_10156201v010301p.pdf.

44. International Electrotechnical Commission (IEC), Power line communication system or power utility applications – Part 1: Planning of analog and digital power line carrier systems operating over EHV/HV/MV electricity grids, September 2012.

45. ETSI Technical Committee PowerLine Telecommunication (PLT), PowerLine Telecommunication (PLT); Basic data relating to LVDN measurements in the 3 MHz to 100 MHz frequency range, Technical Report TR 102 370 V1.1.1, November 2004. [Online] Available: http://www.etsi.org/deliver/etsi_tr/102300_102399/102370/01.01.01_60/.

46. European Telecommunication Standards Institute (ETSI), Powerline Telecommunications (PLT); MIMO PLT Universal Coupler, Operating Instructions – Description, May 2011. [Online] Available: http://www.etsi.org/deliver/etsi_tr/101500_101599/101562/01.01.01_60/tr_101562v010101p.pdf.

47. A. Schwager, Powerline communications: Significant technologies to become ready for integration, Dissertation, Universität Duisburg-Essen, Fakultät für Ingenieurwissenschaften, Duisburg-Essen, Germany, 2010. [Online] Available: http://duepublico.uni-duisburg-essen.de/servlets/DerivateServlet/Derivate-24381/Schwager_Andreas_Diss.pdf.

48. T. Banwell and S. Galli, A novel approach to the modeling of the indoor power line channel part I: Circuit analysis and companion model, *IEEE Transactions on Power Delivery*, 20(2), 655–663, April 2005.

49. M. Ishihara, D. Umehara and Y. Morihiro, The correlation between radiated emissions and power line network components on indoor power line communications, in *IEEE International Symposium on Power Line Communications and Its Applications*, Orlando, FL, 2006, pp. 314–318.

50. M. Tanaka, High frequency noise power spectrum, impedance and transmission loss of power line in Japan on intrabuilding power line communications, *IEEE Transactions on Consumer Electronics*, (CE-34), 321–326, May 1988.

51. S. Tsuzuki, S. Yamamoto, T. Takamatsu and Y. Yamada, Measurement of Japanese indoor power-line channel, in *Fifth International Symposium Power-Line Communications*, Malmö, Sweden, 2001, pp. 79–84.

52. J. A. Malack and J. R. Engstrom, RF impedance of United States and European power lines, *IEEE Transactions on Electromagnetic Compatibility*, EMC-18, 36–38, February 1976.

53. J. R. Nicholson and J. A. Malack, RF impedance of power lines and line impedance stabilization networks in conducted interference measurements, *IEEE Transactions on Electromagnetic Compatibility*, EMC-15, 84–86, May 1973.

54. Ministry of Internal Affairs and Communications (MIC), Report of the CISPR committee, the information and communication council, June 2006, available only in Japanese. [Online] Available: http://www.soumu.go.jp/joho_tsusin/policyreports/joho_tsusin/bunkakai/pdf/060629_3_1-2.pdf.

55. M. Hove, T. O. Sanya, A. J. Snyders, I. R. Jandrell and H. C. Ferreira, The effect of type of transient voltage suppressor on the signal response of a coupling circuit for power line communications, in *IEEE AFRICON*, Livingstone, Zambia, 2011, pp. 1–6.

56. G. R. and S. K. Das, Power line transient interference and mitigation techniques, in *IEEE INCEMIC*, Chennai, India, 2003, pp. 147–154.

57. N. Mungkung, S. Wongcharoen, C. Sukkongwari and S. Arunrungrasmi, Design of AC electronics load surge protection, *International Journal of Electrical, Computer and Systems Engineering*, 1(2), 126–131, 2007.

58. Littlefuse, Combining GDTs and MOVs for surge protection of AC power lines, application Note EC640. [Online] Available: www.littelfuse.com.

59. ETSI Technical Committee PowerLine Telecommunication (PLT), Power line telecommunications (PLT); Channel characterization and measurement methods, Technical Report TR 102 175 V1.1.1, 2003. [Online] Available: http://www.etsi.org/deliver/etsi_tr/102100_102199/102175/01.01.01_60/tr_102175v010101p.pdf.

60. D. M. Pozar, *Microwave Engineering*, 3rd edn. John Wiley & Sons, New York, 2005.

61. S. Galli, A simplified model for the indoor power line channel, in *IEEE International Symposium on Power Line Communications and Its Applications (ISPLC)*, Dresden, Germany, March 2009, pp. 13–19.

62. F. J. Cañete Corripio, L. Díez del Río and J. T. Entrambasaguas Muñoz, A time variant model for indoor power-line channels, in *International Symposium on Power Line Communications (ISPLC)*, Malmö, Sweden, March 2001, pp. 85–90.

63. F. J. Cañete, L. Díez, J. A. Cortés and J. T. Entrambasaguas, Broadband modelling of indoor power-line channels, *IEEE Transactions on Consumer Electronics*, 48(1), 175–183, February 2002.

64. T. Esmailian, F. R. Kschischang and P. G. Gulak, An in-building power line channel simulator, in *International Symposium on Power Line Communications and Its Applications (ISPLC)*, Athens, Greece, March 2002.

65. F. J. Cañete, J. A. Cortés, L. Díez and J. T. Entrambasaguas, Modeling and evaluation of the indoor power line transmission medium, *IEEE Communications Magazine*, 41(4), 41–47, April 2003.

66. S. Galli and T. Banwell, A novel approach to the modeling of the indoor power line channel – Part II: Transfer function and its properties, *IEEE Transactions on Power Delivery*, 20(3), 1869–1878, July 2005.

67. T. Sartenaer and P. Delogne, Deterministic modeling of the (shielded) outdoor power line channel based on the multiconductor transmission line equations, *IEEE Journal on Selected Areas in Communications*, 24(7), 1277–1291, July 2006.

68. H. Philipps, Development of a statistical model for powerline communication channels, in *International Symposium on Power Line Communications (ISPLC)*, Limerick, Ireland, April 2000, pp. 153–160.

69. J.-H. Lee, J.-H. Park, H.-S. Lee, G.-W. Lee and S.-C. Kim, Measurement, modelling and simulation of power line channel for indoor high-speed data communications, in *International Symposium on Power Line Communications (ISPLC)*, Malmö, Sweden, March 2001, pp. 143–148.

70. S. Barmada, A. Musolino and M. Raugi, Innovative model for time-varying power line communication channel response evaluation, *IEEE Journal on Selected Areas in Communications*, 7(24), 1317–1326, July 2006.

71. S. Galli and T. C. Banwell, A deterministic frequency-domain model for the indoor power line transfer function, *IEEE Journal on Selected Areas in Communications*, 24(7), 1304–1316, July 2006.

72. J. Anatory, N. Theethayi and R. Thottappillil, Power-line communication channel model for interconnected networks – Part I: Two-conductor system, *IEEE Transactions on Power Delivery*, 24(1), 118–123, January 2009.

73. J. Anatory, N. Theethayi and R. Thottappillil, Power-line communication channel model for interconnected networks – Part II: Multiconductor system, *IEEE Transactions on Power Delivery*, 24, 124–128, January 2009.

74. G. Moreno-Rodríguez and L. T. Berger, An IIR-filter approach to time variant PLC-channel modelling, in *IEEE International Symposium on Power Line Communications and Its Applications (ISPLC)*, Jeju, South Korea, April 2008, pp. 87–92.

75. A. M. Tonello and F. Versolatto, Bottom-up statistical PLC channel modeling – Part II: Inferring the statistics, *IEEE Transactions on Power Delivery*, 25(4), 2356–2363, October 2010.

76. A. M. Tonello and F. Versolatto, Bottom-up statistical PLC channel modeling – Part I: Random topology model and efficient transfer function computation, *IEEE Transactions on Power Delivery*, 26(2), 891–898, April 2011.

77. M. Zimmermann and K. Dostert, A multi-path signal propagation model for the power line channel in the high frequency range, in *International Symposium on Power-Line Communications and its Applications (ISPLC)*, Lancaster, U.K., April 1999, pp. 45–51.

78. M. Tlich, A. Zeddam, F. Moulin and F. Gauthier, Indoor power line communications channel characterization up to 100 MHz – Part I: One-parameter deterministic model, *IEEE Transactions on Power Delivery*, 23(3), 1392–1401, July 2008.

79. M. Tlich, A. Zeddam, F. Moulin and F. Gauthier, Indoor power line communications channel characterization up to 100 MHz – Part II: Time-frequency analysis, *IEEE Transactions on Power Delivery*, 23(3), 1402–1409, July 2008.

80. J. P. Kermoal, L. Schumacher, K. I. Pedersen, P. E. Mogensen and F. Frederiksen, A stochastic MIMO radio channel model with experimental validation, *IEEE Journal on Selected Areas in Communications*, 20, (6), 1211–1226, August 2002.

81. C. S. Cowies and J. P. Leveille, Modal transmission method and apparatus for multi-conductor wireline cables, Patent EP 0 352 869 A2, January, 1990. [Online] Available: http://worldwide. espacenet.com/espacenetDocument.pdf?flavour=trueFull&locale=es_LP&FT=D&date=19900 131&CC=EP&NR=0352869A2&KC=A2&popup=true, accessed October 2012.

82. J. Dagher, System and method for transporting high-bandwidth signals over electrically conducting transmission lines, Patent US 5 553 097 A, September 1996. [Online] Available: http:// worldwide.espacenet.com/espacenetDocument.pdf?flavour=trueFull&locale=es_LP&FT=D& date=19960903&CC=US&NR=5553097A&KC=A&popup=true, accessed October 2012.

83. D. C. Mansur, Eigen-mode encoding of signals in a data group, Patent US 6 226 330 B1, May 2001. [Online] Available: http://worldwide.espacenet.com/espacenetDocument. pdf?flavour=trueFull&locale=es_LP&FT=D&date=20010501&CC=US&NR=6226330B1&KC= B1&popup=true, accessed October 2012.

84. B. Honary, J. Yazdani and P. A. Brown, Space time coded data transmission via inductive effect between adjacent power lines, Patent GB2 383 724, December 2001. [Online] Available: http://worldwide.espacenet.com/publicationDetails/biblio?DB=EPODOC& II=0&ND=3&adjacent=true&locale=en_EP&FT=D&date=20030702&CC=GB&NR=2383724 A&KC=A, accessed October 2012.

85. C. L. Giovaneli, J. Yazdani, P. Farrell and B. Honary, Application of space-time diversity/coding for power line channels, in *International Symposium on Power Line Communications and Its Applications (ISPLC)*, Athens, Greece, March 2002.

86. L. Stadelmeier, D. Schill, A. Schwager, D. Schneider and J. Speidel, MIMO for inhome power line communications, in *Seventh International ITG Conference on Source and Channel Coding (SCC)*, Ulm, Germany, January 2008.

87. R. Hashmat, P. Pagani and T. Chonavel, MIMO capacity of inhome PLC links up to 100 MHz, in *Third Workshop on Power Line Communications*, Udine, Italy, October 2009.

88. R. Hashmat, P. Pagani and T. Chonavel, MIMO communications for inhome PLC networks: Measurements and results up to 100 MHz, in *IEEE International Symposium on Power Line Communications and Its Applications*, Rio de Janeiro, Brazil, March 2010.

89. R. Hashmat, P. Pagani, A. Zeddam and T. Chonavel, Measurement and analysis of inhome MIMO PLC channel noise, in *Fourth Workshop on Power Line Communications*, Boppard, Germany, September 2010.

90. D. Veronesi, R. Riva, P. Bisaglia, F. Osnato, K. Afkhamie, A. Nayagam, D. Rende and L. Yonge, Characterization of in-home MIMO power line channels, in *2011 IEEE International Symposium on Power Line Communications and Its Applications (ISPLC)*, April 2011, pp. 42–47.

91. D. Rende, A. Nayagam, K. Afkhamie, L. Yonge, R. Riva, D. Veronesi, F. Osnato and P. Bisaglia, Noise correlation and its effects on capacity of inhome MIMO power line channels, in *IEEE International Symposium on Power Line Communications*, Udine, Italy, April 2011, pp. 60–65.

92. D. Schneider, A. Schwager, W. Baschlin and P. Pagani, European MIMO PLC field measurements: Channel analysis, in *2012 16th IEEE International Symposium on Power Line Communications and Its Applications (ISPLC)*, Beijing, China, March 2012, pp. 304–309.

93. F. Versolatto and A. M. Tonello, MIMO PLC random channel generator and capacity analysis, in *IEEE International Symposium on Power Line Communications and Its Applications*, Udine, Italy, April 2011, pp. 66–71.

94. T. Banwell, Accurate indoor residential PLC model suitable for channel and EMC estimation, in *IEEE 6th Workshop on Signal Processing Advances in Wireless Communications*, New York, June 2005, pp. 985–990.

95. F. Versolatto and A. M. Tonello, An MTL theory approach for the simulation of MIMO power-line communication channels, *IEEE Transactions on Power Delivery*, 26(3), 1710–1717, July 2011.

96. A. Canova, N. Benvenuto and P. Bisaglia, Receivers for MIMO-PLC channels: Throughput comparison, in *IEEE International Symposium on Power Line Communications and Its Applications*, Rio, Brazil, March 2010.

97. R. Hashmat, P. Pagani, A. Zeddam and T. Chonave, A channel model for multiple input multiple output in-home power line networks, in *2011 IEEE International Symposium on Power Line Communications and Its Applications (ISPLC)*, Udine, Italy, April 2011, pp. 35–41.

98. A. M. Tonello, Wideband impulse modulation and receiver algorithms for multiuser power line communications, *EURASIP Journal on Advances in Signal Processing*, 1–14, 2007.

99. M. Tlich, P. Pagani, G. Avril, F. Gauthier, A. Zeddam, A. Kartit, O. Isson et al., PLC channel characterization and modelling, OMEGA, European Union Project Deliverable D3.2 v.1.2 IST Integrated Project No ICT-213311, February 2011. [Online] Available: http://www.ict-omega.eu/publications/deliverables.html (accessed April 2013).

100. A. Tomasoni, R. Riva and S. Bellini, Spatial correlation analysis and model for in-home MIMO power line channels, in *IEEE International Symposium on Power Line Communications and Its Applications*, Beijing, China, April 2012.

101. M. Zimmermann and K. Dostert, Analysis and modeling of impulsive noise in broad-band powerline communications, *IEEE Transactions on Electromagnetic Compatibility*, 44(1), 249–258, February 2002.

102. V. Degardin, M. Lienard, A. Zeddam, F. Gauthier and P. Degauque, Classification and characterization of impulsive noise on indoor power lines used for data communications, *IEEE Transactions on Consumer Electronics*, 48(4), 913–918, November 2002.

103. Y.-T. Ma, K.-H. Liu, Z.-J. Zhang, J.-X. Yu and X.-L. Gong, Modeling the colored background noise of power line communication channel based on artificial neural network, in *Wireless and Optical Communications Conference*, Shanghai, China, May 2010.

104. J. A. Cortés, L. Díez, F. J. Cañete and J. J. Sánchez-Martínez, Analysis of the indoor broadband power line noise scenario, *IEEE Transactions on Electromagnetic Compatibility*, 52(4), 849–858, November 2010.

105. P. Pagani, R. Hashmat, A. Schwager, D. Schneider and W. Baschlin, European MIMO PLC field measurements: Noise analysis, in *2012 16th IEEE International Symposium on Power Line Communications and Its Applications (ISPLC)*, Beijing, China, March 2012, pp. 310–315.

106. R. Hashmat, P. Pagani, T. Chonavel and A. Zeddam, Analysis and modeling of background noise for inhome MIMO PLC channels, in *IEEE International Symposium on Power Line Communications and its Applications*, Beijing, China, March 2012.
107. R. Hashmat, P. Pagani, T. Chonavel and A. Zeddam, A time domain model of background noise for inhome MIMO PLC networks, *IEEE Transactions on Power Delivery*, 27(4), 2082–2089, October 2012.
108. F. J. Cañete Corripio, Caracterizacion y modelado de redes electricas interiorescomo medio de transmision de banda ancha, Dissertation, Universidad de Malaga, Escuela Tecnica Superior de Ingenierya de Telecomunicacion, Malaga, Spain, 2006.
109. T. Sartenaer, Multiuser communications over frequency selective wired channels and applications to the powerline access network, PhD dissertation, Faculty of Applied Sciences of the Université Catholique de Louvain, Louvain-la-Neuve, Belgium, September 2004.
110. Ansoft Corporation, Maxwell 2D, web page, accessed April 2013. [Online] Available: http://www.ansys.com/Products/Simulation+Technology/Electromagnetics/Electromechanical+Design/ANSYS+Maxwell.
111. T. Hubeny, Line parameters simulator of symmetric lines, web page, http://matlab.feld.cvut.cz/en/view.php?cisloclanku = 2006011801 (accessed April 2013).
112. T. Bostoen and O. Van de Wiel, Modelling the low-voltage power distribution network in the frequency band from 0.5 MHz to 30 MHz for broadband powerline communications (PLC), in *International Zurich Seminar on Broadband Communications*, Zurich, Switzerland, 2000, pp. 171–178.

2

Narrowband Characterisation in an Office Environment

Klaus Dostert, Martin Sigle and Wenqing Liu

CONTENTS

Reliable communication plays a key role in Smart Grid applications like value-added services, distribution automation, advanced meter reading, load control and remote diagnostics. PLC applications in the frequency range up to 500 kHz, the so-called narrowband PLC (NB-PLC), are becoming more and more popular. NB-PLC is frequently used within Smart Grids due to its relatively large coverage. Unfortunately, the NB power line channel exhibits highly dynamic unpredictable and irreproducible characteristics. Furthermore, it is featured by frequency-selective attenuation and pretty complex noise scenarios. These channel conditions make even low-speed data transmission quite challenging.

This chapter will provide an approach for determining the link quality in a low-voltage grid under the influence of the time-variant features of the transmission channel. Beyond noise scenario and channel *transfer function* (TF), there will be a focus on the access impedance of the low-voltage grid as a crucial factor. Analysis methods as well as implementation aspects will be discussed. Typical results obtained from measurements on the low-voltage grid of a small university campus will be presented. The topology of the low-voltage grid

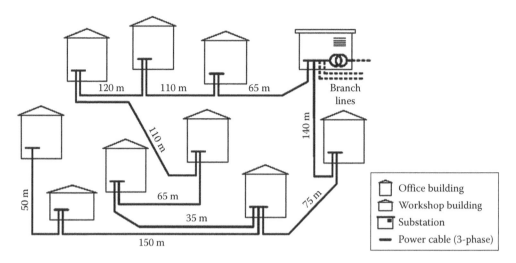

120 m   110 m   65 m   Branch
                                  lines

110 m

140 m

50 m   65 m   75 m

35 m

150 m

Office building
Workshop building
Substation
Power cable (3-phase)

FIGURE 2.1
Topology of the low-voltage grid.

is depicted in Figure 2.1. It consists of a 630 kVA transformer that feeds several office build-ings via underground cables. The connections between the buildings and its length are indicated in the figure. All cables consist of four wires (three phases plus neutral), some of them are additionally shielded. There are two other cables connected to the busbar of the substation feeding two other buildings via branch lines, which are not illustrated in the diagram.

At first, a model for NB-PLC channels is given and a measurement methodology for obtaining the channel characteristics is described. In Sections 2.3 through 2.6, results of the evaluation of the essential channel properties, noise scenario, attenuation, *signal-to-noise ratio* (SNR) and access impedance are presented.

2.1 Modelling NB-PLC Channels

In order to describe the power line channel between a PLC *transmitter* (Tx) and a PLC *receiver* (Rx) within power line networks, we employ an expanded one-way linear time-varying model. As shown in Figure 2.2, it consists of mains access impedance, a linear filter and an additive noise source. t and f indicate the time and the frequency domains, respectively. It is used to describe the time-variant feature. The time-varying access impedance $\underline{Z}_A(t,f)$ is the equivalent impedance seen into a power outlet. Together with the equivalent time-invariant output impedance $\underline{Z}_T(f)$ of the transmitter, it determines the signal level that can be injected into the power line network. Let $x(t)$ and $s_A(t)$ denote the original and the injected signals, respectively. $X(t,f)$ and $S_A(t,f)$ are their spectral components, respectively. The signal level ratio of $s_A(t)$ to $x(t)$ can be obtained by

$$H_A(t,f) = \frac{S_A(t,f)}{X(t,f)} = \frac{\underline{Z}_A(t,f)}{\underline{Z}_A(t,f) + \underline{Z}_T(f)}. \tag{2.1}$$

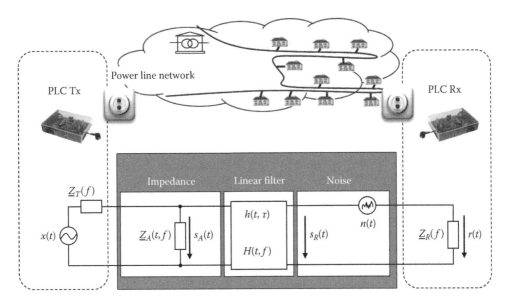

FIGURE 2.2
Equivalent electrical channel model.

$H_A(t,f)$ is usually frequency selective. The magnitude of $H_A(t,f)$ is usually smaller than 1 at most frequencies. This frequency-selective attenuation is caused by an impedance mismatching and called coupling loss [1]. Many efforts have been made to compensate for the coupling loss. For example, adaptive algorithms and special coupling circuitry have been developed to match the impedance during the transmission [2]. The coupling loss can also be reduced by decreasing $\underline{Z}_T(f)$. For systems with these features, the influence of $\underline{Z}_A(t,f)$ can be neglected. The received signal $s_R(t)$ is usually the attenuated and distorted counterpart of $s_A(t)$. For a linear system, the relation between $s_R(t)$ and $s_A(t)$ can be described by a TF $H(t,f)$ in the frequency domain or by an impulse response $h(t,\tau)$ in the time domain. Generally speaking, $h(t,\tau)$ gives the response at time t caused by an impulse excitation applied at time $t-\tau$. t and τ are also called the time of the observation instant and the time of excitation, respectively [3]. $s_R(t)$ is related to $s_A(t)$ by

$$s_R(t) = \int_{-\infty}^{\infty} h(t,\tau) \cdot s_A(t-\tau)d\tau. \tag{2.2}$$

$h(t,\tau)$ and $H(t,f)$ fulfil the relation:

$$H(t,f) = \int_{-\infty}^{\infty} h(t,\tau) \cdot e^{-j2\pi f\tau} d\tau. \tag{2.3}$$

The additive noise $n(t)$ is a collection of interference seen at the receiver. With respect to the time variation, it is believed that the equivalent input impedance of many connected electrical appliances vary periodically and synchronous to mains frequency or to its harmonics. This phenomenon causes periodic frequency-selective fading both for transmit signals and interferences [4]. At the same time, many appliances themselves generate noise synchronously to the mains voltage [5,6].

From a signal processing point of view, the received signal $r(t)$ is obtained by

$$r(t) = x(t) \cdot h_A(t, \tau) \cdot h(t, \tau) + n(t) \tag{2.4}$$

in the time domain or

$$R(t, f) = X(t, f) \cdot H_A(t, f) \cdot H(t, f) + N(t, f) \tag{2.5}$$

in the frequency domain. The channel TF $H(t,f)$ describes the ratio (magnitude and phase) between the voltage level seen at the receiver and the voltage level injected into the grid (at a certain frequency and at a certain instant of time). In general, the channel TF is not symmetric for NB-PLC channels.

2.2 Measurement of Channel Parameters

The main emphases of this section are practical methods for the determination of noise, attenuation and impedance. Applicability for the variety of scenarios observed in reality is a key point for the selected approaches. Representative results will be presented and discussed.

Noise and access impedance have a cyclostationary behaviour and therefore the channel TF and the SNR are cyclostationary too. For this reason, the measurements applied were synchronised with the mains zero crossings. Due to the time variance of the channel, all parameters have to be determined simultaneously or at least within a short time. Multiple distributed measurement devices are involved in the procedure so that an overall picture of the whole grid can be obtained at one time. The following steps are performed periodically:

- *Investigation of the idle channel*: At the beginning of every cycle, the noise is recorded.
- *Transmission/capturing of a test sequence*: Every device (one after another) transmits a test sequence in a predefined order and synchronised with the mains. The sequence consists of sinusoidal signals at discrete frequencies.

The analysis of the captured data is done offline after the measurement. Every mains period is segmented into 20 intervals and statistics of selected parameters are calculated. Analysis of these statistics yields the short-time behaviour (within one measurement cycle) as well as the long-time behaviour of the channel.

In order to obtain all relevant parameters, a systematic measurement has to be carried out. Three highly flexible measurement and communication platforms [7] have been deployed for investigating low-voltage grids.

Each platform allows coupling arbitrary test signals into the power line as well as receiving and capturing signals. The received signal is streamed in realtime to a PC via USB interface. Therefore, signals of arbitrary length can be captured. For transmitting and receiving, a sample rate of 1.333 MSamples/s is used, which allows observations in the low-frequency

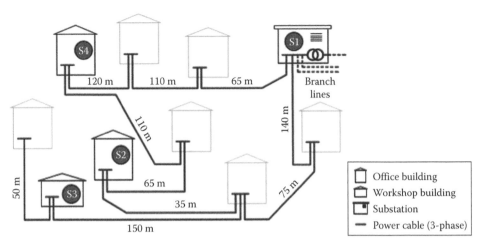

FIGURE 2.3
Measurement locations.

range up to 500 kHz. The platform is synchronised with mains frequency. In addition, the device can also be configured as an *orthogonal frequency-division multiplexing* (OFDM) modem for data transmissions. The measurement campaign is divided into multiple measurement cycles. Totally, three steps are carried out during each cycle. In the first step, all three platforms record the local noise at the same time for 30 s. The second step measures the TFs from one location to another. The channel transfer characteristics feature usually no symmetry (see Section 2.4). Thus, each link has to be measured for both directions. As a result, there are totally six directions to investigate using three measurement platforms. One of the three platforms is allowed to transmit sounding signals while the other two must record the distorted signals for the estimation of channel distortions. In this way, always two directions are investigated at the same time. When one platform finishes the transmission, it switches into the receiving mode, and another platform starts to transmit sounding signals. This process continues until the attenuation values are captured for all six directions.

Several measurement campaigns have been carried out in a small low-voltage grid. The measurement results presented in this chapter are mainly based on two measurement campaigns: a 24 h measurement comprising three locations marked as S1, S2 and S3 in Figure 2.3 and another campaign also comprising three locations marked as S1, S3 and S4 in Figure 2.3. In addition, impedance measurements have also been taken in a suburban and a rural low-voltage grid for the investigation of the access impedance in Section 2.6.

2.3 Noise Scenario

Noise plays a significant role in the channel impairment. On the one hand, the transmit signal is largely attenuated; thus, the communication quality is mostly determined by the different noise scenarios seen at the receiver. On the other hand, the noise scenario deviates from the traditional AWGN model. It has more complicated spectral characteristics and features time-varying behaviours. It is necessary to analyse the noise in detail and deploy realistic noise scenarios to evaluate NB-PLC systems. This part describes the analysis of the noise scenario recorded during the measurement campaigns. At first an overview of the behaviour of the noise is given. This includes the short-time behaviour within one mains period as well as the long-time characteristics within hours or days. In the subsequent sections, a detailed analysis of the different kinds of noise is done. The typical noise scenario is divided into three classes: narrowband interferers, impulsive noise and coloured background noise. In addition, a new subclass in the narrowband noise class will be discussed. Interferers in this subclass have each a swept frequency. Due to this spectral feature, it is also named *swept-frequency noise* (SFN) in the following.

2.3.1 Overview

For the in-depth analysis, N_c mains cycles of the noise are captured and each mains cycle is segmented into 20 intervals (Figure 2.4), which leads to a time duration of about 1 ms (in a 50 Hz environment) and a length of L samples (about 1333 samples at 1.333 MSamples/s). Depending on the actual period length of the mains cycle, this value may vary slightly.

At first, the total noise power of each mains cycle segment is calculated (for each of the N_c mains cycles). For each of the intervals, the power spectral density is calculated by using the spectral estimator

$$\hat{S}_{n,k,i} = DFT\{\tilde{x}_{n,i,l} w_L\}, \tag{2.6}$$

where

$\tilde{x}_{n,i,l}$, $n = 1, \ldots, N_c$, $l = 1, \ldots, L$, $i = 1, \ldots, 20$ denotes the lth sample value within the ith interval and the nth mains cycle

w_L is a Blackman–Harris window with a length equal to the length of one interval

Mean value and variance in both time and frequency domains for every mains cycle interval are evaluated.

Due to the varying impedance of the low-voltage grid at frequencies up to 500 kHz, not power but voltage amplitude has been examined and therefore the results are indicated in dBV^2 or dBV^2/Hz, respectively.

Figure 2.5a shows a result of the typical noise observed close to the consumer unit, that is, at the location of the electricity meter, of a building. Statistics of the total noise signal power within a mains cycle interval are shown in the upper part.

The signal energy calculated for intervals of 1 ms per mains cycle varies strongly within one mains cycle, but is periodic with 10 ms. Between different intervals within one mains cycle,

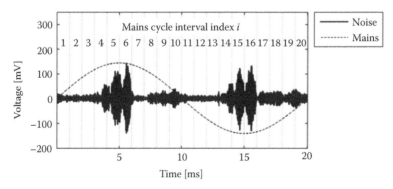

FIGURE 2.4
Mains cycle segmentation.

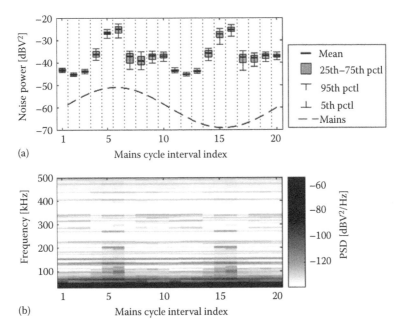

FIGURE 2.5
Received noise power over mains cycle interval. (a) Statistics of overall noise power and (b) noise power distribution over frequency (mean value).

deviations of >15 dB can be observed. The noise energy is mainly concentrated at the maximum and the minimum of the mains voltage.

In Figure 2.5b, the frequency spectrum of one mains cycle is shown. The noise power is concentrated to its largest extent on frequencies below 70 kHz. The noise bandwidth exceeds this frequency band at the same time intervals as the noise power per cycle part increases in the time domain. This is due to the mostly impulse-like characteristics of the time signal at these instants of time. Although the noise bandwidth is increased, however, most of the additional power introduced by impulses is still located below 70 kHz. Furthermore, narrowband noise components become apparent at various frequencies. The respective powers of these narrowband interferers vary only slightly over in-cycle time and in general strongly depend on the particular location of the measurement.

For the long-term analysis, we concatenate the courses of the median noise signal power over the mains cycle obtained from the short-term analyses. Figure 2.6 depicts the results for a 24 h measurement. The characteristic pattern of the noise power over the mains cycle remains almost unchanged over several hours. A clear distinction can be made between daytime and night-time. Noise power reduces significantly during night-time.

In summary of the overview of the noise scenario, the median noise signals are found to be approximately cyclostationary for relatively long periods of time. The course of the median noise power within a mains cycle shows characteristic patterns that remain constant for interval lengths in the order of hours. Depending on the instant of time within the mains cycle, the noise power distribution (noise power spectral density) over frequency varies in accordance with the noise power over time, since noise power maxima in the time domain coincide temporally with increased noise bandwidth.

In the following sections, a detailed analysis of the different noise classes is done.

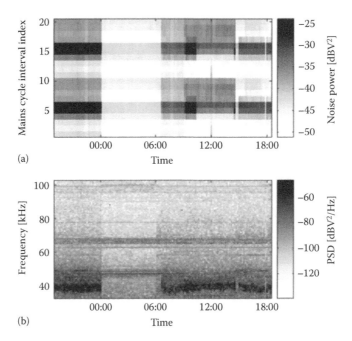

FIGURE 2.6
Noise power distribution over 24 h: (a) overall noise power within each mains cycle interval and (b) noise power distribution over frequency (up to 100 kHz).

2.3.2 Narrowband Interferers

This interference class is characterised by a significant noise level in the frequency domain compared to the background noise [1]. Disturbances at frequencies such as 25, 30, 49, 55, 75 and 82 kHz have been measured. They could be probably caused by switched power supplies. It has also been reported that the narrowband noise occurs mostly at frequencies below 140 kHz or above 410 kHz. The average bandwidth is about 3 kHz [8]. The narrowband interferers with time-invariant amplitude level have been well investigated. Modelling and emulating this kind of noise is also straightforward. Due to the time-varying nature of the power line network, the envelopes of a number of narrowband interferers also exhibit dynamics. The following parts introduce an approach for the estimation of the time-varying envelopes. Finally, a simplified model will be proposed for describing these envelopes.

2.3.2.1 Estimation of the Noise Powers

Figure 2.7a shows a segment of a noise acquired in a university laboratory. This segment lasts for 40 ms. The noise waveform in the time domain is dominated by impulsive disturbances. Furthermore, the overall envelope of the non-impulsive components changes with time. As shown in the spectrogram of the *short-time Fourier transform* (STFT) in plot (Figure 2.7b), there is a significant spectral component at around 64 kHz with a bandwidth of 4 kHz. The spectral density changes periodically. The local maxima are synchronised to the peak of the mains voltage. It is a good example of the narrowband interference with the cyclostationary features. In order to estimate the time dependence of its envelope, the influence of the other noise types such as the coloured background noise and the impulsive noise should be reduced as much as possible.

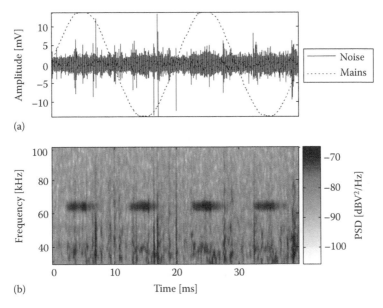

(a)

(b)

FIGURE 2.7
Narrowband noise with cyclostationary feature: (a) waveforms of noise and scaled mains in the time domain and (b) STFT spectrogram of noise (Blackman window, window length: 500 µs, overlap ratio: 83.33%). Both (a) and (b) share the same abscissa.

Figure 2.8 shows the STFT of the noise segment from another point of view. b_M denotes the frequency band that contains the most significant noise power. b_L and b_R are the direct neighbouring bands to the left and right of b_M, respectively. b_L and b_R have the same bandwidths as b_M. Gaps are inserted between b_L and b_M as well as between b_M and b_R so that the tails of each band will not interfere the other two bands.

The PSD of the background noise is relatively small compared to that of the narrowband noise. Therefore, for simplicity it is assumed that the background noise has the same noise power spectral density in all three bands. The spectra of short impulses usually exhibit wideband character. The PSD values are decreasing over frequency. Since these three bands are relatively short in comparison with the bandwidth of the impulsive noise, and they are close to each other, it is assumed that the power of the impulsive noise is

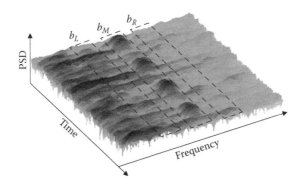

FIGURE 2.8
STFT of noise segment. b_M is the middle frequency band. b_L and b_R are the left and right neighbouring bands, respectively.

a linear function of the frequency. Therefore, the power falling into b_M is approximated by the mean value of the noise powers in b_L and b_R for this impulsive noise:

$$P_M(t)\big|_{\mathrm{dB}} = \frac{P_L(t)\big|_{\mathrm{dB}} + P_R(t)\big|_{\mathrm{dB}}}{2},$$

(2.7)

where $P_M(t)|_{\mathrm{dB}}$, $P_L(t)|_{\mathrm{dB}}$ and $P_R(t)|_{\mathrm{dB}}$ denote the power values of impulsive noise in logarithmic scale in b_M, b_L and b_R, respectively. The linear-scale noise power in the middle band can be estimated using

$$P_M = 10^{\left\lfloor \log_{10}(P_L) + \log_{10}(P_R) \right\rfloor / 2}.$$

(2.8)

2.3.2.2 Reconstruction of Noise Waveforms

Figure 2.9 shows a flow chart for the steps to reconstruct the narrowband interferer $n_{nbn}(t)$ in the time domain. The basic idea is to estimate the envelope $A_M(t)$ and the oscillation waveform $\check{n}_M(t)$ separately and then modulate $\check{n}_M(t)$ using $A_M(t)$. Three FIR band-pass filters are used to obtain the waveform of the noise components located in each band. f_M, f_L and f_R denote the middle frequencies of b_M, b_L and b_R, respectively. Each filter is followed by a square operator and a low-pass filter with cut-off frequency f_E. These two components are used to estimate the time-variant envelope of the total noise power falling in each frequency band. $P_{total}(t)$, $P_L(t)$ and $P_R(t)$ denote the envelopes for the middle, the left and the right band, respectively. Figure 2.10a shows an example of the estimated envelopes for $P_L(t)$, $P_{total}(t)$ and $P_R(t)$, respectively. The zones A_1 to A_5 cover the impulses that superimpose the narrowband interferer, while B_1 and B_2 cover the impulses appearing in the intervals between two narrowband interferers. $n_M(t)$ in Figure 2.10b is the filtered waveform corresponding to the narrowband interferer superimposed by impulsive and background noise. $P_L(t)$ and $P_R(t)$ are used to estimate the non-narrowband noise power $P_M(t)$ for the middle band.

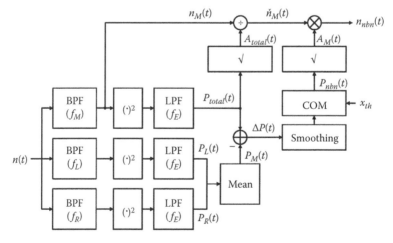

FIGURE 2.9
Flow chart for detecting and extracting a single-frequency narrowband interferer.

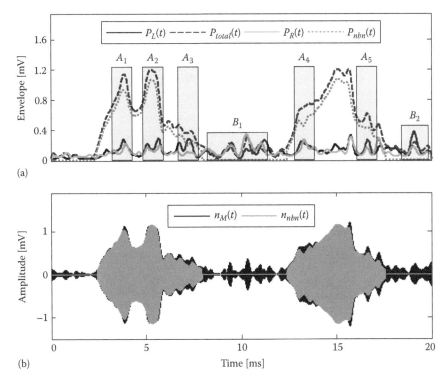

FIGURE 2.10
(a) Estimating envelopes for all three sub-bands and (b) extracting the envelope of the narrowband interferer. Both (a) and (b) share the same abscissa.

In the next step, $P_M(t)$ is subtracted from $P_{total}(t)$ and the difference $\Delta P(t)$ contains mainly the narrowband interferer. The error caused by the background noise can be further reduced by smoothing $\Delta P(t)$ properly. A comparator compares the smoothed $\Delta P(t)$ with a threshold x_{th}. A value greater than x_{th} indicates a valid envelope for the narrowband noise; otherwise, it is considered to be invalid and is forced to be zero. In this way, the power envelope of the narrowband interferer $P_{nbn}(t)$ can be estimated. Its square root $A_M(t)$ is the expected envelope of $n_{nbn}(t)$.

The key aspect in the extraction of $\check{n}_M(t)$ is to keep $\check{n}_M(t)$ in phase with $n_M(t)$. The first idea is to estimate the phase of $n_M(t)$ and use it to synthesise a sinusoidal waveform. The phase estimation algorithm can be very simple if the frequency does not change over time; otherwise, a sophisticated frequency tracking strategy must be implemented. An alternative method is to estimate and compensate the fluctuation in the envelope of $n_M(t)$. The spectral feature is not affected in this way; therefore, it can also be applied even if $n_M(t)$ exhibits a time-varying frequency. This method is implemented by simply dividing $n_M(t)$ by $A_{total}(t)$, which denotes the square root of $P_{total}(t)$.

In the last step, $n_{nbn}(t)$ is obtained by multiplying $\check{n}_M(t)$ by $A_M(t)$. After having estimated the envelope and synthesised the narrowband interferer, $n_{nbn}(t)$ is removed from the original noise $n(t)$ in the time domain. Figure 2.11 shows the remaining noise waveform in Figure 2.11a and the spectrogram of the STFT in Figure 2.11b. The investigated narrowband interferer disappears from the noise segment, while the background noise and the impulsive noise are not affected in comparison with those in Figure 2.7.

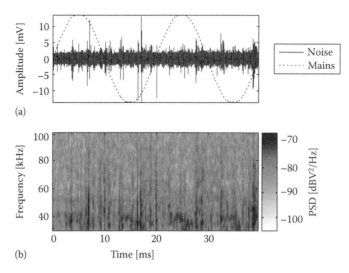

FIGURE 2.11
Waveform and STFT of residual noise; the cyclostationary narrowband disturbance has been removed.
(a) Waveforms of modified noise and scaled mains in the time domain and (b) STFT of modified noise (Blackman window, window length: 500 μs, overlap ratio: 83.33%). Both (a) and (b) share the same abscissa.

2.3.2.3 Modelling Noise Envelope

The individual envelope can be modelled by one or more unsymmetrical triangular functions. The shape and the location of the normalised peak is determined by

$$y(t) = \begin{cases} 0, & t \le t_1, \\ \dfrac{t - t_1}{t_2 - t_1}, & t_1 \le t < t_2, \\ \dfrac{t_3 - t}{t_3 - t_2}, & t_2 \le t < t_3, \\ 0, & t_3 \le t, \end{cases} \tag{2.9}$$

where t_1, t_2 and t_3 are time points for the beginning, peak and end of the triangular curve, respectively. Figure 2.12 shows a normalised measured envelope and the reconstructed artificial envelope in plot (Figure 2.12a). The simulation is a sum of three fundamental shapes $y_1(t)$, $y_2(t)$ and $y_3(t)$.

2.3.3 Swept-Frequency Noise

In addition to the typical narrowband interferers, a subclass of interferers with time-varying frequencies has been observed in both indoor channel and the access domain. Due to the feature in the frequency domain, it is called SFN in the following parts.

2.3.3.1 Typical Waveform and STFT

Figures 2.13 through 2.15 show some measured SFNs. A measurement was made in a university laboratory where several PCs, fluorescent lamps, an *uninterruptible power supply* (UPS),

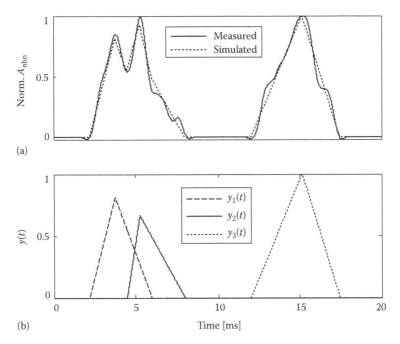

FIGURE 2.12
Measured and simulated envelope of narrowband interferers.

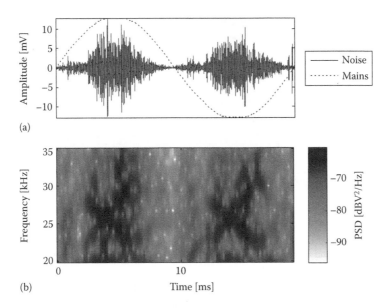

FIGURE 2.13
Measured periodic noise with rising and falling swept frequencies at the same time: (a) waveform in time domain; the envelope fluctuates periodically and synchronised by mains voltage and (b) spectrogram of STFT applied to the waveform.

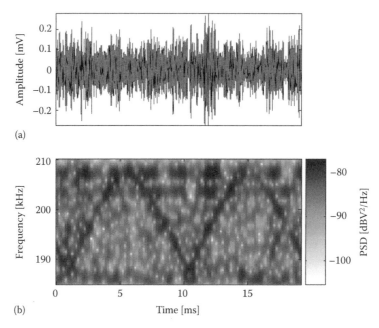

FIGURE 2.14
Measured periodic noise with sequential rising and falling swept frequencies: (a) waveform in time domain and (b) STFT.

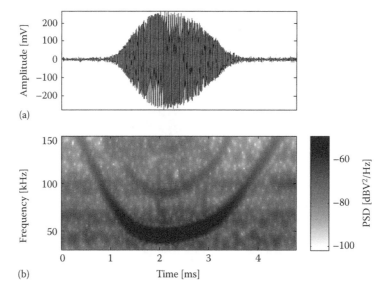

FIGURE 2.15
Measured noise with rising and falling swept frequencies at substation: (a) waveform in time domain and (b) STFT.

a printer and several measurement devices form the appliance scenario. The noise waveform shown in Figure 2.13a is filtered by a band-pass FIR filter with a passband between 22 and 35 kHz. The scaled mains voltage is shown in the same plot to illustrate the synchronisation of the noise envelope with the mains frequency. Obviously, the noise envelope reaches the maximum at the same time as the mains voltage reaches the absolute peak value. The spectrogram of the STFT in Figure 2.13b shows two intersecting periodic traces. The first trace starts with 22 kHz at about 2 ms and increases to 32 kHz linearly within 5 ms. The second frequency trace decreases from 30 to 22 kHz in the same time interval.

Figure 2.14 gives another example obtained at a transformer substation. Again, two periodic frequency traces can be observed. Similar to those traces in Figure 2.13, the frequencies here also change linearly and periodically. Nevertheless, the frequencies sweep in a much higher range, between 185 and 210 kHz. There is no overlapped area, and these two traces appear one after the other. In addition, the noise level does not show a significant fluctuation over time.

Figure 2.15 shows a third example. The noise level is much higher than the first two examples. The waveform looks like a damped oscillation with duration of about 2 ms. The spectrogram of the STFT illustrates a more complicated pattern. Obviously, the frequency does not change linearly over time. Instead, two convex-shaped traces can be observed. The longer trace starts with 140 kHz and decreases to 45 kHz. At the same time, the PSD increases to its maximum at 45 kHz. In the second part, the frequency increases until it reaches 140 kHz again. The second trace seems to be the first harmonic.

This waveform appears as a single noise event in our measurement. However, similar patterns can also occur periodically with a period of 10 ms [9] has reported multiple periodic noise of this kind and has named them recurrent oscillations. An example is shown in Figure 2.16 for a quick comparison. The noise has been recorded in the direct vicinity of a fluorescent lamp after this lamp had been turned on. The filtered noise waveform reaches almost 2 V. The individual oscillations have quite similar envelope as the one shown in Figure 2.15a. The spectral patterns between 30 and 140 kHz in the spectrogram also match the convex shape shown in Figure 2.15b very well.

All examples have some points in common. In the frequency domain, their instantaneous frequencies have small bandwidths but change with time, either linearly as shown in Figures 2.13 and 2.14, or nonlinearly such as the pattern in Figures 2.15 and 2.16.

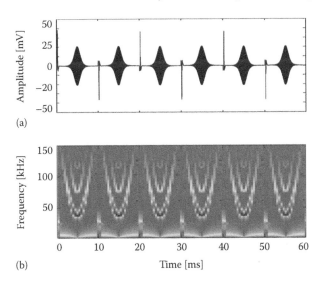

(a)

(b)

Time [ms]

FIGURE 2.16
Periodic damped oscillation reported in [9]: (a) periodic waveform in time domain and (b) spectrogram of STFT, the scale of the colour map is ignored and only the spectral patterns are shown for simplicity. Both plots share the same abscissa.

The sweeping bandwidth can range from tens to hundreds of kHz. Furthermore, almost all NB frequency bands could be disturbed. In the time domain, their envelopes can be periodic and synchronous to the mains frequency, or relatively constant. They can even be aperiodic and appear as individual lobes with high noise levels.

2.3.3.2 Origin of SFN

One main source of this noise class are active *power factor correction* (PFC) circuits in power supply units of many end-user appliances, such as fluorescent lamps and PCs. Figure 2.17 shows a simplified circuit of a *switch mode power supply* (SMPS) with an inserted active PFC module.

In an SMPS without the PFC circuit, the input capacitor C is placed directly behind the rectifier diodes D_1 through D_4. Current i_M is drawn from mains to charge C. As shown in Figure 2.18, C will only be charged when the rectified voltage u_1 exceeds the voltage u_C across C. As a result, i_M has large spikes during the charging of the capacitor and is zero otherwise. This kind of waveform contains large amount of harmonics. At the same time,

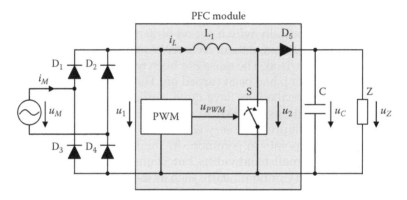

FIGURE 2.17
Simplified circuit of PFC boost pre-regulator. (From Fairchild semiconductor, Power factor correction (PFC) basis, Application note 42047, Rev.0.9.0, August 19, 2004.)

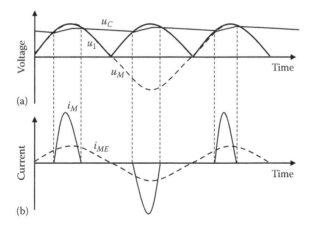

FIGURE 2.18
Voltage and current waveforms in an SMPS without any PFC module [11]: (a) voltage waveforms and (b) current waveforms, i_{ME} is the expected current waveform for a unity power factor.

the power factor – defined as the ratio of the real power to the apparent power – is very low. A low power factor burdens the power utilities because they have to deliver more power than necessary. Meanwhile, the harmonic distortions degrade the power quality and causes *electromagnetic compatibility* (EMC) problems.

The active PFC module is used to keep the current in phase with the mains voltage and to minimise the input current distortion so that the power factor can be raised. As shown in Figure 2.17, the PFC circuit is actually a boost converter that is mainly composed of an inductor (L), a *pulse-width modulation* (PWM) and a power MOSFET as a *switch* (S). The boost converter will be able to provide a higher voltage u_2 than the peak value of u_1 at its output. Simultaneously, i_L must be well controlled so that i_M is proportional to u_M at any given instant. Implementation details of the control unit can be found in [10,11]. Figure 2.19 shows waveforms of i_M, i_L and the PWM signal u_{PWM} when the PFC operates in continuous mode. Suppose the inductor L is uncharged initially. When the switch closes at t_0, u_{PWM} becomes logic high. The inductor current i_L increases linearly:

$$i_L(t) = i_{MIN}(t_0) + \frac{1}{L} \cdot u_L \cdot (t - t_0), \qquad t_0 < t \leq t_{ON} + t_0, \tag{2.10}$$

where u_L is the voltage across L and can be approximated as a constant within t_{ON}:

$$u_L = u_1, \qquad t_0 < t \leq t_{ON} + t_0. \tag{2.11}$$

The switch opens (low level of u_{PWM} within time window t_{OFF}) as soon as $i_L(t)$ reaches $i_{MAX}(t)$ at t_1 and the inductor starts to discharge:

$$i_L(t) = i_{MAX}(t_1) + \frac{1}{L} \cdot u_L \cdot (t - t_1), \qquad t_1 < t \leq t_{OFF} + t_1, \tag{2.12}$$

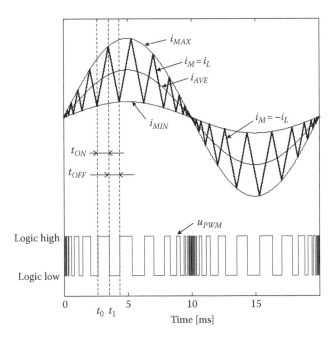

FIGURE 2.19
Waveforms of current and PWM control signals generated by PFC circuit in continuous mode of operation.

where u_L is now the voltage difference between u_1 and u_C:

$$u_L = u_1 - u_C, \quad t_1 < t \leq t_{OFF} + t_1. \tag{2.13}$$

Due to the feature of the boost converter, u_C will be greater than u_1; thus, u_L is now a negative value. Therefore, i_L maintains its direction and decreases linearly. As soon as i_L drops to i_{MIN}, the switch closes again. In this way, the current drawn from mains i_M is kept within the area defined by i_{MAX} and i_{MIN}. Its average value i_{AVE} follows u_m and therefore is in phase with the mains voltage. Depending on the value of $i_{MIN}(t)$, the PFC module can operate in either discontinuous or continuous modes. In the first mode, $i_{MIN}(t)$ is zero over the entire mains cycle. i_L can reach zero, and the current waveform swings between 0 and i_{MAX}. If $i_{MIN}(t)$ is greater than zero and is synchronised with $i_{MAX}(t)$, such as the one in Figure 2.19, i_L can never reach zero during the switching cycle.

The HF components in i_M can be converted to voltage by any connected impedance. Although many SMPS have EMI filters inserted between their rectifier bridges and the power plugs, these filters are usually less effective to reduce the differential-mode noise for the frequency range up to 150 kHz. Therefore, most noise spectral components can still appear in mains and can be coupled into NB-PLC systems. Figure 2.20a shows a band-pass filtered waveform of i_M. The spectrogram of the STFT is shown in Figure 2.20b. Similarities can be observed in both the noise waveform and the spectral characteristics between the synthesised noise and the measured noise shown in Figures 2.15 and 2.16.

More and more active PFC modules are being applied to reduce the harmonics emitted by end-user devices and to comply with international standards such as IEC 61000-3-2. Furthermore, active PFC circuits are the most favourable solutions to limit harmonics for lighting equipment with HF-ballast [9]. Therefore, the influence of the SFN on the system

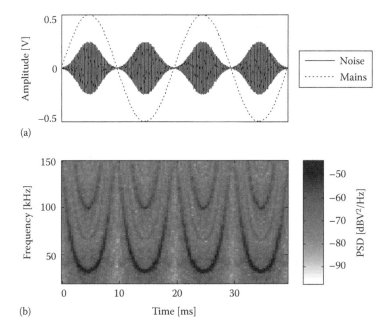

FIGURE 2.20
HF components of i_M obtained by applying a band-pass filter. Passband: 20–150 kHz: (a) filtered waveform and (b) spectrogram of STFT.

performance of NB-PLC will become larger. It is important to improve the robustness of communication systems against the SFN. Meanwhile, it is necessary to add this noise class to the noise scenario and to emulate it as accurately as possible.

2.3.4 Impulsive Noise

Impulsive noise is usually grouped into three different classes: *periodic impulsive noise asynchronous to the mains frequency* mainly caused by switched power supplies, *periodic impulsive noise synchronous to the mains frequency* caused by switching of rectifier diodes of power supplies and *asynchronous impulsive noise* originated from transients caused by switching events. Setting aside the time behaviours such as inter-arrival time, individual impulses have some features in common. In the time domain, the impulses are peaks with short duration and high magnitude levels. Instead of random waveforms of the background noise, the most impulsive noise have deterministic appearance patterns, such sharp rising edges followed by damped oscillations, low-level oscillations terminated by sharp endings [12], impulse chains either equal or not equal spaced [13]. In the frequency domain, the impulsive noise can be distinguished from the background noise by raised wideband PSD. Most measured impulses exceed the background noise spectral density for at least 10–15 dB within most portions of the frequency range. These common features apply not only for broadband PLC but also for NB-PLC. Figure 2.21a shows a noise segment

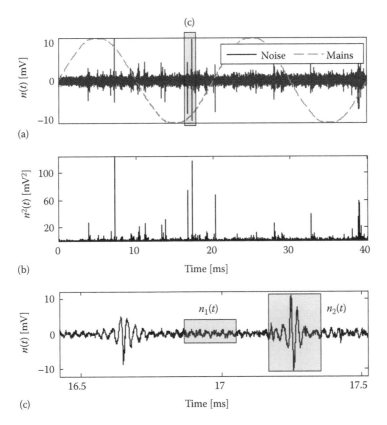

FIGURE 2.21
Measured noise in time domain: (a) noise waveform $n(t)$, (b) $n^2(t)$ and (c) detailed view of $n(t)$, corresponding to (c) in plot (a).

dominated by rich impulsive noise and the background noise. All the narrowband interferers have been removed using the method introduced in Section 2.3.2.1. Figure 2.21b
shows its instantaneous power and Figure 2.21c is a detailed view of the part between 16
and 17.5 ms. The difference between the level of the background noise $n_1(t)$ and the $n_{nbn}(t)$
of the impulse $n_2(t)$ is very large.

2.3.5 Coloured Background Noise

The coloured background noise is a collection of low-level noise from all possible sources.
Its average power level is dependent on the number and type of connected and active
electrical devices. Therefore, it can also be considered to have cyclostationary characteristics [6,13]. In order to investigate the time variance, it is necessary to divide the noise
waveform into multiple segments and to estimate the instantaneous PSD of each segment.
For this purpose, STFT is performed on the remaining noise waveform from which the
narrowband and the impulsive noise have been removed. In addition to the variance in
the time domain, the coloured background noise is supposed to have a smooth spectrum,
the power spectral density is a decreasing function of the frequency [1,8]. Therefore, the
STFT result is smoothed in the time and the frequency domains, respectively. Figure 2.22a
shows the overlapped PSDs of all noise segments. The averaged PSD can be approximated
by the sum of two exponential functions:

$$\hat{P}_{BGN}(f) = a \cdot e^{b \cdot f} + c \cdot e^{d \cdot f}, \tag{2.14}$$

where f is the frequency in kHz. Table 2.1 shows a set of coefficients that fits the averaged PSD.

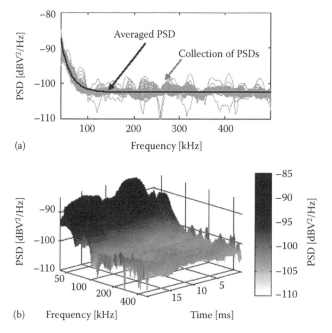

FIGURE 2.22
PSD of background noise. (a) PSD overlapped over time axis and (b) time–frequency view of PSD.

TABLE 2.1

Coefficient of Polynomial

a	b	c	d
0.4413	−0.12	3.132×10^{-4}	-1.32×10^{-4}

TABLE 2.2

Recommended Parameters

A_0	A_1	b	c	d
200	1	−0.12	3×10^{-4}	-3×10^{-4}

Figure 2.22b shows the smoothed STFT of the remaining noise. The PSD fluctuation over time can clearly be recognised in the frequency range below 100 kHz. The maximum of the noise level is synchronous with the peak of the mains voltage. Perturbations can also be observed at higher frequencies. Since the noise level is comparatively low, these perturbations are ignored for simplicity.

Let $m(t)$ denote the mains voltage, the parameter $a(t)$ can be obtained by

$$a(t) = \frac{|m(t)|}{A_0} + A_1, \tag{2.15}$$

where A_0 and A_1 determine the scaling factor and the minimum level of the fluctuation, respectively. Table 2.2 lists a set of recommended parameters for modelling the time-variant PSD of the background noise.

2.4 Channel Attenuation

The channel attenuation is evaluated at discrete frequencies. For this purpose, a test sequence synchronised with the mains is transmitted. The sequence consists of sinusoidal signals at discrete frequencies f_k at time index k consecutively by every device like illustrated in Figure 2.23a. The signals are captured by the receiver. By monitoring the injected test signal at the transmitter, any influence of the coupling circuits (assuming linear behaviour of the components) and the variation of the impedance at the transmitter can be compensated. The signal flow is depicted in Figure 2.23b.

With the Fourier transform of the transmitted signal $S(f)$, the channel TF $H(f)$ and of the noise $N(f)$ assuming that

$$|S(f)C_T(f)H(f)| \gg |N(f)|, \tag{2.16}$$

the modulus of the channel TF at discrete frequencies f_k can be estimated in the frequency domain by evaluating

$$|\hat{H}(f_k)| = \frac{|R(f_k)|}{|S_R(f_k)|}. \tag{2.17}$$

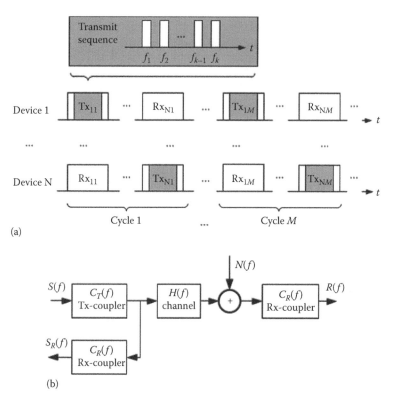

FIGURE 2.23
Evaluation of the attenuation: (a) test sequence flow and (b) signal path.

It is important to note that $|\hat{H}(f_k)|$ gives the ratio between the received voltage level and the transmitted voltage level for a certain frequency. Its attenuation is highly influenced by the particular appliances connected close to the particular receiver location. Therefore, the TF is not symmetric in general and has to be observed in both directions.

Mean value and variance for every mains cycle interval is calculated, as in the case of the noise scenario analysis. Figure 2.24 depicts exemplarily the resulting channel attenuation (up to 100 kHz) between two measurement locations over 24 h.

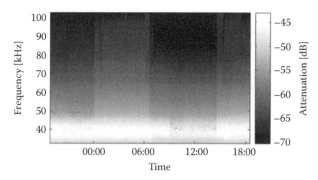

FIGURE 2.24
Variation of the attenuation over 24 h (S1 → S3).

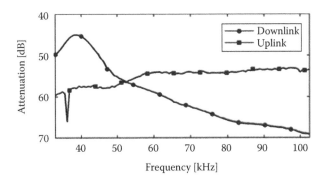

FIGURE 2.25
Channel attenuation for downlink and uplink (S1 → S3).

Obviously, the TF varies over time, but the variation remains comparably small and the overall characteristics of the TF do not change during any of the observed intervals. Contrary to this, the differences between uplink and downlink can be significant. An example is shown in Figure 2.25.

The notch in the uplink is caused by narrowband interferers; due to low SNR, the signal attenuation cannot be calculated reliably for this frequency.

2.5 Signal-to-Noise Ratio

Having performed the analysis of the TF, the received signal level is known at each frequency. The next step is to obtain an estimate of the noise level. It is assumed that the behaviour of the noise energy at a certain frequency within a short period of time and within a small bandwidth is approximately constant. Thus, for time and frequency index k (see also Section 2.4), the Fourier transform of the noise estimate $\hat{N}_k(f_k)$ can be obtained by taking the mean value of the received noise in the pause prior to and after the test sequence and the energy at frequencies f_{k-4} and f_{k+4} while the signal f_k is being transmitted:

$$\left|\hat{N}_k(f_k)\right| = \frac{1}{4}\left(\left|R_k(f_{k-4})\right| + \left|R_k(f_{k-4})\right| + \left|R_{k-1}(f_k)\right| + \left|R_{k+1}(f)\right|\right). \tag{2.18}$$

The modulus of the SNR can then easily be obtained by

$$\text{SNR}_k\left(f_k\right) = \frac{\left|R_k(f_k)\right|}{\left|N_k(f_k)\right|}. \tag{2.19}$$

This estimation only holds true for assumption (Equation 2.16) in the evaluation of the TF. The proposed method is a pragmatic approach, since it is restricted to SNRs well above 0 dB. This is considered sufficient for our investigations because of the high energy of the transmitted test sequence. In fact, the energy of the test signal sequence is much higher than would be reasonable for any real implementation of a communication system.

Similar to the analysis of the noise scenario at first, the short-time behaviour within one mains cycle is investigated. The SNR for one frequency varies within one mains cycle. An example is shown in Figure 2.26. There is a clear variation within one mains cycle and an even stronger deviation relative to the median within the same mains cycle interval.

Considering the mean value of the SNR over one mains cycle and investigated for the whole time of 24 h and for all test frequencies basically reflects the results of the preceding analysis of noise scenario and TF and is depicted in Figure 2.27: the SNR remains constant for long periods of time.

In addition, Figure 2.28 displays statistics of the SNR at a receiver input. The overall SNR appears to be very high. However, these values should be put into perspective by the fact

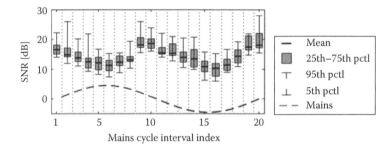

FIGURE 2.26
Statistics of the SNR at 32.5 kHz within one mains cycle (S1 → S4).

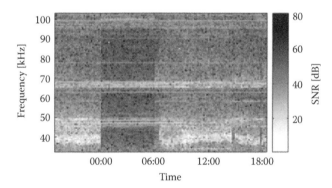

FIGURE 2.27
Variation of the SNR over 24 h (S1 → S3).

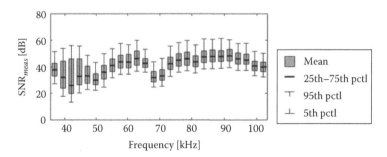

FIGURE 2.28
Statistics of the SNR for different frequencies (S1 → S3).

that the transmitter level is $U_{meas} = 2.8$ V and the observation window for each transmitted sinusoidal test signal is 20 ms. In other words, the SNR depicted would be valid for a single carrier system with a symbol rate of $r_{meas} = 50$ Bd. For simplification, we assume a Gaussian distribution for each frequency slot. A rough estimate of the actual SNR value for a communication system operating at a symbol rate of r_{com} and a transmit amplitude of U_{com} can then be obtained from the measured SNR values by

$$\text{SNR}_{com}\left(f_k\right) = \text{SNR}_{meas}\left(f_k\right) - 20\log\left(\frac{U_{meas}}{U_{com}}\right) - 10\log\left(\frac{r_{com}}{r_{meas}}\right). \tag{2.20}$$

For a symbol rate of $r_{com} = 10$ kBd and a transmit amplitude of $U_{com} = 1$ V, this leads to a reduction of the SNR of about 32 dB compared to the measurement results. In this case, the SNR between about 67 and 72 kHz given in Figure 2.28 would be reduced to about 0 dB. Obviously using a communication system utilising this frequency band would not make much sense in this environment.

For a multi-carrier system, the higher peak-to-average ratio of the transmit signal (compared to a single carrier system) in conjunction with a fixed maximum output amplitude of the transmitter can reduce this value additionally.

2.6 Access Impedance

Exact knowledge of the impedance is a key aspect for power line modem development as well as for modelling channel transfer characteristics. Hence, this aspect is described explicitly.

The impedance seen at a particular point in a low-voltage grid is a superposition of any appliances online and its connecting lines. Although a lot of results regarding access impedance in the low-frequency range have been reported in the last few years – for example, [14–16] – updated measurements are needed due to the replacement of many appliances connected to the grid. Especially in the last years, more and more devices use SMPA.

Usually, a *vector network analyser* (VNA) is used for determination of scattering parameters. However, the expected impedance in the frequency ranging up to 500 kHz is very low and therefore there is a huge difference between the impedance measured and the internal impedance of the VNA, which is usually 50 Ω. In any case, a coupling circuit is needed to connect the VNA to the mains. This coupling circuit needs to be characterised in detail and its influence has to be compensated properly. Furthermore, the noise level in the live mains is high; therefore, for measuring very low impedance with an absolute value below 1 Ω, high power is needed to obtain accurate results. Hence, we choose direct measurement of current and voltage at the primary side of a coupling circuit for our measurements. Our test set-up for determination of the channel attenuation (see Section 2.4) can be used to evaluate the access impedance Z_A simultaneously. This is illustrated in Figure 2.29. A shunt for measuring the current was inserted in the transmit path. The waveform generator includes a powerful front-end, which is able to deliver high currents to measure even very low impedance down to 0 Ω. A schematic of the set-up is depicted in Figure 2.29.

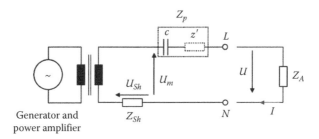

FIGURE 2.29
Impedance measurement set-up.

Every measurement is synchronised to the zero crossings of the mains cycle to allow an analysis of the cyclostationary behaviour of the power line channel.

To compensate for parasitic effects of the shunt resistor and the series impedance of the circuit, two measurements are done for the calibration: one with the output shorted (U_{m1}/U_{Sh1}) and one with a defined reference load of $Z_{Ref} = 1\,\Omega$ (U_{m2}/U_{Sh2}):

$$\frac{U_{m1}}{U_{Sh1}} Z_{Sh} = Z_p, \tag{2.21}$$

$$\frac{U_{m2}}{U_{Sh2}} Z_{Sh} = Z_p + Z_{Ref}. \tag{2.22}$$

The impedance of the shunt Z_{Sh} and the parasitic series impedance Z_p can be estimated by

$$Z_{Sh} = \frac{Z_{Ref}}{\left(\dfrac{U_{m2}}{U_{Sh2}} - \dfrac{U_{m1}}{U_{Sh1}} \right)}, \tag{2.23}$$

$$Z_p = \frac{U_{m1}}{U_{Sh1}} Z_{Sh}. \tag{2.24}$$

With Equations 2.21 through 2.24, the access impedance can be expressed as

$$Z_A = \frac{U_m \cdot Z_{Sh}}{U_{Sh}} - Z_p. \tag{2.25}$$

This is evaluated 20 times within each mains cycle to account for its variance. Measurements have been executed on different three-phase low-voltage distribution grids. The measurement equipment was connected as close as possible to the consumer unit. Three representative results are shown in Figure 2.30. In small buildings where all appliances are usually very close to the house connection box, the impedance is usually very low and irregular. Figure 2.30 shows a snapshot taken in a detached house. The dashed line is a snapshot taken at the same place but with a single additional *energy-saving lamp* (ESL) switched on. The example illustrates that the access impedance is mainly dominated by appliances with low input impedance and that are not very distant from the measurement location. This is caused by the fact that the modified access impedance can be seen as a parallel circuit

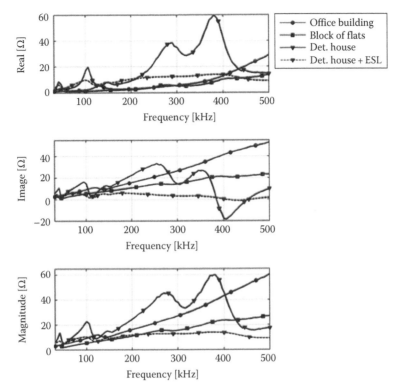

FIGURE 2.30
Access impedance measured at different premises (mean value).

of the input impedance of the appliance and the originally measured access impedance without the additional device. For low frequencies, the decoupling influence of the wires can be observed only for very long distances.

At bigger buildings (block of flats and office building), the general behaviour is more regular. The resistance, as well as the (inductive) reactance, increases towards higher frequencies. The loads are farther away from the house connection point, which leads to an overall higher impedance. The absolute value of the impedance up to 100 kHz is usually below 10 Ω; towards lower frequencies, values below 1 Ω could be observed.

The impedance is stable over long periods of time, that is, in the order of hours. However, abrupt changes can occur by switching operations. The cyclic variation within one mains cycle (20 ms at 50 Hz) depends mainly on the nearby appliances. The variation for the three examples is depicted in Figure 2.31.

Contrary to the detached house, nearly no variation can be observed at the office building or the block of flats. This holds also true for a long-time observation of about 36 h, which is depicted in Figure 2.32. Despite the impedance drop at the beginning and at the end of the measurement – which was caused by switching on a fluorescent lamp very close to the measurement location – nearly no variation of the impedance can be observed at all.

The results obtained from different measurements have shown that for buildings with no appliances close to the house connection box, the impedance is inductive in general and increasing towards higher frequencies. For small buildings, where the length of the cables is shorter in general, the behaviour of the impedance is not predictable and depends mainly on the properties of the devices close to the consumer unit.

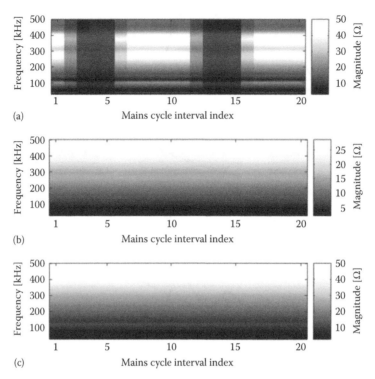

(a)

(b)

(c)

FIGURE 2.31
Variation of the access impedance within one mains cycle: (a) detached house (close to an ESL), (b) block of flats and (c) office building.

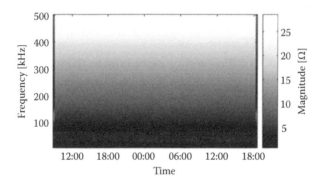

FIGURE 2.32
Long-time behaviour of the access impedance.

2.7 Conclusion

By means of the proposed measurement procedure, it is possible to investigate real-world power line channels comprehensively with respect to properties of the low-voltage mains grid relevant to communication systems. The suggested procedure allows for both short-term and long-term evaluation of different links of the grid at the same time. The measurement procedure has successfully been implemented and yielded valuable information on the channel properties in a small low-voltage network supplied by a single transformer substation. The noise signal power distribution over time and over frequency is periodic with the mains cycle period and its distribution patterns remain stable for periods of time mainly in the order of hours. Its overall characteristics depend strongly on location and on time of day in most cases.

Beside the well-investigated noise classes background and impulsive noise, a detailed look on the noise scenario identified two steady appearing noise classes: narrowband interferers with time-variant amplitude and switched frequency noise. Both have been investigated in detail and appropriate models have been proposed.

The influence of SFN on NB-PLC systems will be an important aspect in the near future due to the increasing number of SMPS in low-voltage grids. Our proposed model for SFN provides a good basis for comprehensive investigations.

The conclusions drawn from the measurement results presented earlier show that the channel TF is stable over long periods of time, but its overall characteristics depend on location. Its characteristics need not necessarily be symmetric. In conjunction with the noise observed at a receiver, this leads to a high variation of the SNR at the receiver, which is the crucial factor for reliable data transmission.

In general, reliable communication in the LF range is feasible for moderate data rates. However, a fixed communication set-up may fail due to the varying conditions. In fact, an adaptive communication approach is required for reliable NB-PLC systems.

Finally, a field study of an NB channel and noise emulator can be found in Chapter 21.

References

1. O. G. Hooijen, A channel model for the residential power circuit used as a digital communications medium, *IEEE Trans. Electromagn. Compat.*, 40(4), November 1998, 331–336.
2. W. Choi and C. Park, A simple line coupler with adaptive impedance matching for power line communication, *IEEE International Symposium Power Line Communications and its Applications*, Pisa, Italy, 2007, pp. 187–191.
3. T. A. C. M. Claasen and W. F. G. Mecklenbraeuker, On stationary linear time-varying systems, *IEEE Trans. Circuits Syst.*, 29(3), March 1982, 169–184.
4. M. H. L. Chan and R. W. Donaldson, Attenuation of communication signals on residential and commercial intrabuilding power-distribution circuits, *IEEE Trans. Electromagn. Compat.*, EMC-28(4), November 1986, 220–230.
5. F. J. Cañete, J. A. Cortés, L. Díez and J. T. Entrambasaguas, Analysis of the cyclic short-term variation of indoor power line channels, *IEEE J. Sel. Areas Commun.*, 24(7), July 2006, 1327–1338.
6. M. Katayama, T. Yamazato and H. Okada, A mathematical model of noise in narrowband power line communication systems, *IEEE J. Sel. Areas Commun.*, 24(7), July 2006, 1267–1276.
7. M. Sigle, M. Bauer, W. Liu and K. Dostert, Transmission channel properties of the low voltage grid for narrowband power line communication, in *IEEE International Symposium Power Line Communication and Application*, Udine, Italy, 2011, pp. 289–294.
8. J. Bausch, T. Kistner, M. Babic and K. Dostert, Characteristics of indoor power line channels in the frequency range 50–500 kHz, *IEEE International Symposium on Power Line Communications and its Applications*, Orlando, FL, March 2006, pp. 86–91.
9. A. Larsson, On high-frequency distortion in low-voltage power systems, Doctoral thesis, Universitetstryckeriet, Luleå, Sweden, 2011.
10. Fairchild semiconductor, Power factor correction (PFC) basics, Application note 42047, Rev.0.9.0, August 19, 2004.

11. U. Tietze, C. Schenck and E. Gamm, *Electronic Circuits: Handbook for Design and Application*, 2nd edn, Spinger, New York, 2008.

12. M. Zimmermann and K. Dostert, Analysis and modeling of impulsive noise in broad-band powerline communications, *IEEE Trans. Electromagn. Compat.*, 44(1), February 2002, 249–258.

13. J. A. Cortes, L. Diez, F. J. Canete and J. J. Sanchez-Martinez, Analysis of the indoor broadband power-line noise scenario, *IEEE Trans. Electromagn. Compat.*, 52(4), November 2010, 849–858.

14. M. Arzberger, K. Dostert, T. Waldeck and M. Zimmermann, Fundamental properties of the low voltage power distribution grid, *International Symposium on Power Line Communications (ISPLC), Proceedings*, Essen, Germany, 1997.

15. O. G. Hooijen, A channel model for the low-voltage power-line channel: Measurement- and simulation results, *Power Line Communications and Its Applications, Proceedings*, Essen, Germany, 1997.

16. M. Katayama, S. Itou, T. Yamazato and A. Ogawa, A simple model of cyclostationary power-line noise for communication systems, *International Symposium on Power Line Communications (ISPLC), Proceedings*, Tokyo, Japan, 1998.

3

Narrowband Measurements in Domestic Access Networks

Weilin Liu, Guangbin Chu and Jianqi Li

CONTENTS

3.1 Introduction

Narrowband power line communications (NB PLC) technology for frequency between 9 and 500 kHz has been widely used for *automatic metering infrastructure* (AMI) [1–4]. It is well known that NB PLC for frequency 9–500 kHz subjects to hostile channel conditions. Power line channel at low frequencies is characterised by time-selective and frequency-selective interferences, low access impedances and frequency-selective attenuation [2–8]. Due to negligible cable loss in the low frequencies, the power line channel of access network greatly depends on electrical properties of loads at the network and in particular at customer premises. This dependency is a major source of considerable time variation of power line channel, which makes design of a reliable NB PLC system a challenging task. In China, the influence of electrical appliances in the distribution network and at the customer premises on the performance of NB PLC system is even more evident, partly due to non-strict compliance of electrical appliances with the *electromagnetic compatibility* (EMC) regulation. To improve the reliability of NB PLC and to enhance its capability to support advanced Smart Grid services, there is still a great need for better understanding of power line channel characteristics in the low frequencies bands.

In China, there are three major types of *low-voltage* (LV) access network: urban network with high-riser buildings, urban/suburban network with low-riser apartment buildings and rural network with scattered houses. The urban LV access network with high-riser buildings usually uses a dedicated underground power cable for each building, and the cable distance is relatively short. This type of network is considered to be less critical for NB PLC. This chapter addresses power line channel of LV access network in urban/suburban area with low-riser apartment buildings.

Noise and attenuation are two major power line channel characteristics. To compare different channels, the so-called *link quality index* (LQI) is proposed, which combines both noise and attenuation and describes the quality of a particular link.

A proper coupling mode is essential for the performance of NB PLC. Both single-phase and three-phase couplings are possible at the distribution transformer. There are pros and cons between these two approaches. Assisted by measurement results, comparison between these two coupling modes is given.

This chapter is organised as follows. In Section 3.2, an approach to determine the access impedance of single-phase and three-phase coupling is described. In Section 3.3, the three types of LV access network and the network site for the channel measurement are described. Channel measurement results are presented and analysed in Section 3.4. Section 3.5 concludes this chapter.

3.2 Impedance Determination

Access impedance depends on coupling mode. While at the customer side phase to neutral coupling (single-phase coupling) is the default mode, at the transformer, both phase to neutral and three phases to neutral couplings (three-phase coupling) are possible. As will be shown, the access impedance of single-phase coupling at low frequency is already very small. A parallel connection of three phases will further lower the access impedance and hence may lead to additional reduction of effective signal level injected into the grid.

In [7], a voltage/current-based approach is described for the determination of impedance at access point. This approach is adopted here. The measurement set-up is shown in Figure 3.1 with the corresponding coupling network. R_{sh} is a shunt resistor for the current measurement [7]. C denotes the coupling capacitor which has a capacitance of 1 µF.

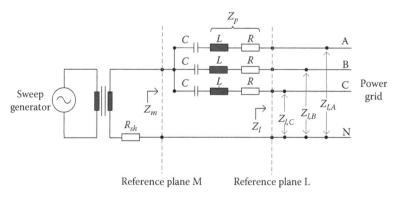

FIGURE 3.1
Impedance measurement set-up.

Z_p denotes the parasitic impedance of the entire coupling network, which includes the coupling circuit, coupling leads and fuses. For low frequencies, Z_p can be described by an inductance L and a resistor R with $L \approx 6\ \mu H$ and $R \approx 0.6\ \Omega$ for the used coupling network. Depending on whether a single phase or three phases of the coupling network are actually connected to the power grid, impedance of single-phase coupling, denoted by $Z_{m,A}$, $Z_{m,B}$, $Z_{m,C}$, and of three-phase coupling, denoted by $Z_{m,ABC}$, at the reference plane M can be measured. The access impedances at the reference plane L can be determined from, $Z_{m,A}$, $Z_{m,B}$, $Z_{m,C}$, $Z_{m,ABC}$, with corresponding calibrations which take into account the coupling network itself. Let $Z_{l,A}$, $Z_{l,B}$, $Z_{l,C}$ denote the access impedance of single-phase coupling and $Z_{l,ABC}$ the access impedance of three-phase coupling at the reference plane L [7]. Then $Z_{l,A}$, $Z_{l,B}$, $Z_{l,C}$ can be derived by ($X = A, B, C$) [7]:

$$Z_{l,X} = Z_{m,X} - \left(R + j\omega L + \frac{1}{j\omega C} \right). \tag{3.1}$$

The access impedance $Z_{l,ABC}$ for the three-phase coupling is defined as

$$Z_{l,ABC} := Z_{m,ABC} - Z_{calibration}. \tag{3.2}$$

In a first approximation, the calibration impedance $Z_{calibration}$ is determined to be the impedance Z_m by a short cut of all three phases of the coupling network against the neutral. Due to mutual magnetic coupling between the conductors of the coupling network, $Z_{calibration}$ is different from $(R + j\omega L + 1/j\omega C)/3$. Given the access impedance of single-phase coupling $Z_{l,A}$, $Z_{l,B}$, $Z_{l,C}$, a theoretical value of the access impedance of the three-phase coupling, denoted by $Z'_{l,ABC}$, can be obtained from the parallel connection of $Z_{l,A}$, $Z_{l,B}$, $Z_{l,C}$ as

$$Z'_{l,ABC} = \frac{1}{(1/Z_{l,A} + 1/Z_{l,B} + 1/Z_{l,C})}. \tag{3.3}$$

$Z'_{l,ABC}$ is a theoretical value as it neglects the effect of mutual magnetic coupling between the parallel phases. The effect of the mutual magnetic coupling leads to an equivalent higher inductance at each phase compared to the inductance of single-phase coupling. A higher inductance at each phase will lead to an overall higher magnitude of the access impedance of three-phase coupling. Hence, the magnitude $Z_{l,ABC}$ is larger than the magnitude of $Z'_{l,ABC}$. In other words, $|Z'_{l,ABC}|$ is a lower bound of $Z_{l,ABC}$.

3.3 LV Network Topology and Measurement Site

The majority of the LV power distribution network in China has a radiation topology. There are three major network topologies.

3.3.1 High-Riser Residential Network

In this network, one transformer supplies power to 1–5 high-riser buildings, with each building containing 10–30 floors. There is a direct underground power cable to each building with a cable distance <150 m. This network can be found typically in big cities.

3.3.2 Low-Riser Residential Network

In this network, one transformer supplies power to dozens of low-riser buildings with less than eight floors of each building. Typically, one underground power cable feeds one or a few buildings. The power line distance from the transformer to the main switch of buildings is usually <300 m. This network is quite common in urban and suburban areas.

3.3.3 Scattered Residential Network

In this network, one transformer supplies power to dozens to hundreds of scattered households. The feed line from the transformer to households is usually cascaded through overhead lines. The maximum power line distance is in the range between 500 and 2000 m. This network topology can be found in rural areas.

3.3.4 Description of the Measurement Site

Channel measurement was carried out in a typical urban residential area in north China. Figure 3.2 illustrates the network with three buildings #1, #2 and #3 chosen for the measurement. Each building is supplied by a dedicated underground power cable. The distance between the transformer and *house access point* (HAP) of the three buildings is about 50, 250 and 350 m, respectively. Buildings #1 and #2 have three blocks; each has six floors and a meter panel with 12 single-phase meters. Building #3 has one block of eight floors and a meter panel with 16 single-phase meters. The distance between the meter panel and the HAP is about 5–20 m.

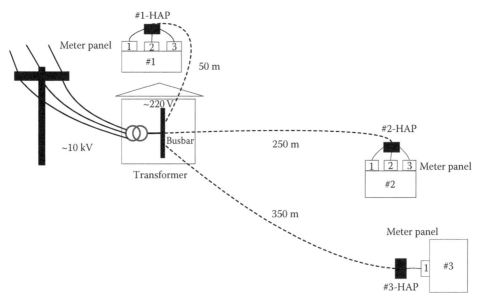

FIGURE 3.2
Measurement site.

3.4 Measurement Results and Discussion

3.4.1 Impedance

Impedance Z_m of single-phase and three-phase coupling was measured and the corresponding access impedance Z_l was computed with calibrations (3.1) and (3.2) (see Figure 3.1).

Figure 3.3 shows an example of access impedances (magnitude and phase) at the transformer. It confirms the observations from other measurements and reports [7,8] that the

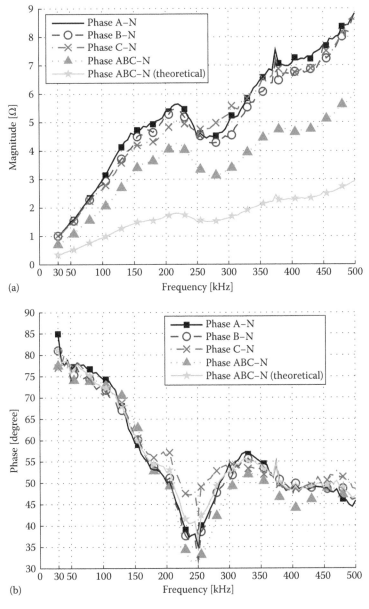

FIGURE 3.3
Impedance at the transformer: (a) magnitude and (b) phase.

access impedance is mainly inductive and resistive. The magnitude of the access impedances with single-phase coupling grows from 1 Ω at 30 kHz to 9 Ω at 500 kHz. In this example, the difference of access impedance between the different single-phase coupling modes is small. This is likely due to similar wiring structure and line properties of each phase, and there is no dominant and noticeable load at a particular phase. As expected, the magnitude of the access impedance of three-phase coupling $Z_{l,ABC}$ is smaller than the corresponding value of single-phase coupling. It is below 1 Ω for frequency below 50 kHz and increases to 6 Ω at 500 kHz. For comparison, the theoretical access impedance of three-phase coupling $|Z'_{l,ABC}|$ determined from Equation 3.3 is also depicted. As it can be seen, $|Z'_{l,ABC}|$ is noticeably smaller than $|Z_{l,ABC}|$. As described in Section 3.2, $Z'_{l,ABC}$ corresponds to the ideal case which assumes that there is no mutual magnetic coupling between the phases. Hence, $|Z'_{l,ABC}|$ is a low bound of $|Z_{l,ABC}|$.

Figure 3.4 shows an example of impedances (magnitude and phase) at the HAP of building #2. Here, the access impedance is largely inductive. The magnitude of the impedance grows steadily from a value below 1 Ω at 30 kHz to 9 Ω at 500 kHz. The difference of the access impedance between different phases is slightly larger than the difference at the transformer. Figure 3.5 shows an example of access impedances at a meter panel of building #2, which is about 5 m apart from the HAP. In general, the access impedance at the meter panel shows a similar behaviour as at the HAP. However, since the meter panel is closer to customer premises, special properties of electrical loads may have more visible impact. In this example, a strong local maximum of the magnitude of the access impedance around frequency 40–70 kHz and at phase B is observed. The magnitude has a peak value of 7 Ω for this low frequency range and the phase of the access impedance is changed by >90°. The load becomes for a short frequency range capacitive. This phenomenon may be caused by a parallel resonance circuit of a load at customer premise.

Figure 3.6 shows a collection of magnitude of access impedances of single-phase coupling measured at different HAPs and different phases. It confirms the trend that the magnitude of the impedances increases from small value in the range of 1–2 Ω at low frequency around 30 kHz to about 8–9 Ω at high frequency around 500 kHz. Local maxima may exist at particular phases depending on particular loads at the network. Impedance value may change with the time due to possible change of grid and/or of electrical loads at the network.

Figure 3.7 gives a comparison of the access impedance (magnitude) measured at phase A–N of the transformer but at two different times. One was measured on 16 September 2012 and the other on 8 November 2012, hence a time gap of nearly 2 months. In this example, the magnitude of the impedance is almost the same after 2 months.

3.4.2 Noise

Noise is a major concern for NB PLC. There are different kinds of noises: background noise, periodic impulsive noises, asynchronous impulsive noise, frequency-selective narrowband interferences, etc. (see Chapter 5) [2–4]. Figure 3.8 shows examples of noise waveform at phase A of the transformer. The noise waveform was recorded at two different times, one at night around 0:00 (Figure 3.8a) and one at noon time (Figure 3.8b). The *alternate current* (AC) mains cycle is also shown as a time reference. Figure 3.8a shows that the noise at night is small and there are no considerable impulsive noises, and the noise level at noon (Figure 3.8b) is significantly higher and excessive periodic impulsive noises can be observed. This may be explained by the increasing usage of electrical appliances at customer premises at noon time.

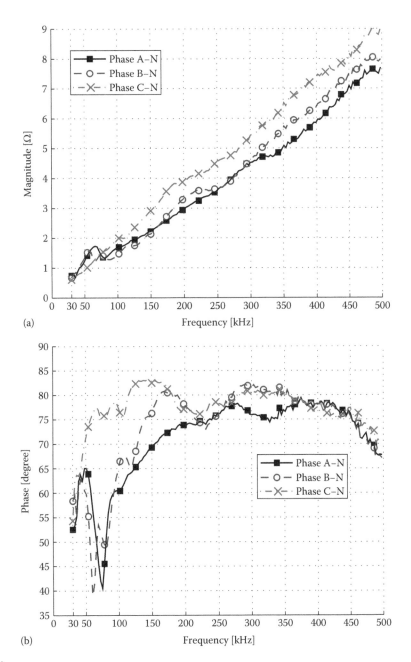

FIGURE 3.4
Impedance at the HAP of building #2: (a) magnitude and (b) phase.

Figure 3.9 shows a noise waveform recorded at the HAP of building #1 at evening 19:03. It also shows a periodic behaviour related to the AC mains cycle and the noise level is at a minimum at AC zero-crossing points. However, in this example, the noise level in general and the peak of impulsive noises in particular are considerably smaller compared to the value measured at the transformer at the noon time.

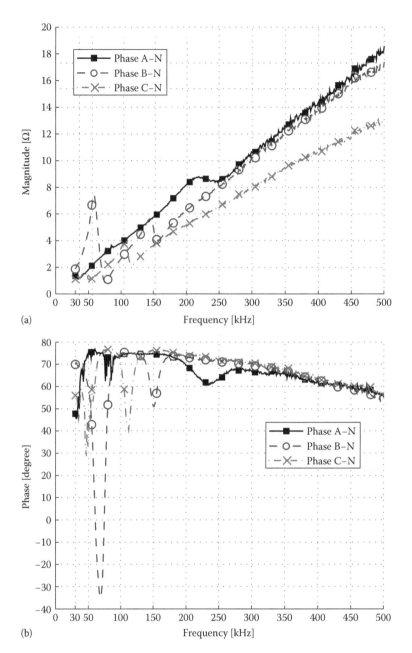

FIGURE 3.5
Impedance at a meter panel of building #2: (a) magnitude and (b) phase.

Calculation of autocorrelation of noise waveform over 1 s (50 AC mains cycle) shows that the noise has a wide-sense cyclostationary behaviour as reported in [9,10]. The period corresponds to the half AC mains cycle, that is, 10 ms. Figure 3.10 shows the corresponding standard deviation of noise variance over a period of 10 ms based on recorded noise waveform over 1 s. In contrast to the noise at the night, there are significant variations of noise variance at noon. Two maxima can be observed: one is about 5 ms and the other is about 7.5 ms from the AC zero-crossing points, which coincide with the periodic impulsive

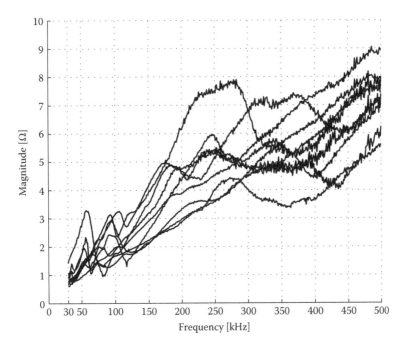

FIGURE 3.6
Collection of the magnitude of access impedance measured at different HAPs and different phases, and with single-phase coupling.

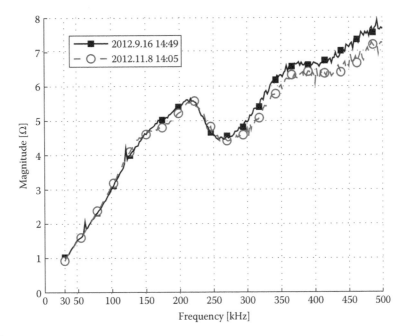

FIGURE 3.7
Impedance (magnitude) measured at the same place (phase A–N of transformer) but at two different times, one on 16 September 2012 and the other on 8 November 2012.

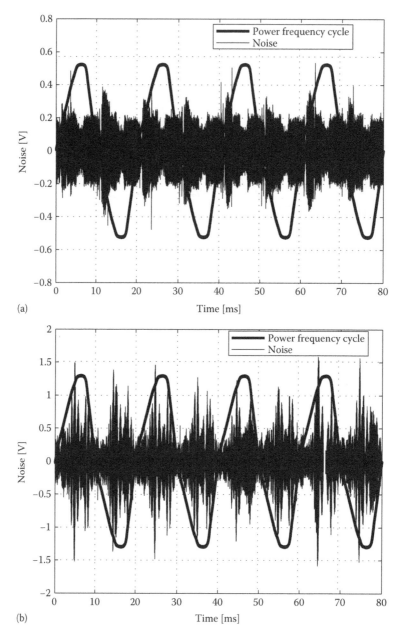

FIGURE 3.8
Examples of noise waveform at transformer recorded at night 0:20 (a) and at noon 12:00 (b).

noises visible at Figure 3.8b. The difference between the maximum and the minimum of the noise variance is nearly 17 dB ($20 \times \log_{10}(0.7 \text{ V}/0.1 \text{ V})$). The noise variance at HAP also shows a maximum at 7.8 ms from the AC zero crossing. But it has a much smaller magnitude compared with the maximum value at the transformer.

The noise power in the frequency domain is captured with a spectrum analyser with logarithmic video signal averaging and with *root mean square* (RMS) detector. The noise resolution bandwidth is 10 kHz.

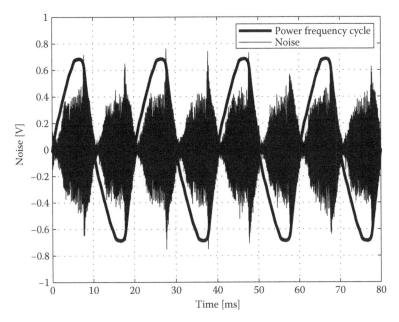

FIGURE 3.9
Example of noise waveform at HAP of building #1, recorded at evening 19:03.

Figure 3.11 shows an example of noise power spectrum at the transformer with the single-phase coupling. The noise power decreases with the frequency. The largest decrease is observed for the phase A, where the noise power is reduced from 80 dBμV at 50 kHz to 33 dBμV at 500 kHz. This is a reduction of >45 dB. At the phases B and C, the reduction of the noise power from 50 to 500 kHz is about 30 dB. The difference of noise power spectrum between

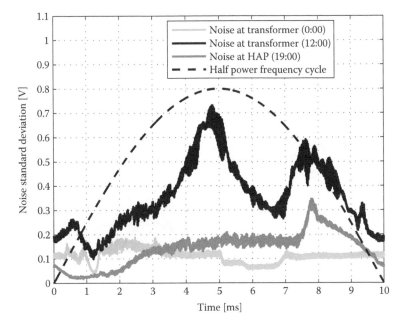

FIGURE 3.10
Standard deviation of noise variance over half power frequency cycle.

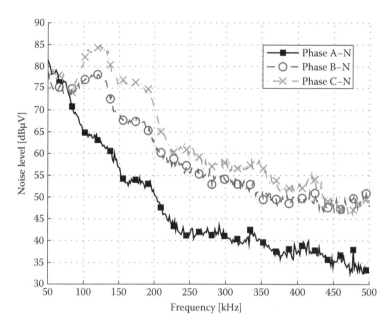

FIGURE 3.11
Noise power spectra at the transformer.

phase A and the other two phases B and C is between 10 and 15 dB. The shape of the noise power spectrum at different phases is quite similar indicating a coupling effect among phases.

Figure 3.12 shows a collection of noise power spectrums captured at different HAPs and at different phases and with different coupling modes. It provides an impression of variability of noise. It confirms that the noise at low frequencies is significantly stronger than

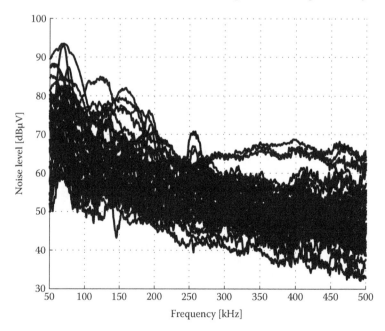

FIGURE 3.12
Collection of noise power spectrum at different HAPs, different phases and with different coupling modes.

the noise at higher frequencies. The most noise powers are within 50–80 dBμV between 50 and 100 kHz and within 30–60 dBμV between 400 and 500 kHz. The differences can be as large as 30–40 dB. NB PLC operating at low frequency range such at CENELEC A band may face considerably stronger noise.

3.4.3 Attenuation

Attenuation is measured between the transformer and HAP of the three buildings. A sweep generator is used. The transmitter is able to inject high signal level for impedances with a magnitude below 1 Ω. The attenuation includes the coupling at transmitter and receiver.

Figure 3.13 shows an example of attenuation between the transformer and the HAPs of buildings #1, #2 and #3. Three-phase coupling is used at the transformer. The receiver uses single-phase coupling between phase B and neutral N. The attenuation varies between 30 and 70 dB. It is well known that unlike for broadband power line (BPL), for NB PLC at LV access network, there is not necessarily a strong correlation between the attenuation and frequencies, nor between the attenuation and distances. This can be observed in Figure 3.13. Among the three links, the longest link with 350 m has the smallest attenuation between 150 and 250 kHz, while the shortest link with 50 m only has the smallest attenuation for frequency above 425 kHz.

Three-phase coupling at the transformer is widely used, as this coupling may have better chance to cover most PLC nodes at customers directly. However, three-phase coupling has two effects. First, the parallel connection of three phases will result in smaller access impedance, which makes an efficient injection of signal difficult, and second, the signal will further split into the three phases. These together may lead to a remarkable reduction

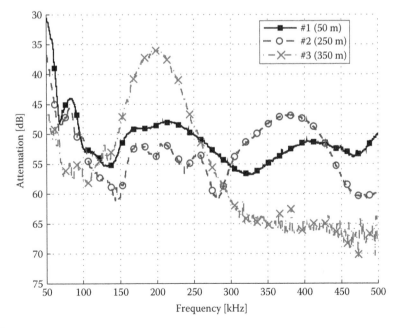

FIGURE 3.13
Attenuation for the three downlinks.

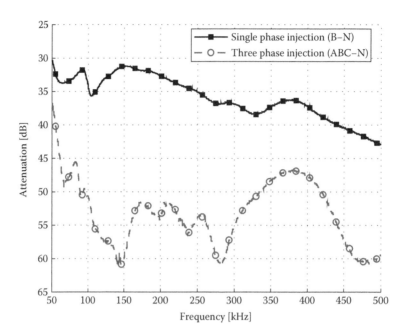

FIGURE 3.14
Attenuation from the transformer to the HAP of building #2 with single-phase and three-phase coupling at the transformer.

of the signal level at the receiver compared with the case when the signal is only injected into one and the same phase as the receiver.

As an example, Figure 3.14 compares attenuation from the transformer to the HAP of building #2 between single- and three-phase coupling at the transformer. For the single-phase coupling, the same phase is used at both sides. In this example, the attenuation with the single-phase coupling is considerably lower than the attenuation with the three-phase coupling. The difference of attenuation value at 150 kHz is nearly 30 dB. Obviously, single-phase coupling is a preferred coupling mode for PLC nodes at the same phase. There are different ways to connect PLC nodes at different phases. An expensive one is to use three transmitters to inject the same signal into the three phases simultaneously. Another possibility is via crosstalk. Many papers, for example, [7,8] report considerable crosstalk between different phases in the considered frequency band. Due to this crosstalk, a PLC node may still be able to directly receive a signal with a sufficient level which is sent at a different phase. It has also been observed that there is significant crosstalk between different phases at the same meter panel. So, PLC nodes coupled at the same phase as the transmitter at the transformer may act as repeater for PLC nodes which are coupled at different phases.

Figure 3.15 shows a collection of attenuation, which are measured between the transformer and different HAPs and with different coupling modes. It provides an impression of variability of attenuations. The attenuation has a large spread which is between 28 and 70 dB. The most attenuation values are in the range 35–60 dB. For the frequency above 250 kHz, the attenuation increases slightly with the frequency.

3.4.4 Link Quality Index

To evaluate power line channel quality and to consider the impact of two important channel characteristics, namely, the noise and the attenuation, the so-called link quality index

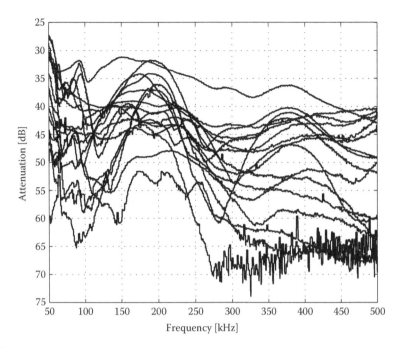

FIGURE 3.15
Collection of attenuations between the transformer and different HAPs.

LQI(f,t) is introduced. LQI(f,t) is defined as the sum of the noise and the attenuation (loss) as follows:

$$\mathrm{LQI}(f,t) := \mathrm{Noise}(f,t) + \mathrm{Loss}(f,t), \tag{3.4}$$

where
  Loss(f,t) is given as a positive attenuation in dB
  Noise(f,t) is the noise power given, for example, in dBμV or dBm measured at frequency
    f and time t and for a given equivalent noise bandwidth

Hence, LQI(f,t) has the same dimension as Noise(f,t). LQI(f,t) of a link is equivalent to the signal power to be sent at the transmitter to get an equivalent 0 dB *signal-to-noise ratio* (SNR) at the receiver. Given LQI(f,t) of a link and a signal power at the transmitter Tx(f,t) in dBμV or dBm, the corresponding SNR at the receiver of the link is simply given by

$$\mathrm{SNR}(f,t) = \mathrm{Tx}(f,t) - \mathrm{LQI}(f,t). \tag{3.5}$$

Obviously, the smaller LQI(f,t) is, the less signal power is needed for a required SNR.

  Figure 3.16 shows example of LQI for the three downlinks from the transformer to the HAPs of the three buildings. The noise bandwidth is 10 kHz. Three-phase coupling is used at the transformer. The downlink to the building #2 has a minimum LQI value of 85 dBμV at 375 kHz and a maximum LQI value of 109 dBμV at 300 kHz. The downlink to the building #3 has, overall, the highest LQI value and hence requires, on average, the highest signal power than the other two downlinks to obtain the same SNR. Interesting to note is that this

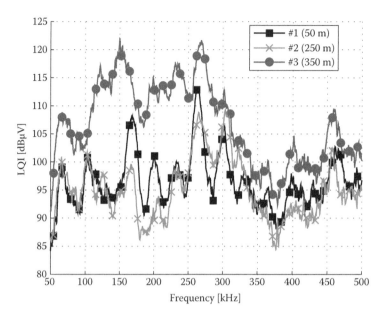

FIGURE 3.16
LQI for the three downlinks between the transformer and the three buildings (HAPs).

link has a quite large distance, which is 350 m. Nevertheless, the smallest LQI of that link is not measured at low frequencies, but rather at higher frequencies between 350 and 500 kHz.

Figure 3.17 shows a collection of LQI from different measurements. The noise bandwidth is 10 kHz. The smallest LQI value is about 80 dBµV and the largest LQI value is about 137 dBµV. Most links have an LQI value between 85 and 115 dBµV. A weak tendency can be observed that the LQI decreases with the frequency.

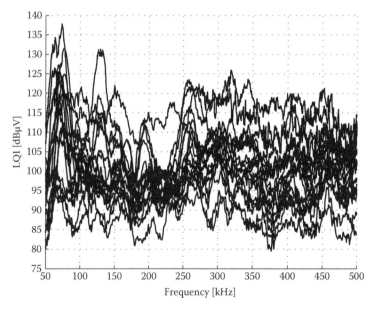

FIGURE 3.17
Collection of LQI from different measurements.

3.4.5 SNR Examples for NB PLC in CENELEC A Band

It is interesting to know what SNR will be available for the LQI values presented in Figure 3.17 if CENELEC A band signal is employed which uses the signal level defined in EN50065-1 [11].

For this purpose, an *orthogonal frequency division multiplex* (OFDM) signal between 40 and 90 kHz with a peak-to-RMS ratio of 8 dB is assumed. For wideband signal (>5 kHz), EN50065-1 specifies a limit of the peak signal of 140 dBµV for three-phase coupling measured with an *artificial mains network* (AMN) which has an input impedance between 5 and 20 Ω [11]. This means an RMS limit of 132 dBµV for the assumed OFDM signal. Since the LQI value in Figure 3.17 includes the coupling network in Figure 3.1, to determine available SNR, the RMS of transmitter signal at the input of the coupling network needs to be derived. Since the magnitude of the impedance of the coupling network for frequency between 40 and 90 kHz is in the range of 0.3–1.3 Ω, which is very small compared to the input impedance of 5–20 Ω as defined in [11], the signal loss due to the coupling network can be neglected. Hence, the OFDM signal has approximately an RMS of 132 dBµV between 40 and 90 kHz at the input of the coupling network. The available SNR for LQI from Figure 3.17 with 10 kHz equivalent noise bandwidth can be calculated as

$$\text{SNR}(f) = 132 \text{ dBµV} - 10\log_{10}\left(50 \text{ kHz}/10 \text{ kHz}\right) - \text{LQI}(f) = 125 \text{ dBµV} - \text{LQI}(f). \quad (3.6)$$

The SNR value of Equation 3.6 is depicted in Figure 3.18 which shows that in this case, the most available SNR values are between 10 and 40 dB. It can also be observed that the SNR of a particular link may have a large variation with the frequency. The lowest SNR values are found between 60 and 75 kHz with some SNR values far below 0 dB.

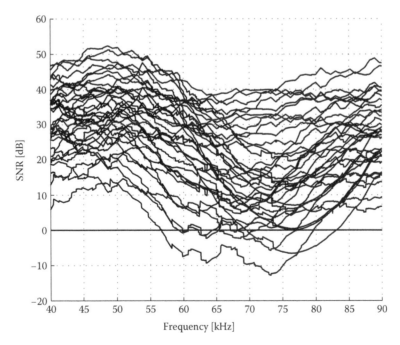

FIGURE 3.18
Available SNR for various links with LQI value in Figure 3.17 and for CENELEC A band OFDM signal between 40 and 90 kHz and with a maximum permitted signal limit according to [11].

Actual PLC systems may not exploit the permitted maximum signal limit. *Power line intelligent metering evolution* (PRIME) specifies OFDM signal between 42 and 89 kHz with a minimum RMS of transmitter signal of 114 dBμV for the three-phase coupling and for the input impedance of 1 Ω [12]. This impedance value is comparable with the value of the coupling network used for the measurement. Hence, for a first approximation, a 6 dB signal loss due to the coupling network is assumed, which leads to an equivalent RMS of transmitter signal of 120 dBμV (114 dBμV + 6 dB) at the input of the coupling network. In this case, PRIME signal will have ~12 dB less SNR compared with the earlier described OFDM scheme which transmits at maximum permitted signal level [11]. An overall reduction of SNR values in Figure 3.18 by 12 dB means a considerable deterioration of link reliability for PRIME for the various links with LQI value in Figure 3.18.

3.5 Conclusion

Narrowband power line channel characteristics in typical underground LV access network in China are evaluated based on measurement results. A three-phase coupling at transformer has very low impedance. This low impedance together with a signal splitting into the three phases may result in considerable signal loss compared to a single-phase coupling. The attenuation has a broad range between 20 and 70 dB. In contrast to BPL, for NB PLC there is not always a clear trend for the attenuation to increase with the frequency and with the distance. On the other side, the noise power decreases with the frequency significantly. Measurements show a reduction of average noise power of 10 kHz noise bandwidth from a range between 50 and 90 dBμV at 50 kHz to a range between 30 and 60 dBμV at 500 kHz. A considerable difference of noise level between night and day is observable. The noise also exhibits cyclostationary behaviour with a period which corresponds to the half cycle of AC mains. The *link quality index* (LQI), taking into account both noise and attenuation proved, is a possible measure to compare different channels. An LQI analysis shows that for a given link, there is a strong variation of LQI values along the frequency axis. On average, there is a weak trend in favour of higher frequencies.

Acknowledgements

This work was funded by the project 'The New Generation Smart PLC Key Technologies Research' of *State Grid Corporation of China* (SGCC). The authors wish to thank Prof. Dostert and his assistants for the support in measurement tools.

References

1. Liu, J., Zhao, B., Geng, L. et al., Current situations and future developments of PLC technology in China, *Proceedings of IEEE International Symposium Power Line Communications and Its Applications*, Beijing, China, 28–30 March 2012.
2. Galli, S., Scagllone, A. and Wang, Z., For the grid and through the grid: The role of power line communications in the smart grids, *Proceedings of the IEEE*, 99(6), 998–1027, June 2011.

3. Dostert, K., *Power Line Communications*. Prentice Hall, Upper Saddle River, NJ, 2001.

4. Ferreira, H. C., Lampe, L., Newbury, J. et al., *Power Line Communications*. John Wiley & Sons, West Sussex, UK, 2010.

5. Sigle, M., Bauer, M., Liu, W. et al., Transmission channel properties of the low voltage grid for narrowband power line communication, *Proceedings of IEEE International Symposium Power Line Communications and Its Applications*, Udine, Italy, 3–6 April 2011, pp. 289–294.

6. Varadarajan, B., Kim, I., Dabak, A. et al., Empirical measurements of the low-frequency power-line communications channel in Rural North America, *Proceedings of IEEE International Symposium Power Line Communications and Its Applications*, Udine, Italy, 3–6 April 2011, pp. 463–467.

7. Sigle, M., Liu, W. and Dostert, K., On the impedance of the low-voltage distribution grid at frequencies up to 500 kHz, *Proceedings of IEEE International Symposium Power Line Communications and Its Applications*, Beijing, China, 28–30 March 2012.

8. Arzberger, M., Dostert, K., Waldeck, T. et al., Fundamental properties of the low voltage power distribution grid, *Proceedings of IEEE International Symposium Power Line Communications and Its Applications*, Essen, Germany, 2–4 April 1997, pp. 45–50.

9. Katayama, M., Yamazato, T. and Okada, T., A mathematic model of noise in narrowband power line communications systems, *IEEE Journal on Selected Areas in Communications*, 24(7), 1267–1276, July 2006.

10. Katayama, M., Itou, S., Yamazato, T. et al., Modeling of cyclostationary and frequency dependent power-line channels for communications, *Proceedings of IEEE International Symposium Power Line Communications and Its Applications*, Ireland, April 2000, pp. 123–127.

11. CENELEC, EN 50065-1 Signaling on low-voltage electrical installations in the frequency range 3 kHz to 148.5 kHz – Part 1: General requirements, frequency bands and electromagnetic disturbances, April 2011.

12. PRIME Alliance, Power Line Intelligent Metering Evolution (PRIME) Specification, v1.3E, 2010.

4

Broadband In-Home Characterisation and Correlation-Based Modelling

Kaywan Afkhamie, Paola Bisaglia, Arun Nayagam, Fabio Osnato,
Deniz Rende, Raffaele Riva and Larry Yonge

CONTENTS

4.1 Introduction*

The in-home network of the future will be a hybrid network where multimedia contents must be provisioned over a stable broadband backbone (as described in Chapter 15). In agreement with this vision, in the past years, the HomePlug Alliance had released the *HomePlug AV* 1.1 (HPAV 1.1) specification [4]. This solution allows achieving a maximum throughput of 200 Mbit/s over the *power line* (PL; see also Chapter 13). Today, the market demand requires a migration to a much higher performance in order to support applications like high-definition multimedia contents and gaming. To accommodate the request for this increase in capacity, reliability and coverage, the HomePlug Alliance has

---

* Permission to re-use part of the material in [1–3] in this and in all subsequent editions, revisions and derivative works in English and in foreign translations, in all formats, including electronic media, has been granted to the authors by the *Institute of Electrical and Electronics Engineers, Incorporated* (the 'IEEE').

defined the specification for the next generation of PL technology, namely, *HomePlug AV 2.0* (HPAV 2.0) (see Chapter 14). One of the major enhancements related to this technology is the introduction of *multiple-input multiple-output* (MIMO) technique. This technique has already attracted attention in wireless communications (see, e.g. the standards IEEE 802.11n [5] and 3GPP LTE [6]). MIMO applied to PL, as already analysed in the literature, addresses either the channel capacity [7,8] or the performance of different MIMO schemes [9,10] (see Chapter 9). Moreover, MIMO PL channels are also addressed in [11–18]. An introduction to PL channel and noise characterisation is provided in Chapter 1, where the most representative in-home PL models that can be found in recent literature are summarised. More details on the characteristics of the MIMO channel are presented in this chapter and in Chapter 5. In the literature, the characteristics of the PL noise have not been widely discussed; in many contributions, it is simply assumed to be *additive white Gaussian noise* (AWGN). The main results suggested that the in-home PL channel capacity can be increased by a factor of around 2 when MIMO techniques are applied.

This chapter addresses the statistical characterisation of MIMO PL channels and noise based on experimental measurements collected in the United States. It presents a MIMO channel modelling approach based on correlation matrices. Statistics of the in-home channel and noise characterisation in Europe is presented in Chapter 5, along with a stochastic channel model based on experimental databases. Analysing a wide set of measurements, the statistical distribution of the most important channel and noise parameters is reported. The considered set of in-home PL channels contains 92 transfer functions measured in the 1.8–88 MHz range, in five North American houses.

This chapter is organised as follows. Section 4.2 details the measurement setup and the procedure to estimate the coefficients of the MIMO PL channels. Section 4.3 presents a statistical analysis of the collected data and a comparison with previous available literature analysis. Based on these results, Section 4.4 proposes a MIMO PL channel model with particular attention to the spatial correlation properties. Section 4.5 concentrates on the characterisation of MIMO PL noise and its effect on channel capacity. Finally, in Section 4.6, conclusions are reported.

The following notation will be used. Vectors and matrices are in bold face. The symbols $(\cdot)^T$, $(\cdot)^{-1}$, $(\cdot)^H$, $(\cdot)^*$, $E[\cdot]$ denote the transposition, the inverse, the Hermitian (conjugate transposition), the conjugate and the expectation, respectively.

4.2 Description of the MIMO PL Field Test System

The apparatus and signalling used to characterise the MIMO PL channel is described in this section. In details, the characterisation of the transfer function of the channel is needed, as well as the noise that is present on the different receiver ports. A typical approach that is used to observe the transfer function of a channel involves the use of a *vector network analyser* (VNA). The VNA sounds the channel using narrowband tones and computes the various *S*-parameters. This results in an accurate measurement of the channel transfer function. However, time-domain phenomena cannot be observed using a VNA.

In order to observe time-domain behaviour and to have the ability to collect noise samples, a time-domain characterisation approach has been chosen. A block diagram of the system is presented in Figure 4.1. Although three ports are available for signalling, only two out of the three ports are useful for transmitting simultaneously. This is because

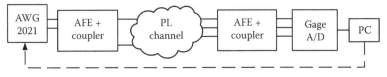

GPIB back-channel to program AWG2021

FIGURE 4.1
The MIMO power line field test system. (From Rende, D., Nayagam, A., Afkhamie, K., Yonge, L., Riva, R., Veronesi, D., Osnato, F., and Bisaglia, P., Noise correlation and its effects on capacity of inhome MIMO power line channels, in *Power Line Communications and Its Applications (ISPLC), IEEE International Symposium on*, April 2011. Copyright © 2011 IEEE. With permission.)

on the transmitter side, the voltage on the third port is a linear combination of the voltages on the other two ports (Kirchhoff's law). In this effort, *line–neutral* (L-N) and *line–protective earth* (L-PE) were used as the transmitting wire pairs. On the transmitter side, two independent signals are generated by a Tektronix AWG2021 *arbitrary waveform generator* (AWG) and passed through an *analogue front end* (AFE). The two-port AFE amplifies the signals and provides the coupling circuitry to inject the signal onto the electrical wires. The two time-domain waveforms are packets composed of two portions: the preamble and the payload. Preamble is dedicated to automatic gain control tuning, frame detection and frame synchronisation. The payload portion is *orthogonal frequency-division multiplexing* (OFDM) modulated, with a carrier spacing of 24.414 kHz (as in the HPAV 1.1 specification [4]) and 4096 carriers. In order to facilitate the estimation of the MIMO channel coefficients, two orthogonal codes are applied to the transmitted signals; hence, during the payload, each OFDM symbol is repeated twice, with the sign fixed by the orthogonal codes, that is, [+1, +1] code applied to the signal transmitted from one port and [+1, −1] code applied to the signal transmitted from the other port. Considering a sampling frequency of 200 MHz, the two signals occupy the frequency range from 0 to 100 MHz. The actual starting frequency is 1.8 MHz, as in HPAV 1.1, and to avoid the noise interference from the FM band, the frequencies above 88 MHz have been masked. Each packet consists of five orthogonal symbol pairs.

To improve the received signal quality, all three of the ports can be used on the receive side, that is, L-N, L-PE and N-PE. In fact, a fourth port is also available if the common-mode signal can be captured [9]. The receiver in the field test system consists of a three-port receiver AFE and coupler module followed by a 200 MSamples/s, 4-channel 16-bit digitizer (CS16200 manufactured by Gage-Applied Inc.). Note that coupling on both the transmitter and receiver sides is capacitive and designed to minimise crosstalk between the individual ports of the AFE. The digitizer is connected to a personal computer. The *General Purpose Interface Bus* (GPIB) back-channel link is used to load the AWG with the two transmit waveforms that are generated by the computer. The receiver AFE filters and amplifies signals that are coupled from the PL on the L-N, L-PE and N-PE wire pairs, respectively.

The signalling structure is shown in Figure 4.2. The AWG transmits a repeating sequence of three packets. As illustrated in Figure 4.2, the transmitter first performs a MIMO transmission using two transmitter ports, then a *single-input single-output* (SISO) transmission using L-N wire pairs only and then a SISO transmission using L-PE wire pairs only. This sequence is repeated continuously, and received signals are captured to collect about 16 repetitions of MIMO, SISO L-N and SISO L-PE transmissions. During the L-N- and L-PE-only transmissions, the transmit power is increased by 3 dB to keep the overall transmit

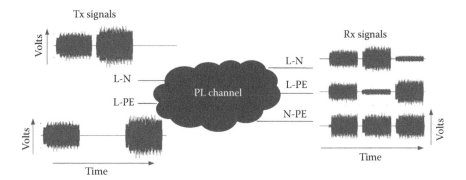

FIGURE 4.2
Signalling used in the PL MIMO field test system. (From Rende, D., Nayagam, A., Afkhamie, K., Yonge, L., Riva, R., Veronesi, D., Osnato, F., and Bisaglia, P., Noise correlation and its effects on capacity of inhome MIMO power line channels, in *Power Line Communications and Its Applications (ISPLC), IEEE International Symposium on*, April 2011. Copyright © 2011 IEEE. With permission.)

power spectral densities (PSDs) even between the MIMO and SISO transmissions. At the output of the transmit coupler, the PSD for each transmit stream in the 1.8–88 MHz band is −67 dBm/Hz during the MIMO packet transmission (first packet in the sequence) and −64 dBm/Hz during the SISO transmissions (second and third packets in the sequence).

As seen in Figure 4.2, there are gaps between the different packets. The received waveform samples in the 'transmission-free' gaps between the packets are samples of noise on the PL, and these samples are used to characterise the PL noise.

This test setup was used to collect data from 92 paths distributed equally among five homes in North America. All homes are detached single family homes varying in age from 5 to 25 years and in size from ∼180 to ∼320 m². The extraction of relevant channel and noise parameters from the waveforms collected in the field is described next.

During MIMO transmission, the received samples in the frequency domain on carrier k are given by

$$x_k = H_k a_k + n_k,\qquad(4.1)$$

where
$a_k = \left[a_k^{(1)}, a_k^{(2)} \right]^T$ are the symbols transmitted from the two transmitter ports
$x_k = \left[x_k^{(1)}, x_k^{(2)}, x_k^{(3)} \right]^T$ are the samples received at the three receiver ports
$n_k = \left[n_k^{(1)}, n_k^{(2)}, n_k^{(3)} \right]^T$ are the noise samples at the three receiver ports
H_k is the 3×2 channel matrix on carrier k

$$H_k = \begin{bmatrix} H_k^{(1,1)} & H_k^{(1,2)} \\ H_k^{(2,1)} & H_k^{(2,2)} \\ H_k^{(3,1)} & H_k^{(3,2)} \end{bmatrix}.\qquad(4.2)$$

$H_k^{(r,t)}$, in the channel matrix, represents the channel from the transmitter port t to receiver port r on carrier k. Note that on the PL the noise is not generally AWGN, that is, n_k is a Gaussian random vector with zero mean and covariance matrix $R_{n,k} \neq I$, where I is the identity matrix ($n_k \in \mathcal{N}_3(0, R_{n,k})$, with $0 = [0,0,0]^T$). The noise covariance matrix is defined as $R_{n,k} = E\left[n_k n_k^H\right]$.

The linear model in Equation 4.1 can be whitened by pre-multiplying by $(R_{n,k})^{-1/2}$. This pre-whitening approach is often used in signal processing to extend results on AWGN to coloured noise. Pre-whitening Equation 4.1 yields

$$(R_{n,k})^{-1/2} x_k = (R_{n,k})^{-1/2} H_k a_k + w_k, \tag{4.3}$$

where $w_k \in \mathcal{N}_3(0, I)$. Henceforth, the pre-whitened channel $H_{w,k} = (R_{n,k})^{-1/2} H_k$ will be referred to as the composite channel (the effect of noise correlation and channel attenuation is combined into one matrix).

Let $a_k^{(t)}[m]$ be the symbol loaded on carrier k with $k = 1, 2, \ldots, 4096$ on the transmitter port t with $t = 1, 2$ during the mth OFDM symbol of the payload, with $m = 1, 2, \ldots, N$. Let $x_k^{(r)}[m]$ be the received sample on the carrier k, on the receiver port r with $r = 1, 2, 3$, during the mth OFDM symbol. Given these definitions and recalling the fact that two orthogonal codes are applied to the transmitted signals, the estimates of the channel coefficients are defined by the following equation:

$$H_k^{(r,t)} = \begin{cases} \dfrac{1}{N/2} \displaystyle\sum_{m=1}^{N/2} \dfrac{x_k^{(r)}[2m] + x_k^{(r)}[2m-1]}{2 \cdot a_k^{(1)}[2m]} & \text{when } t = 1, \\[4mm] \dfrac{1}{N/2} \displaystyle\sum_{m=1}^{N/2} \dfrac{x_k^{(r)}[2m] - x_k^{(r)}[2m-1]}{2 \cdot a_k^{(2)}[2m]} & \text{when } t = 2. \end{cases} \tag{4.4}$$

By using Equation 4.4, the assumption made is that the channel does not change from the mth OFDM symbol to the $(m+1)$th. In our study, the number of transmitted OFDM symbols N is 10. For given values of r and t, by computing the *inverse fast Fourier transform* (IFFT) of $H_k^{(r,t)}$, the channel impulse responses $h_u^{(r,t)}$ are obtained with $u = 1, 2, \ldots, L_H$. Choosing $L_H = 2000$, at least 99% of the channel energy is preserved. Considering a sampling frequency of 200 MHz, the sampling time is equal to 5 ns. Thus, the channel impulse response $L_H = 2000$ is equivalent to 10 μs. In the next section, the channels that were estimated on the 92 paths are analysed.

For the noise samples on a certain carrier, $n_k = \left[n_k^{(1)}, n_k^{(2)}, n_k^{(3)}\right]^T$, the noise covariance matrix $R_{n,k}$ can be empirically obtained as an ensemble average from the noise samples that are captured using the field test system.

4.3 Channel Statistical Analysis

In this section, the estimated channels are analysed in order to provide a statistical description of the most relevant parameters useful to characterise MIMO PL channel responses. As an example, Figure 4.3 shows the time impulse response and the frequency response of a generic SISO channel extracted from one of the MIMO channel captures. In other words, the time and frequency characterisation for one element of the H matrix is shown in Figure 4.3. The dashed line of Figure 4.3b refers to a channel realisation obtained as detailed in Section 4.4.

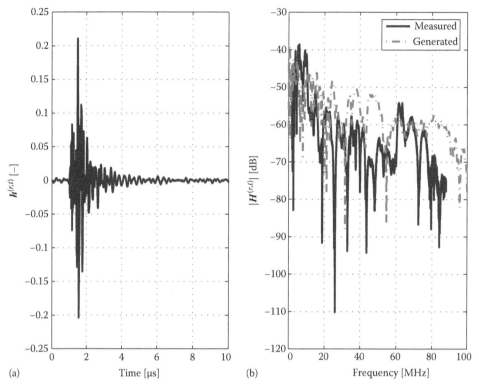

(a)   Time [μs]   (b)   Frequency [MHz]

FIGURE 4.3
Example of a channel realisation: (a) time impulse response and (b) frequency response. (From Veronesi, D., Riva, R., Bisaglia, P., Osnato, F., Afkhamie, K., Nayagam, A., Rende, D., and Yonge, L., Characterization of in-home MIMO power line channels, in *Power Line Communications and Its Applications (ISPLC), IEEE International Symposium on*, April 2011. Copyright © 2011 IEEE. With permission.)

4.3.1 Average Attenuation versus Frequency

In this section, each MIMO channel is considered as a collection of six SISO channels and the energy of each channel is normalised as follows:

$$\bar{H}_k^{(r,t)} = \frac{H_k^{(r,t)}}{\sqrt{\sum_{k=1}^{4096} |H_k^{(r,t)}|^2}}, \tag{4.5}$$

where $\bar{H}_k^{(r,t)}$ denotes the coefficients of the normalised channel matrix $\bar{H}_k$. In Figure 4.4a, the channel frequency response is reported, for all the normalised channels. The super-imposed bold black curve represents the average value. It can be observed that the PL channel gain linearly depends on the frequency with a negative slope. The linear approximation of the average channel gain can be defined as

$$|H(f)|_{dB} = A \cdot f + B \tag{4.6}$$

with $A = -1.9819 \times 10^{-7}$ [1/Hz] and $B = 1.2578$. This linear approximation is shown in Figure 4.4a with the bold straight grey line. It is also interesting to observe how the captured noise

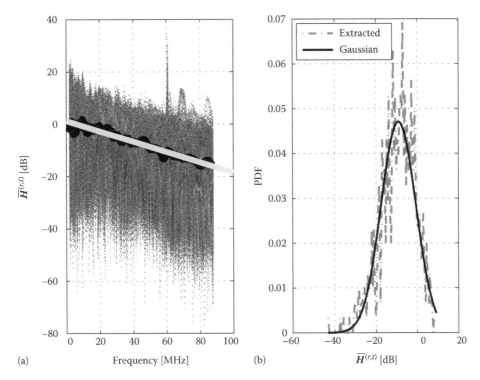

FIGURE 4.4

(a) Channel frequency response for the normalised channels and (b) PDF of the amplitude on the 2048th carrier. (From Veronesi, D., Riva, R., Bisaglia, P., Osnato, F., Afkhamie, K., Nayagam, A., Rende, D., and Yonge, L., Characterization of in-home MIMO power line channels, in *Power Line Communications and Its Applications (ISPLC), IEEE International Symposium on*, April 2011. Copyright © 2011 IEEE. With permission.)

can affect the channel measurements. This phenomenon could be clearly seen from one measurement in the figure showing a spike around 65 MHz. In Figure 4.4b, the *probability density function* (PDF) of the channel gain on a generic carrier is shown. The mean and the standard deviation of this PDF are extracted and then they are used to superimpose the Gaussian distribution. On this carrier, it appears that the truncated Gaussian distribution (shown with the dashed line) fits the PDF of the channel gain. This consideration holds true for all the carriers. Other possible distributions for the observed experimental data would be the Rayleigh or Weibull distributions.

4.3.2 Distribution of the Channel Coefficients

To facilitate the evaluation of the statistical properties of the channel impulse response, in our analysis, all the time-domain profiles are aligned on their maximum value. In other words, the channel impulse response is time shifted such that its maximum absolute value appears always at a given time instant. Let us identify this time instant with the label *MaxPos*. Figure 4.5 reports the superposition of all the estimated channel impulse responses, where the maximum absolute value of the channel impulse response is placed at the time instant *MaxPos* = 300.

Analysing the distribution of the real-valued channel coefficient at a given time instant, three statistical distributions could characterise all the channel impulse responses. For the time instants belonging to the range $\chi_L = [1, MaxPos - 2]$, the channel coefficients have

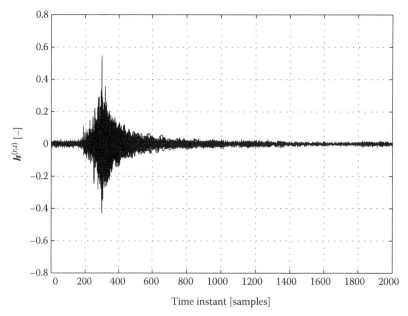

FIGURE 4.5
Superposition of all the estimated channel impulse responses. (From Veronesi, D., Riva, R., Bisaglia, P., Osnato, F., Afkhamie, K., Nayagam, A., Rende, D., and Yonge, L., Characterization of in-home MIMO power line channels, in *Power Line Communications and Its Applications (ISPLC), IEEE International Symposium on*, April 2011. Copyright © 2011 IEEE. With permission.)

a random sign and amplitude values described by a Weibull distribution. For the time instants belonging to the range $\chi_C = [MaxPos - 1, MaxPos + 1]$, the channel coefficients have a random sign and amplitude values described by a Gaussian distribution with non-zero mean. For the time instants belonging to the range $\chi_R = [MaxPos + 2, L_H]$, the channel coefficients have a Gaussian distribution with zero mean. As an example, Figure 4.6 reports the measured PDFs with a black line and the proposed distributions with a marked grey line for the time instants: (a) $MaxPos - 5$, (b) $MaxPos$ and (c) $MaxPos + 5$.

The PDF of a Weibull random variable x is completely characterised by two parameters – the shape parameter k and the scale parameter λ – as

$$f(x,\lambda,k) = \begin{cases} \dfrac{k}{\lambda}\left(\dfrac{x}{\lambda}\right)^{k-1} e^{-(x/\lambda)^k}, & x \geq 0, \\ 0, & x < 0. \end{cases} \tag{4.7}$$

The shape parameter k is assumed to be a constant value equal to 3/4 and the scale parameter λ is expressed as a function of the time instant:

$$\lambda_p = e^{C_0 + C_1 \cdot p + C_2 \cdot p^2 + C_3 \cdot p^3 + C_4 \cdot p^4 + C_5 \cdot p^5} \tag{4.8}$$

with $C_0 = -6.83$, $C_1 = 9.5 \times 10^{-3}$, $C_2 = -2.25 \times 10^{-4}$, $C_3 = 2.07 \times 10^{-6}$, $C_4 = -6.98 \times 10^{-9}$ and $C_5 = 9.15 \times 10^{-12}$. Figure 4.7a shows the value of the scale parameter extracted from the captures with a dashed line and Equation 4.8 with markers, as a function of the time instant.

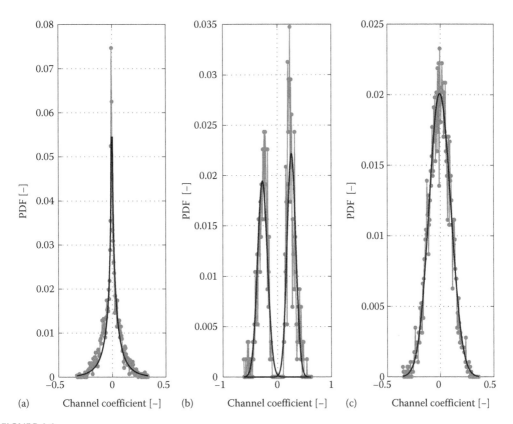

FIGURE 4.6

PDF of the channel impulse response at three different time instants: (a) *MaxPos* − 5, (b) *MaxPos* and (c) *MaxPos* + 5. (From Veronesi, D., Riva, R., Bisaglia, P., Osnato, F., Afkhamie, K., Nayagam, A., Rende, D., and Yonge, L., Characterization of in-home MIMO power line channels, in *Power Line Communications and Its Applications (ISPLC), IEEE International Symposium on*, April 2011. Copyright © 2011 IEEE. With permission.)

The Gaussian distribution with non-zero mean is characterised by the mean and standard deviation values reported in Table 4.1. These parameters are extracted from the captures for the time instants belonging to the range [*MaxPos* − 1, *MaxPos* + 1].

The Gaussian PDF having zero mean is completely characterised by the standard deviation which can be expressed as a function of the time instant:

$$\sigma_p = e^{C_0 + C_1 \cdot p + C_2 \cdot p^2 + C_3 \cdot p^3 + C_4 \cdot p^4 + C_5 \cdot p^5} \tag{4.9}$$

with $C_0 = 1.1$, $C_1 = -1.59 \times 10^{-2}$, $C_2 = 1.84 \times 10^{-5}$, $C_3 = -1.37 \times 10^{-8}$, $C_4 = 5.62 \times 10^{-12}$ and $C_5 = -9.22 \times 10^{-16}$. Figure 4.7b reports the value of the standard deviation extracted from the captures with a dashed line and Equation 4.9 with markers, as a function of the time instant.

4.3.3 RMS-DS versus the Attenuation

In [19], the relation between the *root mean square of the delay spread* (RMS-DS) and the average gain of the PL channel is proposed.

Figure 4.8 shows a scatter plot of the PL measures together with the trend lines, calculated using the least squares algorithm. Also in this figure, the MIMO channels are

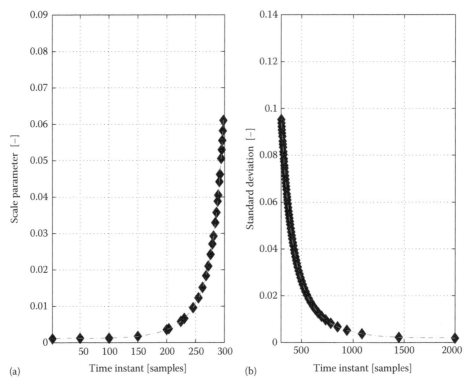

(a) Time instant [samples]   (b) Time instant [samples]

FIGURE 4.7

(a) Scale parameter versus the time instant, expressed in samples, and (b) standard deviation versus the time instant, expressed in samples. (From Veronesi, D., Riva, R., Bisaglia, P., Osnato, F., Afkhamie, K., Nayagam, A., Rende, D., and Yonge, L., Characterization of in-home MIMO power line channels, in *Power Line Communications and Its Applications (ISPLC), IEEE International Symposium on*, April 2011. Copyright © 2011 IEEE. With permission.)

TABLE 4.1

Mean and Standard Deviation Values for the Gaussian
Distribution with Non-Zero Mean

	Time Instant Samples		
	MaxPos − 1	*MaxPos*	*MaxPos* + 1
Mean	1.45×10^{-1}	2.56×10^{-1}	1.47×10^{-1}
Standard deviation	6.96×10^{-2}	8.16×10^{-2}	7.49×10^{-2}

Source: Veronesi, D., Riva, R., Bisaglia, P., Osnato, F., Afkhamie, K., Nayagam, A., Rende, D., and Yonge, L., Characterization of in-home MIMO power line channels, in *Power Line Communications and Its Applications (ISPLC), IEEE International Symposium on*, April 2011. Copyright © 2011 IEEE. With permission.

considered as the composition of six independent SISO channels, and they are depicted with circles. In the figure, two distinct 'clouds' of attenuation values could be seen. In general, higher attenuation values correspond to longer physical paths. One aspect to be noticed is that the experimental measurements belong to a finite set. It could be expected that increasing the number of collected measurements, the average attenuation could span all the range of values roughly from −70 to −10 dB. In Figure 4.8, the same relation as in [19]

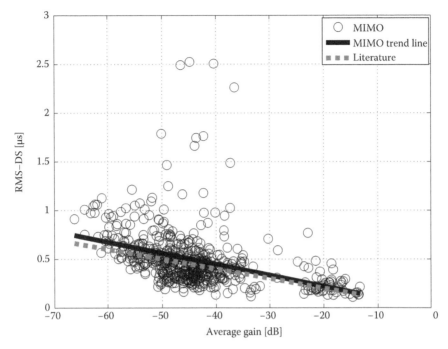

FIGURE 4.8
RMS-DS versus the average channel gain. (From Veronesi, D., Riva, R., Bisaglia, P., Osnato, F., Afkhamie, K., Nayagam, A., Rende, D., and Yonge, L., Characterization of in-home MIMO power line channels, in *Power Line Communications and Its Applications (ISPLC), IEEE International Symposium on*, April 2011. Copyright © 2011 IEEE. With permission.)

(labelled 'literature'), that is, $\sigma_{\tau,\mu s} = -0.01 \cdot \bar{G}_{dB}$ (with $\sigma_{\tau,\mu s}$ being the RMS-DS and $\bar{G}_{dB}$ being the average channel attenuation), and the equivalent results obtained from the MIMO PL measures (labelled 'MIMO trend line') are reported. Note that the measurements analysed in [19] also belong to a set of channel measurements collected in the past years by the HomePlug Alliance.

Based on these results, the relationship between the RMS-DS and the average gain proposed in [19] holds true also for the other five SISO channels that constitute the composite 3×2 MIMO channel.

4.3.4 MIMO Channel Correlation

Let us decompose the channel matrix on a given carrier k with the singular value decomposition as $H_k = U_k S_k V_k^H$. The matrix S_k has only two non-zero entries on the major diagonal. These entries $\lambda_{H_k}^{(i)}$ with $i = 1, 2$ are the singular values of H_k. Similarly to [8], a correlation factor among the channel coefficients on carrier k is introduced:

$$\kappa_k = \left(\frac{\min\{\lambda_{H_k}^{(1)}, \lambda_{H_k}^{(2)}\}}{\max\{\lambda_{H_k}^{(1)}, \lambda_{H_k}^{(2)}\}} \right)^2. \tag{4.10}$$

By definition, when the channel is completely correlated, $\kappa_k = 0$; while, when the channel is completely uncorrelated, $\kappa_k = 1$. To analyse this factor, Figure 4.9a reports the *cumulative distribution function* (CDF) of κ_k considering all the measurements and all the frequencies

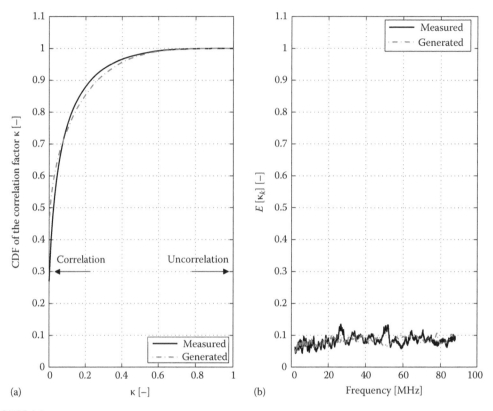

FIGURE 4.9
(a) CDF of the channel correlation and (b) mean value of the channel correlation versus the frequency. (From Veronesi, D., Riva, R., Bisaglia, P., Osnato, F., Afkhamie, K., Nayagam, A., Rende, D., and Yonge, L., Characterization of in-home MIMO power line channels, in *Power Line Communications and Its Applications (ISPLC), IEEE International Symposium on*, April 2011. Copyright © 2011 IEEE. With permission.)

(labelled 'measured'), while Figure 4.9b reports the average value of κ_k, of all the measurements, versus the frequency (labelled 'measured'). The dashed line of Figure 4.9 refers to channel realisations obtained as detailed in Section 4.4.

From Figure 4.9a, it is clear that the estimated channels have a correlation factor κ_k that is not uniformly distributed in the range [0, 1], and in 90% of the cases, there is a correlation factor lower than 0.2. For each carrier, the correlation factor has the same PDF (it is obtained considering all the estimated channels). To prove this observation, Figure 4.9b reports the average value of the correlation factor versus the frequency. It is quite evident that the average correlation is almost constant with respect to the frequency. As a consequence, the proposed channel model of Section 4.4 introduces a unique correlation factor for all the carriers.

4.3.4.1 Spatial Correlation Analysis

In this section, the 'spatial' properties of MIMO PL channels are investigated, where the term 'spatial' refers to the MIMO ports. The generic carrier k is considered, and for simplicity, the index k is omitted where not necessary.

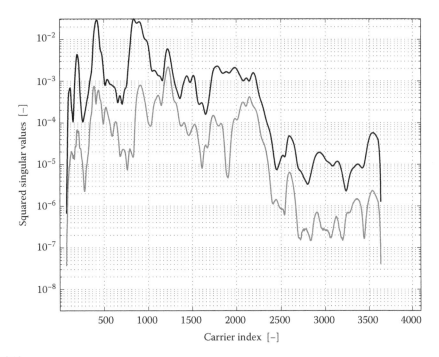

FIGURE 4.10
An example of channel spectrum from one field measurement. The two lines represent the squared singular values of *H* in logarithmic scale for each carrier. (From Tomasoni, A., Riva, R., and Bellini, S., Spatial correlation analysis and model for in-home MIMO power line channels, in *Power Line Communications and Its Applications (ISPLC), IEEE International Symposium on*, March 2012. Copyright © 2012 IEEE. With permission.)

The first step for the analysis of the spatial correlation is to evaluate the singular values of the channel coefficients in the frequency domain. As an example, Figure 4.10 plots the squared singular values, as defined in Section 4.3.4, of the channel *H* in logarithmic scale, from one field measurement. It is clear from the figure that the channel strongly fades at high frequencies (the Nyquist frequency is 100 MHz, and the channel is measured in the frequency range 1.8–88 MHz with a carrier spacing equal to 24.414 kHz).

The channel taps $h_u^{(r,t)}$ have null mean (as shown in the previous sections):

$$E[h_u^{(r,t)}] = 0. \tag{4.11}$$

The second-order statistic is more complicated. Indeed, channel taps cannot be considered independent. This can be easily understood thinking on the topology of the network: wires have almost the same length and are likely to cover similar paths. Besides, Kirchhoff's law must hold at the receiver side. All these effects can be captured by a channel covariance matrix:

$$\boldsymbol{R}_h = E\left[vect(\boldsymbol{H}) vect(\boldsymbol{H})^H \right] \tag{4.12}$$

with dimensions $N_t\,N_r \times N_t\,N_r$ providing information about all couples of channel taps, where N_t and N_r stand for the number of transmitter and receiver ports, respectively. The operator vect(·) aligns the columns of a matrix with size $N_r \times N_t$ to form a column vector of length $N_r\,N_t$.

The channel covariance matrix can be written as

$$R_h = G^2 \cdot R_t \otimes R_r, \tag{4.13}$$

where
  G is a constant introducing an overall channel gain
  R_t and R_r are the transmitter and receiver correlation matrices
  $\otimes$ is the Kronecker product and the normalisation chosen is as follows

$$\mathrm{tr}(R_t) = N_t, \tag{4.14}$$

$$\mathrm{tr}(R_r) = N_r, \tag{4.15}$$

where the operator $\mathrm{tr}(\cdot)$ denotes the trace.

The computation of the covariance matrix of the transmitter and receiver ports from the field measurements is an interesting aspect in the channel analysis. The average of the covariance matrices along the used carriers is not suitable due to the high dynamic range shown in Figure 4.10. The lower frequencies tend to experience lower attenuation than the higher frequencies, therefore, covariance matrices are better calculated as a weighted average.

Although the study of the channel attenuation is mainly useful to model the spectral properties of MIMO PL channels, it also helps to characterise the spatial properties of the same channels. First, the average power of the samples of H is calculated as

$$P_h(k) = \frac{1}{N_t N_r} \mathrm{tr}(H^H H) = \frac{1}{N_t N_r} \sum_{r=1}^{N_r} \sum_{t=1}^{N_t} \left| H_k^{(r,t)} \right|^2. \tag{4.16}$$

Once $P_h(k)$ has been obtained for each carrier, it is used to calculate from the measurements the correlation matrices $\hat{R}_t$ and $\hat{R}_r$ at the transmitter and at the receiver, respectively, as

$$\hat{R}_t = G_t \cdot \sum_k \frac{1}{P_h(k)} H_k^T H_k^*, \tag{4.17}$$

$$\hat{R}_r = G_r \cdot \sum_k \frac{1}{P_h(k)} H_k^* H_k^T, \tag{4.18}$$

where G_t and G_r are normalising factors, necessary to fulfil Equations 4.14 and 4.15.

Figure 4.11 depicts an example extracted from one field measurement. Spatial correlation matrices highlight pronounced values for the diagonal elements and lower values for the off-diagonal elements. This behaviour should be taken into account during the modelling phase. The behaviours of the eigenvalues and of the diagonal elements of the measured $\hat{R}_t$ and $\hat{R}_r$ are reported in Figures 4.12 and 4.13 with dashed lines (solid lines represent synthetic values, described in the following analysis). Each point on the x-axis represents one of the $N = 92$ measurements. Horizontal lines are the average values of the corresponding curves.

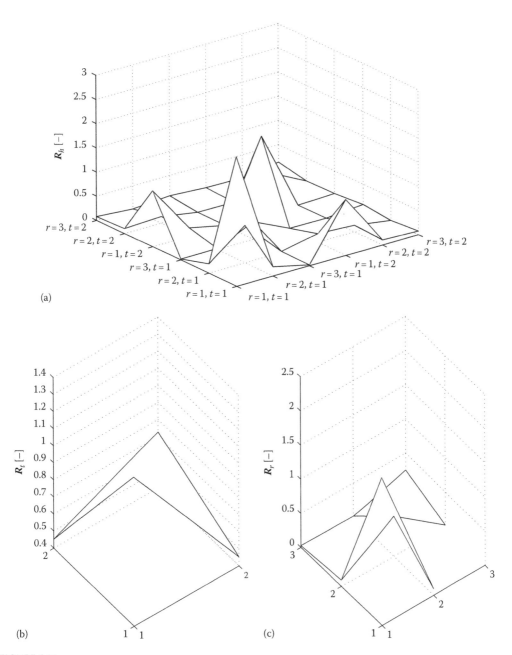

FIGURE 4.11
An example of channel spatial covariance matrices from one field measurement, obtained with the weighted average over the frequency of Equations 4.17 and 4.18 R_h (a), R_t (b) and R_r (c). (From Tomasoni, A., Riva, R., and Bellini, S., Spatial correlation analysis and model for in-home MIMO power line channels, in *Power Line Communications and Its Applications (ISPLC), IEEE International Symposium on*, March 2012. Copyright © 2012 IEEE. With permission.)

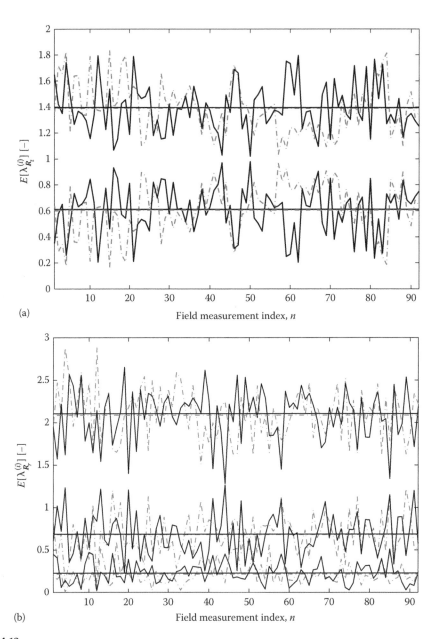

FIGURE 4.12
Eigenvalues of the channel covariance matrices. R_t (a) and R_r (b). Dashed lines are the actual measurements. Solid lines are the results obtained through model simulations. Horizontal lines represent the average value over the $N = 92$ field measurements. (From Tomasoni, A., Riva, R., and Bellini, S., Spatial correlation analysis and model for in-home MIMO power line channels, in *Power Line Communications and Its Applications (ISPLC), IEEE International Symposium on*, March 2012. Copyright © 2012 IEEE. With permission.)

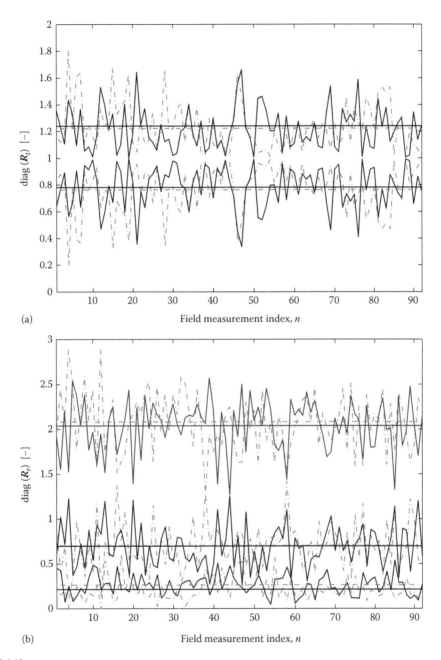

(a)

(b)

FIGURE 4.13
Diagonal elements of the channel covariance matrices. (a) $\mathbf{R}_t^{(i,i)}$ and (b) $\mathbf{R}_r^{(i,i)}$. Dashed lines are the actual measurements. Solid lines are the results obtained through model simulations. Horizontal lines represent the average value over the $N = 92$ field measurements. (From Tomasoni, A., Riva, R., and Bellini, S., Spatial correlation analysis and model for in-home MIMO power line channels, in *Power Line Communications and Its Applications (ISPLC), IEEE International Symposium on*, March 2012. Copyright © 2012 IEEE. With permission.)

4.4 Proposed PL Channel Model

4.4.1 SISO Channel Model

The PL SISO channel model is obtained by generating each channel impulse response coefficient accordingly to the distributions observed in Section 4.3. Note that until now the results do not take into account the presence of correlation among contiguous samples. In order to introduce a correlation among the generated impulse response coefficients, it should be considered that not all the channel coefficients in the range $[1, \ldots, L_H]$ should be randomly generated. Some of them should be obtained as a combination of the generated ones. Considering the channel coefficients in the range χ_R, it can be observed that samples with large standard deviation values correspond to samples with low time correlation. On the other hand, samples with small standard deviation values correspond to samples with high time correlation. Based on these facts, a suitable set of time instants named χ_L^* has been defined and a random variable for each element in this set has been generated; the elements of χ_L^* are shown in Figure 4.7a with diamonds. The same considerations hold true for the samples in the range χ_R, considering the scale parameter, instead of the standard deviation. In this case, the new subset is labelled with χ_R^* and its elements are shown in Figure 4.7b with diamonds. Finally, for the range χ_C, all the time samples have been randomly generated, that is, $\chi_C^* = \chi_C$. Using the earlier definitions, a realisation of the SISO PL channel is obtained with the following procedure:

- For each time instant in χ_L^*, generate a random variable accordingly to the Weibull distribution with scaling factor given by Equation 4.8 and shape parameter equal to 3/4. Then, randomly change the sign of each random variable.

- For each time instant in χ_C^*, generate a random variable accordingly to the Gaussian distribution with mean and variance given in Table 4.1. Then, randomly change the sign of each random variable.

- For each time instant in χ_R^*, generate a random variable accordingly to the Gaussian distribution with zero mean and variance given in Equation 4.9.

- The generated channel must have its maximum absolute value at the time instant *MaxPos*. Let it be ξ; all the samples that have an absolute value higher than ξ are removed from the subsets χ_L^*, χ_C^* or χ_R^*.

- Interpolate, with a linear function, the channel impulse response on all the time instants χ_L, χ_C and χ_R (the removed samples need to be obtained by interpolation).

- Evaluate the RMS-DS of the generated SISO PL channel.

- Introduce an average attenuation using the relation between the RMS-DS and the average attenuation proposed in [19] and shown in Figure 4.8.

Let $H_{SISO,k}$ be the channel frequency response. An example is reported in Figure 4.3b with the dashed line.

4.4.2 MIMO Channel Model

Let $\tilde{H}_{MIMO,k}$ be a $N_r \times N_t$ matrix that denotes the MIMO channel coefficients on carrier k. The three receiving signals are not linearly dependent; hence, the coefficients $\tilde{H}_{MIMO,k}^{(r,t)}$ could be generated as follows:

$$\tilde{H}_{MIMO,k}^{(r,t)} = H_{SISO,k}, \tag{4.19}$$

where each (r,t) refers to a different SISO channel realisation. As a consequence, by construction, the coefficients of $\tilde{H}_{MIMO,k}$ are statistically uncorrelated. However, as shown in Figure 4.9, the coefficients of $\tilde{H}_{MIMO,k}$ must be correlated.

Due to this observation, the channel correlation is modelled based on the assumption that the phenomena correlating the channel taps act only at the two link extremities (as it happens in wireless channels [20,21]), that is, for the generic kth carrier, the correlated MIMO channel coefficients are as follows:

$$H = G \cdot R_r^{1/2} H' R_t^{1/2}, \tag{4.20}$$

where the $N_t N_r$ channel taps $h'^{(r,t)}_u \in \mathcal{N}(0,1)$ are independent and identically distributed (i.i.d.) variables. Because of G, the two correlation matrices R_t and R_r can be arbitrarily scaled.

Therefore, the problem is to derive a synthetic model to generate such matrices as close as possible to the observed ones.

In the following sections, it will be analysed how to introduce a statistical correlation.

4.4.2.1 Spatial Correlation Model

Now, both R_t and R_r must be modelled. Being interested not only in the covariance matrices but also in their eigenvalues, the considered eigenvector–eigenvalue decompositions are:

$$R_t = U_t \cdot D_t \cdot U_t^H, \tag{4.21}$$

$$R_r = U_r \cdot D_r \cdot U_r^H, \tag{4.22}$$

and their eigenvectors and eigenvalues are drawn separately, respecting the channel properties (as estimated in Section 4.3.4.1).

The eigenvalues are modelled as uniform random variables (as suggested by Figure 4.12, dashed lines), with average and standard deviation values optimised evaluating the measurements. However, one should remember that covariance matrices have been normalised. Since the trace of any square matrix equals the sum of its eigenvalues, this constraint is inherited by eigenvalues, too, and can be satisfied, for example, eventually normalising them. For the time being, consider a trace normalisation to 1. In order to model the distribution of the eigenvalues of R_t, two variables uniformly ($\mathcal{U}(\cdot)$) distributed x_1 and x_2 are introduced such that $\bar{x}_1$ is obtained as

$$\bar{x}_1 = \frac{x_1}{x_1 + x_2} \tag{4.23}$$

with $x_i \in \mathcal{U}(x_{i,min}, x_{i,max})$. Similarly, for R_r,

$$\overline{x}_1 = \frac{x_1}{x_1 + x_2 + x_3}. \tag{4.24}$$

First, notice that the earlier two cases can be considered as special cases of

$$\overline{x} = \frac{x}{x + y} \tag{4.25}$$

with $x \in \mathcal{U}(x_{min}, x_{max})$ and y having any PDF $f_Y(y)$. In the former case, $y = x_2$ has uniform distribution as well. In the latter, $y = x_2 + x_3$ will exhibit a triangular PDF. In the same way, higher-order convolutions of uniform variables are easy to manage and integrate. The PDF of the normalised variable (Equation 4.25) is computed as follows:

$$f_{\overline{X}}(\overline{x}) = \int_{x=x_{min}}^{x_{max}} \frac{1}{\Delta_x} \int_{y_{min}}^{y_{max}} f_Y(y)\delta\left(\overline{x} = \frac{x}{x+y}\right) dy\, dx,$$

$$= \frac{1}{\Delta_x}\max\left(\int_{\max(x_{min},\frac{\overline{x}}{1-\overline{x}}y_{min})}^{\min(x_{max},\frac{\overline{x}}{1-\overline{x}}y_{max})} \frac{x}{\overline{x}^2}\underbrace{f_Y\left(x\frac{1-\overline{x}}{\overline{x}}\right)dx}_{@z}, 0\right),$$

$$= \frac{1}{\Delta_x(1-\overline{x})^2}\max\left(\int_{\max(x_{min}\frac{1-\overline{x}}{\overline{x}},y_{min})}^{\min(x_{max}\frac{1-\overline{x}}{\overline{x}},y_{max})} z f_Y(z)dz, 0\right), \tag{4.26}$$

$\Delta_x = x_{max} - x_{min}$ and $\delta(\cdot)$ being the impulsive (Dirac) generalised function. Notice that the integral in Equation 4.26 can be simply interpreted as a truncated expectation of y. Finally, if the rescaled version of $\overline{x}$ is considered, that is, the actual synthetic eigenvalues, the following equations are obtained:

$$\lambda = \alpha\overline{x} \tag{4.27}$$

$$f_\Lambda(\lambda) = \frac{f_{\overline{X}}(\lambda/\alpha)}{\alpha} \tag{4.28}$$

with $\alpha = N_t$ or N_r, respectively. By means of the earlier PDF, the mean and variance of the $N_t = 2$ and $N_r = 3$ uniform eigenvalues have been iteratively optimised, so that the mean and variance of their normalised versions are as close as possible to the measured eigenvalues, along all the experiments. Table 4.2 reports these optimisation results.

Figure 4.12a and b compares measured eigenvalues with those of the earlier model. Dashed lines are the actual eigenvalues; solid lines are the synthetic eigenvalues, obtained with the optimised parameters; as already stated, horizontal lines represent their average values over the $N = 92$ channel measurements. We expect this to match very well.

TABLE 4.2

Upper and Lower Bounds, Mean and Standard
Deviation for the Uniform Random Variables x_i
of the Eigenvalues Model

	$x_1^{(t)}$	$x_2^{(t)}$	$x_1^{(r)}$	$x_2^{(r)}$	$x_3^{(r)}$
Max	1.33	0.75	1.6	0.52	0.19
Min	0.67	0.13	0.4	0.13	0.01
Mean	1	0.44	1	0.325	0.1
Standard deviation	0.1905	0.1791	0.3464	0.1126	0.052

Source: Tomasoni, A., Riva, R., and Bellini, S., Spatial correlation analysis and model for in-home MIMO power line channels, in *Power Line Communications and Its Applications (ISPLC), IEEE International Symposium on*, March 2012. Copyright © 2012 IEEE. With permission.

The eigenvector choice is now considered. Our focus is on the main diagonal elements of R_t and R_r that must be as close as possible to the actual ones. If the eigenvectors were circularly distributed (i.e. with uniform PDF on the unitary hypersphere), one could choose among many strategies to draw them, for example, singular value decomposition, eigenvalue–eigenvector decomposition or QR decomposition of a square matrix of i.i.d., zero-mean Gaussian samples. Actually, this is not the case, since Figure 4.11 shows that covariance matrices have pronounced diagonal elements, compared to off-diagonal ones. Anyway, the QR decomposition of a square matrix with Gaussian samples can still be exploited, yet biasing the mean of the elements on its main diagonal, to emphasise them:

$$QT = aI + W, \tag{4.29}$$

where
$w^{(i,j)} \in \mathcal{N}(0, 1)$
Q is unitary ($QQ^H = Q^HQ = I$)
T is upper triangular

In the limit, when $a \to 0$, Q is circularly distributed; conversely, when $a \to \infty$, $Q \to I$; therefore, also covariance matrices would have a diagonal structure.

To avoid dependency of the eigenvectors on the column position w.r.t. Q (the chosen QR is based on a successive Gram–Schmidt orthonormalisation process), the columns and rows of Q are randomly permuted by a permutation matrix Π, obtaining

$$U = \Pi Q \Pi^H. \tag{4.30}$$

Also the biasing parameters a_t and a_r have been optimised numerically, to find the best match with R_t and R_r, respectively. The optimisation ended with $a_t = 1.25$ and $a_r = 8$. Figure 4.13 compares measured diagonal elements with the synthetic ones. Again, dashed lines represent the actual values, while solid lines are the diagonal entries of Equations 4.21 and 4.22, assuming the earlier two models for the eigenvalues and the eigenvectors.

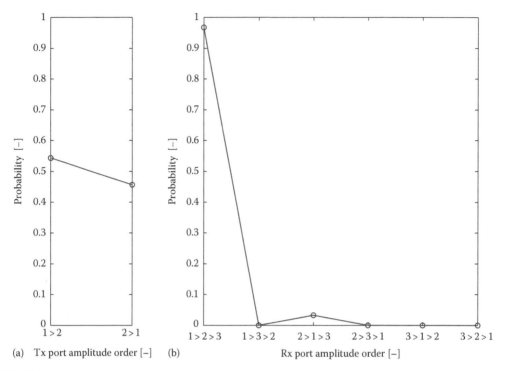

FIGURE 4.14
Rank statistics of the diagonal elements of R_t and R_r. (a) Transmitter ports 1 and 2 correspond to the couples L-N and L-PE, respectively. (b) Receiver ports 1, 2 and 3 correspond to the couples L-N, L-PE and N-PE, respectively. (From Tomasoni, A., Riva, R., and Bellini, S., Spatial correlation analysis and model for in-home MIMO power line channels, in *Power Line Communications and Its Applications (ISPLC), IEEE International Symposium on*, March 2012. Copyright © 2012 IEEE. With permission.)

Horizontal lines represent their average values over the $N = 92$ channel measurements. Again, the match between the model and the actual measurements is very good.*

Finally, the rank statistics of the covariance matrices diagonal elements is studied. Indeed, the mapping between the elements of Figure 4.13 and the transmitter and receiver ports is not uniform. Actually, some ports are likely to have gains larger than others and must most frequently be associated to the highest eigenvalues. These statistics are reported in Figure 4.14.

4.4.2.2 Simpler Correlation Model

In this section, a simpler spatial correlation model is introduced. A spatial correlation between each SISO link can be introduced for each carrier based on an approach

---

* Of course, the eigenvectors model can be further refined to find a better match between synthetic and measured off-diagonal covariance matrix samples. For example, one could run a multivariate optimisation of a biasing matrix A rather than of the diagonal one aI that we have chosen for simplicity. In our opinion, the measurement campaign at the moment is too small to justify such a precision.

documented in several publications on MIMO channel modelling (e.g. [22]). The transmitter and receiver correlation matrices, based on the definition in [23], are explicitly stated as

$$R_t = \begin{bmatrix} 1 & \rho_t \\ \rho_t & 1 \end{bmatrix}, \quad R_r = \begin{bmatrix} 1 & \rho_r & \rho_r^2 \\ \rho_r & 1 & \rho_r \\ \rho_r^2 & \rho_r & 1 \end{bmatrix}. \tag{4.31}$$

Successively, the correlated channel matrix H for each carrier is obtained using Equation 4.13.
The eigenvalues of the proposed matrices in Equation 4.31 obtained resolving the characteristic equation are

$$\begin{cases} \lambda_{R_t}^{(1)} = 1 + \rho_t \\ \lambda_{R_t}^{(2)} = 1 - \rho_t \end{cases}, \tag{4.32}$$

and

$$\begin{cases} \lambda_{R_r}^{(1)} = 1 - \rho_r^2, \\ \lambda_{R_r}^{(2)} = \frac{1}{2}\left(2 + \rho_r^2 + \sqrt{8 + \rho_r^2}\right), \\ \lambda_{R_r}^{(3)} = \frac{1}{2}\left(2 + \rho_r^2 - \sqrt{8 + \rho_r^2}\right). \end{cases} \tag{4.33}$$

Considering the results obtained in Section 4.3.4.1 on the average values of measured eigenvalues of R_t and R_r, the values of $\rho_t = 0.4$ and $\rho_r = 0.6$ guarantee similar behaviour for the eigenvalues of the models in Equation 4.31. Thus, it is recommended to adopt these last values for the model parameters, in order to obtain simulated channel models close to the field measurements in terms of channel capacity. The elements on the diagonal of R_t and R_r still remain equal to 1, thus losing insight on some characteristics observed in Figures 4.11 and 4.14. Nevertheless, this simpler model allows to study MIMO PL channels either in terms of spatial correlation or in terms of channel capacity.

4.5 Characterisation of Noise in MIMO PL Channels

In contrast to MIMO channel modelling on the PL, the analysis of MIMO PL noise is relatively unexplored to date. In many previous treatments of the PL channel, noise is either not addressed or often simply assumed to have an independent and Gaussian distribution. Pagani et al. [15] uses the ETSI STF410 measurements focusing on the characterisation of noise power with respect to frequency and different geographies. Hashmat et al. [16,17] provide two statistical models for the background noise found in MIMO PL channels. In addition to [15–17], which deal specifically with PL noise, a number of contributions also treat the case of correlated noise in MIMO communication systems in [24–27].

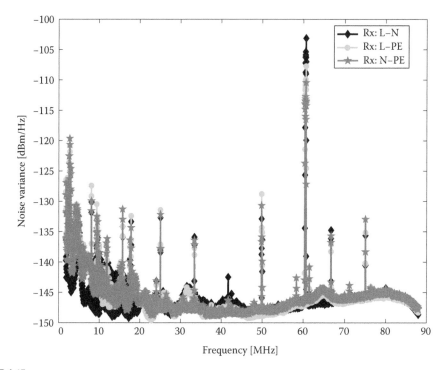

FIGURE 4.15
Variation of noise power across frequency (average over all measured paths). (From Rende, D., Nayagam, A., Afkhamie, K., Yonge, L., Riva, R., Veronesi, D., Osnato, F., and Bisaglia, P., Noise correlation and its effects on capacity of inhome MIMO power line channels, in *Power Line Communications and Its Applications (ISPLC), IEEE International Symposium on*, April 2011. Copyright © 2011 IEEE. With permission.)

Noise characteristics are derived using the captured samples in all three receiver ports during the silent period shown in Figure 4.2. Noise samples are used to calculate the noise covariance matrix $R_{n,k}$, defined in Section 4.2, for every carrier. Noise power on all three receiver ports as well as the correlation between them is derived from the estimated noise covariance matrix.

The noise power on each receiver port corresponds to the diagonal elements of $R_{n,k}$. The variation of noise power across frequency is shown in Figure 4.15. The noise power is averaged across all measured paths for each frequency. It is seen that noise tends to be higher in the 1.8–30 MHz frequency band as compared to frequencies greater than 30 MHz. It is also seen that the average noise power is similar across all three of the receiver ports. Thus, from a noise power perspective, a certain receiver port is not better than any of the others.

For every carrier k, the correlation coefficient between noise in receiver ports i and j is defined in terms of elements of $R_{n,k}$ as

$$C_k^{(i,j)} = \frac{\left|R_{n,k}^{(i,j)}\right|}{\sqrt{R_{n,k}^{(i,i)}R_{n,k}^{(j,j)}}}, \quad i,j = 1,\ldots,N_r. \tag{4.34}$$

Note that the off-diagonal elements of $R_{n,k}$ are complex since the noise n_k is complex, and hence, the correlation is defined in terms of the magnitude of $R_{n,k}^{(i,j)}$ in Equation 4.34. When the noise is identical on ports i and j, then $C_k^{(i,j)} = 1$. When the noise is completely uncorrelated, then the correlation coefficient becomes 0. Correlation coefficients for MIMO PL

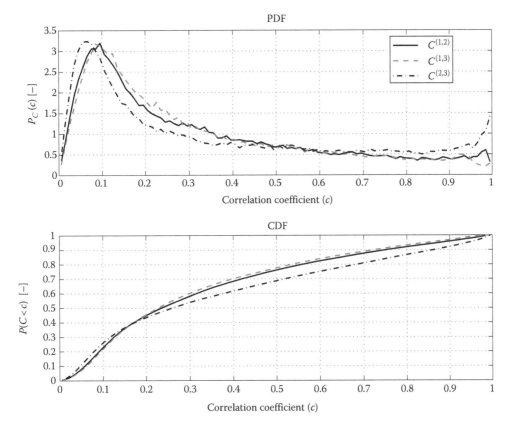

FIGURE 4.16
Density and distribution functions of the correlation coefficients of MIMO PL noise. (From Rende, D., Nayagam, A., Afkhamie, K., Yonge, L., Riva, R., Veronesi, D., Osnato, F., and Bisaglia, P., Noise correlation and its effects on capacity of inhome MIMO power line channels, in *Power Line Communications and Its Applications (ISPLC), IEEE International Symposium on,* April 2011. Copyright © 2011 IEEE. With permission.)

noise for all carriers from all measured paths are collected to obtain the PDF and CDF that are shown in Figure 4.16. Note that the density function is heavy tailed, that is, noise with high correlation is not unlikely on the PL. Also, the CDF indicates that there is a higher correlation on L-PE and N-PE wire pairs. For instance, there is a 20% chance that $C_k^{(2,3)}$ is greater than 0.7 and only a 15% chance that $C_k^{(1,2)}$ or $C_k^{(1,3)}$ is greater than 0.7.

The frequency dependence of the correlation coefficients of PL noise is shown in Figure 4.17. It is seen that noise is more correlated on the lower frequencies as compared to the higher frequencies. On the lower frequencies, it is also seen that the correlation is higher between the L-PE and N-PE wire pairs as compared to the other pairs of receiver ports.

4.5.1 Eigenspread Analysis of MIMO PL Channels

In this section, results on eigenspread of the MIMO PL channel matrix are presented. For a given channel matrix H, the eigenspread is defined as the ratio of the square root of the largest eigenvalue of $H^H H$ to the square root of the lowest eigenvalue [28]. For square H matrices, the eigenspread is the ratio of the eigenvalues of H, and for rectangular H matrices, it is the ratio of the singular values of H. The eigenspread is also referred to as the condition number of a matrix. In our case, the channel is a 3×2 matrix and hence the condition

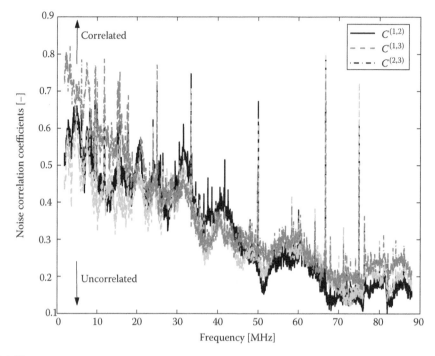

FIGURE 4.17
Variation of noise correlation coefficients across frequency. (From Rende, D., Nayagam, A., Afkhamie, K., Yonge, L., Riva, R., Veronesi, D., Osnato, F., and Bisaglia, P., Noise correlation and its effects on capacity of inhome MIMO power line channels, in *Power Line Communications and Its Applications (ISPLC), IEEE International Symposium on*, April 2011. Copyright © 2011 IEEE. With permission.)

number is defined as $\psi = \lambda_H^{(1)}/\lambda_H^{(2)}$, where $\lambda_H^{(1)}$ is the largest singular value of H and $\lambda_H^{(2)}$ is the smallest singular value. A large eigenspread means that there is more correlation in the channel (the channel is more ill conditioned), whereas a small eigenspread means that the channels are less correlated. The *signal to noise ratios* (SNRs) on the two transmit streams are proportional to $\left(\lambda_H^{(1)}\right)^2$ and $\left(\lambda_H^{(2)}\right)^2$ [9]. A large eigenspread implies that the SNR of the first stream that corresponds to the largest eigenvalue is significantly better than the other stream. Since the SNRs are proportional to the eigenvalues, the capacity of the MIMO channel is also dependent on the eigenvalues [9]. Thus, in order to understand how the noise correlation affects capacity, it is also important to understand how the noise correlation affects the eigenvalues and eigenspread.

To isolate the effect of the noise correlation, the eigenspread is calculated for three different cases:

1. For the composite channel, $H_w = (R_{n,k})^{-1/2}H$. The eigenspread in this case affects the capacity of a PL channel with correlated noise.

2. For the raw PL channel, H. The eigenspread in this case affects the capacity of a PL channel with i.i.d. noise.

3. For a fictitious channel that has independent but nonidentically distributed (i.n.i.d.) noise, $H_D = (R_{ndiag})^{-1/2}H$. In this case, R_{ndiag} is a diagonal matrix with different entries along the diagonal elements. This fictitious channel does not have any correlation between the noise received in different MIMO parts. It provides

a mechanism to study channel and noise conditions from the noise correlation perspective and enables direct comparison between correlated and independent noise cases. The eigenspread in this case affects the capacity of a PL channel with i.n.i.d. noise. In order to keep the comparison to the eigenspread of the composite channel fair, the elements of R_{ndiag} are generated such that the Frobenius norms of R_{ndiag} and R_n are equal, that is, $\|R_{ndiag}\| = \|R_n\|$. This is accomplished by computing each diagonal entry of R_{ndiag} as

$$R_{ndiag}^{(v,v)} = \sqrt{\sum_{r=1}^{N_r} \left| R_n^{(v,r)} \right|^2}, \quad v = 1, \ldots, N_r. \tag{4.35}$$

The normalisation earlier preserves the noise power on each receiver port in R_n and R_{ndiag}. Thus, a comparison of capacities or eigenspreads on H_w and H_D is fair because it compares uncorrelated and correlated noise with equal effective powers.

The variation of the eigenspread of the MIMO PL channel across frequency is shown in Figure 4.18. Eigenspread for a given carrier is obtained by averaging the eigenspread for that carrier across all the paths that were measured. It is seen that eigenspread for the raw PL channel and the fictitious channel with *i.n.i.d.* noise is nearly constant across frequency. Eigenspread of the composite MIMO channel H_w shows more variation across frequency with higher eigenspread in low frequencies.

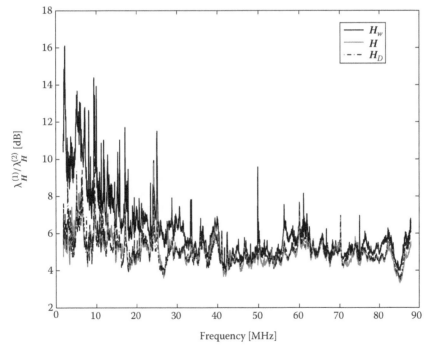

FIGURE 4.18
Frequency dependence of eigenspread of the MIMO PL channel when averaged over all homes. (From Rende, D., Nayagam, A., Afkhamie, K., Yonge, L., Riva, R., Veronesi, D., Osnato, F., and Bisaglia, P., Noise correlation and its effects on capacity of inhome MIMO power line channels, in *Power Line Communications and Its Applications (ISPLC), IEEE International Symposium on*, April 2011. Copyright © 2011 IEEE. With permission.)

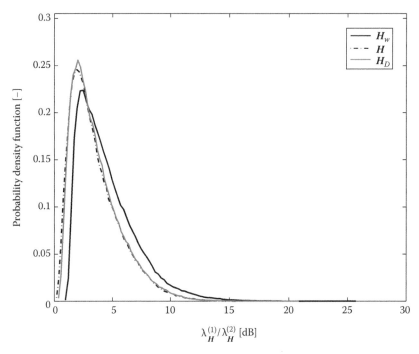

FIGURE 4.19

PDF of the eigenspread of the composite MIMO PL channel under different noise assumptions. (From Rende, D., Nayagam, A., Afkhamie, K., Yonge, L., Riva, R., Veronesi, D., Osnato, F., and Bisaglia, P., Noise correlation and its effects on capacity of inhome MIMO power line channels, in *Power Line Communications and Its Applications (ISPLC), IEEE International Symposium on*, April 2011. Copyright © 2011 IEEE. With permission.)

The PDF of the eigenspread of the MIMO PL channel is shown in Figure 4.19. The PDF is obtained by looking the corresponding channel matrices over all the carriers on all paths that were measured. It is observed that eigenspread is in general not very high for the three different cases.

The PDF can be approximated by the Rayleigh distribution and this fact has also been observed for the eigenspread of random matrices in [29].

It is also seen that eigenspreads for the raw PL channel and the fictitious channel with i.n.i.d noise have almost identical distributions. Thus, as long as the noise is independent, the actual noise powers on the three different ports do not have a significant influence on the eigenspread. In this case, the eigenspreads are dominated by the correlation in the channel only. The eigenspread for the composite channel is higher compared to the independent noise cases. This suggests that noise correlation increases the eigenspread of the composite channel. Since eigenspread is an indication of increased channel correlation, this in turn implies that noise correlation causes a reduction in capacity compared to the case of having uncorrelated noise.

However, channel correlation or eigenspread is not the only factor that determines capacity. In fact, the capacity is determined by the individual eigenvalues. The two eigenvalues for the composite PL MIMO channel for R_n and R_{ndiag} are shown in Figure 4.20. Note that the two eigenvalues are closer together when the noise is i.n.i.d. and the spread in eigenvalues is larger when noise is correlated. When the noise is correlated, each of the eigenvalues increases. Since SNRs are proportional to eigenvalues, the capacity will actually increase when the noise is correlated. So, even though the noise correlation increases

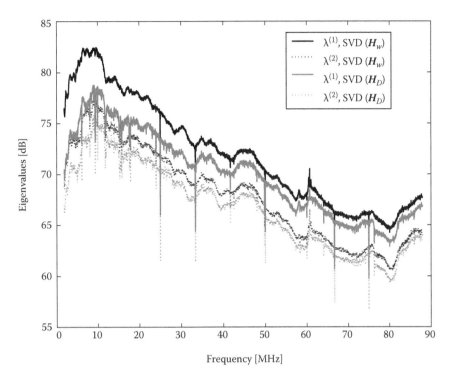

FIGURE 4.20
Variation of eigenvalues with frequency for the MIMO PL channel (averaged over all measured paths). (From Rende, D., Nayagam, A., Afkhamie, K., Yonge, L., Riva, R., Veronesi, D., Osnato, F., and Bisaglia, P., Noise correlation and its effects on capacity of inhome MIMO power line channels, in *Power Line Communications and Its Applications (ISPLC), IEEE International Symposium on*, April 2011. Copyright © 2011 IEEE. With permission.)

the spread in eigenvalues (as seen in Figure 4.20), this is more than compensated for by the increase in the individual eigenvalues. The fact that capacity increases when the noise is correlated is also observed in [24] and will be illustrated for MIMO PL channels in the next section.

4.5.2 Impact of Noise Correlation on the MIMO PL System

The effect of the noise correlation on the capacity of a MIMO PL system with two or three receiver ports is studied in this section. The channel capacity for a single carrier in a MIMO system with correlated noise is given in [24]:

$$C = BW \log\left(\left(I + R_n^{-1}HFH^H\right)\right). \tag{4.36}$$

The system capacity is obtained by summing the capacity over all the carriers. It is easy to derive the expression for capacity in Equation 4.36 by starting with the pre-whitened linear model given in Equation 4.3. In Equation 4.36, BW is the carrier bandwidth. F is found through the water-filling solution after applying singular value decomposition on the composite MIMO channel $H_w = R_n^{-1/2}H$.

It was shown that noise in MIMO PL channels is correlated. In order to study the effect of noise correlation on capacity, the capacity is studied with three different noise

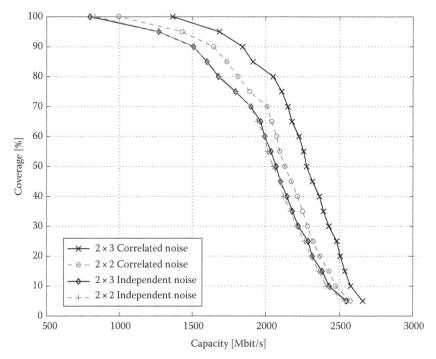

FIGURE 4.21
Capacity of MIMO PL channels with correlated and independent noise. (From Rende, D., Nayagam, A., Afkhamie, K., Yonge, L., Riva, R., Veronesi, D., Osnato, F., and Bisaglia, P., Noise correlation and its effects on capacity of inhome MIMO power line channels, in *Power Line Communications and Its Applications (ISPLC), IEEE International Symposium on*, April 2011. Copyright © 2011 IEEE. With permission.)

assumptions as described in the beginning of Section 4.5, namely, correlated noise, i.i.d. noise and i.n.i.d. noise. The capacity of 2×2 and 2×3 MIMO PL channels with correlated and independent noise is shown in Figure 4.21. The results illustrate that noise correlation actually improves channel capacity. This observation can also be found in [24–27]. It is seen that 2×2 and 2×3 MIMO capacities with correlated noise are higher than the capacities with independent noise.

Note also that 2×3 MIMO shows a larger difference between capacities with correlated and independent noise as compared to 2×2 MIMO. The additional receiver port in the 2×3 system when compared to the 2×2 MIMO system is the N-PE wire pair. It is mentioned that the correlation between L-PE and N-PE is higher than the other wire pairs (see Figure 4.16). Thus, by including the N-PE wire pair, the correlation among the noise samples goes up thereby increasing the capacity more for the 2×3 case when compared to the 2×2 case.

The results on capacity are obtained by summing the capacity of each individual carrier. It is possible that the capacity could be dominated by a few carriers that exhibit low noise correlation, and this is clouding the results. To establish that this is not the case, capacity as a function of frequency for R_n and R_{ndiag} is plotted in Figure 4.22.

The capacity for a single carrier is obtained as the average of the capacity on that carrier in each path that is tested. It can be seen that noise correlation improves capacity on all frequencies. The benefit is higher on lower frequencies because the noise correlation is actually higher on lower frequencies as shown in Figure 4.17.

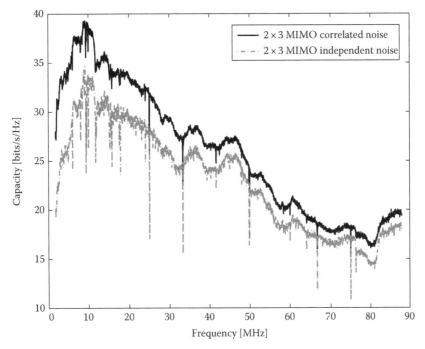

FIGURE 4.22
MIMO capacity as a function of frequency for correlated and independent noise (averaged over all measured paths). (From Rende, D., Nayagam, A., Afkhamie, K., Yonge, L., Riva, R., Veronesi, D., Osnato, F., and Bisaglia, P., Noise correlation and its effects on capacity of inhome MIMO power line channels, in *Power Line Communications and Its Applications (ISPLC), IEEE International Symposium on*, April 2011. Copyright © 2011 IEEE. With permission.)

4.6 Conclusions

In this chapter, a statistical description of PL channels were provided and the noise in the 1.8–88 MHz band was analysed from a MIMO perspective. Channel and noise samples collected on 92 paths in five different homes in North America formed the basis of the analysis.

This chapter began reporting properties that provide insight into important physical characteristics of MIMO PL channels and compares them to previous studies on SISO PL channels finding a good match. A MIMO PL channel model was proposed, based on field measurements and on the extracted statistics related to meaningful physical parameters. As a first step, each single communication link within the MIMO channel matrix was completely characterised through the collected statistics. Then, the MIMO PL channel model is defined, introducing a representation of the correlation that is verified on the field. The channel model proposed here was suited to perform system-level simulations in a realistic scenario.

The last portion of this chapter was devoted to MIMO PL noise characterisation; in particular, it has been shown that the noise is correlated on the L-N, L-PE and N-PE receiver ports. The strongest correlation is measured between the L-PE and N-PE receiver ports. Moreover, the correlation is stronger on lower frequencies when compared to higher frequencies. The effect of the noise correlation on the capacity of a MIMO PL system with two transmit ports and three receiver ports is studied, and it is observed that noise correlation indeed helps to increase the MIMO channel capacity.

The analysis and the models described so far can be conveniently leveraged to simulate and compare the performance of different MIMO PL communication systems.

References

1. D. Veronesi, R. Riva, P. Bisaglia, F. Osnato, K. Afkhamie, A. Nayagam, D. Rende and L. Yonge, Characterization of in-home MIMO power line channels, in *Power Line Communications and Its Applications (ISPLC), IEEE International Symposium on*, Udine, Italy, April 2011, pp. 42–47.

2. D. Rende, A. Nayagam, K. Afkhamie, L. Yonge, R. Riva, D. Veronesi, F. Osnato and P. Bisaglia, Noise correlation and its effects on capacity of inhome MIMO power line channels, in *Power Line Communications and Its Applications (ISPLC), IEEE International Symposium on*, Udine, Italy, April 2011, pp. 60–65.

3. A. Tomasoni, R. Riva and S. Bellini, Spatial correlation analysis and model for in-home MIMO power line channels, in *Power Line Communications and Its Applications (ISPLC), IEEE International Symposium on*, Beijing, China, March 2012, pp. 286–291.

4. HomePlug AV Specification, Version 1.1, 21 May 2007, HomePlug PowerLine Alliance. http://www.homeplug.org.

5. Standard for Local and Metropolitan Area Networks – Part 11: Wireless LAN MAC and PHY Specifications. Amendment 5: Enhancements for Higher Throughput, http://standards.ieee.org/findstds/standard/802.11n-2009.html, IEEE 802.11n.

6. 3G Release 8. See version 8 of *Technical Specifications and Technical Reports for a UTRAN-based 3GPP system*, 3GPP TR 21.101, http://www.3gpp.org/article/lte, accessed 17 October 2013.

7. R. Hashmat, P. Pagani and T. Chonavel, MIMO capacity of inhome PLC links up to 100 MHz, in *Workshop on Power Line Communications (WSPLC)*, Udine, Italy, October 2009, pp. 4–6.

8. R. Hashmat, P. Pagani, A. Zeddam and T. Chonavel, MIMO communications for inhome PLC networks: Measurements and results up to 100 MHz, in *Power Line Communications and Its Applications (ISPLC), IEEE International Symposium on*, Rio de Janerio, Brazil, March 2010, pp. 120–124.

9. L. Stadelmeier, D. Schill, A. Schwager, D. Schneider and J. Speidel, MIMO for inhome power line communications, in *International ITG Conference on Source and Channel Coding (SCC)*, Ulm, Germany, January 2008, pp. 1–6.

10. C. L. Giovaneli, B. Honary and P. G. Farrell, Space–time coding for power line communications, in *Communication Theory and Applications (ISCTA), International Symposium on*, Ambleside, Lake Districk, UK, July 2003, pp. 162–169.

11. R. Hashmat, P. Pagani, A. Zeddam and T. Chonavel, A channel model for multiple input multiple output in-home power line networks, in *Power Line Communications and Its Applications (ISPLC), IEEE International Symposium on*, Udine, Italy, April 2011, pp. 35–41.

12. F. Versolatto and A. M. Tonello, A MIMO PLC random channel generator and capacity analysis, in *Power Line Communications and Its Applications (ISPLC), IEEE International Symposium on*, Udine, Italy, April 2011, pp. 66–71.

13. A. Schwager, W. Bäschlin, H. Hirsch, P. Pagani, N. Weling, J. L. González Moreno and H. Milleret, European MIMO PLT field measurements: Overview of the ETSI STF410 Campaign & EMI analysis, in *Power Line Communications and Its Applications (ISPLC), IEEE International Symposium on*, Beijing, China, April 2012, pp. 298–303.

14. D. M. Schneider, A. Schwager, W. Bäschlin and P. Pagani, European MIMO PLC field measurements: Channel analysis, in *Power Line Communications and Its Applications (ISPLC), IEEE International Symposium on*, Beijing, China, April 2012, pp. 304–309.

15. P. Pagani, R. Hashmat, A. Schwager, D. M. Schneider and W. Bäschlin, European MIMO PLC field measurements: Noise analysis, in *Power Line Communications and Its Applications (ISPLC), IEEE International Symposium on*, Beijing, China, April 2012, pp. 310–315.

16. R. Hashmat, P. Pagani, T. Chonavel and A. Zeddam, Analysis and modeling of background noise for inhome MIMO PLC channels, in *Power Line Communications and Its Applications (ISPLC), IEEE International Symposium on*, Beijing, China, April 2012, pp. 316–321.

17. R. Hashmat, P. Pagani, T. Chonavel and A. Zeddam, A time domain model of background noise for inhome MIMO PLC networks, *IEEE Trans. Power Deliv.*, 27(4), 2082–2089, October 2012.

18. F. Versolatto and A. M. Tonello, An MTL theory approach for the simulation of MIMO power-line communication channels, *IEEE Trans. Power Deliv.*, 26(3), 1710–1717, July 2011.

19. S. Galli, A simplified model for the indoor power line channel, in *Power Line Communications and Its Applications (ISPLC), IEEE International Symposium on*, Dresden, Germany, March 2009, pp. 13–19.

20. D. McNamara, M. Beach and P. Fletcher, Spatial correlation in indoor MIMO channels, in *Proceedings of IEEE International Symposium on Personal, Indoor and Mobile Radio Communications*, Pavilho Atlantico, Lisbon, Portugal, September 2002, pp. 290–294.

21. K. Yu, M. Bengtsson, B. Ottersten, D. McNamara, P. Karlsson and M. Beach, Second order statistics of NLOS indoor MIMO channels based on 5.2 GHz measurements, in *Proceedings of IEEE Global Telecommunications Conference*, San Antonio, TX, November 2001, pp. 156–160.

22. J. P. Kermoal, L. Schumacher, K. I. Pedersen, P. E. Mogensen and F. Frederiksen, A stochastic MIMO radio channel model with experimental validation, *IEEE J. Sel. Areas Commun.*, 20(6), 1211–1226, August 2002.

23. S. L. Loyka, Channel capacity of MIMO architecture using the exponential correlation matrix, *IEEE Commun. Lett.*, 5(9), 369–371, September 2001.

24. S. Krusevac, P. Rapajic and R. A. Kennedy, Channel capacity estimation for MIMO systems with correlated noise, *Proceedings of IEEE GLOBECOM '05*, St. Louis, MO, December 2005, pp. 2812–2816.

25. Y. Dong, C. P. Domizioli and B. L. Hughes, Effects of mutual coupling and noise correlation on downlink coordinated beamforming with limited feedback, *EURASIP J. Adv. Signal Process.*, 2009, Article ID 807830, 2009.

26. C. P. Domizioli, B. L. Hughes, K. G. Gard and G. Lazzi, Receive diversity revisited: Correlation, coupling and noise, in *Proceedings of the IEEE Global Telecommunications Conference (GLOBECOM '07)*, Washington, DC, November 2007, pp. 3601–3606.

27. J. W. Wallace and M. A. Jensen, Mutual coupling in MIMO wireless systems: A rigorous network theory analysis, *IEEE Trans. Wireless Commun.*, 3(4), 1317–1325, 2004.

28. V. Madisetti and D. B. Williams, *The Digital Signal Processing: Handbook*, CRC Press, Boca Raton, FL, 1997.

29. A. Edelman, Eigenvalues and condition numbers of random matrices, PhD dissertation, Department of Mathematics, Massachusetts Institute of Technology, Cambridge, MA, 1989.

5

Broadband In-Home Statistics and Stochastic Modelling

Andreas Schwager, Pascal Pagani, Daniel M. Schneider,
Rehan Hashmat and Thierry Chonavel

CONTENTS

5.1 Motivation

PLC systems available today use only one transmission path between two outlets. It is the *differential-mode* (DM) channel between the phase (or live) and neutral contact of the mains. These systems are called *single input single output* (SISO). In contrast, *multiple-input multiple-output* (MIMO) PLC systems make use of the third wire, *protective earth* (PE), which provides several transmission combinations for feeding and receiving signals into and from the *low-voltage distribution network* (LVDN). Various research publications, for example, [1–3] describe up to eight transmission paths that might be used simultaneously in the home.

Channel measurements, as described in this publication, are verified by the *European Telecommunications Standards Institute* (ETSI). A Special Task Force was set up [4] in 2010 to record MIMO channel, noise and *electromagnetic interference* (EMI) properties in many European countries.

As MIMO PLC modems also utilise the protective earth, they are able to alternately feed from *live to neutral* (L-N), *live to protective earth* (L-PE) and *neutral to protective earth* (N-PE). The protective earth may be grounded inside (e.g. at the foundations) or outside (at the transformer station) the building and provides low impedance for the 50 Hz AC power. However, high-frequency signal measurements show the PE wire to be a rather excellent communication path which by no means represents a ground. This is due to the inductivity of the long PE wires.

This chapter provides a statistical evaluation of these properties plus the spatial correlation of the MIMO paths. In addition, the noise received in a MIMO PLC system is analysed. Information on the presence of the protective earth wire, on measurement methods, and on MIMO topologies and corresponding coupling devices are described in detail in Chapter 1. Finally, this chapter presents new models of the MIMO *channel transfer function* (CTF) and multiple-output noise, based on the observed statistics.

5.2 Statistical Evaluation of Results

5.2.1 CTF and Attenuation

S21 of MIMO PLC channels was measured in 36 buildings. Records of 4684 sweeps in total were collected using a *network analyser* (NWA). Each sweep recorded the attenuation of 1601 frequency values which resulted in 7,499,084 measures for the statistical analysis later.

Figure 5.1 depicts an example of individual sweeps of all MIMO paths between two outlets. There are 12 sweeps in total transmitted at the delta ports D1, D2 and D3 and received at the star ports S1, S2, S3 and the CM port S4. The ports described here and the PLC coupler used for the measurements are described in Chapter 1. The black area in Figure 5.1 shows the distance from the minimum attenuation of all sweeps to the maximum values. Additionally, three individual records are given. The area between the paths is 15 dB up to 35 dB wide depending on the frequency. An increasing attenuation towards the higher frequencies is visible. Chapter 9 shows that the more deviation a link provides between the individual paths, the more gain will be achieved when utilising MIMO technology. *Frequency-modulated* (FM) radio reception was excellent in the location where Figure 5.1 was recorded. Some ingress of FM radio stations onto the NWA data is also visible in the graphs. Electromagnetic compatibility between PLC and FM radio reception will be discussed in Chapter 7.

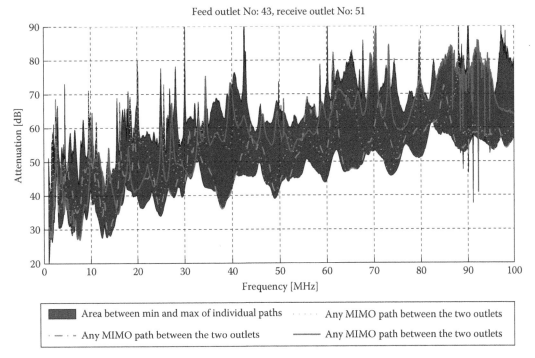

FIGURE 5.1
All MIMO paths between two outlets.

Figure 5.2 represents the cumulative probability of measuring such as attenuation. Usually, a stretched S-style line from the bottom-left to top-right corner is depicted in the graph. These figures may be read in the following way:

- Top-right corner, 100% or 'always' point: The maximum recorded value in Figure 5.2 is 100 dB attenuation.

- 80% point at light grey line (Δ marker) in Figure 5.2: 80% of all measured values provide less attenuation than 66 dB when feeding style D1 (SISO feeding between live and neutral) is selected.

- Median or 50% point: Every second measured value provides higher or lesser attenuation than ~53 dB in Figure 5.2.

- Lower left corner, the 0% or 'never' point: The attenuation was never smaller than 5 dB (minimum point outside Figure 5.2).

All attenuation measurements – independent of location, country or frequency – are separated into individual feeding possibilities: D1, D2 and D3 on delta coupler and T1 and T2 at T-style coupler. Their statistical distribution is visualised in Figure 5.2. A zoom into the high attenuated records (lower-right figure in Figure 5.2) shows that the T-style coupler provides better PLC coverage. They are less attenuated than other feeding styles. The two wires used for traditional PLC modems (live–neutral, D1) show the worst coverage in this case. This is probably caused by multiple consumers connected to live and neutral inside homes.

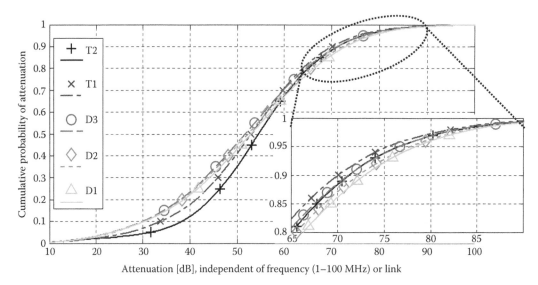

FIGURE 5.2
Statistical distribution of attenuation for all feeding styles.

NOTES: The unused ports at the coupler were terminated with 50 Ω in these measurements. The classical SISO style modems operate by sending and receiving signals symmetrically via P and N. Unused feeding combinations in SISO are either open or unterminated. Termination of unused ports, when identical energy is fed into the couplers, theoretically consumes 1.96 dB of signal energy from mains wires compared to an unterminated coupler where no loss takes place.

Results for the possibility of receiving – identical analytical process as applied for the transmission possibilities – are displayed in Figure 5.3. In the lower right corner is a close up of the high attenuated values. These statistics are derived from identical measurements

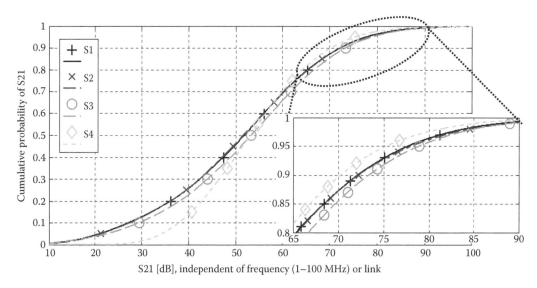

FIGURE 5.3
Cumulative probability of attenuation (S21) of all receiving styles.

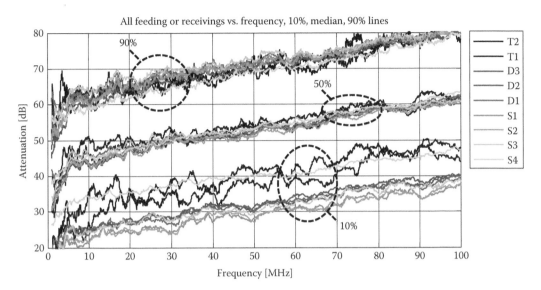

FIGURE 5.4
Attenuation (S21) depending on frequency.

already presented in Figure 5.2 but sorted by individual receiving styles. Of the receiving possibilities displayed, CM reception on the coupler's port S4 (◊) provides the best coverage (less attenuation at high attenuated channels).

NOTES: The coupler itself is considered to be part of the channel in these statistics. ETSI TR 101562 in clause 7.1.6 [5] shows that CM reception with the STF 410 coupler increases attenuation by 3–4 dB compared to reception on S1, S2 or S3. If someone is interested in the channel attenuation without the influence of the STF 410 coupler, the coupler verification values from ETSI TR 101562 [5–7]* might be considered causing the S4 (CM) line to move to lower attenuated records. The gap between the CM line and the others becomes larger at high attenuated values.

The 10% values, median values and 90% values at each frequency, for all feeding and receiving styles, are shown in Figure 5.4. A bunch of curves for each feeding and receiving port is shown three times in Figure 5.4. The top bunch increasing over frequency from 60 to 80 dB represents the 90% values of all attenuations over the frequency. These are the high attenuated channels out of the top-right corners of Figures 5.2 and 5.3. The median values out of all sweeps are shown by the middle bunch of Figure 5.4. The 10% values have a higher standard variation and are the lower bunch in Figure 5.4. A constant trend from 5 to 100 MHz can be observed, where higher frequencies are more attenuated than lower frequencies. This trend is independent of the feeding or receiving port. This slope is 0.2 dB/MHz. This frequency trend is valid at high as well as at low attenuated channels.

Figure 5.5 shows that the CM path (received on port S4 and marked with ◊) provides less attenuation in higher attenuated channels. Figure 5.5 shows a graph of the top 90% of attenuated values depending on frequency. The phenomenon of the CM having less attenuation than the DM signals or single-ended lines (S1, S2 or S3) exists at all frequencies.

---

* Members to STF 410: Andreas Schwager (STF Leader), Sony, Germany; Werner Bäschlin, JobAssist, Switzerland; Holger Hirsch, University of Duisburg-Essen, Germany; Pascal Pagani, France Telecom, France; Nico Weling, Devolo, Germany; Jose Luis Gonzalez Moreno, Marvell, Spain; Hervé Milleret, Spidcom, France. STF 410 produced following technical reports: [ETSI TR 101 562-1], [ETSI TR 101 562-2] and [ETSI TR 101 562-3].

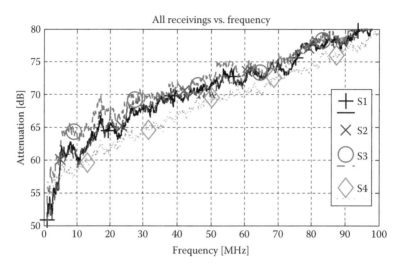

FIGURE 5.5
Statistical 90% value of attenuation at higher attenuated records. Depending on frequency and receiving port.

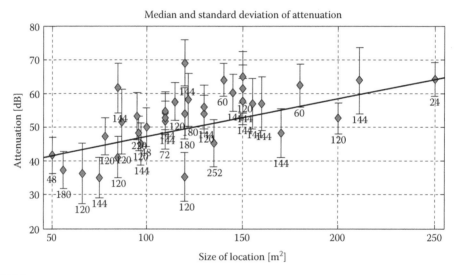

FIGURE 5.6
Size of location, median attenuation, standard deviation and number of sweeps.

It is interesting to compare the median values and the standard deviation of attenuation at each location. The number of sweeps at each location gives an indication of the size and maturity of the statistics. Figure 5.6 shows median attenuation using the diamonds; the standard deviation is indicated by the error bars. The number of sweeps is listed below each error bar.

Figure 5.6 displays the relationship between the size of the location under measurement and the attenuation between two outlets. Assuming a linear relationship between these two parameters, a fitting line can be drafted into Figure 5.6. The formula of this line is

$$\text{Attenuation (dB)} = 0.11 \text{ dB/m}^2 * \text{size (m}^2) + 36.07 \text{ dB}.$$

Figure 5.7 shows the median attenuation of all attenuation records of each country. Highest attenuation is found in Germany. This is caused by the three-phase installations

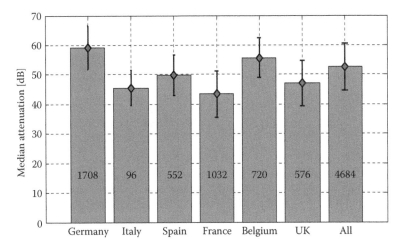

FIGURE 5.7
Median attenuation of each country.

in German homes. Additionally, the error bars give the standard deviation. It is interesting to see that the standard deviation is almost identical in all countries. The results of an individual record at an outlet vary heavily to the other outlet or location. However, comparing a statistical mass of data between two locations shows always identical results. The numbers given on each bar indicate the number of sweeps recorded in this country.

The mains installation in the United Kingdom and the Commonwealth of Nations uses a ring wire on each housing level. The outlets are daisy chained along the ring. As a result each outlet is connected to two sets of wires, each going in a different direction in the room. Electrical installations in the rest of the world follow a combination of star (at the fuse cabinet) and tree style (branches into rooms, outlets, light switches, etc.). Even the ring-wire installations used in the United Kingdom do not show any major difference results here.

5.2.2 Reflection (S_{11})

Reflection measurements were conducted in 33 locations in Belgium, France, Italy, Germany, Spain and the United Kingdom. In total, 661 frequency sweeps have been recorded with 1601 points in each frequency domain. This results in a statistical compilation of 1,058,261 values.

Figure 5.8 presents a sweep recording the reflection parameter at the star coupler. The three single-ended ports S1 ($\Diamond$), S2 (o) and S3 (×) show above 10 MHz a very similar reflection characteristic compared to the attenuation measurements shown earlier. The CM termination (S4, +) is different to the others and provides less fading over the frequency.

Figure 5.9 shows an overview of the probability of measuring a reflection parameter of all S_{11} measurements. Indoor power line networks show weak impedance conditions. Due to the high variations of S_{11} parameters, it is difficult to implement impedance matching couplers. Time-, frequency- and location-dependent characteristics influence the coupler's feeding or receiving properties. A S_{11} parameter of zero or negative infinite in the logarithm representation shows optimal termination, nothing is reflected. If this parameter equals 1 in linear representation or 0 dB, the full wave is reflected. If the S_{11} parameter is worse than −6 dB (−6 dB < S_{11} < 0 dB), more than half of the fed signals are reflected back to the coupler at the connected outlet. This is the case for >60% of S_{11} measurements

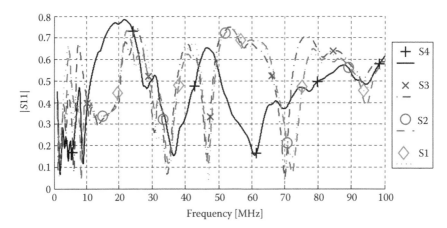

FIGURE 5.8
S_{11} at the star-style coupler at a typical outlet.

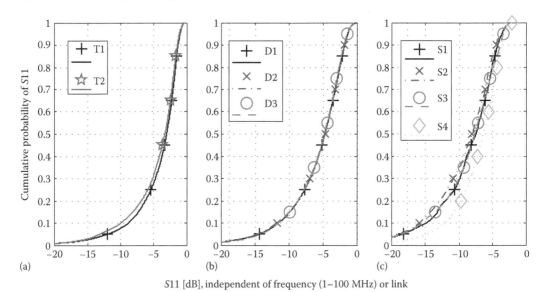

$S11$ [dB], independent of frequency (1–100 MHz) or link

FIGURE 5.9
CDF of magnitude of S_{11} from T-style (a), delta-style (b) and star-style (c). Independent of location or frequency.

using the delta- or T-style coupler. Star-style ports offer better termination than the others; T-style ports are the worst.

The impedances of the delta ports D1 (o), D2 (×) and D3 (+) at a typical outlet are shown per frequency in Figure 5.10. It is interesting to see the variance of the impedance values in between the individual ports to be small. This phenomenon was also found at the single-ended star ports in Figure 5.8. Obviously, if a MIMO PLC coupler should become impedance matched to the mains, all MIMO paths could be matched identically.

The ring-wire mains installation in the United Kingdom might provide half the impedance of a continental outlet, because the outlet is connected to two sets of wires. Table 5.1 compares the median impedances of UK installations with all measurements recorded on the European continent. As expected, the UK installations show lower impedance but not half the value than found on the continent.

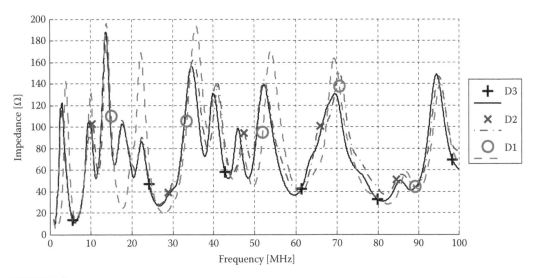

FIGURE 5.10
Impedance Z (delta ports) of a typical outlet.

TABLE 5.1

Median Impedance for each Feeding/Receiving Style

Feeding/ Receiving Style	Impedance in Ω, All Locations	Impedance in Ω, All Locations Except United Kingdom	Impedance in Ω, Locations in United Kingdom Only
D1	86.04	86.86	77.14
D2	87.41	88.27	78.31
D3	88.94	89.73	81.96

5.2.3 Noise

Experimental measurements of the MIMO PLC noise were conducted in 31 different dwelling units in five European countries, including Belgium, France, Germany, Spain and the United Kingdom. Measurements were taken in flats and houses of different sizes. On average, four electrical plugs were considered in each location. Considering that the measurement process included four different signals (live, protective earth, neutral and common-mode), and that each signal was recorded using four different filters, the statistical set was constituted of 1928 records measured in the time domain over a duration of 20 ms each.

5.2.3.1 Frequency Domain Results

The first analysis of the collected noise is conducted in the frequency domain up to a maximum frequency of 100 MHz. For this purpose, the *power spectral density* (PSD) of the recorded noise was estimated using the method of Welch [8]. The final *resolution bandwidth* (ResBw) of the frequency domain data is 122 kHz, which is comparable to the resolutions commonly used in EMI testing. The ResBw of 122 kHz = 50.9 dB (Hz) was subtracted from the noise records to visualise the noise PSD in Figures 5.11 through 5.16.

First statistical results are depicted in Figure 5.11, where different percentiles of the noise PSD recorded during the experiment over the full frequency band (Band 1) are shown.

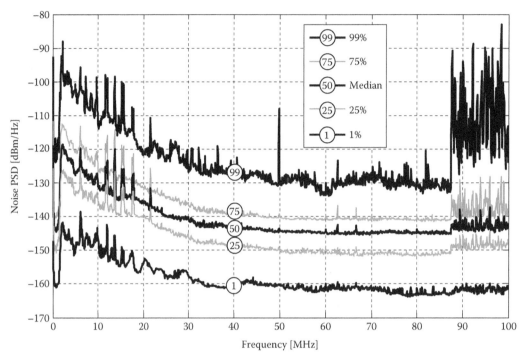

FIGURE 5.11
Statistics of the noise PSD measured on Band 1: 1%, 25%, median, 75% and 99%.

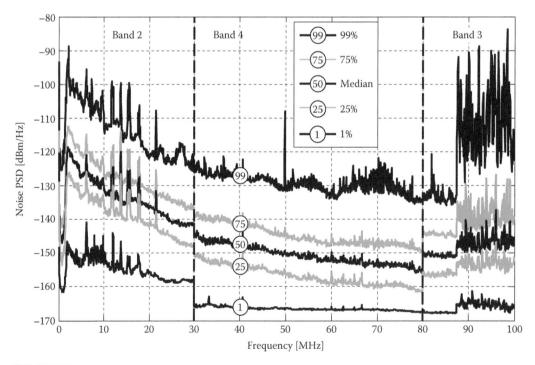

FIGURE 5.12
Statistics of the noise PSD measured on the combined band: 1%, 25%, median, 75% and 99%.

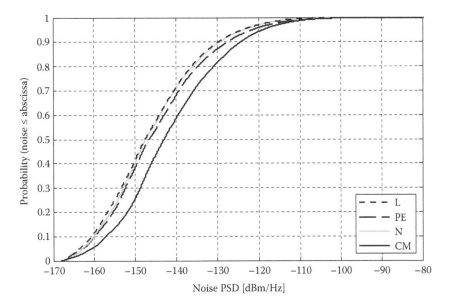

FIGURE 5.13
Statistics of the noise PSD for different reception methods.

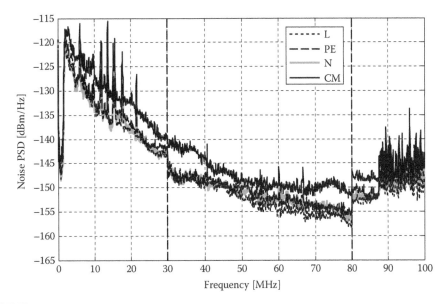

FIGURE 5.14
Median noise PSD measured on the combined band for different reception methods.

The considered data set includes all measurements, regardless of the location or the *receiver* (Rx) port. For each frequency, the graph represents the 1% and 25% percentiles, the median value and the 75% and 99% percentiles. The observed PSD lies within minimum values around −160 dBm/Hz and maximum values around −80 dBm/Hz. The presence of *short wave* (SW) broadcast frequencies is clearly noticeable at 6, 7.3, 9.5, 12, 13.5, 15.5, 17.5 and 21 MHz (metre bands of 49, 41, 31, 25, 22, 19, 16 and 13 m, respectively). At the highest frequencies above 87 MHz, strong noise peaks reveal the presence of broadcast FM signals.

In order to be able to observe low noise levels, one needs to make sure that the signal is not contaminated by quantisation effects due to the *digital sampling oscilloscope* (DSO) limitations. To avoid this problem, noise was also recorded using filters over narrower bands. In the experimental set-up, Band 2 rejected FM signals, Band 3 rejected SW signals and Band 4 rejected both. From these measurements, a combined band was defined by aggregating results from Band 2 over the (0–30 MHz) range, Band 4 over the (30–80 MHz) range and Band 3 over the (80–100 MHz) range.

Figure 5.12 reproduces similar results as in Figure 5.11, based on records from the combined band. Using the combined band mainly improves the sensitivity for the lowest noise values. Considering the 1% percentile, lowest values now reduce to −168 dBm/Hz in the (30–87 MHz) range.

It is now interesting to observe the influence of the reception method on the level of the recorded noise. Figure 5.13 presents the *cumulative distribution function* (CDF) of the noise PSD for each of the L wire, PE wire, N wire and CM signal. A first observation is that the CM signal is subject to stronger noise than the signals transmitted over the L, PE or N wire. This difference of 5 dB can be explained by the higher sensitivity of the CM signal to interference from external sources, such as radio broadcasting. The L, PE and N ports present similar noise statistics. However, when considering large noise records (5% percentile), one can observe that the PE port is more sensitive to noise by ∼2 dB than the N or L ports. Similarly, for low noise records (95% percentile), the L port is less sensitive to noise by ∼1 dB than the N or PE ports.

Figure 5.14 presents the median value of the noise PSD measured on the combined band for different reception methods. This frequency domain observation confirms that the CM noise is stronger than the noise collected over each of the L, PE and N wires separately. This difference is particularly noticeable in the frequency range from 1 to 45 MHz. This leads to the assumption that the ingress of noise signals to the mains grid is responsible for the background noise and not conducted noise sources. Another interpretation of this phenomenon is that conducted noise sources mainly generate CM noise instead of DM noise. On the statistical median, the noise observed on the L, PE and N ports appears to take similar values over the whole combined band.

This noise measurement campaign performed in 35 locations in different European countries was the occasion to observe the possible relation between the level of the noise PSD and the size of the location. Without further knowledge, one could assume that larger houses with a larger number of connected electrical appliances would lead to higher observed noise values. Figure 5.15 presents statistics of the recorded noise PSD across the location size. The location size ranges from 50 to 200 m², and noise statistics include the 1% and 25% percentiles, the median and the 75% and 99% percentiles. Statistically, it can be expected that larger homes have more electrical appliances installed. However, one can clearly observe that there is no correlation between the surface of the house and the statistics of the noise PSD level. Using other words, the noise accumulation is independent of the location's size. One can conclude that the PLC modems will be mostly affected by interferers in their close vicinity or by external

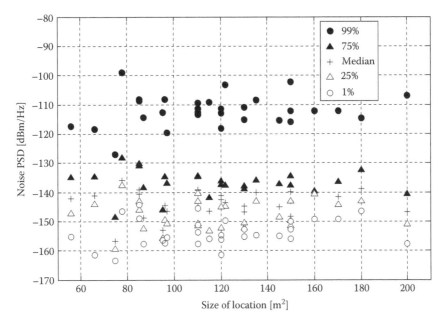

FIGURE 5.15
Influence of the size of the location on the noise statistics.

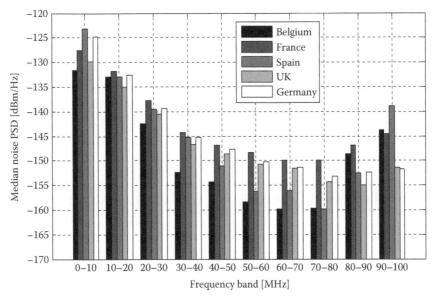

FIGURE 5.16
Median of the noise PSD received over frequency bands of 10 MHz each, separated by country.

broadcast signals. Distant electromagnetic interferers on the same electrical network do not seem to have a noticeable impact on the received noise.

Finally, the analysis compares the level of noise received in different countries. Figure 5.16 presents the median value of the noise PSD collected over 10 frequency bands of 10 MHz each, for each of the following countries: Belgium, France, Spain, United Kingdom and Germany. The aggregated data has been collected by considering measurements over the combined band. The level of recorded noise is comparable in Germany, France and United Kingdom, except for frequencies above 70 MHz, where France presents larger values. In Spain, measurements below 10 MHz and above 90 MHz led to the highest noise records, but in the range (40–80 MHz), the values found for this country are among the lowest ones. From measurements taken in Belgium, it can be concluded that the noise level is lower than for the other countries for frequencies below 80 MHz. Nevertheless, in the FM band, strong interference can be observed in Belgian records.

5.2.3.2 Time Domain Results

In the following, the characteristics of the noise are analysed in the time domain. Out of the 1928 records of 20 ms duration each, four typical structures could be identified:

1. *Stationary noise*: Most of the records present a flat temporal structure, without noticeable fluctuation. An example of such record is given in Figure 5.17, which is representative of approximately two-thirds of our observations.

2. *Periodical 50 Hz synchronous noise*: In some records, a clear periodical structure with a period of 10 ms is observable. This is the case in the example given in Figure 5.18. The period of such a noise structure corresponds to the half of the period of the 50 Hz mains. Such impulsive noise can be generated by, for example, switched mode power supplies or light dimmers. Such structure occurred between 10% and 15% of our observations.

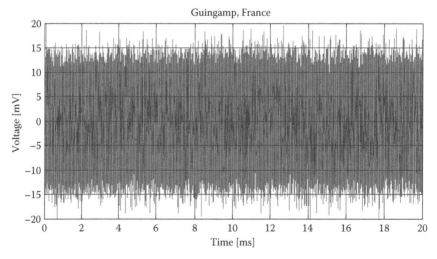

FIGURE 5.17
20 ms noise sample recorded in the time domain, stationary noise.

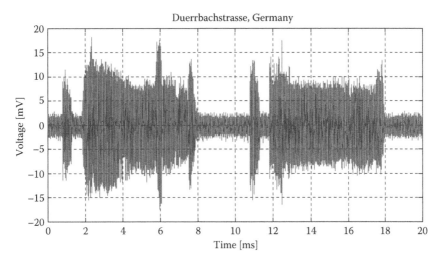

FIGURE 5.18
20 ms noise sample recorded in the time domain, periodical 50 Hz synchronous structure.

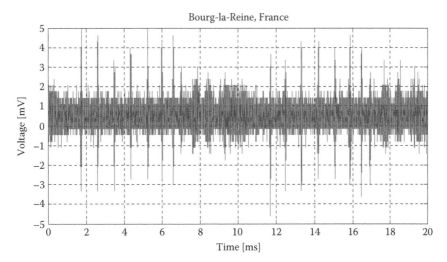

FIGURE 5.19
20 ms noise sample recorded in the time domain, periodical asynchronous structure.

3. *Periodical asynchronous structure*: Other records showed a typical periodical structure, but without being synchronised with the 50 Hz mains. Figure 5.19 presents such a periodical, asynchronous noise, with impulses reproduced at a rate of 1.3 kHz. This type of noise could be observed in 10% to 15% of our records, and can be generated by, for example, electronic devices or compact fluorescent lamps.

4. *Strong impulsive structure*: In the database, some noise records presented larger voltage values and an impulsive structure, like the sample depicted in Figure 5.20. This strong impulsive structure is observed in 10%–15% of the cases.

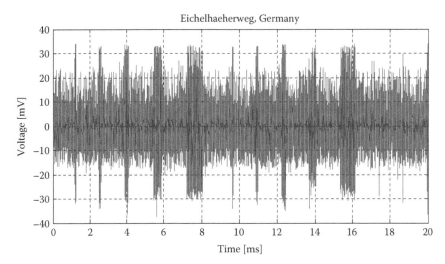

FIGURE 5.20
20 ms noise sample recorded in the time domain, strong impulsive structure.

Finally, the correlation of the noise signals received via the L, PE, N and CM ports was investigated. For this purpose, the statistical correlation $\rho_{k,l}$ between the noise records $n_k(t)$ and $n_l(t)$ was computed using the following equation:

$$\rho_{k,l} = \frac{\overline{n_k(t)\cdot n_l(t)} - \overline{n_k(t)}\cdot\overline{n_l(t)}}{\sqrt{\left(\overline{\left|n_k(t)\right|^2} - \left|\overline{n_k(t)}\right|^2\right)\left(\overline{\left|n_l(t)\right|^2} - \left|\overline{n_l(t)}\right|^2\right)}}, \tag{5.1}$$

where $\overline{n_k(t)}$ represents the average of variable $n_k(t)$ in the time domain.

The first observation when computing this correlation coefficient is that all possible values of correlation were found between 0 (i.e. completely decorrelated) and 1 (i.e. completely correlated). Such observation was already revealed by previous analyses in the literature [9].

Figure 5.21 represents the median noise correlation $|\rho|$ between any possible combination of two Rx ports, for Band 1 (left part) and Band 4 (right part). In Band 1, corresponding to a single measurement in the (2–100 MHz) frequency range, the highest median correlation is obtained between the L and PE ports. The lowest median correlation is observed between the PE and CM ports. When considering Band 4 (i.e. measurements excluding the FM and SW bands), the median correlation is reduced by 0.1. It can be concluded that the broadcast interference generated in the FM and SW bands leads to an increase of the correlation between noise samples received on different ports.

5.2.4 Singular Values and Spatial Correlation

The used input and output ports define the MIMO PLC channel. Assume for example that the delta-style coupler is used at the *transmitter* (Tx). Up to 2 out of the 3 ports (D1, D2 and D3) may be used according to Kirchhoff's law (see details on MIMO coupling in Chapter 1); both ports of the T-style coupler may be used to feed the signals to the wires. At the receiver, all four ports of the star-style coupler may be used. Let N_T be the number

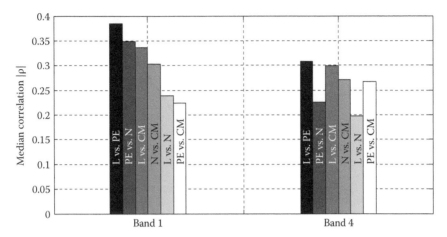

FIGURE 5.21
Median noise correlation between different Rx ports, for Band 1 and Band 4.

of transmit ports, N_R the number of the receive ports and h_{nm} the complex channel coefficient from transmit port m ($m = 1, ..., N_T$) to receive port n ($n = 1, ..., N_R$) for each frequency measurement point (1601 frequency points from 1 to 100 MHz). Then, the channel matrix **H** for each frequency measurement point can be defined as

$$\mathbf{H} = \begin{pmatrix} h_{11} & \cdots & h_{1N_T} \\ \vdots & \ddots & \vdots \\ h_{N_R1} & \cdots & h_{N_RN_T} \end{pmatrix}. \tag{5.2}$$

The *singular value decomposition* (SVD), which was already introduced in Chapter 4, decomposes the channel matrix **H** into three matrices:

$$\mathbf{H} = \mathbf{UDV}^H, \tag{5.3}$$

where

^{H} is the Hermitian operator (transpose and conjugate complex)

D is a diagonal matrix containing the singular values $\sqrt{\lambda_j}$ ($j = 1, ..., r$) of the channel matrix **H** in decreasing order; r is the rank of **H**

The number of non-zero singular values is derived from the number of transmit and receive ports as $r = \min(N_T, N_R)$ if **H** has full rank. In practice, this condition is always fulfilled (see the results for the spatial correlation later). **U** and **V** are unitary matrices, that is, $\mathbf{U}^{-1} = \mathbf{U}^H$ and $\mathbf{V}^{-1} = \mathbf{V}^H$. The SVD decomposes the channel matrix into parallel and independent SISO branches as denoted by the diagonal structure of **D**.

Since **U** and **V** are unitary matrices, multiplication by these matrices preserves the total energy. The singular values $\sqrt{\lambda_j}$ of **D** describe the attenuation of the decomposed SISO branches (note that the channel matrix **H** was not normalised before calculating the SVD). The decomposition into parallel and independent spatial streams by the SVD is very useful in MIMO theory and will be used and discussed later in more detail, for example, in Chapters 8 and 9.

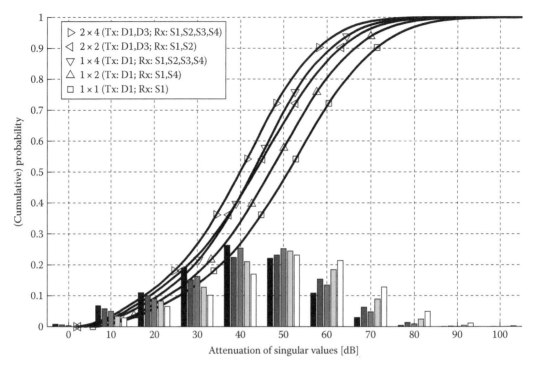

FIGURE 5.22
Attenuation of the singular values of the first MIMO stream.

Figure 5.22 shows the probability (bar graphs) and the cumulative probability (line graphs) of the attenuation of the first singular value $-20 * \log_{10}(\sqrt{\lambda_1})$ (first MIMO stream) for different MIMO configurations. The underlying data of this plot are all the MIMO measurement channels (346 in total) of all buildings, countries and all frequencies. Each bar graph from left to right shows the MIMO configurations depicted in the legend (from top to down). As explained earlier, the number of the used transmit and receive ports define the number of MIMO streams: One spatial stream is available if only one transmit port is used; two spatial streams can be used for transmission if two transmit ports and more than two receive ports are used. Figure 5.23 shows a zoom plot of Figure 5.22 at the high coverage area. Figure 5.24 illustrates the corresponding attenuation of the second singular value where a second spatial stream is available, that is, for the 2 × 2 (◁) and 2 × 4 (▷) configuration. Increasing the number of transmit and receive ports results in a lower attenuation of the spatial streams. A comparison between the 2 × 4 MIMO configuration (▷) and SISO (1 × 1 configuration, □) shows the MIMO gain. The first 2 × 4 stream is 11 dB less attenuated at the 50% point (median) than the SISO path. Additionally, the second stream supports data transmission. The attenuation of the second stream (2 × 4 (▷) in Figure 5.24) is 3 dB higher compared to SISO (□ in Figure 5.22). The zoom at high attenuations reveals even a higher gain of MIMO. Considering the 90% percentile the first stream is 13 dB less attenuated than the SISO path while the attenuation of the second stream is only 1.5 dB higher compared to SISO. Thus, MIMO improves in particular the performance of highly attenuated channels which are most important to meet coverage requirements.

The relation between the MIMO paths can be attributed by the spatial correlation which is an important measure of the MIMO channel. The spatial correlation affects not only the channel capacity (see Chapter 9) but also the performance of different MIMO schemes

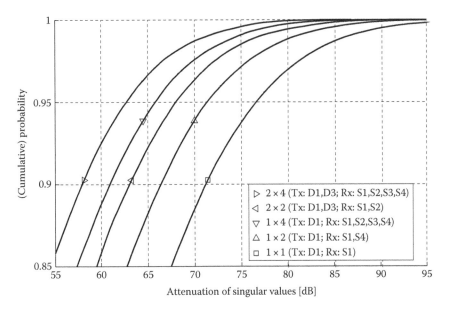

FIGURE 5.23
Attenuation of the singular values of the first MIMO stream: zoom.

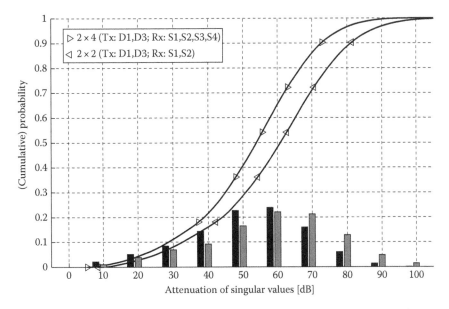

FIGURE 5.24
Attenuation of the singular values of the second MIMO stream.

(see Chapters 8 and 9). Note that a discussion on the spatial correlation of the MIMO PLC channel can be also found in Chapter 4. The condition number of a matrix can be used as a measure of the spatial correlation. The condition number is defined as the ratio of the largest singular value to the smallest singular value:

$$20 * \log_{10}\left(\frac{\sqrt{\lambda_1}}{\sqrt{\lambda_2}}\right). \tag{5.4}$$

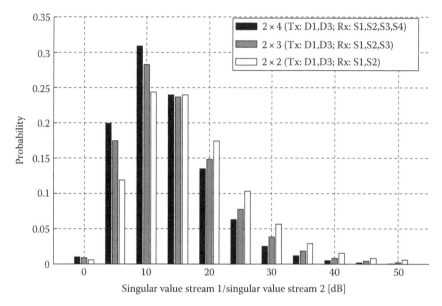

FIGURE 5.25
Spatial correlation.

As explained earlier, the singular values describe the attenuation of the MIMO streams. The attenuation fraction of the channel is reduced by calculating the ratio according to Equation 5.4. The higher the condition number according to Equation 5.4, the higher the attenuation of the second stream compared to the first stream and the higher the spatial correlation. The definition of the spatial correlation is illustrated by the extreme case of a fully correlated MIMO channel where all the channel coefficients of the channel matrix **H** have the same value. As a consequence, the rank of this channel matrix is one, the second singular value is equal to zero, and thus the spatial correlation according to Equation 5.4 approaches infinity. Figure 5.25 shows the probability of the spatial correlation according to the definition of Equation 5.4 for the 2 × 2, 2 × 3 and 2 × 4 MIMO configurations. The rare occurrence of high values of the condition number indicates a rather high spatial correlation of the MIMO PLC channel. The 2 × 4 MIMO configuration comprises in 60% of the channels and frequencies a spatial correlation of >10 dB. Although the second stream is significantly attenuated compared to the first stream in this case, it has to be kept in mind that the first stream is significantly less attenuated compared to SISO. The MIMO performance is gained by the combination of both spatial streams and not only by the availability of the second stream (see also the channel capacity results in Chapter 9).

5.3 Statistical Modelling of the PLC MIMO Channel Based on Experimental Data

5.3.1 MIMO CTF Modelling

In this section, a step-by-step description is given for the statistical modelling of the PLC MIMO *channel transfer function* (CTF) based on experimental measurements. The purpose of the model is to capture the main physical features of the transmission channel, including its median attenuation, its frequency domain path loss characteristics, its multi-path and frequency fading structure and the correlation between different MIMO paths. These parameters were characterised from experimental measurements in Section 5.2.1. The proposed model is statistical in nature, and will be developed following a top-down approach. Therefore, its purpose is not to reproduce exactly the CTF measured at a given measurement location, but rather to enable the generation of random CTF exhibiting similar statistics in line with actual channel measurements.

The philosophy used in the modelling process has been first presented in [10]. In this first proposal, the model parameters were extracted from a limited set of measurements performed in five houses in France, leading to a collection of 42 CTF matrices. Measurements were taken using differential ports at both the Tx and Rx. Hence, 3 × 3 MIMO channel matrices were generated by the model.

In the present study, a larger set of data is considered, as measurements have been taken in 36 buildings in six different countries: Belgium, France, Germany, Italy, Spain and United Kingdom. Differently from the analysis in [10], the couplers used for this measurement series used differential ports at the Tx (D1: line–neutral port, D2: neutral–protective earth port and D3: protective earth–line port), and star-style ports at the Rx (S1: line port, S2: neutral port and S3: protective earth port). In addition, a fourth Rx port (S4) was available to receive CM signals flowing through the three wires. More details on MIMO coupling can be found in Chapter 1. The corresponding MIMO configuration is called 3 × 4 MIMO, and therefore, each measured CTF between any two outlets consists of a 4 × 3 matrix, as shown in Figure 5.26. The overall measurement set comprises 351 4 × 3 CTF matrices. More details about the measurement campaign and the corresponding channel characterisation analysis can be found in Section 5.2.

As explained in Section 5.2.1, no significant difference between the channels measured in different countries was observed. Highly attenuated channels as well as channels with better propagation conditions could be found in each country. Therefore, the following model will capture the global measurement statistics independently from the country where the measurement was taken. The output will be representative of the PLC channel transmissions conditions in Europe.

5.3.1.1 Median Channel Gain Modelling

A first important feature representative of the PLC CTF is the median channel gain, since it governs the overall signal attenuation to be expected by a transmission system. In this

$$
\begin{array}{cccc}
\text{Input} & \text{L-N} & \text{N-PE} & \text{PE-L} \\
\text{port} & \text{D1} & \text{D2} & \text{D3}
\end{array}
$$

$$
H(f) = \begin{bmatrix} h_{S1,D1}(f) & h_{S1,D2}(f) & h_{S1,D3}(f) \\ h_{S2,D1}(f) & h_{S2,D2}(f) & h_{S2,D3}(f) \\ h_{S3,D1}(f) & h_{S3,D2}(f) & h_{S3,D3}(f) \\ h_{S4,D1}(f) & h_{S4,D2}(f) & h_{S4,D3}(f) \end{bmatrix} \begin{array}{cc} \text{S1} & \text{L} \\ \text{S2} & \text{N} \\ \text{S3} & \text{PE} \\ \text{S4} & \text{CM} \end{array}
$$

$$
\begin{array}{c}
\text{Output} \\
\text{port}
\end{array}
$$

FIGURE 5.26
Description of the measured and modelled channel matrix: input and output ports.

analysis, the median gain of the CTF is defined as a matrix A providing the median value across frequency for each element of the channel matrix $H(f)$:

$$A = \begin{bmatrix} A_{S1,D1} & A_{S1,D2} & A_{S1,D3} \\ A_{S2,D1} & A_{S2,D2} & A_{S2,D3} \\ A_{S3,D1} & A_{S3,D2} & A_{S3,D3} \\ A_{S4,D1} & A_{S4,D1} & A_{S4,D3} \end{bmatrix} = med\left(\|H(f)\|\right), \tag{5.5}$$

where $med()$ represents the element-wise median operator. Complementarily, the representation of the median channel gain in dB will be denoted A_{dB}:

$$A_{dB} = 20 \cdot \log_{10}(A) = med\left(10 \cdot \log_{10}\left(\|H(f)\|^2\right)\right). \tag{5.6}$$

As a first step, the statistics of the median gain of the D1–S1 link, $A_{S1,D1}$, is investigated. The solid line in Figure 5.27 represents the CDF of this parameter in dB. One can observe that this median gain varies between −80 dB for the most attenuated channels and −10 dB for weakly attenuated channels. This observation is comparable to the results in Figure 5.2, where the CDF of the channel attenuation over all frequencies for a given feeding style is provided. In addition, the shape of the CDF shows that the distribution of parameter $A_{dB\,S1,D1}$ is close to a Gaussian distribution. The dashed curve represents the fitted Gaussian approximation, with mean $\mu_A = -50.1$ dB and standard deviation $\sigma_A = 15.6$ dB. Therefore, the first parameter of the model is $A_{dB\,S1,D1}$, and will be randomly selected according to the following normal distribution:

$$A_{dB\,S1,D1} = \mathcal{N}(\mu_A, \sigma_A), \tag{5.7}$$

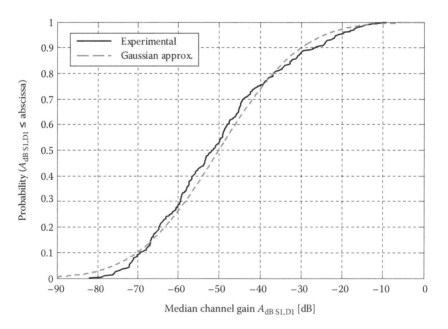

FIGURE 5.27
CDF of the median gain of the D1–S1 link $A_{dB\,S1,D1}$.

where the *probability density function* (PDF) of the normal distribution $\mathcal{N}(\mu, \sigma)$ is given by

$$p_X(x) = \frac{1}{\sqrt{2\pi}\sigma} \exp\left(-\frac{(x-\mu)^2}{2\sigma^2}\right). \tag{5.8}$$

In order to characterise the other elements in matrix A_{dB}, that is, the median gain of the other links in the MIMO channel matrix, it is interesting to analyse how they are related to parameter $A_{dB\,S1,D1}$. For this purpose, Figure 5.28 provides scatter plots of the observed values of $A_{dB\,S2,D1}$ and $A_{dB\,S3,D1}$, respectively, versus the value of $A_{dB\,S1,D1}$, for the 351 measured channel matrices. As a reminder, $A_{dB\,S2,D1}$ provides the median channel gain for the D1–S2 link and $A_{dB\,S3,D1}$ provides the median channel gain for the D1–S3 link. As can be seen on these graphs, the median gain parameters are highly similar, as the plots approximately align along the $y = x$ line (dotted line). A similar observation was made in the characterisation phase: in Figure 5.3, it is shown that the channel attenuation is similar when considering either the S1, S2 or S3 receive ports. In Figure 5.28, the difference between the plotted parameters can be seen as a deviation with respect to perfect equality. By comparing Figure 5.28a and b, one can conclude that the deviation between $A_{dB\,S3,D1}$ and $A_{dB\,S1,D1}$ is larger than the deviation between $A_{dB\,S2,D1}$ and $A_{dB\,S1,D1}$. This can be physically interpreted, since the output ports S1 (L wire) and S2 (N wire) both comprise a wired connection to the differential input port D1 (L-N), and are symmetrical with respect to each other. On the contrary, the output port S3 corresponds to the PE wire which is not wired to the input port D1. This can explain the larger deviation.

Similar characteristics were observed for all input ports D1, D2 and D3, and for the three output ports S1, S2 and S3. Therefore, it is proposed to model parameters $A_{dB\,Sm,Dn}$ for $m \in [1, 2, 3]$ and $n \in [1, 2, 3]$ as

$$A_{dB\,Sm,Dn} = A_{dB\,S1,D1} + \mathcal{N}(0, \sigma_{Sm,Dn}), \tag{5.9}$$

where the deviation between parameter $A_{dB\,S1,D1}$ and parameter $A_{dB\,Sm,Dn}$ is modelled as a Gaussian distribution with mean 0 and standard deviation $\sigma_{Sm,Dn}$. This last parameter can be easily derived from experimental measurements by computing the standard deviation of the difference between $A_{dB\,Sm,Dn}$ and $A_{dB\,S1,D1}$, as will be done in the following paragraphs.

A closer look needs now to be taken at the median channel gain parameter $A_{dB\,S4,Dn}$ when the output port S4 is involved, that is, using CM reception. Figure 5.29 provides similar scatter plots as Figure 5.28 for parameters $A_{dB\,S4,D1}$ and $A_{dB\,S4,D3}$, respectively.

In both cases, one can observe that the approximation given in Equation 5.9 is no longer valid. Indeed, the scatter plot is no longer aligned with the dotted $y = x$ line. For highly attenuated channels (bottom-left corner of the graph), reception using the CM tends to generate a lower attenuation, while for low attenuated channels (top-right corner of the graph), the attenuation is stronger when receiving using the CM (S4 port). This particularity was already observed when statistically analysing the experimental measurements (see Figures 5.3 and 5.5). A linear regression analysis showed that these scatter plots are much better aligned with the line given by $y = 0.5x{-}30$ (solid line). The same result holds for parameter $A_{dB\,S4,D3}$ (data not shown). Therefore, a refined model will be used when considering MIMO links involving CM reception on port S4, as follows ($n \in [1{-}3]$):

$$A_{dB\,S4,Dn} = 0.5 \times A_{dB\,S1,D1} - 30 + \mathcal{N}(0, \sigma_{S4,Dn}), \tag{5.10}$$

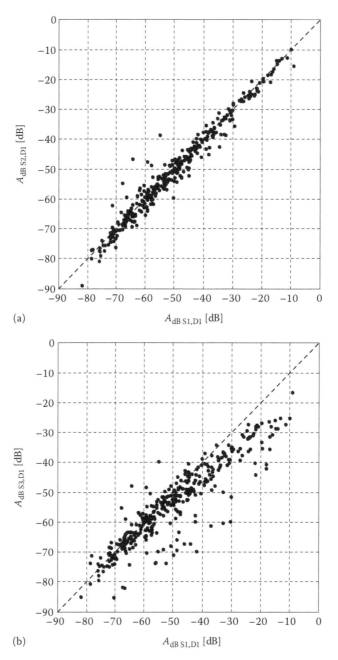

(a)

(b)

FIGURE 5.28
Scatter plots of median channel gains: (a) $A_{dB\ S2,D1}$ versus $A_{dB\ S1,D1}$ and (b) $A_{dB\ S3,D1}$ versus $A_{dB\ S1,D1}$.

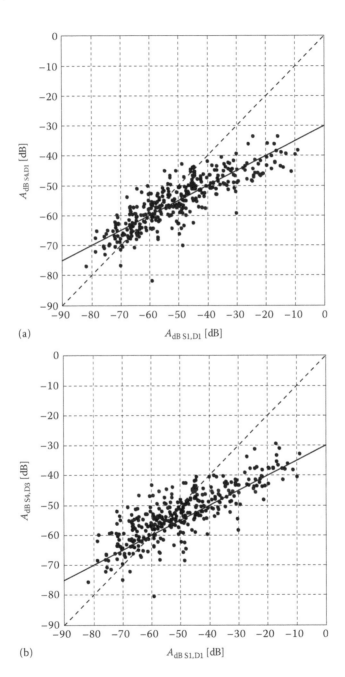

(a)

(b)

FIGURE 5.29
Scatter plots of median channel gains: (a) $A_{dB\ S4,D1}$ versus $A_{dB\ S1,D1}$ and (b) $A_{dB\ S4,D3}$ versus $A_{dB\ S1,D1}$.

where values 0.5 and −30, respectively, represent a slope and an y-intercept accounting for the particular shape of the scatter plot observed in Figure 5.29. The deviation from this linear model is given by a Gaussian distribution with mean 0 and standard deviation $\sigma_{S4,Dn}$. The values of $\sigma_{S4,Dn}$ can be easily derived from observations of experimental measurements. All values of the standard deviations $\sigma_{Sn,Dn}$ in dB are grouped in a single matrix as follows:

$$[\sigma_{Sm,Dn}] = \begin{bmatrix} 0 & 5.1 & 3.8 \\ 2.9 & 5.7 & 5.2 \\ 6.6 & 7.8 & 6.9 \\ 4.6 & 5.9 & 5.1 \end{bmatrix} \text{ in dB.} \tag{5.11}$$

5.3.1.2 Frequency Domain Path Loss Modelling

Once the median channel gain is settled, the focus is now placed on the modelling of the frequency domain path loss. For this purpose, the proposed MIMO channel model builds on a widely accepted SISO PLC channel model first developed by Zimmerman [11]. This model accounts for both the multi-path characteristics of the PLC channel and its particular frequency domain power decay. The same formalism was used within the ICT OMEGA Project [12]. This CTF model $H(f)$ is mathematically described as follows:

$$H(f) = A \sum_{p=1}^{N_p} g_p e^{-j(2\pi d_p/v)f} e^{-(a_0 + a_1 f^K)d_p}, \tag{5.12}$$

where
  v represents the speed of the electromagnetic wave in the copper wire
  d_p and g_p, respectively, represent the length and gain of the propagation path
  N_p represents the number of propagation paths
  The other parameters a_0, a_1, K and A are attenuation factors

An interesting statistical extension of this model was proposed by Tonello in [13] and further developed in [12]. This extension assumes that the multiple paths are generated by a Poisson arrival process with intensity Λ, and L_{max} denotes the maximum length of the signal paths. In addition, the gains g_p of the propagation paths are considered to be uniformly distributed in the range [−1, 1]. Under these assumptions, Tonello computed the statistical expectation of the frequency domain path loss as

$$PL(f) = A^2 \frac{\Lambda}{3} \frac{1 - e^{-2L_{max}(a_0 + a_1 f^K)}}{(2a_0 + 2a_1 f^K)(1 - e^{-\Lambda L_{max}})}. \tag{5.13}$$

Note that in this model, parameter A governs an arbitrary attenuation of the channel. Therefore in the following, we will consider it as the experimental median gain of the measured channels as described in Section 5.3.1.1. The expression of the expected path loss provided by Equation 5.13 is useful to derive parameters a_0, a_1 and K from experimental measurements. For this purpose, parameter A was taken as the median gain for each measurement, and parameters L_{max} and Λ were arbitrarily fixed to 800 m and 0.2 m^{-1} as it is done in [12]. By selecting L_{max} = 800 m, one considers that the signals travelling along propagation paths (including multiple reflections) longer than 800 m do not play a significant role in the structure of the path loss. This assumption is reasonable due to the strong attenuation of signals with increasing distance. The choice of Λ = 0.2 m^{-1} implies that the average excess length between any two successive propagation paths is 5 m. This means that the average distance between any two successive junctions or outlets in the electrical network is 2.5 m. This value is in line with observed wiring practices for indoor electrical networks.

The values of a_0, a_1 and K can then directly be obtained for each measurement using a simulated annealing procedure. This method requires input parameters, and the values defined in the OMEGA Project were used, namely $a_{0,init}$ = 3 × 10^{-3}, $a_{1,init}$ = 4 × 10^{-10} and K_{init} = 1. Figure 5.30 presents an example of such a fitting, performed on an experimental CTF measured for the D1–S1 link in a German house. For this particular example, the simulated annealing procedure resulted in the following path loss parameters: a_0 = 3.6 × 10^{-4}, a_1 = 4 × 10^{-10} and K = 1.04.

Path loss parameters were obtained for the 12 possible links for each of the 351 measured MIMO channel matrices. As a result, 4212 values were available for the statistical evaluation of each of the path loss parameters. As a first observation, no particular dependency of the path loss parameters with the particular link considered within the MIMO matrix was detected. Therefore, the statistical distribution of each parameter was considered globally from the 4212 samples. Figure 5.31 provides the experimental estimate of the PDF for parameter a_0, obtained using a normalised histogram. It can be observed that the PDF of a_0 follows an exponential decay with negative minimum values. Hence, it is proposed to

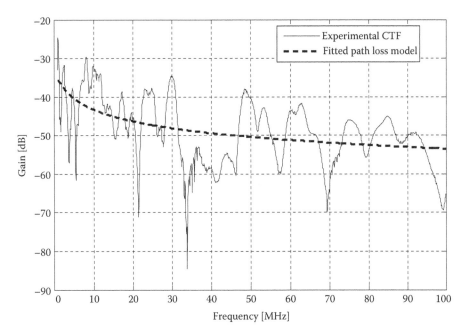

FIGURE 5.30
PLC CTF measured in a German house, and fitted path loss model.

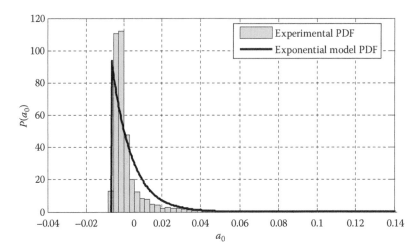

FIGURE 5.31
Experimental PDF versus shifted exponential model PDF for variable a_0.

model this parameter using a shifted exponential distribution $\mathcal{E}_{shift}(\mu_{a0}, \delta_{a0})$. The PDF of the shifted exponential distribution $\mathcal{E}_{shift}(\mu, \delta)$ is defined as

$$p_X(x) = \frac{1}{\gamma}\exp\left(-\frac{x-\delta}{\gamma}\right).\qquad(5.14)$$

Figure 5.31 also represents the shifted exponential model PDF for variable a_0, where the best fitting parameters are given by $\mu_{a0} = 1.04 \times 10^{-2}$ and $\delta_{a0} = -6.7 \times 10^{-3}$. These parameters will be used to randomly generate realistic values for parameter a_0. The same statistical distribution was already observed in [10], with similar parameter values.

Parameter a_1 will be much more easily modelled. Indeed, for 97% of the measured channels, the estimated value of a_1 stayed equal to the input value of $a_{1,init} = 4 \times 10^{-10}$. Therefore, this value will be constant in the proposed model, as was already done in [10].

Finally, Figure 5.32 represents the experimental estimate of the PDF for parameter K, in the form of a normalised histogram. This experimental PDF takes a slightly asymmetric shape, which is in general well represented using a heavy-tailed distribution. The Weibull distribution $\mathcal{W}(\alpha, \beta)$ is an example of such heavy-tailed distributions that can be adapted to experimental data by fitting its parameters α and β. The Weibull distribution PDF is given by

$$p_X(x) = \alpha\beta \cdot x^{\beta-1}\exp(-\alpha \cdot x^\beta).\qquad(5.15)$$

The Weibull distribution best fitting the experimental distribution of parameter K was obtained with parameters $\alpha_K = 5.7 \times 10^{-2}$ and $\beta_K = 57.7$. Its theoretical PDF is represented in Figure 5.32, and will be used to randomly generate values of parameter K for the MIMO channel model. It can be noted that in [10], a normal distribution was selected, with a similar mean value. In the present work, the larger dataset allowed to refine this estimate, in particular by accounting for the distribution asymmetry. Table 5.2 summarises the statistical models of the path loss parameters.

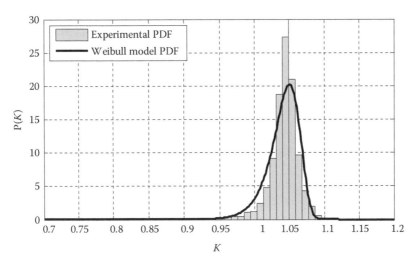

FIGURE 5.32
Experimental PDF versus Weibull model PDF for variable K.

TABLE 5.2

Statistical Models of the Path Loss Parameters

Path Loss Parameter	Model	Parameters
$A_{\text{dB S1,D1}}$	$\mathcal{N}(\mu_A, \sigma_A)$	$\mu_A = -50.1$ dB $\sigma_A = 15.6$ dB
$A_{\text{dB Sm,Dn}}$ ($m \in [1\text{–}3]$)	$A_{\text{dB Sm,Dn}} = A_{\text{dB S1,D1}} + \mathcal{N}(0, \sigma_{Sm,Dn})$	
$A_{\text{dB S4,Dn}}$	$A_{\text{dB S4,Dn}} = 0.5 \times A_{\text{dB S1,D1}} - 30 + \mathcal{N}(0, \sigma_{S4,Dn})$	$[\sigma_{Sm,Dn}] = \begin{bmatrix} 0 & 5.1 & 3.8 \\ 2.9 & 5.7 & 5.2 \\ 6.6 & 7.8 & 6.9 \\ 4.6 & 5.9 & 5.1 \end{bmatrix}$ dB
a_0	$\mathcal{E}_{shift}(\mu_{a0}, \delta_{a0})$	$\mu_{a0} = 1.04 \times 10^{-2}$ $\delta_{a0} = -6.7 \times 10^{-3}$
a_1	Constant	$a_1 = 4 \times 10^{-10}$
K	$\mathcal{W}(\alpha_K, \beta_K)$	$\alpha_K = 5.7 \times 10^{-2}$ $\beta_K = 57.7$
L_{max}	Constant	$L_{max} = 800$ m
Λ	Constant	$\Lambda = 0.2$ m^{-1}

Note: Parameters a_0, a_1 and K are dimensionless.

5.3.1.3 Multi-Path Model for the SISO Line–Neutral (D1) to Line (S1) Path

Once the path loss parameters of the MIMO channel model are settled, the modelling procedure focuses on the multi-path and frequency fading structure of the PLC channel. Statistical channel models dedicated to SISO links already exist, where this specific feature is well captured. In particular, the proposed MIMO channel model builds on the SISO channel framework developed by Zimmermann and Tonello in [11,14]. More specifically, the first step in the MIMO channel model consists in emulating the line–neutral to line path (D1–S1 link) using Equation 5.12, where the generic parameter A is replaced with $A_{S1,D1}$, and the path loss parameters $A_{S1,D1}$, a_0, a_1 and K are selected according to the statistical distributions specified in Table 5.2. Following [12], the path lengths d_p are randomly generated as events of a Poisson arrival process $\mathcal{P}(\Lambda)$ characterised by its intensity Λ.

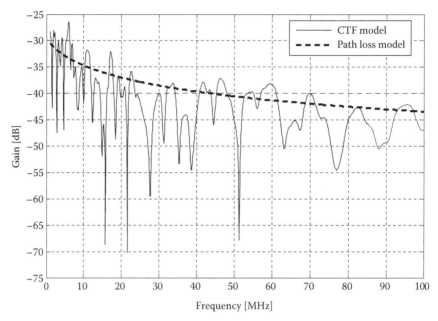

FIGURE 5.33
PLC CTF generated using the SISO model, and corresponding path loss model.

Using a maximum path length L_{max}, the Poisson process also leads to a finite number of paths N_p. Note that the values of Λ and L_{max} are fixed and provided in Table 5.2. Finally, the gains g_p of the propagation paths are generated using a uniform statistical distribution $\mathcal{U}(g_{min}, g_{max})$ within the range [−1, 1].

Figure 5.33 presents a CTF of a D1–S1 link generated using the proposed channel model. The corresponding path loss model is also represented. One can notice its similarity with measured CTF such as the one presented in Figure 5.30.

5.3.1.4 Extension of the Multi-Path Model to the MIMO Channel Matrix

In order to extend the SISO channel model to the full MIMO matrix, it is first necessary to investigate the characteristics of the last feature that the model intends to capture, namely the correlation between different MIMO links. As explained in Section 5.2.4, the degree of correlation within the MIMO channel matrix has an impact on the channel capacity and on the performance of MIMO signal processing techniques (see also Chapter 9). In order to capture the frequency fading correlation between individual channels in the MIMO matrix, this study uses the Pearson correlation coefficient ρ already introduced in [10]. It is recalled that the computation of this coefficient requires evaluating the statistical expectation of the channel gain. As we are interested only in the correlation of the fast fading behaviour of the CTF, the Pearson correlation coefficient ρ will be computed on a normalised CTF defined as follows:

$$\tilde{h}_{Sm,Dn}(f) = \frac{h_{Sn,Dm}(f)}{\sqrt{PL_{Sn,Dm}(f)}}, \tag{5.16}$$

where

$h_{Sn,Dm}(f)$ represents the CTF between input port Dm and output port Sn
$PL_{Sn,Dm}(f)$ represents its expected path loss computed using Equation 5.13

This normalised CTF is Wide Sense Stationary and thus the statistical expectations used in the computation of the correlation coefficient can be replaced by the frequency domain average. The complex correlation coefficient between channel $h_{Sn,Dm}(f)$ and channel $h_{Si,Dj}(f)$ in the MIMO matrix is finally computed using the following equation:

$$\rho_{SnDm,SiDj} = \frac{\left\langle \tilde{h}_{Sn,Dm}(f)\tilde{h}^*_{Si,Dj}(f) \right\rangle - \left\langle \tilde{h}_{Sn,Dm}(f) \right\rangle \left\langle \tilde{h}^*_{Si,Dj}(f) \right\rangle}{\sqrt{\left(\left\langle \left| \tilde{h}_{Sn,Dm}(f) \right|^2 \right\rangle - \left| \left\langle \tilde{h}_{Sn,Dm}(f) \right\rangle \right|^2 \right) \left(\left\langle \left| \tilde{h}_{Si,Dj}(f) \right|^2 \right\rangle - \left| \left\langle \tilde{h}_{Si,Dj}(f) \right\rangle \right|^2 \right)}}, \tag{5.17}$$

where
$\langle \, \rangle$ denotes the frequency domain average
* denotes the complex conjugate operator

Note that Equation 5.17 provides a correlation coefficient independent of frequency, by comparing the variations of any two channels across their measurement frequency range. Computation of frequency-dependent correlation factors is also possible but requires considering all channels in a measurement matrix simultaneously (see, for instance, the study presented in Chapter 4).

The correlation coefficient $\rho_{SnDm,SiDj}$ was computed for each of the 351 measured channel matrices, and for each possible pair of MIMO links. As the measurements were performed using three possible Tx ports (D1 to D3) and four possible Rx ports (S1 to S4), the MIMO channel matrix contains 12 possible links, and thus 66 pairs of distinct channels can be considered. Figure 5.34 provides the statistical CDF of coefficient ρ for a number of specific cases. As a first observation, the statistical distribution of coefficient ρ can be analysed over

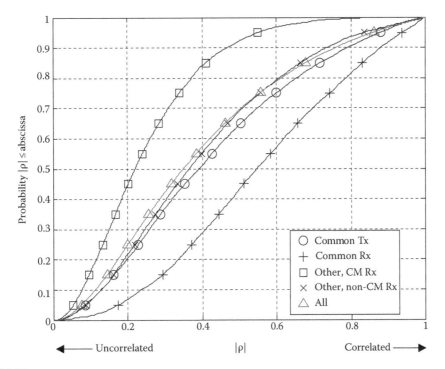

FIGURE 5.34
Statistical CDF of the correlation coefficient ρ obtained from experimental measurements.

the whole set of channel pairs (Δ marker). One can notice that the correlation coefficient is ranging all values between 0 and 1, with a quasi-uniform distribution. Hence, both correlated channels ($|\rho| > 0.8$) and uncorrelated channels ($|\rho| < 0.2$) could be measured. In order to get more insights in the nature of the correlation between MIMO links, Figure 5.34 also represents the correlation coefficient CDF for subsets of the measurement ensemble, namely

- All channel pairs using the same Tx port (o marker)
- All channel pairs using the same Rx port (+ marker)
- All channel pairs with different Tx and Rx port, where one channel uses the CM (S4) Rx port (□ marker)
- All channel pairs with different Tx and Rx port, where the CM (S4) Rx port is not used (× marker)

Observations of the resulting CDF reveal that using the same Tx port does not increase the correlation between channels significantly. However, two channels using the same Rx port are more correlated than average, as the corresponding CDF curve moves towards the right-hand side of the figure. When considering channel pairs that do not share the same Tx port nor the same Rx port, it appears that the correlation coefficient is similar to the average when the CM port is not used. Nevertheless, employing the CM Rx port provides a channel that is less correlated to any other channel using different ports. This could be expected from physical considerations, as CM reception involves unique transmission mechanisms that can differ from classical wire reception.

From this evaluation of the correlation characteristics of the MIMO channel matrix, a full MIMO PLC channel model can be developed. The SISO D1–S1 link is already modelled using the procedure described in Section 5.3.1.3. All other links in the MIMO channel matrix will be modelled using an extended CTF definition as follows:

$$H(f) = A \sum_{p=1}^{N_p} g_p e^{-j\varphi_p} e^{-j(2\pi d_p/v)f} e^{-(a_0+a_1 f^K)d_p}. \tag{5.18}$$

In this CTF equation, each path was assigned an arbitrary additional phase φ_p. The idea originally developed in [10] is to randomly select the value φ_p for each path using a uniform distribution $\mathcal{U}(-\Delta\varphi/2, \Delta\varphi/2)$ within a given range $[-\Delta\varphi/2, \Delta\varphi/2]$. The larger the value of $\Delta\varphi$, the less correlated the resulting channel will be with respect to the line–neutral to line SISO path. For an extreme value of $\Delta\varphi = 2\pi$, a low level of correlation is expected for the corresponding channel. On the contrary, selecting a value of $\Delta\varphi = 0$ will produce a channel identical to the reference channel, with a correlation coefficient $\rho = 1$. In order to keep the degree of correlation observed for channels involving the L, N or PE Rx ports, it is proposed to modify only two parameters when generating different links within the MIMO matrix: the median gain A and the vector of arbitrary phases $\{\varphi_p, p = 1, ..., N_p\}$. This choice is consistent with the physical observation that the topology of the electrical network is the same, independently from the considered MIMO link. Therefore, it seems reasonable to keep parameters such as the distribution of path lengths $\{d_p, p = 1, ..., N_p\}$ or path gains $\{g_p, p = 1, ..., N_p\}$ constant when considering L, N or PE reception. An exception is made for CM reception (port S4): as observed in Figure 5.34, those channels present a lower degree of correlation with channels using different Rx ports. Therefore, for CM reception, a specific distribution of path lengths $\{d_{S4,p}, p = 1, ..., N_{p,S4}\}$ will be generated using the same Poisson

TABLE 5.3

Statistical Models of the MIMO Multi-Path Parameters

MIMO Multi-Path Parameter	Model	Parameters
$\{d_p, p = 1 \dots N_p\}$	$\mathcal{P}(\Lambda)$	$\Lambda = 0.2 \text{ m}^{-1}$ $L_{max} = 800 \text{ m}$
$\{g_p, p = 1 \dots N_p\}$	$\mathcal{U}(g_{min}, g_{max})$	$g_{min} = -1 \text{ V}$ $g_{max} = 1 \text{ V}$
$\{d_{S4p}, p = 1 \dots N_{p, S4}\}$	$\mathcal{P}(\Lambda)$	$\Lambda = 0.2 \text{ m}^{-1}$ $L_{max} = 800 \text{ m}$
$\{g_{S4p}, p = 1 \dots N_{p, S4}\}$	$\mathcal{U}(g_{min}, g_{max})$	$g_{min} = -1 \text{ V}$ $g_{max} = 1 \text{ V}$
v	Constant	$v = 2 \times 10^8 \text{ m/s}$
$\{\varphi_p, p = 1 \dots N_p\}$	$\mathcal{U}(-\Delta\varphi/2, \Delta\varphi/2)$	$\Delta\varphi = 2\pi$ rad to generate links D1–S2 and D1–S3 from link D1–S1 $\Delta\varphi = 4\pi/3$ rad to generate links D2–Sm and D1–Sm from links D1–Sm ($m = 1, \dots, 4$)

Note: Parameters g_{min} and g_{max} are normalised voltages.

arrival process $\mathcal{P}(\Lambda)$ as for the LN-N path. Complementarily, a new set of path gains $\{g_{S4,p}, p = 1, \dots, N_{p,S4}\}$ will be drawn from a uniform distribution $\mathcal{U}(g_{min}, g_{max})$ within the range $[-1, 1]$.

In order to select the appropriate values of $\Delta\varphi$, Monte Carlo simulations were run and the resulting channel correlation factors were compared to the experimental observations. The following values are finally recommended:

- To generate link D1–S2 from link D1–S1, and link D1–S3 from link D1–S1, select $\Delta\varphi = 2\pi$.

- For $m = 1, \dots, 4$, to generate link D2–Sm from link D1–Sm, and link D3–Sm from link D1–Sm, select $\Delta\varphi = 4\pi/3$.

Table 5.3 summarises all necessary parameters for the simulation of random 3×4 MIMO channel matrices.

5.3.1.5 Evaluation of the Channel Model

Figure 5.35 provides an example of generated CTF using the proposed statistical MIMO PLC channel model. The channel model generally produces 3×4 channel matrices, but for the sake of legibility, only the four channels corresponding to the line–neutral differential port (D1) at Tx are represented. One can note that the model produces random MIMO CTFs that faithfully replicates the multi-path structure observed for experimental measurements (see Section 5.2.2). By construction, the channel model also reproduces the statistics observed for the main path loss parameters, namely the median channel gain A and the frequency domain power decay parameters a_0, a_1 and K.

Finally, it is interesting to analyse how the proposed MIMO channel model renders the correlation between different links within the MIMO matrix. For this purpose, a random draw of 351 3×4 MIMO channel matrices was generated. From this synthetic data set, the correlation coefficient ρ between any two links within each MIMO matrix was computed using the same method already described in Section 5.3.1.4. The CDF of the magnitude of

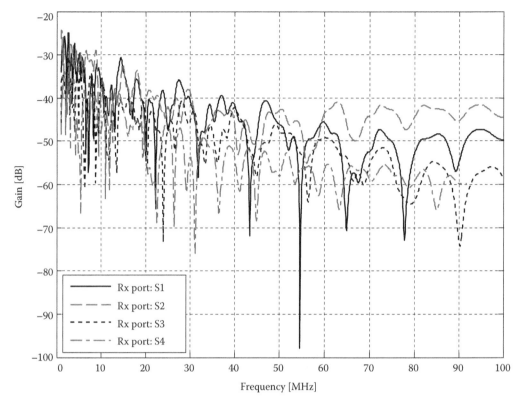

FIGURE 5.35
Example of PLC CTF generated using the proposed MIMO model (Tx port: D1 only).

ρ is displayed in Figure 5.36 for the same specific cases as for the experimental data set: for channels sharing the same Tx port, for channels sharing the same Rx port, for channels without any port in common and using the CM Rx port, and for channels without any port in common and using the L, N or PE port at Rx. By comparing Figures 5.34 and 5.36, one can observe similar correlation trends between the proposed model and the experimental measurements. In particular, the larger correlation between channels sharing the same Rx port is well reproduced. Second, the low correlation between channels involving the CM port and other channels is well captured. Finally, all other channels present an average correlation, spanning the [0, 1] interval in a quasi-uniform way.

5.3.2 Multiple-Output Noise Modelling

5.3.2.1 Multivariate Time Series

A complete and detailed noise model can be achieved through both frequency domain and time domain modelling techniques. However, in a multiple-output noise model, one of the most important features one wishes to capture is the correlation between the signals received at different ports. Characteristics of the experimentally observed noise correlation are exposed in Section 5.2.3.2. On the other hand, the multi-path structure of the received signals, that produce fading notches in the frequency domain, can also be seen as the correlation between successive samples in the time domain. Hence, modelling of the multiple-output noise will present less complexity and more effectiveness in the time domain.

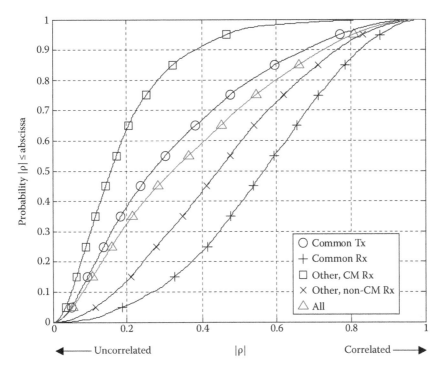

FIGURE 5.36
Statistical CDF of the correlation coefficient ρ obtained from model simulations.

Multivariate time series (MTS) analysis is a powerful mathematical technique used for the modelling and forecasting of the future values of time sequences. The mathematical framework usually employed for the modelling of MTS is the *vector autoregressive* (VAR) model. The suitability and utility of MTS modelling in the context of MIMO PLC noise has been first demonstrated in [15] where it was applied to a small series of noise measurements conducted in French homes. The method is described in the following sections through its application to the ETSI noise measurements presented in Section 5.2.3.

The time domain noise, as mentioned in Section 5.2.3, is measured on four ports simultaneously: S1, S2, S3 and S4. This noise can be visualised as a four-variate time series which is a special case of MTS. A MTS is an ensemble of individual time sequences measured at fixed and equally spaced time intervals Δt. A MTS consists of simultaneous observations of several variables. Consider a MTS $\{N_t\}$ with n variables such that $N_t = (x_{1t}, ..., x_{nt})^T$, where T stands for transpose. A four-variate MTS consists of $\{x_{1t}\}$, $\{x_{2t}\}$, $\{x_{3t}\}$ and $\{x_{4t}\}$. The measured noise can be characterised as an MTS with $n = 4$ and a sampling period $\Delta t = 2$ ns. It should be noted that $\{x_{1t}\}$, $\{x_{2t}\}$, $\{x_{3t}\}$ and $\{x_{4t}\}$ represent S1, S2, S3 and S4 noise, respectively. For analysis, we consider the duration of the vector as 46.52 μs which is equal to the duration of an OFDM symbol in the HomePlug AV2 standard for commercial PLC devices (see Chapter 14).

5.3.2.2 Figures of Merit for Model Validation

Each noise measurement is an ensemble of four noise series; therefore, their spectra need to be observed individually. The overall spectral form of the noise is of a decreasing exponential function as shown in Figure 5.11. At this point, we define a parameter $\Psi_{i,j}$, shown in Equation 5.19, which denotes frequency domain cross-correlation between the measured noise PSD and the modelled noise PSD (for any of S1, S2, S3 and S4 measurements). The parameter $\Psi_{i,j}$ serves as a figure of merit of spectral resemblance between the measured and modelled noise:

$$\psi_{i,j} = \frac{\overline{N_i(f)N_j^*(f)} - \overline{N_i(f)}\,\overline{N_j^*(f)}}{\sqrt{\left(\overline{N_i(f)^2} - \overline{N_i(f)}^2\right)\left(\overline{N_j(f)^2} - \overline{N_j(f)}^2\right)}}, \tag{5.19}$$

where
  $N(f)$ represents the envelope of the noise PSD in dBm/Hz
  Subscripts i and j stand for the measured and modelled noise, respectively
  $\overline{(.)}$ denotes the frequency domain average

Naturally, high values of $\Psi_{i,j}$ are desirable as it would suggest that the PSD of the modelled noise has a close resemblance to the PSD of the measured noise.

The root mean square error is also commonly used to analyse the degree of closeness or match between two datasets. In our case, we use the frequency domain *root mean square error* (RMSE) as a metric of spectral resemblance between the measured and the modelled noise. The frequency domain RMSE, denoted $\varepsilon_{rms,f}$ is defined as

$$\varepsilon_{rms,f} = \sqrt{\frac{1}{K}\sum_{k=1}^{K}(N_i(f_k) - N_j(f_k))^2}, \tag{5.20}$$

where
  $N_i(f_k)$ and $N_j(f_k)$ represent the spectrum in dBm/Hz of a given MIMO PLC noise measurement and its VAR realisation, respectively, measured at a given frequency f_k
  K is the length of noise sequences in the frequency domain

The frequency domain RMSE, like $\Psi_{i,j}$, is a parameter used to estimate the accuracy of the VAR model. Lower values of RMSE indicate a more accurate model and vice versa.

5.3.2.3 Vector Autoregressive Model

In the earlier sections, field measurement and mathematical characterisation of noise are described. In this section, we present a noise model and verify its correctness and efficiency.

A VAR model is a two-dimensional extension of basic *autoregressive* (AR) model. VAR models are used to model time sequences by exploiting their prediction property. A well-known notation of VAR models is VAR(p) where p stands for the model order. A VAR(p) model for a MTS with m variables takes the following mathematical form:

$$x_t = w + \sum_{l=1}^{p} A_l x_{t-l} + \varepsilon_t, \operatorname{cov}(\varepsilon_t) = C, \tag{5.21}$$

where
  $x_t \in \mathfrak{R}^m$ are the vectors representing the MTS at a given time instant t
  ε_t are zero-mean, uncorrelated random noise vectors
  $C \in \mathfrak{R}^{m \times m}$ is the noise covariance matrix
  $A_1, ..., A_p \in \mathfrak{R}^{m \times m}$ are model coefficient matrices

The vector $w \in \mathfrak{R}^m$ serves to introduce mean value if the MTS has non-zero mean [16].

For given values of m and p, Equation 5.21 is used to extract the parameters of the VAR model from the measured noise data at a given socket. The parameters of the VAR model consist of $A_1, \ldots, A_p, C$ and w. We denote this model by VAR$(p)_{socket}$ as this model is associated to a given socket for a given order p. Once VAR$(p)_{socket}$ is obtained, MTS noise can be statistically generated through computer simulations by using Equation 5.21.

5.3.2.4 Order Selection and Extraction of the VAR Model

As mentioned in Section 5.3.2.2, the frequency domain cross-correlation $\Psi_{i,j}$ and the frequency domain RMSE denoted by $\varepsilon_{rms,f}$ are used for assessing the accuracy of the noise model. These two parameters can be effectively used for studying the spectral resemblance between measured noise and its VAR realisation VAR$(p)_{socket}$. For this purpose, one needs to calculate the RMSE between the spectra of a given measured noise and its VAR$(p)_{socket}$ realisation for different p. Similar calculation is performed for $\Psi_{i,j}$ as well. Figure 5.37 shows the dependence of RMSE on the VAR order averaged over 116 measurements. Similar to the observations in [15], the RMSE initially decreases sharply as the VAR order increases, but later it starts settling at a floor value of 4 dB. As for the frequency domain cross-correlation, Figure 5.38 shows the dependence of $\Psi_{i,j}$ on VAR order averaged over 116 measurements. The value of $\Psi_{i,j}$ increases sharply as the VAR order increases but later settles around 0.93. We select a value of $p = 50$, that is, VAR(50) model as a compromise between complexity and accuracy.

Selection of the order of the VAR model is the first step; in our case, it is fixed at $p = 50$. The next step in the modelling process is the extraction of the model parameters from the measured data. To simplify things, we observe in Figures 5.17 through 5.20 that the measured data has zero mean; therefore, w can be set as a 1×4 null vector. At this point, it should be noted that our four-variate MTS will require matrices A and C which are 4×200

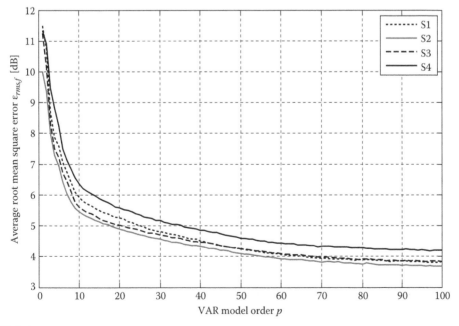

FIGURE 5.37
Dependence of RMSE on the VAR order.

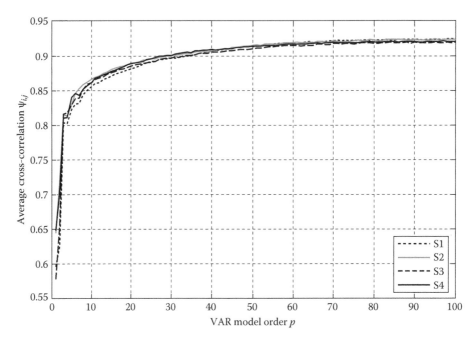

FIGURE 5.38
Dependence of $\Psi_{i,j}$ on the VAR order.

and 4×4 matrices, respectively. The estimation methodology of A and C has been elaborated in [16]. It should be noted here that matrices A and C are real valued. Since we had measured data at 116 PLC sockets, we end up with 116 A and 116 C matrices.

5.3.2.5 Results and Model Validation

Let us first consider the spectrum of a typical measured noise shown in Figure 5.39. It can be observed that it is a typical coloured noise with a high PSD towards the low frequencies. As the frequency increases, the noise PSD decreases. The presence of FM noise PSD is also visible. This noise measurement sample is representative of the statistical noise characteristics detailed in Section 5.2.3.1.

Now we study the PSD of the noise obtained from the VAR(50) simulation. It is clear from Figure 5.40 that the modelled noise successfully captures the coloured nature of the measured noise. The FM noise PSD is also successfully reserved by the model. Although the exact peak-by-peak match is not achieved the measured noise and the modelled noise, the VAR(50) model confidently reflects the overall inverse relationship between the frequency and the noise PSD.

Now we turn to the statistical analysis of the VAR model efficiency. First we consider the frequency domain cross-correlation between the measured noise and its VAR(50) model, as shown in the CDF of Figure 5.41. The CDF represents the dataset of 116 measurements. Figure 5.41 shows that for 90% of the measurements the VAR(50) model achieves a correlation of 0.85–0.98. The typical midrange value of the correlation is around 0.93. These high correlation values suggest that the VAR(50) model efficiently captures and generates the spectral characteristics of the measured noise.

The RMSE $\varepsilon_{rms,f}$ between a given noise measurement and its VAR(50) model ranges from 2 to 7 dB, as shown in Figure 5.42. The typical midrange value of RMSE is around 4.5.

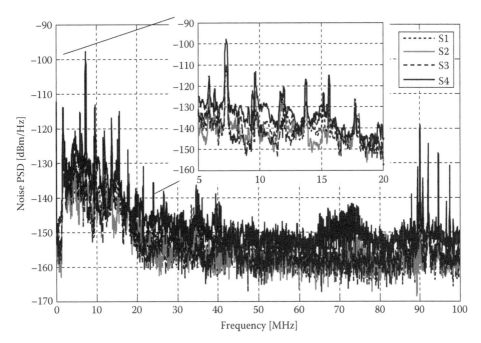

FIGURE 5.39
PSD of a typical measured noise.

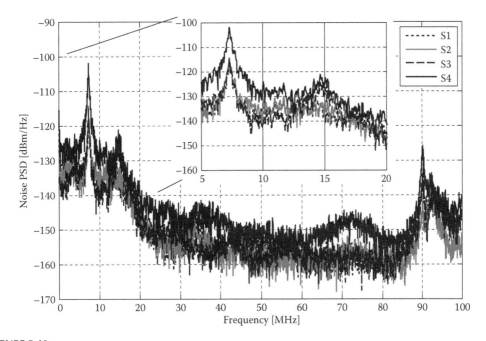

FIGURE 5.40
PSD of the VAR(50) modelled noise corresponding to the measured noise of Figure 5.39.

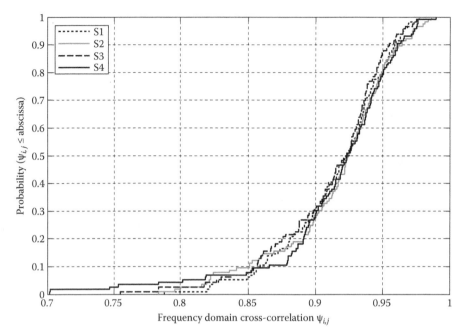

FIGURE 5.41
CDF: Correlation between the spectrum of the measured noise and its VAR(50) model.

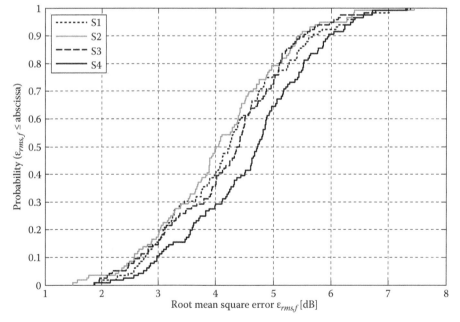

FIGURE 5.42
CDF: RMSE between the spectrum of the measured noise and its VAR(50) model.

We note that for 90% of the cases the RMSE value is below 6 dB, that is, ±3 dB in absolute terms. These statistics affirm that the VAR(50) model successfully regenerates the spectral characteristics of the measured noise.

5.4 Conclusion

Extensive field measurements recoding MIMO PLC properties in private homes were conducted in seven European countries with an identical measurement set-up according to a harmonised measurement specification. This generated comparable results among all locations and a huge statistical mass of data.

Following highlights were found when interpreting the data:

- T-style coupler provides less attenuation at the high attenuated channels (important to achieve maximum coverage).
- CM reception also provides a lower attenuation at the bad links.
- 55 dB is the median attenuation of the main grid in the frequency range 1 MHz < f < 100 MHz, where the attenuation raises with 0.2 dB/MHz.
- The median attenuation within a private home of any size is 0.11 dB/m² × size (m²) +36.07 dB.
- The three-phase installations in Germany causes the highest attenuation among the visited European countries.
- Impedance matching at outlets is quite low, where the star-style coupler shows best and the T-style coupler worst matching properties.
- Usually the impedance over frequency is identical among the wires at one outlet.
- Ring wires in the United Kingdom only show minimal lower input impedance than the European continent.
- The spatial correlation of the MIMO PLC channel is relatively high. For the 2 × 4 MIMO configuration, 60% of the channels and frequencies comprise a spatial correlation of >10 dB, that is, the second spatial stream is >10 dB attenuated compared to the first spatial stream.
- The background noise is the lowest in the frequency range 30 MHz < f < 87 MHz, where the median PSD is −150 dBm/Hz.
- The noise level is independent of the size of a home.
- In the time domain, impulsive noise is present in one-third of the observations, due to the presence of electronic devices like switched mode power supplies or fluorescent lamps.

Furthermore, this chapter presented a new statistical channel model for the MIMO CTF matrix, based on the measurements described. This model allows the generation of random CTFs faithfully reproducing the path loss statistics, multi-path structure and channel correlation coefficients statistically observed within the field data base. Such a model will prove useful to evaluate the performance of existing MIMO PLC transmission systems and to develop new signal processing strategies optimally exploiting the channel characteristics. Modelling of multiple-output noise has been proposed by fitting series of noise measurements to a MTS with a limited set of parameters. This formalism allows reproducing both the noise multi-path structure and the correlation between output ports. In order to improve such studies, it will become important in the future to better understand the noise component of the MIMO PLC channel and to develop stochastic models statistically representative of the variety of noise environments.

Acknowledgements

The authors wish to thank Jose Luis Gonzalez Moreno from Marvell (Spain), Holger Hirsch from the University of Duisburg-Essen (Germany), Hervé Milleret from MStar Semiconductor (France), Andrea Tonello from the University of Udine (Italy) and Nico Weling from Devolo (Germany) for their help in conducting field measurements, in the framework of the ETSI Specialist Task Force (STF) 410. Special thanks goes to Werner Bäschlin from JobAssist (Switzerland) for his excellent support, development and production of the STF 410 couplers which had been used by all teams in the field measurements.

References

1. R. Hashmat, P. Pagani, A. Zeddam and T. Chonavel, MIMO communications for inhome PLC networks: Measurements and results up to 100 MHz, *International Symposium on Power Line Communications and Its Applications (ISPLC)*, Rio de Janerio, Brazil, 2010.
2. A. Schwager, Powerline communications: Significant technologies to become ready for integration, PhD dissertation, Universität Duisburg-Essen, Duisburg, Germany, 2010. http://duepublico.uni-duisburg-essen.de/servlets/DerivateServlet/Derivate-24381/Schwager_Andreas_Diss.pdf.
3. L. Stadelmeier, D. Schneider, D. Schill, A. Schwager and J. Speidel, MIMO for inhome power line communications, *International Conference on Source and Channel Coding (SCC)*, Ulm, Germany, 2008.
4. Special Task Force 410. http://portal.etsi.org/stfs/ToR/Archive/ToR410v11_PLT_MIMO_measurem.doc.
5. ETSI TR 101 562-1 V1.3.1, PowerLine Telecommunications (PLT); MIMO PLT; Part 1: Measurement Methods of MIMO PLT.
6. ETSI TR 101 562-2 V1.2.1, PowerLine Telecommunications (PLT); MIMO PLT; Part 2: Setup and Statistical Results of MIMO PLT EMI Measurements.
7. ETSI TR 101 562-3 V1.1.1, PowerLine Telecommunications (PLT); MIMO PLT; Part 3: Setup and Statistical Results of MIMO PLT Channel and Noise Measurements.
8. P. D. Welch, The use of Fast Fourier transform for the estimation of power spectra: A method based on time averaging over short, modified periodograms, *IEEE Trans. Audio Electroacoust.*, 15, 70–73, June 1967.
9. D. Rende, A. Nayagam, K. Afkhamie, L. Yonge, R. Riva, D. Veronesi, F. Osnato and P. Bisaglia, Noise correlation and its effects on capacity of inhome MIMO power line channels, *IEEE International Symposium on Power Line Communications, ISPLC 2011*, Udine, Italy, April 2011.

10. R. Hashmat, P. Pagani, A. Zeddam and T. Chonavel, A channel model for multiple input multiple output inhome powerline networks, *IEEE International Symposium on Power Line Communications and Its Applications (ISPLC)*, Udine, Italy, April 2011.
11. M. Zimmermann and K. Dostert, A multipath model for the power line channel, *IEEE Trans. Commun.*, 50(4), 553–559, April 2002.
12. Seventh Framework Programme: Theme 3 ICT-213311 OMEGA, Deliverable D3.2, v.1.2, PLC Channel Characterization and Modelling, February 2011 (http://www.ict-omega.eu/publications/deliverables.html).
13. A. M. Tonello and F. Versolatto, New results on top-down and bottom-up statistical PLC channel modeling, *Third Workshop on Power Line Communications*, Udine, Italy, October 2009.
14. A. M. Tonello, Wideband impulse modulation and receiver algorithms for multiuser power line communications, *EURASIP J. Adv. Signal Process.*, 2007, 1–14.
15. R. Hashmat, P. Pagani, T. Chonavel and A. Zeddam, A time domain model of background noise for inhome MIMO PLC networks, *IEEE Transactions on Power Delivery*, 27(4), 2082–2089, October 2012.
16. A. Neumaier and T. Schneider, Estimation of parameters and eigenmodes of multivariate autoregressive models. *ACM Trans. Math. Softwares* 2001, 27(1), 27–57.

Part II

Regulations, Electromagnetic Compatibility and MIMO Capacity

6

Power Line Communication Electromagnetic Compatibility Regulations

Andreas Schwager and Lars T. Berger

CONTENTS

This chapter introduces *power line communication* (PLC)-related *electromagnetic compatibility* (EMC) regulations worldwide. In particular, it introduces the feeding levels used in the PLC capacity and throughput analysis found in Chapter 9.

6.1 Historical Overview

Regulations are one of the fundamental means to enable or simplify business and the exchange of products. Early civilisations in China, Egypt, Greece, India and the Roman Empire had established regulations. Weights and measures were regulated first; currency was probably the next issue where people felt harmonisation to be mandatory. Historically, only non-competitive producers created trade barriers to protect their market. Furthermore, non-transparent regulations tend to favour cheats. With the formation of the EU, there is a movement to deregulate national laws to enable seamless transit of people, products and services. Deregulating national laws requires agreement on an international level, which is not easily negotiated in most cases. Globally operating organisations, like the *International Organization for Standardization* (ISO), the *International Electrotechnical Commission* (IEC) and the *International Telecommunication Union* (ITU), create standards to harmonise products. ISO focuses on mechanical engineering, IEC focuses on electric and electronic technologies and ITU addresses various forms of telecommunication. Each international organisation is mirrored by local committees. For example, the *Comité Européen de Normalisation* (CEN, in English: European Committee for Standardization) mirrors ISO. The *Comité Européen de Normalisation Électrotechnique* (CENELEC, in English: European Committee for Electrotechnical Standardization) mirrors IEC and the *European Telecommunications Standards Institute* (ETSI) mirrors ITU. In addition, there are national organisations like the *Deutsches Institut für Normung* (DIN) and the *Deutsche Kommission Elektrotechnik* (DKE) in Germany that delegate their experts to the international committees, often also submitting proven national proposals to a larger audience.

Market introduction of a new product requires a certification process. In the United States, this is called the 'FCC marking' (FCC for *Federal Communications Commission*). In Europe, this process is called 'CE marking' (CE for *Conformité Européenne*). Generally, the manufacturer declares that the product conforms to the essential applicable requirements, for example, *European Commission* (EC) directives. In addition to several other tests, EMC is verified, which examines the immunity and level of interference caused by a device. An immunity test, for example, of a TV, ensures the safety of the users in the room, should the rooftop antenna be struck by lightning. Emission tests basically guarantee that all devices in the vicinity of the product operate as intended. EMC tests are standardised by the *Comité International Spécial des Perturbations Radioélectriques* (CISPR, in English: Special International Committee on Radio Interference) which is part of the IEC. Additionally, the ITU can issue EMC-related recommendations.* Usually, all nations follow the EMC standards from CISPR. Of course, there are exceptions: for example, in the case of PLC EMC in the United States and JP, national authorities specified individual processes.

Historically, shortly after S. Morse sent his message over a telegraph line between Washington and Baltimore in 1844, the demand for borderless communications arose and the ITU, founded in 1865 – at that time called the *International Telegraph Union* – was the first international organisation to fill that gap. Standardisation is an obviously successful practice because – we can remember – 15 agreements were required in Prussia to link Berlin and frontier localities by telegraph. Since the invention of information transmission, engineers have continuously been challenged to find new, creative means for enhancing generation and transmission of power. The foundation was laid by W. von Siemens in 1866 with the invention of the dynamo. It was soon found in engines, trains, elevators, etc.

Communication (*low-voltage* [LV] signalling) and high-power applications have the potential to cause interferences if EMC is not regulated. The first 'EMC regulation' was established in Prussia in 1892, according to [1]. It clarified that establishing and operating telegraph equipment is only allowed by authorities, and in the case of interference, placed the cost of eliminating interference on the owner of the most recent installation. In 1908, the law was amended to also regulate wireless transmissions. The first interference complaints occurred in innovative cities in the 1890s, where tramcars and telegraph links had been installed. The next complaints arose when railway signalling lines were influenced by power transmissions. Soon railway operators, energy utilities and telecommunication

---

* Here, the authors frequently have the impression that there is not much communication between IEC and ITU, even though they are located in Geneva across the road. It should be noted that ITU recommendations are not mandatory for product certification.

suppliers were seeking regulations. Medical radiation therapy equipment emerged as a source and sink of EMC issues. The world's first commercial radio station was created by the Westinghouse Electric Corporation in 1920 in Pittsburgh, Pennsylvania, United States. Since then, the *electromagnetic interference* (EMI) of tramcars frequently interrupted radio broadcast services. Also in the 1920s, rotary dial phones converted an EMC victim to an EMC source. In these early days, power transmission lines were also used for transmitting radio programmes. In Norway, this technique was called *Linjesender*, and the radio programme was fed into the lines by special transformers. Filters for carrier frequencies used by transmission systems were installed in substations and at line branches, in order to prevent uncontrolled propagation. It is probable that the EMC of electric engines stopped this technology. The *short-wave* (SW) radio receivers use the mains as counterpoise to the monopole whip antenna. When utilising one and the same media for various applications, of course EMC has to be considered with great care. In *World War II*, another EMC source and sink found widespread installation: the radar. The transistor – developed by the *Bell Labs* in 1947 – launched the success of microelectronics and all kinds of EMC sources and victims. One of the most dramatic interference sources was found in 1964, when the lights in Hawaii, 1445 km away, had been switched off due to a high-altitude nuclear test called *Starfish Prime* [1].

Aside from PLC, the 'Digital Dividend' (new frequency split between mobile Internet service providers and TV broadcast) and emissions from *light-emitting diode* (LED) illuminations keep EMC experts busy today. The end user detects an obvious EMC issue when a mobile phone operates in the vicinity of a speaker device. The interruptions in the mobile signal transmission fall into the audible frequency range and are detectable by human ears. Here, market surveillance authorities do not take initiative, possibly because the issue might be solved by placing the mobile phone in a different location.

6.2 Creation of EMC Regulations

Whenever there is a new EMC issue, regulatory committees might be requested, through a *New Work Item Proposal* (NWP), to solve the problem. A national mirror committee has to submit the NWP, for example, to the IEC, and other nations are needed to support the submission. If the NWP is accepted, issues of political interest are represented in a newly founded group. Typically, there are industry representatives who source the interference, device manufacturers or service providers of the electromagnetic sinks, national authorities, measurement device manufacturers, universities (in case of academic interest) and many consultants who seem to monitor the process. Finding a compromise is usually as difficult as mediating a married couple. There is always the question whether the immunity threshold of one side is too low or the emission level of the other is too high and adaptive/cognitive compromises might, in some cases, be the way forward. In the human case, such adaptations are possible in the time or location domain. Technically, the frequency domain is also usable. An example of adaptations already approved by ETSI and CENELEC can be studied in Chapter 22, which deals with dynamically notching parts of the PLC frequency spectrum.

The classical concept of EMC requires constant emission and immunity limits against *high-frequency* (HF) signals. The emission limit and their own immunity threshold define the operating range of all devices. Devices working within this range operate without producing any serious interference in their environment. This classical concept of guaranteeing EMC

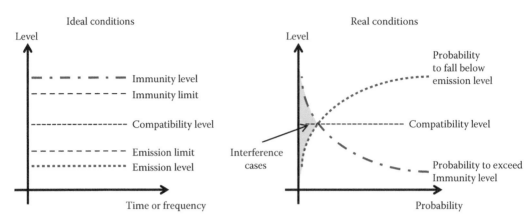

FIGURE 6.1
EMC coordination, ideal and real conditions.

as shown on the left side of Figure 6.1 has the drawback that some resources are left unused. Furthermore, devices need costly shielding or isolations, even though there are often no signals causing disturbance. The device is shielded against a broad range of frequencies independent of where and when the device is operated. In short, the resources are not used efficiently.

In some cases, the reception of a low-power signal is disturbed despite the fact that surrounding devices comply with the relevant EMC standards. Unfortunately, this is the actual situation, shown on the right side in Figure 6.1. In rare cases, interference issues might occur when the immunity level is below the emission level. From both an economic and a technical point of view, this is not satisfactory. In this case, stricter limits should have been chosen. Alternatively, these rare cases can be avoided with adaptive behaviour at the device and it is expected that more adaptive approaches will be developed in the near future.

Whenever a proposal, which is likely to be accepted, is made to the EMC committees, there is the question how to integrate it into the existing world of EMC documents. An engineer would describe the new solution technically in an innovative way. Due to the profound approval process of EMC standards (national entities, who frequently do not join the discussions in the EMC committee, have to agree), innovative solutions are difficult to explain to all decision makers. This is why minor or minimal progress is made, as the new version of a document boasts only minute amendments. Many 'gives' and 'gets' need to be exchanged enriching the final document by multiple extras, initially unplanned and in the worst case technically ambiguous. The base of regulation documents has also historically grown with the risk of including inconsistencies. For instance, CISPR 16 [2] specifies the *common-mode* (CM) voltage (V_{cm}) in an electric network to be half the vector sum of V_a and V_b, that is, $(V_a + V_b)/2$, where V_a is the vector voltage between one of the mains terminals and earth and V_b is the vector voltage between the other mains terminal and earth. CISPR 22 [3] specifies V_{cm} to be the voltage appearing across the CM impedance presented to the *equipment under test* (EUT) by an *impedance stabilisation network* (ISN). If the EUT is a PLC modem, the ISN is intentionally not implemented fully symmetrically, resulting in the two definitions from CISPR 16 and CISPER 22 not matching. Nevertheless, both standards have to be adhered to at the time of product certification for information technology equipment connected to and powered by the electrical mains network.

When modifying the existing documents, it has to be considered that the collection of current EMC regulations dates back to the days of our grandfathers. Changing any value or measurement method usually has several side effects where even the experienced delegates no longer remember the foundation of the status quo. Of course, this does not simplify compromise. For instance, the approach enhancing CISPR 22 by PLC in [4] was one of the main challenges of the CISPR *project team power line telecommunications* (PT PLT). However, CENELEC found faster agreements after deciding to draft a dedicated product standard for PLC.

6.3 PLC-Specific Regulations*

Power line cables were not designed to carry communication signals and, hence, give rise to conducted emission, as well as radiated emission that may interfere, for example, with Amateur Radio or radio broadcast receivers. When looking at power line EMC regulations, one may distinguish between regulations for *narrowband* (NB)-PLC and *broadband* (BB)-PLC.

The NB-PLC regulations deal with a spectrum from 3 kHz up to around 500 kHz. Important NB-PLC regulations are listed in Table 6.1. As a subset of all other bands, CENELEC bands are the only ones available on a global basis. Four CENELEC bands are defined as A (9–95 kHz), B (95–125 kHz), C (125–140 kHz) and D (140–148.5 kHz) [5]. Aside from specifying transmission limits and their measurement procedures, the CENELEC standard also mandates that the A-band only be used by electricity suppliers and their licensees, while the other bands may be used by consumers. Further, devices operating on the C-band have to comply with a *carrier sense medium access/collision avoidance* (CSMA/CA) protocol that allows a maximum channel holding period of 1 s, a minimum of 125 ms between channel uses by the same device and a minimum of 85 ms before declaring the channel idle. In the United States, there are currently ongoing efforts [6] to specify the band from 9 to 534 kHz for NB-PLC operations with a mandatory CSMA/CA protocol compliant with CENELEC EN 50065-1 [5]. The advantages are that equipment manufacturers would easily be able to adapt their NB-PLC products to the EU and US markets and to many other markets that follow these standards. Specific details on current ITU and IEEE NB-PLC standard-related band plans and power masks may be found in Chapter 11.

TABLE 6.1

Important Regulations Related to NB-PLC

States	Frequency [kHz]	Institution	Reference
EU	3–148.5	CENELEC	[5]
United States	10–490	FCC	[7]
JP	10–450	*Association of Radio Industries and Businesses* (ARIB)	[8]

Source: Berger, L.T. and Iniewski, K., Eds., *Smart Grid – Applications, Communications and Security*, John Wiley & Sons, Hoboken, NJ, April 2012, ch. 7, ISBN: 978-1-1180-0439-5. Copyright © 2012 John Wiley & Sons. With permission.

---

* Section in parts based on Berger, L.T. and Iniewski, K., Eds., *Smart Grid – Applications, Communications and Security*, John Wiley & Sons, Hoboken, NJ, April 2012, ch. 7, ISBN: 978-1-1180-0439-5, reproduced with permission of John Wiley & Sons, and Berger, L.T. et al., Hindawi Publishing Corporation, *J. Electric. Comput. Eng.*, 2013, [Online] http://www.hindawi.com/journals/jece/aip/712376/ (accessed February 2013) [10], reproduced with permission from the copyright holders.

Looking at BB-PLC, one may again distinguish between two frequency ranges, that is, 1–30 MHz, where conducted emission is the focus of regulation (except for the United States), and 30–100 MHz, where the focus shifts to radiated emission. A technical reason for establishing this division at 30 MHz is the size limitation inside anechoic chambers. Toward lower frequencies, the wavelength gets so long that *far-field* (FF) conditions can no more be realised. Emission tests record below 30 MHz only the magnetic component or H-field. Above 30 MHz, the electrical wave or E-field is considered. Surprising is the 30 MHz threshold to be this sharp without any transition period. It is interesting to note that for immunity testing, the threshold is at 80 MHz.

The goal of the conducted emission test setup is to represent the properties of any field network in a static and reproducible way in the test lab. Therefore, ISN [2] and an *Artificial Mains Network* (AMN) [3] are specified in the CISPR documents. The level of CM signals inside the network should be verified when assessing the EMI potential of a device. An ISN creates a defined amount of CM signal which should be of similar magnitude as the CM signals in the field. Figure 6.2 illustrates the idea behind the ISN. PLC modems create a differential signal feed to the mains. Even minimal parasitic elements (denoted Z_{Para} in Figure 6.2) generate a CM signal launched by the PLC modem. The ISN provides a specified *differential-mode* (DM) impedance (Z_{DM}), CM impedance (Z_{CM}) and the level of asymmetry called *longitudinal conversion loss* (LCL) [11]. The LCL is the ratio between the symmetrical and asymmetrical signals in a network. This LCL level can be adjusted by selecting the impedance of the series or shunt asymmetry resistors (denoted $Z_{Asym\ series}$ and $Z_{Asym\ shunt}$ in Figure 6.2). The internal realisation of an ISN is implementation specific. CISPR 22 only describes the ISN's outside parameters like impedances and LCL. Implementing a $Z_{Asym\ series}$ allows a higher feeding level at identical ISN outside parameters than the usage of $Z_{Asym\ shunt}$. The CM choke prevents the CM signal from being influenced by the outside and the DM choke blocks DM signals not being shortcut by Z_{CM}. The CM signal generated by the ISN from the injected DM signal is called the 'converted' CM current. For PLC, the converted CM signals are dominant compared to the launched CM signals. Signals passing Z_{CM} toward ground are used for assessing the interference potential of the device under examination. Components needed for safety or blocking the AC mains signal are not shown in Figure 6.2.

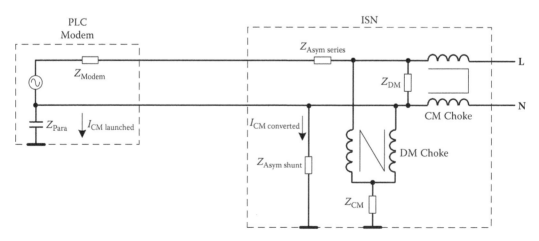

FIGURE 6.2
Principle of an ISN-based conducted emission test setup.

CISPR I PT PLT [4] was making efforts to regulate BB-PLC-generated interference. At the beginning, CISPR 22 [3] defined two sets of limits and measurement methods for conducted emissions of telecommunication equipment. One set was defined for the telecommunication port and the other was defined for the mains port. For PLC modems, it is not defined whether the PLC signal port, which at the same time is used for power supply, is considered a mains or a telecommunication port. The early versions of CISPR 22 were written without PLC applications in mind. The method for telecommunication ports respects the symmetry properties of an attached cable.

Asymmetries like an open light switch or asymmetrical parasitic capacitances convert the symmetrically fed signals into CM signals (see Chapter 1 for details). However, the method used for devices connected to the mains, specified in CISPR 16 [2], is based on measuring the asymmetric voltage level of either the phase or neutral wire to the ground. From the perspective of a PLC modem, this is the worst case, because this voltage consists of asymmetric and symmetric voltage, that is, not only the emission and there with the interference relevant part, but also the desired signal is measured and compared against limits.

IEC CISPR/I/89/CD [12] tried to clarify the situation by interpreting PLC as an application following the telecommunication limits of CIPSR 22. Therefore, the LCL parameter was used in an identical way as, for example, in the testing of *digital subscriber line* (DSL) equipment.

The benefit of using the LCL parameter is the simplicity of measuring it. It is a reflection parameter whose measurement requires equipment to be connected to only one location of the grid. Other options for verifying the potential of interference would have been to place antennas and measure the field generated by the fed PLC signal as done, for example, during the FCC marking process in the United States. However, this would make a field measurement setup significantly more cumbersome and difficult to reproduce. Nevertheless, with respect to highly attenuated wires like the power lines the simple LCL measurement of a reflection parameter as in CISPR 22 [3] only describes the local situation instead of giving detailed insight of what is happening if a PLC signal travels deeply into the electrical network. Here, the *Transverse Conversion Transfer Loss* (TCTL) [11] would have been the better approach from an engineering point of view but the LCL was already introduced to the CISPR EMC universe and, hence, this measure was adopted by CISPR I PT PLT [4]. In 2008, CISPR/I/257/CD [13] was published with an LCL parameter reduced by 6 dB. Simultaneously, CISPR/I/258/DC [14] indicated that mitigation techniques like cognitive notching for PLC modems [15,16] and dynamic transmit power management [17] are the compromises to solve the never-ending EMC discussions, and new adaptive approaches started to become integrated into EMC documents.

CIS/I/301/CD [18] answered the question whether PLC is connected to the telecommunication or the mains port, specifically defining a PLT port. In this document, the adaptive EMI mitigation techniques are specified as normative. However, CISPR never proposed a *committee draft for voting* (CDV). This is why CENELEC became active after the lifetime of the CISPR committee, to find at least in Europe, an acceptable solution. The result, FprEN 50561-1:2012 [17], was approved in November 2012 with national committees' acceptance of 91%. The document specifies the following:

- An emission measurement procedure at the PLT port while no communication takes place.
- A second measurement procedure at the PLT port when communication takes place. Here the signal level fed by a PLC modem is recorded and the symmetrically injected signal level is verified. Signals converted to CM using any LCL parameter as done by previous CISPR documents are no longer relevant for the assessment of the PLC signal level.

- Permanent notching of certain parts of the radio spectrum, that is, related to amateur radio and aeronautical bands.

- A procedure for adaptive notching, meaning that the PLC equipment senses the presence of radio services and notches the affected frequencies for its own operation (see also Chapter 22).

- A procedure of adaptive transmit power management, meaning that the transmitting equipment limits its transmit power as a function of channel attenuation and noise to a level below the allowed maximum, that is just sufficient to achieve the required data rate.

The approval of EN50561-1:2012 finally gives certainty to PLC stakeholders of European interference limits. In the United States, the FCC is in charge of regulating electromagnetic emissions. In general, all digital equipment has to comply with the FCC part 15 standard (47 CFR §15) [7]. Specifically, Access PLC systems over *medium-voltage* (MV) and LV power lines, and for a frequency range from 1.705 to 80 MHz, are treated in the standard's Section G. Conducted emission limits are explicitly not applicable but radiated emission limits are imposed through a transmit *power spectral density* (PSD) mask. Additionally, PLC systems have to be able to notch certain frequencies that might be used by other services. Further, the FCC defines excluded bands where no PLC signal will be injected, as well as geographical exclusion zones where no Access PLC systems may be deployed. Further, procedures in which service providers inform about prospective PLC Access deployments as well as complaint handling procedures are a requirement. More details on PLC EMC regulations as well as conducted and radiated interference measurement results can be found in [19,20]. Also, the 'IEEE Standard for Power Line Communication Equipment – *Electromagnetic Compatibility* (EMC) Requirements – Testing and Measurement Methods' [21] was released, intending to provide an internationally recognised EMC measurement and testing methodology. It endorses, among others, CISPR 22 and FCC part 15 as normative references, but does not establish any emission limits itself. Looking at the developments in CISPR/CENELEC, as well as at FCC part 15, it becomes clear that the next generation of PLC equipment will have to be highly configurable to apply PSD shaping masks, as well as adaptive notching.

6.4 Broadband PLC Feeding Levels

Frequently, the feeding level of a PLC modem is given as PSD, for example, in V^2/Hz as in Chapter 2 or dBm/Hz as in Chapter 9. The power unit is used, because modems are connected to weak impedance networks. Chapter 5 shows the input impedance at an outlet to vary between about 10 Ω and hundreds of Ω. If at some frequencies a current and at other frequencies voltage is fed into the grid, power is the best unit to describe both situations. If the PLC modem is connected to a reproducible environment as described in CISPR or CENELEC documents, where an ISN provides a given impedance, then the specification of current or voltage would be free of ambiguity. On the median, the symmetrical input impedance Z_{DM} of a pair of wires at an outlet is around 100 Ω [15] in the frequency range 1 MHz $< f <$ 30 MHz. In view of a *multiple-input multiple-output* (MIMO) PLC modem, this impedance exists on all three wire pair combinations (see Chapter 5).

The conversion factor $CF_{dBm2dB\mu V}$ from dBm to dBµV is given in logarithmic representation as

$$CF_{dBm2dB\mu V} = (V/\mu V) - Z_{DM} - (mW/W),$$

$$= 20 * \log_{10}(1e6) - 20 * \log_{10}(100\ \Omega) - 10 * \log_{10}(1e-3),$$

$$= 110\ dB\ (\mu V/mW). \tag{6.1}$$

In the time domain, PLC modems send bursts of data with an unknown duty cycle. Spectrum analysers offer a wide selection of detectors to record time variant data: *peak*, *quasi-peak*, *root mean square* (RMS), *average* and *sample*. The selection of the detector influences the measured results.

Using the sample detector, the result is hardly reproducible because it could be any value between the min and max values which is captured. The peak detector is the most conservative as it only records the signal peak value, ignoring any quiet signals within the sample interval. However, this detector is also the one delivering most reproducible results, because parameters like the actual transmitted data payload do not influence results, which conversely is the case when using alternative detectors. The average detector weights periods with signals and without signals equally. Finally, quasi-peak and RMS are in between peak and average with a weighting factor specified in the standards for measurement receivers and spectrum analysers CISPR 16-1-1 [2].

All the CISPR approaches [11,13,18] use average and quasi-peak detectors. This is owed to the history of CISPR documents. FprEN 50561-1:2012 [17] introduced the peak detector for PLC measurements for the first time.

In analogue receivers, the psychophysical annoyance of the interference interpreted by humans gives parameters to define a weighting factor. The quasi-peak detector was invented to cover this subjective effect for *amplitude-modulated* (AM) services. Nevertheless, the subjective effect depends on whether the receiver is acoustic or visual.

In digital receivers, the interference is measured by the *bit error rate* (BER), which is minimised by error correction codecs. Both noise and signal levels are compared in order to calculate the theoretical channel capacity as in Chapter 9. Here, a digital communication signal protected with *forward error correction* (FEC) is compared to a noise signal recorded with a spectrum analyser. None of the available detectors weights both levels correctly. The average detector was chosen for these calculations because of its balanced weighting of periods with and without signals. Compared to quasi-peak, the average detector is also fast and useful for sweep measurements when, for example, recording the noise. In the following, the conversion from several regulatory limits specified with quasi-peak detector to average value is done by subtracting 8 dB. This value can be found in an experimental setup using a spectrum analyser.

The level of a signal with the property of a rectangular spectral shape like in *orthogonal frequency division multiplexing* (OFDM) is frequently described using a spectral density. If it is a power signal, it is the PSD in dBm/Hz. Technically, a PSD in dBm/Hz cannot be measured (even if many spectrum analysers provide results using this unit) because the PSD is the power in an infinitely small *bandwidth* (BW), that is, the derivation $\Delta P/\Delta BW$. If the BW becomes infinitely small, the question of which measurement detector is applied becomes obsolete, as there is no more variance in the signal. However, simultaneously the measurement time goes to infinity.

For simplification, the PSD is calculated by subtracting the *resolution BW* (ResBw) from the power without applying any weighting window. For instance, for a measured power $P = 1$ mW using a resolution BW ResBw = 9 kHz, the PSD is given by

$$\text{PSD} = P - \text{ResBw} = 10*\log_{10}(1 \text{ mW}) - 10*\log_{10}(9 \text{ kHz}) = -39.5 \text{ dBm/Hz}. \quad (6.2)$$

In the frequency range below 30 MHz, a resolution BW of 9 kHz is used CISPR 22 [3]. Still this calculation is a theoretical one, because the measurement device has to use a detector and a detector with a weighting window with vertical slopes in the frequency domain cannot be implemented.

6.4.1 US Feeding Levels

In the United States, [22,23] specify how emissions from PLC devices are evaluated. The documents call PLC a *broadband over power line* (BPL) system, a new type of carrier current technology. The emission limits (the numbers can be found in FCC 15 [24] §15.209) are given in 1.705 MHz $< f <$ 30 MHz in a radiated field strength of $E_{f < 30 \text{ MHz} @ 30 \text{ m}} = 30 \text{ μV/m}$ (= 29.5 dBμV/m, quasi-peak, ResBw = 9 kHz) at a distance of 30 m from the exterior wall of the building. In the band 30–88 MHz, the radiated field is $E_{f > 30 \text{ MHz} @ 3 \text{ m}} = 100 \text{ μV/m}$ (= 40 dBμV/m, quasi-peak, ResBw = 120 kHz) at a 3 m distance (for *Class B* devices). The reader has to check FCC 15 [24] §15.31(f) for measurable emissions at distances other than the distances specified. There is a distance *extrapolation factor* (EF) of $\text{EF}_{\text{NF}} = 40$ dB/decade for frequencies where the antenna is located in the *near field* (NF) below 30 MHz and $\text{EF}_{\text{FF}} = 20$ dB/decade in the FF above 30 MHz.

The limit drop when passing the 30 MHz mark can be calculated as

$$\text{Drop}_{@ 30 \text{ MHz}} = E_{f < 30 \text{ MHz} @ 30 \text{ m}} + 10 * \log_{10}\left(120 \text{ kHz}/9 \text{ kHz}\right) - E_{f > 30 \text{ MHz} @ 3 \text{ m}} + \text{EF},$$

$$= 29.5 \text{ dBμV/m} + 11 \text{ dB } (\text{Hz}) - 40 \text{ dBμV/m} + \text{EF},$$

$$= 0.5 \text{ dB} + \text{EF}. \quad (6.3)$$

The question remains, which EF (EF_{NF} or EF_{FF}) should be used when dropping over the 30 MHz line? Looking at the PLC interoperability standards IEEE 1901 [25] and ITU-T G.9964 [26], one can see that the compromise of 30 dB between both EFs was selected without specifying further details.

Class A refers to devices intended to operate in an industrial environment, whereas Class B devices are for use in private homes. Further, FCC specifies Class A limits to apply on MV wires and Class B on LV wires. Usually, Class A devices have limits increased by 10 dB.

Additionally, notches are required to protect *aeronautical, mobile* and *radionavigation services*, in some geographical zones extra frequencies have to be excluded, and care must be taken for public safety services. Adaptive interference mitigation techniques are also described by the FCC. For better comparison among individual feeding PSD, these additional notches are not applied to the PSD mask used for the capacity calculations in Chapter 9.

The verification of emissions is done in the field at three typical buildings. The antenna is placed at various locations in the garden. The antenna's height can be adjusted between 1 and 4 m where the maximum reading has to be recorded or 5 dB added if only the convenient 1 m high is measured.

The PLC modem feeding level may be derived using the coupling factor. The coupling factor is the radiated E-field level of the building minus the level of the symmetrical fed signal into the mains. In Europe, ETSI published several coupling factor measurements [27–29]. In the United States, no such exhaustive measurement campaign has been published. This is why the capacity calculations presented in Chapter 9 assume a PSD_{US} of −50 dBm/Hz (quasi-peak) or −58 dBm/Hz (average).

A wrap-up of regulations on RF emissions from PLC systems in the United States can be found in ITU's recommendation ITU SM.1879-1 [30].

6.4.2 European Feeding Levels

FprEN 50561-1:2012 [17] limits the maximum PLC transmit signal level between 1.6065 and 30 MHz using a setup with >40 dB attenuation between EUT and the auxiliary equipment to 95 dBμV (average detector). Using Equations 6.1 and 6.2, the modem's PSD can be calculated as

$$PSD_{Europe} = 95 \text{ dBμV} - 110 \text{ dB (μV/mW)} - 39.5 \text{ dBm/Hz} = -55 \text{ dBm/Hz}. \qquad (6.4)$$

Furthermore, in Europe, CENELEC aims to draft an EMC standard for frequencies above 30 MHz, which was decided on the TC 210 [31] meeting in December 2012. The document is not finalised so far. The frequency spectrum below the so-called Band II (the *frequency-modulated* (FM) radio band, 87–108 MHz; the exact numbers vary from country to country) contains fewer sensitive services for which a protection through frequency exclusion is required. Even so, the number of services to be protected is low and the standardisation work at CENELEC has not been completed. For purposes of the calculation here, a 30 dB reduction is selected for feeding levels from 30 to 86 MHz. However, there still is the open question of justifying this 30 dB step at 30 MHz. When measuring the radiation inside or outside buildings as, for example, shown in Chapter 7, an increase of the interference potential of 30 dB at 30 MHz has never been observed.

As mentioned previously, in all upcoming calculations – for a better comparison of the individual feeding PSD – mandatory notches of aeronautical, amateur radio, emergency services, radio broadcast and the influence of a deep notch for the protection of FM radio are not considered.

6.4.3 JP Feeding Levels

The JP limits for PLC transmissions in the HF band apply to the CM current measured at the mains port of a PLC modem. The specified measurement methods used in JP are similar to the concept of CISPR 22. According to ITU SM.1879-1 [30], a JP ISN has the properties of LCL = 16 dB, $Z_{DM} = 100 \ \Omega$ and $Z_{CM} = 25 \ \Omega$. The modem's communication signals are assessed by measuring the CM current converted by the ISN from the symmetrically fed levels. However, these properties are not typical for JP buildings. Statistically, the median values found in JP houses are LCL = 35.5 dB, $Z_{DM} = 83.4 \ \Omega$ and $Z_{CM} = 240.1 \ \Omega$ [32].

The chosen LCL value of 16 dB is the 99% worst-case value of the cumulative distribution. Such a low LCL value inside the JP ISN generates a high CM signal. This generated CM signal is by far higher (usually >1000 times) than any CM signal launched by PLC modems under normal deployment, as PLC modems usually benefit using a highly symmetrical implementation. Additionally, the low Z_{CM} value leads to a high CM current in the test setup, much higher than would be the case in any private home. As a result of the selected measurement procedure, the maximum allowed feeding level is reduced.

Furthermore, the limits of the frequencies below and above 15 MHz differ by 10 dB in JP. The CM current I_{CM} must not exceed $I_{CM f < 15 MHz} = 20$ dBμA (average) at frequencies smaller than 15 MHz and the limit is $I_{CM f > 15 MHz} = 10$ dBμA above 15 MHz.

The resulting PSD for JP on the test bed (using the JP ISN) depends on the CM impedance of the PLC modem. A typical PLC modem with the size of a human fist might inject

$$\text{PSD}_{\text{Japan } f < 15\,\text{MHz}} = -71 \text{ dBm/Hz} \quad \text{and} \quad \text{PSD}_{\text{Japan } f > 15\,\text{MHz}} = -81 \text{ dBm/Hz (average)}. \quad (6.5)$$

However, the uncertainty of this measurement method is about 14 dB. This is due to the modem's CM impedance, which is usually dominated by the size of the modem.

Kitagawa measured the differentially injected current of PLC modems in JP buildings [33]. The atypical parameters selected for the JP ISN and the uncertainty of the measurements cause the CM current in the field to exceed the CM limits at the modems' verification setup by about 40 dB.

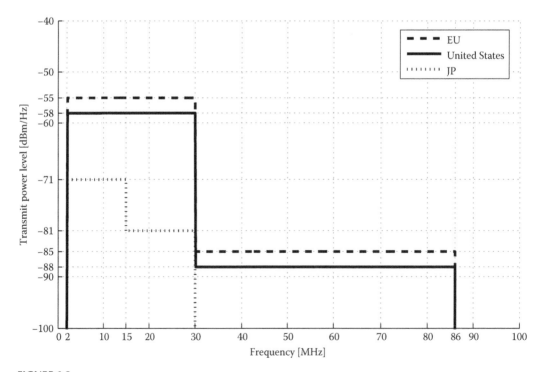

FIGURE 6.3
Transmit PSD masks. The EU limit above 30 MHz is subject to pending regulation.

Also, modems in JP omit frequencies of radio amateurs and some JP HF radio broadcast stations. However, as before, additional notches or extra frequency exclusions are not considered for further calculations. Any PLC transmissions above 30 MHz are also not considered.

6.4.4 Feeding Level Comparison

Based on the derivations and considerations in the previous sections, Figure 6.3 displays the US, EU and JP PSD masks used for the capacity calculations in Chapter 9.

6.5 PSD Masks in BB-PLC Specifications

The PSD masks introduced in the PLC communication standards are only relevant for interoperability, not for product certification. This implies, for example, a receiver must not become overloaded when an adjacent transmitter feeds signals with the given power. The receiver is also prepared not to find any of the information in the notched carriers previously specified in the interoperability standard. Here is a collection of EMC-related parameters of individual standards.

6.5.1 ITU-T G.hn Mask

The ITU-T G.hn group specified a PSD mask in ITU-T G.9964 [25], which is shown in Chapter 12. The max PSD has its upper bound at −55 dBm/Hz. The PSD is supposed to be verified with an RMS detector with a 'maximum hold' function making it difficult to compare the result against the regulatory rules using average, quasi-peak or peak detectors. However, the 'maximum hold' function enhances the reproducibility when measuring the pulse pause–interrupted PLC signals, after waiting several sweeps. The recommendation specifies a network termination of 100 Ω and a minimum receiver input impedance of 40 Ω in 1.8 MHz $< f <$ 50 MHz. It does not recommend any ISN on which the PLC modem, its communication partner device and the measurement receiver are connected. Moreover, ITU-T. 2010. G.9964 [25] allows dynamic reduction of the transmit power and the additional exclusion of frequencies (notches).

Toward the frequencies above 30 MHz, there is a 30 dB step specifying a PSD of −85 dBm/Hz up to 100 MHz. ITU-T. 2010. G.9964 [25] specifies subcarriers above 80 MHz will be masked (zero power transmitted), but without describing any consequences to the PSD mask. Considering that FM radio transmissions in *Band II* overlap with this definition and taking the FM interference threshold recorded in Chapter 7 into account, there is the risk that PLC modems applying maximal PSD will jam FM radio reception in 25% of the cases.

6.5.2 IEEE 1901 Mask

IEEE standardises the transmit spectrum masks PSD, in IEEE 1901 [25] (see Chapter 13), to be −55 dBm/Hz below 30 MHz and −85 dBm/Hz above 30 MHz. In contrast to

ITU-T G.hn, IEEE 1901 applies the quasi-peak detector and limits the transmission up to 50 MHz, thereby providing better FM radio protection. The termination impedance is also 100 Ω. The standard includes a note that these limits apply for Class B devices, whereas access modems might use Class A levels increased by the 10 dB.

Power back-off and notching are also encouraged in IEEE 1901 where the wavelet part optionally specifies a 'stand-alone dynamic notching' as described in FprEN 50561-1:2012 [16] and Chapter 22.

6.5.3 HomePlug AV2 Mask

The HomePlug AV2 [34] specification (see also Chapter 14) lists a PSD of −50 dBm/Hz recorded using a quasi-peak detector below 30 MHz. The level above 30 MHz is marked as 'to be defined' where a special protection for FM radio is foreseen, but not specified in detail. Today, it is the only specification providing a test circuit to measure the individual MIMO feeding levels. The signals of the three wires are recorded sequentially using a 50 Ω measurement receiver. The total PSD is calculated by summing up the individual levels. The unused wires are also terminated via 50 Ω to ground while measuring on one line. This results in a 100 Ω termination from one wire to the other. The auxiliary equipment is connected through a channel attenuation network. The throughput rates of HomePlug AV2 modems even benefit when reducing the max transmission level, as explained in Chapter 14. Moreover, the specification allows very flexible notching by digital filtering with minimal loss of transmission resources.

6.6 Conclusions and Outlook

This chapter provided an overview of PLC regulation. The recent development of the EMC standards specifying the signal level to be injected by PLC modems shows the trend to adaptive design concepts resolving interferences at the location, the frequency and the time where the modem is operated. The next discussion on PLC EMC regulation is related to feeding levels in the frequencies above 30 MHz. Here, again, today's specifications are drafted without PLC in mind, which enables ambiguities when assessing the levels. Protection of FM radio services in the *very-high-frequency* (VHF) band demands lower thresholds than applied today for AM services in the HF band. Furthermore, impedance modulating devices – appliances that toggle their input impedance synchronously with the AC line cycle [16] – may cause intermodulations of the PLC signal, which might fill up the deep notch originally established for FM protection. Here, another adaptation could solve the problem. Specifically, the PLC modems could pause their transmission during the time when the impedance toggling happens. Today's PLC modems apply adaptive constellations (see Chapter 9) for their carriers depending on the available, frequency-dependent *signal-to-noise ratio* (SNR). The estimation of the channel is done continuously over the line cycle and several tone maps (number of bits modulated on each carrier) are applied periodically within a line cycle. Due to these adaptations already implemented, avoidance of the intermodulations could be realised via minimal add-ons.

Acknowledgements

The work has been partially supported by the Spanish Ministry of Science and Innovation (MICINN) Program INNCORPORA-Torres Quevedo 2011.

References

1. Anton Kohling, EMV: Umsetzung der technischen und gesetzlichen Anforderungen an Anlagen und Gebäude sowie CE-Kennzeichnung von Geräten, ISBN-10: 3800730944, VDE-Verlag, 2012.
2. CISPR 16-1-1, Specification for radio disturbance and immunity measuring apparatus and methods – Part 1-1: Radio disturbance and immunity measuring apparatus – Measuring apparatus.
3. CISPR, Information technology equipment; radio disturbance characteristics; limits and methods of measurement, ICS CISPR, International Standard Norme CISPR 22:1997, 1997.
   Was updated later by
   CISPR, Specification for radio disturbance and immunity measuring apparatus and methods. Part 1-1: Radio disturbance and immunity measuring apparatus – Measuring apparatus, 2003. and
   CISPR, Amendment to CISPR 22: Clarification of its application to telecommunication system on the method of disturbance measurement at ports used for PLC, 2003.
4. Project Team of Power line Telecommunication at CISPR, Limits and method of measurement of broadband telecommunication equipment over power lines, http://www.iec.ch/dyn/www/f?p=103:14:0::::FSP_ORG_ID,FSP_LANG_ID:3204,25.
5. European Committee for Electrotechnical Standardization (CENELEC), Signalling on low-voltage electrical installations in the frequency range 3 kHz to 148.5 kHz – Part 1: General requirements, frequency bands and electromagnetic disturbances, Standard EN 50065-1, September 2010.
6. National Institute of Standards and Technology (NIST), Priority Action Plan 15 (PAP15), Harmonize Power Line Carrier Standards for Appliance Communications in the Home, Coexistence of narrow band power line communication technologies in the unlicensed FCC band, April 2010, http://collaborate.nist.gov/twiki-sggrid/pub/SmartGrid/PAP15PLCForLowBitRates/NB_PLC_coexistence_paper_rev3.doc (accessed December 2010).
7. FCC, Title 47 of the code of federal regulations (CFR), Federal Communications Commission, Technical Report 47 CFR §15, July 2008, http://www.fcc.gov/oet/info/rules/part15/PART15 07-10-08.pdf (accessed February 2009).
8. Association of Radio Industries and Businesses (ARIB), Power line communication equipment (10 kHz–450 kHz), November 2002, STD-T84, Version 1.0 (in Japanese), http://www.arib.or.jp/english/html/overview/doc/1-STD-T84v1_0.pdf (accessed April 2013).
9. L. T. Berger and K. Iniewski, Eds., Wireline communications in smart grids, in *Smart Grid – Applications, Communications and Security*, Hoboken, NJ: John Wiley & Sons, April 2012, ch. 7, ISBN: 978-1-1180-0439-5.
10. L. T. Berger, A. Schwager and J. J. Escudero-Garzás, Power line communications for smart grid applications, Hindawi Publishing Corporation, *Journal of Electrical and Computer Engineering*, 2013, [Online] http://www.hindawi.com/journals/jece/aip/712376/ (accessed February 2013).
11. ETSI TR 102 175 V1.1.1 (2003–03); Power Line Telecommunications (PLT); Channel characterization and measurement methods.

12. CISPR/I/89/CD: Amendment to CISPR 22: Clarification of its application to telecommunication system on the method of disturbance measurement at port used for PLC (power line communication), IEC, November 2003.
13. CISPR, CISPR 22 am3 f1 ed. 5.0, Limits and method of measurement of broadband telecommunication equipment over power lines, February 2008.
14. CISPR, Report on mitigation factors and methods for power line telecommunications, February 2008.
15. European Telecommunication Standards Institute, PowerLine Telecommunications (PLT); Coexistence between PLT modems and short wave radio broadcasting services, August 2008.
16. A. Schwager, Powerline communications: Significant technologies to become ready for integration, Dissertation, Universität Duisburg-Essen, Fakultät für Ingenieurwissenschaften, Duisburg-Essen, Germany, 2010, http://duepublico.uni-duisburg-essen.de/servlets/DerivateServlet/Derivate-24381/Schwager_Andreas_Diss.pdf.
17. CENELEC, Power line communication apparatus used in low-voltage installations – Radio disturbance characteristics – Limits and methods of measurement – Part 1: Apparatus for in-home use, November 2012.
18. CISPR, Amendment 1 to CISPR 22 ed. 6.0, Addition of limits and methods of measurement for conformance testing of power line telecommunication ports intended for the connection to the mains, July 2009.
19. The OPEN meter Consortium, Description of current state-of-the-art of technology and protocols description of state-of-the-art of PLC-based access technology, European Union Project Deliverable FP7-ICT-2226369, March 2009, d 2.1 Part 2, Version 2.3, http://www.openmeter.com/files/deliverables/OPEN-Meter%20WP2%20D2.1%20part2%20v2.3.pdf (accessed April 2011).
20. R. Razafferson, P. Pagani, A. Zeddam, B. Praho, M. Tlich, J.-Y. Baudais, A. Maiga, O. Isson, G. Mijic, K. Kriznar and S. Drakul, Report on electromagnetic compatibility of power line communications, OMEGA, European Union Project Deliverable D3.3 v3.0, IST Integrated Project No ICT-213311, April 2010, [Online] http://www.ictomega.eu/publications/deliverables.html (accessed December 2010).
21. Institute of Electrical and Electronics Engineers, IEEE standard for power line communication equipment – Electromagnetic compatibility (EMC) requirements – Testing and measurement methods, January 2011.
22. Amendment of Part 15 regarding new requirements and measurement guidelines for access broadband over power line systems, Report and Order in ET Docket No. 04-37, FCC 04-245, released 28 October 2004, http://hraunfoss.fcc.gov/edocs_public/attachmatch/FCC-04-245A1.pdf.
23. Amendment of Part 15 regarding new requirements and measurement guidelines for access broadband over power line systems; carrier current systems, including broadband over power line systems Memorandum Opinion and Order in ET Docket No. 04-37, FCC-06-113 released 07/08/2006, http://hraunfoss.fcc.gov/edocs_public/attachmatch/FCC-06-113A1.pdf.
24. FCC PART 15 – RADIO FREQUENCY DEVICES, http://www.gpo.gov/fdsys/pkg/CFR-2009-title47-vol1/pdf/CFR-2009-title47-vol1-part15.pdf (accessed March 2013).
25. Institute of Electrical and Electronics Engineers, IEEE Std 1901–2010: IEEE Standard for Broadband over Power Line Networks: Medium Access Control and Physical Layer Specifications, IEEE Standards Association, 2010.
26. ITU-T. 2010. G.9964, Unified high-speed wireline-based home networking transceivers – Power spectral density specification.
27. ETSI TR 102 259 V1.1.1 (2003–09); Power Line Telecommunications (PLT); EMI review and statistical analysis.
28. ETSI TR 102 370 V1.1.1 (2004–11); PowerLine Telecommunications (PLT); Basic data relating to LVDN measurements in the 3 MHz to 100 MHz frequency range.
29. ETSI TR 101 562-2 V1.2.1 (2012–02); PowerLine Telecommunications (PLT); MIMO PLT; Part 2: Setup and statistical results of MIMO PLT EMI measurements.

30. Recommendation SM.1879-1 (09.11), The impact of power line high data rate telecommunication systems on radiocommunication systems below 30 MHz and between 80 and 470 MHz.
31. CLC/TC 210 Technical Committee on Electromagnetic Compatibility, http://www.cenelec.eu/dyn/www/f?p=104:7:4171075401399912::::FSP_ORG_ID,FSP_LANG_ID:814,25 (accessed in March 2013).
32. Ministry of Internal Affairs and Communications (MIC), Report of the CISPR Committee, the Information and Communication Council, June 2006, available only in Japanese, http://www.soumu.go.jp/joho_tsusin/policyreports/joho_tsusin/bunkakai/pdf/060629_3_1-2.pdf (accessed 2006).
33. M. Kitagawa and M. Ohishi, Measurements of the Radiated Electric Field and the Common Mode Current from the In-house Broadband Power Line Communications in Residential Environment, EMC Europe, September 2008.
34. HomePlug Alliance, HomePlug AV Specification Version 2.0, HomePlug Alliance, January 2012.

7

MIMO PLC Electromagnetic Compatibility Statistical Analysis

Andreas Schwager

CONTENTS

7.1 Motivation

This chapter describes a field measurement campaign performed under the umbrella of ETSI [1] and STF410 [2–5]. The radiated E-field was recorded from *power line communication* (PLC) signals injected into the mains grid within private homes. Due to the special interest in *multiple-input multiple-output* (MIMO) PLC, the radiations for all potential feeding possibilities were compared. Furthermore, interference tests were conducted with *frequency-modulated* (FM) radio receivers.

7.2 Measurement Description

7.2.1 Introduction

Electromagnetic interference (EMI) properties of the *low-voltage distribution network* (LVDN) can be recorded in the *time domain* (TD) or in the *frequency domain* (FD). An example of a record in the TD is a continuous transmission of a pseudo-random sequence where the receiver calculates the correlation to derive the channel's impulse response. An FD measurement is typically done by *network analysers* (NWAs) when sweeping a carrier over the frequency range of interest. A receiver monitors the channel's modification of the carrier in amplitude and phase. The pros and cons of each measurement method were evaluated at the beginning of the measurement campaign. It was concluded that the FD approach is better suited for the following reasons:

- Most of the earlier PLC *electromagnetic compatibility* (EMC) measurements were performed in FD. Thus, a comparison of the results obtained by previous measurements and this campaign is facilitated.

- The human ear is vitally an FD analyser.

- Interferences assessed by human ears like the *signal, interference, noise, propagation and overall* (SINPO) measurements use consumer electronic devices like *amplitude-modulated* (AM) or FM radio receivers. Such measurements were performed in ETSI TR 102 616 [6] and ITU-R [7] earlier. Test signals are fed to all *transmitter* (Tx) paths simultaneously or sequentially. These investigations are conducted with a pulsed signal to allow recognition by the human ear–brain chain.

- Field levels are monitored with a calibrated antenna, which is straightforward to process in FD. EMI measurements in TD have the risk that periodicities in the transmitted PN sequence may cause additional spurs. Furthermore, the measurement dynamics do not seem to be adequate.

- FD measurements can be done using a comb generator and spectrum (or EMI) analyser. This set-up has the benefit that transmitter and receiver do not need to be synchronised. On the other hand, the dynamic range or frequency resolution is limited as feeding energy from the comb generator has to be shared among all signal carriers.

Alternatively, a sweeping source like an NWA might be used. Special care must be taken with signals received by the antenna, as they can be influenced by additional signals being picked up through the long cables connecting the antenna to the NWA. To minimise this effect, double-shielded cables, *common-mode absorption devices* (CMADs) and ferrites have to be installed (see Chapter 1). This measurement method was selected for faster recoding of frequency sweeps and the high dynamic range.

Measurement campaigns were conducted in Germany, Switzerland, Belgium, France and Spain. To guarantee comparability of the data recorded in each country, all teams were equipped with identical probes or PLC couplers. The antenna was shipped to each team in turn. The actual measurements were performed with a general purpose NWA.

A commercially available, small biconical antenna (with built-in amplifier) was used because of its frequency range of up to 100 MHz. In one location, the loop antenna (limited to frequencies up to 30 MHz) was used in order to have a comparison for the results obtained by earlier measurement campaigns. Figure 7.1 shows the measurement set-up used for the EMI radiation measurements.

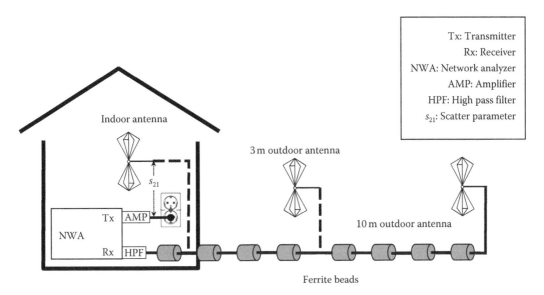

FIGURE 7.1
Measurement set-up arranged in private home.

7.2.2 General Requirements for the Measurements

The test signals for all EMI measurements are fed using MIMO PLC couplers. These couplers as well as the preparation of the power supply for the measurement equipment is described in Chapter 1.

7.3 Radiation Measurements

The radiation measurement set-up basically consists of an NWA connected with a MIMO PLC coupler to the mains. To enhance the dynamic range of the set-up, the NWA is connected to an *amplifier* (AMP) and the amplified signal is fed into the MIMO coupler. The signal has to be at least 10 dB above the noise floor, that is, the signal indicated by the NWA without the signal injection connected. Care should be taken that the output power does not exceed 1 W to avoid damaging the couplers and disturbing the appliances connected to the mains grid. On the other side, the antenna is connected through a cable with ferrites as shown in Figure 7.1 to a *high-pass filter* (HPF) and the receiving end of the NWA. The HPF attenuates signals below 2 MHz, which were identified in a few cases reducing the dynamic range of the NWA.

In the past, NWA *k*-factor measurements using coaxial cables to connect the couplers were considered unacceptable, as results may be influenced by signals picked up by the cable. Therefore, the coaxial measurement set-up (including the ferrite beads) described herein was validated by comparative measurements using fibre-optical link equipment between the antenna and the NWA. No difference could be detected. Thus, the fibre-optical link was not further used, because of its reduced dynamic range, higher noise floor and its more cumbersome installation.

The outlets where PLC signals were fed were arbitrarily selected from within the building. The antenna is positioned at a distance of 10 or 3 m from the exterior wall outside the building (see Figure 7.1).

7.3.1 Calibration of the NWA

The NWA needs to be calibrated in order to eliminate the effects caused by using long cables in the building. A response ('through') calibration is done by shortcutting the endings of both coaxial cables. A conventional adapter (BNC female to BNC female) is used as calibration kit. The MIMO coupler is considered to be part of the PLC channel.

7.3.2 Calculation of the Radiation Measure (*k*-Factor)

The coupling factor or *k*-factor is the ratio between the electrical field caused by signal radiation and the signal power fed into the main grid. The *k*-factor was used first in ETSI TR 102 175 [8].

The radiation of buildings is evaluated by

$$k_{E,H} = E_{antenna} - P_{max,feed},$$

$$= U_{Receiver} + AF - P_{max,amp\_output} + A_{PLC\_Coupler},$$

$$= P_{Receiver} + Conv_{dBm2dB\mu V} + AF - P_{max,amp\_output} + A_{PLC\_Coupler},$$

$$= s_{21} + Conv_{dBm2dB\mu V} + AF + A_{PLC\_Coupler}, \tag{7.1}$$

where

$k_{E,H}$ is *k*-factor in dB(μV/m)–dBm with regard to the electric component (k_E) or in dB(μA/m)–dBm with regard to the magnetic field component (k_H)

$E_{antenna}$ is the electrical field strength in dB(μV/m) received at the location of the antenna

$P_{max,feed}$ is the signal power in dBm at the output of the PLC coupler (in case of terminated output)

$U_{Receiver}$ is the voltage in dB(μV) at the output of the antenna

AF is antenna factor in dB(1/m) of the antenna

P_{max,amp_output} is signal power in dBm at the output of the amplifier (in case of termination)

$A_{PLC\_Coupler}$ is the attenuation of the PLC coupler in dB ETSI TR 101 562-2 [4]

$P_{Receiver}$ is the power from in dBm the output of the antenna

$Conv_{dBm2dB\mu V}$ is the conversion factor from dBm to dBμV in a 50 Ω environment: 107 dBμV–dBm (see Chapter 6)

s_{21} is the scattering parameter in dB as measured by the calibrated network analyser (as shown in Figure 7.1)

The literal meaning of the formula earlier is as follows: If a signal is fed with 0 dBm into the mains of a building, then an electrical field of $E_{antenna}$ dBμV/m is recorded inside as well as outside the building.

From the recorded values s_{21} of the NWA, the *k*-factor can be derived using Equation 7.1.

The combinations of different antenna polarisations or orientations are antenna dependent. The calculations presented in Table 7.1 apply to derive a single *k*-factor for each injection, plug and antenna location combination. Loop antennas capture signals only from one orientation or direction. The circumpolar symmetry of the biconical consequences signals

TABLE 7.1

Calculation of Resulting k-Factor in Dependence of Antenna Type

Antenna Type	Calculation of the Resulting k-Factor
Biconical	$k_{res} = \max(k_{horizontal}, k_{vertical})$
Loop	$k_{res} = \sqrt{|k_x|^2 + |k_y|^2 + |k_z|^2}$, where x, y and z are the three orientations

to be picked up from two directions. The first orientation is the one where the antenna spots to where the second one is the horizontal or vertical orientation (as set up by the operator). The resulting field is derived after calculating the maximum of the two biconical records or the vector sum of the three dimensions individually measured using the loop antenna.

These calculations are performed individually for each frequency in each record.

7.4 Statistical Evaluation of k-Factor Results

The k-factor was measured at 15 locations in Spain, France and Germany.

A typical sweep from 1 to 100 MHz of any k-factor measurement is shown in Figure 7.2. Fading characterises the shape of a k-factor sweep. The legend of Figure 7.2 says that the signals are fed to outlet No. '01' for this record. The feeding port was 'LNNT', which is the D1 of the MIMO PLC coupler displayed in Chapter 1. The abbreviation represents 'live to neutral'; other ports are 'not terminated', which is the *single-input single-output* (SISO) feeding case; and 'A03_D0' is the antenna position number 3 at a distance of 10 m from the outside wall of the building (symbolised by 'D0').

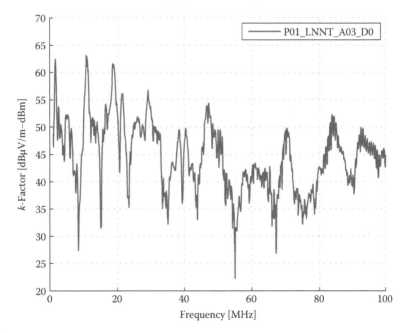

FIGURE 7.2
Typical sweep of a k-factor measurement outdoors at 10 m distance.

In total, 1294 such sweeps were recorded in this measurement campaign.

Figures 7.3 and 7.4 show the median of all data separated into the individual feeding possibilities. The median value for each measured frequency and feeding style is calculated individually. The values here are derived from data received from all antenna locations indoors and outdoors at 3 and 10 m distance from the building. For MIMO PLC,

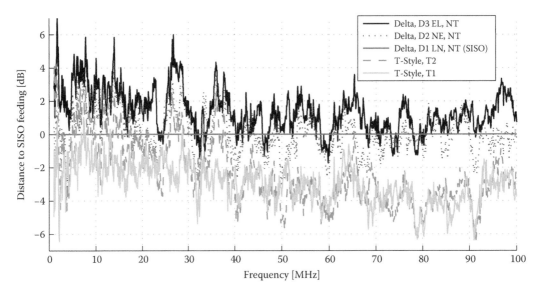

FIGURE 7.3
Median values for non-terminated delta- and T-style feeding possibilities. Outlet or antenna position always matches at comparisons relative to SISO.

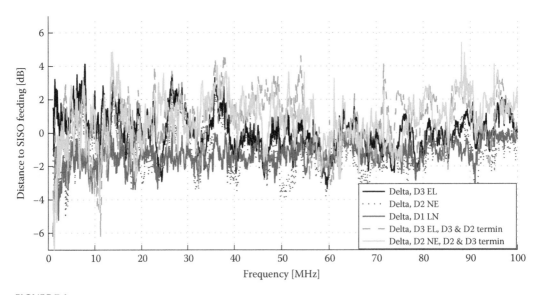

FIGURE 7.4
Median values for delta feeding possibilities. Outlet or antenna position always matches at comparisons relative to SISO.

a comparison to the traditional SISO feeding style is interesting. All median values in Figures 7.3 and 7.4 are shown relative to SISO.

Figure 7.3 presents the unterminated delta- and T-style feeding. The SISO case (from *live neutral* [LN] and *not terminated* [NT]) is one of them. It is the straight 0 dB line. Figure 7.4 depicts the terminated delta feedings. Either all delta ports are terminated or only 2 out of the 3 are terminated, as indicated in the legend. The three-port termination (D1, D2, D3) feeds 1.3 dB less energy into the mains after the PLC coupler compared to one-port termination (D1 NT, D2 NT, D3 NT on the left side). The feeding style causing the highest E-fields is the unterminated D3. The reason could be the absence of any devices connected to the mains between protective earth and live. There are no resistors present consuming the HF power between these ports. The two T-style couplings show the lowest potential for interference. This is valid for the full monitored frequency range. Apparently, they meet the best symmetry in the mains grid.

The median *k*-factor value of each location is given in Table 7.2 for all antenna positions indoors and at 3 and 10 m distance from the outside wall of the building.

Due to the high variance of the results between the individual locations and the low number of locations surveyed, a statistical evaluation of the *k*-factor for each country has not been calculated. Furthermore, the number of records from each location is unique and given in Table 7.2 in brackets. The number of antenna positions was selected according to the size of the location, size of the garden and accessibility to each location. In total, in all frequencies and feeding possibilities, 771,682 values (482 sweeps) have been recorded from a 10 m distance, 650,006 values (406 sweeps) from 3 m and 441,876 values (276 sweeps) from indoors. An explanation why there is such a high spread in the median values might be due to local conditions surrounding the building and inside the flat. In most measurements conducted, the area around the building was flat and the outdoor antenna positions

TABLE 7.2

Median Coupling Factors of Each Location

Location	Country	Median *k*-Factor Indoor in dBμV/m–dBm (No. of Records)	Median *k*-Factor from 3 m Distance in dBμV/m–dBm (No. of Records)	Median *k*-Factor from 10 m Distance in dBμV/m–dBm (No. of Records)
Duerrbachstr	Germany	72.60 (38,425)	Not measured	45.63 (76,849)
ImGeiger	Germany	69.17 (51,233)	Not measured	44.77 (102,465)
Nauheimerstr	Germany	73.48 (51,233)	Not measured	43.24 (102,465)
Rothaldenweg	Germany	73.60 (12,809)	Not measured	51.54 (102,465)
Schlossbergstr	Germany	68.46 (38,425)	57.10 (115,273)	44.74 (76,849)
VickiBaumWeg	Germany	61.19 (51,233)	Not measured	47.70 (102,465)
Boenen	Germany	71.88 (38,425)	62.38 (25,617)	Not measured
Universitaet	Germany	Not measured	55.58 (12,809)	49.84 (12,809)
Voerde	Germany	Not measured	69.53 (60,839)	59.76 (60,839)
El_Puig	Spain	55.89 (48,031)	40.04 (144,091)	Not measured
Sant_Sperit	Spain	Not measured	49.51 (144,091)	44.87 (48,031)
Torre_en_Conill	Spain	57.83 (48,031)	45.12 (96,061)	31.87 (48,031)
Guingamp	France	72.92 (25,617)	59.52 (25,617)	54.22 (12,809)
RueBunuel	France	69.61 (12,809)	Not measured	52.67 (12,809)
RueDepasse	France	70.67 (25,617)	62.89 (25,617)	50.69 (12,809)
All locations		67.98	51.3	46.96

were located on the same level as the ground floor. If the residential unit was a flat located on the second floor of a building or a multi-level house, then some of the feeding outlets had an additional vertical distance to the antenna. For example, the *k*-factor measurements in France and the location Voerde in Germany were recorded where all feeding outlets were located on the ground floor and the area around the building is flat land. This is why the outdoor *k*-factors at these locations tend to be higher than at others.

The *complementary cumulative distribution* (C-CDF) of the *k*-factor at a location where all three antenna positions were recorded is depicted in Figure 7.5.

The *k*-factor records presented earlier were obtained using an E-field antenna, because magnetic EMC antennas are not available for frequency ranges up to 100 MHz. Furthermore, consumer electronic devices in private homes use an E-field antenna (stick or whip) in the HF and VHF bands. The marker in Figure 7.5 is to be interpreted as 'in 20% of all records in a 3 m distance from the house, the *k*-factor was larger than 51.6 dBµV/m–dBm'.

Comparisons between the magnetic field (H-field) and electric field (E-field) were recorded at the location in Voerde, Germany. Radiation measurements were recorded from the building using an E-field biconical antenna [9] and H-field ring antenna [10], from 3 and 10 m away, with the antennas in the same fixed position for each reading. In order to compare H- and E-field values, the magnetic fields – recorded in dBµA/m – need to be converted into electric fields with a free space wave impedance of 377 Ω = 51.5 dBΩ. Figure 7.6 shows low correlation between H- and E-fields at 3 m distance. Obviously, a distance of 3 m may still be in the near field, where the free space wave impedance of 377 Ω cannot be applied. At 10 m from the building, the E- and H-fields display a similar pattern, as expected in the far field (see Figure 7.7), especially for frequencies higher than 15 MHz. The graphs stop at 30 MHz because the loop antenna [10] is only specified up to this frequency. At frequencies above 30 MHz, it is expected that far-field radiation conditions from a building are valid at closer distances or even indoors.

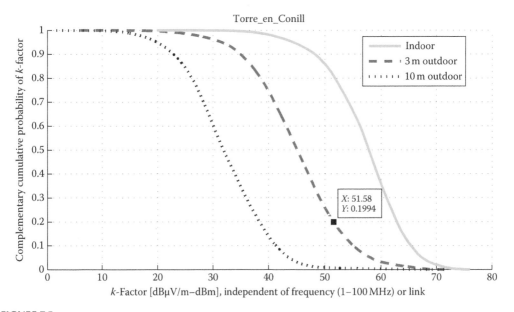

FIGURE 7.5
C-CDF of *k*-factor of a location. Note: Number of sweeps indoor, 33; in 3 m, 66; and in 10 m, 33.

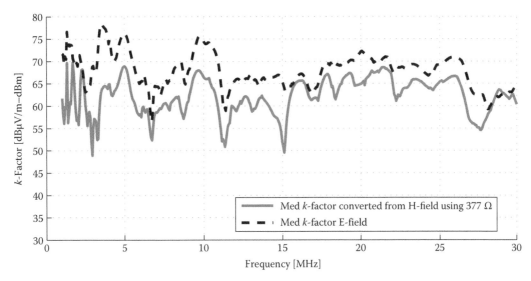

FIGURE 7.6
k-Factor measured with biconical and loop antenna at 3 m distance at one location.

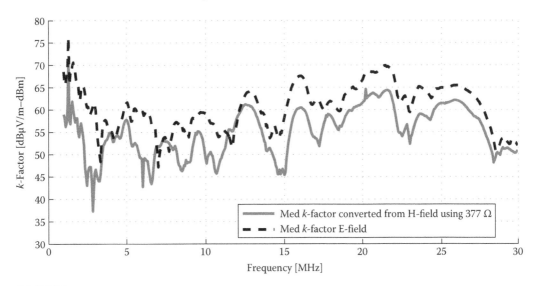

FIGURE 7.7
k-Factor measured with biconical and loop antenna at 10 m distance.

7.5 Subjective Evaluation of the Interference to Radio Broadcast

Subjective evaluations of interference to AM radio reception in the *high-frequency* (HF) bands were performed in the past in TD ETSI TR 102 616 [6]. Performing identical tests with all MIMO feeding possibilities would deliver unstable results, because the variance of received signal level (fading in TD) is more dynamic than an operator might be able to test. During a MIMO test, the interference from all MIMO feeding possibilities should be compared. The signal level is usually never stable in HF bands. Schwager [11] describes dynamic changes in the received HF signal level caused by reflections on the ionosphere. Broadcasting conditions in the *very high-frequency* (VHF) band are by far more stable over time, allowing a comparison of levels recorded over a period of a few minutes.

The source of the signal is a broadband noise generator and a frequency modulator shifting the noise to the desired frequency. The generator can be switched on and off, in order to distinguish the disturbed and undisturbed states. The FM-modulated signal is injected as an interfering signal to the mains using all possible feeding styles of the MIMO coupler shown in Chapter 1.

7.5.1 Verification and Calibration

Prior to the test, the disturbance signal has to be analysed with a measurement receiver or spectrum analyser to document the 3 dB bandwidth. As a first step, the amplification between the generator output level and the signal injected into the Bayonet BNC plug of the MIMO PLC coupler has to be determined. The feeding level of the signal generator is U_{max_feed}.

7.5.2 Measurement Procedure

At each frequency of a selected radio station, the level of the RF generator is adjusted to a lower level where no disturbance at the radio receiver is recognised. From that value, the generator level is increased until a disturbance can hardly be recognised. After that the level of interference signal, U_{gen} in dBµV is verified and documented.

This procedure is repeated for each

- Coupling type
- Selected frequency
- Feeding outlet (injection point)
- Radio receiver
- Radio receiver location

Furthermore, the measurements have to be done when the radio receiver is battery driven and mains powered.

7.6 Interference Threshold of FM Radio Broadcasts

The level of interference in FM radio broadcasts was tested at 10 different locations. The field test set-up is shown in Figure 7.8. In total, 1179 subjective evaluations of PLC interference, which can be detected by human ears, in FM radio were conducted. This includes nine feeding styles times 131 radio stations with various radio devices in a range of different positions. The threshold at which human ears are able to detect interference is noted in the measurement protocol.

Figure 7.9 shows the C-CDF of the threshold of all radio services at all locations independent of the radio device or power supply used. The x-axis is the feeding PSD injected to a power outlet. All MIMO feeding styles are within the black area, and the three delta feedings are explicitly given. The FM services received with a low signal level are interfered also with low PLC feeding levels. As shown in the top left corner of Figure 7.9, there is only

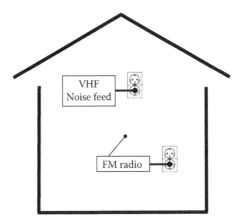

FIGURE 7.8
Basic set-up for FM interference tests.

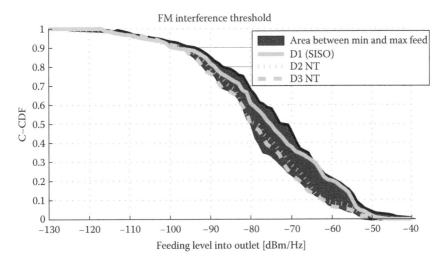

FIGURE 7.9
C-CDF of threshold when interference is noticeable by human ears.

a small variation between the individual feeding styles. Sensitive FM stations do not show higher or reduced interference potential when another feeding style is selected.

When performing similar tests in the HF range with AM radio stations, a significant difference was found if the radio was battery driven or mains powered [9]. HF radio devices also used the mains grid for communications (like PLC modems). They use the mains as a counterpoise to the whip antenna resulting in a bipolar reception instead of only a monopole. This phenomenon could not be verified in the VHF; the conducted coupling path was not dominant.

Or in other words, the FM radio receivers used in this test are sufficiently isolated from the mains interferer. There was never any influence whether the FM radio was mains powered or battery driven.

The FM radio immunity field tests were performed using radio receivers operated by the habitants of the flats. The immunity of all radio devices was pretty much identical. No dependency among device manufacturers or HIFI radio systems versus kitchen radios could be found.

7.7 Conclusions

EMI field tests were conducted measuring the PLC's radiation from buildings as well as the interference to FM radio receivers. The results show all three wires connected to power outlets in private homes to have a similar potential for interference. No significant difference was found when injecting signals into the mains in any MIMO feeding style compared to the traditional SISO style. Chapter 16 shows further experiments of the influence of beamforming on the radiated fields.

References

1. European Telecommunications Standards Institute, http://www.etsi.org, accessed April 2013.
2. Special Task Force 410. http://portal.etsi.org/STFs%5CToR%5CToR410v11_PLT_MIMO_measurem.doc Members to STF410: Andreas Schwager (STF Leader), Sony, Germany; Werner Bäschlin, JobAssist, Switzerland; Holger Hirsch, University of Duisburg-Essen, Germany; Pascal Pagani, France Telecom, France; Nico Weling, Devolo, Germany; Jose Luis Gonzalez Moreno, Marvell, Spain; Hervé Milleret, Spidcom, France. STF410 produced following technical reports: [ETSI TR 101 562-1]; [ETSI TR 101 562-2]; [ETSI TR 101 562-3].
3. ETSI TR 101 562-1 V1.3.1 (2012-02), PowerLine Telecommunications (PLT); MIMO PLT; Part 1: Measurement methods of MIMO PLT.
4. ETSI TR 101 562-2 V1.3.1 (2012-10), PowerLine Telecommunications (PLT); MIMO PLT; Part 2: Setup and statistical results of MIMO PLT EMI measurements.
5. ETSI TR 101 562-3 V1.1.1 (2012-02), PowerLine Telecommunications (PLT); MIMO PLT; Part 3: Setup and statistical results of MIMO PLT channel and noise measurements.
6. ETSI TR 102 616 V1.1.1 (2008-03): PowerLine Telecommunications (PLT); Report from Plugtests™ 2007 on coexistence between PLT and short wave radio broadcast; Test cases and results.
7. ITU-R Recommendation BS.1284: General methods for the subjective assessment of sound quality.
8. ETSI TR 102 175 V1.1.1 (2003-03): PowerLine Telecommunications (PLT); Channel characterization and measurement methods.
9. SCHWARZBECK MESS – ELEKTRONIK; EFS 9218: Active electric field probe with biconical elements and built-in amplifier 9 kHz … 300 MHz.
10. R&S®HFH2-Z2: Loop Antenna Broadband active loop antenna for measuring the magnetic field-strength; 9 kHz–30 MHz.
11. A. Schwager, Powerline communications: Significant technologies to become ready for integration, PhD dissertation, Universität Duisburg-Essen, 2010, http://duepublico.uni-duisburg-essen.de/servlets/DerivateServlet/Derivate-24381/Schwager_Andreas_Diss.pdf.

8

MIMO PLC Signal Processing Theory

Daniel M. Schneider and Andreas Schwager

CONTENTS

8.1 Introduction

The idea of *multiple-input multiple-output* (MIMO) [1–3] started many research activities, mainly for wireless applications. MIMO technology has recently been successfully introduced by several wireless standards such as IEEE 802.11n, WiMAX and LTE [4]. Compared to their basic *single-input single-output* (SISO) solutions, they offer a fundamental increase in data rate without the demand of higher transmit power or bandwidth. By using more than two wires, for example, by using the *protective earth* (PE) wire in addition to the *live* (L) and *neutral* (N) wire in in-home *power line communications* (PLC) (see Chapter 1), the MIMO principles can be applied to PLC as well. The properties of the MIMO PLC channel are discussed in detail in Part I. This chapter deals with the MIMO signal processing theory and the application to MIMO PLC. Before pointing out the outline of this chapter, a brief overview of literature on the application of MIMO schemes to PLC is discussed.

References [5–9] apply space–time codes to access PLC systems and assume perfect isolation between different phase wires. This means that signals fed into one transmit port are only visible at the same receive port. Based on the same channel assumptions [10], proposes an *orthogonal frequency division multiplex* (OFDM)–based space–time MIMO system for PLC. Reference [11] considers a 2 × 2 MIMO system for in-home applications based on space–frequency-coded multi-tone M-ary frequency shift keying. The wires are assumed to be uncoupled, that is, also no crosstalk between the wires is assumed. Reference [12] considers for the first time a coupled access MIMO PLC channel but also restricts itself to space–time and space–frequency codes. Reference [13] investigates a *multiple-input single-output* (MISO) system for in-home PLC, based on OFDM and space–frequency codes. In JP, two-wire installations are used. However, in the fusion box, three wires are available which make MISO possible if the transmitter is located in the fuse box. The crosstalk between the wires is considered. Reference [14] investigates a MISO access PLC system based on single carrier modulation for narrowband applications. The crosstalk between the wires is considered here as well. The idea of space–time coding is applied to PLC by several authors even if no MIMO channel is assumed. Reference [15] applies space–time codes for coding across OFDM subcarriers to use the diversity across the subcarriers for the frequency-selective PLC channels. Distributed space–time codes are applied in [16] in the context of multi-hop transmission, in which each network node acts as a potential repeater. The use of distributed space–time codes allows an efficient combination at the receiver when different repeater nodes transmit simultaneously. References [17–20] analyse different MIMO schemes and system proposals for in-home broadband PLC. These algorithms are discussed in the following sections.

The outline of this chapter is as follows. In the first step, the concept of the MIMO channel matrix and its decomposition into parallel and independent MIMO streams based on the *singular value decomposition* (SVD) is discussed (Section 8.2). Next, this chapter focuses on the comparison of several MIMO schemes and their suitability for broadband PLC, taking into account the characteristics of the MIMO PLC channel.

In order to deal with the frequency-selective characteristic of the PLC channel, today's SISO-PLC systems use multi-carrier modulation like OFDM. Section 8.3 introduces the basic MIMO-OFDM system including the system assumptions used in this chapter. In the next step, several basic MIMO schemes are introduced and discussed. The different MIMO schemes may be divided into two groups aiming for different goals. MIMO offers spatial diversity to combat fading. The spatial diversity, which is defined by the number of available MIMO paths, is fully exploited by sending the signals over different MIMO paths. These MIMO schemes are called space–time–frequency coded and are introduced in Section 8.4 with the main focus on the Alamouti scheme. Typically, these schemes do not use *channel state information* (CSI) at the transmitter. The second group of MIMO schemes aims to maximise the throughput by sending different streams via the available transmit ports. These *spatial multiplexing* (SMX) schemes are discussed in Section 8.5, including precoding at the transmitter where the precoding exploits CSI at the transmitter.

In particular, the *signal-to-noise ratio* (SNR) after MIMO detection is calculated. The measured MIMO PLC channels obtained during the European measurement campaign (ETSI STF410, see Chapter 5) form the basis of a detailed comparison of the investigated MIMO schemes with respect to the SNR after MIMO detection (see Section 8.6). The basic principles of MIMO signal processing discussed in this chapter form the basis for many of the following chapters in this book dealing with MIMO aspects. The decomposition of the SVD is useful to derive the theoretical channel capacity and the SNR after MIMO decoding and is used to compute the throughput of MIMO PLC systems with adaptive modulation in Chapter 9. This chapter also provides the background knowledge for the MIMO algorithms incorporated in the next-generation broadband PLC specifications (see Chapters 12 and 14).

8.2 MIMO Channel Matrix and Its Decomposition in Eigenmodes

8.2.1 MIMO Channel Matrix

Chapter 1 showed how the PE wire can be used for MIMO communications. The chapter also discussed the coupling of the MIMO signals into the PLC channel. The used input and output ports define the MIMO PLC channel. Figure 8.1a is the schematic measurement set-up with all available MIMO paths. When the delta-style coupler according to Chapter 1 is attached to the MIMO paths, three ports for feeding are available (according to Kirchhoff's law, up to two might be used simultaneously). The use of the star-style coupler allows up to four receive options. Assume, for example, that the input ports D_1 and D_3 and the output ports S_1, S_2 and S_4 should be used for communication (as indicated by the bold arrows in Figure 8.1a). Then, the resulting MIMO PLC channel is shown in Figure 8.1b with $N_T = 2$ transmit and $N_R = 3$ receive ports, resulting in overall six transmission paths.

Figure 8.2 illustrates the magnitude response of the transfer functions of a typical measurement recorded in the ETSI MIMO PLC channel measurement campaign (see Chapter 5) with the delta-style coupler at the transmitter (the ports D_1 and D_3 are used) and the star-style coupler at the receiver (all four ports S_1–S_4 are used). The frequency range goes up to 100 MHz. Since $N_T = 2$ transmit and $N_R = 4$ receive ports are used, overall eight transfer functions are available. The figure aims to illustrate the relationship between the MIMO paths and the same MIMO PLC channel is used in Section 8.2.2 to illustrate the decomposition in eigenmodes. The example shows that the signal, which is fed into one port, is not only visible at the same receive port as the feeding port, but also visible at all the other receive ports. The crosstalk caused by the coupling of the wires results in the presence of all possible MIMO paths. The similar shape of the transfer functions (especially for low frequencies) is due to the same underlying topology of the different MIMO paths. Usually, the wires are run in parallel within the walls facing a similar multi-path propagation. The magnitude responses show the typical frequency-selective behaviour of PLC channels. For details on the MIMO PLC channel properties, refer to Chapters 4 and 5.

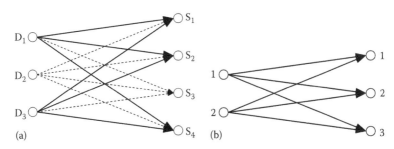

FIGURE 8.1
Constructing the MIMO channel matrix (b) from measured MIMO paths (a) by using the input ports D_1 and D_3 and output ports S_1, S_2 and S_4.

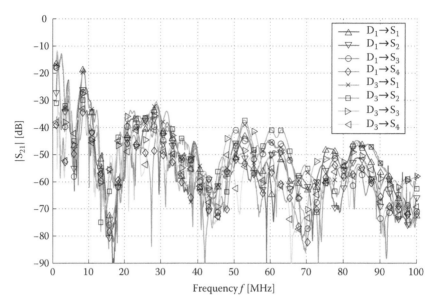

FIGURE 8.2
Magnitude responses of a typical MIMO PLC channel, 0–100 MHz. MIMO configuration: D_1 and D_3 as feeding and S_1–S_4 as receiving ports. SISO configuration: D_1 as feeding and S_1 as receiving port.

With $h_{ml}(f)$, the complex frequency response of frequency f from feeding port l ($l = 1, \ldots, N_T$) to receiving port m ($m = 1, \ldots, N_R$), the MIMO channel matrix of each frequency f is defined as

$$
\mathbf{H}(f) = \begin{bmatrix} h_{11}(f) & h_{12}(f) & \cdots & h_{1N_T}(f) \\ h_{21}(f) & h_{22}(f) & \cdots & h_{2N_T}(f) \\ \vdots & \vdots & \ddots & \vdots \\ h_{N_R1}(f) & h_{N_R2}(f) & \cdots & h_{N_RN_T}(f) \end{bmatrix}.
\tag{8.1}
$$

Note that $h_{ml}(f)$ are complex numbers and thus $\mathbf{H}(f)$ is a complex matrix. $h_{ml}(f)$ can be related to the complex transfer function $H_{ml}(n\Delta f)$ ($n = 1, \ldots, N$) as measured by the network analyser (e.g. for $N = 1601$ measurement points in Chapter 5), that is, the channel is divided into the narrow subchannels measured by the network analyser. For the example shown in Figure 8.1 with the 2×3 MIMO set-up, the dimensions of the matrix in Equation 8.1 are 3×2. Since the maximum number of feeding and receive ports is $N_T = 2$ and $N_R = 4$, up to 2×4 MIMO is possible.

8.2.2 Decomposition in Eigenmodes

To simplify the notation and for better legibility, the dependency of f of the channel matrix $\mathbf{H}(f)$ will be dropped in the following section. The SVD [21] decomposes the channel matrix $\mathbf{H}$ into three matrices:

$$
\mathbf{H} = \mathbf{U}\mathbf{D}\mathbf{V}^H.
\tag{8.2}
$$

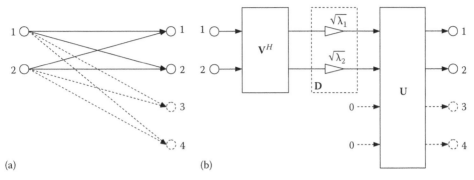

FIGURE 8.3
Decomposition (b) of the MIMO channel (a) into two parallel SISO channels or spatial streams for 2 × 2 MIMO (solid ports and paths) and 2 × 4 MIMO (adding the dashed ports and paths).

$\mathbf{D}$ is a diagonal matrix of dimensions $R \times R$ containing the real-valued singular values* $\sqrt{\lambda}_p$, $p = 1, \ldots, R$ of the channel matrix $\mathbf{H}$:

$$\mathbf{D} = \mathrm{diag}\left(\sqrt{\lambda}_1, \ldots, \sqrt{\lambda}_R\right) = \begin{bmatrix} \sqrt{\lambda}_1 & 0 & 0 \\ 0 & \ddots & 0 \\ 0 & 0 & \sqrt{\lambda}_R \end{bmatrix}. \tag{8.3}$$

R is defined as the minimum of the number of transmit and receive ports†, that is, $R = \min(N_T, N_R)$. Note that the singular values are real and defined in decreasing order without loss of generality [21]. $\mathbf{U}$ and $\mathbf{V}$ are unitary matrices, that is, $\mathbf{U}^{-1} = \mathbf{U}^H$ and $\mathbf{V}^{-1} = \mathbf{V}^H$. $(\cdot)^H$ is the Hermitian operator, that is, transpose $(\cdot)^T$ and conjugate complex $(\cdot)^*$. $(\cdot)^{-1}$ is the matrix inverse operator. The dimensions of the complex matrices $\mathbf{U}$ and $\mathbf{V}$ are $N_R \times R$ and $N_T \times R$, respectively. The SVD decomposes the channel matrix into parallel and independent SISO branches or streams [3]. Figure 8.3 visualises the decomposition for $N_T = 2$ transmit ports and $N_R = 2$ and $N_R = 4$ receive ports, respectively. For the example of two transmit ports, at least two receive ports are needed to support $R = 2$ spatial streams. The full channel matrix in Figure 8.3a (ports drawn with solid lines: 2 × 2 MIMO and including the dashed drawn receive ports: 2 × 4 MIMO) is decomposed into two SISO branches (see Figure 8.3b).

Since $\mathbf{U}$ and $\mathbf{V}$ are unitary matrices, multiplication by these matrices preserves the total energy. The singular values $\sqrt{\lambda}_p$ of $\mathbf{D}$ describe the attenuation of the decomposed SISO branches.

Figure 8.4 shows an example of the attenuation of the decomposed MIMO channel. The same channel measurement results are taken as in Figure 8.2. The transmit ports D_1 and D_3 are used as feeding ports and S_1–S_4 as receive ports. The number of transmit ports is $N_T = 2$ and the number of receive ports is $N_R = 4$, that is, $R = \min(N_T, N_R) = 2$ singular values $\sqrt{\lambda}_1$ and $\sqrt{\lambda}_2$ are obtained. The two singular values are compared to the attenuation of the SISO channel (D_1 is used for feeding and S_1 for receiving). The first stream is less attenuated compared to SISO, while the second stream shows higher and lower attenuation depending on

---

* The singular values $\sqrt{\lambda}_p$ are derived from the eigenvalues λ_p of $\mathbf{H}^H\mathbf{H}$.
† R is not necessarily the rank R' of $\mathbf{H}$, $R' \leq R$. If $\mathbf{H}$ has full rank, the rank of $\mathbf{H}$ is $R' = R$. If $\mathbf{H}$ does not have full rank $R' < R$, some of the singular values $\sqrt{\lambda}_p$ are 0.

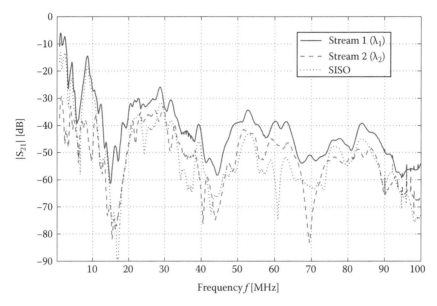

FIGURE 8.4
Comparison of magnitude response between decomposed MIMO streams (singular values) and SISO (same measurement as in Figure 8.2). MIMO configuration: D_1 and D_3 as feeding and S_1–S_4 as receiving ports. SISO configuration: D_1 as feeding and S_1 as receiving port.

the frequency compared to SISO. The fact that the singular values are in decreasing order results in the first stream being stronger than the second stream. The singular values show the gain of MIMO. Instead of one transmission stream for SISO, two MIMO streams are available. The decomposed streams are still very frequency selective with deep fades. This can be explained by recalling Figure 8.2 around 15 MHz, for example. At this frequency, all MIMO paths experience high attenuation. The singular values of the MIMO channel at this frequency are therefore highly attenuated as well (see frequencies around 15 MHz in Figure 8.4). The attenuation of the singular values not only reflects the attenuation of the MIMO paths but also indicates the spatial correlation between the MIMO paths (see Chapters 4 and 5).

8.3 MIMO-OFDM System

The basic MIMO system introduced in this chapter is based on OFDM. OFDM divides the frequency-selective channel into parallel and orthogonal sub-bands or subcarriers [22]. If the number of OFDM subcarriers is large, each sub-band is so narrow that the corresponding magnitude response is flat. Or in other words, the subcarrier spacing must be smaller than the channel coherence bandwidth. The equalisation is simplified and reduced to normalisation by a complex scalar. Each subcarrier is modulated separately which makes OFDM robust against strong fading of certain frequencies or noise in specific sub-bands (narrowband interference). This makes OFDM a suitable modulation scheme for PLC [23]. In order to eliminate *intersymbol interference* (ISI), OFDM requires a guard interval.

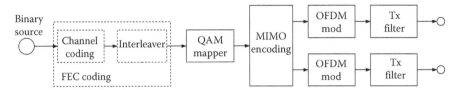

FIGURE 8.5
Block diagram of the MIMO transmitter.

Its length depends on the duration of the impulse response. The insertion of the guard interval reduces the bandwidth efficiency of OFDM.

Figure 8.5 shows the basic transmitter structure of the OFDM-based MIMO system with two transmit ports. Typically, the bits of the binary source are encoded by *forward error correction* (FEC). The FEC is drawn by dashed lines since this chapter focuses on the following blocks and does not consider FEC. The encoded bits are modulated by the *quadrature amplitude modulation* (QAM) block. The QAM may comprise different modulation orders depending on the subcarrier index (see Section 9.3.1). Next, the complex QAM symbols are processed by the MIMO encoding block. The MIMO encoding may comprise a *serial-to-parallel* (S/P) conversion to split the symbols to the different transmit ports and may comprise precoding (see Section 8.5). In the case of space–time–frequency coding (see Section 8.4), the symbols are rearranged within this block. The complex symbols of each MIMO stream are OFDM-modulated and the digital time-domain signal is obtained. The block *transmitter* (Tx) filter comprises impulse shaping, frequency up conversion*, transmit filtering and *digital-to-analogue* (D/A) conversion. Finally, the analogue signal is transmitted via the two transmit ports to the MIMO PLC channel.

Figure 8.6 shows the block diagram of the MIMO receiver with four receive ports. The four receiver front-ends (*receiver* [Rx] filters) receive the signals from the four receive ports and include receive filters, *analogue-to-digital* (A/D) converters and the frequency down-conversion. Next, each signal is OFDM demodulated. The MIMO processing at the receiver depends on the deployed MIMO scheme and is discussed in detail in the following sections.

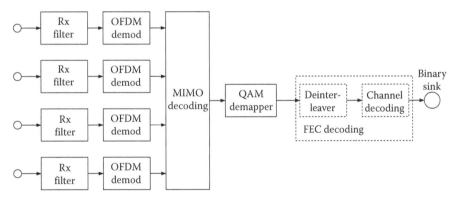

FIGURE 8.6
Block diagram of the MIMO receiver.

---

* Alternative implementations use an *inverse fast Fourier transform* (IFFT) of double size with complex conjugate input to obtain the real-valued signal which is often referred to as *discrete multi-tone modulation* (DMT). In this case, frequency up-conversion is not needed.

In case of SMX schemes, the MIMO decoding block contains an equaliser or detection to obtain the two logical MIMO streams and a *parallel-to-serial* (P/S) conversion to obtain the complex QAM symbols. The QAM symbols are further processed by the QAM demapper. Finally, the demodulated bits are decoded by the FEC decoding to obtain estimates of the transmitted bits.

The FEC shown in Figures 8.5 and 8.6 illustrates an example containing channel coding and interleaving. However, different architectures are possible, for example, an outer and inner code with several interleavers, turbo codes or *low-density parity check* (LDPC) codes (see the PLC systems described in Part III of this book).

The MIMO system introduced here works on the level of QAM symbols; the MIMO encoding follows the QAM modulation in the transmitter and precedes QAM demodulation in the receiver. The FEC processes a single bit stream. This coding is often called vertical coding since the streams of the multiple transmit ports are encoded together. On the other hand, horizontal coding handles each MIMO stream separately. Starting from a SISO system, vertical coding allows an easy evolution without changing the encoding and decoding stages. This may be preferable for the further development of a SISO system toward a MIMO communications system to keep the downward compatibility, for example, of a given standard and to reduce complexity.

The receiver in Figure 8.6 does not contain synchronisation or channel estimation. Perfect synchronisation and channel estimation are assumed. Furthermore, this chapter assumes in the following an uncoded system without FEC.

As explained earlier, OFDM decomposes the bandwidth of the channel into small intervals given by multiple subcarriers. In the following, N is the number of subcarriers and n ($n = 1, \ldots, N$) is the subcarrier index. The channel matrix $\mathbf{H}(n)$ of each subcarrier n describes the system between OFDM modulation at the transmitter and OFDM demodulation at the receiver including the couplers. The channel matrix $\mathbf{H}(n)$ includes not only the MIMO PLC channel but also includes all filters in the transmitter and receiver. $\mathbf{s}_k(n)$ describes the $N_T \times 1$ symbol vector containing the symbols transmitted via the N_T transmit ports and $\mathbf{r}_k(n)$ the $N_R \times 1$ received symbol vector of the N_R receive ports after OFDM demodulation of carrier n and time k. The transmit symbols $s_{l,k}(n)$ ($l = 1, \ldots, N_T$) have the average power P_T/N_T, where P_T is the total transmit power. The system between the input of the OFDM modulator and the output of the OFDM demodulator is described by

$$\mathbf{r}_k(n) = \mathbf{H}(n)\mathbf{s}_k(n) + \mathbf{n}_k(n), \tag{8.4}$$

where $\mathbf{n}_k(n)$ is the symbol vector containing the noise samples of subcarrier n and time k. The noise samples are assumed to follow a zero mean Gaussian distribution with variance N_0 and are independent for each subcarrier and receive port. The channel is assumed to not change during several OFDM symbols. Typically, the in-home PLC channel experiences long-term time variation, for example, when the network topology changes (toggling of a light switch or a new device is plugged to the mains grid). Additionally, there are short-term and typically line cycle–dependent time variations caused by impedance-modulating devices such as small-sized power supplies. In the following, a quasi-static channel is assumed. Thus, $\mathbf{H}(n)$ does not include the time index k. The block diagram of the MIMO-OFDM system is shown in Figure 8.7 for $N_T = 2$ and $N_R = 4$.

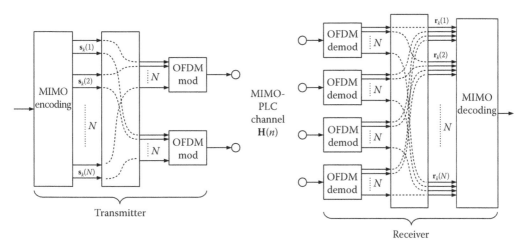

FIGURE 8.7
Block diagram of the MIMO-OFDM system for $N_T = 2$ and $N_R = 4$.

8.4 Space–Time–Frequency Coding

The number of available MIMO paths defines the spatial diversity. A MIMO system with N_T transmit ports and N_R receive ports provides the maximal spatial diversity of order $N_T N_R$. Space–time–frequency codes aim to transmit the signals via each available MIMO path to achieve the full spatial diversity. Space–time block codes achieve this goal by transmitting replicas of each symbol via the different transmit ports at several time instances. The transmission of replicas makes transmission more reliable. However, the bitrate is not maximised. Orthogonal space–time block codes are a subclass of these codes, allowing easy decoding. For the special case of two transmit ports as deployed here for in-home PLC, the orthogonal space–time block codes lead to the well-known Alamouti scheme [24] which is introduced in Section 8.4.1. Several other space–time–frequency codes are discussed briefly in Section 8.4.2.

8.4.1 Alamouti Scheme

The Alamouti scheme encodes two QAM symbols. At the first time instance, k, the two symbols $b_{1,k}(n)$ and $b_{2,k}(n)$ are transmitted on subcarrier n via transmit port 1 and 2, respectively, forming the transmit symbol vector $\mathbf{s}_k(n) = \begin{bmatrix} b_{1,k}(n) \\ b_{2,k}(n) \end{bmatrix}$. At the second time instance, $(k+1)$, the symbols $-b_{2,k}^*(n)$ and $b_{1,k}^*(n)$ are transmitted via transmit port 1 and 2, respectively. Note that $(\cdot)^*$ is the conjugate complex operator. The transmit symbol vector is $\mathbf{s}_{k+1}(n) = \begin{bmatrix} -b_{2,k}^*(n) \\ b_{1,k}^*(n) \end{bmatrix}$.

Two QAM symbols are transmitted during two time instances or OFDM symbols over the two transmit ports. Thus, the spatial code rate is $r_S = 1$. Each symbol is transmitted via each transmit port and reaches the receiver on every possible MIMO path. Therefore, the full spatial diversity $N_T N_R$ is achieved. The Alamouti scheme works independently of the

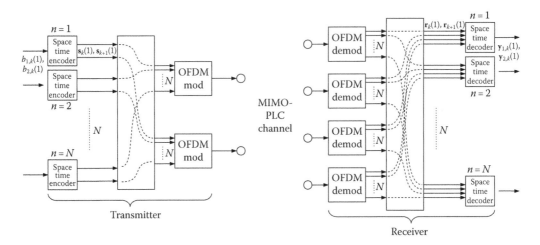

FIGURE 8.8
Block diagram of the MIMO-OFDM Alamouti system, the space–time processing is shown, using the first sub-carrier as an example, $N_T = 2$ and $N_R = 4$.

number of receive ports. MISO is possible as well. Figure 8.8 shows the block diagram of the Alamouti scheme. The space–time encoding is illustrated, using the first subcarrier as an example.

The two transmit symbol vectors are summarised in the space–time coding matrix:

$$\mathbf{S}(n) = [\mathbf{s}_k(n) \quad \mathbf{s}_{k+1}(n)] = \begin{bmatrix} b_{1,k}(n) & -b_{2,k}(n)^* \\ b_{2,k}(n) & b_{1,k}(n)^* \end{bmatrix}. \tag{8.5}$$

The two columns of $\mathbf{S}(n)$ are orthogonal, allowing simple decoding at the receiver. According to Equations 8.4 and 8.5, the received symbol vectors at the two time instances are

$$\mathbf{r}_k(n) = \mathbf{H}(n)\mathbf{s}_k(n) + \mathbf{n}_k(n), \tag{8.6}$$

$$\mathbf{r}_{k+1}(n) = \mathbf{H}(n)\mathbf{s}_{k+1}(n) + \mathbf{n}_{k+1}(n). \tag{8.7}$$

Note that it is assumed that the channel does not change during the two time instances, that is, $\mathbf{H}_k(n) = \mathbf{H}_{k+1}(n) = \mathbf{H}(n)$.

Taking the conjugate complex of Equation 8.7 and rearranging the channel matrix

$$\mathbf{H}(n) = [\mathbf{h}_1(n) \quad \mathbf{h}_2(n)] \tag{8.8}$$

with the two column vectors $\mathbf{h}_1(n)$ and $\mathbf{h}_2(n)$ into a new channel matrix

$$\tilde{\mathbf{H}}(n) = \left[\mathbf{h}_2^*(n) \quad -\mathbf{h}_1^*(n)\right] \tag{8.9}$$

results in

$$\mathbf{r}_{k+1}^{*}(n) = \breve{\mathbf{H}}(n)\mathbf{s}_k(n) + \mathbf{n}_{k+1}^{*}(n). \tag{8.10}$$

Writing Equations 8.6 and 8.10 in matrix notation provides

$$\underbrace{\begin{bmatrix} \mathbf{r}_k(n) \\ \mathbf{r}_{k+1}^{*}(n) \end{bmatrix} = \begin{bmatrix} \mathbf{H}(n) \\ \breve{\mathbf{H}}(n) \end{bmatrix}}_{G(n)} \mathbf{s}_k(n) + \begin{bmatrix} \mathbf{n}_k(n) \\ \mathbf{n}_{k+1}^{*}(n) \end{bmatrix}, \tag{8.11}$$

where $\mathbf{G}(n)$ is the concatenated channel matrix. Since the columns of $\mathbf{S}(n)$ are orthogonal, the columns of $\mathbf{G}(n)$ are also orthogonal, that is,

$$\mathbf{G}^{H}(n)\mathbf{G}(n) = \beta(n)\mathbf{I}_2 \tag{8.12}$$

with $\beta(n) = \|\mathbf{H}(n)\|_{F}^{2}$, where $\|\mathbf{H}(n)\|_{F} = \sqrt{\sum_{l=1}^{N_T}\sum_{m=1}^{N_R}|h_{ml}(n)|^{2}}$ is the Frobenius norm of the channel matrix $\mathbf{H}(n)$ and $h_{ml}(n)$ is the matrix element m, l of the channel matrix $\mathbf{H}(n)$.

Left multiplying $\dfrac{1}{\beta(n)}\mathbf{G}^{H}(n)$ on Equation 8.11 yields

$$\mathbf{y}_k(n) = \begin{bmatrix} y_{1,k}(n) \\ y_{2,k}(n) \end{bmatrix} = \frac{1}{\beta(n)}\mathbf{G}^{H}(n)\begin{bmatrix} \mathbf{r}_k(n) \\ \mathbf{r}_{k+1}^{*}(n) \end{bmatrix},$$

$$= \mathbf{s}_k(n) + \frac{1}{\beta(n)}\mathbf{G}^{H}(n)\begin{bmatrix} \mathbf{n}_k(n) \\ \mathbf{n}_{k+1}^{*}(n) \end{bmatrix},$$

$$= \begin{bmatrix} b_{1,k}(n) \\ b_{2,k}(n) \end{bmatrix} + \frac{1}{\beta(n)}\mathbf{G}^{H}(n)\begin{bmatrix} \mathbf{n}_k(n) \\ \mathbf{n}_{k+1}^{*}(n) \end{bmatrix}, \tag{8.13}$$

where $\mathbf{y}_k(n)$ contains estimates of the transmit symbol vector $\mathbf{s}_k(n) = \begin{bmatrix} b_{1,k}(n) \\ b_{2,k}(n) \end{bmatrix}$ which are superimposed with the filtered noise. Equation 8.13 shows the processing by the Alamouti scheme. In a first step, the first received symbol vector $\mathbf{r}_k(n)$ and the conjugate complex of the following received symbol vector $\mathbf{r}_{k+1}^{*}(n)$ are stacked into one vector. Next, $\dfrac{1}{\beta(n)}\mathbf{G}^{H}(n)$ is formed based on the channel matrix $\mathbf{H}(n)$. The multiplication of the stacked receive vector by $\dfrac{1}{\beta(n)}\mathbf{G}^{H}(n)$ leads to estimates of the two transmitted symbols corrupted by noise. Also, according to Equation 8.13, the signal energy of the symbols after Alamouti processing is $E\{|b_p|^2\} = \dfrac{P_T}{2}$ ($p = 1, 2$) and the variance of the filtered noise is

$$\left[\frac{1}{\beta(n)} \mathbf{G}^H(n) \frac{1}{\beta(n)} \mathbf{G}(n) \right]_{pp} N_0 = \frac{N_0}{\beta(n)} \ (p = 1, 2) \text{ with } [\cdot]_{ml} \text{ the matrix element with position } (m, l).$$

Hence, the SNR after Alamouti processing of subcarrier n is calculated as

$$\Lambda(n) = \frac{\left(\dfrac{P_T}{2} \right)}{\left(\dfrac{N_0}{\beta(n)} \right)} = \| \mathbf{H}(n) \|_F^2 \frac{\rho}{2} \tag{8.14}$$

with $\rho = P_T / N_0$. The factor $1/2$ results from the distribution of the total transmit power over the two transmit ports. The SNR is the same for both symbols since each symbol is transmitted via each MIMO path. For the same reason, the Alamouti scheme does not benefit from any unequal *power allocation* (PA) to the two transmit ports.

The Alamouti coding shown here is applied in space and time. A space–frequency coding is possible as well. In this case, the two transmit symbols are not assigned to two time instances but are assigned to two adjacent subcarriers at the same time instant. Correct decoding requires that the channel matrices of the two adjacent subcarriers are identical. This assumption is usually fulfilled for a sufficient large number of subcarriers. However, only small differences in the channel matrices of adjacent subcarriers may lead to distortions. This can cause errors in detection, especially for high QAM constellations, such as 1024-QAM or 4096-QAM.

8.4.2 General Space–Time–Frequency Codes

As explained in the previous subsection, the Alamouti scheme is an orthogonal space–time or space–frequency block code for two transmit ports. There are also orthogonal space–time or space–frequency block codes for more than two transmit ports. In order to keep the orthogonal property for simple decoding, the spatial code rate r_S is decreased compared to two transmit ports with $r_S < 1$ [25]. Non-orthogonal space–time–frequency block codes achieve higher spatial code rates while still achieving full spatial diversity. Reference [26] proposes a full diversity space–time code for two transmit ports, which achieves a spatial code rate of $r_S = 2$. This code was further developed in [27] for an arbitrary number of transmit ports. Reference [20] and references therein discuss space–time–frequency block codes for MIMO PLC. For PLC, space–frequency codes are preferred to space–time codes since the quasi-static PLC channel offers no temporal diversity. Note that the Alamouti scheme requires that the channel does not change over two consecutive OFDM symbols. Therefore, no temporal diversity is exploited by the Alamouti scheme; the encoding of the Alamouti scheme is designed to achieve the full spatial diversity. High-rate space–frequency codes require computationally complex *maximum likelihood* (ML) decoding. In ML decoding, an estimate of the transmitted code word is found by comparing the received code word to all possible code words and finding the most likely one. The complexity increases exponentially with the number of transmit ports and the block length (number of bundled subcarriers) to the basis of the modulation order. The computational complexity may be too high for implementation, especially for high QAM constellations, such as 1024-QAM. Sphere decoding [28] alleviates the burden of ML decoding. However, the application of adaptive modulation is difficult because calculation of the carrier specific SNR after decoding is not possible. Adaptive modulation (see later) already adapts to the frequency fading which makes space–frequency codes less effective for PLC (see also Section 9.3.2).

8.5 (Precoded) Spatial Multiplexing

SMX aims to increase the bitrate by transmitting different streams via the different transmit ports. The number of MIMO streams or spatial code rate r_S is limited by the minimum of transmit and receive ports*:

$$r_S \leq \min(N_T, N_R). \tag{8.15}$$

The spatial diversity depends on the detection algorithm and the application of precoding. Figure 8.9 illustrates the basic block diagram of the OFDM-based SMX system. Two complex QAM symbols are assigned to a vector $\mathbf{b}_k(n) = \begin{bmatrix} b_{1,k}(n) \\ b_{2,k}(n) \end{bmatrix}$ for each subcarrier n. $\mathbf{F}(n)$ is an optional precoding matrix. If no precoding is considered, $\mathbf{F}(n) = \mathbf{I}_{N_T}$ and therefore $\mathbf{s}_k(n) = \mathbf{b}_k(n)$. The received symbol vector $\mathbf{r}_k(n)$ is the superposition of both transmitted symbols weighted by the channel matrix $\mathbf{H}(n)$ according to Equation 8.4. MIMO detection aims to recover the two transmitted streams. In this figure, linear detection is considered, as described by the detection matrix $\mathbf{W}(n)$.

Section 8.5.1 describes different detection algorithms. In a next step, precoding is introduced. The optimum precoding matrix can be separated into unitary precoding (see Section 8.5.2) and PA (see Section 8.5.3). Finally, Section 8.5.4 explains how to combine precoding and an optional noise whitening filter together with the channel matrix to obtain an equivalent channel. This equivalent channel can be used to apply the detection algorithms introduced in Section 8.5.1. For convenience and better legibility, the subcarrier index n and time index k are dropped in the following. However, it has to be kept in mind that the following vector and matrix operations are applied for each subcarrier separately if not stated otherwise.

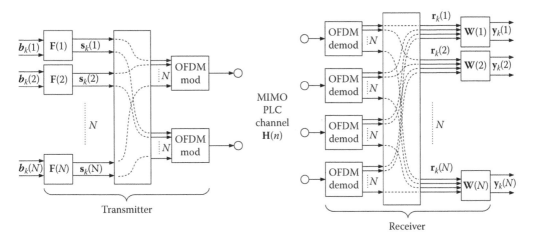

FIGURE 8.9
Block diagram of the MIMO-OFDM spatial multiplexing system with precoding, $N_T = 2$ and $N_R = 4$.

---

* This is only true if the channel matrix has full rank; otherwise, the number of spatial streams is limited by the rank R' of the channel matrix $r_S \leq R'$.

8.5.1 Detection

The detection or equaliser algorithms can be separated into linear and non-linear equalisers. In the following, different detection algorithms are briefly introduced. Details may be found, for example, in [3,29].

8.5.1.1 Linear Equalisers: Zero-Forcing and Minimum Mean-Squared Error

Referring to Figure 8.9, the detection matrix $\mathbf{W}$ is applied on the received symbol vector $\mathbf{r}$ to obtain estimates of the transmit symbol vector $\mathbf{s}$. With $\mathbf{r} = \mathbf{Hs} + \mathbf{n}$ according to Equation 8.4, we obtain

$$\mathbf{y} = \mathbf{Wr} = \mathbf{WHs} + \mathbf{Wn}. \tag{8.16}$$

Assume that the total transmit power P_T is equally distributed among the N_T transmit ports and that the noise power N_0 is the same on all N_R receive ports. According to Equation 8.16, the pth symbol y_p (pth row of $\mathbf{y}$, $p = 1, \ldots, r_S$) contains the pth transmit symbol s_p of $\mathbf{s}$. s_p is weighted by the diagonal entry $[\mathbf{WH}]_{pp}$ in Equation 8.16; the signal energy is therefore $\left|[\mathbf{WH}]_{pp}\right|^2 \frac{P_T}{N_T}$. y_p is also disturbed by the interchannel interference from the other transmit symbols s_i ($i = 1, \ldots, r_S, i \neq p$) which is described by the off-diagonal elements of $\mathbf{WH}$ in Equation 8.16. The energy of the interference is then $\frac{P_T}{N_T} \sum_{i=1, i \neq p}^{r_S} \left|[\mathbf{WH}]_{pi}\right|^2$. Additionally, y_p is corrupted by noise. The noise variance is given by $[\mathbf{WW}^H]_{pp} N_0$. The *signal to interference plus noise ratio* (SINR) after linear detection of the streams $p = 1, \ldots, r_S$ is then

$$\Lambda_p = \frac{\left|[\mathbf{WH}]_{pp}\right|^2 \dfrac{P_T}{N_T}}{\sum_{i=1, i \neq p}^{r_S} \left|[\mathbf{WH}]_{pp}\right|^2 \dfrac{P_T}{N_T} + \left[\mathbf{WW}^H\right]_{pp} N_0} = \frac{\left|[\mathbf{WH}]_{pp}\right|^2}{\sum_{i=1, i \neq p}^{r_S} \left|[\mathbf{WH}]_{pi}\right|^2 + \left[\mathbf{WW}^H\right]_{pp} \dfrac{N_T}{\rho}}, \tag{8.17}$$

with $\rho = P_T / N_0$. Note that the interference term $\sum_{i=1, i \neq p}^{r_S} \left|[\mathbf{WH}]_{pi}\right|^2$ in Equation 8.17 is equal to 0 if $r_S = 1$.

Basically, there are two types of linear receivers: *zero-forcing* (ZF) and *minimum mean-squared error* (MMSE) detection. ZF detection aims to remove the interchannel interference and does not take the noise into account. MMSE detection minimises both the noise and interchannel interference. The diversity order of the linear SMX receivers is $N_R - N_T + 1$ ($N_R \geq N_T$) [29].

8.5.1.1.1 Zero-Forcing

The detection matrix $\mathbf{W}$ of ZF is the pseudoinverse $[\cdot]^\dagger$ of the channel matrix $\mathbf{H}$:

$$\mathbf{W} = \mathbf{W}_{ZF} = \mathbf{H}^\dagger = (\mathbf{H}^H \mathbf{H})^{-1} \mathbf{H}^H. \tag{8.18}$$

Applying $\mathbf{W} = \mathbf{W}_{ZF}$ according to Equation 8.18 to the received vector $\mathbf{r}$ according to Equation 8.4 results in

$$\mathbf{y} = \mathbf{Wr} = \mathbf{W}(\mathbf{Hs} + \mathbf{n}) = \mathbf{H}^\dagger \mathbf{Hs} + \mathbf{H}^\dagger \mathbf{n},$$

$$= (\mathbf{H}^H \mathbf{H})^{-1} \mathbf{H}^H \mathbf{Hs} + \mathbf{H}^\dagger \mathbf{n} = \mathbf{s} + \mathbf{H}^\dagger \mathbf{n}. \tag{8.19}$$

Equation 8.19 shows again the design criterion. If the noise is zero, the transmit symbol vector is recovered and the interchannel interference is removed completely. However, if there is noise, the variance of $\mathbf{H}^\dagger\mathbf{n}$ after detection might be increased compared to the variance of $\mathbf{n}$.

Replacing Equation 8.18 in Equation 8.17 results in the SINR after detection

$$\Lambda_p = \frac{1}{\left[\mathbf{WW}^H\right]_{pp}}\frac{\rho}{N_T} = \frac{1}{\|\mathbf{w}_p\|^2}\frac{\rho}{N_T} = \frac{1}{\left[(\mathbf{H}^H\mathbf{H})^{-1}\right]_{pp}}\frac{\rho}{N_T} \tag{8.20}$$

with $\mathbf{w}_p$ the pth row of the detection matrix $\mathbf{W}$.

8.5.1.1.2 Minimum Mean-Squared Error

The detection matrix $\mathbf{W}$ of MMSE is defined as

$$\mathbf{W} = \mathbf{W}_{\mathrm{MMSE}} = \left(\mathbf{H}^H\mathbf{H} + \frac{N_T}{\rho}\mathbf{I}_{N_T}\right)^{-1}\mathbf{H}^H. \tag{8.21}$$

Again, applying the detection matrix $\mathbf{W} = \mathbf{W}_{\mathrm{MMSE}}$ from Equation 8.21 to the received vector $\mathbf{r}$ according to Equation 8.4 results in

$$\mathbf{y} = \mathbf{Wr} = \mathbf{W}(\mathbf{Hs}+\mathbf{n}) = \underbrace{\left(\mathbf{H}^H\mathbf{H} + \frac{N_T}{\rho}\mathbf{I}_{N_T}\right)^{-1}\mathbf{H}^H\mathbf{H}}_{\mathbf{J}}\mathbf{s} + \mathbf{Wn}. \tag{8.22}$$

$\mathbf{J}$ in Equation 8.22 describes the remaining interchannel interference. There is a trade-off between removing interchannel interference and minimising noise enhancement. For $\rho \to \infty$, that is, for diminishing noise, Equation 8.22 falls back to ZF detection with $\mathbf{J} = \mathbf{I}$. The SINR after detection can be calculated by replacing Equation 8.21 in Equation 8.17.

8.5.1.2 Non-Linear Equalisers: (Ordered) Successive Interference Cancellation and Maximum Likelihood

8.5.1.2.1 Successive Interference Cancellation

The linear equalisers described in the previous section detect the symbols of the different streams in parallel. The application of the detection matrix $\mathbf{W}$ in Equation 8.16 results in estimates of all transmitted symbols. The *successive interference cancellation* (SIC) algorithm detects the symbols sequentially. The basic idea of this algorithm is similar to the Gaussian elimination and is explained for the example of $N_T = 2$ transmit ports and $N_R = 4$ receive ports. Equation 8.4 can be expanded to

$$\underbrace{\begin{bmatrix} h_{11} & h_{12} \\ h_{21} & h_{22} \\ h_{31} & h_{32} \\ h_{41} & h_{42} \end{bmatrix}}_{\mathbf{H}=[\mathbf{h}_1 \quad \mathbf{h}_2]} \cdot \underbrace{\begin{bmatrix} s_1 \\ s_2 \end{bmatrix}}_{\mathbf{s}} + \underbrace{\begin{bmatrix} n_1 \\ n_2 \\ n_3 \\ n_4 \end{bmatrix}}_{\mathbf{n}} = \underbrace{\begin{bmatrix} r_1 \\ r_2 \\ r_3 \\ r_4 \end{bmatrix}}_{\mathbf{r}}. \tag{8.23}$$

The superscript index (p) with $p = 1, 2$ indicates the iteration to detect the two symbols s_1 and s_2. The first iteration $p = 1$ starts with $\mathbf{H}^{(1)} = \mathbf{H}$ and $\mathbf{r}^{(1)} = \mathbf{r}$. The detection matrix of the first iteration is

$$\mathbf{W}^{(1)} = \mathbf{W} \tag{8.24}$$

with either $\mathbf{W} = \mathbf{W}_{\mathrm{ZF}}$ or $\mathbf{W} = \mathbf{W}_{\mathrm{MMSE}}$ according to Equation 8.18 or Equation 8.21, respectively. Assume that the first symbol is detected first

$$y_1 = \mathbf{w}_1^{(1)}\mathbf{r}^{(1)} \tag{8.25}$$

with $\mathbf{w}_1^{(1)}$ the first row of $\mathbf{W}^{(1)} = \mathbf{W}$. The decision leads to an estimate of the first transmitted symbol:

$$\check{s}_1 = Q(y_1), \tag{8.26}$$

where $Q(\cdot)$ is the decision operator. The step of the decision in Equation 8.26 makes this method a non-linear detection algorithm although all the other steps comprise only linear operations.

Assume that the decision is correct, that is, $\check{s}_1 = s_1$. The influence of the first symbol is subtracted from the received vector $\mathbf{r}$ according to Equation 8.23 as

$$\underbrace{\mathbf{h}_2}_{\mathbf{H}^{(2)}} s_2 + \mathbf{n} = \underbrace{\mathbf{r}^{(1)} - \mathbf{h}_1 \check{s}_1}_{\mathbf{r}^{(2)}}, \tag{8.27}$$

$$\mathbf{H}^{(2)}s_2 + \mathbf{n} = \mathbf{r}^{(2)}.$$

Equation 8.27 is similar to Equation 8.23 and the matrix operation simplifies to a vector operation. The new detection matrix $\mathbf{W}^{(2)} = \mathbf{w}_1^{(2)}$ is derived from the new channel matrix $\mathbf{H}^{(2)}$ and the second symbol is detected

$$y_2 = \mathbf{w}_1^{(2)}\mathbf{r}^{(2)} \tag{8.28}$$

and decided

$$\check{s}_2 = Q(y_2). \tag{8.29}$$

The post-detection SINR of the two MIMO streams based on Equation 8.17 is then

$$\Lambda_1 = \frac{|[\mathbf{W}^{(1)}\mathbf{H}]_{11}|^2}{|[\mathbf{W}^{(1)}\mathbf{H}]_{12}|^2 + [\mathbf{W}^{(1)}(\mathbf{W}^{(1)})^H]_{11}\dfrac{2}{\rho}} \tag{8.30}$$

and

$$\Lambda_2 = \frac{|\mathbf{W}^{(2)}\mathbf{H}^{(2)}|^2}{\mathbf{W}^{(2)}(\mathbf{W}^{(2)})^H\dfrac{2}{\rho}}. \tag{8.31}$$

The SINR of the first stream has not changed compared to the parallel linear detection, and the SINR of the second stream is improved. Equation 8.31 assumes that the first symbol is decoded correctly and that the channel estimate is noise-free. This simplification is not true in reality, where error propagation has to be considered as well. Reference [30] describes how to incorporate the error propagation into the post-detection SINR.

8.5.1.2.2 Ordered SIC

The first symbol was detected first in the previous section. *Ordered successive interference cancellation* (OSIC) takes the optimum order of the detection process into account. To find the optimum detection order, the post-detection SINR of each symbol is evaluated and the symbol with the highest SINR is decoded first. The SINR of the first symbol is given by Equation 8.30, while the SINR of the second symbol is given by

$$\Lambda_2 = \frac{|[\mathbf{W}^{(1)}\mathbf{H}]_{22}|^2}{\left|[\mathbf{W}^{(1)}\mathbf{H}]_{21}\right|^2 + \left[\mathbf{W}^{(1)}(\mathbf{W}^{(1)})^H\right]_{22}\frac{2}{\rho}}. \tag{8.32}$$

If the SINR of the first symbol according to Equation 8.30 is higher than the SINR of the second symbol according to Equation 8.32, then the first symbol is decoded first. This results in the SINR of the two symbols according to Equations 8.30 and 8.31. Otherwise, the second symbol is decoded first, resulting in the SINR of the two symbols according to Equation 8.32 and

$$\Lambda_1 = \frac{|\mathbf{W}^{(2)}\mathbf{H}^{(2)}|^2}{\mathbf{W}^{(2)}(\mathbf{W}^{(2)})^H\frac{2}{\rho}} \tag{8.33}$$

with $\mathbf{H}^{(2)} = \mathbf{h}_1$ and $\mathbf{W}^{(2)}$ derived based on the new channel $\mathbf{H}^{(2)}$ depending on the detection scheme (ZF or MMSE).

8.5.1.2.3 Maximum Likelihood

ML decoding is the optimum decoder. ML decoding compares the received vector $\mathbf{y}$ with all possible transmit vectors $\mathbf{s}$ and finds the most likely one. The statistical estimation problem leads to the geometrical task of solving [3]:

$$\mathbf{y} = \arg\min_{\mathbf{s}} \| \mathbf{r} - \mathbf{H}\mathbf{s} \|^2. \tag{8.34}$$

The ML receiver searches through all possible transmit symbol vectors. The complexity grows exponentially with the number of transmit ports N_T. With 1024-QAM and two transmit ports, a brute force implementation requires the search through 1024^2 vector symbols. As already explained in Section 8.4.2, it is difficult to estimate the SINR after ML detection. The SINR after detection is needed for adaptive modulation.

8.5.2 Eigenbeamforming

Incorporating the precoding matrix $\mathbf{F}$ in Equation 8.16 by replacing the transmit symbol vector $\mathbf{s} = \mathbf{F}\mathbf{b}$ results in

$$\mathbf{y} = \mathbf{W}\mathbf{r} = \mathbf{W}\mathbf{H}\mathbf{F}\mathbf{b} + \mathbf{W}\mathbf{n}. \tag{8.35}$$

The optimum linear precoding matrix $\mathbf{F}$ for precoded SMX systems can be factored into two matrices $\mathbf{V}$ and $\mathbf{P}$ [31]*:

$$\mathbf{F} = \mathbf{VP}. \tag{8.37}$$

$\mathbf{P}$ is a diagonal matrix, which describes the PA of the total transmit power to each of the transmit streams. PA is considered in Section 8.5.3 in more detail. $\mathbf{V}$ is the right-hand unitary matrix of the SVD of the channel matrix $\mathbf{H} = \mathbf{UDV}^H$ (see Section 8.2.2), $\mathbf{U}$ is the left hand unitary matrix and $\mathbf{D}$ is a diagonal matrix containing the singular values of the channel matrix $\mathbf{H}$. Since $\mathbf{V}$ is a unitary matrix, the average signal power is not affected by the precoding matrix.

Precoding by just the unitary matrix $\mathbf{F} = \mathbf{V}$ is often referred to as unitary precoding or *eigenbeamforming* (EBF). Substituting the detection matrix $\check{\mathbf{W}} = \mathbf{U}^H$ in Equation 8.35 results in

$$\mathbf{y}' = \check{\mathbf{W}}\,\mathbf{HVb} + \check{\mathbf{W}}\mathbf{n} = \mathbf{U}^H \underbrace{\mathbf{H}}_{\mathbf{UDV}^H} \mathbf{Vb} + \check{\mathbf{W}}\mathbf{n} = \mathbf{Db} + \mathbf{U}^H\mathbf{n}. \tag{8.38}$$

The combination of precoding, channel and detection matrix according to Equation 8.38 decomposes the equivalent channel into parallel streams described by the diagonal matrix $\mathbf{D}$ (see Figure 8.10).

The normalisation by $\mathbf{D}^{-1}$ leads to estimates of the transmit symbol vector $\mathbf{b}$:

$$\mathbf{y} = \mathbf{D}^{-1}\mathbf{y}' = \mathbf{b} + \mathbf{D}^{-1}\mathbf{U}^H\mathbf{n}. \tag{8.39}$$

Consequently, the final detection matrix is obtained by $\mathbf{W} = \mathbf{D}^{-1}\mathbf{U}^H$.

If only one spatial stream carries information, that is, $\mathbf{b} = \begin{bmatrix} b_1 \\ 0 \end{bmatrix}$, the precoding simplifies to

$$\mathbf{s} = \mathbf{Vb} = \begin{bmatrix} \mathbf{v}_1 & \mathbf{v}_2 \end{bmatrix} \begin{bmatrix} b_1 \\ 0 \end{bmatrix} = \mathbf{v}_1 b_1 \tag{8.40}$$

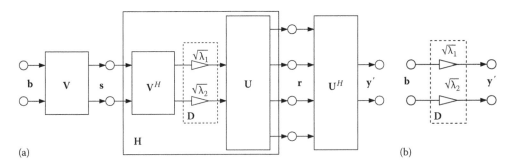

(a)   (b)

FIGURE 8.10
EBF: The unitary precoding and detection (a) decomposes the equivalent channel into parallel streams (b).

---

* The optimum precoder is derived in [31] by minimising the MSE matrix:

$$E\{(\mathbf{y}-\mathbf{b})(\mathbf{y}-\mathbf{b})^H\} = \mathrm{MSE}(\mathbf{F},\mathbf{W}) = (\mathbf{WHF}-\mathbf{I})\mathbf{R}_{bb}(\mathbf{WHF}-\mathbf{I})^H + \mathbf{WR}_{nn}\mathbf{W}^H \tag{8.36}$$

with $\mathbf{y}$, $\mathbf{b}$, $\mathbf{F}$, $\mathbf{W}$ and $\mathbf{H}$ from Equation 8.35 and the covariance matrices $\mathbf{R}_{bb} = E\{\mathbf{bb}^H\}$ and $\mathbf{R}_{nn} = E\{\mathbf{nn}^H\}$.

with $\mathbf{v}_1$ the first column of the precoding matrix $\mathbf{V}$. This one-stream *beamforming* (BF) is also called spot-BF. Even though only one logical MIMO stream carries information, both transmit ports are used since $\mathbf{s}$ is a 2×1 vector for two transmit ports. One-stream BF might be used if only one receive port is available (MISO) or if the receiver supports only the decoding of one stream (full MIMO with using only one stream). In this case, detection is simplified to a vector multiplication

$$\mathbf{W} = \mathbf{w} = \frac{1}{\sqrt{\lambda_1}} \mathbf{u}_1^H \qquad (8.41)$$

with $\mathbf{u}_1$ as the first column of $\mathbf{U}$. In the case of a low SINR channel, spot-BF improves the coverage.

Let us go back to the general case $b_1 \neq 0$ and $b_2 \neq 0$. The precoding matrix $\mathbf{V}$ superimposes the symbols of $\mathbf{b}$, and thus the transmit symbol vector $\mathbf{s}$ contains all symbols of $\mathbf{b}$, that is, each symbol is transmitted via each MIMO path. Thus, the full spatial diversity $N_T N_R$ is achieved.

$\mathbf{U}^H$ is a unitary matrix, that is, the noise variance of $\mathbf{n}$ in Equation 8.38 is not affected by the multiplication of $\mathbf{U}^H$ if the noise power is the same for all receive ports. $\mathbf{D}$ contains the singular values $\sqrt{\lambda}_p$ on the diagonal, and, from Equation 8.39, it follows that the post-detection SINR of the two streams is

$$\Lambda_p = \frac{\rho}{N_T} \lambda_p, \quad p = 1, 2. \qquad (8.42)$$

If each MIMO stream is modulated independently, the signals of the r_S MIMO streams are uncorrelated

$$E\{\mathbf{b}\mathbf{b}^H\} = \mathbf{I}_{r_S}. \qquad (8.43)$$

Thus, in the case that no precoding is applied, $\mathbf{s} = \mathbf{b}$, the transmitted signals are also uncorrelated $E\{\mathbf{s}\mathbf{s}^H\} = E\{\mathbf{b}\mathbf{b}^H\} = \mathbf{I}_{N_T}$ ($r_S = N_T$). If precoding is applied and the entries of $\mathbf{b}$ are uncorrelated, the correlation matrix results in

$$E\{\mathbf{s}\mathbf{s}^H\} = E\{(\mathbf{F}\mathbf{b})(\mathbf{F}\mathbf{b})^H\} = \mathbf{F}E\{\mathbf{b}\mathbf{b}^H\}\mathbf{F}^H = \mathbf{F}\mathbf{F}^H. \qquad (8.44)$$

In case of EBF with $\mathbf{F} = \mathbf{V}$, the unitary property of $\mathbf{V}$ simplifies Equation 8.44 to $\mathbf{V}\mathbf{V}^H = \mathbf{I}_{N_T}$, and the transmit signals are still uncorrelated. If one-stream BF or PA (see next subsection) is used, the simplification is not valid. This results in correlation of the transmit signals. The effect of MIMO transmission on EMI is discussed in Chapters 7 and 16 in detail.

8.5.3 Power Allocation for Spatial Multiplexing Streams

There are basically two options to allocate the total transmit power for MIMO-OFDM systems. The power can be allocated across subcarriers and the power can be distributed among the available MIMO streams. In case of PLC, the first option, the PA across subcarriers, is only possible within the resolution bandwidth used for regulatory assessment (9 kHz for frequencies below 30 MHz and 120 kHz above 30 MHz [32]). EMC regulation accepts a flat power spectral density as maximal feeding level across the subcarriers. Shifting energy between carriers is only possible if subcarrier spacing is smaller than the resolution *bandwidth* (BW) used for regulatory assessment.

In the following, the second option, the distribution of the total transmit power among the available MIMO streams, is discussed. *Water filling* (WF) is the optimum MIMO PA to maximise the system capacity [3]. In case that Gaussian distributed input signals are present, the system capacity is maximal. However, if the input signals are taken from a finite set of symbols like QAM, WF is not the optimum solution. Reference [33] derives the optimum PA for arbitrary input distributions and for parallel channels corrupted by *additive white Gaussian noise* (AWGN). This algorithm is called *mercury water filling* (MWF). A detailed description of this algorithm goes beyond the scope of this chapter; for details, refer to [33,34]. Only a few key observations are discussed here. The PA coefficients of the MWF algorithm depend on the SNR and modulation scheme of each MIMO stream. The PA coefficients a_1 and a_2 of the two MIMO streams have to meet the constraint:

$$a_1 + a_2 = 2. \tag{8.45}$$

Note that the total transmit power P_T is assumed to be constant and that the transmit symbols have average power $P_T/2$ (see Section 8.3). Equation 8.45 ensures that the total power $(a_1 + a_2)P_T/2 = P_T$ is not exceeded.

The PA matrix is constructed by

$$\mathbf{P} = \begin{bmatrix} \sqrt{a_1} & 0 \\ 0 & \sqrt{a_2} \end{bmatrix}. \tag{8.46}$$

Figure 8.11 shows the MWF PA coefficient a_1 (illustrated by the contour lines) depending on the SNR of each MIMO stream and the modulation scheme. Figure 8.11a illustrates *binary phase-shift keying* (BPSK) and *quadrature phase-shift keying* (QPSK) modulation on the

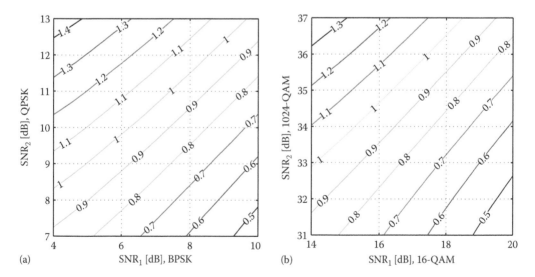

FIGURE 8.11
PA coefficients of *mercury water filling* (MWF); the contour lines represent the PA coefficient a_1 of the first stream with (a) BPSK and QPSK modulation and (b) 16-QAM and 1024-QAM modulation on the two streams.

two MIMO streams while Figure 8.11b shows 16-QAM and 1024-QAM on streams 1 and 2, respectively. According to the WF algorithm, most power is allocated to the MIMO stream with the highest SNR [3]. However, under some circumstances, the MWF algorithm mandates the opposite, as illustrated in Figure 8.11b, for example. Assume that the first stream has an SNR of 15 dB and the second stream an SNR equal to 36 dB. According to Figure 8.11b, more power is assigned to the weaker stream 1 (a_1 = 1.2) than to the stronger stream 2 (a_2 = 0.8).

A simplified PA algorithm may be used in conjunction with adaptive modulation [18,20]. Only three PA coefficients are used: 0, 1 and $\sqrt{2}$. If one stream of one subcarrier is not allocated any information because of insufficient SNR for the lowest possible constellation, the power is added to the other stream. This results in a 3 dB higher SNR of the boosted stream. If both streams of one subcarrier carry information, the power is equally allocated to the two streams ($a_1 = a_2$ = 1). The transmitter concludes from the adaptive modulation feedback from the receiver the PA coefficients. Thus, there is no additional feedback of the PA coefficients required. The performance of this simplified PA algorithm is close to optimum MWF (see Section 9.3).

8.5.4 Equivalent Channel and Detection

8.5.4.1 Equivalent Channel Including Precoding

An equivalent channel matrix can be defined, which combines a general precoding matrix $\mathbf{F}$ and the channel matrix $\mathbf{H}$ as

$$\tilde{\mathbf{H}}_{\text{pre}} = \mathbf{HF}. \tag{8.47}$$

The detection algorithms introduced in Section 8.5.1 can be applied to the equivalent channel matrix by substituting $\mathbf{H}$ by $\tilde{\mathbf{H}}_{\text{pre}}$.

For the optimum unitary precoding matrix $\mathbf{F} = \mathbf{V}$, the different detection algorithms according to Section 8.5.1 yield all the same SINR after detection (see [20]). Thus, if EBF is used, the most simple equaliser (the ZF equaliser) is sufficient. In fact, the ZF equaliser results in $\mathbf{W} = \mathbf{D}^{-1}\mathbf{U}^H$, which is the same detection matrix as obtained in Section 8.5.2 (see [20]). In case of quantisation of the optimum precoding matrix $\mathbf{F} = \tilde{\mathbf{V}}$ with $\tilde{\mathbf{V}}$ the quantised precoding matrix $\mathbf{V}$, the precoding is no more optimal and more sophisticated equalisers might be used. However, the performance gain of the more sophisticated equalisers compared to ZF detection is very small for a sufficient fine quantisation.

8.5.4.2 Whitening Correlated Noise

Correlation of the noise results in a noise covariance matrix $\mathbf{N}_c = E\{\mathbf{nn}^H\}$ unequal to the scaled identity matrix. The noise can be whitened by multiplying the received vector $\mathbf{r}$ by $\mathbf{N}_c^{-\frac{1}{2}}$

$$\tilde{\mathbf{r}} = \mathbf{N}_c^{-\frac{1}{2}}\mathbf{Hs} + \mathbf{N}_c^{-\frac{1}{2}}\mathbf{n}, \tag{8.48}$$

where $\mathbf{N}_c^{-\frac{1}{2}}$ is defined such that

$$\mathbf{N}_c^{-\frac{1}{2}}\mathbf{N}_c^{-\frac{1}{2}} = \mathbf{N}_c^{-1} \tag{8.49}$$

is fulfilled. $\mathbf{N}_c$ is a Hermitian matrix, that is, $\mathbf{N}_c = \mathbf{N}_c^H$ since $E\{\mathbf{nn}^H\}^H = E\{\mathbf{nn}^H\}$. A Hermitian matrix $\mathbf{N}_c$ can be diagonalised by a unitary matrix $\mathbf{U}_c$ [21]:

$$\mathbf{N}_c = \mathbf{U}_c \mathbf{D}_c \mathbf{U}_c^H, \tag{8.50}$$

where $\mathbf{D}_c$ is a diagonal matrix with real elements. Using $\mathbf{N}_c^{-1} = \mathbf{U}_c \mathbf{D}_c^{-1} \mathbf{U}_c^H$, $\mathbf{N}_c^{-\frac{1}{2}}$ is defined as

$$\mathbf{N}_c^{-\frac{1}{2}} = \mathbf{U}_c \mathbf{D}_c^{-\frac{1}{2}} \mathbf{U}_c^H \tag{8.51}$$

with $\mathbf{D}_c^{-\frac{1}{2}}$ containing the inverse elements of the square root of the elements of $\mathbf{D}_c$. Equation 8.51 fulfils Equation 8.49 since $\mathbf{N}_c^{-\frac{1}{2}} \mathbf{N}_c^{-\frac{1}{2}} = \mathbf{U}_c \mathbf{D}_c^{-\frac{1}{2}} \mathbf{U}_c^H \mathbf{U}_c \mathbf{D}_c^{-\frac{1}{2}} \mathbf{U}_c^H = \mathbf{U}_c \mathbf{D}_c^{-1} \mathbf{U}_c^H = \mathbf{N}_c^{-1}$.

The filtered noise $\tilde{\mathbf{n}} = \mathbf{N}_c^{-\frac{1}{2}} \mathbf{n}$ is uncorrelated with

$$E\{\tilde{\mathbf{n}}\tilde{\mathbf{n}}^H\} = \mathbf{N}_c^{-\frac{1}{2}} \underbrace{E\{\mathbf{nn}^H\}}_{\mathbf{N}_c} \left(\mathbf{N}_c^{-\frac{1}{2}} \right)^H \overset{(0.51)}{=} \mathbf{D}_c^{-1} \mathbf{D}_c = \mathbf{I}_{N_R}. \tag{8.52}$$

The equivalent channel matrix

$$\tilde{\mathbf{H}}_{\text{white}} = \mathbf{N}_c^{-\frac{1}{2}} \mathbf{H} \tag{8.53}$$

can be used to determine the detection matrix and optimum precoding matrix by replacing $\mathbf{H}$ by $\tilde{\mathbf{H}}_{\text{white}}$.

8.6 Simulation Results

The MIMO-OFDM system described in Section 8.3 forms the basis of the system simulations. 1296 subcarriers are used in the frequency range from 4 to 30 MHz*.

The noise is modelled by AWGN with zero mean, and it is assumed that the noise is uncorrelated and that the noise power is the same for all receive ports. The transmit power to noise power level is assumed to be $\rho = 65$ dB. This value corresponds to transmit power spectral density of −55 dBm/Hz (see Chapter 6) and an average noise power spectral density of −120 dBm/Hz (this corresponds to the 90% point of CDF of the noise according to Chapter 5). Impulsive noise is not considered. The focus is on the comparison between MIMO and SISO schemes. It is expected that impulsive noise will influence all receive ports in a similar way. Thus, mitigation techniques known from SISO systems can be applied [23,35]. The measured MIMO PLC channels obtained during the European

---

* The system uses a 2048 OFDM from 0 to 40 MHz with 1296 active subcarriers in the frequency range from 4 to 30 MHz. The other subcarriers are not active.

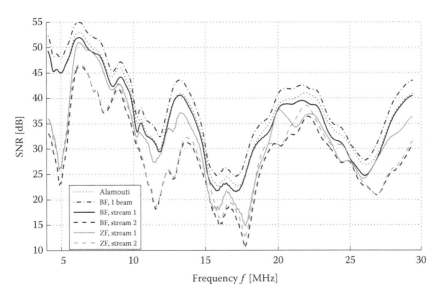

FIGURE 8.12
Frequency-dependent SNR for multiplexing streams 1 and 2 of different MIMO schemes, $\rho = 65\,\text{dB}$, $N_T = 2$, $N_R = 4$.

measurement campaign (ETSI STF410, see Chapter 5) are used in the system simulations. In case of MIMO, the two feeding ports D_1 and D_3 (i.e. L-N and L-PE; see Chapter 5) and all four receive ports (S_1, S_2, S_3, S_4) are used; in case of SISO, the D_1 (L-N) port is used at the transmitter and the S_1 port (L) at the receiver. It was observed that using the S_2 port (N) at the receiver yields the same performance as using the S_1 port. The corresponding SNR is calculated based on the channel matrix of each subcarrier (channel estimation is assumed to be perfect), depending on the MIMO scheme as shown in the previous sections.

Figure 8.12 compares the SNR after detection for different MIMO schemes: the Alamouti scheme, SMX with ZF detection and BF with one and two streams. The precoding matrix in case of BF is assumed to be perfect. A typical MIMO channel is shown in the figure. No PA is applied for the schemes with two streams.

The SNR after detection is still very frequency selective for all the MIMO schemes shown. BF with one stream (spot-BF) and the Alamouti scheme achieve the highest SNR. However, these two schemes support only one MIMO stream. This has to be kept in mind with respect to the achieved bitrate (see Section 9.3). The SMX schemes provide two MIMO streams. For SMX without precoding (ZF in the figure), there is no preferred spatial stream; the two streams range in the same order of SNR. Consider the frequencies between 20 and 25 MHz: For some frequencies, the first stream offers higher SNR than the second one and vice versa. Precoding (BF) enforces one of the streams and weakens the other stream. However, the gain of stream 1 is much higher compared to the loss of the second stream. This behaviour is especially visible for the low frequencies of this channel. This results in higher bitrates for precoding (see Chapter 9). The SNR curve of one-stream BF is shifted by 3 dB compared to the first stream of EBF. The second stream is not used, and therefore the power is assigned completely to the first stream.

Figure 8.13 presents the SNR after detection of different SMX schemes. The two MIMO streams are shown in Figure 8.13a and b for the first and second streams, respectively. To show the advantage of MMSE, the same channel is used as in Figure 8.12 with a lower transmit to noise power level of $\rho = 44\,\text{dB}$ to make the effects of very low SNR more visible. The gain of BF is again visible for the first stream, as already discussed earlier.

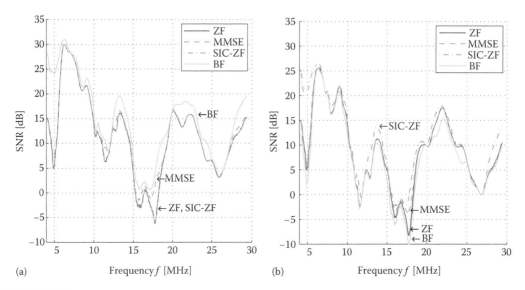

FIGURE 8.13
Frequency-dependent SNR of stream 1 (a) and stream 2 (b) of different spatial multiplexing MIMO schemes, same channel as in Figure 8.12, $\rho = 44$ dB, $N_T = 2$, $N_R = 4$.

ZF detection and SIC yield the same SNR for the first stream since the detection of the first stream is performed in the same way (see Section 8.5.1). The gain of SIC becomes visible for the second stream. MMSE detection improves the SNR only in very poor SNR regions (see frequencies around 17 MHz). For high SNR regions, the MMSE gain diminishes compared to ZF. The gain of OSIC (not shown in the figures) is only marginal compared to SIC.

Figure 8.14 compares the SNR of PA for BF and SIC. The top row of the figure illustrates the SNR; the lower row shows the PA coefficients of MWF. For frequencies with very low SNR on the second stream, no information is assigned by the adaptive modulation to the second stream. The complete power is assigned to the first stream. The PA coefficient of the first stream becomes 2 while the PA coefficient of the second stream becomes 0 (see Figure 8.14c). For these frequencies, two-stream BF falls back to one-stream BF (see Figure 8.14a). The SNR of the two streams of SIC is closer together compared to BF (see Figure 8.14b). Thus, the variation of the PA coefficients is not that pronounced and the extreme PA coefficients 0 and 2 do not appear that often compared to BF. The gain of PA for EBF is higher, compared to SMX without precoding. The same behaviour is observed for the other SMX schemes (ZF and MMSE, data not shown). The gain of PA becomes smaller for higher SNR.

Figure 8.15 shows the *cumulative distribution function* (CDF) of the SNR of all MIMO PLC channel measurements. No PA is applied. The spatial streams are drawn separately. One spatial stream is available for SISO and the Alamouti scheme, while the SMX schemes with ZF, SIC-ZF and EBF have two spatial streams. The Alamouti scheme and the first spatial stream of EBF give the highest SNR. Note that Alamouti uses only one spatial stream, while EBF offers a second spatial stream which can be used for communication. The second spatial stream has the lowest SNR in Figure 8.15. EBF maximises the first spatial stream. Although the second stream is weakened, the sum of both streams offers the highest performance, which is shown for the throughput results in Chapter 9 the SNR gain of EBF is most visible for the low SNR region which is most important to increase the coverage. The SNR values of the two streams of *SMX with ZF detection* (SMX-ZF) are in the

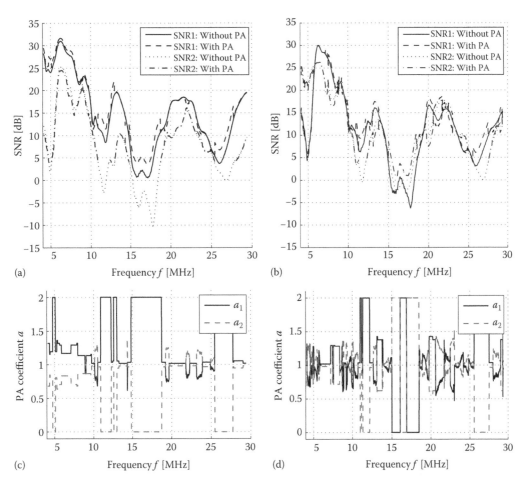

FIGURE 8.14
SNR for spatial multiplexing streams 1 and 2 for BF (a) and SIC (b), and corresponding *power allocation* (PA) coefficients for BF (c) and SIC (d) as a function of frequency, same channel as in Figure 8.12, $\rho = 44$ dB, $N_T = 2$, $N_R = 4$.

same order of magnitude. This is due to the fact that SMX has no preference with respect to the spatial streams. At the median point, the SNR is about 3 dB lower compared to SISO. The SNR of the first spatial stream of SIC-ZF is exactly the same as ZF since the first stream is detected in the same way. However, the SNR of the second spatial stream is increased by 7 dB in median compared to the second stream of ZF.

Figure 8.16 shows the CDF for the SMX schemes with different detection algorithms and EBF. Additionally to ZF, SIC-ZF and EBF as already illustrated in Figure 8.15, MMSE and SIC-MMSE are shown. As for ZF and SIC-ZF, the SNR of the first spatial stream of MMSE and SIC-MMSE are identical. MMSE offers only an increased SNR for low SNR regions (gain of 4 dB at the 10% point), while for moderate toward high SNR, ZF and MMSE converge to the same performance. Comparing the best detection scheme which is SIC-MMSE to EBF leads to the following conclusions. At the high SNR region, the gain of the first stream of EBF compared to the stronger stream of SIC-MMSE is similar to the loss of EBF's second stream compared to the weaker stream of SIC-MMSE. This indicates that the same throughput could be expected for these two schemes for low attenuated channels (see also the Section 9.3) when neglecting any error propagation in

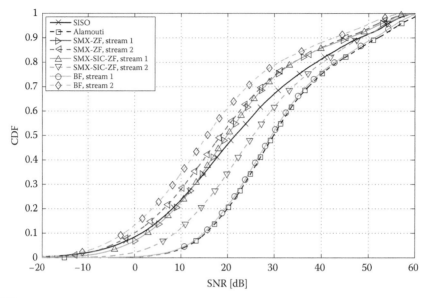

FIGURE 8.15
Cumulative distribution function (CDF) of the SNR for multiplexing streams 1 and 2 of different MIMO schemes and SISO, $\rho = 65$ dB, $N_T = 2$, $N_R = 4$.

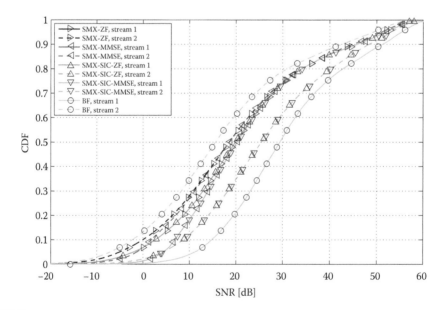

FIGURE 8.16
Cumulative distribution function (CDF) of the SNR for multiplexing streams 1 and 2 of different two-stream MIMO schemes, $\rho = 65$ dB, $N_T = 2$, $N_R = 4$.

TABLE 8.1

Qualitative Comparison of Different MIMO Schemes for PLC

MIMO Scheme	Minimum Required MIMO Configuration	High Coverage Performance (Low SNR Region)	Maximum Throughput Performance (High SNR Region)	Feedback Required	Transmitter Complexity	Receiver Complexity
SISO	1×1	o	o	No	Very low	Very low
SISO (SD)	1×1^a	+	o	No	Very low	Low
Alamouti	2×1	++	o	No	Low	Medium
SMX-ZF	2×2	−	+	No	Low	Medium
SMX-MMSE	2×2	o	+	No	Low	Medium–high
SMX-(O)SIC-ZF/ MMSE	2×2	o	++	No	Low	High
One-stream BF (spot BF)	2×1	++	+	Yes	Medium	Low–medium
Two-stream BF (EBF)	2×2	++	++	Yes	Medium	Medium

a SISO with SD requires several transmit ports or several receive ports; however, only one stream is processed.

SIC-MMSE due to wrongly decided symbols in the first iteration of the SIC algorithm. However, at the low SNR region, due to the fact that EBF maximises the first stream, the first EBF stream might carry information while the SNR of the two streams of SIC-MMSE might be too low for communication (again refer to Section 9.3).

Table 8.1 shows a qualitative comparison and summary of the different MIMO schemes and illustrates the pros and cons with respect to MIMO PLC. The MIMO schemes discussed in this chapter are shown in the first column. The second column shows the minimum required configuration of transmit and receive ports. SISO is considered as a special MIMO configuration with one transmit and one receive port. SISO with *selection diversity* (SD) uses several transmit or receive ports at the coupler. However, only one of the transmit or receive streams is processed which makes the signal processing a SISO scheme. The Alamouti scheme requires two transmit ports and at least one receive port, that is, a MISO configuration is sufficient. The SMX schemes use two spatial modes, that is, at least two transmit and two receive ports are needed, the same holds for EBF. Spot BF utilises only one spatial stream. Thus, a MISO configuration is sufficient, at least two transmit ports are required to form the beam. The third and fourth columns show the achievable performance in the high coverage area (for channels with low SNR) and in the low coverage area where the maximum throughput could be achieved (for channels with high SNR), respectively. Compared to the SISO reference, the Alamouti scheme achieves very good performance for low SNR channels; however, no multiplexing gain is achieved for channels with high SNR due to the transmission of only one spatial stream. The SMX schemes use two spatial streams and therefore good performance is achieved for high SNR channels. The performance for low SNR channels depends on the MIMO detection algorithm or equaliser. The most simple detection algorithm is ZF which results in poor SNR after detection for highly correlated channels. The more sophisticated detection algorithms MMSE, SIC-ZF, SIC-MMSE, OSIC-ZF and OSIC-MMSE increase the performance if real-world imperfections are not considered. The BF schemes achieve the highest performance for all channel conditions since the transmission is optimally adopted to the channel by the precoding. The first spatial stream is maximised, which gives the best performance for low SNR channels. For high attenuated channel, the second spatial stream has only low SNR after detection. Thus, one-stream and two-stream BF achieve almost the same performance in the high coverage area. For low attenuated channels, one-stream BF does not achieve the multiplexing gain and two-stream BF outperforms one-stream BF. The last three columns compare the MIMO schemes in terms of complexity. The BF schemes usually require feedback from the receiver to the transmitter. This feedback is no significant drawback for PLC as will be discussed in the conclusions. With respect to the implementation complexity at the transmitter and receiver, the SISO complexity serves as reference with very low complexity. The Alamouti scheme and SMX schemes without precoding require a de-multiplexing to split the incoming bits or QAM symbols to the two transmit ports. BF applies precoding at the transmitter which is a matrix multiplication. At the receiver, the Alamouti decoding is basically a matrix multiplication making use of the orthogonal property of the Alamouti code. The detection complexity of the SMX schemes increases with the more sophisticated equaliser algorithms. For BF, the most simple detection algorithm is sufficient which is ZF detection. For one-stream BF, the complexity is further reduced since only one spatial stream has to be decoded.

8.7 Conclusions

This chapter introduced MIMO signal processing theory and its application to MIMO PLC. The comparison of several MIMO schemes aimed at answering the question which MIMO scheme is suited best for in-home PLC. Although the investigations focused on in-home PLC, many aspects and conclusions may also apply to access PLC. The SNR after MIMO processing at the receiver was investigated. The Alamouti scheme improves the SNR compared to SISO. However, no multiplexing gain is achieved because of the transmission of replicas of each symbol. SMX uses two spatial streams and different detection algorithms were investigated. ZF detection is the most simple detection scheme. However, ZF detection fails for a high correlation of the channel which results in low SNR after detection, which might be even below the SNR of SISO for high spatial correlation. More complex detection algorithms (like MMSE and SIC) improve the performance of the SMX schemes. In particular, the SIC receiver approaches the BF performance for channels with low attenuation. However, the general problem for correlated channels remains.

BF offers the highest SNR under all channel conditions by adapting the transmission to the eigenmodes of the channel. The full spatial diversity gain is achieved for highly attenuated channels while the SMX gain is achieved for channels with low attenuation. The most simple detection algorithm, which is ZF detection, can be used in case of BF since the performance of BF is independent of the detection algorithm. Spatial correlation of the transmit signals may cause higher radiation emission of the power lines. However, the unitary precoding matrix of EBF does not introduce any correlation of the transmit signals if the two streams before precoding are uncorrelated. Chapter 16 investigates the influence of spot BF on the EMI. BF offers flexibility with respect to the receiver configuration. Only one spatial stream may be activated by the transmitter if only one receive port is available, that is, if the outlet is not equipped with the third wire or if a simplified receiver implementation is used which supports only one spatial stream. Since BF aims to maximise one MIMO stream, the performance loss of not utilising the second stream is relatively small compared to the SMX schemes without precoding. This is especially true for highly attenuated and correlated channels, where the SNR of the second MIMO stream is very low and no information might be carried on this stream. These channels are most critical for PLC coverage and adequate MIMO schemes are important. BF uses CSI at the transmitter. Usually, the CSI in terms of the precoding matrix is fed back from the receiver to the transmitter. Due to this feedback, BF is also called a closed-loop MIMO scheme in contrast to the open-loop MIMO schemes which do not use CSI at the transmitter. The feedback creates an additional overhead. However, the feedback of the precoding matrix can be combined with the feedback of the tone maps for adaptive modulation (see Chapter 9). Additionally, the update rate of the feedback is low since the time variations of the PLC channel are low (for changes in the network topology) or periodic within a line cycle. The amount of feedback to quantise the precoding matrix without performance loss is investigated in [18,20].

The throughput results of the different MIMO schemes are investigated in Chapter 9. In particular, the SNR after detection is used to apply adaptive modulation and to calculate the achieved bitrate. HomePlug and ITU-T G.hn adopted MIMO and precoded SMX or BF in their next-generation specifications HomePlug AV2 (see Chapter 14) and G.hn/G.9963 (see Chapter 12), respectively.

References

1. G. J. Foschini, Layered space–time architecture for wireless communication in a fading environment when using multi-element antennas, *Bell Labs Tech. J.*, 1(2), 41–59, 1996.
2. I. E. Telatar, Capacity of multi-antenna Gaussian channels, *Eur. Trans. Telecom.*, 10, 585–595, 1999.
3. A. Paulraj, R. Nabar and D. Gore, *Introduction to Space–Time Wireless Communications*. Cambridge University Press, Cambridge, U.K., 2003.
4. L. Schumacher, L. T. Berger and J. Ramiro Moreno, Recent advances in propagation characterisation and multiple antenna processing in the 3 GPP framework, in *XXVIth URSI General Assembly*, Maastricht, the Netherlands, August 2002, session C2. [Online] Available: http://www.ursi.org/Proceedings/ProcGA02/papers/p0563.pdf.
5. C. L. Giovaneli, P. F. J. Yazdani and B. Honary, Application of space–time diversity/coding for power line channels, in *International Symposium on Power Line Communications and Its Applications*, Athens, Greece, 2002, pp. 101–105.
6. C. L. Giovaneli, P. G. Farrell and B. Honary, Improved space–time coding applications for power line channels, in *International Symposium on Power Line Communications and Its Applications*, Kyoto, Japan, 2003, pp. 50–55.
7. C. L. Giovaneli, B. Honary and P. G. Farrell, Optimum space-diversity receiver for Class A noise channels, in *International Symposium on Power Line Communications and Its Applications*, Zaragoza, Spain, 2004, pp. 189–194.
8. A. Papaioannou, G. D. Papadopoulos and F.-N. Pavlidou, Performance of space–time block coding over the power line channel in comparison with the wireless channel, in *International Symposium on Power Line Communications and Its Applications*, Zaragoza, Spain, 2004, pp. 362–366.
9. A. Papaioannou, G. D. Papadopoulos and F.-N. Pavlidou, Performance of space–time block coding in powerline and satellite communications, *J. Commun. Inf. Syst.*, 20(3), 174–181, 2005.
10. C. Giovaneli, B. Honary and P. Farrell, Space–frequency coded OFDM system for multi-wire power line communications, in *International Symposium on Power Line Communications and Its Applications*, Vancouver, British Columbia, Canada, April 2005, pp. 191–195.

11. B. Adebisi, S. Ali and B. Honary, Multi-emitting/multi-receiving points MMFSK for power-line communications, in *International Symposium on Power Line Communications and Its Applications*, Dresden, Germany, 2009, pp. 239–243.

12. L. Hao and J. Guo, A MIMO-OFDM scheme over coupled multi-conductor power-line communication channel, in *International Symposium on Power Line Communications and Its Applications*, Pisa, Italy, 2007, pp. 198–203.

13. H. Furukawa, H. Okada, T. Yamazato and M. Katayama, Signaling methods for broadcast transmission in power-line communication systems, in *International Symposium on Power Line Communications and Its Applications*, Kyoto, Japan, 2003.

14. F. de Campos, R. Machado, M. Ribeiro and M. de Campos, MISO single-carrier system with feedback channel information for narrowband PLC applications, in *International Symposium on Power Line Communications and Its Applications*, Dresden, Germany, 2009, pp. 301–306.

15. M. Kuhn, D. Benyoucef and A. Wittneben, Linear block codes for frequency selective PLC channels with colored noise and multiple narrowband interference, in *Vehicular Technology Conference, VTC Spring 2002*, vol. 4, Birmingham, AL, 2002, pp. 1756–1760.

16. L. Lampe, R. Schober and S. Yiu, Distributed space–time coding for multihop transmission in power line communication networks, *IEEE J. Sel. Areas Commun.*, 24(7), 1389–1400, 2006.

17. L. Stadelmeier, D. Schneider, D. Schill, A. Schwager and J. Speidel, MIMO for inhome power line communications, in *International Conference on Source and Channel Coding (SCC), ITG Fachberichte*, Ulm, Germany, 2008.

18. D. Schneider, J. Speidel, L. Stadelmeier and D. Schill, Precoded spatial multiplexing MIMO for inhome power line communications, in *Global Telecommunications Conference, IEEE GLOBECOM*, New Orleans, LA, 2008.

19. A. Canova, N. Benvenuto and P. Bisaglia, Receivers for MIMO-PLC channels: Throughput comparison, in *International Symposium on Power Line Communications and Its Applications*, Rio de Janeiro, Brazil, 2010, pp. 114–119.

20. D. Schneider, Inhome power line communications using multiple input multiple output principles, Dr.-Ing. dissertation, Verlag Dr. Hut, January 2012.

21. R. A. Horn and C. R. Johnson, *Matrix Analysis*. Cambridge University Press, Cambridge, U.K., 1985.

22. J. Proakis, *Digital Communications*, 4th edn. McGraw-Hill Book Company, New York, 2001.

23. E. Biglieri, Coding and modulation for a horrible channel, *IEEE Commun. Mag.*, 41(5), 92–98, 2003.

24. S. Alamouti, A simple transmit diversity technique for wireless communications, *IEEE J. Sel. Areas Commun.*, 16(8), 1451–1458, 1998.

25. B. Vucetic and J. Yuan, *Space–Time Coding*. John Wiley, Chichester, U.K., 2003.

26. W. Zhang, X.-G. Xia, P. Ching and H. Wang, Rate two full-diversity space–frequency code design for MIMO-OFDM, in *Workshop on Signal Processing Advances in Wireless Communications*, New York, June 2005, pp. 303–307.

27. W. Zhang, X.-G. Xia and P. C. Ching, High-rate full-diversity space–time–frequency codes for broadband MIMO block-fading channels, *IEEE Trans. Commun.*, 55(1), 25–34, January 2007.

28. E. Viterbo and J. Boutros, A universal lattice code decoder for fading channels, *IEEE Trans. Inf. Theory*, 45(5), 1639–1642, July 1999.

29. A. Paulraj, D. Gore, R. Nabar and H. Bolcskei, An overview of MIMO communications – A key to gigabit wireless, *Proc. IEEE*, 92(2), 198–218, February 2004.

30. A. Boronka, Verfahren mit adaptiver Symbolauslschung zur iterativen Detektion codierter MIMO-Signale, Dr.-Ing. dissertation, Institute of Telecommunications, University of Stuttgart, November 2004.

31. A. Scaglione, P. Stoica, S. Barbarossa, G. Giannakis and H. Sampath, Optimal designs for space–time linear precoders and decoders, *IEEE Trans. Signal Process.*, 50(5), 1051–1064, May 2002.

32. CISPR 16-1:1999: Specification for radio disturbance and immunity measuring apparatus and methods. Radio disturbance and immunity measuring apparatus, CISPR Std.

33. A. Lozano, A. Tulino and S. Verdu, Mercury/waterfilling: Optimum power allocation with arbitrary input constellations, in *International Symposium on Information Theory*, Adelaide, South Australia, Australia, 2005, pp. 1773–1777.
34. A. Lozano, A. Tulino and S. Verdu, Optimum power allocation for parallel Gaussian channels with arbitrary input distributions, *IEEE Trans. Inf. Theory*, 52(7), 3033–3051, 2006.
35. D. Fertonani and G. Colavolpe, On reliable communications over channels impaired by bursty impulse noise, *IEEE Trans. Commun.*, 57(7), 2024–2030, 2009.

9

MIMO PLC Capacity and Throughput Analysis

Daniel M. Schneider, Pascal Pagani and Andreas Schwager

CONTENTS

9.1 Introduction

Multiple-input multiple-output (MIMO) systems have been used for many years in the field of wireless communications [1,2]. The huge increases in coverage and capacity offered by MIMO technology are the key benefits of using multiple sensors at the *transmitter* (Tx) and *receiver* (Rx). For a single-user transmission, and under the assumption that the channel information is perfectly known at both the Tx and Rx, it has been demonstrated that the capacity increases linearly with the number of antennas. However, in more realistic wireless scenarios, the capacity of a MIMO system depends on a number of practical considerations including channel estimation in a time-varying environment, spatial correlation induced by the sensors and the value of the *signal-to-noise ratio* (SNR) available at the Rx [3]. Recently, MIMO technology has been applied in the context of *power line communications* (PLCs), with the aim of offering higher channel capacity and therefore larger system coverage, by including the use of the *protective earth* (PE) wire in addition to the *line* (L) and *neutral* (N) wires [4–8] (see Chapter 1). This new application for MIMO technology offers different characteristics as compared to wireless communications, which can in turn effect the capacity gain achieved. On the one hand, the number of input and output ports of a PLC channel is much more constrained than for a radio channel. Due to Kirchhoff's law, only two differential input ports can be used simultaneously, in the three possible combinations (L-N, N-PE and PE-L). At the Rx, three different signals can be monitored, either on a wire-to-wire basis or using differential reception between two wires. In addition, the common-mode (CM) signal generated by asymmetries in the transmission medium can be measured at the Rx, which provides a fourth output of the MIMO system. As a result, a MIMO PLC transmission up to a 2×4 configuration is implementable. More details about coupling for MIMO signal injection and reception are given in Chapter 1. SNR values observed in typical PLC scenarios can be much higher than in the case of a classical wireless communication. This high SNR condition is beneficial for MIMO transmission, as it ensures a high capacity gain with respect to *single-input single-output* (SISO) transmission, even if the channel presents a high degree of spatial correlation.

This chapter is divided into two main sections. First, the PLC channel capacity offered by MIMO technology is analysed in detail (see Section 9.2). The channel capacity provides an upper limit for achievable throughput and does not take system implementation or a particular MIMO scheme into consideration. The achievable throughput for different MIMO schemes in an OFDM system with adaptive modulation is investigated in Section 9.3. The channel capacity analysis in Section 9.2 is elaborated in the following subsections. The mathematical framework used for capacity computation is first presented in Section 9.2.1. The capacity results presented in the literature for different scenarios are discussed in Section 9.2.2. Finally, Section 9.2.3 presents a statistical analysis of the MIMO channel capacity based on the experimental measurement campaign ETSI STF410 presented in Chapter 5. The throughput analysis in Section 9.3 is organised as follows: First, adaptive modulation is applied to the MIMO-OFDM systems introduced in Chapter 8 (Section 9.3.1). As it was shown in Chapter 8, the SNR after MIMO detection/equalisation of the different investigated MIMO schemes is still very frequency-selective. Thus, adaptive modulation specific to the subcarrier is the method of choice for dealing with this frequency selectivity. The bitrate achieved by the different MIMO PLC systems is investigated in Section 9.3.2 for the same set of MIMO PLC channels as used in Section 9.2 for the channel capacity analysis.

9.2 MIMO PLC Channel Capacity

9.2.1 Theoretical Background

In the following, a wideband signal transmission is considered, where the transmitted signal $s(f)$ is defined for a set of frequencies f in the range $[f_{\min}, f_{\max}]$. In general, multi-carrier transmission schemes, such as *orthogonal frequency division multiplexing* (OFDM), are used to convey wideband signals without suffering from *inter-symbol interference* (ISI) due to the frequency-selective nature of the channel. More information about MIMO-OFDM systems is detailed in Chapter 8. For transmission over a SISO channel, involving one Tx port and one Rx port, the relation between the received signal $r(f)$ and the transmitted signal $s(f)$ is given by

$$r(f) = h(f)s(f) + n(f), \tag{9.1}$$

where
$h(f)$ represents the SISO *channel transfer function* (CTF) defined for all frequencies f
$n(f)$ denotes the received noise

The concept of channel capacity was developed by Shannon in [9]. According to information theory, data transmission can occur at an arbitrary low error probability, provided that the data rate is lower than the maximum channel capacity. Channel capacity is thus a measure of the maximum transmission rate that can be theoretically obtained over a given channel. For a single-carrier SISO channel, the channel capacity C_{SISO} is given as

$$C_{SISO} = B \cdot \log_2(1 + \Lambda) \; [\text{bit/s}], \tag{9.2}$$

where
  B represents the signal bandwidth in Hz
  Λ represents the SNR at the receiver

For a multi-carrier transmission scheme with L carriers defined at frequencies f_1 to f_L with an inter-carrier spacing Δf, Equation 9.2 translates to

$$C_{SISO} = \Delta f \cdot \sum_{n=1}^{L} \log_2(1 + \Lambda(f_n)) \; [\text{bit/s}]. \tag{9.3}$$

For a given signal *power spectral density* (PSD) $P(f)$ defined in W/Hz at the Tx injection point and denoting $N(f)$ the noise PSD at the Rx in W/Hz, Equation 9.3 can be further detailed as

$$C_{SISO} = \Delta f \cdot \sum_{n=1}^{L} \log_2 \left(1 + \frac{P(f_n)|h(f_n)|^2}{N(f_n)} \right) \; [\text{bit/s}]. \tag{9.4}$$

In the case of a MIMO transmission involving N_T Tx ports and N_R Rx ports, the transmitted signal $\mathbf{s}(f)$ is represented as a $N_T \times 1$ symbol vector and the received signal $\mathbf{r}(f)$ is represented as a $N_R \times 1$ symbol vector. Their relation is given by the following equation (see also Section 8.2):

$$\mathbf{r}(f) = \mathbf{H}(f)\mathbf{s}(f) + \mathbf{n}(f), \tag{9.5}$$

where
  $\mathbf{n}(f)$ is the $N_R \times 1$ symbol vector representing the noise received at the N_R Rx ports
  $\mathbf{H}(f)$ is the $N_R \times N_T$ MIMO CTF matrix given by

$$\mathbf{H}(f) = \begin{bmatrix} h_{11}(f) & h_{12}(f) & \cdots & h_{1N_T}(f) \\ h_{21}(f) & h_{22}(f) & \cdots & h_{2N_T}(f) \\ \vdots & \vdots & \ddots & \vdots \\ h_{N_R 1}(f) & h_{N_R 2}(f) & \cdots & h_{N_R N_T}(f) \end{bmatrix}, \tag{9.6}$$

where $h_{ml}(f)$ represents the CTF between input port l ($l = 1, \ldots, N_T$) and receiving port m ($m = 1, \ldots, N_R$).

As shown in Section 8.2, the channel matrix $\mathbf{H}(f)$ can be decomposed into $R = \min(N_T, N_R)$ parallel streams where the attenuation of the streams is described by the singular values $\sqrt{\lambda_p(f)}$, $p = 1, \ldots, R)$ of $\mathbf{H}(f)$.

The channel capacity formula of Equation 9.4 can be extended to the sum of the channel capacities of the R independent SISO streams as follows [10]:

$$C_{\mathrm{MIMO}} = \Delta f \cdot \sum_{n=1}^{L} \sum_{p=1}^{R} \log_2 \left(1 + \frac{P(f_n)\lambda_p(f_n)}{R \cdot N(f_n)} \right) \, [\mathrm{bit/s}]. \tag{9.7}$$

In Equation 9.7, it is assumed that the noise of the N_R receive ports is uncorrelated and that the noise power is the same for all receive ports. It can be noted in Equation 9.7 that the signal PSD $P(f)$ is now divided by $R = \min(N_T, N_R)$ as the available power is shared between the R parallel and independent streams. Note that this assumption may be considered a worst-case assumption. The MIMO PLC signal PSD is not limited by the total power but rather constraint by the EMI properties of the transmission. The discussion on EMI properties of the MIMO PLC transmission can be found in Chapter 7. The analysis in Chapter 7 suggests that the reduction of the transmit power per transmit port may be <3 dB in the case of two transmit ports. The analysis of Chapter 7 is further developed in Chapter 16 in the context of *beamforming* (BF).

The MIMO channel capacity formula according to Equation 9.7 can be further elaborated by taking the correlated noise at the receiver into account. Assuming that the noise is correlated with the noise covariance matrix

$$\mathbf{N}_c(f) = E\{\mathbf{n}(f)\mathbf{n}^H(f)\} \tag{9.8}$$

(see also Chapters 4 and 5), where $\mathbf{N}_c(f)$ is of dimensions $N_R \times N_R$. As shown in Section 8.5.4, a noise whitening filter $\mathbf{N}_c^{-\frac{1}{2}}(f)$ can be applied at the receiver. The filtered noise $\tilde{\mathbf{n}}(f) = \mathbf{N}_c^{-\frac{1}{2}}(f)\mathbf{n}(f)$ is then uncorrelated with

$$E\{\tilde{\mathbf{n}}(f)\tilde{\mathbf{n}}^H(f)\} = \mathbf{I}_{N_R}. \tag{9.9}$$

The noise whitening filter $\mathbf{N}_c^{-\frac{1}{2}}(f)$ and the channel matrix $\mathbf{H}(f)$ can be combined to form an equivalent channel:

$$\tilde{\mathbf{H}}(f) = \mathbf{N}_c^{-\frac{1}{2}}(f)\mathbf{H}(f). \tag{9.10}$$

Applying the SVD to the equivalent channel gives the singular values $\sqrt{\tilde{\lambda}_p(f)}$. Using the new equivalent singular values $\sqrt{\tilde{\lambda}_p(f)}$, the channel capacity of Equation 9.7 is extended to

$$C_{\mathrm{MIMO}} = \Delta f \cdot \sum_{n=1}^{L} \sum_{p=1}^{R} \log_2 \left(1 + \frac{P(f_n)\tilde{\lambda}_p(f_n)}{R} \right) \, [\mathrm{bit/s}]. \tag{9.11}$$

Note that the term $N(f_n)$ is removed in Equation 9.11 since the noise power is already considered via the noise whitening filter $\mathbf{N}_c^{-\frac{1}{2}}(f)$ in $\tilde{\lambda}_p(f)$ and therefore the noise power is equal to 1 according to Equation 9.9. MIMO transmission offers another degree of freedom, namely, the allocation of the total transmit power to the R MIMO streams. This *power allocation* (PA) can be incorporated in Equation 9.11 by a factor a_p for each transmit stream with the constraint of $\sum_{p=1}^{N_T} a_p = N_T$. The optimum PA with respect to the channel capacity is achieved by the *water filling* (WF) algorithm (see [10]). The channel capacity of MIMO PLC using WF was investigated in [8]. It was shown that WF improves the channel capacity for links with very low SNR, while the channel capacity of links with medium to high SNR was only marginally increased when WF was applied. In conclusion, links with low SNR (which are most important to reach coverage goals) may benefit from PA. The application of PA to MIMO PLC systems is discussed in Chapter 8. The simulation results in this chapter do not consider WF.

9.2.2 Review of MIMO Channel Capacity Computations from the Literature

The MIMO PLC channel capacity was investigated for the first time in [4] based on in-home PLC measurements in German houses and flats. These investigations were further elaborated in [8]. The authors found that the MIMO channel capacity is on average double that of SISO. These results were confirmed and extended to a frequency range of up to 100 MHz in [6,11] for measurements in France. In [12], the throughput of different MIMO PLC schemes is compared on the basis of a theoretical MIMO channel model and *additive white Gaussian noise* (AWGN). Rende et al. [13] investigated the MIMO PLC channel and channel capacity based on measurements in North America, focusing on the influence of the noise correlation on the channel capacity. Versolatto and Tonello [14] derived a bottom-up MIMO PLC channel model and compared, among other features, the MIMO PLC channel capacity of the proposed model to the MIMO channel capacity based on the earlier mentioned measurements. Schneider et al. [15] analysed the MIMO PLC channel and computed MIMO channel capacity based on the ETSI MIMO channel measurement campaign of STF410 (see Chapter 5, [16–18]). However, the noise was assumed to be white and uncorrelated in this analysis.

9.2.3 Statistical Analysis of the MIMO Channel Capacity from European Field Measurements

The statistical analysis of the channel capacity is based on the MIMO PLC channels obtained in the European (EU) field measurement campaign of ETSI STF410. This measurement campaign recorded the complex S_{21} scattering parameter (among other channel and EMI features) between hundreds of outlets for different MIMO feeding and receiving options. Additionally, the noise at the outlets was recorded for all MIMO ports simultaneously. This allows the derivation of the noise correlation among the receive ports. The noise correlation is considered in the analysis of the channel capacity below. The frequency range of the measurement campaign goes up to 100 MHz. The measurements were performed in Germany, Italy, Spain, France, Belgium and in the United Kingdom. Details about the measurement campaign may be found in Chapter 5. In total, 285 channels are used in the statistical analysis later. Note that reciprocal channel measurements were removed from the data set.

The transmit power is considered as a parameter in the following analysis to reflect different regulatory constraints in different parts of the world. The transmit PSD masks introduced in Chapter 6 are used. In particular, the EU, US and JP transmit masks according to Figure 6.3 are assumed.

The channel capacity was calculated for each link according to the channel capacity of Equation 9.11. Figures 9.1 through 9.3 show the complementary cumulative distribution function (C-CDF) of the channel capacity for different transmit power masks, namely, the ones in Europe, the United States and JP, respectively. The C-CDF figures may be read to obtain different coverage values, that is, with which probability a certain bitrate is exceeded. The channel capacity is derived for different MIMO configurations. The MIMO configurations from left to right in Figures 9.1 through 9.3 are summarised in Table 9.1. The first configuration is SISO, where the D1 port of the delta-style coupler (differential feeding between L and N) was used at the transmitter and the S1 port (L) of the star-style coupler was used as the receive port. Note that the SISO configuration with feeding on D1 and reception on S2 (N) yields the same performance as the SISO configuration with reception on D1 (L). Therefore, this second SISO configuration is not shown in Figures 9.1 through 9.3. Following are three SIMO configurations, each with feeding on D1 port (L-N) and with an increasing number of receive ports. The 1×2 configuration might be used in homes where the third wire is not present; 1×4 might be used if the transmitter is a legacy (non MIMO) modem and the receiver has full MIMO capabilities. The three MIMO configurations use the same receive ports as the SIMO configurations and use feeding on D1 (L-N) and D3 (L-PE). Details about the couplers and port definitions may be found in Chapter 1. The 2×2 configuration might be the most advantageous as coupler resources are used symmetrically when transmitting or receiving. 2×4 is today's maximum configuration on a three-wire network.

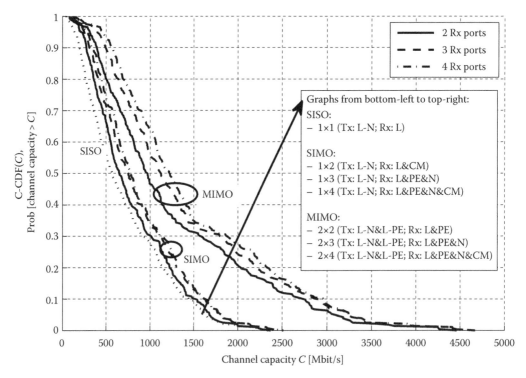

FIGURE 9.1
Channel capacity for the EU transmit power mask.

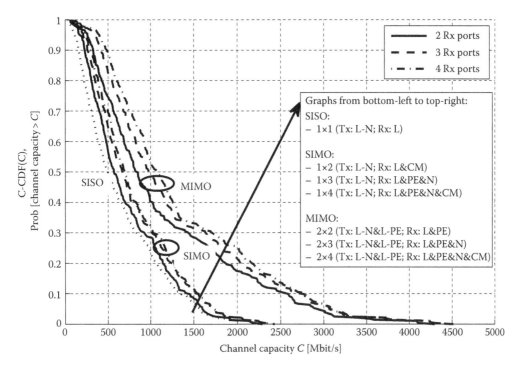

FIGURE 9.2
Channel capacity for the US transmit power mask.

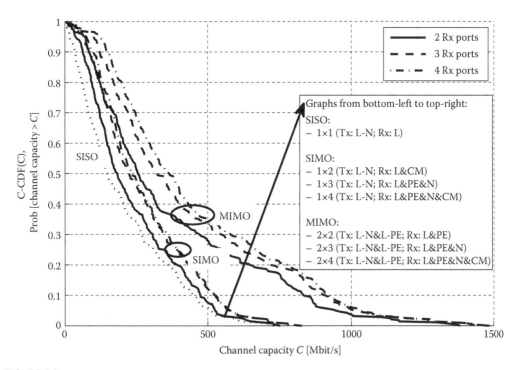

FIGURE 9.3
Channel capacity for the JP transmit power mask.

TABLE 9.1

MIMO Configurations

MIMO Configuration	Tx Ports	Rx Ports
1 × 1	D1 (L-N)	S1 (L)
1 × 2	D1 (L-N)	S1 (L) and S4 (CM)
1 × 3	D1 (L-N)	S1 (L), S2 (N) and S3 (PE)
1 × 4	D1 (L-N)	S1 (L), S2 (N), S3 (PE) and S4 (CM)
2 × 2	D1 (L-N) and D3 (L-PE)	S1 (L) and S3 (PE)
2 × 3	D1 (L-N) and D3 (L-PE)	S1 (L), S2 (N) and S3 (PE)
2 × 4	D1 (L-N) and D3 (L-PE)	S1 (L), S2 (N), S3 (PE) and S4 (CM)

Table 9.2 summarises the median value (50% point) and the 98% point (high coverage point) of the C-CDF shown in the Figures 9.1 through 9.3. The following observations may be drawn from the figures and the table. First, consider the median values:

- The SIMO configurations with only one transmit port already offer a gain compared to SISO. The channel capacity of the best SIMO scheme, 1 × 4, is increased by 37% (EU mask), 39% (US mask) and 56% (JP mask) compared to SISO. The increasing relative gain starting with the EU mask, followed by the US mask, and ending with the JP mask, is explained considering absolute bitrates. The higher the absolute bitrate, the lower the relative SIMO gain. And vice versa, channels supporting only lower absolute bitrates offer the highest gain relative to SISO. The same tendency is observed for MIMO in the high coverage area. The overall lower bitrates of the US and JP masks compared to the EU mask are derived from the transmit masks: the US mask is 5 dB below the EU mask, while the JP mask does not even allow any transmission above 30 MHz.

- Using the second transmit port results in full MIMO configurations and provides a significant increase in bitrate compared to SISO, reflected by 71% (EU and US masks) and 72% (JP mask) for the 2 × 2 scheme and 116% (EU and US masks) and 146% (JP mask) for the 2 × 4 MIMO configuration. (The bitrate increase when applying the JP power transmit mask is hypothetical, because the third wire only rarely exists in JP outlets.)

TABLE 9.2

Channel Capacity and Gain Compared to SISO for Different Transmit Power Masks: Median Values and 98% Coverage Point

MIMO Config- uration	EU Mask				US Mask				JP Mask			
	Median		98%		Median		98%		Median		98%	
	Mbit/s	Gain	Mbit/s	Gain	Mbit/s	Gain	Mbit/s	Gain	Mbit/s	Gain	Mbit/s	Gain
1 × 1	568		82		499		62		149		6	
1 × 2	651	1.15	126	1.55	571	1.14	103	1.65	184	1.24	23	3.63
1 × 3	751	1.32	154	1.88	670	1.34	127	2.04	226	1.51	31	4.81
1 × 4	777	1.37	173	2.12	694	1.39	143	2.30	233	1.56	34	5.37
2 × 2	971	1.71	153	1.87	851	1.71	121	1.94	257	1.72	23	3.52
2 × 3	1126	1.98	201	2.46	984	1.97	160	2.57	323	2.17	35	5.48
2 × 4	1227	2.16	235	2.88	1077	2.16	190	3.05	367	2.46	41	6.35

In conclusion, the MIMO channel capacity of the full (2 × 4) MIMO configuration is on average more than double the SISO capacity.

Next, consider the high coverage area (98% point):

- At the high coverage area, the SIMO configurations already offer a significant gain compared to SISO: factors of 2.12 (EU mask), 2.03 (US mask) and even 5.37 (JP mask) are observed for the 1 × 4 configuration.
- In contrast to 1 × 4 MIMO, the 2 × 2 configuration provides less gain: 1.87, 1.94 and 3.52 for the EU, US and JP masks, respectively. The second, weaker stream (eigenmode) does not contribute much for low SNR channels. It is more important to collect all the available signal energy at the receiver. This is reflected by the number of receive ports.
- Combining the use of the maximum number of receive ports and the use of two streams for the 2 × 4 configuration shows the highest gain compared to SISO: 2.88, 3.05 and 6.35 for the EU, US and JP masks, respectively.

The MIMO gain in the high coverage area is even higher compared to the median values. Thus, MIMO especially improves the difficult links with high attenuation and therefore low SNR at the receiver, making MIMO a promising method for meeting ambitious coverage requirements. Note that the presented results are the theoretical channel capacity. The choice of the implemented MIMO scheme and the system parameters for modulation, coding and implementation aspects and limitations influence the achievable throughput in real modem implementations (e.g. a 10 bit *analogue-to-digital converter* (ADC) might not utilise the full SNR available at the channel).

9.3 MIMO PLC Throughput Analysis

The aim of this section is to verify the MIMO channel capacity gains of the previous section for different MIMO PLC systems. In particular, the OFDM-based MIMO PLC schemes introduced in Chapter 8 are investigated with respect to the achieved bitrate. Adaptive modulation is applied to the SNR after detection (see Section 9.3.1). As for SISO, the SNR after MIMO processing is still very frequency-selective. This makes subcarrier-specific adaptive modulation a good choice for maximising the PLC throughput. The achieved bitrate is analysed in Section 9.3.2 for a large set of MIMO PLC channels.

9.3.1 Adaptive Modulation

The frequency-selective PLC channel leads to high SNR variations for different OFDM subcarriers. To overcome this problem, adaptive modulation is applied. Each subcarrier is bit loaded and modulated according to the corresponding SNR. The higher the SNR, the higher the *quadrature amplitude modulation* (QAM) constellation. Figure 9.4 shows an example of a typical SISO PLC channel. The SNR, depending on the frequency from 4 to 30 MHz, illustrates the frequency-selective channel. The QAM constellations with modulation order M used here are *binary phase-shift keying* (BPSK) ($M = 2$) and the even or square QAM constellations from *quadrature phase-shift keying* (QPSK) to 4096-QAM ($M = 4, 16, 64, 256, 1024, 4096$). The number of bits per QAM symbol is $\log_2(M)$. The SNR thresholds θ_M

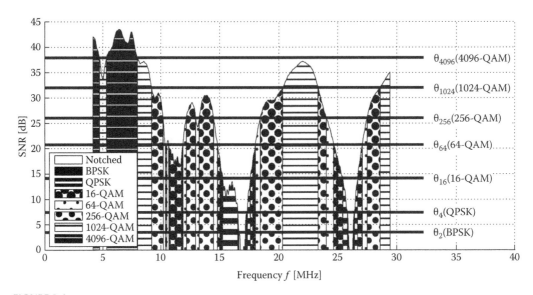

FIGURE 9.4
Adaptive modulation: available SNR depending on the frequency, and application of QAM constellations depending on the available SNR, SISO, transmit power to noise power level $\rho = 65$ dB.

for each modulation order M are illustrated by the horizontal lines in Figure 9.4 for each modulation scheme. These thresholds define the assignment of the QAM constellation depending on the SNR of each subcarrier. A subcarrier can also be omitted (notched) if the SNR is too low to carry any information.

If $\Lambda(n)$ is the SNR of subcarrier n ($1 \leq n \leq N$), the assignment to modulation order $M(n)$ of subcarrier n depends on the thresholds θ_M of each modulation order M:

$$M(n) = \begin{cases} 1, & \Lambda(n) < \theta_2 \text{ (not modulated).} \\ 2, & \theta_2 \leq \Lambda(n) < \theta_4 \text{ (BPSK).} \\ 4, & \theta_4 \leq \Lambda(n) < \theta_{16} \text{ (QPSK).} \\ 16, & \theta_{16} \leq \Lambda(n) < \theta_{64} \text{ (16-QAM).} \\ 64, & \theta_{64} \leq \Lambda(n) < \theta_{256} \text{ (64-QAM).} \\ 256, & \theta_{256} \leq \Lambda(n) < \theta_{1024} \text{ (256-QAM).} \\ 1024, & \theta_{1024} \leq \Lambda(n) < \theta_{4096} \text{ (1024-QAM).} \\ 4096, & \theta_{4096} \leq \Lambda(n) \text{ (4096-QAM).} \end{cases} \tag{9.12}$$

In Equation 9.12, the modulation order $M = 1$ indicates that no information is assigned to the subcarrier, that is, $\log_2(1) = 0$ bit is assigned. The choice of the SNR thresholds θ_M influences both the achieved bitrate and the *bit error ratio* (BER).

The bitrate D is the sum of the bits assigned to each subcarrier divided by the OFDM symbol length T_u:

$$D = \frac{\sum_{n=1}^{L} \log_2(M(n))}{T_u}. \tag{9.13}$$

The BER depends on the SNR Λ and the modulation order M. For an AWGN channel, the BER is given by the tight approximation [19,20]:

$$P_b(M,\Lambda) = \begin{cases} \dfrac{1}{2} \cdot \text{erfc}\left(\sqrt{\Lambda}\right), & M = 2, \\[2ex] \dfrac{2}{\log_2(M)} \cdot \left(1 - \dfrac{1}{\sqrt{M}}\right) \cdot \text{erfc}\left(\sqrt{\dfrac{3\Lambda}{2(M-1)}}\right), & \log_2(M) \text{ even} \end{cases} \tag{9.14}$$

with $\text{erfc}(x) = \dfrac{2}{\sqrt{\pi}} \displaystyle\int_x^\infty \exp(-t^2)dt$ being the complementary error function. Binary reflected Gray bit labelling is assumed for the even constellations ($\log_2(M)$ even) in Equation 9.14.

With $M(n)$ the modulation order and $\Lambda(n)$ the SNR of subcarrier n ($1 \leq n \leq N$), the BER of each subcarrier $P_b(M(n),\Lambda(n))$ can be calculated according to Equation 9.14. The overall average BER $\bar{P}_b$ is given as the average number of bits in error divided by the total number of transmitted bits [20] and can be calculated as

$$\bar{P}_b = \frac{\text{Number of errors}}{\text{Number of bits}} = \frac{\sum_{n=1}^{N} P_b(M(n),\Lambda(n)) \cdot \log_2(M(n))}{\sum_{n=1}^{N} \log_2(M(n))}. \tag{9.15}$$

Recall that the SNR thresholds θ_M determine the modulation order $M(n)$ of each subcarrier according to Equation 9.12 and thus influence both the bitrate according to Equation 9.13 and the average BER according to Equation 9.14. The design of the SNR thresholds θ_M can be optimised with respect to two different criteria:

- Minimising the BER for a fixed bitrate
- Maximising the bitrate for a fixed BER

The second criterion is considered here. A simple algorithm that guarantees a certain target BER P_b' with $\bar{P}_b \leq P_b'$ uses fixed SNR thresholds. For a given target BER P_b' and modulation order M, Equation 9.14 can be solved for Λ which is then used as SNR threshold θ_M. Figure 9.5 shows the BER depending on the SNR for different modulation orders. The figure also provides the SNR thresholds for a target BER of $P_b' = 10^{-3}$. The BER value of 10^{-3} may be sufficient for the raw physical layer since additional *forward error correction* (FEC) will improve the BER. This algorithm guarantees that the average BER $\bar{P}_b$ does not exceed the target BER P_b' in every case. Usually, the average BER $\bar{P}_b$ will be lower than the target BER P_b' since many subcarriers have higher SNR than the SNR thresholds.

Schneider et al. [5,8] propose an algorithm which maximises the bitrate D, for a desired target BER P_b'. The algorithm takes the SNR distribution of the given channel into account, that is, the algorithm adapts the SNR thresholds to the current channel conditions.

The SNR thresholds shown in Figure 9.4 are obtained according to the algorithm described earlier for a target BER of 10^{-3}. Note that the SNR thresholds are lower compared to the fixed SNR thresholds shown in Figure 9.5. Thus, the bitrate of the proposed algorithm is higher compared to the algorithm with fixed thresholds.

The MIMO scheme and detection algorithm determine the SNR of the MIMO streams as shown in Chapter 8. Adaptive modulation is applied for each MIMO stream separately. Figure 9.6a shows the block diagram of the adaptive modulation algorithm for MIMO.

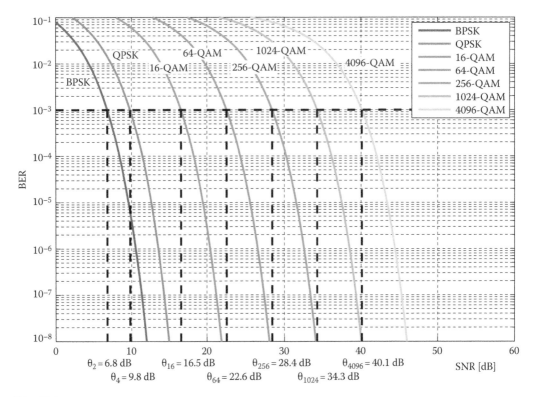

FIGURE 9.5
BER depending on SNR for different QAM orders and SNR values required for a target BER of 10^{-3}.

The SNR thresholds θ_M are calculated according to the SNR of the subcarriers after MIMO detection. These thresholds are used to assign the QAM constellation to each subcarrier. Note that the blocks *adaptive QAM pattern* and *adaptive thresholds* in Figure 9.6 require knowledge of the SNR for all subcarriers; thus, these blocks operate in parallel. The calculation of the PA coefficients is based on the SNR and the QAM constellations (see Section 8.3). This results in a dependency between the PA and the adaptive modulation as highlighted in Figure 9.6b. The combination of adaptive modulation and PA has to be calculated iteratively in this case.

9.3.2 Simulation Results

The MIMO-OFDM system described in Chapter 8 forms the basis of the system simulations. 1296 subcarriers are used in the frequency range from 4 to 30 MHz. Each subcarrier is adaptively modulated, according to the adaptive modulation algorithm described in Section 9.3.1. The target average BER of the uncoded system is adjusted to 10^{-3}. An additional FEC might easily reduce this BER. The bitrate is obtained as the sum of the number of bits assigned to all subcarriers divided by the OFDM symbol length. This bitrate describes the raw physical layer bitrate without considering the guard interval length, training data or FEC overhead. The basic system parameters are summarised in Table 9.3.

The noise is modelled by AWGN with zero mean, and it is assumed that the noise is uncorrelated and that the noise power is the same for all receive ports. The transmit power to noise power level is assumed to be $\rho = 65$ dB. This value corresponds to a transmit

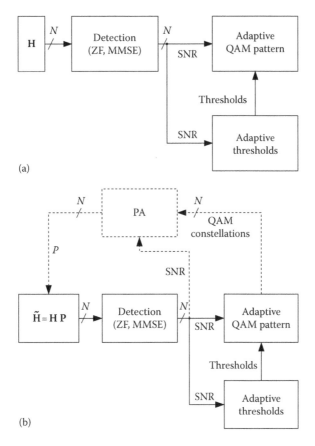

FIGURE 9.6
Block diagram of the adaptive modulation algorithm, without PA (a) and with PA (b).

TABLE 9.3

Basic System Parameters

FFT points	2048
Nyquist frequency (MHz)	40
Frequency band (MHz)	4–30
Number of active subcarriers (4–30 MHz)	1296
Carrier spacing (kHz)	19.53
Symbol length (μs)	51.2
Modulation (per subcarrier)	BPSK, QPSK, 16-, 64-, 256-, 1024-, 4096-QAM
Uncoded target BER	10^{-3}

PSD of −55 dBm/Hz (see Chapter 6) and an average noise PSD of −120 dBm/Hz (this corresponds to the 90% point of CDF of the noise according to Chapter 5). Impulsive noise is not considered. The focus is on the comparison between MIMO and SISO schemes. It is expected that impulsive noise will influence all receive ports in a similar way. Thus, mitigation techniques known from SISO systems can be applied [21,22]. The measured MIMO PLC channels obtained during the European measurement campaign (ETSI STF410, see Chapter 5) are used in the system simulations. In the case of MIMO, the two feeding ports D1 and D3 (i.e. L-N and L-PE, see Chapter 1) and all four receive ports (S1, S2, S3 and S4) are used; in the case of SISO, the D1 (L-N) port is used at the transmitter and the S1 port (L) at the receiver. It was observed that using the S2 (N) at the receiver yields the same performance as using the S1 port. The corresponding SNR is calculated based on the channel matrix of each subcarrier (channel estimation is assumed to be perfect), depending on the MIMO scheme as shown in Chapter 8. Then, the derived SNR is used with the adaptive modulation algorithm to determine the subcarrier's constellations. Adaptive modulation is applied to each of the MIMO schemes in this section.

Figure 9.7 compares the C-CDF of the bitrate at ρ = 44 dB for different MIMO schemes, namely, SISO, the Alamouti scheme and *spatial multiplexing* (SMX) with different detection algorithms (see Chapter 8) and BF. The measured MIMO PLC channels form the basis of the comparison. No PA is applied here. SISO is expected to offer the lowest bitrate. However, SMX with *zero-forcing* (ZF) detection performs about the same or even worse, compared to SISO for most channels and bitrates up to about 40 Mbit/s. The high correlation of the power line channels results in high values of the detection matrix entries, leading to an amplification of the noise (refer also to Section 8.6). This effect is mitigated using more advanced detection algorithms. The bitrate is increased, as seen in Figure 9.7 for *minimum mean squared error* (MMSE), *successive interference cancellation* (SIC)-ZF and SIC-MMSE. *Ordered SIC* (OSIC) receivers are not shown in Figure 9.7, because their performance improvement compared to SIC is only marginal. *Eigenbeamforming* (EBF) achieves the highest bitrate. The Alamouti scheme performs almost as well as EBF, especially for low bitrates or the high coverage point. The bitrate gain of MIMO compared to SISO is highest for the low bitrate region in Figure 9.7, that is, for channels with high attenuation.

Figure 9.8 is similar to Figure 9.7 with a higher transmit signal to noise power level of ρ = 65 dB. Here, SMX with ZF detection overrides SISO in contrast to Figure 9.7.

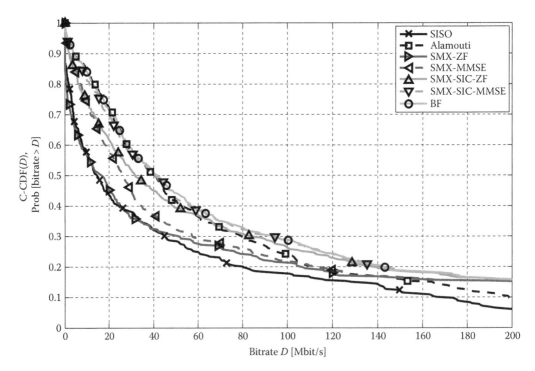

FIGURE 9.7
C-CDF of the bitrate for different MIMO schemes, ρ = 44 dB, no PA, N_T = 2, N_R = 4.

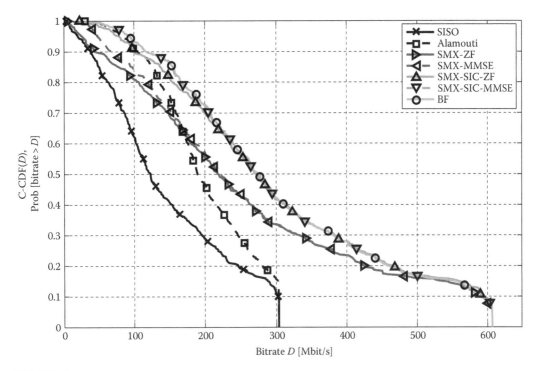

FIGURE 9.8
C-CDF of the bitrate for different MIMO schemes, $\rho = 65$ dB, no PA, $N_T = 2$, $N_R = 4$.

The gain of MMSE versus ZF becomes smaller compared to the lower transmit to noise power level in Figure 9.7. The SIC receivers reach the performance of EBF. However, it has to be kept in mind that no PA and no error propagation of the SIC receiver are considered here. The Alamouti scheme performs well for high values of the C-CDF, that is, for channels with low bitrate due to high attenuation and correlation. However, due to the transmission of replicas, no multiplexing gain is achieved for channels with high bitrate (as can be seen by the low values of the C-CDF in Figure 9.8 where the line of the Alamouti scheme reaches the SISO's line).

Figure 9.9 compares the PA for SMX and EBF for $\rho = 44$ dB and $\rho = 65$ dB. There is only a marginal performance improvement of *mercury water filling* (MWF) compared to the simplified PA (as introduced in Chapter 8). The gain of PA is most visible for low ρ. As explained also by the SNR results in Chapter 8 (see Figure 8.14), EBF benefits most from PA. The PA's gain of SMX (SMX-ZF) is relatively small. This is observed for all different receivers for SMX.

Figure 9.10 is similar to Figure 9.7. However, it includes PA. Additionally, one-stream BF is shown where the total power is assigned to the first stream. Figure 9.10a shows the complete coverage range, Figure 9.10b shows the median coverage point and Figure 9.10c shows the high coverage point. Most carriers of the EBF's second stream turn out to carry no information for highly attenuated channels. The shift of the second stream's power to the first stream of EBF results in 3 dB higher SNR of the first stream if none of the carriers of the second stream are carrying any information. Obviously, the performance of one-stream BF is close to two-stream BF because as only a few carriers of the second stream contribute to the bitrate. The superior performance of BF over the Alamouti scheme is

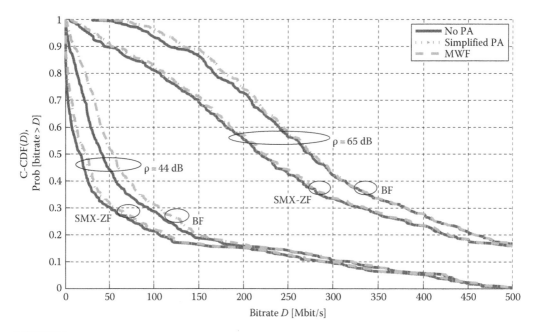

FIGURE 9.9
C-CDF of the bitrate for *spatial multiplexing* (SMX-ZF), *beamforming* (BF) and different PA schemes: no PA, simplified PA and MWF, ρ = 44 dB and ρ = 65 dB, $N_T = 2$, $N_R = 4$.

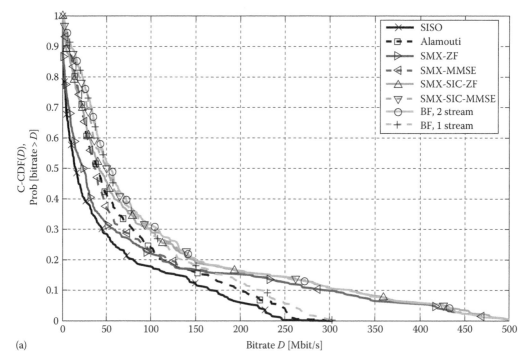

FIGURE 9.10
(a) C-CDF of the bitrate for different MIMO schemes with PA, ρ = 44 dB, $N_T = 2$, $N_R = 4$.

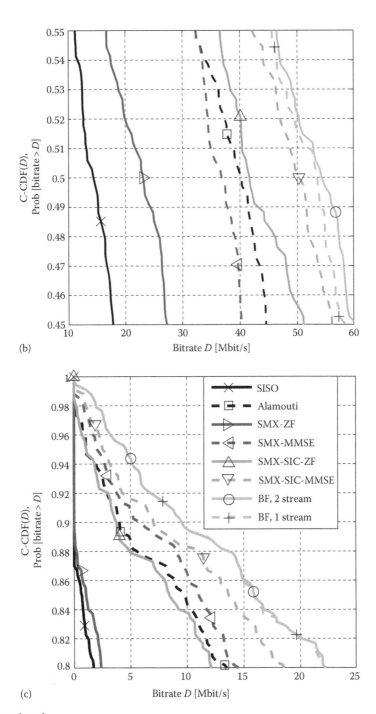

FIGURE 9.10 (continued)
(b,c) C-CDF of the bitrate for different MIMO schemes with PA, $\rho = 44$ dB, $N_T = 2$, $N_R = 4$.

TABLE 9.4

Mean Values of the Bitrates for Different MIMO and Power Allocation Schemes, 2 × 4 MIMO Configuration

| MIMO Scheme | ρ = 44 dB | | | | | | ρ = 65 dB | | | | | |
| | No PA | | Simplified PA | | MWF | | No PA | | Simplified PA | | MWF | |
	Mbit/s	Gain	Mbit/s	Gain	Mbit/s	Gain	Mbit/s	Gain	Mbit/s	Gain	Mbit/s	Gain
SISO	55	1	PA cannot be used				147	1	PA cannot be used			
Alamouti	79	1.4	PA cannot be used				201	1.4	PA cannot be used			
One-stream BF	93	1.7	PA cannot be used				217	1.5	PA cannot be used			
SMX-ZF	85	1.6	88	1.6	91	1.7	264	1.8	266	1.8	269	1.8
SMX-MMSE	92	1.7	98	1.8	101	1.8	270	1.8	273	1.9	276	1.9
SMX-SIC-ZF	105	1.9	110	2.0	113	2.1	306	2.1	309	2.1	313	2.1
SMX-SIC-MMSE	109	2.0	114	2.1	117	2.1	310	2.1	313	2.1	316	2.1
Two-stream BF	110	2.0	119	2.2	120	2.2	311	2.1	315	2.1	317	2.2

more visible compared to no PA in Figure 9.7. Please note that PA cannot be used for the Alamouti scheme since each symbol is transmitted via each transmit port.

Table 9.4 summarises the mean values of the bitrates for the different MIMO schemes. The two transmit to noise power levels of ρ = 44 dB and 65 dB are considered. The gain of a MIMO scheme is defined as the ratio of the MIMO bitrate to the SISO bitrate.

9.4 Conclusions

This chapter analysed the theoretical MIMO channel capacity based on an extensive measurement set of MIMO PLC channels obtained during the ETSI measurement campaign of STF410. The channel capacity was computed under different regulatory constraints, that is, different transmit power masks. Not only were the measured MIMO CTF taken into account but also the measured noise statistics, which incorporate the spatial correlation of the noise. The MIMO channel capacity is, on average, doubled compared to SISO. In particular, highly attenuated channels benefit most from the application of MIMO which makes MIMO a promising method for improving PLC coverage.

In a next step, the gain of throughput was verified for different MIMO PLC systems. Adaptive modulation was applied to the SNR after MIMO detection of the MIMO PLC schemes introduced in Chapter 8, and the achieved bitrate was analysed. Similar conclusions as for the SNR analysis in Chapter 8 can be drawn for the throughput analysis. Generally, there is a significant increase of bitrate for all MIMO schemes compared to SISO transmission. This confirms the channel capacity gain found in the first part of this chapter. The Alamouti scheme improves the bitrate compared to SISO and showed good performance for highly attenuated channels. However, no multiplexing gain is achieved because of the transmission of replicas of each symbol. Adaptive modulation already adapts to the frequency-selective channel, making the property of the Alamouti scheme to combat fading counterproductive. SMX increases the bitrate compared to the Alamouti scheme for low attenuated channels. Care has to be taken with respect to which MIMO detection algorithm is used. ZF detection fails for a high correlation of the channel, and more complex detection algorithms (like MMSE and SIC) are suggested to increase the performance. The earlier MIMO schemes are open-loop MIMO schemes which require no channel state information at the transmitter. An additional performance gain is achieved by closed-loop MIMO schemes like BF which use channel state information at the transmitter. BF offers the highest bitrate in all scenarios and comes closest to the channel capacity by adapting the transmission to the eigenmodes of the channel. The full spatial diversity gain is achieved for highly attenuated channels and maximum bitrate gain is achieved for channels with low attenuation.

BF requires knowledge about the channel state information at the transmitter. Usually, only the receiver has channel state information. Thus, information about the precoding matrix has to be fed back from the receiver to the transmitter. The application of adaptive modulation already requires feedback about the constellation maps from the receiver. This feedback path might also be used to return the BF information to the transmitter. The feedback rate can be kept low because the in-home PLC channel is less time varying compared to, for example, a mobile channel. Schneider et al. [5,8] investigated the amount of feedback overhead needed to feed back the information about the precoding matrices and showed that the required feedback for the precoding matrices lies in the same order of magnitude as the feedback required for adaptive modulation. For these reasons, a BF-based MIMO-OFDM system with adaptive modulation is a well-suited MIMO system for PLC. The adoption of MIMO and precoded SMX or BF to the latest PLC specifications is discussed in detail in Chapter 12 for G.hn/G.9963 and in Chapter 14 for HomePlug AV2. A study of a MIMO PLC hardware implementation with BF can be found in Chapter 24.

References

1. A. Paulraj, D. Gore, R. Nabar and H. Bolcskei, An overview of MIMO communications – A key to gigabit wireless, *Proceedings of the IEEE*, 92(2), 198–218, February 2004.
2. L. Schumacher, L. T. Berger and J. Ramiro Moreno, Recent advances in propagation characterisation and multiple antenna processing in the 3 GPP framework, in *XXVIth URSI General Assembly*, Maastricht, the Netherlands, August 2002, session C2. [Online] Available: http://www.ursi.org/Proceedings/ProcGA02/papers/p0563.pdf.
3. A. Goldsmith, S. Jafar, N. Jindal and S. Vishwanath, Capacity limits of MIMO channels, *Selected Areas in Communications, IEEE Journal on*, 21(5), 684–702, 2003.
4. L. Stadelmeier, D. Schneider, D. Schill, A. Schwager and J. Speidel, MIMO for inhome power line communications, in *International Conference on Source and Channel Coding (SCC), ITG Fachberichte*, Ulm, Germany, 2008.
5. D. Schneider, J. Speidel, L. Stadelmeier and D. Schill, Precoded spatial multiplexing MIMO for inhome power line communications, in *Global Telecommunications Conference, IEEE GLOBECOM*, New Orleans, LA, 2008.
6. R. Hashmat, P. Pagani, A. Zeddam and T. Chonavel, MIMO communications for inhome PLC networks: Measurements and results up to 100 MHz, in *International Symposium on Power Line Communications and Its Applications*, Rio de Janeiro, Brazil, 2010, pp. 120–124.

7. A. Schwager, Powerline communications: Significant technologies to become ready for integration, Dr.-Ing. dissertation, Universität Duisburg-Essen, Germany, May 2010.
8. D. Schneider, Inhome power line communications using multiple input multiple output principles, Dr.-Ing. dissertation, Verlag Dr. Hut, Germany, January 2012.
9. C. Shannon, Communication in the presence of noise, *Proceedings of the IRE*, 37(1), 10–21, 1949.
10. A. Paulraj, R. Nabar and D. Gore, *Introduction to Space–Time Wireless Communications*. Cambridge University Press, New York, 2003.
11. R. Hashmat, P. Pagani and T. Chonavel, MIMO capacity of inhome PLC links up to 100 MHz, in *Workshop on Power Line Communications*, Udine, Italy, 2009.
12. A. Canova, N. Benvenuto and P. Bisaglia, Receivers for MIMO-PLC channels: Throughput comparison, in *International Symposium on Power Line Communications and Its Applications*, Rio de Janeiro, Brazil, 2010, pp. 114–119.
13. D. Rende, A. Nayagam, K. Afkhamie, L. Yonge, R. Riva, D. Veronesi, F. Osnato and P. Bisaglia, Noise correlation and its effect on capacity of inhome MIMO power line channels, in *International Symposium on Power Line Communications and Its Applications*, Udine, Italy, 2011, pp. 60–65.
14. F. Versolatto and A. Tonello, A MIMO PLC random channel generator and capacity analysis, in *International Symposium on Power Line Communications and Its Applications*, Udine, Italy, 2011, pp. 66–71.
15. D. Schneider, A. Schwager, W. Bäschlin and P. Pagani, European MIMO PLC field measurements: Channel analysis, in *International Symposium on Power Line Communications and Its Applications*, Beijing, China, 2012, pp. 304–309.
16. ETSI, TR 101 562-1 v1.3.1, Powerline telecommunications (PLT), MIMO PLT, part 1: Measurement methods of MIMO PLT, Technical Report, 2012.
17. ETSI, TR 101 562-2 v1.2.1, Powerline telecommunications (PLT), MIMO PLT, part 2: Setup and statistical results of MIMO PLT EMI measurements, Technical Report, 2012.
18. ETSI, TR 101 562-3 v1.1.1, Powerline telecommunications (PLT), MIMO PLT, part 3: Setup and statistical results of MIMO PLT channel and noise measurements, Technical Report, 2012.
19. S. T. Chung and A. Goldsmith, Degrees of freedom in adaptive modulation: A unified view, *IEEE Transactions on Communications*, 49(9), 1561–1571, September 2001.
20. J. Proakis, *Digital Communications*, 4th ed. McGraw-Hill Book Company, New York, 2001.
21. E. Biglieri, Coding and modulation for a horrible channel, *IEEE Communications Magazine*, 41(5), 92–98, 2003.
22. D. Fertonani and G. Colavolpe, On reliable communications over channels impaired by bursty impulse noise, *IEEE Transactions on Communications*, 57(7), 2024–2030, 2009.

Part III

Current PLC Systems and Their Evolution

10

Current Power Line Communication Systems: A Survey

Lars T. Berger, Andreas Schwager, Stefano Galli,
Pascal Pagani, Daniel M. Schneider and Hidayat Lioe

CONTENTS

10.1 Introduction*

The idea of using power lines also for communication purposes has already been around since the beginning of the last century [6,7]. It is now broadly referred to as *power line communications* (PLCs). The obvious advantage is the widespread availability of electrical infrastructure, so that theoretically, deployment costs are confined to connecting modems to the existing electrical grid. Following the nomenclature introduced in [8], power line technologies can be grouped into the following:

1. *Ultra narrowband* (UNB) technologies operating at very low data rate in the *ultra-low-frequency* band (ULF, 0.3–3 kHz) or in the upper part of the *super-low-frequency band* (SLF, 30–300 Hz). Examples of UNB-PLC are *ripple carrier signalling* (RCS) [6], the *turtle system* [9] and the more recent *two-way automatic communications system* (TWACS) [10,11]. Especially, *automated meter reading* (AMR) systems frequently used UNB-PLC technologies to gain access and in parts control over the energy meters within private homes. UNB-PLC systems are usually designed to communicate

---

* Chapter in parts based on [1–5] with permission of the copyright holders.

over long distances with their signals passing through low-voltage/medium-voltage transformers. This helps to keep the amount of required modems and repeaters to a minimum. Drawbacks are low data rates, for example, in the order of 0.001 bit/s (Turtle) and 2 bits per mains frequency cycle (TWACS)*. Additionally, these systems are sometimes limited to unidirectional communications.

2. *Narrowband* (NB) technologies operate in the *very-low-*, the *low-* and in parts of the *medium-frequency* (VLF/LF/MF) bands, which include the European *Comité Européen de Normalisation Électrotechnique* (CENELEC) bands (3–148.5 kHz), the *US Federal Communications Commission* (US FCC) band (10–490 kHz), the JP *Association of Radio Industries and Businesses* (ARIB) band (10–450 kHz) and the Chinese band (3–500 kHz). Within this class of NB-PLC, one may further subdivide into the following:

   a. *Low data rate* (LDR), which refers to technologies capable of data rates of a few kbit/s. These technologies are usually based on single-carrier or spread-spectrum modulation. Typical examples of LDR NB-PLC technologies are devices conforming to the recommendations: ISO/IEC 14908-3 (*LonWorks*), ISO/IEC 14543-3-5 (*KNX*), CEA-600.31 (*CEBus*), IEC 61334-3-1, IEC 61334-5 (*FSK and spread-FSK*). They are backed by *Standard Development Organisations* (SDOs), precisely by the *International Electrotechnical Commission* (IEC) and the *International Organization for Standardization* (ISO). Additional non-SDO-based examples are *Insteon, X10, HomePlug C&C, SITRED, Ariane Controls* and *BacNet*. LDR NB-PLC technologies have also been referred to as *distribution line carrier* or *power line carrier*.

   b. *High data rate* (HDR) refers to technologies capable of data rates ranging between tens of kbit/s and around 500 kbit/s. Today, HDR technologies are based on *orthogonal frequency division multiplexing* (OFDM) [12]. Typical examples of HDR NB-PLC technologies are those included in the family of approved *International Telecommunications Union – Telecommunication Standardization Sector* (ITU-T) NB-PLC Recommendations [13–15] and the ongoing *Institute of Electrical and Electronics Engineers* (IEEE) P1901.2 project [16]. Original non-SDO-based examples are the industry specifications G3-PLC and *Powerline-Related Intelligent Metering Evolution* (PRIME), which have recently become ITU-T Recommendations G.9903 and G.9904, respectively.

3. *Broadband* (BB) technologies operate in the *medium-*, *high-* or *very-high-frequency* (MF/HF/VHF) bands (1.8–250 MHz) and have a *physical layer* (PHY) rate ranging from several Mbit/s to several hundred Mbit/s. Typical examples of BB-PLC technologies are devices conforming to TIA-1113 (*HomePlug 1.0*), IEEE 1901 and ITU-T G.hn (G.9960–G.9964) standards. Additional non-SDO-based examples are *HomePlug AV2, HomePlug Green PHY, UPA Powermax* and *Gigle MediaXtreme*. BB-PLC technologies devoted to 'last mile' and access applications have also sometimes been referred to as *broadband over power lines* (BPL).

An overview of UNB-, NB- and BB-PLC specifications and standards is presented in Figure 10.1. Apart from the systems in Figure 10.1, the company Watteco [17] developed a technology called *Watt pulse communication* (WPC), which does not easily fit into the aforementioned UNB/NB/BB categorisation. WPC roughly occupies the band from 500 kHz up to 7 MHz while at the same time transmitting LDR around 10 up to 50 kbit/s.

---

* Data rate can be increased by conveying up to six TWACS channel per mains phase using *code division multiple access* (CDMA) with orthogonal (Hadamard) codes.

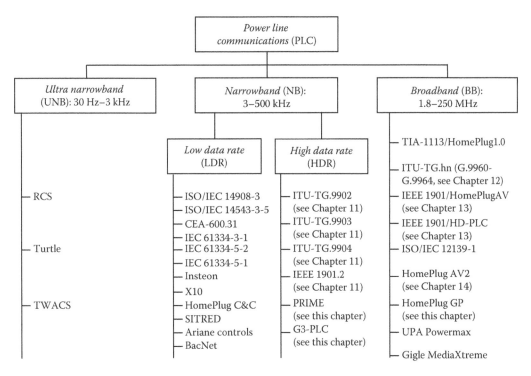

FIGURE 10.1
Overview of UNB-, NB- and BB-PLC specifications and standards.

The transmitter shortcuts the AC with a defined impedance which produces spikes on the mains. The information to be broadcasted is given in the timing relation between individual spikes. Systems can be realised with very low component cost, as the transmitter does not require a *digital-to-analogue converter* (DAC) or an amplifier [18]. Although interesting for a plurality of *command and controls* (C&C), and *Smart Home* applications, controlling *electromagnetic compatibility* (EMC) or implementing frequency notches (see, Chapter 22) is very difficult for such WPC-PLC systems.

Leaving UNB-PLC and WPC at the sideline, this chapter provides an overview of current NB-PLC and BB-PLC systems in Sections 10.2 and 10.3, respectively. Further details on the NB-PLC systems IEEE 1901.2 and the family of ITU-T NB-PLC Recommendations are presented in Chapter 11. Details on ITU-T G.hn, IEEE 1901 and HomePlug AV2 may be found in Chapters 12 through 14, respectively.

10.2 NB-PLC

NB-PLC systems usually operate in the frequency range from 3 to 500 kHz, that is, the CENELEC/ARIB/FCC bands. A pioneering LDR NB-PLC system is definitely *LonWorks*. While today's PLC standards only specify one or two of the *ISO-Open Systems Interconnection* (ISO-OSI) layers, the LonWorks system includes all seven tiers from programming networked applications down to the physical media power line, twisted pair, RF and fibre optics. After becoming an ANSI standard, international organisations agreed to approve it as ISO/IEC 14908-3 [19,20]. LonWorks PLC transceivers are designed to operate in one of two frequency ranges depending on the end application. When configured for use in electric utility applications, the CENELEC A-band is used, whereas in-home/commercial/ industrial applications use the C-band. Achievable data rates are in the order of few kbit/s.

Another LDR NB-PLC standard is ISO/IEC 14543-3-5 (KNX, EN 50090) [21]. It spans all layers of the OSI model and can, besides over power line, also be used over other media such as twisted pair and in some cases even wirelessly.

The most widespread NB-PLC technologies deployed today are based on *frequency shift keying* (FSK) or spread-FSK as specified in the IEC 61334-5-2 [22] and IEC 61334-5-1 [23] standards, respectively. The availability of standards for these technologies goes from the stack of PHY communications protocols up to the application layer (IEC 62056-53 for *Companion Specification for Energy Metering* [COSEM]) [24], thus facilitating the development of interoperable solutions. Such *advanced metering infrastructure* (AMI) solutions are now provided by a number of companies and are widely and successfully applied by utilities.

Due to the increasing desire for higher data rates and the growing attention that smart metering projects are getting in various national efforts directed at modernising the aging power grid toward a *Smart Grid* [2,8], there is today a renewed interest in HDR NB-PLC solutions operating in the CENELEC/FCC/ARIB bands. This section on NB-PLC will primarily focus on current HDR-PLC systems. As in many other communication systems, OFDM has emerged as the modulation scheme of choice. Examples of HDR NB-PLC systems are G3-PLC [25] and PRIME [26] that were initially designed as proprietary technologies within industry alliances, specifically, the *G3-PLC Alliance* and the *PRIME Alliance*, but have recently become International standards as ITU-T Recommendation G.9903 [14] and ITU-T Recommendation G.9904 [15], respectively. The processing power requirements of HDR NB-PLC are moderate, so that implementation of multiple technologies in the same device is possible, for example, via *digital signal processing* (DSP) platforms. This allows for upgradeability via software updates, and can especially be an advantage when considering that Smart Grid devices have a very long obsolescence horizon and utilities rely on equipment in the field for a few decades [8]. The technically challenging bits are related to higher layers where hundreds or even thousands of, for example, smart energy meters are connected to a single PLC network. With respect to coexistence, the US *National Institute of Standards and Technology* (NIST) *priority action plan 15* (PAP15) working group* recently approved that newly developed NB-PLC standards shall all implement a coexistence protocol as to 'have minimal performance impact on the existing deployed devices', that is, devices using ISO/ IEC 14543-3-5 or a standard from the IEC 61334 family [27]. Furthermore, the PAP15 working group also agreed to mandate a single NB-PLC coexistence protocol for the newer OFDM-based technologies. This NB-PLC coexistence protocol is being specified in IEEE P1901.2.

In the next subsections, we give an overview of PRIME and G3-PLC, whereas more details on IEEE P1901.2 [16] and the family of NB-PLC ITU-T Recommendations [13–15] will be given in Chapter 11. PRIME and G3-PLC were both designed initially as proprietary technologies by the PRIME Alliance and the G3-PLC Alliance, respectively, and are mentioned here for historical reasons since they were the first OFDM-based HDR NB-PLC to be developed.

10.2.1 PRIME

PRIME was developed within the PRIME Alliance, with its steering committee chaired by the Spanish utility Iberdrola [28]. In 2012, PRIME version 1.3.6 became an international standard – see ITU-T Recommendation G.9904 [15].

---

* Working group title: 'Harmonize Power Line Carrier Standards for Appliance Communications in the Home'.

The PRIME system uses a total of 96 ODFM subcarriers over the frequencies from 42 to 89 kHz, that is, within the CENELEC A-band. Further, it deploys differential *binary, quaternary* and *eight phase shift keying* (BPSK, QPSK and 8PSK) and an optional 1/2 rate convolutional code. Therewith, it is able to achieve a PHY peak data rate of 128.6 kbit/s [29]. The OFDM symbol interval is of 2240 µs including a 192 µs cyclic prefix which suffices to deal with most common power line delay spreads. In order to deal with unpredictable impulsive noise, PRIME offers the option to implement *automatic retransmission request* (ARQ), based on the *selective repeat* mechanism [30].

Turning to the system architecture, PRIME is forming subnetworks, where each subnetwork has one *base node* and several *service nodes*. The base node is the 'master' that manages the subnetwork's resources and connections using a periodically sent *beacon* signal. The base node is further responsible for PLC channel access arbitration. A contention-free and a contention-based access mechanism exists, whose usage time and duration are decided by the base node. Within the contention-free *time division multiplex* (TDM) channel access period, the base node assigns the channel to only one node at a time. The contention-based access uses *carrier sense multiple access with collision avoidance* (CSMA/CA) [26,29].

To interface *medium access control* (MAC) and *application layer*, PRIME defines a *convergence layer* (CL) between the two. The CL can be split into a *common part convergence sublayer* (CPCS) and a *service-specific convergence sublayer* (SSCS). The CPCS performs tasks of data segmentation and reassembling and is adjusted to the specific application. Three SSCS are currently defined: The '*NULL Convergence Sublayer* provides the MAC layer with a transparent way to the application, being as simple as possible and minimising the overhead. It is intended for applications that do not need any special convergence capability'. 'The *IPv4 Convergence Layer* provides an efficient method for transferring IPv4 packets over the PRIME network'.

Finally, the *IEC 61334-4-32 CL* 'supports the same primitives as the IEC 61334-4-32 standard' [31], making it easy, for example, to support advanced metering applications that make use of the standardised data models of IEC 62056-62 [32]. PRIME could therefore also be used to replace the aging PHY and MAC layer of the single-carrier power line standard IEC 61334-5-1 [23], also known as S-FSK.

10.2.2 G3-PLC

The other OFDM-based HDR NB-PLC specification G3-PLC [33–35] was published in August 2009. In 2012, members of the G3-PLC Alliance brought the original G3-PLC specs into the ITU-T where they were enhanced and additional functionalities were added. The enhanced G3-PLC became an International Standard in 2012 as ITU-T Recommendation G.9903 [14].

G3-PLC can be configured to operate in the internationally accepted bands from 10 to 490 kHz (FCC, CENELEC, ARIB). Using *differential* BPSK, QPSK and 8PSK for constellation mapping, time–frequency interleaving, concatenated convolutional Reed–Solomon *forward error correction* (FEC) and the option of robust modes (ROBO, for *robust OFDM*), it reaches PHY peak data rates close to 300 kbit/s. Peak and typical data rates for various frequency bands have been reported in [36]. A comparison of the rates achievable by all the HDR NB-PLC technologies considered here is also given in Tables 11.2 and 11.3.

The MAC layer is based on IEEE 802.15.4-2006 [37]. 6LoWPAN [38] is used to adapt the IEEE 802.15.4-2006 MAC to IPv6 [39]. This allows the application layer to comply with ANSI C12.19/C12.22 [40], or IEC 62056-61/62 (DLMS/COSEM) [41,42] to run standard Internet services. A detailed description of the additional features that were introduced in G.9903 is given in Chapter 11.

A comparison between PRIME and G3-PLC, mainly focusing on PHY aspects, is presented in [43]. There it is found that G3-PLC performs much more robustly than PRIME, while PRIME's design paradigm is low complexity. Additional considerations on the comparison of the various NB-PLC technologies are also made in Section 11.7.

10.3 BB-PLC

In the last decade, BB-PLC chips by semiconductor vendors, such as Intellon*, DS2†, Gigle‡ and Panasonic, came to market that operate in the band from around 2 to 86 MHz and, in some cases, optionally up to 300 MHz. The chips are mainly based on three consortia backed specifications developed within the frameworks of the *HomePlug Powerline Alliance* (HomePlug), the *Universal Powerline Association* (UPA) and the *High-Definition Power Line Communication* (HD-PLC) Alliance. Related products allow data rates around 200 Mbit/s and are not interoperable.

However, to make BB-PLC systems a broad success, internationally adopted standards became essential. The ITU-T, as well as the IEEE, commenced work on such next-generation standards, namely, *ITU-T G.hn* [44] and *IEEE 1901* [45].

ITU-T G.hn is not only applicable to power lines but also to phone lines and coaxial cables, therewith for the first time defining a single standard for all major wireline communications media. At the end of 2008, the PHY layer and the overall architecture were consented in ITU-T Recommendation G.9960 [46]. The *data link layer* (DLL) Recommendation G.9961 [47] was approved in June 2010, and a *multiple-input multiple-output* (MIMO) transceiver extension G.9963 was consented in September 2011 [48]. Alongside, the *HomeGrid Forum* was founded to promote the ITU-T G.hn standard and to address certification and interoperability issues [49]. Details on ITU-T G.hn may be found in Chapter 12.

Simultaneously, IEEE P1901 [50] was working on the 'Standard for Broadband over Power Line Networks: Medium Access Control and Physical Layer Specifications' [51]. It covers the aspects access, in-home, as well as coexistence of access–in-home and in-home–in-home networks, and the official IEEE standard 1901-2010 was published in December 2010. To assure a broad industrial backing of IEEE 1901, two optional PHY technologies, namely, *FFT-PHY* (based on HomePlug AV) and *wavelet-PHY* (based on HD-PLC), were included in the standard. The two resulting PHY layers are not interoperable, but a mandatory coexistence protocol, *inter-system protocol* (ISP), was included to assure their coexistence. The HomePlug Powerline Alliance [52] serves as the certifying body for IEEE 1901 FFT-PHY compliant products, whereas the HD-PLC Alliance serves as the certifying body for IEEE 1901 wavelet-PHY compliant products. While IEEE 1901 wavelet-PHY/HD-PLC is presently mainly used on the JP market, IEEE 1901 FFT-PHY/HomePlug AV is used in many countries around the globe, with products of the HomePlug family currently possibly being the most deployed BB-PLC technology worldwide. Details on IEEE 1901 may be found in Chapter 13.

In analogy to the introduction of MIMO to ITU-T G.hn, the HomePlug Alliance introduced the HomePlug AV2 specification in January 2012. The HomePlug AV2

---

* In 2009 acquired by Atheros; Atheros in 2011 acquired by Qualcomm.
† In 2010 acquired by Marvell.
‡ In 2010 acquired by Broadcom.

specification includes features like MIMO with beamforming, an extended frequency range of up to 86 MHz, efficient notching, several transmit power optimisation techniques, 4096-*quadrature amplitude modulation* (QAM), power save modes, short delimiter and delayed acknowledgement, boosting the maximum PHY rate to around 2 Gbit/s (see Chapter 14 for details). Further, to cover multiple home networking media under one umbrella, IEEE P1905.1 devised a 'Standard for a Convergent Digital Home Network for Heterogeneous Technologies' [53]. It defines an abstraction layer for multiple home networking technologies like IEEE 1901, IEEE 802.11 (Wi-Fi), IEEE 802.3 (Ethernet) and MoCA 1.1 and is extendable to work with other home networking technologies. Further details on IEEE 1905.1 may be found in Chapter 15.

10.3.1 IEEE 1901 and ITU-T G.hn

IEEE 1901 uses the band from 2 MHz up to 50 MHz with services above 30 MHz being optional. ITU-T G.hn (G.9960/G.9961) operates from 2 up to 100 MHz using bandwidth scalability, with three distinct and interoperable bands defined as 2–25, 2–50 and 2–100 MHz. The architectures defined by IEEE 1901 and ITU-T G.hn (G.9960/G.9961) are similar in several aspects. In ITU-T G.hn, one refers to a subnetwork as *domain*. Operation and communication are organised by the *domain master* who communicates with various *nodes*. Similarly, the subnetwork in IEEE 1901 is referred to as *basic service set* (BSS). The equivalent to the domain master is the *BSS manager,* which connects to the so-called stations. These network items with their system-specific terminology are summarised in Table 10.1.

Even if many features appear to be individually developed by ITU-T and IEEE, several are actually identical. The fact that ITU-T G.hn and IEEE 1901 largely agree on channel coherence time, coherence bandwidth, guard interval, roll-off window timings, etc. shows that the BB-PLC channel is analysed similarly and that channel difference for comparable topologies is not very different around the globe. Similarities continue with PHY-frame header settings making use of QPSK, FEC code rate 1/2 and repetition codes. The segmentation process of embedding the application data into PLC convenient packets is similar and data are in both cases encrypted using AES-128 [54]. The MAC cycle or beacon period is selected to be two line cycles. The bit loading of carriers can be line cycle–dependent and immediate, and delayed acknowledgments are possible.

If ITU-T G.hn and IEEE 1901 modems are installed in the same home, one transmission is an unintended interferer for the other one. Confusion is avoided due to different preambles that are sent at the beginning of a frame. A correlator at the receiver is able to identify precise timing information of a frame by detecting the inverted transmitted preamble symbols. A PLC transmission is identified by verifying the timing results of the correlated received signals with an ITU-T G.hn or an IEEE 1901 mask. A multi-mode receiver could identify both correlation masks and could forward the subsequent data signals to the corresponding decoding chain.

Another likely scenario is that same technology networks exist in close proximity, with the risk of the so-called *neighbouring network* interference. To deal with neighbouring network interference, ITU-T G.hn uses different preamble symbol seeds in each network. Therewith, ITU-T G.hn networks are able to coexist and communicate simultaneously, that is, not using time division. Instead, link adaptation procedures adjust the throughput to cope with degraded *signal to interference plus noise ratios* (SINR). In many cases, the throughput will be throttled only slightly allowing ITU-T G.hn networks to coexist nearly unimpeded. On the other hand, IEEE 1901 relies on a CSMA/CA medium access strategy,

TABLE 10.1

Synopsis of Terms used in the BB-PLC Standards ITU-T G.hn and IEEE 1901

Item	ITU-T G.hn	IEEE 1901
Subnetwork	Domain	BSS
Transceiver	Node	*Station* (STA)
Subnetwork controller	*Domain master* (DM)	BSS manager/*central coordinator* (CCo)
Layer 2 (L2) of the OSI model	*Data link layer* (DLL) including application, protocol convergence	*Medium access control* (MAC) with isolated convergence layer
Relaying transceiver	Relay (L2)	Repeater (L2)
Network controller proxy	Relay (assigned as a proxy)	Proxy BSS manager
Allocated frequencies	Frequency bandplan	Spectrum mask
Time frame	MAC cycle	Beacon interval
Time between frames	Inter-frame gap	Inter-frame space
Synchronisation and training symbol	Preamble	Preamble
Information broadcasted at the beginning	PHY-frame header (168 bit)	Frame control (128 bit)
Robust transmission	*Robust communication mode* (RCM)	*Robust OFDM* (ROBO) mode
SINR estimation signals	Probe	Sound
SINR feedback info	*Bit allocation table* (BAT)	Tone map
Smallest data packet	*Logical link control* (LLC) *protocol data unit* (LPDU)	PHY block
Encryption responsible	LLC	BSS manager
Link set-up and QoS responsible	DLL management	Connection manager
Access method	CSMA/CA, TDMA, STXOP (*shared transmission opportunities*)	CSMA/CA, TDMA
Access control schedule	*Media access plan* (MAP)	Beacon
Interface to higher layers	A-Interface	H1 Interface

Source: Extended from Table 7.7 in Berger, L.T., *Smart Grid – Applications, Communications and Security*, Berger, L.T. and Iniewski, K., Eds., John Wiley & Sons, Hoboken, NJ, April 2012, ch. 7. Copyright © 2012 John Wiley & Sons. With permission.

which may lead to an increased number of collisions. As countermeasure, IEEE 1901 introduces a *coordinated mode* that allows neighbouring networks to allocate times over the shared medium for specific communications. This coordinated *time division multiple access* (TDMA) mode enables traffic to get through unimpeded albeit at the price of time division (orthogonal throughput sharing).

Set aside these minor differences, two very different FECs, that is, *low-density parity-check code* (LDPC) in ITU-T G.hn and Turbo Code in IEEE 1901, were chosen – see [55] for a comparative analysis. Some have argued that this makes it more difficult (or costly) to implement both standards in a single chip, as the FEC part is up to the present day a non-negligible cost/space factor when manufacturing wafers. However, dual mode devices have already started to appear on the market.

In terms of data rate and silicon cost, the full-fletched ITU-T G.hn and IEEE 1901 systems are targeting primarily in-home data delivery, web browsing as well as audio and video distribution. To offer an alternative to automation and energy management tasks with respect to HDR NB-PLC as introduced in Section 10.2, ITU-T G.hn includes a *low complexity profile* (LCP), while HomePlug developed on the basis of IEEE 1901 the *HomePlug Green PHY* specification. These are outlined in the following two subsections, respectively.

10.3.2 ITU-T G.hn LCP

It is envisioned that G.hn nodes are in the future embedded into *Smart Grid home* (SGH) *area network* devices. SGH nodes will typically make use of the ITU-T G.hn LCP, operating in the frequency range from 2 to 25 MHz while still being interoperable with the full G.hn profile. This allows for reduced component cost and power consumption. Example SGH nodes could be heating and air-conditioning appliances, as well as *plug-in electric vehicles* (PEVs) and *electric vehicle supply equipment* (EVSE). Together they form a multi-domain *home area network* (HAN).

The SGHs interact with the *utility's access network* (UAN) and its AMI through an *energy service interface* (ESI). The AMI domain comprises *AMI meters* (AMs), *AMI submeters* (ASMs), as well as an AMI *head end* (HE). The HE is a local hub (concentrator) that controls all meters downstream from it and interfaces to the utility's wide area/backhaul network upstream from it. Each AMI HE supports up to 250 AM and/or ASM nodes forming an AMI domain (in dense urban areas, 150–200 m is a frequently encountered maximum). Further, a network supports up to 16 AMI domains, delivering support for up to $16 \times 250 = 4000$ AMI devices. The ability to support 16 domains with 250 nodes each is a general property of G.hn not limited to Smart Grid/AMI applications. Domains may be formed over any kind of wiring. The nodes within a domain are grouped into SGH and non-SG nodes. For security reasons, non-SG nodes are logically separated from SGH nodes using a secure upper-layer protocol.

In every domain, there is a domain master that coordinates operation of all nodes. G.hn nodes of different domains communicate with each other via *inter-domain bridges* (IDBs). IDBs are simple data communications bridges on OSI Layer 3 and above, enabling a node in one domain to pass data to a node in another domain. In a multi-domain situation, a *global master* (GM) provides coordination of resources, priorities and operational characteristics between G.hn domains. Besides, ITU-T G.hn domains can be bridged to alien (non-G.hn) domains, for example, to IEEE 1901/1901.2 and wireless technologies. For example, besides the UAN/AMI connection through the ESI, the HAN might be connected to the outside world via a digital subscriber line or cable modem gateway communicating with the ITU-T G.hn HAN via an alien domain bridge.

10.3.3 HomePlug GreenPhy

In analogy to the ITU-T G.hn LCP, the HomePlug Powerline Alliance has released the *HomePlug GreenPhy* (Home Plug GP) specification. HomePlug GP is a subset of HomePlug AV that is intended for use within Smart Grid applications. It was developed specifically to support applications on the HAN within the customer premises. A means of reducing cost and power consumption has to be found while maintaining HomePlug AV/IEEE 1901 interoperability, as well as preserving reliability and coverage. Optimised for low-power applications and costs, HomePlug GP uses the most robust communication mode of HomePlug AV technology. OFDM carrier spacing, preamble, frame control and FEC are identical to HomePlug AV/IEEE 1901 resulting in identical coverage and reliability. CSMA/CA is used as channel access scheme. Further, nodes may use long power save periods if a higher latency is acceptable. In the sleep state, modems have only a 3% power consumption compared to the awake time resulting in an average power reduction of >90% with respect to standard HomePlug AV products.

One of the biggest differences between HomePlug AV and HomePlug GP is the peak PHY rate. HomePlug AV supports a peak PHY rate of 200 Mbit/s, which is simply not required for Smart Grid applications. Based on extensive discussion with the utility industry, it was learned that coverage and reliability were paramount considerations. Peak data rate could be reduced, which in turn would allow for reductions in both cost and power consumption. HomePlug GP supports a peak PHY rate of 10 Mbit/s, which is the result of two key simplifications:

1. Restriction of OFDM subcarrier constellation mapping to QPSK
2. Restriction to data rates supported by ROBO (*robust OFDM*) modes, thereby eliminating the need for adaptive bit loading and management of tone maps

By making exclusive use of QPSK, the analogue front-end and line driver requirements (linearity and converter resolution) are less stringent. As a result, HomePlug GP devices should be able to achieve a higher level of integration, including single-chip architectures, which will help reduce cost and footprint.

The HomePlug GP MAC shares the same CSMA and priority resolution mechanisms as HomePlug AV. It does not, however, support the optional TDMA mechanism.

These measures were essential to enable low-cost, low-power devices that interoperate with HomePlug AV/IEEE 1901 while maintaining the same robust coverage and reliability. A comparative overview on the differences between HomePlug AV and HomePlug GP is presented in Table 10.2.

There are some features that are unique to the HomePlug GP MAC. In order to ensure that HomePlug GP devices would not adversely affect network throughput of HomePlug

TABLE 10.2

HomePlug AV and HomePlug GreenPhy Comparison

Parameter/Function	HomePlug AV	HomePlug GP
Frequency spectrum	2–30 MHz	2–30 MHz
Frequency division multiplexing	OFDM	OFDM
Number of subcarriers	1155	1155
Subcarrier spacing	24.414 kHz	24.414 kHz
Bit-loading constellation mapping	BPSK, QPSK, 16 QAM, 64 QAM, 256 QAM, 1024 QAM	QPSK
FEC type	Turbo code	Turbo code
FEC rate	1/2 and 16/21 (punctured)	1/2
Robust mode data rate	4–10 Mbit/s	4, 5 and 10 Mbit/s
Adaptive bit-loading data rate	20–200 Mbit/s via pre-negotiated tone maps	No (instead use of mini ROBO 3.8 Mbit/s, standard ROBO 4.9 Mbit/s and high-speed ROBO 9.8 Mbit/s)
Channel access	CSMA/CA with optional TDMA	CSMA/CA
Central coordinator capability	Yes	Yes (limited modes)
Power save mode	No	Yes (identical to HomePlug AV2)
Bandwidth sharing	No (channel access CSMA/CA, TDMA)	*Distributed bandwidth control* (DBC)

Source: Based on HomePlug Powerline Alliance, HomePlug Green PHY 1.1 – The standard for In-Home Smart Grid Powerline Communications: An application and technology overview, HomePlug Powerline Alliance, Technical Report, October 2012, white Paper, Version 1.02, http://www.homeplug.org/tech/whitepapers/HomePlug_Green_PHY_whitepaper_121003.pdf (accessed February 2013). Copyright © 2012 HomePlug Powerline Alliance, Inc. With permission.

AV devices, an ad hoc bandwidth sharing algorithm that limits HomePlug GP's *time on wire* (ToW) was included in the HomePlug GP specifications. In addition, a specific routing protocol was implemented allowing for repeater functionality. A novel power-saving mechanism was also added. Finally, a method for characterising signal-level attenuation was included to facilitate association and binding of electric vehicles with charging equipment in public parking areas.

If the slower HomePlug GP devices operating in the presence of heavy HomePlug AV voice or video traffic were able to access the medium in an unconstrained manner, it is possible that HomePlug AV throughput could be adversely affected. As a countermeasure, *distributed bandwidth control* (DBC) was developed. When traffic at various *channel access priorities* (CAP) is detected, DBC will limit aggregate HomePlug GP channel access time, or ToW to ~7%. This corresponds to an effective PHY rate of 700 kbit/s (7% ToW at 10 Mbit/s) and a MAC throughput rate of 400–500 kbit/s, which provides ample capacity for Smart Grid applications. HomePlug GP clients monitor all HomePlug GP transmissions in a two line cycle sliding window that immediately precedes any attempt to contend for channel access. This is possible because HomePlug GP packets have a special flag in the *start of frame* (SoF) delimiter. If the pending packet will cause aggregate HomePlug GP ToW to exceed 7%, the HomePlug GP client cannot contend for channel access at a CAP3 (highest) priority and must wait for an ensuing channel access opportunity or the next window. In most instances, the local medium will not be completely occupied. HomePlug GP devices may exploit unused ToW by contending for channel access at CAP0 (lowest priority). HomePlug GP equipment installed in locations in which HomePlug AV equipment is not present may occupy up to 100% ToW.

The routing and repeating feature implemented in HomePlug GP is interoperable with the repeating functionality specified in IEEE 1901 (see Chapter 13). The main purpose of the repeating functionality is to provide an extension of the HomePlug network coverage when some stations in the network are too distant to allow error-free communication. To support repeating, each HomePlug GP station maintains a *local routing table* (LRT), containing routing information for every associated station in the network. The table includes information such as the identifier of the next station through which to route, the *route data rate* (RDR) and the *route number of hops* (RNH). Regularly, a specific procedure updates the routing tables. New routes can be selected, depending on the computed RDR and RNH for the candidate routes.

As already introduced in connection with ITU-T G.hn LCP, PEV charging is a major Smart Grid application, with upcoming need to provide charging facilities at home, at work and in public parking areas. In order to ensure error-free billing, it is necessary to unambiguously resolve which PEV is physically connected to a charging spot, that is, an EVSE. In order to reliably perform PEV/EVSE association, a special feature called *signal-level attenuation characterisation* (SLAC) was designed into HomePlug GP. A PEV invokes SLAC broadcasting a message. Any available EVSE within hearing range computes the PEV's signal strength and reports back. The EVSE with the highest received signal strength is identified as the correct EVSE, and the two devices set-up a private network for the duration of the charging session. Details of how to use HomePlug GP SLAC in automotive applications can be found in ISO/IEC 15118-3 [56].

Finally, reduced power consumption is a critical factor for Smart Grid applications. A special power save mode has hence been implemented within HomePlug GP. This power save mode is also an integral part of HomePlug AV2 with details described in Chapter 14.

10.3.4 BB-PLC Coexistence and Interoperability

Despite the similarities between the BB-PLC systems, one should note that G.hn defines a PHY/DLL used for operation over any wireline medium. The OFDM parameters are adjusted to account for different medium-dependent channel and noise characteristics and to allow for straightforward scaling when adjusting the parameters between one and another medium. On the contrary, IEEE 1901 defines two disparate PHY/MAC technologies based on HomePlug AV and HD-PLC. One of the key differences is their frequency division multiplexing scheme. The HomePlug AV-based version uses the *fast Fourier transform* (FFT), while the HD-PLC-based version uses wavelets. Hence, they are sometimes also referred to as FFT-PHY and wavelet-PHY, respectively. A special coexistence mechanism has to be used when operating IEEE 1901 devices from both PHYs on the same power line which is standardised within IEEE 1901 as ISP (see also [57,58]). A nearly identical mechanism was standardised by ITU-T in Recommendation G.9972 [59], also known as G.cx. Technical contributions to ITU-T and IEEE from members of the NIST PAP15 assured the alignment of both standards. As a result, the NIST PAP 15 recommended to mandate that all BB-PLC technologies implement Recommendation ITU-T G.9972 or ISP [60] (see also [8] Sect. III.E.).

The ISP protocol allows a TDM scheme to be implemented between coexisting in-home systems and between coexisting in-home and access systems. Each of the PLC system categories is allocated a particular ISP window in a round robin fashion. The allocation is determined by (1) the number of systems on the power line, (2) the type of the systems present and (3) the systems' bandwidth requests. The TDM synchronisation period for the in-home and access systems is defined with the parameter T_H in Figure 10.2. There are four ISP time slots (T_{ISP}) within a single T_H period, one for *access modems* (ACC), *in-home wavelet* (IH-W), *in-home OFDM* (IH-O) and in-home G.hn (IH-G). Each ISP time slot is further divided into three *TDM units* (TDMUs), leading to a total of 12 TDMUs in each T_H period. Each TDMU is further divided into eight *TDM slots* (TDMSs), labelled TDMS#0 through TDMS#7. Figure 10.2 also illustrates the TDM partitioning relative to the AC line cycles. The ISP window is used to generate and detect the ISP signal that is allocated within the first TDMS#0 in TMDU#0, TMDU#3, TMDU#6 and TDMU#9. The duration of a TDMU is equivalent to the *beacon interval/MAC cycle*, that is, two AC line cycles.

Coexistence signalling is carried out by the use of periodically repeating ISP signals within the ISP time slots. The phase of the transmitted ISP signal conveys the coexistence information. This set of instantaneous information is termed the *network status* that defines the allocation of resources to each coexisting system. By monitoring the ISP signal transmitted within the ISP time slots allocated to other systems, a coexisting system is able to determine the number and type of coexisting systems present on the line and their resource requirements. Similarly, by monitoring the signal within its own ISP time slot, a coexisting system is able to detect a resynchronisation request from one of the other coexisting systems. The ISP signal consists of 16 consecutive short OFDM symbols, each of T_s duration. Each symbol is formed by a set of 'all-one' BPSK sequences. The 16 symbols are multiplied by a window function with the length of T_W to reduce out-of-band energy complying with the transmit spectrum requirement. Since all devices send the signal simultaneously, the ISP signal must be sent with 8 dB less power than the normal transmission.

The TDM synchronisation scheme is used such that each PLC system shares the medium without interfering with others. However, it is possible that two or more systems are synchronised to two or more different, mutually visible ISP sequences [51, Annex R]. In such cases, in order to prevent mutual interference, it is important that they resynchronise to the same ISP sequence. In other words, whenever a BB-PLC device starts up or restarts,

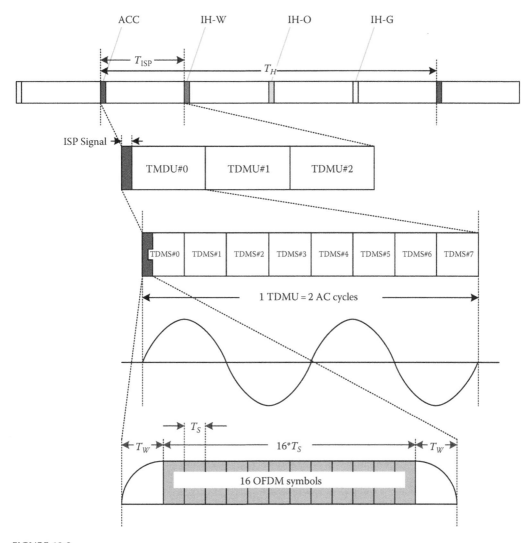

FIGURE 10.2
Time division multiplexing on ISP and timing parameters. (From Yonge, L. et al., *Hindawi J. Electric. Comput. Eng.*, Article ID 892628, 20, 2013, http://downloads.hindawi.com/journals/jece/2013/892628.pdf. Copyright © 2013.)

it needs to be aware of the presence of any other systems with which it is able to coexist. Accordingly, a start-up and resynchronisation procedure is defined within the ISP.

10.4 Conclusions

This chapter provided an overview of UNB-, NB- and BB-PLC systems and standards. It should be noted that more in-depth information on the ITU-T family of NB-PLC Recommendations and IEEE 1901.2 is presented in Chapter 11. Further, Chapters 12 through 14 present details on ITU-T G.hn, IEEE 1901 and HomePlug AV2, respectively. Finally, IEEE 1905.1 enriches home networking by creating a common layer on top of PLC,

IEEE 802.11 (Wi-Fi), IEEE 802.3 (Ethernet) and MoCA technologies. This allows seamless hybrid networking increasing the coverage to transport, for example, several high-definition video streams to remote locations within a home (for details, see Chapter 15).

Acknowledgements

The work has been partially supported by the Spanish Ministry of Science and Innovation (MICINN) Program INNCORPORA-Torres Quevedo 2011.

References

1. L. T. Berger, Broadband powerline communications, in *Convergence of Mobile and Stationary Next Generation Networks*, K. Iniewski, Ed. Hoboken, NJ: John Wiley & Sons, 2010, ch. 10, pp. 289–316.

2. L. T. Berger, Wireline communications in smart grids, in *Smart Grid – Applications, Communications and Security*, L. T. Berger and K. Iniewski, Eds. Hoboken, NJ: John Wiley & Sons, April 2012, ch. 7.

3. HomePlug Powerline Alliance, HomePlug Green PHY 1.1 – The standard for In-Home Smart Grid Powerline Communications: An application and technology overview, HomePlug Powerline Alliance, Technical Report, October 2012, White Paper, Version 1.02, http://www. homeplug.org/tech/whitepapers/HomePlug_Green_PHY_whitepaper_121003.pdf (accessed February 2013).

4. L. T. Berger, A. Schwager and J. J. Escudero-Garzás, Power line communications for Smart Grid applications, *Hindawi Publishing Corporation Journal of Electrical and Computer Engineering*, Article ID 712376, 1–16, 2013, received 3 August 2012; accepted 29 December 2012, Academic Editor: Ahmed Zeddam, http://www.hindawi.com/journals/jece/aip/712376/.

5. L. Yonge, J. Abad, K. Afkhamie, L. Guerrieri, S. Katar, H. Lioe, P. Pagani, R. Riva, D. Schneider and A. Schwager, An overview of the HomePlug AV2 technology, *Hindawi Journal of Electrical and Computer Engineering*, Article ID 892628, 20, 2013, http://downloads.hindawi.com/journals/jece/2013/892628.pdf.

6. K. Dostert, Telecommunications over the power distribution grid – Possibilities and limitations, in *International Symposium on Power Line Communications and Its Applications (ISPLC)*, Essen, Germany, April 1997, pp. 1–9.

7. P. A. Brown, Power line communications – Past present and future, in *International Symposium on Power Line Communications and Its Applications (ISPLC)*, Lancaster, UK, September 1999, pp. 1–8.

8. S. Galli, A. Scaglione and Z. Wang, For the grid and through the grid: The role of power line communications in the smart grid, *Proceedings of the IEEE*, 99(6), 998–1027, June 2011.

9. D. Nordell, Communication systems for distribution automation, in *IEEE Transmission and Distribution Conference and Exposition*, Bogota, Colombia, April 2008, pp. 1–14.

10. S. Mak and D. Reed, TWACS, a new viable two-way automatic communication system for distribution networks. Part I: Outbound communication, *IEEE Trans. Power App. Syst.*, 101(8), 2941–2949, August 1982.

11. S. Mak and T. Moore, TWACS, a new viable two-way automatic communication system for distribution networks. Part II: Inbound communication, *IEEE Trans. Power App. Syst.*, 103(8), 2141–2147, August 1984.

12. R. van Nee and R. Prasad, *OFDM for Wireless Multimedia Communications*, ser. Universal personal communication. Artech House Publishers, London, U.K., 2000.

13. International Telecommunications Union (ITU) – Telecommunication Standardization Sector STUDY GROUP 15, Narrowband orthogonal frequency division multiplexing power line communication transceivers for G.hnem networks, Recommendation ITU-T G.9902, October 2012.
14. International Telecommunications Union (ITU) – Telecommunication Standardization Sector STUDY GROUP 15, Narrowband orthogonal frequency division multiplexing power line communication transceivers for G3-PLC networks, Recommendation ITU-T G.9903, October 2012.
15. International Telecommunications Union (ITU) – Telecommunication Standardization Sector STUDY GROUP 15, Narrowband orthogonal frequency division multiplexing power line communication transceivers for PRIME networks, Recommendation ITU-T G.9904, October 2012.
16. IEEE 1901.2: Draft standard for low frequency (less than 500 kHz) narrow band power line communications for Smart Grid applications, http://grouper.ieee.org/groups/1901/2/.
17. Watteco, Next generation wireless IP sensors for the Internet of things, http://www.watteco.com/(accessed April 2013).
18. P. Bertrand, O. Pavie and C. Ripoll, Watteco's WPC: Smart, safe, reliable and low power automation for home, White Paper, La Garde, France, December 2008.
19. American National Standards Institute/Electronic Industries Association (ANSI/EIA), Control network power line (PL) channel specification, September 2006, ANSI/CEA-709.2-A.
20. International Organization for Standardization, Interconnection of information technology equipment – Control network protocol – Part 3: Power line channel specification, January 2011, International Standard ISO/IEC 14908-3, Revision 11.
21. International Organization for Standardization, Information technology – Home electronic system (HES) architecture – Part 3–5: Media and media dependent layers – Powerline for network based control of HES class 1, May 2007, international standard ISO/IEC 14543-3-5, First edition.
22. International Electrotechnical Commission (IEC), Distribution automation using distribution line carrier systems – Part 5-2: Lower layer profiles – Frequency shift keying (FSK) profile, Geneva, Switzerland, Standard IEC 61334-5-2, Ed. 1.0, 1998.
23. International Electrotechnical Commission (IEC), Distribution automation using distribution line carrier systems – Part 5-1: Lower layer profiles – The spread frequency shift keying (S-FSK) profile, Standard IEC 61334-5-1, Ed. 2.0, 2001.
24. International Electrotechnical Commission (IEC),Electricity metering – Data exchange for meter reading, tariff and load control – Part 53: COSEM application layer, Standard IEC 62056-53, Ed. 2, December 2006.
25. Électricité Réseau Distribution France, G3-PLC: Open standard for Smart Grid implementation, http://www.maxim-ic.com/products/powerline/g3-plc/ (accessed April 2013).
26. PRIME Alliance, Draft standard for PoweRline Intelligent Metering Evolution, 2010, www.prime-alliance.org/Docs/Ref/PRIME-Spec_v1.3.6.pdf (accessed March 2013).
27. NIST Priority Action Plan 15, Narrowband PLC coexistence requirement, http://collaborate.nist.gov/twiki-sggrid/pub/SmartGrid/PAP15PLCForLowBitRates/Requirements_on_NB_PLC_coexistence_Final_Oct_11r1.xls (accessed November 2011).
28. PRIME Alliance, Powerline related intelligent metering evolution (PRIME), http://www.prime-alliance.org (accessed March 2013).
29. I. Berganza, A. Sendin and J. Arriola, PRIME: Powerline intelligent metering evolution, in *CIRED Seminar 2008: SmartGrids for Distribution*. Frankfurt, Germany: CIRED, June 2008, pp. 1–3.
30. A. S. Tannenbaum, *Computer Networks*, 4th edn. Englewood Cliffs, NJ: Prentice Hall International, 2003.
31. International Electrotechnical Commission (IEC), Distribution automation using distribution line carrier systems – Part 4: Data communication protocols – Section 32: Data link layer – Logical link control (LLC), November 1997.
32. International Electrotechnical Commission (IEC), Electricity metering – Data exchange for meter reading, tariff and load control – Part 62: Interface classes, Standard IEC 62056-62, Ed. 2, November 2006.

33. Electricité Réseau Distribution France (ERDF), G3-PLC physical layer specification, August 2009, http://www.maxim-ic.com/products/powerline/pdfs/G3-PLC-Physical-Layer-Specification.pdf (accessed February 2011).

34. Electricité Réseau Distribution France (ERDF), G3-PLC MAC layer specification, August 2009, http://www.maxim-ic.com/products/powerline/pdfs/G3-PLC-MAC-Layer-Specification.pdf (accessed February 2011).

35. Electricité Réseau Distribution France (ERDF), G3-PLC profile specification, August 2009, http://www.maxim-ic.com/products/powerline/pdfs/G3-PLC-Profile-Specification.pdf (accessed February 2011).

36. K. Razazian, G3-PLC provides an ideal communication platform for the smart gird, in *IEEE International Symposium on Power Line Communications and Its Applications (ISPLC)*, Rio de Janeiro, Brazil, March 2010, keynote Presentation, http://ewh.ieee.org/conf/isplc/2010/KeynoteAndPanelFiles/9-40-KAVEH.pdf (accessed December 2012).

37. Institute of Electrical and Electronics Engineers, Local and metropolitan area networks – Specific requirements part 15.4: Wireless medium access control (MAC) and physical layer (PHY) specifications for low-rate wireless personal area networks (WPANs), September 2006, standard for Information Technology – Telecommunications and Information Exchange Between Systems.

38. Z. Shelby and C. Bormann, *6LoWPAN: The Wireless Embedded Internet.* Chichester, U.K.: John Wiley & Sons, November 2009.

39. S. Deering and R. Hinden, Internet protocol, version 6 (IPv6) specification, RFC 2460, December 1998, http://tools.ietf.org/html/rfc2460 (accessed February 2011).

40. American National Standards Institute (ANSI), Utility industry end device data tables, ANSI Standard C12.19, 2008.

41. International Electrotechnical Commission (IEC), Electricity metering – Data exchange for meter reading, tariff and load control – Part 61: Object identification system (OBIS), November 2006, International Standard IEC 62056-61, second edition.

42. International Electrotechnical Commission (IEC), Electricity metering – Data exchange for meter reading, tariff and load control – Part 62: Interface classes, November 2006, International Standard IEC 62056-62, second edition.

43. M. Hoch, Comparison of G3 PLC and PRIME, in *IEEE International Symposium on Power Line Communications and Its Applications (ISPLC)*, Udine, Italy, April 2011, pp. 165–169.

44. V. Oksman and S. Galli, G.hn: The new ITU-T home networking standard, *IEEE Commun. Mag.*, 47(10), 138–145, October 2009.

45. S. Galli and O. Logvinov, Recent developments in the standardization of power line communications within the IEEE, *IEEE Commun. Mag.*, 46(7), 64–71, July 2008.

46. International Telecommunications Union (ITU), ITU-T Recommendation G.9960, Unified high-speed wire-line based home networking transceivers – Foundation, August 2009.

47. International Telecommunications Union (ITU), ITU-T Recommendation G.9961, Data link layer (DLL) for unified high-speed wire-line based home networking transceivers, June 2010.

48. International Telecommunications Union (ITU), ITU-T Recommendation G.9963, Unified high-speed wire-line based home networking transceivers – Multiple input/multiple output (MIMO), September 2011 (ex G.hn-MIMO).

49. HomeGrid Forum, For any wire, anywhere in your home, http://www.homegridforum.org/ (accessed February 2011).

50. Institute of Electrical and Electronic Engineers (IEEE), Standards Association, Working group P1901, IEEE standard for broadband over power line networks: Medium access control and physical layer specifications, http://grouper.ieee.org/groups/1901/ (accessed February 2011).

51. Institute of Electrical and Electronics Engineers (IEEE) Standards Association, P1901 working group, IEEE standard for broadband over power line networks: Medium access control and physical layer specification, December 2010, http://standards.ieee.org/findstds/standard/1901-2010.html.

52. HomePlug Powerline Alliance, About us, http://www.homeplug.org/home (accessed February 2011).

53. Institute of Electrical and Electronics Engineers, Standards Association, Working Group P1905.1, IEEE standard for a convergent digital home network for heterogeneous technologies, April 2013, http://standards.ieee.org/findstds/standard/1905.1-2013.html (accessed April 2013).

54. National Institute of Standards and Technology (NIST), U.S. Department of Commerce, Specification for the advanced encryption standard (AES), Federal Information Processing Standards Publication 197, November 2001.

55 S. Galli, On the fair comparison of FEC schemes, in *IEEE International Conference on Communication (ICC)*, Cape Town, South Africa, 23–27 May 2010.

56. International Organization for Standardization (ISO), Road vehicles – Vehicle to grid communication interface – Part 3: Physical and data link layer requirements, 2013, International Standard ISO/DIS 15118-3, under development.

57. S. Galli, A. Kurobe and M. Ohura, The inter-PHY protocol (IPP): A simple coexistence protocol for shared media, in *IEEE International Symposium on Power Line Communications and Its Applications (ISPLC)*, Dresden, Germany, March 2009, pp. 194–200.

58. S. Galli, M. Koch, H. Latchman, S. Lee and V. Oksman, Chap. 7: Industrial and International Standards on PLC-based networking technologies, in *Power Line Communications*, 1st edn., H. Ferreira, L. Lampe, J. Newbury and T. Swart, Eds. New York: John Wiley & Sons, 2010, ch. 7.

59. International Telecommunications Union (ITU), ITU-T Recommendation G.9972, Coexistence mechanism for wireline home networking transceivers, June 2010.

60. D. Su and S. Galli, PAP 15 recommendations to SGIP on broadband PLC coexistence, December 2010, http://collaborate.nist.gov/twiki-sggrid/pub/SmartGrid/PAP15PLCForLowBitRates/PAP15_-_Recommendation_to_SGIP_BB_CX_-_Final_-_APPROVED_2010-12-02.pdf (accessed February 2011).

11

Narrowband Power Line Standards

Stefano Galli and James Le Clare

CONTENTS

In this chapter, we will give an overview of the latest standardisation efforts in the *Institute of Electrical and Electronics Engineers* (IEEE) and the *Telecommunication Standardization Sector of the International Telecommunication Union* (ITU-T) on *narrowband power line communication* (NB-PLC). The list of acronyms used throughout this chapter can be found in Table 11.1.

11.1 Historical Overview

One of the first *low data rate* (LDR) NB-PLC standards ratified is the *American National Standards Institute* (ANSI)/*Electronic Industries Alliance* (EIA) 709.1 standard, also known as LonWorks. Issued by ANSI in 1999, it became an international standard in 2008 (ISO/IEC 14908) [1]. This seven-layer *Open Systems Interconnection* (OSI) protocol provides a set of services that allow the application program in a device to send and receive messages to/from other devices over twisted pair or PLC. Achievable data rates are in the order of few kbps. The most widespread NB-PLC technologies deployed today are based on *frequency shift keying* (FSK) or spread-FSK as specified in the IEC 61334-5-2 [2] and IEC 61334-5-1 [3] standards, respectively. The availability of standards for these technologies goes from recommendations that specify the stack of communications protocols from the physical up to the application layer (IEC 62056-53 for COSEM) thus facilitating the development of interoperable solutions.

The increasing desire for higher data rates and the growing attention that smart metering projects are getting in various national efforts directed at modernising the aging power grid has led to a renewed interest in *high data rate* (HDR) NB-PLC solutions operating in the *Comité Européen de Normalisation Électrotechnique* (CENELEC)/*Federal Communications Commission* (FCC)/*Association of Radio Industries and Businesses* (ARIB) bands which are able to provide higher data rates than LDR NB-PLC [4]. For example, the *Powerline Related Intelligent Metering Evolution* (PRIME) initiative has gained industry support in Europe and has specified an HDR NB-PLC solution based on *orthogonal frequency division multiplexing* (OFDM)

TABLE 11.1

List of Acronyms Used in This Chapter

Acronym	Meaning	Acronym	Meaning
6LoWPAN	IPv6 over Low-power Wireless Personal Area Networks	IoAC	Interleave over AC cycle
		IoF	Interleave over fragment
AC	Alternate current	ISI	Inter-Symbol Interference
ACK	Acknowledgement frame	ISO	International Organization for Standardization
AES	Advanced Encryption Standard		
		ITU-T	International Telecommunication Union – Telecommunication Standardization Sector
AKM	Authentication and Key Management		
AMI	Advanced Metering Infrastructure	LDR	Low data rate
		LF	Low frequency
ANSI	American National Standards Institute	LLC	Logical link control
		LOAD/ LOADng	Lightweight On-demand Ad hoc Distance-vector Routing Protocol – Next Generation
AODV	Ad hoc On-Demand Distance-Vector Routing		
ARIB	Association of Radio Industries and Businesses	LPTV	Linearly and periodically time varying
AWGN	Additive white Gaussian noise	LV	Low voltage
		MAC	Medium access control
BPSK	Binary phase shift keying	MP2P	Multi-point to point
CCM	Counter Cipher Mode	MV	Medium voltage
CENELEC	Comité Européen de Normalisation Électrotechnique	NACK	Negative acknowledgement
		NB-PLC	Narrowband PLC
		NIST	US National Institute of Standards and Technology
CES	Channel estimation symbol		
CFP	Contention-free period	NPCW	Normal-priority contention window
CFS	Contention-free slot		
CP	Contention period	OFDM	Orthogonal frequency division multiplexing
CPCS	Common part convergence sublayer		
		OSI	Open Systems Interconnection
CRC	Cyclic redundancy check	P2MP	Point to multi-point
CSMA/CA	Carrier sense multiple access/ collision avoidance	P2P	Point to Point
		PAN	Personal area network
CW	Contention window	PAP	Priority action plan
DLL	Data link layer	PEV	Plug-in electric vehicle
DM	Domain master	PFH	PHY frame header
DPSK	Differential phase shift keying	PHY	Physical layer
EAP-PSK	Extensible authentication protocol–pre-shared key	PLC	Power line communications
		PRIME	Powerline Related Intelligent Metering Evolution
EIA	Electronic industries alliance		
EMC	Electromagnetic compatibility	PSD	Power spectral density
ERM	Extremely robust mode	PSK	Phase shift keying
EUI	Extended unique identifier	QAM	Quadrature amplitude modulation
FCC	Federal Communications Commission	RCM	Robust communication mode
		RERR	Route Error
FCH	Frame control header		

(continued)

TABLE 11.1 (continued)

List of Acronyms Used in This Chapter

Acronym	Meaning	Acronym	Meaning
FEC	Forward error correction	RPL	Routing protocol for low-power and Lossy networks
FSK	Frequency shift keying		
HDR	High data rate	RREP	Route Reply
HNEM	Home network energy management	RREQ	Route Request
		RS	Reed–Solomon
HPCW	High-priority contention window	SAE	Society of Automotive Engineers
		SCP	Shared contention period
IEC	International Electrotechnical Commission	SDO	Standards Developing Organization
		SGIP	Smart Grid interoperability panel
IEEE	Institute of Electrical and Electronics Engineers	SNR	Signal-to-noise ratio
		SSCS	Service-specific convergence sublayer
IFS	Interframe spacing		

and operating in the CENELEC-A band [5]. A similar initiative, G3-PLC, was initiated in 2008 driven by Maxim Integrated Products, Sagemcom and ERDF who developed the G3-PLC specifications and brought them to the ITU-T. The G3-PLC Alliance was then formed few years later in 2011 [6]. G3-PLC is an OFDM-based HDR NB-PLC technology that can operate in the CENELEC and FCC bands, with many features aimed at improving reliability of communications, for example, adaptive tone mapping, *Advanced Encryption Standard* (AES)-128, concatenated coding, robust modes, and priority-based medium access. Both G3-PLC and PRIME were designed to be open public specifications. Several papers about PRIME and G3-PLC performance have appeared in the past few years [7–12].

Recognising the importance of standardising a next-generation NB-PLC technology, the IEEE Standards Association and the ITU-T started in 2010 the standardisation of next-generation OFDM-based NB-PLC technologies launching the IEEE P1901.2 [13] and ITU-T G.hnem (Home Networking Energy Management) projects, respectively. An effort was made to bring to official standardisation both PRIME and G3-PLC technologies and this was successfully accomplished in the ITU-T Study Group 15 Question 15 (Q15/15) [14]. As a consequence, the original G.hnem project was widened to include also G3-PLC and PRIME. The technical approach initially followed by G3-PLC was later adopted also in IEEE P1901.2 and ITU-T Q15/15, which have both inherited many of the technical features of G3-PLC.

The formation of the IEEE P1901.2 work group started in 2010 with PLC discussions among several companies attending automotive standards meetings. The discussions centred on how to standardise a sub-500 kHz power line communication solution that would meet the automotive specifications in the upcoming SAE J2931/3 [15] and the issued ISO/IEC 15118-3 [16]. At that time, there was no standardisation effort for PLC solutions above the CENELEC band and in the low-frequency (FCC and lower) range. IEEE P1901.2 came to life in fall 2009 with the IEEE Communications Society agreeing to serve as sponsor and the IEEE P1901.2 started its work in 2010. The IEEE P1901.2 Draft is still under study; it passed sponsor ballot and it should be ratified and published in 2014.

The first standards on next-generation OFDM-based NB-PLC to be approved were ITU-T Recommendations G.9955 [17] and G.9956 [18]. These two recommendations contain the *physical layer* (PHY) and *data link layer* (DLL) specifications, respectively, of three NB-PLC technologies: G.hnem, G3-PLC and PRIME:

1. *G.hnem*: A new NB-PLC technology developed by ITU-T members based on G3-PLC and PRIME. The PHY/*medium access control* (MAC) are specified in the main body of the G.9955/G.9956 Recommendations, and the solution operates over CENELEC-A–D bands and the US FCC band.

2. *G3-PLC*: An established and field-proven NB-PLC technology contributed by ITU-T members of the G3-PLC Alliance. The PHY is specified in Annex A (CENELEC-A band) and Annex D (FCC band) of G.9955; the DLL is specified in Normative Annex A of G.9956. The G3-PLC Annexes are normative and stand alone, that is, they can be implemented independently from the main body and the other annexes.

3. *PRIME*: An established and field-proven NB-PLC technology contributed by ITU-T members of the PRIME Alliance. The PHY/MAC are specified in Annex B and the solution operates over CENELEC-A band. The PRIME Annexes are normative and stand alone, that is, they can be implemented independently from the main body and the other annexes.

Basically, Recommendations ITU-T G.9955 and G.9956 define a family of three stand-alone international next-generation NB-PLC standards. These technologies are not interoperable with each other. To reduce confusion in the industry and allow better visibility to the each of the NB-PLC technologies, the three NB-PLC technologies mentioned above and specified in G.9955/G.9956 have been repackaged into three separate recommendations plus a fourth one containing the material with regulatory implication. Recommendations ITU-T G.9955/G.9956 were thus split into the following four separate ITU-T Recommendations which have received final approval in late 2012 and now supersede G.9955/G.9956:

1. G.9901 [19] 'Narrowband OFDM Power Line Communication Transceivers–Power Spectral Density (PSD) Specification'. This recommendation contains all the material in G.9955 with regulatory relevance and implications, such as OFDM control parameters that determine spectral content, PSD mask requirements and the set of tools to support reduction of the transmit PSD.

2. G.9902 (G.hnem) [20] 'Narrowband OFDM Power Line Communication Transceivers for ITU-T G.hnem Networks'. This recommendation contains the PHY and the DLL specifications for G.hnem NB-PLC transceivers. It uses the material in G.9955 and G.9956 as is, specifically using material in its main body and the Annexes that pertain to the main body. This recommendation normatively references G.9901.

3. G.9903 (G3-PLC) [21] 'Narrowband OFDM Power Line Communication Transceivers for G3-PLC Networks'. This recommendation contains PHY and the DLL specifications for G3-PLC NB-PLC transceivers. It uses the material in G.9955 and G.9956 as is, specifically using material in Annexes A and D of G.9955 and Annex A of G.9956. This recommendation normatively references G.9901.

4. G.9904 (PRIME) [22] 'Narrowband OFDM Power Line Communication Transceivers for PRIME Networks'. This recommendation contains the PHY and the DLL specifications for PRIME NB-PLC transceivers. It uses the material in G.9955, G.9956 and G.9956 Amd1 as is, specifically using material in Annex B of G.9955, Annex B of G.9956 and G.9956 Amd1. This recommendation normatively references G.9901.

G3-PLC has evolved since its initial version and this evolution has been included in the Recommendation ITU-T G.9903 (2012) and its 2013 Revision. For example, G3-PLC as defined now in G.9903 has a coherent option in addition to its original mandatory differential modulation, higher-order modulations, and additional bandplans were added (CENELEC-B and ARIB).

11.2 IEEE P1901.2 Working Group Draft

The scope of the IEEE P1901.2 effort as outlined in its *Project Authorisation Request* (PAR) is the following [13]:

> This standard specifies communications for low-frequency (less than 500 kHz) narrowband powerline devices via alternating current and direct current electric powerlines. This standard supports indoor and outdoor communications over a low-voltage line (line between transformer and meter, less than 1000 V), through a transformer low-voltage to medium-voltage (1000 V up to 72 kV), and through transformer medium-voltage to low-voltage powerlines in both urban and in long-distance (multikilometer) rural communications. The standard uses transmission frequencies less than 500 kHz. Data rates will be scalable to 500 kbps depending on the application requirements. This standard addresses grid-to-utility meter, electric vehicle-to-charging station, and within home area networking communications scenarios. Lighting and solar-panel powerline communications are also potential uses of this communications standard. This standard focuses on the balanced and efficient use of the powerline communications channel by all classes of low-frequency narrowband (LF NB) devices, defining detailed mechanisms for coexistence between different LF NB standards developing organizations (SDO) technologies, assuring that desired bandwidth may be delivered. It also ensures coexistence with broadband powerline (BPL) devices by minimizing out-of-band emissions in frequencies greater than 500 kHz. The standard addresses the necessary security requirements that assure communication privacy and allow use for security sensitive services. This standard defines the physical layer and the medium access sublayer of the data link layer, as defined by the International Organization for Standardization (ISO) Open Systems Interconnection (OSI) Basic Reference Model.

The IEEE P1901.2 Working Group is constituted by several subgroups to address solutions for various key areas. These areas included harmonisation technologies operating in low-frequency band; robustness for through-transformer communication; defining limits and testing for *electromagnetic compatibility* (EMC); defining complete coexistence mechanism with existing SDO technologies; and prioritising IP addressing. The EMC subgroup was formed to define EMC limits in areas where regulations are missing as for the FCC band, whereas limits exist for the CENELEC and ARIB bands. The EMC subgroup was additionally tasked to develop test criteria to meet these limits. A NB-PLC coexistence subgroup was also formed to address challenges to manage a simple and fair mechanism that could be globally applicable. To address global regulations, three main bands were defined: the CENELEC band (Europe, CENELEC bands A–D) which has an upper limit of 148.5 kHz, the ARIB band (JP) which has an upper limit of 450 kHz, and the FCC band (multiple countries) which has an upper limit of ~490 kHz. Although these bands have definitive upper limits, it is customary to define subbands within these limits to maximise system parameters for optimal performance in varying conditions and to maximise shared bandwidth. An example of this would be an FCC subband which has a start frequency above the CENELEC bands at 154.7 kHz and a stop frequency at 488.3 kHz. Because of the inherently low EMC emissions (and, consequently, limited emissions), properly defined NB-PLC solutions can transmit in frequency bands with a relatively small guard band without disturbance-related issues.

Each subband is defined with a start and stop frequency and with a specific number of subcarriers (tones) per band. Once the number of carriers is defined, a table is generated that indicates the phase vector definition per carrier. With the known number of carriers per symbol, along with the number of symbols per PHY frame and the number of parity bits added by *forward error correction* (FEC) blocks, the PHY data rate can then be calculated. The number of symbols in each PHY frame is selected based on two parameters: the required data rate and the acceptable robustness.

As mentioned earlier, the IEEE P1901.2 is still under study and is not available publicly so only limited information is publicly available (see [23]). Since the P1901.2 Draft is still under study, the technical information reported here may not be correct and is also subject to change.

11.2.1 Physical Layer

The PHY layer of the IEEE P1901.2 follows closely the PHY layers of G3-PLC. There are some differences that are related to the number of *Reed–Solomon* (RS) code words per frame, the *frame control header* (FCH), the interleaver, and the FCC tone spacing for the FCC bandplan. As work on IEEE P1901.2 is still ongoing, it is not possible at this time to state whether IEEE P1901.2 is interoperable or not with G.9903 (G3-PLC).

11.2.2 Medium Access Control

The MAC layer is an interface between the *logical link control* (LLC) and the PHY. The MAC layer regulates access to the medium using *carrier sense multiple access with collision avoidance* (CSMA/CA). It provides feedback to upper layers via the use of positive (ACK) and negative (NACK) acknowledgement frames, and also performs packet fragmentation and reassembly. Packet encryption/decryption is carried out by MAC as well. The primary areas developed in the MAC include the MAC sublayer service specification, MAC frame formats, MAC command frames, MAC constants, attributes such as *interframe spacing* (IFS), MAC functional description and MAC security-suite specifications.

The IEEE P1901.2 Draft focuses on OSI Layer 1(L1) and parts of OSI Layer 2 (L2) and does not normatively specify any L2 or Layer 3 (L3) routing mechanism. IEEE P1901.2 is agnostic to routing mechanism. In particular, a candidate for L2 routing is the Lightweight *On-demand Ad hoc Distance-vector Routing Protocol – Next Generation* (LOADng) [24]; for the L3 routing, a candidate is the *Routing Protocol for Low-Power and Lossy Networks* (RPL) [25]. LOADng and RPL are further discussed in Sections 11.4.2.2, 11.4.2.3 and 11.7.2.

A unique feature of IEEE P1901.2 that is not present in the ITU-T Recommendations is that it provides an optional adaptive multi-tone mask for the preamble and header.

This allows for PHY/MAC protocols that facilitate additional reliable communication for *low-voltage* (LV)/*medium-voltage* (MV) crossing in the US grids [23].

11.3 Recommendation ITU-T G.9902: G.hnem

G.9902 targets all main Smart Grid applications: *Advanced Metering Infrastructure* (AMI) for residential and business sites, in-home energy management, including *demand-response* (DR) programs and smart appliances, home automation and *plug-in electric vehicle* (PEV) charging. The default G.9902 network layer protocol is IPv6, while others can also be supported using an appropriate convergence sublayer.

A G.9902 network consists of one or more logical domains, and a domain is constituted by all nodes registered in that domain (see Figure 11.1). Each node is identified by its domain ID and node ID. One node in the domain is assigned as a *domain master* (DM). The DM controls operation of all other nodes and performs admission, resignation and other domain-wide management operations. Domains of the same network are connected by *interdomain bridges* (IDB), allowing nodes in different domains to communicate. Any domain may also be bridged to a non-G.9902 (alien) domain.

The PHY can be programmed to operate in different bandplans over CENELEC, FCC and ARIB bands, over different types of power line wiring, such as MV, LV, in-home wiring and *alternate current* (AC) and pilot wires of PEV cables. Both synchronous beacons (sent periodically) and asynchronous beacons (sent at the discretion of the DM) are defined. If a node operating as a DM fails, the DM function is automatically passed to another node of the domain.

The NB-PLC technology specified in G.9902 is based on G3-PLC and PRIME, but is non-interoperable with the other technologies specified in IEEE and ITU-T. There are a few published papers on G.hnem (see [26,27]).

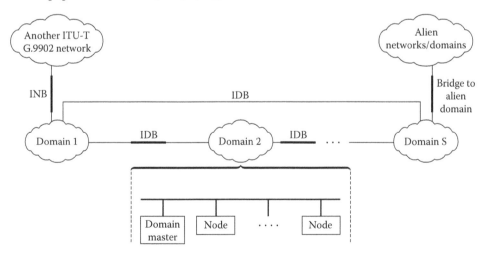

FIGURE 11.1
Generic network architecture. (From Narrowband orthogonal frequency division multiplexing power line communication transceivers for ITU-T G.hnem networks, ITU-T Rec. G.9902, October 2012. [Online] Available: http://www.itu.int/rec/T-REC-G.9902 (Fig. 5-1). With permission.)

11.3.1 Physical Layer

11.3.1.1 Bandplans

G.9902 defines several bandplans. Support of at least one bandplan is mandatory. The CENELEC band (3–148.5 kHz) is divided into three bandplans: CENELEC-A (35.9375–90.625 kHz), CENELEC-B (98.4375–120.3125 kHz) and CENELEC-CD (125–143.75 kHz). Three bandplans are currently defined over the FCC band (9–490 kHz). Those are FCC (34.375–478.125 kHz), FCC-1 (34.375–137.5 kHz) and FCC-2 (150–478.125 kHz). The ARIB bandplan uses the 154.7–403.1 kHz range.

11.3.1.2 Modulation

G.9902 uses windowed OFDM with the following set of programmable parameters:

- Number of carriers
  - CENELEC: 128, of which 36, 15 and 13 carriers are active in the CENELEC-A, CENELEC-B and CENELEC-CD bandplans, respectively.
  - FCC bandplan: 256, of which 143, 34 and 106 carriers are active in the FCC, FCC-1 and FCC-2 bandplans, respectively.
- Carrier spacing
  - CENELEC bandplans: 1.5625 kHz
  - FCC bandplan: 3.125 kHz
- Carrier modulation: *Quadrature amplitude modulation* (QAM) with 1, 2, 3 or 4 bits per carrier
- Guard interval: 0 for the frame header and*
  - CENELEC bandplans: 60 and 120 μs for the payload
  - FCC bandplans: 30 and 60 μs for the payload
- Size of transmitter windowing (PSD shaping)
  - CENELEC bandplans: 8 samples
  - FCC bandplan: 16 samples

11.3.1.3 FEC and Interleaver

The FEC encoder consists of an inner convolutional encoder with rate 1/2 and constraint length $L = 7$ and an outer RS encoder. A (2,1,6) mother convolutional code with rate 1/2 is used with the octal encoding matrix $G = [171;133]$; an additional rate of 2/3 is obtained by puncturing. The RS encoder uses input blocks up to 239 bytes and can be shortened to a 25-byte input block. The concatenated scheme is mandatory for payload data, while the RS encoder is bypassed for the *PHY frame header* (PFH). The interleaver is designed to combat both frequency domain and time domain erasures, including repetitive erasures with a period of 1/2 AC cycle and duration up to 1/4 AC cycle. For the payload, the interleaver

---

* For payload symbols, two guard intervals are defined. The longest guard interval is used for 16-QAM modulation due to its higher sensitivity to *Inter-Symbol Interference* (ISI).

first splits the payload into multiple fragments and then each fragment may be repeated for 2, 4, 6 or 12 times to increase robustness. Two modes of interleaving are defined:

1. *Interleave over fragment* (IoF)
2. *Interleave over AC cycle* (IoAC)

If the IoF mode is set, each fragment is interleaved separately. If the IoAC mode is set, each repeated fragment is further padded by additional repetitions up to the closest multiple of half AC cycle. The IoAC mode is used to handle channels with severe periodic erasures.

G.9902 allows the transmission of multiple RS code words per PHY frame. More in detail, G.9902 allows transmitting at most 64 segments in an LLC frame or having at most a 250 ms time on wire. This allows the transmission of up to 300 OFDM symbols in a single PHY frame.

11.3.1.4 Tone Mapping

In order to maximise throughput under varying channel conditions, G.9902 employs tone mapping. Tone mapping allows loading a specified number of bits on carriers based on the *signal-to-noise ratio* (SNR) per carrier. The frame header always uses QPSK on all carriers of the used bandplan, and the tone mapping used for the payload is indicated in the frame header. Only flat bit loading of 1, 2 and 4 bits/carrier is specified in G.9902, that is, all carriers are loaded with the same constellation.

11.3.1.5 Channel Estimation and Pilot Tones

Since G.9902 requires a coherent receiver, it is necessary to provide accurate synchronisation and channel estimation. This is accomplished by the use of a combination of *channel estimation symbols* (CES) and pilot carriers added at pre-defined locations in both the header and the payload. The CES are transmitted inside or right after the PFH; the modulation parameters of the CES, including transmitter windowing, are the same as for the preamble symbols. The position of pilot tones in each OFDM symbol is shifted by three carriers from the previous symbol to reduce interpolation errors. Pilot tones are also essential to improve reception in time-varying channels like the PLC channel.

11.3.1.6 Robust Modes

For challenging links where node connectivity is compromised, G.9902 specifies a *robust communication mode* (RCM). Payload transmission in RCM uses a uniform loading of 1 bit per carrier (binary phase shift keying [BPSK]), with the possibility of repetition encoding with 2, 4, 6 or 12 repetitions. Repetitions are applied before interleaving. Protection of the PFH is critical since its loss implies the loss of the whole frame; thus, the PFH uses the rate 1/2 convolutional encoder plus 12 repetitions and an additional 12 bit *cyclic redundancy check* (CRC). An *extremely robust mode* (ERM) is specified optional in Annex A of G.9902. In ERM, repetitions can be set to 32, 64 or 128.

11.3.2 Data Link Layer

11.3.2.1 Medium Access Control

An ITU-T G.9902 node of a standard profile supports prioritised contention-based medium access with four priorities; nodes of low complexity profile support two priorities.

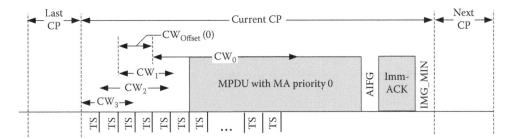

FIGURE 11.2
Example of contention period. (From Narrowband orthogonal frequency division multiplexing power line communication transceivers for ITU-T G.hnem networks, ITU-T Rec. G.9902, October 2012. [Online] Available: http://www.itu.int/rec/T-REC-G.9902 (Fig. 9-10). With permission.)

A contention-based prioritised CSMA/CA is specified. The three lower priorities are intended for user data frames, and the fourth priority is for frames carrying emergency signalling. All management frames are granted the third priority, to make network management more dynamic.

The prioritised contention-based medium access is defined in terms of *contention periods* (CP) (see Figure 11.2). The contention process starts at the beginning of the CP. The CP ends T_{IFG_MIN} after the node that won the contention completes the transmission of the frame sequence, which includes the transmitted frame and the ACK frame (if required). A new CP starts immediately after the end of the previous CP. Each CSMA/CA CP starts with priority resolution. The CP will consist of four priority resolution periods that may also overlap. Each such period is associated with a *contention window* (CW) with adjustable size and position. CWs of different priorities may overlap. Each node picks randomly the transmission slot in the corresponding CW and then monitors the medium using physical carrier sensing via preamble detection. If the medium becomes busy prior to the chosen transmission slot, the node refrains from transmitting. Synchronised medium access is optional. If enabled, the DM coordinates nodes by periodically sending (synchronous) beacons.

11.3.2.2 Link Layer Control

Long packets are segmented so that the size of a segment fits actual channel conditions. For higher efficiency, a G.9902 frame may carry multiple segments, each encapsulated into a link protocol data unit and protected by its own CRC. Transmission of frames containing multiple segments is supported by a selective acknowledgement scheme. The receiver checks the CRC of each segment and acknowledges all corrupted segments, which are then retransmitted.

Similarly to IEEE P1901.2, G.9902 is agnostic to routing mechanisms. G.9902 allows performing mesh networking by using either L2 routing (by relaying the LLC frames) or L3 routing (by relaying the protocol data units above the A interface). Prior to initialisation, each domain will be set into a particular routing mode that either allows intradomain L2 routing (L2 relaying mode) or that disallows it (no L2 relaying is allowed for any node of the domain). If L2 routing is set for the domain, the DM assigns one or more nodes as relays.

11.3.2.3 Application Protocol Convergence

Although IPv6 is the default protocol for G.9902, it is also possible to support other networking protocols. Each protocol is supported by the corresponding *application protocol convergence* (APC) function. The type of APC used by the transmitter is indicated in the packet header.

11.3.2.4 Security

Security is provided by encryption and authentication of the relevant data and management frames communicated between the nodes of the domain. Encryption and authentication of data and relevant management messages use the *Counter with Cipher Block Chaining-Message Authentication Code* (CCM) algorithm based on 128-bit AES-128. Node authentication, generation and distribution of encryption keys between nodes, encryption key updates and node authentication updates are provided by a set of *Authentication and Key Management* (AKM) procedures. The AKM procedure can establish both group keys (i.e. a unique set of keys for a particular group of nodes) and pairwise keys (i.e. a unique set of keys per every pair of communicating nodes).

11.4 Recommendation ITU-T G.9903: G3-PLC

The G3-PLC technology was originally conceived by Maxim Integrated Products, Sagemcom and ERDF, who brought the G3-PLC specs into the ITU-T where they were enhanced and additional functionalities were added. In parallel, companies from different areas (utilities, system integrators, meter manufacturers and silicon vendors) founded G3-PLC Alliance in 2011 with the objective of refining and enhancing the specifications, set up certification, and promote the technology. Several published papers on G3-PLC PHY performance and field trials are available (see [8,11,12] and references therein).

A G.9903 network consists of one or more domains (called a personal area network or PAN), and a domain is constituted by all nodes registered in that domain. Each node is identified by its PAN ID and short address (16 bit ID). One node in the domain is assigned as a DM (or PAN-coordinator). The coordinator controls operation of all other nodes and performs admission, resignation and other domain-wide management operations, in addition to connectivity to other domains or WAN.

The specification details given in the next sections pertain to the approved G.9903 Recommendation [21] and its Revision which was approved in May 2013. The following enhancements were made in the 2013 Revision of G.9903:

- Support for a new bandplan in CENELEC-B
- A Regional Annex for JP, with an ARIB bandplan (Annex K).
- The support for an optional coherent mode for use in CENELEC bandplans
- The update of RFC4944 by RFC6282 for header compression
- A new routing algorithm (LOADng) [24] that replaces the one used previously (LOAD) [28]
- A clause on coexistence with other NB-PLC technologies that also specifies when it is mandatory or optional

11.4.1 Physical Layer

11.4.1.1 Bandplans

G.9903 specifies a CENELEC-A bandplan, three FCC bandplans, and an ARIB bandplan. When operating in the CENELEC-A bandplan, G.9903 uses the frequencies in the 35.938–90.625 kHz range with carrier spacing of 1.5625 kHz. The three FCC bandplans are FCC-1

(154.6875–487.5 kHz), FCC-1.a (154.687–262.5 kHz) and FCC-1.b (304.687–487.5 kHz). The ARIB bandplan uses the 154.7–403.1 kHz range. Tone masking allows the network manager to limit the bandwidth of the active band or to divide the band into smaller sub-bands to handle several domains. An Amendment currently under development is revising the FCC bandplans.

11.4.1.2 Modulation

ITU-T G.9903 uses windowed OFDM with the following set of programmable parameters:

- Number of carriers
  - CENELEC-A bandplan: 128, of which 36 are active
  - FCC bandplan: 128, of which 72, 24 and 40 carriers are active in the FCC-1, FCC-1.a and FCC-1.b bandplans, respectively
- Carrier spacing
  - CENELEC bandplans: 1.5625 kHz
  - FCC bandplan: 4.6875 kHz
- Carrier modulation
  - Mandatory *differential phase shift keying* (DPSK) with 1, 2 or 3 bits per carrier. This is the classical 'time' differential modulation where the information is encoded in the phase difference between two consecutive symbols. This is different from G.9904 where frequency domain differential modulation is employed.
  - The optional coherent mode specifies *phase shift keying* (PSK) with 1, 2 and 3 bits per carrier and QAM modulation with 4 bits per carrier (16 QAM).
- Guard interval: 0 for the frame header
  - CENELEC bandplans: 55 µs for the payload
  - FCC bandplan: 18.3 µs for the payload
- Size of transmitter windowing (PSD shaping): 8 samples

11.4.1.3 FEC and Interleaver

ITU-T G.9903 uses an inner convolutional encoder with rate 1/2 and constraint length $L = 7$ and an outer RS encoder. The inner convolutional code is the same one used in G.9902. An RS (255, 239) is specified where the number of correctable symbol errors is 8. The concatenated scheme is mandatory for payload data, while the RS encoder is bypassed for the FCH.

ITU-T G.9903 specifies a channel interleaver designed to provide protection against two sources of errors:

1. A burst error that corrupts a few consecutive OFDM symbols
2. A frequency deep fade that corrupts a few adjacent tones for a large number of OFDM symbols

To fight both sources of errors at the same time, interleaving is done in two steps. In the first step, each column is circularly shifted a different number of times. Therefore, a corrupted OFDM symbol is spread over different symbols. In the second step, each row is circularly shifted a different number of times, which prevents a deep frequency fade from disrupting the whole column.

The first version of G.9903 allowed the transmission of a single RS code word per PHY frame, thus allowing at most to transmit around 20–40 OFDM symbols per packet. This reduces transmission efficiency but protects from impulsive noise. A new amendment to G.9903 is being finalized and two RS code words per PHY frame can be used in the FCC bandplans.

11.4.1.4 Tone Mapping

Also, G.9903 adaptively selects the usable tones and optimum modulation and code rate to ensure reliable communication over the power line channel. The modulation and coding selection is based on the estimated SNR per subcarriers. Only flat bit loading is specified in G.9903.

11.4.1.5 Channel Estimation and Pilot Tones

Since G.9903 defines a mandatory non-coherent scheme, CES and pilot tones are not specified. However, for the optional coherent mode, six pilots are defined to help with clock recovery and channel estimation.

11.4.1.6 Robust Modes

Two robust modes are specified in G.9903: a robust mode and a super-robust mode. In robust mode, every bit at the output of the convolutional encoder is repeated four times and then passed as input to the interleaver. In the super-robust mode, six repetitions are used. Both robust modes bypass the RS encoder. The Frame Control Header uses the super-robust mode.

11.4.2 Data Link Layer

The ITU-T G.9903 DLL specifications comprise two sublayers:

1. MAC sublayer based on the IEEE 802.15.4 standard [29]
2. IPv6 adaptation sublayer based on a modified version of 6LoWPAN (IPv6 over Low-power Wireless Personal Area Networks) [30]

11.4.2.1 Medium Access Control Sublayer

The channel access is accomplished by using *carrier sense multiple access with collision avoidance* (CSMA/CA) mechanism with a random back-off time. The random back-off mechanism spreads the time over which stations attempt to transmit, thereby reducing the probability of collision. Each time a device wishes to transmit data frames, it will wait for a CP according to the packet's priority and then start the random period. If the channel is found to be idle following the random back off, the device will transmit its data. If the channel is found to be busy following the random back off, the device will wait for the next CP according to the packet's priority and then start another random period before trying to access the channel again.

G.9903 supports unslotted CSMA/CA for non-beacon PANs, as described in IEEE 802.15.4 [29]. The random back-off mechanism spreads the time over which stations attempt to transmit, thereby reducing the probability of collision, using a truncated binary exponential back-off mechanism. The CSMA/CA algorithm will be used before the transmission of data or MAC command frames. The algorithm is implemented using units of time called back-off periods, where one back-off period will be equal to the duration of the contention slot time (two data symbols).

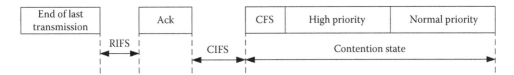

FIGURE 11.3
Priority contention windows. (From Narrowband orthogonal frequency division multiplexing power line communication transceivers for G3-PLC networks, ITU-T Rec. G.9903, October 2012. [Online] Available: http://www.itu.int/rec/T-REC-G.9903 (Fig. C-3). With permission.)

Prioritised access to the channel is specified in G.9903 with two levels of priority (high and normal). Priority resolution is implemented by using two contention time windows during the contention state as shown in Figure 11.3.

The first slot of the CW is called the *contention-free slot* (CFS). The CFS will be used for transmission of subsequent segments of a MAC packet without the back-off procedure to prevent possible interruption from other nodes and to simplify the MAC packet reassembly procedure for a receiver. In this case, only the first segment is sent using either a *normal-priority* CW (NPCW) or *high-priority* CW (HPCW) and the rest are sent using the CFS. The high- and normal-priority stations will compete for channels during the HPCW and normal NPCW correspondingly. Since the HPCW is located before the NPCW, high-priority stations will get access to the channel before the station with normal priority.

Automatic repeat request is implemented based on acknowledged and unacknowledged retransmission. The ACK is a positive acknowledgement that when received allows the transmitter to assume successful delivery of the frame. The *negative acknowledgement* (NACK) is used to inform a packet originator that the receiver received the packet but it was corrupted. If the originator does not receive an acknowledgement after a waiting period, it assumes that the transmission was unsuccessful and retries the frame transmission in the correct CP. If an acknowledgement is still not received after several retries, the originator can choose either to terminate the transaction or to try again.

11.4.2.2 IPv6 Adaptation Sublayer

G.9903 uses IPv6 as an interconnection scheme at the network layer and adopted 6LoWPAN to facilitate IPv6 interaction at low-rate networks. 6LoWPAN allows power line networks to incorporate IPv6 in embedded equipment and seamless connection of MAC layer and network layer to achieve the header compression, routing, fragmentation and assembly.

To maintain good performance despite the hostile PLC environment, it is essential to adopt an efficient routing protocol over 6LoWPAN which can quickly adapt to varying network topologies and link conditions. G.9903 relies on the *6LoWPAN Ad hoc On-Demand Distance Vector Routing* (LOAD) L2 routing protocol defined in [28]. LOAD is a simplified on-demand routing protocol based on *Ad hoc on Demand distance Vector routing* (AODV) [31]. It is a reactive protocol, and operates creating a mesh network topology underneath the IPv6 network layer.

The route to a destination is found by broadcasting *Route Requests* (RREQs) per demand, and each node learns about its neighbour node through which it can connect to the destination. Upon reception of RREQ at a destination, the destination node will unicast a *Route Reply* (RREP) back to the source node using the reverse path characterized by the lowest route cost. Moreover, should a node detect a broken link a *Route Error* (RERR) is broadcasted to initiate a new source discovery process.

Since LOAD was originally designed for wireless networks, a number of enhancements are specified in G.9903 to improve LOAD's performance in a PLC network. For example,

- *Asymmetrical routes*: In PLC, the link between two nodes can be highly asymmetrical, meaning that the quality of the channel is significantly different depending on the communication direction. This is due to the possible proximity of noise sources or mismatch between transmitter and receiver impedances. There is no provision in LOAD to allow communicating the reverse channel quality to an intermediate node so that it can be taken into account in its link cost calculation. Furthermore, LOAD does not specifically propose any method to calculate the link cost between nodes. G.9903 solves this by defining mathematically the link cost on the basis of several parameters and also defines a neighbouring table with entries that include the forward and reverse link cost.

- *Forwarding multiple RREQs*: Sometimes RREQs can be delayed due to processing load at a node, so that it is not necessarily true that the first RREQ to arrive at an intermediate node has the best route. However, LOAD does not consider this case and subsequent RREQs are not rebroadcasted. This issue is addressed in G.9903 by rebroadcasting late RREQs if their associated route cost is lower than the last rebroadcasted one.

- *Minimising sending RREPs*: The destination node generates a RREP in response to every RREQ if its route cost is lower than the previously received RREQs from the same source with same RREQ ID. This results in sending multiple RREP through multiple routes. As a consequence, the source needs to wait for some time to make sure that there is no better RREP in transit. In order to reduce the RREP traffic, G.9903 moved the waiting period for collecting all RREPs at the source to the destination. After receiving first RREQ, the destination will wait for a defined period to make sure no other RREQ with a better route cost is in transit. At the end of this wait period, the best route is chosen and a RREP associated with that route is generated and sent toward the source.

- *Orphan node*: An orphan node is defined as a device that is associated to a network, and, although it had at some time a very good link with the concentrator, it now has a very bad link with the concentrator. In this scenario, the orphan node erroneously believes it has a good route to the data concentrator and this condition may exist for some time before the node or concentrator notices the failure in the link because in AMI applications, there is infrequent access to the meter. Now, since the orphan node always responds to any beacon request from a new associating device, a series of repeated failures of the association procedure for that device can occur. To alleviate this problem, G.9903 specifies a route cost to coordinator field which is added to the payload of the beacon response. Once the first association through an orphan node fails, a very high value of the route cost to coordinator is reported in any future beacon responses of the orphan node. The device seeking to join the network can use this information to avoid selecting the orphan node for the next association attempt.

11.4.2.3 The 2013 G.9903 Revision and LOADng

As mentioned in Section 11.4, the ITU-T approved in May 2013 a Revision of G.9903. One of the main changes in this Revision is the replacement of the routing algorithm LOAD with LOADng [24].

LOADng is designed to further enhance LOAD with additional features and capabilities. Derived from AODV [31], the basic operation of LOADng is similar to LOAD and includes generation of RREQs by a LOADng router, forwarding them until they reach the destination, generation of RREPs upon receipt of an RREQ by the indicated destination and hop-by-hop forwarding of these unicast RREPs toward the originator. It also employs RERR message to report a broken link if a data packet cannot be forwarded. Compared to AODV, LOADng contains both extensions and simplification and it also includes features that are useful in a PLC environment.

Here are some of additional features of LOADng:

- *Blacklisting*: The blacklisted neighbour set maintains the address of nodes to which the connection is detected to be unidirectional. More specifically, if a node sends a RREP to a neighbour from which a RREQ has been received and the RREP fails, then the neighbour is added to the blacklist. When a neighbour is blacklisted, any RREQ received from that node is dropped.

- *Separate forward and reverse route*: LOADng introduces provisions to establish different routes between two nodes based on the initiator of the route discovery and hence to create and use two separate routes between two nodes, A and B depending on the direction of communication. This allows the discovery of an optimised route for sending packets from A to B while having a viable link from B to A; and a different optimised route for sending packets from B to A is also found, while having a viable link from A to B.

- *Extension for route cost calculation*: LOADng allows 16-bit values to be used for route metrics while LOAD only allocates 8 bits. This allows a better resolution when comparing route costs and it also allows for a higher number of hops to be used when accumulating the link costs along a route.

- Optimized flooding is supported, reducing the overhead incurred by RREQ generation and flooding. When a node receives an RREQ looking for a path to a destination address already contained in its routing table, it may propagate the RREQ in an unicast manner to the final destination node avoiding unnecessary flooding.

A preliminary performance evaluation has been conducted in [32], and it has shown that LOADng yields comparative performance to that of AODV except that LOADng incurs a substantially lower control traffic overhead.

11.4.2.4 Security

An end device may not access the network without a preliminary identification and authentication which are based on two parameters that are unique to the end device:

- An *Extended Unique Identifier* (EUI-48) MAC address. This address may be easily converted into an EUI-64, if requested.

- A 128-bit shared secret (also known as pre-shared key) used as a credential during the authentication process. It is shared by the ED itself (also known as peer) and an authentication server. The mutual authentication is based on proof that the other party knows the pre-shared key.

The authentication process is *extensible authentication protocol–pre-shared key* (EAP-PSK) method in place. Confidentiality and integrity services are ensured at different levels at the MAC level and at the EAP-PSK level. At the MAC level, a CCM type of ciphering is delivered to every frame transmitted between nodes in the network. The MAC frames are encrypted and decrypted at every hop. To support this service, all the nodes in the network receive the same group master key which is individually and securely distributed to every node by using the EAP-PSK secure channel.

11.5 Recommendation ITU-T G.9904: PRIME

The PRIME technology was originally conceived within the PRIME Alliance [5]. ITU-T Members of the PRIME Alliance brought the PRIME specifications v. 1.3.6 into the ITU-T, and the specifications were ratified as Recommendation ITU-T G.9904. There are several published papers on PRIME (see [7,9–11].

The reference model of G.9904 is shown in Figure 11.4.

The *convergence layer* (CL) classifies traffic associating it with its proper MAC connection. This layer performs the mapping of any kind of traffic to be properly included in *MAC service data units* (MSDUs). It may also include compression functions. Several SSCSs are defined to accommodate different kinds of traffic into MSDUs. The MAC layer provides core MAC functionalities of system access, bandwidth allocation, connection establishment/maintenance and topology resolution. The PHY layer transmits and receives MPDUs between neighbour nodes.

A G.9904 system is composed of subnetworks, each of them defined in the context of a transformer station. A G.9904 subnetwork can be logically seen as a tree structure with two types of nodes: the base node (master) and service nodes (slaves).

A base node is at the root of the tree structure and acts as a master node that provides all network elements with connectivity. It manages the network resources and connections. There is only one base node per G.9904 network.

The service nodes are either leaves or branch points of the tree structure. They are initially in a disconnected functional state and follow the registration process to become part of the network. Service nodes have two functions in the network: keeping connectivity to the other nodes in the network for their application layers and switching other nodes' data to propagate connectivity.

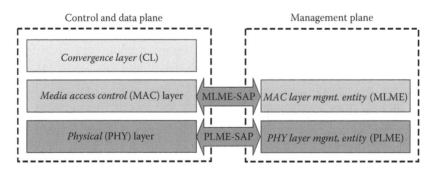

FIGURE 11.4
Reference model of G.9904 protocol layers. (From Narrowband orthogonal frequency division multiplexing power line communication transceivers for PRIME networks, ITU-T Rec. G.9904, October 2012. [Online] Available: http://www.itu.int/rec/T-REC-G.9904 (Fig. 6-1). With permission.)

11.5.1 Physical Layer

11.5.1.1 Bandplans

G.9904 specifies only a CENELEC-A bandplan. This bandplan specifies the use of the frequencies in the 41,992–88,867 kHz range, with carrier spacing of 488.28125 Hz. The three FCC bandplans are FCC-1 (154.6875–487.5 kHz), FCC-1.a (154.687–262.5 kHz) and FCC-1.b (304.687–487.5 kHz), all with carrier spacing of 4.6875 kHz.

11.5.1.2 Modulation

G.9904 uses OFDM without any transmitter windowing and with the following set of programmable parameters:

- Number of carriers: 256, of which 96 carry data and 1 is a pilot
- Carrier spacing: 488.28125 Hz
- Carrier modulation: DPSK with 1, 2 or 3 bits per carrier. As opposed to G.9903, G.9904 specifies differential modulation in the frequency domain, that is, the information is encoded in the phase difference between adjacent subcarriers
- Guard interval: 0 for the frame header and 192 μs for the payload
- Size of transmitter windowing: 0 samples

Performing differential encoding in the frequency domain as opposed to in the classical time domain allows better resiliency to impulse noise but also presents other challenges due to the necessity of placing a reference symbol in every first subcarrier of each OFDM symbol. For example, the necessity of placing a reference symbol after every notched carrier introduces overhead reducing transmission efficiency and G.9904 deals with this by not allowing to turn off subcarriers. This is a limitation when dealing with channels affected by narrowband interferers.

G.9904 allows the transmission of a variable number of OFDM symbols per PHY frame, up to 63.

11.5.1.3 FEC and Interleaver

G.9904 specifies a convolutional encoder with rate 1/2 and constraint length $L = 7$, which is the same as the one used in G.9902 and G.9903. There is no RS encoder and the convolutional encoder can be turned off. G.9904 specifies a channel interleaver but, differently from G.9903, interleaving is done within a single OFDM symbol.

11.5.1.4 Tone Mapping

Similarly to G.9902 and G.9903, G.9904 specifies the adaptive selection of usable tones and optimum modulation and code rate to maximise throughput.

11.5.1.5 Channel Estimation and Pilot Tones

In a differential scheme, channel estimation is not needed. Nevertheless, G.9904 specifies 13 pilots in the header to facilitate the estimation of the frequency offset and one pilot to provide a phase reference for frequency domain DPSK demodulation.

11.5.1.6 Robust Modes

No robust mode is defined in G.9904.

11.5.2 Data Link Layer

11.5.2.1 Medium Access Control

G.9904 devices access the channel based on dividing time into composite units of abstraction for channel usage, called MAC frames. The service nodes and base node on a network can access the channel in the *shared contention period* (SCP) or request a dedicated *contention-free period* (CFP). CFP channel access needs devices to request allocation from the base node. Depending on channel usage status, the base node may grant access to the requesting device for a specific duration or deny the request. SCP channel access does not require any arbitration. However, the transmitting devices need to respect the SCP timing boundaries in a MAC frame. The composition of a MAC frame in terms of SCP and CFP is communicated in every frame as part of the beacon. A MAC frame is comprised of one or more beacons, one SCP and zero or one CFP (see Figure 11.5).

G.9904 also specifies a L2 mechanism for routing within a G.9904 network. If the base node cannot communicate with a node directly, switch nodes relay traffic to/from the base node so that every node on the network is effectively able to communicate with the base node. Switch nodes selectively forward traffic that originates from or is destined to one of the service nodes in its control hierarchy. All other traffic is discarded by switches, thus reducing traffic flow on the network. Switch nodes do not necessarily need to connect directly to the base node. They may attach to other switch nodes and form a cascaded chain. There is no limitation to the number of switch nodes that may connect to a switch node down the cascaded chain, thus contributing significantly to range extension and scalability.

11.5.2.2 Convergence Layers

The CL classifies traffic associating it with its proper MAC connection. The CL is separated into two sublayers:

1. The *common part convergence sublayer* (CPCS) provides a set of generic services.
2. The *service-specific convergence sublayer* (SSCS) contains services that are specific to one application layer.

There may be several SSCS, but only one common part CPCS. Several CLs are defined in order to accommodate different kinds of traffic: IPv4 and IPv6 CL and IEC 61334-4-32 [33] as a link toward metering systems.

11.5.2.3 Security

G.9904 security provides privacy, authentication and data integrity to the MAC layer through a secure connection method and a key management policy.

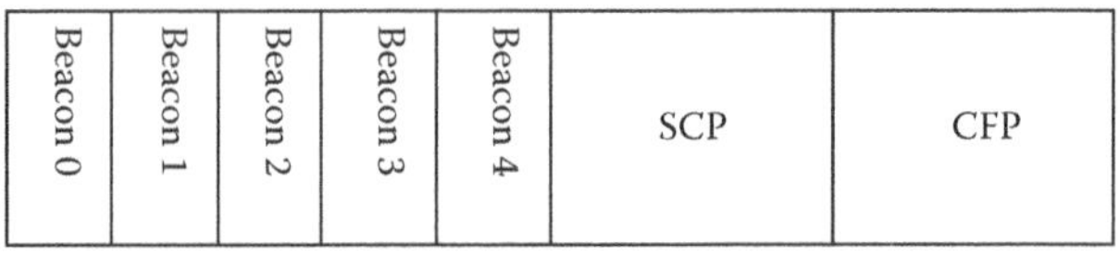

Beacon 0	Beacon 1	Beacon 2	Beacon 3	Beacon 4	SCP	CFP

FIGURE 11.5
Structure of a MAC frame. (From Narrowband orthogonal frequency division multiplexing power line communication transceivers for PRIME networks, ITU-T Rec. G.9904, October 2012. [Online] Available: http://www.itu.int/rec/T-REC-G.9904 (Fig. 8-7). With permission.)

While devices may choose not to encrypt data traffic, it is mandatory for all MAC control messages to be encrypted with a specific security profile. Several security profiles are provided to manage different security needs, which can arise in different network environments. Current G.9904 version specifies two security profiles.

Authentication is guaranteed by the fact that each node has its own secret key known only by the node itself and the base node. Data integrity is guaranteed by the fact that the payload CRC itself is encrypted. Communications having Security Profile 0 are based on transmission of MAC frames without any encryption. This profile will be used by communication that does not have strict requirements on privacy, authentication or data integrity. Security Profile 1 is based on 128 bit AES encryption of data and its associated CRC.

11.6 NB-PLC Coexistence

An additional critical effort taken by the IEEE P1901.2 Draft is the definition of a robust and reliable coexistence mechanism. Coexistence is meant to address the issue of interference between non-interoperable devices that share the same power line cable. Power line cables connect LV transformers to a set of individual homes or set of multiple dwelling units without isolation. Signals generated within the premises interfere among each other, and with signals generated outside the premises. As the interference increases, both from indoors and outdoors sources, PLC stations will experience a decrease in data rate as packet collisions increase, or even complete service interruption. Hence, PL cables are a shared medium (like coax and wireless) and do not provide links dedicated exclusively to a particular subscriber. As a consequence, the PLC channel is interference limited, and approaches based on frequency division multiplexing as in Wi-Fi or coax are not always suitable because only a relatively small band is available in PLC. As a consequence, it is necessary to devise mechanisms to limit the harmful interference caused by non-interoperable neighbouring devices. Note that similar considerations can be made about the interference-limited nature of many wireless networks, for example, Wi-Fi, WiMAX, Zigbee, Bluetooth, and Z-Wave (as treated in [34]).

The issue of PLC coexistence was first raised two decades ago in CENELEC. Since CENELEC did not mandate the use of a specific PHY/MAC, it was necessary to provide a fair channel access mechanism that avoided channel capture and collisions when non-interoperable devices operated on the same wires. In fact, if non-interoperable devices access the medium, then native CSMA and virtual carrier sensing do not work and a common medium access mechanism must be defined. CENELEC mandates a CSMA/CA mechanism only for the C-band [35] where a single frequency (132.5 kHz) is used to inform that the channel is in use.

The use of orthogonal bandplans enables one form of coexistence (frequency separation). Another coexistence mechanism utilises a notching technique, also referred to as tone masking. Notching is a method to avoid certain frequencies that are reserved by power line regulatory bodies for other applications; it also allows for cohabitation with single-carrier NB-PLC [2,3] systems and cohabitation with other potential systems operating over the power line. However, to meet the IEEE 1901.2 objective of developing a complete coexistence mechanism, additional mechanisms are required.

11.6.1 IEEE P1901.2 Preamble-Based Coexistence

The requirement of defining a NB-PLC coexistence between IEEE P1901.2 and the three technologies specified in ITU-T (G.hnem, G3-PLC, PRIME) was raised in the SGIP/US

National Institute of Standards and Technology (NIST) *Priority Action Plan* (PAP) 15 Group [36]. It was also agreed in PAP 15 that the NB-PLC mechanism should be future proof, so that future technologies could also coexist with today's. The IEEE P1901.2 Working Group took the responsibility to work on a preamble-based coexistence mechanism tailored for NB-PLC OFDM-based solutions. This preamble-based coexistence mechanism allows different NB-PLC solutions to coexist with fairness and minimal disruption of service.

A preamble-based CSMA coexistence mechanism employs a fixed number of coexistence preamble symbols at a specific frequency, or multiples of a specific frequency depending on the bandplan. Fairness can be achieved through the appropriate sequence of repeated coexistence preambles over a given time duration, and through a defined duty cycle which enables a defined time to occupy the channel (whether same technology or different NB-PLC solutions).

To implement a coexistence procedure, several configuration parameters need to be set. These parameters determine channel access and the amount of time that a channel can be controlled. Additionally, coexistence behaviour/fairness attributes need to be defined with type, range, description and default values. The use of a coexistence mechanism is dependent on the technology type and region of deployment. Control attributes are defined to set default values that enable or disable the preamble-based coexistence mechanism. For example, an IEEE P1901.2 solution implementing only a CENELEC-A bandplan in a region where energy providers control the CENELEC-A frequency band could most likely not implement the preamble-based coexistence mechanism because the probability that the same utility deploys non-interoperable NB-PLC devices is negligible. Instead, the solution would rely on existing frequency separation or notching techniques.

G.9902 supports three types of coexistence:

1. *Frequency division*: Allows the suppression of interference from G.9902 into a particular frequency band or bands by using non-overlapping ITU-T G.9902 bandplans.

2. *Frequency notching*: Allows to suppress interference from G.9902 into a particular (relatively narrow) frequency range by notching out one or more subcarriers; frequency notching allows G.9902 to coexist with the existing narrowband FSK/PSK systems operating over the same frequency band.

3. *Preamble-based*: Allows to fairly share the medium with other types of PLC technologies operating over the same frequency band (and utilising this coexistence mechanism).

G.9902 states that support for coexistence is mandatory, while G.9903 states that a preamble-based coexistence mechanism is mandatory with the exception of when the network is operated in frequency bands restricted to monitoring or controlling the operation of the grid (e.g. CENELEC-A band), in which case coexistence is optional. G.9904 contains no language on coexistence.

11.7 A Qualitative Comparison of NB-PLC Technologies

The four HDR NB-PLC standards described in this chapter present some similarities but also marked differences. While at the MAC level they all use some variant of CSMA/CA, the PHY exhibits many differences in the design choices. Also the support for routing at L2 or L3 is a philosophical divide that puts on one side G.9903 and G.9904 in support of L2 routing and on the other side IEEE P1901.2 and G.9902 agnostic with respect to routing and capable of supporting it at both L2 and L3.

In the following sections, the major differences between these four standards will be highlighted and discussed qualitatively.

11.7.1 Physical Layer

The main PHY parameters of the four HDR NB-PLC discussed earlier are listed in Tables 11.2 and 11.3 for the CENELEC-A and FCC bandplans, respectively. The maximum data rates take into account of *cyclic prefix* (CP), FEC and the overhead of FCH, preambles, CES and pilots. Similar values for these data rates are also reported in [23]. Note that the data rates shown in the tables are the maximum ones achievable over an ideal channel. Actual data rates will be lower and depend on line conditions.

11.7.1.1 NB-PLC Channel Assumptions

When designing transceivers, one of the first things one has to have is knowledge about the physical channel. When comparing these four technologies, the first difference that can be observed is that some choices indicate that designers made very different assumptions about the NB-PLC channel. For example, the guard interval ranges from 55 to 192 µs in CENELEC-A and from 18.3 to 60 µs for the FCC band; this clearly shows that the various designers had very different assumptions about the frequency selectivity of the NB-PLC channel. Making a mistake on the level of ISI introduced by the channel can cause a reduction of the available SNR, something that may be critical in certain scenarios characterised already by low SNR or that may prevent the use of higher-order constellations.

TABLE 11.2

PHY Parameters of NB-PLC Solutions and Their Maximum Data Rate – CENELEC-A Bandplan

	ITU-T G.9902 (G.hnem)	ITU-T G.9903 (G3-PLC)	ITU-T G.9904 (PRIME)	IEEE P1901.2[a]
Frequency range (kHz)	35.9–90.6	35.9–90.6	42–89	35.9–90.6
Sampling frequency (kHz)	200	400	250	400
Number of carriers	128	128	256	128
Cyclic prefix (µs)	100/160	75	192	75
Guard interval (µs)	60/120	55	192	55
Window size (samples)	8	8	0	8
Subcarrier spacing (Hz)	1562.5	1562.5	488.28125	1562.5
OFDM symbol duration (µs)	700/760	695	2240	695
Modulation	M-QAM	M-DPSK[b]	M-DPSK	M-DPSK
FEC	Conv+RS	Conv+RS	Conv[c]	Conv+RS
Interleaving over	Fragment/AC cycle	Packet	OFDM symbol	Packet
Robust modes	Yes	Yes	No	Yes
PHY frame efficiency	Multiple RS	Single RS	≤63 symbols	Single RS
Max PHY rate for $M = 2$ (kbps)	25.3	20.3	20.5	20.0
Max PHY rate for $M = 4$ (kbps)	50.6	34.9	41.0	34.1
Max PHY rate for $M = 8$ (kbps)	76.0	46.0	61.4	44.6
Max PHY rate for $M = 16$ (kbps)	101.3	N/A	N/A	N/A

[a] At the time of writing, specifications can change before approval.
[b] Differential modulation is mandatory, and an optional coherent mode has been specified in Revised G.9903.
[c] The convolutional code in PRIME is optional.

TABLE 11.3

PHY Parameters of NB-PLC Solutions and Their Maximum Data Rate – FCC Bandplan

	ITU-T G.9902 (G.hnem)	ITU-T G.9903 (G3-PLC)	IEEE P1901.2[a]
Frequency range (kHz)	34.4–478.1/150–478.1[b]	154.7–487.5	154.7–487.5
Sampling frequency	800 kHz	1.2 MHz	1.2 MHz
Number of carriers	256	128	128
Cyclic prefix (μs)	50/80	25	25
Guard interval (μs)	30/60	18.3	18.3
Window size (samples)	16	8	8
Subcarrier spacing (Hz)	3125	4687.5	4687.5
OFDM symbol duration (μs)	350/380	231.7	231.7
Modulation	M-QAM	M-DPSK/M-QAM[c]	M-DPSK/M-QAM
FEC	Conv+RS	Conv+RS	Conv+RS
Interleaving over	Fragment/AC cycle	Packet	Packet
Robust modes	Yes	Yes	Yes
PHY frame efficiency	Multiple RS	Single/Two RS[d]	Single/Two RS[d]
Max PHY rate for $M = 2$ (kbps)	210.2/150.8	106.2	106.2
Max PHY rate for $M = 4$ (kbps)	417.4/301.6	166.5[c]	166.5[c]
Max PHY rate for $M = 8$ (kbps)	616.5/448.1	207.6[c]	207.6[c]
Max PHY rate for $M = 16$ (kbps)	809.5/591.0	233.5[c]	233.5[c]

[a] At the time of writing, specifications can change before approval.

[b] Data rates values are shown for the FCC and FCC-2 bandplans, respectively.

[c] Differential modulation is mandatory, but an optional coherent mode is specified in G.9903.

[d] For the data rate calculation, it is assumed that only a single RS code word can be transmitted in a frame – as in the current Revised G.9903. Both the current IEEE P1901.2 Draft and an Amendment of G.9903 under development specify the use of two RS code words per frame in the FCC bandplan.

Also, choices about robustness against channel noise denote a really diverse set of assumptions on the channel. For example, we go from the G.9904 (PRIME) 'optimism' where a simple convolutional encoder is used (and can also be turned off), no robust mode is specified and interleaving is made within a single OFDM symbol, to the much more conservative choices of G.9903 (G3-PLC) where a concatenated coding scheme is employed, interleaving is done over the whole packet and two robust modes are specified. Furthermore, while G.9902 and G.9904 allow to transmit frames constituted of multiple RS code words or multiple OFDM symbols, G3-PLC designers decided to improve impulse noise protection by constraining transmission efficiency and allowing the transmission of a single RS code word per packet only.

Another difference worth mentioning is that the differential modulation specified in G.9903 is different from the differential modulation specified in G.9904. The former is the classical *'time-differential' phase modulation* (*t*-DPSK), that is, the information is encoded in the phase difference between two consecutive symbols; the latter is a *'frequency-differential' phase modulation* (*f*-DPSK) where the information is encoded in the phase difference between adjacent subcarriers, respectively. These two differential modulations yield the same performance in *additive white Gaussian noise* (AWGN) or block fading channels but behave differently in more realistic channels. In fact, erasures in time like impulse noise affect *t*-DPSK more than *f*-DPSK, but the opposite holds for erasures in the frequency domain like narrowband interferers. Furthermore, different interleaver and FEC design will also contribute to the overall PHY performance.

Hoch compared the PHY performance of G.9903 and G.9904 and reported the following conclusions [11]:

- In frequency flat channels affected by coloured and periodic impulsive noise, G.9903 performs better than G.9904 because of the additional RS coding. For example, at a *frame error rate* (FER) of 10^{-4}, the SNR gain of G.9903 is around 6 dB.
- In frequency-selective channels and AWGN, G.9903 performs again better than G.9904 for DBPSK, while for higher constellations, performances tend to be similar. This may be due to the fact that bad channels (high ISI) affect higher-order constellations of the two systems in a similar way.
- In frequency flat channels affected by AWGN and a narrowband interferer, G.9903 outperforms G.9904 and in some cases G.9904 is degraded so much that communications is not possible.

Finally, while it has been shown that G3-PLC is able to penetrate LV/MV transformers [12] thus ensuring LV to MV and MV to LV connectivity, here are no publicly available reports that confirm that PRIME signals can pass the distribution transformer. The capability of passing LV/MV transformers offers more degrees of freedom in network design. The main architectural consequence of MV/LV connectivity is that many more meters could be handled by a single concentrator located on the MV side. This concentrator node would then send the aggregated data from many meters back to the utility using either PLC or any other networking technology available in situ. This capability also heavily impacts the business case when there is a very different number of customers per MV/LV transformer: in North America, the majority of transformers serves less than 10 customers; in Europe, the majority of transformers serves 200 customers or more. Thus, especially in the United States, it is economically advantageous to avoid coupler installation and resort to technologies that allow connectivity between the MV and LV sides – and possibly also between meters served by different distribution transformers (LV/MV/LV links). When there are very few end points (meters) per distribution transformer as in the United States, it is convenient to push the concentrator up along the MV side (and even up to the substation) and handle multiple LV sections so that more end points can be handled per concentrator. On the other hand, the large number of end points per transformers in Europe does not really require to locate the concentrator up in the substation or on the MV side as it can be conveniently located on the LV section of the grid. Thus, the capability of a NB-PLC technology to pass through distribution transformers is very appealing in areas with low density of population (meters).

11.7.1.2 Coherent versus Differential Modulation

Another interesting difference is given by the choice of differential modulations or PSK/QAM modulation. G.9902 (G.hnem) specifies a mandatory PSK/QAM modulation, while G.9903 (G3-PLC), G.9904 (PRIME) and IEEE P1901.2 specify a mandatory (time or frequency) differential modulation. The choice of specifying PSK/QAM modulations for G.hnem influenced also the other groups, although no public results of the performance of these modulations in the field are today available. For example, while G3-PLC was being discussed in ITU-T, an additional optional coherent modulation was specified for both CENELEC-A and FCC bandplans. Similarly, also the IEEE P1901.2 Working Group that originally specified only the differential modulation used in G3-PLC, decided to add an optional coherent mode.

The basic principle of differential modulation (which is a particular case of modulation with memory) is to use the phase of the previous symbol as reference of the current symbol. In AWGN, the SNR gain of coherent versus incoherent reception is nearly 3 dB, while for Rayleigh slow fading channels it is a little bit <3 dB (at high SNR). Recent work has shown that there are non-coherent decoding schemes that can be applied to almost all types of coded modulation and can approach the performance of coherent detection over AWGN as the observation length grows (see [37]). However, this property does not hold true for channels affected by non-Gaussian noise like the PLC one which is often modelled as an erasure channel. In fact, as the PLC channel is affected by periodic impulse noise, letting the observation length grow to improve non-coherent detection raises also the probability of occurrence of an erasure that will then cause a loss in the memory of the differential modulation. Since an erasure may destroy the memory required for differential decoding, a loss of the reference phase in the previous symbol may occur thus leading to the incorrect detection of the current symbol. Furthermore, even if differential modulation copes well with random shifts of the carrier phase (thus allowing non-coherent demodulation), it does not perform well when the channel has a non-zero Doppler spread* as in the case of PLC [38]. In fact, in this case, the signal phase can lose its correlation between consecutive symbols making the previous symbol a noisy phase reference. Thus, in a severely impulsive channel, the SNR gain of coherent reception over non-coherent may grow beyond the classical 3 dB in AWGN as the erasure probability of the channel grows.

These theoretical considerations can justify the preference for a coherent scheme, and they are also consistent with the results reported in [40], where it was shown that non-coherent schemes suffer performance degradation when the channel is time varying and affected by impulsive noise. The issue of how much gain coherent reception can offer with respect to incoherent reception over the PLC channel and for the many NB-PLC application scenarios still requires further analysis and field testing, and this can explain why certain groups have decided not to specify at all a coherent mode (G.9904) or to specify it just as an option (G.9903 and IEEE P1901.2). Nevertheless, it is an interesting problem that the scientific community should tackle. Of course, it is also true that doing so is problematic because there is no commonly agreed upon NB-PLC channel model.

11.7.2 'Route-Over' or 'Mesh-Under'?

As mentioned earlier, the support in NB-PLC standards for routing at L2 (mesh-under) or L3 (route-over) is still an open problem, and it is not clear yet what the best approach is for the various Smart Grid applications. A mesh-under approach places routing functions in

---

* Cañete et al. [38] measured in indoor PLC channels a median Doppler spread of 100 Hz, a 90% percentile of 400 Hz and Doppler components up to 1750 Hz; thus, the coherence time of the PLC channel is around 600 μs. Interestingly, all observed Doppler components were quantised at multiples of the fundamental frequency of the mains AC cycle, thus suggesting that the PLC channel behaves like a *linear and periodically time-varying* (LPTV) channel [39]. This LPTV behaviour is due to the fact that the electrical devices plugged in outlets (loads) contain non-linear elements such as diodes and transistors that, relative to the small and rapidly changing communication signals, appear as a resistance biased by the AC mains voltage. In fact, the periodically changing AC signal swings the devices over different regions of their non-linear I/V curve and this induces a periodically time-varying change of their resistance. The overall impedance appears as a shunt impedance across the 'hot' and 'return' wires and, since its time variability is due to the periodic AC mains waveform, it is naturally periodic. A time-varying (and sometimes even LPTV) behaviour can also be encountered in the LV/MV outdoor PLC channel.

the link layer to emulate a single broadcast domain where all devices appear as immediate neighbours to the network layer (e.g. for G.9903 devices, every node is at one IPv6 hop from each other), even if this is not the case and multi-hop communication still occurs. In contrast, a route-over approach places all routing functions at the network layer thus following the classical IP architecture.

In choosing between L2 and L3 routing, scalability is the first issue. In L3 routing with IP, there is aggregation at the IP prefix level, and, therefore, solutions are scalable. On the other hand, one can design a L2 addressing and routing scheme that is very much similar to an L3 one, the only difference will be in address assignment and whether it is possible to aggregate addresses and create routing tables that scale well. Another aspect to consider is overhead: IP packets add a 40 byte header to a L2 packet, and this can be as high as 20% of the data payload carried by a G.9903 or IEEE P1901.2 packet. Answering this question in general is not easy as the assumed topological characteristics of the network and the traffic flowing through it can yield different conclusions. As more experimental validation is certainly needed, here we will review the recent literature on this topic and point out what is important with respect to AMI applications over PLC. We will also restrict our overview to the comparison of LOADng [24] and RPL [25], which are the main candidates proposed for Smart Grid applications.

LOADng and RPL are two good solutions for routing, but perhaps they solve two different problems [41]. RPL is highly optimised for specific topologies where *multi-point-to-point* (MP2P) traffic dominates (e.g. wireless sensor networks) and a central controller handles topology formation and maintenance. On the other hand, the philosophy underlying LOADng allows a distributed mode of operation where paths are discovered on demand and bidirectional traffic – P2P, MP2P and *point-to-multi-point* (P2MP) – is present. For many Smart Grid applications, including AMI, the assumption of bidirectional traffic is closer to reality*.

The simulation study in [42] evaluates the quality of RPL routes for the case of point-to-point traffic. For this scenario, it was found that the path quality of RPL is fairly close to an optimised shortest path. However, AMI traffic requires both MP2P and P2MP links. The performance of the RPL routing protocol with bidirectional traffic is investigated in [43], and it is reported that LOAD can provide similar data delivery ratios as RPL but with less protocol overhead. Additional experimental results on RPL are reported in [44] for a wireless network, confirming its good performance; and recent experimental results on advanced LOAD that confirm its good performances have been reported in [45].

In the work by Ancillotti et al. [6,7], it is reported that RPL route selection procedures ensures to find dominant routes that are significantly persistent and dominated by a single route, and this may prevent RPL from quickly adapting to link quality variations which is very common scenario in PLC-based networks. It was also found that RPL lacks a complete knowledge of link qualities, and it may sometimes select suboptimal paths with highly unreliable links. The authors also conclude that further research is required to improve the RPL route selection process in order to increase routing reliability.

It is finally important to point out that most results available in the literature pertain to wireless sensor networks and the specific characteristics of PLC links (low throughput, time varying and asymmetric link quality, unidirectional links – see also Section 11.4.2.3)

---

* Typically, AMI applications require handling up to 10,000 m with both MP2P and P2MP types of traffic: MP2P traffic is due to the support of scheduled read reporting, on-demand meter reading and event-driven responses (e.g. in the case of power outages or leak detections); P2MP traffic is due to the support of meter configuration, meter read requests, network wide firmware upgrades, demand-response applications. In the case of distribution automation, it is also necessary to support point-to-point traffic between a few devices.

are not always taken into account. Thus, we remind the reader about the general caveat that generalisation from simulation studies should always be made *cum grano salis*.

11.8 Conclusions

The design intent for G.9902 was to integrate features of G3-PLC and PRIME and add new features for even better performance. However, there are no experimental results on G.9902 today, so it is not possible yet to say whether that goal was achieved or not. G3-PLC was originally designed for improved robustness sometimes at the cost of reduced data rate, which is a reasonable design choice as HDR is not the primary design goal in smart metering applications – see also considerations made in Sect. V.D of [4]. The G.9903 and IEEE P1901.2 approach was to build on top of the original G3-PLC PHY and enhance it. Several G3-PLC field trials have been going on for years, thus allowing a lot of fine-tuning and improvements that have been included in both G.9903 and IEEE P1901.2. However, while the experience gathered with the LOAD and LOADng routing mechanisms has been retained in G.9903, IEEE P1901.2 and G.9902 have not specified any L2 or L3 routing scheme. Finally, G.9904 was originally designed for simplicity, keeping in mind a combination of robustness and data rate.

References

1. Open data communication in building automation, controls and building management – Control network protocol, ISO/IEC Std. DIS 14908, 2008.
2. Distribution automation using distribution line carrier systems – Part 5-2: Lower layer profiles – Frequency shift keying (S-FSK) profile, IEC Std. 61334-5-2, 1998.
3. Distribution automation using distribution line carrier systems – Part 5-1: Lower layer profiles – Spread frequency shift keying (S-FSK) profile, IEC Std. 61334-5-1, 2001.
4. S. Galli, A. Scaglione and Z. Wang, For the grid and through the grid: The role of power line communications in the Smart Grid, *Proc. IEEE*, 99(6), 998–1027, June 2011.
5. Powerline Related Intelligent Metering Evolution (PRIME). [Online] Available: http://www.prime-alliance.org, accessed 13 October 2013.
6. The G3-PLC Alliance. [Online] Available: http://www.g3-plc.com/.
7. I. Berganza, A. Sendin and J. Arriola, Prime: Powerline intelligent metering evolution, in *IET-CIRED – CIRED Seminar, SmartGrids for Distribution*, Frankfurt, Germany, 23–24 June 2008.
8. K. Razazian, M. Umari, A. Kamalizad, V. Loginov and M. Navid, G3-PLC specification for powerline communication: Overview, system simulation and field trial results, in *IEEE International Symposium on Power Line Communications and Its Applications (ISPLC)*, Rio de Janeiro, Brazil, 28–31 March 2010.
9. I. Berganza, A. Sendin, A. Arzuaga, M. Sharmaand and B. Varadarajan, PRIME interoperability tests and results from field, in *IEEE International Conference on Smart Grid Communications (SmartGridComm)*, Gaithersburg, MD, 4–6 October 2010.
10. J. Domingo, S. Alexandres and C. Rodriguez-Morcillo, PRIME performance in power line communication channel, in *IEEE International Symposium on Power Line Communications and Its Applications (ISPLC)*, Udine, Italy, 3–6 April 2011.

11. M. Hoch, Comparison of PLC G3 and PRIME, in *IEEE International Symposium on Power Line Communications and Its Applications (ISPLC)*, Udine, Italy, 3–6 April 2011.

12. K. Razazian, A. Kamalizad, M. Umari, Q. Qu, V. Loginov and M. Navid, G3-PLC field trials in U.S. distribution grid: Initial results and requirements, in *IEEE International Symposium on Power Line Communications and Its Applications (ISPLC)*, Udine, Italy, 3–6 April 2011.

13. IEEE 1901.2: Draft Standard for Low Frequency (less than 500 kHz) Narrow Band Power Line Communications for Smart Grid Applications. [Online] Available: http://grouper.ieee.org/groups/1901/2, accessed 13 October 2013.

14. ITU-T Question 15/15 – Communications for Smart Grid. [Online] Available: http://www.itu.int/en/ITU-T/studygroups/2013-2016/15/Pages/q15.aspx, accessed 13 October 2013.

15. SAE J2931/3 - PLC Communication for Plug-in Electric Vehicles. [Online] Available: http://standards.sae.org/wip/j2931/3, accessed 13 October 2013.

16. Road vehicles – Vehicle to grid Communication Interface – Part 3: Physical and data link layer requirements, ISO/IEC Std. ISO/DIS 15118-3, 2013.

17. Narrowband OFDM power line communication transceivers – Physical layer specification, ITU-T Rec. G.9955, December 2011.

18. Narrowband OFDM power line communication transceivers – Data link layer specification, ITU-T Rec. G.9956, November 2011.

19. Narrowband orthogonal frequency division multiplexing power line communication transceivers – Power spectral density specification, ITU-T Rec. G.9901, November 2012. [Online] Available: http://www.itu.int/rec/T-REC-G.9901, accessed 13 October 2013.

20. Narrowband orthogonal frequency division multiplexing power line communication transceivers for ITU-T G.hnem networks, ITU-T Rec. G.9902, October 2012. [Online] Available: http://www.itu.int/rec/T-REC-G.9902, accessed 13 October 2013.

21. Narrowband orthogonal frequency division multiplexing power line communication transceivers for G3-PLC networks, ITU-T Rec. G.9903, October 2012. [Online] Available: http://www.itu.int/rec/T-REC-G.9903, accessed 13 October 2013.

22. Narrowband orthogonal frequency division multiplexing power line communication transceivers for PRIME networks, ITU-T Rec. G.9904, October 2012. [Online] Available: http://www.itu.int/rec/T-REC-G.9904, accessed 13 October 2013.

23. M. Nassar, J. Lin, Y. Mortazavi, A. Dabak, I. Kim and B. Evans, Local utility power line communications in the 3–500 kHz band: Channel impairments, noise, and standards, *IEEE Signal Process. Mag.*, 29(5), 116–127, September 2012.

24. T. H. Clausen, A. Colin de Verdiere, J. Yi, A. Niktash, Y. Igarashi, H. Satoh, U. Herberg, C. Lavenu, T. Lys, C. E. Perkins and J. Dean, The Lightweight On-demand Ad hoc Distance-vector Routing Protocol – Next Generation (LOADng), IETF Internet-Draft, 7 January 2013. [Online] Available: http://tools.ietf.org/html/draft-clausen-lln-loadng-07, accessed 13 October 2013.

25. T. Winter, P. Thubert, A. Brandt, J. Hui, R. Kelsey, P. Levis, K. Pister, R. Struik, JP. Vasseur and R. Alexander, RPL: IPv6 routing protocol for low-power and Lossy networks, IETF RFC 6550, March 2012.

26. V. Oksman and J. Zhang, G.hnem: The new ITU-T standard on narrowband PLC technology, *IEEE Commun. Mag.*, 49(12), 138–145, December 2011.

27. A. Rossello-Busquet, G.hnem for AMI and DR, in *International Conference on Computing, Networking and Communications (ICNC)*, Maui, HI, 30 January–2 February 2012.

28. K. Kim, S. Daniel Park, G. Montenegro, S. Yoo and N. Kushalnagar, 6LoWPAN Ad Hoc On-Demand Distance Vector Routing (LOAD), IETF Internet-Draft, 19 June 2007. [Online] Available: http://tools.ietf.org/html/draft-daniel-6lowpan-load-adhoc-routing-03, accessed 13 October 2013.

29. IEEE standard for local and metropolitan area networks – Part 15.4: Low-rate wireless personal area networks (LR-WPANs), IEEE Std. 802.15.4, 2006.

30. G. Montenegro, N. Kushalnagar, J. Hui and D. Culler, Transmission of IPv6 Packets over IEEE 802.15.4 Networks, IETF RFC 4944, September 2007. [Online] Available: https://tools.ietf.org/html/rfc4944, accessed 13 October 2013.

31. C. Perkins, E. Belding-Royer and S. Das, Ad hoc on-demand distance vector (AODV) routing, IETF RFC 3561, July 2003. [Online] Available: http://www.ietf.org/rfc/rfc3561.txt, accessed 13 October 2013.

32. T. Clausen, J. Yi and A. de Verdiere, LOADng: Towards AODV version 2, in *IEEE Vehicle Technical Conference (VTC)*, Québec City, Canada, 3–6 September 2012.

33. Distribution automation using distribution line carrier systems Part 4: Data communication protocols Section 32: Data link layer logical link control (LLC), IEC Std. 61334-4-32, 1996.

34. J. J. García Fernández, L. T. Berger, A. Garcia Armada, M. J. Fernández-Getino, V. P. Gil Jiménez and T. B. S orensen, Wireless communications in smart grids, in *Smart Grid – Applications, Communications and Security*, L. T. Berger and K. Iniewski, Eds. Hoboken, NJ: John Wiley & Sons, April 2012, ch. 6.

35. Signaling on low-voltage electrical installations in the frequency range 3 kHz–148.5 kHz – Part 1: General requirements, frequency bands and electromagnetic disturbances., CENELEC Std. EN 50065-1, 2011.

36. Priority Action Plan (PAP-15): Harmonize power line carrier standards for appliance communications in the home. [Online] Available: http://www.sgip.org/pap-15-power-line-communications/#sthash.e6MzyTB6.dpbs, accessed 13 October 2013.

37. D. Raphaeli, Noncoherent coded modulation, *IEEE Trans. Commun.*, 44, 172–183, February 1996.

38. F. Cañete, J. Cortés, L. Díez and J. Entrambasaguas, Analysis of the cyclic short-term variation of indoor power line channels, *IEEE J. Sel. Areas Commun.*, 24(7), 1327–1338, July 2006.

39. S. Galli and A. Scaglione, Discrete-time block models for transmission line channels: Static and doubly selective cases, 2011. [Online] Available: http://arxiv.org/abs/1109.5382, accessed 13 October 2013.

40. D. Umehara, M. Kawai and Y. Morihiro, Performance analysis of noncoherent coded modulation for power line communications, in *International Symposium on Power Line Communications and Its Applications (ISPLC)*, Malmö, Sweden, 4–6 April 2001.

41. M. P. T. Clausen and U. Herberg, A critical evaluation of the IPv6 routing protocol for low power and Lossy networks (RPL), in *IEEE International Conference on Wireless and Mobile Computing, Networking and Communications (WiMob)*, Shanghai, China, 10–12 October 2011.

42. J. Tripathi, J. de Oliveira and J. Vasseur, Applicability study of RPL with local repair in Smart Grid Substation Networks, in *IEEE International Conference on Smart Grid Communications (SmartGridComm)*, Gaithersburg, MD, 4–6 October 2010.

43. U. Herberg and T. Clausen, A comparative performance study of the routing protocols LOAD and RPL with bi-directional traffic in low-power and lossy networks (LLN), in *International Symposium on Performance Evaluation of Wireless Ad Hoc, Sensor, and Ubiquitous Networks (PE-WASUN)*, Miami Beach, FL, 31 October–4 November 2011.

44. J. P. Vasseur, J. Hui, S. Dasgupta and G. Yoon, RPL Deployment Experience in Large Scale Networks, IETF Internet-Draft, 5 July 2012. [Online] Available: http://tools.ietf.org/id/draft-hui-vasseur-roll-rpl-deployment-01.txt, accessed 13 October 2013.

45. K. Razazian, A. Niktash, T. Lys and C. Lavenu, Experimental and field trial results of enhanced routing based on LOAD for G3-PLC, in *International Symposium on Power Line Communications and Its Applications (ISPLC)*, Johannesburg, South Africa, 24–27 March 2013.

46. E. Ancillotti, R. Bruno and M. Conti, RPL routing protocol in Advanced Metering Infrastructures: An analysis of the unreliability problems, in *IFIP Conference on Sustainable Internet and ICT for Sustainability (SUSTAINIT)*, Pisa, Italy, 4–5 October 2012.

47. E. Ancillotti, R. Bruno and M. Conti, The role of the RPL routing protocol for Smart Grid communications, *IEEE Commun. Mag.*, 51(1), 75–83, January 2013.

12

ITU G.hn: Broadband Home Networking

Erez Ben-Tovim

CONTENTS

12.1 Structure of the Chapter on G.hn and G.hn MIMO

This chapter, focused on the G.hn *multiple-input multiple-output* (MIMO) *power line communication* (PLC) technology, is structured in the following manner:

The chapter starts with Section 12.2, giving an introduction to the G.hn family of standards which are the basis for G.hn MIMO. This includes an overview description of G.hn's network architecture and operation, the structure of the (non-MIMO) G.hn transceiver and G.hn's *media access* (MAC) scheme and an overview on G.hn's functionalities and mechanisms.

Section 12.3 provides an introduction to G.hn MIMO, focusing on the MIMO configurations applicable to the MIMO PLC channel. This section also lists the basic requirements which were set for the design of the G.hn MIMO transceivers.

Section 12.4 describes G.hn MIMO with its various elements: First, the frame format used for MIMO transmissions is described, highlighting its design consideration aimed to allow the receiver to achieve gain, timing and frequency synchronisation along with channel estimates of the MIMO PLC channel. Following that, the section describes the G.hn MIMO transceiver structure, with the blocks added/changed compared to that of a non-MIMO G.hn transceiver: the *spatial streams* (SSs) parser, bit loading, Tx port mapper and more. Finally, the section describes the MIMO transmission schemes used for transmitting the payload. G.hn MIMO provides various schemes in which a G.hn MIMO transmitter can communicate with non-MIMO G.hn receivers (with enhanced performance compared to a case of a non-MIMO transmitter) and with other G.hn MIMO receivers. The latter provides three modes: Two modes use precoding in the transmitter which is based on channel state information feedback obtained from the receiver, while one mode does not require such a feedback (trade-offs between these modes are also explained). Emphasis is given to unique features, such as the ability of the receiver to control the transmitted MIMO Tx port mapping and bit loading on a subcarrier basis.

12.2 Introduction to the G.hn Standards

12.2.1 Background

Standardisation work on G.hn started in the *International Telecommunication Union, Telecommunication* (ITU-T) sector in 2006 with a goal to develop a next-generation home networking technology with transceivers capable of operating over all existing types of wires within homes and businesses (i.e. over power lines, phone lines and coaxial cables).

Until the introduction of G.hn, multiple home networking technologies were developed, usually by private consortia and alliances, rather than international *standardisation development organisations* (SDOs), where each of these technologies was usually targeting and optimised for operation over a single in-home wiring type. G.hn, on the other hand, was designed so that the transceivers and the G.hn protocols are specified in a generic way and more importantly are optimised and configurable for operation over all of the mentioned wire types ('media' in the G.hn jargon). The media-dependent aspect of the transceivers and protocols is specified using media-specific parameters: for example, G.hn uses an *orthogonal frequency division multiplexing* (OFDM) modulation, where the OFDM parameters, such as the number of subcarriers and subcarrier spacing, are defined per each media.

This unique approach, of having a unified home networking transceiver, provides multiple benefits to silicon and system vendors developing the home networking equipment and to end users (consumers) and service providers installing the home networks. These benefits include the following:

- Provides a road map/evolution path from other wired home networking technologies to a single technology
- The development of the standard leveraged from accumulated experience of member companies having a knowledge base in a multitude of existing technologies operating over the different wire types
- Simplifies the development and reduces costs for silicon vendors
- Reduced integration cost, effort and risk for systems manufacturers integrating the G.hn silicon in their design
- Simplifies implementations of equipment which allows bridging between domains operating over different media (wire types)
- Enables self-installable networks, flexible and simplified deployment architecture according to the customer's preference. The user does not need to deal with different installation and operation procedures

G.hn MIMO, which is described in this chapter starting at Section 12.3, is adding MIMO capabilities to the G.hn transceiver in order to increase the throughput and coverage when operating over power lines.

The G.hn technology (including G.hn MIMO) is backed by the HomeGrid forum (http://www.homegridforum.org), which was formed in 2008 and, besides marketing activities aimed to promote the G.hn technology and guarantee its market success, provides a certification program for G.hn silicon and products to ensure compliance of these devices to the ITU-T standard and interoperability between equipments of different vendors.

In December 2012, the HomeGrid forum certified the first G.hn chipset (http://www. marvell.com/wireline-networking/ghn/). This chipset, by Marvell, comprises the Marvell 88LX3142 digital baseband processor and the Marvell 88LX2718 baseband *analogue front end* (AFE). This was followed by Sigma Designs with their G.hn chipset composed of a digital baseband chip CG5211 and AFE chip CG5213 (http://www.sigmadesigns.com/ solutions_subcat.php?id=35###). Several other silicon vendors are expected to follow and have their silicon certified, for example Metanoia (http://www.metanoia.com.tw).

12.2.2 G.hn Family of Standards

The G.hn technology is specified in multiple ITU-T standards. These standards belong to the 'G' family of ITU-T standards, specifying 'transmission systems and media, digital systems and networks'. The acronym 'G.hn' ('hn' stands for home networking) was an intermediate name given at the early stages of developing the standards and is still used as the common name for this technology. G.hn includes the following ITU-T standards; each specifies a different part of the technology:

- *G.9960*: Specification of G.hn's *physical layer* (PHY) and architecture. See ITU-T. 2010. G.9960.

- *G.9961*: Specification of G.hn's *data link layer* (DLL) and security protocols. See ITU-T. 2010. G.9961 [1].

- *G.9961 Amendment 1*: Contains a mechanism for mitigating interferences between neighbouring G.hn domain. See ITU-T. 2012. G.9961 Amendment 1 [2].

- *G.9963*: Enhancement of G.hn to support MIMO ('G.hn MIMO') for power lines. This includes modifications to both the PHY and DLL sub-layers. See ITU-T. 2011. G.9963 [3].

- *G.9964*: G.hn's *power spectral density* (PSD) specifications. See ITU-T. 2010. G.9964 [4].

All of the mentioned recommendations were formally approved by the ITU-T. G.9960 was the first to be approved in 2009, followed by G.9961 and G.9964. G.9963 (G.hn MIMO) was approved in December 2011.

12.2.3 G.hn Network Architecture and Topology

The architecture of G.hn home networks is illustrated in Figure 12.1. The basic G.hn network is referred to as a G.hn 'domain'. A domain is composed of nodes connected to the same medium (wire type). In a domain, one of the nodes is acting as a *domain master* (DM), while the other nodes are acting as 'end point nodes'. The DM manages the domain: It is responsible for registering nodes, manages the *transmission opportunities* (TxOPs) of nodes in the domain, maintains the topology and routing information needed for relaying of management and data messages and broadcasts information to the nodes, needed for operation of the domain (e.g. information on regional masks, required frequency notches).

The G.hn domain's architecture has the form of a mesh network, in which nodes within a domain can communicate with other nodes in that domain, either directly or via nodes acting as relay nodes. A DM is capable of supporting at least 32 registered nodes. Each node is capable of supporting simultaneous communication sessions with at least 8 other nodes.

Nodes connected to different domains (whether these domains are connected to the same or different media, i.e. wire types) generally need to communicate via inter-domain

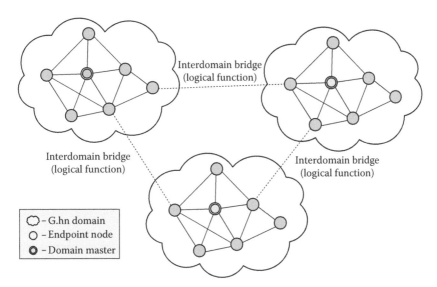

FIGURE 12.1
G.hn home network architecture reference model.

bridges (e.g. L2 or L3 bridging). The operation and specifications of such bridges is outside the scope of the G.hn standards. Nevertheless, G.hn does specify communication between domains in a specific case: G.hn includes a mechanism targeted to mitigate the interferences caused to a G.hn domain connected to power lines, by neighbouring G.hn domains operating over power lines as well. This mechanism specifies communications between the neighbouring domains in order to assist the mitigation process and to coordinate the interferers' transmissions so that the interference is minimised.

12.2.4 G.hn Transceiver

12.2.4.1 General

The G.hn transceiver is based on a burst, windowed-OFDM modem operating with a specific frequency 'bandplan' (a bandplan defines the start/stop frequencies or centre frequency and the bandwidth of the transmitted signal). For each of the supported media, G.hn specifies several possible bandplans. Different OFDM parameters are specified for the different media types. However, for each media, all of the bandplans, specified for this media, are using the same subcarrier spacing. This was done in order to facilitate interoperable operation of devices of different bandplans operating over the same media which belong to the same domain. The bandplans specified in G.hn are as follows:

- For operation over power lines, three baseband bandplans are specified with bandwidth of 25 MHz (1024 subcarriers), 50 MHz (2048 subcarriers) and 100 MHz (4096 subcarriers). The frequency spacing between subcarriers is $F_{SC} = 24.4140625$ kHz.

- For phone lines, two baseband bandplans are defined: 50 MHz (1024 subcarriers) and 100 MHz (2048 subcarriers). $F_{SC} = 48.828125$ kHz.

- For coaxial cables, two baseband bandplans, 50 MHz (256 subcarriers) and 10 MHz (512 subcarriers), and two RF bandplans with bandwidth of 50 MHz (256 subcarriers) and 100 MHz (512 subcarriers). $F_{SC} = 195.3125$ kHz.

G.hn specifies a transmit PSD mask that the transmitter should apply to the signal before transmission. This mask is composed of a limit PSD mask defined in G.hn for each particular medium, notching of international amateur radio bands and other services and any other notching or mask shaping dictated by local regulations.

12.2.4.2 Frames Transmitted and Received by the G.hn Transceiver

The G.hn transceiver, acting in the role of either an end point node or a DM, is transmitting and receiving frames ('PHY frames') with a general structure which consists of a preamble, header and payload. This structure is illustrated in Figure 12.2.

The preamble is a training sequence used by the receiver for detecting the transmitted PHY frame, training the receiver's AGC, acquiring frequency and time synchronisation and performing initial channel estimates.

The *PHY frame header* (PFH) is a field of 168 information bits carrying control information, such as the type of the frame and its length; parameters of the PHY related to the payload, such as the length of the guard interval, identification of the bit-loading vector and parameters of the *forward error correction* (FEC); and also some control information related to the DLL sub-layer mechanisms. In order to guarantee robust transmission of the header, it is transmitted usually over a single OFDM symbol with a FEC code rate of 1/2, bit loading of 2, that is, a constellation of *quadrature phase shift keying* (QPSK) over all of the *active subcarriers* (ASCs) and a strong repetition.

G.hn specifies several PHY frame types used for different purposes. Some of the specified frame types are listed in Table 12.1. As mentioned in this table, some of the frame types carry information only in the header and do not include a payload.

Some of the mechanisms using the previous frame types (e.g. retransmissions and more) are briefly reviewed in Section 12.2.7.

FIGURE 12.2
Format of the G.hn's PHY frame.

TABLE 12.1

PHY Frame Types

Frame Type	Description	Header	Payload
MSG	The basic PHY frame carries user data or management data or both.	√	√
MAP/RMAP	A frame transmitted by the DM, carrying the MAP (scheduling information of the MAC cycle) or relayed MAP.	√	√
ACK	An acknowledgment frame. The relevant ARQ control data are communicated in the header.	√	none
PROBE	A frame carrying probe symbols (for channel and noise estimation) in its payload.	√	√

Source: ITU-T. 2010. G.9961, Unified high-speed wire-line based home networking transceivers – Data link layer specification. With permission.
Note: Few additional PHY frame types are specified in the G.hn specifications.

The payload, which is transmitted over multiple OFDM symbols, delivers data which encapsulate higher-layer information. This encapsulation is described in Section 12.2.4.4. The payload's data are scrambled and coded using a *low-density parity check* (LDPC) FEC code with a flexible code rate, supporting code rates of 1/2, 2/3, 5/6, 16/18 and 20/21 and two possible information block sizes, 120 and 540 bytes. G.hn also specifies flexible loading of bits onto the OFDM symbols. Each subcarrier of a payload OFDM symbol can be loaded with a different number of bits, ranging from 1 to 12 bits (the loading is fixed per subcarrier index along the time axis, i.e. over all OFDM symbols of the payload).

12.2.4.3 Protocol Reference Model of the G.hn Transceiver

The protocol reference model of the G.hn transceiver is presented in Figure 12.3. The G.hn standard specifies the PHY and the DLL, that is, layers 1 and 2 of the *open systems interconnection* (OSI) model.

The DLL is composed of the following sub-layers:

- The *application protocol convergence* (APC) sub-layer provides an interface with the *application entity* (AE), which operates with an application-specific protocol, such as Ethernet. The APC also classifies the incoming data into flows and may provide data rate adaptation between the AE and the home network transceiver in the receive direction.

- The *logical link control* (LLC) sub-layer is responsible for encrypting the transmitted data and handling the security protocols, managing the data path of the retransmissions mechanism, supporting the relaying functionality, managing connections of the node inside the domain and facilitating *quality of service* (QoS) constraints for the connections.

- The MAC sub-layer controls access of the node to the medium using the various specified medium access protocols.

The PHY, as described in the G.hn standard, is also composed of several sub-layers; however, for simplicity reasons, the description hereafter treats the PHY as a single entity.

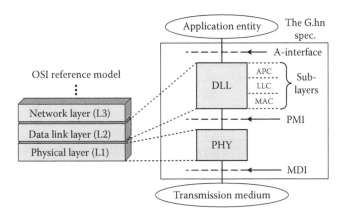

FIGURE 12.3
Protocol reference model of the G.hn home network transceiver. (Based on figure 5-11 from ITU-T. 2011. G.9960, Unified high-speed wireline-based home networking transceivers – System architecture and physical layer specification. With permission.)

The PHY encapsulates the data units obtained from the MAC, into PHY frames, adding a preamble (for various synchronisation and estimation purposes) and a PFH (containing control parameters). The PHY provides *forward error correction* (FEC) of the PHY frame content (header and payload) and loads the coded bits onto the subcarriers of OFDM symbols, according to *bit allocation tables* (BATs). The front-end operations in the PHY include the OFDM modulator, the *inverse discrete Fourier transform* (IDFT), cyclic prefix, the transmit filters (in order to meet regulations) and the AFE. Further description appears in the following sections.

The reference model of Figure 12.3 also identifies the interfaces of the G.hn transceiver. The external interfaces are the A-interface, the interface to the AE, which is user/application dependent (e.g. Ethernet, IP); the *medium-dependent interface* (MDI), which is the physical interface to the medium (i.e. power lines, phone lines and coaxial cables); and the internal interface between the DLL and PHY (the *physical medium-independent interface* [PMI]). The layers above the DLL (above the A-interface) are beyond the scope of the G.hn standard.

12.2.4.4 Data-Plane Processing in the DLL

The overall functional model of the DLL is presented in Figure 12.4.

This section provides a brief overview of the processing of data within the DLL. In addition to this data-plane processing functionality, the DLL provides control-plane functionalities. The interested reader is referred to [1] for further information on these functionalities.

In the transmit direction, *application data primitive* (ADP) sets enter the DLL from the AE across the A-interface. Every incoming ADP meets a format defined by a particular application protocol (e.g. Ethernet frames). Each incoming ADP is converted by the APC sub-layer into *APC protocol data units* (APDUs), which include all parts of the ADP set intended for communication to the destination node(s). The APDUs are transferred to the LLC sub-layer via the x_1-interface. Processing of the data within the G.hn transceiver from this point onward is described in Figure 12.5. In addition to the incoming APDUs, the LLC sub-layer

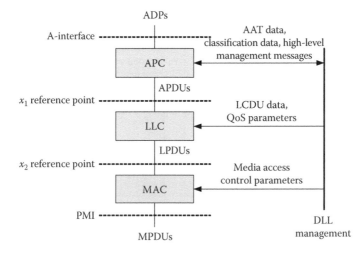

FIGURE 12.4
Functional model of the DLL and partitioning into sub-layers. (Based on figure 8-1 from ITU-T. 2010. G.9961, Unified high-speed wire-line based home networking transceivers – Data link layer specification. With permission.)

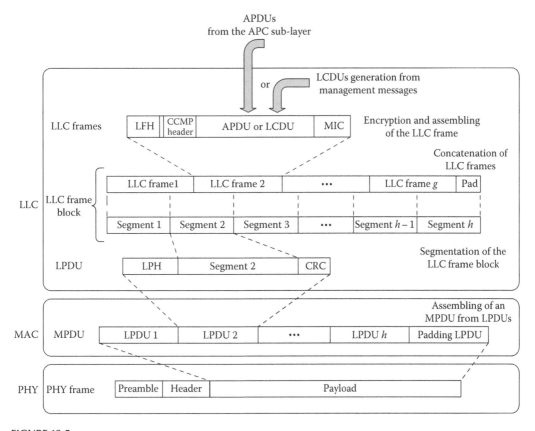

FIGURE 12.5
Processing of data by the G.hn transmitter.

receives management data primitives from the DLL management entity intended for LLC control frames, which are assembled into *link control data units* (LCDUs).

In the LLC, APDUs and LCDUs are converted into LLC frames, by adding an *LLC frame header* (LFH), and may be encrypted using assigned encryption keys. In this case, an additional header, the *CCM encryption protocol* (CCMP) header (CCM is the counter with cipher block chaining message authentication code), and footer (the *message integrity code* [MIC]) information are added. The LLC frame is the basic unit which is subject to the relaying functionality which the LLC manages (G.hn facilitates relaying of both data and management information in layer 2). Several LLC frames are concatenated to form an LLC frame block (this block will eventually fit into the payload part of a single PHY frame).

The LLC frame block is then segmented to segments of equal size (at a previous stage padding, bits are added to the LLC frame block so that its length is an integer multiple of a segment size). Each segment is transformed into an *LLC protocol data unit* (LPDU) by adding an *LPDU header* (LPH) and *cyclic redundancy code* (CRC). The LPDUs are of fixed size of either 540 or 120 bytes. These segments are the basic data unit used for the retransmission mechanism which is controlled by the LLC layer and are also the basic units which in the PHY will be subject to FEC encoding.

The LPDUs are transferred to the MAC sub-layer via the x_2-interface. The only data-plane processing performed in the MAC is assembling of a *MAC protocol data unit* (MPDU) out of the incoming LPDUs. This MPDU is transferred to the PHY via the PMI interface (this MPDU is mapped in the PHY into the payload of a single PHY frame).

12.2.4.5 PHY of the G.hn Transceiver

The functional model of the PHY of the G.hn transmitter is presented in Figure 12.6 (the G.hn standard, as any other standard, specifies the transmitter. The receiver is vendor discretionary).

The following sections describe the transmission flow and the operation of the different PHY building blocks.

12.2.4.5.1 Data Scrambling (for the Header and Payload)

The inputs to the PHY at the PMI reference point are MPDUs coming from the DLL. Each MPDU is mapped to the payload of a single PHY frame. PFH bits are assembled from control information conveyed by the transceiver's management entity (generally, the header carries 168 information bits. There are few exceptions in which the header uses twice this size) and prepended to the payload. This block of bits composed of the header and payload are scrambled (Xor-ed) with a pseudorandom sequence generated by a specific *linear feedback shift register* (LFSR).

12.2.4.5.2 Header and Payload FEC Encoding

After scrambling, both the header and the payload are encoded using a FEC code. The G.hn FEC is using a systematic *quasi-cyclic LDPC block code* (QC LDPC BC). In the process of selecting the coding scheme for G.hn, the LDPC code was evaluated along with a *duo-binary cyclic turbo code* (DB-CTC). While CTC offers good performance at high *block error rates* (BLERs), for example, BLERs of 10^{-2}–10^{-3} which are typical in harsh environments such as power lines, it suffers from a known phenomenon of an error floor in low BLER regions, for example, BLERs of 10^{-6}–10^{-8} which are typical to very benign media-like coax. LDPC, on the other hand, was shown [5] to perform well in a wide range of BLERs (LDPC shows substantial gains over DB-CTC at low BLERs and offers the same or better coding gains at low BLERs), thus proven to be adequate to G.hn intended for multiple media and supporting a wide variety of applications. In addition, LDPC yields itself to very efficient

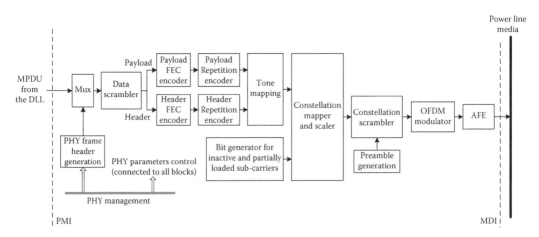

FIGURE 12.6
Functional model of the PHY of a G.hn transceiver.

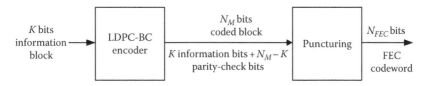

FIGURE 12.7
FEC encoder. (Based on figure 7-7 from ITU-T. 2011. G.9960, Unified high-speed wireline-based home network-ing transceivers – System architecture and physical layer specification. With permission.)

implementations at high rates and is preferred over CTC in many of the recent high-speed communication standards, such as WiFi (IEEE 802.11n and IEEE 802.11ac), DVB-S2/DVB-T2/DVB-C2 (digital video broadcasting, 2nd generation) and 10GBase-TEthernet (IEEE 802.3an).

The structure of the FEC encoder, composed of the LDPC encoder and a puncturing mechanism, is illustrated in Figure 12.7.

In Figure 12.7, K is the size of information block, N_M is the size of coded block (before puncturing), $N_M - K$ is the number of parity check bits and $N_{FEC} \leq N_M$ is the size of the FEC codeword [after puncturing]. The rates related to the coding scheme are $R_M = K/N_M$ (the mother code rate [before puncturing]) and $R = K/N_{FEC}$ (the code rate [after puncturing]).

For the header encoding, the information block size is 168 bits. Header encoding is done using a single mother code with rate $R_M = 1/2$ (corresponding to a mother code matrix, denoted $(1/2)_H$).

For the payload encoding, two possible information block sizes are specified: 960 bits (referred to as 'short' blocks) or 4320 bits (referred to as 'long' blocks). The G.hn payload encoding scheme specifies five code rates 1/2, 2/3, 5/6, 16/18 and 20/21. The encoder is based on mother codes with rates $R_M = 1/2$, $R_M = 2/3$ and $R_M = 5/6$ for each of the informa-tion block sizes (corresponding to mother code matrices denoted $(1/2)_S$, $(1/2)_L$, $(2/3)_S$, $(2/3)_L$, $(5/6)_S$ and $(5/6)_L$, where the matrices with subscript S are used for coding the short blocks and the ones with subscript L are used for the long blocks). From these mother codes, codes with higher code rates are obtained through puncturing, that is, codes with rates 16/18 and 20/21. The codeword at the output of the puncturing block is of size $N_{FEC} \leq N_M$. Table 12.2 summarises all of the coding options.

12.2.4.5.3 Header and Payload Repetition Encoding

In order to improve robustness of transmissions, G.hn specifies a *robust communication mode* (RCM) which is a mode in which encoded data blocks are repeated N_{REP} times. There are two variants for this mode of operation, one specified for the header and for the payload.

The header processing always includes repetition. The coded header block, of size 336 bits (168 bits LDPC coded with rate 1/2), is repeated N_{REP} times:

$$N_{REP} = \text{ceiling } (k_H/N_{FEC}),$$

where k_H is the number of bits to be loaded onto the OFDM symbol carrying the header (equal to 2 bits per subcarrier × the number of *supported subcarriers* [SSCs]) and $N_{FEC} = 336$ bits. The N_{REP} copies are concatenated, where the bits of each copy are cycli-cally shifted by 2 bits within the copy. The bits of this concatenated block are loaded onto subcarriers in the PMD sub-layer.

For the payload, RCM may be applied in MSG frames following a decision of either the receiver or the transmitter: The transmitter may choose to use RCM immediately after

TABLE 12.2

FEC Encoding Parameters

	Code Rate (R)	Information Block Size, K (Bits)	Mother Code Matrix	FEC Codeword Size, N_{FEC} (Bits)
For header	1/2	$PHY_H = 168$	$(1/2)_H$	336
For payload	1/2	960	$(1/2)_S$	1920
	1/2	4320	$(1/2)_L$	8640
	2/3	960	$(2/3)_S$	1440
	2/3	4320	$(2/3)_L$	6480
	5/6	960	$(5/6)_S$	1152
	5/6	4320	$(5/6)_L$	5184
	16/18	960	$(5/6)_S$	1080
	16/18	4320	$(5/6)_L$	4860
	20/21	960	$(5/6)_S$	1008
	20/21	4320	$(5/6)_L$	4536

Source: Based on table 7-56 from ITU-T. 2011. G.9960, Unified high-speed wireline-based home networking transceivers – System architecture and physical layer specification. With permission.

establishing a connection with a receiver and prior to having runtime BATs established; the receiver may choose to request the transmitter to use RCM if channel conditions are very bad. The number of repetitions is determined by the transmitter or receiver based on the needed robustness. G.hn supports payload repetition factors of $N_{REP} = 2, 3, 4, 6, 8$. Without getting into the finer details of the repetition mechanism, it can be said that this mechanism spreads and shuffles the copies of each FEC codeword in both frequency and time (i.e. over multiple OFDM symbols), depending on N_{REP}, N_{FEC} and K_P, which is the number of bits to be loaded onto a payload OFDM symbol (equal to 2 bits per subcarrier × the number of SSCs). This is done to optimise the diversity gained by the repetition.

12.2.4.5.4 Tone Mapping and the Bit Allocation Table

The tone mapper divides the incoming encoded header and payload blocks into groups of bits, according to a BAT and subcarrier grouping being used, and associates each group of bits with specific subcarriers onto which these groups shall be loaded. *The BAT is a vector associating each subcarrier index along the frequency axis with the number of bits to be loaded on it.* A BAT is identified with an identifier called BAT_ID that is indicated in the PFH of each frame transmission.

Two types of BATs are defined:

1. *Predefined BATs*: A set of fixed mappings between subcarrier indices and bit loadings. These mappings are specifying fixed loadings of 1 or 2 bits over all of the (non-masked) subcarriers. These mappings are associated with fixed BAT_IDs.

2. *Runtime BATs*: Mappings that are established for a specific link between a transmitter (source node) and receiver (a destination node), after negotiation and agreement between the two. This negotiation process is part of the 'channel estimation' protocol during which a BAT_ID is dynamically associated to specific runtime BATs.

The PFH is defined with a fixed bit loading of 2 bits on each of the *non-masked subcarriers* (non-MSCs). The payload of the MSG frame can use either the predefined or runtime BATs.

Several different BATs (and BAT_IDs) may be established for a specific link between transmitter and receiver. In addition, BATs may be valid for only a specific portion of the MAC cycle. For example, for a network operating over power lines, in which the MAC cycle is equal to two *alternating current* (AC) cycles, a receiver may divide the AC cycle into intervals and associate each interval of the AC cycle with a different BAT (and BAT_ID). This can be used to cope with the typical cyclic variation of the channel impulse response, noise and interference of the power lines media, synchronised with the AC (mains) cycle.

The establishment of runtime BATs usually involves communicating the explicit BAT (an ordered vector of bit loadings) from the receiver to the transmitter which consumes time on the media. In addition, both the receiver and transmitter need to save the BATs associated with each of their links. In order to reduce the overhead incurred by communication of the BATs over the media and reduce the transmitter and receiver memory requirements, BAT grouping may be used. If grouping is used, all of the subsequent subcarriers of the same group, with G subcarriers, shall use the same bit loading. A G.hn node is capable of supporting grouping of any runtime BAT using grouping of $G = 1$ (no grouping), 2, 4, 8 and 16 subcarriers.

The tone mapper takes into consideration the following types of subcarriers for the purpose of tone mapping:

1. *Masked subcarriers (MSCs)*: Subcarriers on which transmission is not allowed. These subcarriers are not loaded with bits.

2. *Supported subcarriers (SSCs)*: Subcarriers on which transmission is allowed under restrictions of the relevant PSD mask. The number of SSCs is #SSC = N − #MSC. The following types of SSC are defined:

    a. *Active subcarriers (ASCs)*: Subcarriers that have loaded bits ($b \geq 1$) for data transmission. ASCs are subject to constellation point mapping, constellation scaling and constellation scrambling.

    b. *Inactive subcarriers (ISCs)*: Subcarriers that do not have any data bits loaded (e.g. because the SNR is low). The number of ISCs is #ISC = #SSC − #ASC. These subcarriers are loaded with a pseudorandom sequence of constellation points, as described later on. ISCs can be used for measurement purposes.

12.2.4.5.5 Bit Generation and Loading for Inactive and Partially Loaded Subcarriers

There are two cases in which subcarriers of the payload symbols of a MSG frame are loaded or partially loaded with a pseudorandom sequence of bits generated by an LFSR generator, rather than with encoded data bits:

- ISCs with zero data bit loading ($b = 0$). This is usually the result of the receiver (which is usually the node calculating the BATs and communicating them to the transmitter) coming to the conclusion that the SNR on this subcarrier is too low. In this case two bits are taken from the LFSR and loaded on these subcarriers.

- In cases where the number of bits in the encoded payload is not enough to completely fill all subcarriers of the last OFDM symbol in the PHY frame, these last subcarriers will be loaded with bits from the mentioned LFSR. There may be two types of such subcarriers: One subcarrier may be partially loaded with the last few bits from the encoded payload and the rest are taken from the LFSR (this subcarrier is actually an ASC). If there are more subcarriers following this subcarrier in the OFDM symbol, they will be loaded with bits from the LFSR (these last subcarriers are categorised as ISCs). The number of bits loaded from the LFSR on these subcarriers is according to the BAT.

The mentioned LFSR is also used to load the payload symbols of probe frames, also called 'probe symbols'. Both types of symbols are defined as composed of only ISCs. These subcarriers are loaded with 2 bits per subcarrier taken from the LFSR. The probe symbols may be used by receivers for channel and noise estimation purposes.

12.2.4.5.6 Constellation Mapping and Scaling

Constellation mapping associates every group of b bits, $\{d_{b-1}, d_{b-2}, \ldots d_0\}$, loaded onto a subcarrier, with the values of I (in-phase component) and Q (quadrature-phase component) of a constellation diagram.

G.hn specifies constellations with loading of up to 12 bits per subcarrier, as shown in Figure 12.8. All of the even-ordered constellations, that is, loading of 2, 4, 6, 8, 10 and 12 bits per subcarrier, are mandatory for both the transmitter and receiver sides. Theses constellations are square-shaped QAM constellations.

The support of all odd-ordered constellations, that is, loading of 1, 3, 5, 7, 9 and 11 bits per subcarrier, is mandatory for the transmitter. For the receiver, the support of loading of 1 and 3 bits is mandatory, and all other bit loadings with $b \geq 5$ are optional. The constellations with $b \geq 5$ are cross-shaped constellations.

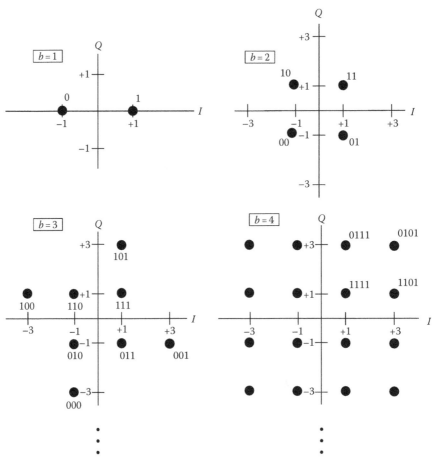

FIGURE 12.8
Constellation mapping ($b = 1, 2, 3, \ldots, 12$).

Mapping of bits into the constellation points is specified in the G.hn standard and basically follows the known Gray-mapping rules (for all even-ordered constellations and most of the odd-ordered constellations).

Each constellation point (I, Q), corresponding to the complex value $I + jQ$ at the output of the constellation mapper, shall be scaled by a power-normalisation factor $\chi(b)$:

$$Z = \chi(b) \times (I + jQ).$$

The normalisation factor, $\chi(b)$, for a subcarrier with b bit loading depends only on the value of b. This factor is calculated so that the values (I, Q) for each constellation point (of each subcarrier) are scaled such that all constellations, regardless of their size, have the same normalised average power equal to 1.

12.2.4.5.7 Constellation Scrambling

An OFDM signal is the weighted sum of N independent narrowband signals (subcarriers). At some time instances, the phases of these independent signals might constructively combine, resulting in a large sum value, while at other times this sum may be small (due to destructive combining). This means that the peak value of the OFDM signal is substantially larger than the average value. This high *peak-to-average power ratio* (PAPR) property is one of the known issues with OFDM.

The phase of constellation points generated by the constellation mapper and scaler is phase shifted (rotated) in accordance with a pseudorandom sequence generated by an LFSR generator. This operation, applied to the subcarriers of the entire frame (including the preamble), reduces the probability of constrictive or destructive combining of the signals constituting the OFDM symbol, thus resulting in lower PAPR of the transmitted OFDM signal. The LFSR is advanced by 2 bits on each subcarrier, and these two output bits are used to determine the phase from the range $\{0, \pi/2, \pi, 3\pi/2\}$. This operation is described as

$$Z_{i,l} = Z_{i,l}^0 \cdot \exp(j\theta) \quad \text{for } l = 0,\dots, M_F - 1, \quad i = 0,\dots, N-1,$$

where

$Z_{i,l}^0$ is the originally mapped constellation point

l is the index of the OFDM symbol within the current frame (M_F denotes the total number of OFDM symbols in the current frame)

i is the index of the subcarrier within the OFDM symbol

θ is the rotation phase

$Z_{i,l}$ is the output of the constellation scrambler (which is fed as input to the IDFT)

12.2.4.5.8 Preamble Generation

The preamble is composed of up to three sections. Each section I comprises N_I repetitions of an OFDM symbol ('mini-symbol') S_I employing subcarrier spacing $k_I \times F_{SC}$, where F_{SC} denotes the subcarrier spacing of the payload OFDM symbols. The third section is only included for the coax media case (and not for power lines or phone lines). The values of k_I can be selected from the set 1, 2, 4 or 8. The subcarriers of section I are spaced such that one subcarrier is included for every k_I subcarriers used for the payload OFDM symbol. The subcarrier spacing of the second section is equal to the subcarrier spacing of the first section ($k_2 = k_1$). The symbols of the second section are an inverted time-domain waveform of the first section ($S_2 = -S_1$). This forms a reference point to detect the start of the received frame. Each preamble section is windowed in order to comply with the PSD mask, as illustrated in Figure 12.9.

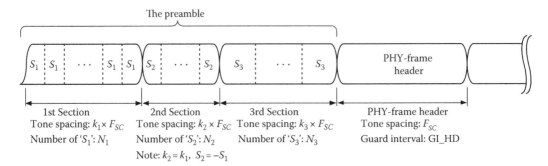

FIGURE 12.9
Structure of the G.hn preamble. (Based on figure 7-25 from ITU-T. 2011. G.9960, Unified high-speed wireline-based home networking transceivers – System architecture and physical layer specification. With permission.)

The non-MSCs of the preamble have a bit sequence of all ones mapped onto them using 1-bit constellations. Following this mapping, constellation phase rotation is applied to the preambles subcarriers using the constellation scrambler described in Section 12.2.4.5.7. The LFSR generator of the constellation scrambler is initialised at the beginning of each one of the used preamble sections to a seed that is section and medium dependent.

As an example, the preamble used for the power lines media comprises only two sections. The first section is composed of 7 mini-symbols and the 2nd of 2 mini-symbols. The two sections use $k_1 = k_2 = 8$.

12.2.4.5.9 OFDM Modulation

The functional diagram of OFDM modulator is presented in Figure 12.10.

The incoming signal to the modulator is a set of N complex values $z_{i,l}$ generated either by the constellation encoder (for symbols of the header and the payload) or by the preamble generator (for symbols of the preamble). The IDFT converts the stream of N complex numbers $z_{i,l}$ into a stream of N complex time-domain samples $x_{n,l}$:

$$x_{n,l} = \sum_{i=0}^{N-1} \exp\left(j \cdot 2\pi \cdot i \frac{n}{N} \right) \cdot z_{i,l} \quad \text{for } n = 0 \text{ to } N-1, \quad l = 0 \text{ to } M_F - 1.$$

Following the IDFT, a cyclic prefix is added by prepending the last $N_{CP}(l)$ samples of the IDFT output to its output N samples. This guard interval is intended to protect against

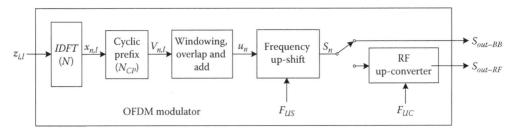

FIGURE 12.10
Functional model of the OFDM modulator. (Based on figure 7-23 from ITU-T. 2011. G.9960, Unified high-speed wireline-based home networking transceivers – System architecture and physical layer specification. With permission.)

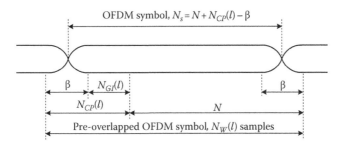

FIGURE 12.11

Structure of an OFDM symbol with cyclic extension and overlapped windowing. (Based on figure 7-24 from ITU-T. 2011. G.9960, Unified high-speed wireline-based home networking transceivers – System architecture and physical layer specification. With permission.)

inter-symbol interference (ISI). The lth OFDM symbol after cyclic prefix addition consists of $N_W(l) = N + N_{CP}(l)$ samples and is described by the following equation:

$$\upsilon_{n,l} = x_{n-N_{CP}(l),l} = \sum_{i=0}^{N-1} z_{i,l} \times \exp\left(j \cdot 2\pi \cdot i \frac{n - N_{CP}(l)}{N} \right) \quad \text{for } n = 0 \text{ to } N_W(l) - 1 = N + N_{CP}(l) - 1.$$

NOTE: The cyclic shift operation does not apply for the preamble.

Following cyclic prefix addition, the time-domain samples are subject to windowing. Windowing shapes the envelope of the transmitted signal to facilitate PSD shaping: it allows sharp PSD roll-offs used to create deep spectral notches and reduction of the out-of-band PSD. The windowing for all OFDM symbols excluding the preamble is done on the first β samples of the cyclic prefix and last β samples of the IDFT output. To reduce the modulation overhead, the windowed samples of adjacent symbols overlap, as shown in Figure 12.11. The value of $N_{CP}(l) - \beta = N_{GI}(l)$ forms the guard interval. The duration of the lth OFDM symbol after overlap is thus $N_S(l) = N + N_{CP}(l) - \beta$. The windowing and the overlap and add operation on the preamble are done on each of the preamble sections separately as can be seen in Figure 12.9.

After applying the windowing and the overlap and add functions, the time-domain samples at the reference point u_n in Figure 12.10 comply with the following equation:

$$u_n = u_n^{(pr)} + \sum_{l=0}^{M_F-1} w(n - M(l), l) \cdot \upsilon_{n-M(l),l} \quad \text{for } n = 0 \text{ to } M(M_F - 1) + N_W(M_F - 1) - 1,$$

where
  $u_n^{(pr)}$ is the nth sample of the preamble (the signal $u_n^{(pr)}$ already includes windowing as necessary)
  $w(n, l)$ is the windowing function and is vendor discretionary
  $M(l)$ is the index of the first sample of the lth symbol

The symbol rate f_{OFDM} (number of symbols per second) and symbol period T_{OFDM} for the given value of N_{CP} and β shall be computed, respectively, as

$$f_{OFDM} = \frac{N \times F_{SC}}{N + N_{CP} - \beta} \quad \text{and} \quad T_{OFDM} = \frac{1}{f_{OFDM}}.$$

The number of IDFT samples, N, and the number of windowed samples, β, are constant over the frame. The value of $N_{CP}(l)$ (and the duration of the pre-overlapped OFDM symbol $N_w(l)$, accordingly) may change during the course of the frame, as follows:

- The header symbol and the first two symbols of the payload use a default value $N_{GI-HD} + \beta$.
- All the rest of the payload symbols use a value of N_{GI} which may be different than the default value, selected from a valid range of values mentioned in Tables 12.3 through 12.5 and indicated in the header.

The frequency up-shift block from Figure 12.10 offsets the spectrum of the transmit signal, shifting it by F_{US} (the value of F_{US} for each bandplan is mentioned later on in Tables 12.3 through 12.5):

$$s_n = u_{n/p} \times \exp\left(j\frac{2\pi mn}{Np} \right) = Re(s_n) + jIm(s_n); \quad \text{for } n = 0 \text{ to } [M(M_F - 1) + N_W(M_F - 1)] \times p - 1,$$

where $u_{n/p}$ is u_n after interpolation with factor p. The interpolation factor p is vendor discretionary and is equal to or higher than 2 (the minimum value of p sufficient to avoid distortions depends on the ratio between the up-shift frequency F_{US} and the bandwidth of the transmit signal $BW = N^*F_{SC}$. It is assumed that an appropriate low-pass filter is included to reduce imaging).

For baseband bandplans, the output signal of the modulator is the real component of s_n:

$$S_{out-BB} = Re(s_n).$$

For the *RF* bandplans (i.e. coax *RF*), the *RF* up-converter produces the following output signal:

$$S_{out-RF}(t) = Re\left[s(t) \times \exp(j2\pi F_{UC}t) \right] = Re[s(t)] \times \cos(2\pi F_{UC}t) - Im[s(t)] \times \sin(2\pi F_{UC}t),$$

TABLE 12.3

OFDM Control Parameters for Power Lines Baseband

Parameter	Bandplan 25 MHz (Notes 1, 2)	50 MHz (Notes 1, 2)	100 MHz (Notes 1, 2)
N	1024	2048	4096
F_{SC}	24.4140625 kHz	24.4140625 kHz	24.4140625 kHz
N_{GI}	$N/32 \times k, k = 1,\dots,8$	$N/32 \times k, k = 1,\dots,8$	$N/32 \times k, k = 1,\dots,8$
N_{GI-DF}	$N/4$	$N/4$	$N/4$
β	$N/8$	$N/8$	$N/8$
F_{US}	12.5 MHz	25 MHz	50 MHz

Source: Based on table 6-4 from ITU-T. 2010. G.9964, Unified high-speed wireline-based home networking transceivers – Power spectral density specification. With permission.

Note 1: The range of subcarrier frequencies is between 0 and $2 \times F_{US}$ MHz.

Note 2: The 25, 50 and 100 MHz bandplans may be used by nodes operating in the same power line domain.

TABLE 12.4

OFDM Control Parameters for Phone Lines (Baseband)

Parameter \ Bandplan	50 MHz (Notes 1, 2)	100 MHz (Notes 1, 2)
N	1024	2048
F_{SC}	48.828125 kHz	48.828125 kHz
N_{GI}	$N/32 \times k, k = 1,\ldots,8$	$N/32 \times k, k = 1,\ldots,8$
$N_{GI\text{-}DF}$	$N/4$	$N/4$
β	$N/32$	$N/32$
F_{US}	25 MHz	50 MHz

Source: Based on table 6-1 from ITU-T. 2010. G.9964, Unified high-speed wireline-based home networking transceivers – Power spectral density specification. With permission.

Note 1: The range of subcarrier frequencies is between 0 and $2 \times F_{US}$ MHz.

Note 2: The 50 and 100 MHz bandplans may be used by nodes operating in the same telephone-line domain.

TABLE 12.5

OFDM Control Parameters for Coax Cables

Parameter \ Bandplan	Coax Baseband		Coax RF	
	50 MHz (Notes 1, 3)	100 MHz (Notes 1, 3)	50 MHz (Notes 2, 3)	100 MHz (Notes 2, 3)
N	256	512	256	512
F_{SC}	195.3125 kHz	195.3125 kHz	195.3125 kHz	195.3125 kHz
N_{GI}	$N/32 \times k, k = 1,\ldots,8$	$N/32 \times k, k = 1,\ldots,8$	$N/32 \times k, k = 1,\ldots,8$	$N/32 \times k, k = 1,\ldots,8$
$N_{GI\text{-}DF}$	$N/4$	$N/4$	$N/4$	$N/4$
β	$N/32$	$N/32$	$N/32$	$N/32$
F_{US}	25 MHz	50 MHz	25 MHz	50 MHz

Source: Based on table 6-6 from ITU-T. 2010. G.9964, Unified high-speed wireline-based home networking transceivers – Power spectral density specification. With permission.

Note 1: The range of subcarrier frequencies is between 0 and $2 \times F_{US}$ MHz.

Note 2: The range of subcarrier frequencies is between F_{UC} and $F_{UC} + 2 \times F_{US}$ MHz.

Note 3: The 50 and 100 MHz baseband bandplans may be used by nodes operating in the same coax baseband domain. The same principle applies to 50 and 100 MHz bandplans defined for the coax RF domain.

where F_{UC} is the frequency shift introduced by the *RF* modulator. After *RF* up-conversion, the centre frequency around which the spectrum of the transmit OFDM signal will be placed is $F_C = F_{UC} + F_{US}$.

The OFDM modulator, as illustrated in Figure 12.10, is described in the G.hn standard in a parameterised way which is media-independent. For each media (power lines, phone lines and coax), a different set of parameters, optimised for the specific media characteristics, is used, as given in Tables 12.3 through 12.5.

Oksman and Galli [5] analysed the OFDM parameters needed for the different wired media types supported by G.hn (power lines, phone lines and coax cables). The reported findings include statistics (scatter plots and trend lines) of the *root mean square delay spread* (RMS-DS) versus the channel gain for the supported wire types. The RMS-DS is a key

metric for optimising the cyclic prefix length (the cyclic prefix length should cover the maximal delay spread to avoid ISI) and the subcarrier spacing (the subcarrier spacing should
be smaller than the coherence bandwidth of the channel, where the coherence bandwidth
is proportional to the inverse of the maximal delay spread of the channel). These statistics show that the different media have distinct RMS-DS and channel gain characteristics,
where the RMS-DSs of the three supported mediums are multiples of each other by a factor that is very close to a power-of-two (the 99% worst case RMS-DS of these media was
found to be 1.75 μs for power lines, 0.39 μs for phone lines and 46 ns for coax). This leads
to a conclusion that a power-of-two scalable OFDM solution is appropriate for all of the
three types of wiring and allows one to calculate the OFDM parameters per each media.
Practically the number of subcarriers is chosen to be higher than the 'theoretical' calculation for higher transmission efficiency, as long as the OFDM symbol length is not too long
(so that the channel does not substantially change during a symbol) and if allowed by
computational complexity and memory constraints.

12.2.4.5.10 *Transmit PSD Mask and AFE*

The signal transmitted by a G.hn transmitter is subject to a transmit PSD mask. The transmit PSD mask is composed of several elements, such as a limit PSD mask, a mechanism
to digitally notch subcarriers and a mechanism for PSD shaping. The limit PSD mask is
an upper bound on the allowed PSD and takes into account currently known regulation
restrictions. As an example, Figure 12.12 shows the G.hn limit PSD specified for power lines.

The notching mechanism allows digital masking of subcarriers in bands that are not
allowed for transmissions (such as bands allocated to amateur radio transmission, radio
broadcasts and other radio services). The existence of these notches is either fixed or
dynamically published by the DM (in the MAP) to all nodes in the domain. A G.hn node
(transmitter) does not load any bits on these MSCs. Another mechanism allows the DM to
publish PSD shaping of the mask in regions where stricter regulations exists compared to
the limit PSD mask.

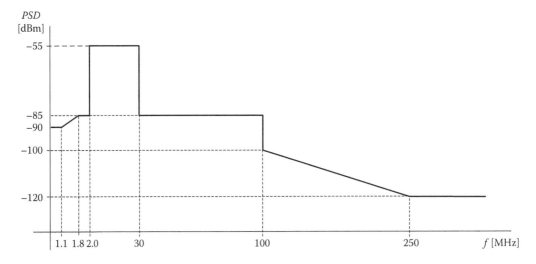

FIGURE 12.12
The limit PSD mask for transmissions over power lines for the 25, 50 and 100 MHz bandplans (notches not
shown). (Based on figure 6-2 from ITU-T. 2010. G.9964, Unified high-speed wireline-based home networking
transceivers – Power spectral density specification. With permission.)

The AFE of the transceiver includes the D/A and A/D converters, analogue filters and the (medium-dependent) line drivers. All of these are vendor discretionary.

12.2.5 G.hn's Media Access

The access of nodes in a domain to the medium, that is, MAC in G.hn is managed by the DM and is synchronised with a cycle called the MAC cycle. This MAC cycle can be synchronised to an external source, and in particular for communication over power lines, the MAC cycle is defined to be synchronised to the mains (AC) cycle, and its length is fixed and equal to 2 AC cycles. This is done to assist in coping with the periodically time-varying nature of the power line channel and noise, which is the result of the power lines topology and electrical devices and appliances connected to the power outlets.

The MAC cycle is divided into time intervals which are assigned as TxOPs for different nodes or groups of nodes in the domain (Figure 12.13).

In each MAC cycle, at least one TxOP is assigned for the DM itself for transmission of the *medium access plan* (MAP). Other TxOPs are assigned by the DM to nodes (or groups of nodes) for transmission of data for various applications. Several types of TxOPs are specified, in which different MAC rules are defined, as explained later on.

Nodes in the domain synchronise with the MAC cycle, decode the MAP and transmit only during TxOPs assigned to them by the DM (transmission in each TxOP is according to the specific MAC rules of that TxOP). Frames transmitted inside a TxOP are separated by *inter-frame gaps* (IFGs). During IFGs the medium is idle.

The DM is responsible for setting up the scheduling in the domain, that is, the length of the MAC cycle (which is fixed for the power lines media, but not for other media types) and the number, timing and parameters of TxOPs in the MAC cycle. This scheduling is set by the DM based on requests for bandwidth from the nodes and may change from one MAC cycle to the other in order to accommodate possible variations of the load in the network, number of registered nodes, changes in the channel conditions, etc.

12.2.5.1 Media Access Plan

The *media access plan* (MAP) transmitted in MAC cycle n contains scheduling information on MAC cycle $n + 1$. The MAP includes the following scheduling information: identification of the boundaries of the MAC cycle, boundaries and content of each one of the TxOPs (the MAC used in the TxOP and its parameters, assignment of nodes or priorities to TxOPs

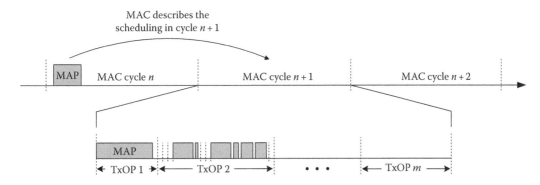

FIGURE 12.13
G.hn's MAP-controlled MAC-cycle-synchronised medium access.

or *time slots* [TSs] within a TxOP and more) and global information needed for the operation of the domain such as regional PSD notches and masks used in the domain. This MAP is broadcasted in the domain by the DM via a MAP PHY frame that is also retransmitted by nodes assigned to be MAP relays, in order to guarantee that the MAP is received by all nodes in the domain.

12.2.5.2 Transmission Opportunities and Time Slots

G.hn specifies several MAC methods with different rules by defining several TxOP types. Each TxOP type specifies a specific MAC method. This was done in order to address different application demands and network conditions (network load, number of nodes, channel characteristics, etc). The DM determines which of these MAC methods to use. It may choose to use a single MAC method or a mix of these methods along the MAC cycle. All G.hn end point nodes mandatorily support all of the MAC methods (to guarantee interoperability). The different TxOP types can be classified into two categories:

1. *Contention-free TxOP* (CFTxOP)
2. *Shared TxOP* (STxOP)

The DM partitions the MAC cycle into CFTxOPs and STxOPs in accordance with service requirements of the domain's nodes and other scheduling issues.

12.2.5.2.1 Contention-Free TxOP

This transmission opportunity is allocated by the DM to a single node, and therefore, transmission in this TxOP is contention-free (i.e. transmissions without collisions). One example for usage of this MAC scheme is the dedicated CFTxOP allocated by the DM in each one of the MAC cycles, for transmission of the MAP itself. This TxOP may also be used for guarantying QoS for certain applications.

12.2.5.2.2 Shared TxOP

An STxOP is divided into a grid of one or more TSs where each TS represents an opportunity to start transmitting for the node or a group of nodes assigned to this TS. The grid starts at the beginning of an STxOP. If a node assigned to the TS gains access to the media, the grid is frozen until the end of the transmission (the TS is virtually expanded). The grid timing is reset after the end of each transmission. A node assigned to the TS may either use the opportunity to start transmitting during the TS or pass on the opportunity to transmit. Transmission rules within each TS depend on the type of the TS. If the node passes on the opportunity, it waits until the next opportunity to transmit in a subsequent TS assigned for this node. The duration of a TS is medium-dependent (e.g. 35.84 µs for power lines). An STxOP can contain the following types of TSs (these are actually two different variants of MAC schemes):

1. *Contention-free TS* (CFTS)
2. *Contention-based TS* (CBTS)

An STxOP can be composed of only CFTSs, only CBTSs or both CFTSs and CBTSs.

A CFTS is a TS assigned for a single node. Other nodes monitor the line (virtual carrier sensing) in order to track the TS grid and be synchronised with the TSs assigned for them for transmission. In this sense, although the STxOP as a whole is shared, the TS MAC itself exhibits a managed carrier-sense, contention-free (i.e. collision avoidance) MAC scheme (i.e. CSMA/CA). This scheme is beneficial for services with flexible bandwidth where QoS is an issue (e.g. VoIP, games, interactive video).

A CBTS is a TS assigned to a group of nodes or to transmissions with a predefined priority or higher (meaning that all nodes which have data of a certain priority or higher can contend for transmission on the line) and therefore exhibits a contention-based MAC (i.e. collisions can occur). This scheme is beneficial for best effort services. There are multiple flavours of CBTS MAC schemes. In the most complex scheme, transmission in a TS is performed in the following way: The node wanting to contend for access to the media transmits an INUSE signal, signalling that it has data to transmit, followed by *priority resolution signals* (PRSs) that indicates the priority of the frame it intends to send. The PRS signalling is built in a way that allows a node to first monitor to see if other nodes have data of higher priority to send before it signals the priority of its planned transmissions. The PRS signalling selects nodes with frames of highest priority: only these nodes are allowed to contend, while all others back off to the next contention period. The probability of collision between the selected nodes is reduced by a random pick of the particular transmission slot inside a contention window. From the beginning of the window, all selected nodes monitor the medium (by carrier sensing). If the medium is inactive at the slot picked by the node, the node transmits its frame; otherwise, it backs off to the next contention period.

12.2.6 Establishing a G.hn Domain and Communication between Nodes of the Domain

Following is a description of how a G.hn domain is established and how nodes in the domain establish communication between each other. Steps 1 through 3 are used to establish a G.hn domain:

1. *Forming a domain (DM selection)*: A domain is started by connecting (and powering up) one or several nodes to the media. These nodes will first look for an existing domain (by trying to detect MAPs from an existing DM) and in the lack of such will undergo a process of selecting a single DM among themselves. The user (or installer of the network) may configure one node to be the DM or it may be automatically selected to be a DM following certain criteria (e.g. if it has the highest visibility of other nodes). Once a node is selected as a DM, it will select a unique domain ID and start transmitting the MAPs conveying the scheduling of the MAC cycle.

2. *Admitting (registration) new nodes into the domain*: End point nodes wishing to join a domain first detect and decode the MAPs and then undergo a process of registration to the domain, during which they convey their capabilities to the DM and receive a unique device ID and other information unique to the domain. After the registration is confirmed, the newly registered node may start communicating with other nodes in the domain (in the TxOPs allocated to it and following the MAC rules of these TxOPs).

3. *Conveying topology (routing) information to end point nodes*: In order to allow communication in the mesh network, the DM gathers information about the connectivity map of nodes in the domain (i.e. a 'routing table' which including all routes from every node to every other node). This routing table is broadcasted to all nodes of the domain and is used by the nodes to know how to 'reach' their destination nodes (see more on 'relaying' in the next section).

The following Steps 4 through 7 are the steps needed for establishing communication between nodes:

4. *Connection setup*: A mechanism initiated by the transmitter during which the transmitter and the receiver exchange information about the connection and establish a one way connection (transmitter to receiver). The information exchanged includes control information of the connection, such as usage of acknowledgments, size of FEC block, type of the connection (unicast/multicast/broadcast) and more. Connections may be released by the transmitter or the receiver.

5. *Communication using RCM*: Once a connection is established, the transmitter can start sending data to the receiver. If the transmitter does not have runtime BATs (bit-loading tables) established with the destination receiver, it will transmit all of the transmissions with a payload transmitted in RCM, that is, using predefined BATs (usually this means loading of 2 bits, i.e. QPSK modulation, uniformly on all SSCs), and repetition.

6. *Channel estimation (establishing runtime BATs)*: This is the process by which a receiver establishes runtime BATs (bit-loading table per subcarrier) with the transmitter. This protocol is very flexible to allow different implementations. For example, the receiver can request the transmitter to transmit probe PHY frames, which are frames in which the entire payload is generated by a pseudorandom sequence generator (each subcarrier is QPSK modulated). Such a receiver will use these probe frames to estimate the channel (and/or noise) and determine the BATs. Another possible receiver implementation will estimate the channel by processing regular data (MSG) PHY frames with their payload transmitted in RCM (e.g. by using channel estimation based on decision-directed processing of the payload). In either case after deriving the BAT, the receiver sends it to the transmitter (along with additional control information such as an identifier of the BAT, called a BAT_ID, the bit-loading grouping) using a special channel estimation management message.

7. *Communication with runtime BATs*: After a transmitter established a runtime BAT with a receiver, it can use it for communication with the receiver. The BAT_ID identifying the BAT used for modulation of the payload is indicated in the PFH. As mentioned earlier, a specific BAT may be associated with an interval of the AC cycle; hence, a transmitter may hold several BATs (and associated BAT_IDs) with every receiver it communicates with. The transmitter will choose the BAT to use according to the time location of the transmitted frame.

A note regarding security mechanisms in G.hn: If the domain is configured to operate in a secure mode, additional security-related procedures will be required along with the previous: This includes authentication, generation and distribution of encryption keys between nodes and periodical key and authentication updates using a set of *authentication and key management* (AKM) procedures. All of the traffic in the domain (all of the data and management communication between nodes in the domain) is encrypted (in the LLC sub-layer) using AES-128 encryption. Several encryption modes are possible, such as a mode in which a single encryption key is used per domain or another mode in which every pair of nodes in unicast and nodes of every multicast group use a unique encryption key.

12.2.7 Overview of Some of G.hn's Mechanisms

This section briefly highlights some of the mechanisms and protocols in G.hn. This is a very brief overview, and the interested reader is referred to the G.hn specifications for further details (see Section 12.2.2 for detailed list of references):

- Acknowledgment and retransmission protocol

  In order to increase robustness of transmissions in noisy media, G.hn specifies a mechanism in which the receiver acknowledges reception of segments (LPDUs) of received frames, allowing the transmitter to retransmit those segments which were not received correctly. G.hn specifies two types of acknowledgments: immediate and delayed. With the first one, a receiver receiving a *message* (MSG) PHY frame replies each frame with an ACK PHY frame (after a predefined IFG). In delayed ACK, the transmission of the ACK frame is deferred to a later TxOP assigned to the receiver. The ACK frame contains information about the segments (LPDUs) which were correctly/incorrectly received (each segment has a CRC attached), in a certain window of segments. The transmitter retransmits segments which were not acknowledged.

- Relaying (at layer 2) of MAPs and data/management frames

  In some media types, some of the nodes in the networks might not be 'visible' to some nodes (these are referred to as 'hidden' nodes) but visible to others. This might be caused by high attenuation in specific links or due to temporary high noise conditions. In order to allow communication between hidden nodes, other nodes must act as relays. G.hn specifies comprehensive relaying mechanisms: For nodes hidden from the DM, MAP relaying is defined in which end point nodes are assigned by the DM to relay the MAPs from time to time (to guarantee that all nodes receive it). In addition, the DM builds and broadcasts the routing table to all nodes of the domain. This table is used by the nodes to know how to 'reach' their destination nodes (e.g. node A will know that if it needs to communicate with nodes B, it needs to send its frames through node C, i.e. node C acts as a relay node).

- Multicast protocols

  G.hn specifies protocols to support efficient transmission of multicast traffic. A PHY level protocol ('multicast binding protocol') enables a transmitter to transmit the same PHY frames to a group of nodes with a common BAT. This protocol also allows member nodes of the multicast group to acknowledge the received data. A DLL level multicast protocol is also defined (allowing multicast traffic to be transmitted over relay nodes).

- PHY frame bursting

  In power lines, the bit-loading allocation table (the BAT) may change along the AC cycle, due to the nature of the channel and typical noise (which is usually periodical with a period of a complete or half AC cycle). This means that for a given link, the receiver (and transmitter) usually will maintain a set of BATs, each to be used on a different time portions (called 'interval') of the AC cycle. The PHY frame bursting mechanism allows to 'break' a long frame into several shorter frames. Each frame in this 'burst' may be transmitted with a different BAT. The frames in this burst are transmitted in succession without relinquishing the medium with very small gaps between them. A single ACK frame is used to acknowledge the status of the LPDUs in all of the frames of the burst, if required.

- Bidirectional transmissions

  Bidirectional transmissions between two nodes may be used to improve throughput and minimise latency of a traffic that is bidirectional in nature, such as TCP traffic with acknowledgments. With this mechanism, a node originating (sourcing) the bidirectional traffic and the destination node exchange special frames: a *bidirectional message* (BMSG) frame and a *bidirectional acknowledgment* (BACK) frame. Both BMSG and BACK carry data and, in the case of acknowledged transmissions, also an acknowledgment on the recently received frame.

- Mitigating interferences from neighbouring (i.e. other G.hn) domains

  G.hn specifies a set of tools to mitigate interferences from neighbouring G.hn domains, for example, domains in adjacent apartments in a *multi-dwelling unit* (MDU) building. Such interference in power lines is common due to inductive propagation between adjacent wirings or due to low attenuation when the networks share common feeder lines. The set of tools include means to quickly detect the existence of neighbouring domains, measure the severity of interferences per node and mitigate the interferences on a node by node basis (some nodes may be interfered and others will not). The mechanism allows the domain to communicate with the other interfering domains and to coordinate transmissions of nodes which strongly interfere with each other.

- Coexistence with alien (non-G.hn) power line networks

  When there is a chance that devices of another non-interoperable PLC technology (i.e. non-G.hn) are simultaneously using the same power line cables in the same frequency range, a coexistence mechanism can be used to mitigate the mutual interferences. This mechanism is specified in ITU-T specification G.9972 [6], also known as 'G.cx', and is also reviewed in Chapter 10.

12.3 Introduction to G.hn MIMO

A detailed introduction to MIMO signal processing in general, as well as specifically with respect to PLC, has been provided in Chapter 8. For convenience, the following reintroduces some of the basic aspects as well as the mathematical notation used throughout the remainder of this chapter.

12.3.1 Received Signal Model

A MIMO system typically consists of N_T transmitters (or transmission ports) and N_R receivers (or reception ports). At a given time instant, the transmitters send dependent or independent data $(x_1, x_2, \ldots, x_{N_T})$ over the channel simultaneously and in the same frequency band through N_T transmitters. The composite channel is characterised by a $N_R \times N_T$ channel transfer matrix H with entries $h_{i,j}$ which stand for the transfer response from transmitter j to receiver i.

In OFDM–MIMO systems, this model is used to characterise each one of the individual subcarriers (tones). The following analysis is per subcarrier of the OFDM signal. Thus, all measures should be functions of the frequency (or subcarrier index), but for simplicity of notation, this dependence is omitted. The following relation results from a transmit vector $x = [x_1, x_2, \ldots, x_{N_T}]^T$, a receive vector $y = [y_1, y_2, \ldots, y_{N_R}]^T$ and noise vector $n = [n_1, n_2, \ldots, n_{N_R}]^T$:

$$y = Hx + n.$$

We will also denote by Λ the cross-correlation matrix of the noise components:

$$\Lambda = E\{n \cdot n^H\},$$

where $E\{\cdot\}$ is the mathematical expectation operator for random variables.

12.3.2 Closed-Loop Transmit Diversity Schemes

Turning to practical considerations in the process of MIMO scheme selection, closed-loop transmit diversity schemes (with 1 SS, 2 Tx ports) appear to be very useful for wireless fading channels [7]. For PLC, where radiation limits apply, these advantages are undermined. For example, for the beamforming configuration (beamforming in the case of 1 SS is sometimes referred to as *spot beamforming*, in contrast to beamforming with 2 SSs which is referred to as *eigenbeamforming* and is a *spatial multiplexing* (SM) scheme), one may not use half the *single-input single-output* (SISO) transmit power (Tx power reduced by 3dB) on each one of the two transmit ports due to the spatial directivity of the radiation pattern: constructive superposition of the signals at the receiver wires may also be achieved on some other point in space. Thus, generally speaking, in transmit beamforming, the restriction of the PSD sum conveyed over the two wire pairs to that of the SISO transmission is no longer satisfactory, and the relations between the MIMO and SISO systems should apply to the transmit voltages instead. This limitation makes the transmit beamforming scheme useless and degenerates the best beamforming configuration (with optimum power allocation) to a simple one-pair transmitter (selecting the best transmitter). Further treatment on beamforming and electromagnetic compatibility may be found in Chapter 16.

12.3.3 Open-Loop Transmit Diversity Schemes

Similarly, Alamouti's space–time code [8] is useful for time-varying channels without channel state information at the transmitter, in which it may exhibit a significant diversity gain. However, assuming that the channel is known and that the transmitter adapts its transmission parameters to the (slowly) varying channel conditions, it is better to deliver all of the allowed power to the best transmitter: Alamouti's space–time code is inferior to a single-port transmission (with per subcarrier Tx selection) with an MRC receiver, assuming an underlying total power constraint.

12.3.4 Spatial Multiplexing MIMO with Precoding (Closed-Loop MIMO)

In a precoding-based SM MIMO system, the receiver estimates the channel response matrix and conveys some sort of channel state information (this is usually called a 'precoding matrix'), derived from this matrix, to the transmitter via a feedback channel. The transmitter uses this channel state information (i.e. the precoding matrix) to adapt ('precode') its transmissions to the varying channel conditions. The scheme, also known as 'closed-loop MIMO' (due to the feedback from the receiver to transmitter), provides a good trade-off between receiver complexity and the ability to nearly achieve the channel capacity, compared to 'open-loop' MIMO schemes.

NOTE: The previous description assumes that the transmitter gains access to the precoding matrix through feedback from the receiver. In a general description of a MIMO precoding scheme, there is another alternative. Theoretically (and under some circumstances), many channels are reciprocal, so the transmitter could estimate the channel on the other direction (based on frames it receives from the other node on the other direction) and use it to precode the data it transmits. This is sometimes referred to as 'open-loop' precoding (do not confuse it with open-loop MIMO which usually refers to SM) or 'implicit feedback'. In practice, channel reciprocity is rather difficult to achieve: it requires some calibrations which might increase the overhead of the transmitted PHY frames. In addition, transmit and receive impedances in PLC are different, and for reciprocity to hold in PLC channels, they must be the same. Therefore, in principle, it seems this method is not viable in PLC.

In order to derive the MIMO precoding scheme, let us take the channel response matrix H and transform it to its *singular value decomposition* (SVD):

$$H = UDV^H,$$

where $U \in C^{N_R \times N_R}$ and $V \in C^{N_T \times N_T}$ are unitary (i.e. $U^{-1} = U^H$, $V^{-1} = V^H$) and $D \in \Re^{N_R \times N_T}$ is non-negative and diagonal. The diagonal entries of the matrix D are the non-negative square roots of the eigenvalues of HH^H.

Using the following (information lossless), orthogonal transformation,

$$\tilde{y} = U^H y, \quad \tilde{x} = V^H x, \quad \tilde{n} = U^H n,$$

on the system model described in Section 12.3.1,

$$y = Hx + n,$$

results in the following equivalent model:

$$\tilde{y} = D\tilde{x} + \tilde{n}.$$

The practical interpretation of the previous transformation and models is as follows:

1. The receiver estimates the channel response matrix H (a process called 'channel estimation').
2. The receiver then calculates the precoding matrix V, for example, by performing the SVD operation on the matrix H ($H = UDV^H$).
3. The receiver sends the precoding matrix V (or some representation of it) to the transmitter via a feedback channel.
4. The transmitter performs precoding of the data, x, it needs to send to the receiver, using the precoding matrix V, that is, performs $x = V\tilde{x}$ and transmits x.
5. At the receiver, the following signal is received: $y = Hx + n$.

6. The receiver applies the matrix U^H on the received signal y:

$$\tilde{y} = U^H y,$$

$$= U^H H x + U^H n,$$

$$= U^H U D V^H V \tilde{x} + U^H n,$$

$$= D \tilde{x} + \tilde{n}.$$

7. Since D is diagonal, both channels are decoupled and the receiver can continue with the decoding process on these two separate channels. Since U is unitary, the noise power is not increased.

A note regarding the additive noise: clearly,

$$E\left\{ \tilde{n} \cdot \tilde{n}^H \right\} = E\left\{ U^H n \cdot n^H U \right\} = U^H \Lambda U.$$

Thus, if $\Lambda = I$, then the transformed noise vector also has this unity covariance matrix. In the general case, the additive noise might not be white; therefore, a whitening filter will be needed as part of the processing. From here onward, we can absorb the whitening filter into the channel matrix response and thus assume that $\Lambda = I$. We henceforth refer to the concatenation of the channel matrix response and the whitening filter as the *equivalent channel response*.

The previous orthogonal transformation decomposes the channel into independent parallel channels:

$$\tilde{y}_i = \lambda_i^{1/2} \tilde{x}_i + \tilde{n}_i, \quad 1 \le i \le \min(N_R, N_T),$$

where

λ_i is an eigenvalue of HH^H

$\min(N_R, N_T)$ is the rank of the channel matrix (assuming it is full rank), which is equal to the maximal number of SSs (which for the case of the MIMO PLC channel equals 2).

The capacity of each of the component channels is given by

$$C_i = \log_2(1 + P_{ii} \lambda_i), \quad P_{ii} \equiv E\left\{ |x_i|^2 \right\}.$$

In order to maximise the total capacity of the channel, we need to maximise the sum mutual information. This maximisation results in the optimum power allocation needed for each of the constituent channels. There are several results known in the literature for this optimisation:

- Assuming that x_i's are chosen to be independent, zero-mean Gaussian variables, the optimum power allocation is obtained by the known 'waterfilling' calculation [9].

- If, however, x_i's are chosen to be discrete signalling constellations such as m-PSK or m-QAM in lieu of the ideal Gaussian signals, the optimum power allocation is obtained by the mercury/waterfilling calculation [10].

Note that experimental treatment on PLC channel capacity has been presented in Chapter 9.

SM MIMO with precoding approaches the channel capacity provided that the channel variations in time are not too rapid. This is indeed the case in PLC channels, as the PLC channel's transfer function is quasi-static; it varies only slowly with time, except for cases in which the power line channel topology is changed (e.g. an electrical device is plugged-in in the vicinity of the considered sockets). In general, calculation of the precoding matrix requires SVD of the equivalent channel matrix response (including the whitening filter). The precoder scheme also facilitates the detection of high-order MIMO schemes. However, its drawbacks are the need to communicate the precoder coefficients to the transmitter (closed-loop MIMO configuration), the rate loss incurred by this communication and its limited capability to cope with channel variations. Likewise, on the practical side, it necessitates a rather large memory at the transmitter end in order to accommodate the precoder coefficients for all of the links from one node to all of the other domain nodes.

12.3.5 Spatial Multiplexing MIMO without Precoding (Open-Loop MIMO)

An alternative to the SM MIMO precoding (closed-loop) scheme is the non-precoded SM MIMO (open-loop) scheme. Such a MIMO scheme was first proposed in the late 1990s and was known as the *Bell Labs Layered Space-Time* (BLAST) scheme. In these SM MIMO schemes, the transmitter has no knowledge on the channel and the transmission is rather simple: independent data streams are transmitted through the multiple Tx ports thus achieving spatial diversity (increased throughput).

Open-loop schemes which use FEC encoding prior to the MIMO mapping include two variants: vertical and horizontal MIMO schemes. Vertical encoding specifies that one FEC block is encoded and multiplexed onto all of the SSs for spatial diversity. Horizontal encoding specifies that separate FEC blocks are encoded for each of the parallel SSs separately. Block diagrams of vertical and horizontal MIMO transmitters are given in Figures 12.14 and 12.15, respectively (**Note:** The 'streams to Tx port mapping' block in the simplest case is an identity mapping, i.e. steam 1 is mapped to Tx port 1).

In open-loop schemes, most of the MIMO processing burden is placed on the receiver side. The receiver usually tries to de-multiplex the SSs in order to detect the transmitted symbols. Simplistically, there are a variety of decoding techniques to achieve this, starting from simple and up to complex ones, to name a few: *zero-forcing* (ZF) that uses simple matrix inversion but results in poor performance when the channel matrix is ill conditioned; *minimum mean square error* (MMSE), which is more robust in that sense but provides limited enhancement if knowledge of the noise/interference is not used; and

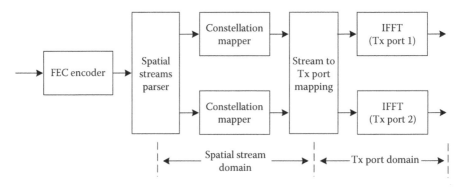

FIGURE 12.14
Vertical SM MIMO (open-loop) transmission.

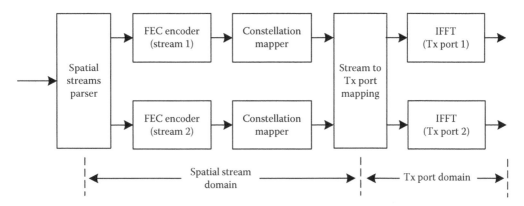

FIGURE 12.15
Horizontal SM MIMO (open-loop) transmission.

maximum likelihood (ML), which is optimal in the sense that it compares all possible combinations of symbols but can be too complex, especially for high-order modulations.

Practically, devising capacity-achieving and non-complex receivers for the open-loops schemes is not a trivial task: By contrast to single-port transmission, the capacity of a (vertical) coded MIMO transmission scheme (with a bit-interleaved coding scheme) cannot be approached by concatenating a MIMO detector that generates the *log-likelihood ratio* (LLR) of the coded bits and a FEC decoder, as the LLRs of the constituent bits of the symbol pair are no longer independent. One alternative solution is to jointly decode the signals from the two wire pairs using a 'turbo equalisation' scheme, where information is iteratively passed between a soft-input-soft-output MIMO detector and a soft-input-soft-output FEC decoder.

As an alternative to the vertical SM MIMO scheme with a turbo equalisation receiver, one can use the horizontal SM MIMO transmission scheme with a *successive interference cancellation* (SIC) receiver: As separate encoding is used at the transmitter, the various codewords in the horizontal MIMO scheme may be decoded successively. A SIC receiver of a horizontally encoded MIMO signal decodes one stream, re-encodes it and subtracts the effect of this codeword from the received signal. The resulting signal, after cancellation (subtraction), is used to detect the second SS. In general, for a higher-order horizontal MIMO system, this procedure is repeated iteratively, where, in each step, the receiver decodes one stream after subtracting the effect of previously decoded streams and treating the effect of all other streams as coloured noise. On the other hand, the horizontal scheme with the SIC receiver suffers from several drawbacks (relative to the vertical one): The scheme requires substantial memory needed to store the received signal points associated with the second decoded stream till the first one is decoded and its effect on the buffered stream is subtracted. Decoding latency is also induced by the SIC policy, which might require a larger IFG period to compensate for the processing latencies in the receiver. It also requires a rather complex scheduling of the FEC decoding.

In conclusion, assuming that the MIMO transmitter is expected to meet the same radiation regulation and that the SISO PSD specified in G.hn was imposed by these requirements (see Section 12.2.4.5.10), the MIMO configuration that seems to hold the greatest promise for power line is SM (with or without precoding) along with the constraint that the sum of PSDs transmitted from the two wire pairs should be equal to the specified SISO PSD. This conclusion is consistent with the findings reported in [11], where the attainable throughputs of the different mentioned MIMO schemes are compared.

12.3.6 Basic Requirements from G.hn MIMO

The following requirements were used to design the various elements of the G.hn MIMO transmission schemes and frame format:

1. *Optimally exploiting the MIMO PLC channel (increasing the throughput and coverage)*: The basic design goal was to design a MIMO transmission scheme which, combined with proper processing in the receiver, will optimally exploit the properties of the MIMO PLC channel, thus providing an increased data rate and enhanced connectivity (i.e. service coverage) of the home network. As previously explained, the spatial diversity of the MIMO PLC channel is 2, meaning that 2 SSs can be transmitted over the channel independently (hence, theoretically doubling the throughput/capacity of the channel compared to single-port transmissions).

2. *Interoperability with legacy, non-MIMO G.hn nodes*: A fundamental assumption made during the development of G.hn MIMO was that a G.hn network can be composed of legacy (native) G.hn nodes, that is, nodes not supporting G.hn MIMO and nodes supporting G.hn MIMO. All of these devices need to interoperate (although clearly the full benefits of the MIMO transmissions will be exhibited when two G.hn MIMO devices are communicating). Two basic requirements result from this assumption:

   a. Native G.hn transceivers (i.e. non-MIMO transceivers complying with G.9960 and G.9961) and G.hn MIMO transceivers (complying with G.9963) need to be able to interoperate when they operate on the same wires and belong to the same domain.

   b. Transmissions from G.hn MIMO transceivers will not degrade the performance of native G.hn transceivers, when operating on the same lines (e.g. allowing non-MIMO G.hn nodes to track the MAC cycle grid in the presence of MIMO transmissions).

   These two requirements had implications on both the selected MIMO frame structure and the various options for transmitting the payload, as explained later on.

3. *Facilitating acquisition of the MIMO channel at the receiver*: Acquisition of PHY frames at the receiver includes several mechanisms, such as detection of the frame (preamble), acquiring gain, frequency and timing synchronisation and obtaining initial channel estimates. A basic requirement related to these acquisition algorithms is that the G.hn MIMO specification should provide all the tools required to perform these acquisitions on a per frame basis (i.e. not relying only on probe frames), in order to cope with rapid variations of the channel response and noise. For the special case of the MIMO PLC channel, care should be taken for the following two acquisition mechanisms:

   a. *Tuning the AGC*: This requires a training sequence which will allow the receiver to set the AGC of the two reception ports, that is, transmission of independent training sequences on the two transmission ports, preparing for receiving the payload which is transmitted over the two transmission ports.

   b. *MIMO channel response estimation*: Providing the tools to allow estimating the full channel response of the MIMO PLC channel (i.e. the four coefficients of the channel matrix), prior to reception of the payload.

12.4 G.hn MIMO

This section begins with describing the road leading to full G.hn MIMO implementations: Although the native G.hn specifications do not mention MIMO transmissions, some implementations which include variants of MIMO schemes are possible and were indeed implemented. Section 12.4.1 describes these possible schemes, paving the way toward full MIMO implementations.

The section continues with describing the various elements of G.hn MIMO: The first element described in Section 12.4.2 is the structure (format) of the PHY frame supporting the MIMO transmissions. The PHY frame for MIMO transmissions was especially designed so that it will accommodate transmissions to both legacy (non-MIMO) G.hn devices and to MIMO G.hn devices, thus allowing for interoperability between these two device types. The description includes some of the criteria used for designing the frame format.

The second element, described in Section 12.4.3, is the structure of the G.hn MIMO transceiver with the new blocks added to enhance the G.hn transceiver to support operation over the MIMO PLC channel.

Finally, Section 12.4.4 describes how the MIMO PHY frame and the transceiver elements are being practically used to implement the different supported transmission schemes used to transmit the payload. A MIMO G.hn node (i.e. a node compliant with the G.9963 specification) is required to be capable of communication with both legacy (non-MIMO) G.hn nodes and MIMO G.hn nodes. This means that a MIMO G.hn node is able to transmit frames using two transmission options: transmissions in which the payload is created as a single SS, for transmissions to legacy (non-MIMO) G.hn nodes (or for cases in which a third conductor is not available), and transmissions in which the payload is created as two SSs, for transmissions to other MIMO G.hn nodes. Both of these options include several possible transmission schemes answering the requirements and allowing different implementations with different complexities, as described in the section.

12.4.1 Road toward a Full G.hn MIMO PLC System

The legacy G.hn standard specifies transmission over a single transmission port (at any given time). The standard does not, however, specifically define any specific mapping of this transmission port to actual wire pairs. The G.hn standard (as any other standard) is focused on specifying the transmitter side and does not directly specify how the receiver should be structured. Receiver structures are left for the implementers to decide on.

Traditional PLC transceivers used only a single transmission port and a single reception port for communication, where these ports are traditionally connected to the *phase-neutral* (P-N) pair of conductors. Such devices can be referred to as SISO devices. Since, as mentioned earlier, the structure of the receiver is not dictated by the standard, a vendor may choose to implement a receiver receiving from multiple ports, either selecting the better port(s) or combining several ports, thus exploiting the receive diversity in the MIMO PLC channel, even when the transmission is done through a single Tx port. Such devices can be referred to as *single-input multiple-output* (SIMO) devices. Other types of devices, still compliant with legacy non-MIMO PLC standards, such as G.hn, may employ different transmit diversity techniques at the transmission ports. For example, a transceiver may be connected to two wire pairs and transmit on one of them at any given time, where the selection may be left for the transmitter or it may be based on some channel quality measures delivered by the receiver to the transmitter (such feedback is proprietary, i.e. not specified in the standard). The receiver can either use a single Rx port (tuned to the same Tx port) or use multiple Rx ports. Topology wise, such schemes can be considered as a 'MIMO' scheme, although they will not achieve any spatial or capacity gain since they are basically a single SS scheme. They will be able to achieve some diversity gain (either transmit or/and receive diversity gains).

All of variants of transceivers mentioned previously are not full MIMO transceivers and will obviously not reach the full potential gain of MIMO systems since inherently any scheme based on the legacy G.hn specifications is limited to being a single SS scheme. However, these variants can be seen as a road paving the way for full G.hn MIMO transceivers, as described in the following sections.

12.4.2 G.hn MIMO PHY Frame

12.4.2.1 Structure of the G.hn MIMO PHY Frame

The general structure of the frames used for MIMO transmissions is illustrated in Figure 12.16.

The MIMO transmission frame format adheres to the following:

- The entire frame, that is, the preamble/header/*additional channel estimation* (ACE) symbol/payload, is transmitted simultaneously on both Tx ports.

- The preamble and the header symbol transmitted on the second Tx port are copies of the preamble and header symbol(s) transmitted on the first Tx port.

- For the case where the payload is created as two SSs, one ACE symbol is added to the frame, after the header. The ACE symbol does not carry data but rather a pseudorandom sequence of constellation points (2 bits per subcarriers), used to assist the receiver in obtaining estimates of the MIMO channel response (as explained later on). It is constructed exactly as probe symbols (the payload of the probe frames). The ACE symbol transmitted on the second Tx port is an inverted version of the ACE symbol transmitted on the first Tx port (i.e. for each time sample of the ACE, $x_2^{ACE} = -x_1^{ACE}$, where the subscripts 1 and 2 denote the SS number).

- The payload may be created as either two SSs (indicated by setting a field called MIMO_IND in the PFH to 1) or a single SS (indicated by setting the field MIMO_IND in the PFH to 0).

- Transmissions on the second Tx port are done with a cyclic shift with respect to the transmission on the first Tx port.

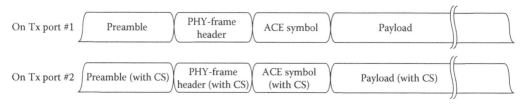

ACE = Additional channel estimation symbol
CS = Cyclic shift

FIGURE 12.16
Format of the G.hn MIMO PHY frame.

12.4.2.1.1 *Purpose of the Cyclic Shift*

The cyclic shift on the second Tx port is needed for AGC setting at the receiver: It is aimed at guarantying that the two versions of the underlying preamble signal transmitted through the two wire pairs are practically uncorrelated so that the received signal power during preamble transmission is the simple sum of the power received from each one of the two transmitting wire pairs (the appealing choice for a MIMO preamble signal in greenfield technologies is based on a combination of independent signals transmitted each through one of the different Tx ports. Albeit, such a selection for the MIMO scheme for G.hn would not be interoperable with legacy G.hn devices which employ detection algorithms based on cross-correlation).

12.4.2.1.2 *Purpose of the Added ACE Symbol*

The addition of this symbol is aimed at allowing the receiver to estimate the full MIMO channel. As mentioned earlier, one of the requirements while designing G.hn MIMO was to allow the receiver to derive MIMO channel estimates on a frame-by-frame basis, which then may be used by the receiver for decoding the payload. On the other hand, in order not to decrease the efficiency of MIMO transmissions, any additional overhead to the PHY frame should be minimised. For this reason the G.hn MIMO spec facilitates the following MIMO channel estimation scheme:

1. The receiver produces an initial channel estimate based on the received header (and may be the last part of the preamble). The received signal at Rx port 1 and 2 is as follows:

$$y_1^{header} = h_{11} \cdot x_1^{header} + h_{12} \cdot x_2^{header} + n_1 = (h_{11} + h_{12}) \cdot x^{header} + n_1,$$

$$y_2^{header} = h_{21} \cdot x_1^{header} + h_{22} \cdot x_2^{header} + n_2 = (h_{21} + h_{22}) \cdot x^{header} + n_2,$$

$$\text{in the previous: } x_1^{header} = x_2^{header} \equiv x^{header},$$

since the same header data are transmitted over SSs 1 and 2. Hence, the receiver can only derive channel estimates for the composite channel $(h_{11} + h_{12})$ from receiving the header at Rx port 1 and $(h_{21} + h_{22})$ from receiving it in Rx port 2.

2. The receiver produces another set of channel estimates based on the received ACE symbol. The received signal at Rx port 1 and 2 is as follows:

$$y_1^{ACE} = h_{11} \cdot x_1^{ACE} + h_{12} \cdot x_2^{ACE} + n_1,$$

$$= h_{11} \cdot x^{ACE} + h_{12} \cdot (-x^{ACE}) + n_1 = (h_{11} - h_{12}) \cdot x^{ACE} + n_1,$$

$$y_2^{ACE} = h_{21} \cdot x_1^{ACE} + h_{22} \cdot x_2^{ACE} + n_2,$$

$$= h_{21} \cdot x^{ACE} + h_{22} \cdot (-x^{ACE}) + n_2 = (h_{21} - h_{22}) \cdot x^{ACE} + n_2,$$

$$\text{in the previous: } x_1^{ACE} \equiv x, \quad x_2^{ACE} \equiv -x^{ACE},$$

since the *ACE* symbol transmitted on SS 2 is an inverted version of the *ACE* symbol transmitted over SSs 1. Hence, the receiver can derive channel estimates for the composite channels, $(h_{11} - h_{12})$ and $(h_{21} - h_{22})$ from receiving the *ACE* at *Rx* port 1 and 2, respectively.

3. The extraction of the individual channel responses for the complete MIMO channel matrix can now be derived directly by subtraction and summation of the two composite responses obtained at steps 1 and 2, for each of the received port.

12.4.2.2 Design Aspects of the G.hn MIMO PHY Frame

This section provides some insights regarding the design criteria and performance merits of the PHY frame structure selected for G.hn MIMO:

Interoperability with legacy G.9960/1 (non-MIMO) devices: The cyclic shift scheme is, by design, fully interoperable with legacy G.hn devices. This is due the fact that the frames start with the legacy G.hn preamble and header. This interoperability is important for the case of a domain which includes both MIMO and non-MIMO G.hn nodes. In such a case when two MIMO nodes communicate with each other using a shared MAC (STxOP), all nodes including non-MIMO nodes need to be able to detect the preamble and decode the header of PHY frames so they will be able to track the MAC cycle grid ('virtual carrier sensing'). Decoding the header is important since it contains the 'duration' field, describing the duration of the PHY frame. The impact of the cyclic shift on the detection performance of legacy G.hn devices (passively listening to MIMO transmissions for virtual carrier sensing) is examined in the following text.

Selecting the value of the cyclic shift applied to the second Tx port: This selection involves two adverse considerations: On the one hand, correct AGC setting for receiving a MIMO payload transmission of two independent SSs is in favour of large values of cyclic shift (larger than the typical delay spread of the power line channels). On the other hand, large values of the cyclic shift widen the duration of the composite channel response as sensed during preamble transmission and blunt the sharp, single peak nature of the cross-correlation measure that is calculated by the frame (preamble) detector in the receiver, thus degrading its performance.

The selection of the cyclic shift value was the result of an evaluation which was based on real-field MIMO channels measurements. This database included 72 MIMO power line channels measured in 13 homes in North America (6 channels measured per home). The evaluation showed the following:

1. The AGC mismatch factor is bounded by around ±1dB for $CS > T_S/8$ (T_S is the time duration of a preamble's mini-symbol, $T_S = 5.12$ µs) and ±0.5 dB for $CS > T_S/4$.

2. The performance degradation of a cross-correlation-based frame (preamble) detector due to the cyclic shift depends on the actual implementation of the detector, its threshold setting and target miss-detection and false alarm probabilities. An evaluation (see the next paragraph) showed that a cyclic shift of $T_S/8$ results in an acceptable degradation. Larger values of the cyclic shift result in relatively significant degradation (and the selection of $CS = T_S/2$ renders the detector useless).

According to these results the cyclic shift in G.hn MIMO was set to a value of $T_S/8$ (= 0.64 µs), which reflects the best compromise between the two conflicting design targets.

What is the performance impact of using the MIMO preamble on a legacy G.hn device that is listening to a MIMO transmission for virtual carrier sensing purposes?

In order to answer this question, real-field MIMO channel measurements (same measurement database mentioned in the previous paragraph) were used to simulate the effect of the cyclic shift. These simulations assumed 3 dB reduction of the transmit power through each wire pair of the MIMO preamble relative to the legacy G.hn preamble so that the total transmit power is preserved. As we are interested in the performance of legacy SISO receivers, the analysis evaluated the performance of a single-port receiver coupled to the phase-neutral (marked as 'P-N' in Figures 12.17 and 12.18) wire pair only.

From the detection performance perspective, the new preamble signal may exhibit either performance gain or loss relative to the legacy preamble. On the one hand, it enjoys diversity and robustness due to the simultaneous transmission through the second wire pair, but on the other hand, it experiences 3 dB power cutback of the legacy transmission through the P-N wire pair.

For each channel, the sensitivity loss of a detector connected to the P-N wire pair, under the MIMO cyclic shift transmission scheme, was checked, relative to the legacy G.hn preamble transmission scheme (i.e. transmission through the P-N wire pair only). The sensitivity loss is defined as the difference in noise levels required to achieve the same detection performance (misdetection and false alarm rates) for the MIMO and legacy preambles using the same cross-correlation-based detector. We note that negative loss means performance gain of the detector when operated with the MIMO preamble relative to its performance when detecting the legacy G.hn preamble. This gain is ascribed to the use of the second wire pair. Figure 12.17 presents the results for all the measured MIMO channels, for the case of cyclic shift of $T_S/8 = 0.64$ μs.

Figure 12.18 presents the same results but with classification of the MIMO channels to one of three groups based on the average attenuation of the legacy channel connecting the P-N wire pair at both ends of the communication channel. This measure is indicative of the SNR of the received legacy preamble.

Figure 12.17 shows that though in some channels the MIMO preamble exhibits better performance, on the average, it is slightly degraded compared to the legacy preamble.

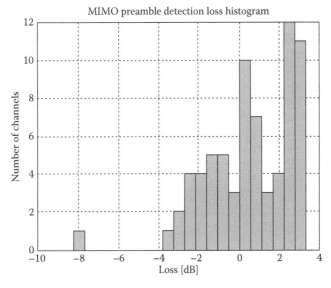

FIGURE 12.17
Detection loss of the MIMO preamble.

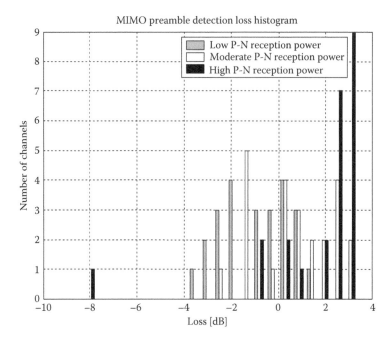

FIGURE 12.18
Detection loss of the MIMO preamble with channel segmentation.

However, looking at Figure 12.18, one can see that there is an outstanding correlation between the performance gain/loss of the frame detector for the MIMO cyclic shift transmission scheme (relative to its performance under the legacy G.hn preamble) and the quality (SNR) of the original P-N wire pair. For the 33% worst channels, the MIMO preamble exhibits 0.9 dB performance gain relative to the detection performance of the legacy preamble over the same channels. On the 33% best channels, the MIMO preamble incurs on the average 1.7 dB performance loss. On the remaining channels (those with moderate reception power), the MIMO preamble incurs an average loss of 0.7 dB. In other words, for poor channels, the frame detector exhibits better performance when operated with the MIMO preamble (with the cyclic shift format) than its performance under processing the legacy G.hn preamble. This means that for these channels, the benefit of the use of the second wire pair is larger than the penalty of the 3 dB power reduction. However, over the good channels, the frame detector shows inferior performance when it processes the proposed MIMO preamble. Yet, over these channels the detector operates very reliably and enjoys a significant noise margin and this loss is not reflected by any practical implication.

To summarise, the results show that the channel coverage of the frame detector for MIMO with the cyclic-shift-based preamble is larger than its counterpart channel coverage for the detection of the legacy preamble. As this is the prominent figure of merit of the preamble signal, the degradation in the SNR experienced by the detector in good channels does not incur any practical loss. It seems that for the challenging (poor) channels, on the average, the proposed MIMO preamble will not degrade the performance of the frame detector but rather offer some gain (better detection performance) despite the 3 dB power reduction.

12.4.3 G.hn MIMO Transceiver

The block diagram of the PHY of the G.hn transceiver (transmitter) is presented in Figure 12.19.

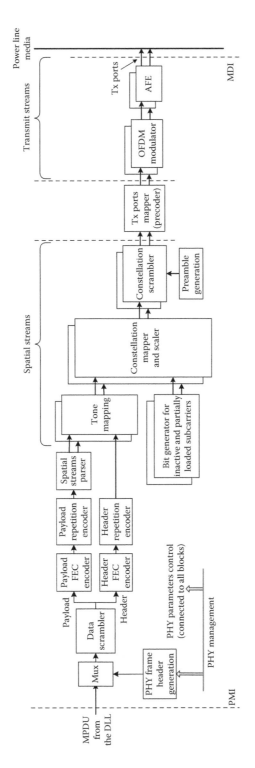

FIGURE 12.19
Functional model of the PHY of a G.hn MIMO transceiver.

The description hereafter will focus on the blocks unique to the MIMO processing and the changes needed in other blocks needed for MIMO operation (i.e. the appropriate extensions needed to have them work on 2 SSs).

12.4.3.1 Data Scrambling and FEC Encoding of the Header and Payload

Both the header and payload data (the MPDU coming from the DLL) are scrambled, and LDPC encoded exactly the same way as in a legacy G.hn transceiver.

12.4.3.2 Spatial Stream Parsing (of the Payload)

The SSs parser works on the output of the FEC LDPC encoder of the payload. There are two cases:

1. Whenever the payload is created as a single SS, the parser does not do anything, that is, delivers the encoded payload block to its output as is.

2. Whenever the payload is created as two SSs, the parser parses the encoded payload block into two SSs in the following way: The parser assigns (groups of) bits alternately to each stream at the subcarrier level according to the bit loading on each stream. Defining $b_i^{(q)}$ as the number of data bits to be loaded to subcarrier i of SS q according to the BAT, the SS parser assigns the first $b_0^{(1)}$ bits at its input to SS 1, the next $b_0^{(2)}$ bits at its input to SS 2, the next $b_1^{(1)}$ bits at its input to SS 1, the next $b_1^{(2)}$ bits at its input to SS 2 and so on (if $b_i^{(q)}$ is zero, no data bits are assigned to subcarrier i of SS q).

12.4.3.3 Tone Mapping

The tone mapper operates independently on each one of the incoming SSs, dividing the incoming streams of bits of the header and payload (of each SS) into groups of bits according to the BAT being used for that SS (i.e. BAT$^{(q)}$ for SS q, $q = 1, 2$) and the subcarrier grouping (the BAT grouping is identical for the two SSs), and associates each group of bits with specific subcarriers onto which these groups shall be loaded. This information is passed to the constellation encoder.

12.4.3.4 Bit Generation for Inactive Subcarriers, Constellation Mapping and Scaling and Constellation Scrambling

For MIMO, the LFSR used to generate bits for inactive and partially loaded subcarriers operates on each SS separately (the specification allows implementing this generator with either two separate LFSRs or a single LFSR).

The constellation mapping and scaling is also performed independently on each SS, associating group of bits of a specific SS, with the I (in-phase component) and Q (quadrature-phase component) values of a scaled constellation diagram (for a specific subcarrier i, different constellations may be used for SS 1 and SS 2).

The operation of the constellation mapper is identical to that used for the legacy G.hn transceiver, where the same phase shift applied to the constellation points of both SSs per the same subcarrier (i.e. the LFSR output per a subcarrier is used to determine the phase shift applied to the constellation points of both SSs associated with the specific subcarrier).

12.4.3.5 Tx Port Mapping (Including Precoding)

The Tx port mapper converts the SSs at its input into transmit streams at its output. The inputs to the mapper are either a single or two SSs coming from the constellation scrambler. The outputs are the transmit streams which are transformed to time-domain samples by the OFDM Modulator and connected to Tx ports. The Tx port mapper operates on a per subcarrier basis. On each subcarrier, it maps a single or a pair of constellation points assigned to the SSs of the subcarrier, to modified pair of signals which are connected (after OFDM modulation, i.e. IDFT) to Tx ports, according to a Tx port *mapping allocation table* (MAT) which the receiver sent. The operation of the Tx port mapper, for subcarrier *i*, is described mathematically as follows:

$$\begin{bmatrix} S_{out,i}^{(1)} \\ S_{out,i}^{(2)} \end{bmatrix} = \begin{bmatrix} TPM_{11,i} & TPM_{12,i} \\ TPM_{21,i} & TPM_{22,i} \end{bmatrix} \cdot \begin{bmatrix} S_{in,i}^{(1)} \\ S_{in,i}^{(2)} \end{bmatrix}, \quad i = 0, \dots, N-1,$$

where

$S_{in,i}^{(q)}$ is the input signal associated with subcarrier *i* of SS *q* (in the case where only a single SS is used $S_{in,i}^{(2)} = 0$)

$S_{out,i}^{(k)}$ is the output signal associated with subcarrier *i* of transmit stream *k*, and the *Tx port mapping matrix* (TPM) for subcarrier *i* denoted TPM_i is

$$TPM_i = \begin{bmatrix} TPM_{11,i} & TPM_{12,i} \\ TPM_{21,i} & TPM_{22,i} \end{bmatrix}, \quad i = 0, \dots, N-1,$$

where its elements $TPM_{kq,i}$ denote the mapping from SS *q* to transmit stream *k* for subcarrier *i*.

12.4.3.5.1 Specific Tx Port Mapping Matrices (per a Single Subcarrier i)

The specific Tx port mapping for the different MIMO schemes and different mapping options, for a specific subcarrier *i*, is described by specific mapping matrices (the vector describing the complete allocation of subcarrier indices and Tx port mappings for all subcarriers in the frequency axis is described by the MAT which is described in Section 12.4.3.6):

- The 'direct' mapping: Used for mapping two inputs, to the two Tx ports (e.g. for *spatial mapping* (SM) without precoding)

$$TPM\#0 = \frac{1}{\sqrt{2}} \begin{bmatrix} 1 & 0 \\ 0 & 1 \end{bmatrix}.$$

- The 'duplication' mapping: Used for copying a single input, to the two Tx ports (e.g. for mapping of the preamble and header)

$$TPM\#1 = \frac{1}{\sqrt{2}} \begin{bmatrix} 1 & 0 \\ 1 & 0 \end{bmatrix}.$$

- The 'duplicate and negate' mapping: Used for copying a single input, to the two Tx ports with the second Tx port inverted (e.g. for mapping of the ACE symbol)

$$TPM\#2 = \frac{1}{\sqrt{2}}\begin{bmatrix} 1 & 0 \\ -1 & 0 \end{bmatrix}.$$

- The 'Tx port 1'/'Tx port 2'mapping: Used for mapping a single input to a single Tx port (SS 1 to Tx port 1 or SS 2 to Tx port 2)

$$TPM\#3 = \begin{bmatrix} 1 & 0 \\ 0 & 0 \end{bmatrix}, \quad TPM\#4 = \begin{bmatrix} 0 & 0 \\ 0 & 1 \end{bmatrix}.$$

- The 'precoding' mapping: Used for SM with precoding

$$TPM\#5 = \frac{1}{\sqrt{2}}\begin{bmatrix} e^{j\varphi}\cos\theta & -e^{j\varphi}\sin\theta \\ \sin\theta & \cos\theta \end{bmatrix}; \quad 0 \le \theta \le \frac{\pi}{2}; \quad 0 \le \varphi < 2\pi.$$

- The 'precoding without SS 2 input'/'precoding without SS 1 input' mapping: Used in one of the modes of SM with precoding when only one input is present

$$TPM\#6 = \begin{bmatrix} e^{j\varphi}\cos\theta & 0 \\ \sin\theta & 0 \end{bmatrix}, \quad TPM\#7 = \begin{bmatrix} 0 & -e^{j\varphi}\sin\theta \\ 0 & \cos\theta \end{bmatrix}; \quad 0 \le \theta \le \frac{\pi}{2}; \quad 0 \le \varphi < 2\pi.$$

12.4.3.6 Bit Allocation and Tx Port Mapping Allocation Table

In G.hn MIMO, the concept of BAT is expanded to include not only the variable bit loading (modulation) per subcarrier but also the Tx port mapping associated with each subcarrier. The *Bit Allocation and Tx Port Mapping Allocation Table* (BMAT) is the combination of the following elements:

1. The BATs for the payload of the PHY frame:
   a. The BAT of SS 1, $BAT^{(1)}$
   b. The BAT of SS 2, $BAT^{(2)}$
2. The MAT for the payload of the PHY frame

The MAT is a vector associating each subcarrier index along the frequency axis with a specific Tx port mapping (i.e. a *TPM* matrix) to be used for this subcarrier. A specific BMAT is associated with an index called BMAT_ID.

The receiver is calculating the BMAT (i.e. the combination of both BATs and MAT) during the process of channel estimation and conveys this information to the transmitter using a specific 'channel estimation' management message. A transmitter may hold several BMATs with every receiver it communicates with. The specific BMAT used by the transmitting node in a particular PHY frame is indicated to the receiving node by a BMAT_ID field in the PFH.

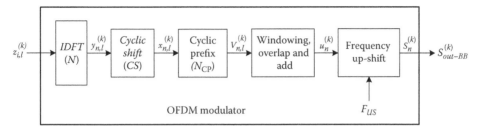

FIGURE 12.20

Block diagram of the MIMO OFDM modulator for transmit stream (k). (Based on figure 7-10 from ITU-T. 2011. G.9963, Unified high-speed wireline-based home networking transceivers – Multiple input/multiple output specification. With permission.)

12.4.3.7 OFDM Modulator (Including Cyclic Shift on the Second Tx Port)

A MIMO transceiver comprises two OFDM modulators, one for each one of the two Tx ports. The OFDM modulator of an individual Tx port is modified compared to that of a legacy G.hn transceiver by introducing a new functional block which performs a cyclic shift on the two transmit streams. The block diagram for the OFDM modulator of an individual Tx port is given in Figure 12.20.

The cyclic shift cyclically shifts the samples of an OFDM symbol at the output of the IDFT, $y_{n,l}^{(k)}$, to generate a shifted version of this sequence, $x_{n,l}^{(k)}$. This shift depends on both the transmit stream index and symbol type (preamble, PFH, ACE and payload). This operation is defined by the following equation:

$$x_{n,l}^{(k)} = y_{\left(n-CS_l^{(q)}\right)\bmod N,l}^{(k)} = \sum_{i=0}^{N-1} z_{i,l}^{(k)} \times \exp\left(j \cdot 2\pi \cdot i \frac{n-CS_l^{(k)}}{N}\right), \quad \text{for } n = 0, 1, \ldots, N-1, \text{ and } k = 1, 2,$$

where $CS_l^{(k)}$ is the cyclic shift used for the lth OFDM symbol of the kth transmit stream. The values of the cyclic shift for the two transmit streams and the different OFDM symbols are listed in Table 12.6.

12.4.3.8 AFEs, Mapping of Tx Ports to Power Lines Conductors and PSD Requirements

The MIMO transceiver comprises two AFEs, one for each one of the two Tx ports. The mapping of transmit streams (Tx ports) to an actual combination of conductors

TABLE 12.6

Cyclic Shift Values

Symbol Type	Number of Subcarriers (N)			Cyclic Shift for Transmit Stream 1 ($k = 1$) [Samples]	Cyclic Shift for Transmit Stream 2 ($k = 2$) [Samples]
	25 MHz	50 MHz	100 MHz		
Preamble	128	256	512	0	N/8
Header	1024	2048	4096	0	N/64
ACE symbol	1024	2048	4096	0	N/64
Payload	1024	2048	4096	0	N/64

Source: Based on table 7-15 from ITU-T. 2011. G.9963, Unified high-speed wireline-based home networking transceivers – Multiple input/multiple output specification. With permission.

(e.g. Tx port 1 connects to the P-N wire pair) is not specified in the G.hn MIMO specifications and is vendor discretionary. However, this mapping cannot change once a node is registered to a domain.

For a G.hn MIMO transceiver, the PSD requirements are that the sum of PSDs of the two transmit signals transmitted from the two Tx ports at any frequency shall never exceed the transmit PSD mask specified for single-port transmissions (see Section 12.2.4.5.10).

12.4.4 G.hn MIMO Payload Transmission Schemes

A G.hn MIMO transceiver is capable of transmitting using two transmission options:

1. Transmissions in which the payload is created as a single SS. This is used for several cases in which the 'full MIMO scheme', that is, the scheme which uses 2 SSs cannot be used. This includes cases in which a G.hn MIMO node transmits (unicast) to a legacy (non-MIMO) G.hn node or transmits to a multicast or broadcast group of nodes which includes such legacy nodes. Another case is a case in which the power lines installation does not include a third conductor (certain geographies do not include the ground/protective earth conductors). There are several variants for performing such transmissions. Further details on this option are given in Section 12.4.4.1.

2. Transmissions in which the payload is created as two SSs. These transmission schemes are used for transmissions between MIMO G.hn nodes and are aimed at exploiting the spatial diversity of the MIMO PLC channel to its full extent. The various possible schemes for such transmissions are described in Section 12.4.4.2.

In general, the selection between the various transmission options (transmission schemes) is usually in the hands of the receiver which is the side that has the knowledge of the MIMO channel response. The receiver performs the selection using management messages, of the 'channel estimation' protocol, which it sends to the transmitter. In the case of transmissions in which the payload is created as a single SS, the transmitter has two transmission options and the selection between the two is in its hands, as described in Section 12.4.4.1 (both options are decodable by legacy G.hn nodes).

12.4.4.1 Payload Transmitted as a Single Spatial Stream
(Transmission to Legacy G.hn Nodes)

Transmissions from a G.hn MIMO node in which the payload is created as a single SS can be done in two possible ways:

1. Transmission as specified in the legacy (non-MIMO) G.hn specifications (although these are single Tx port transmissions, several 'semi-MIMO' options are covered by this, as described in Section 12.4.1). (Clarification note: Although the term 'SS' is not explicitly defined in the legacy G.hn specifications, in essence, the legacy G.hn transmissions are single SS transmissions.)

2. A MIMO transmission (using the MIMO PHY frame) in which the payload is created as a single SS and duplicated for transmission over the two Tx ports simultaneously (with cyclic shift on the second port).

Transmissions according to both of these options will be decodable by legacy (non-MIMO) G.hn nodes. However, since the second option is actually a sort of transmit diversity transmission scheme, using it instead of the previous one (single Tx port transmission) is expected to provide improved performance with increased coverage when a G.hn MIMO node is transmitting to a legacy (non-MIMO) G.hn node.

The G.hn MIMO node needs to select one of the previously mentioned transmission rules to legacy devices, at the time of registration, and does not change this decision as long as it is still registered. A node may select a different transmission rule if it resigns from a domain and reregisters to the domain. The intention behind this is that since the transmission rule for a given legacy node does not change from frame to frame; the channels perceived by other nodes in the domain are consistent from frame to frame.

12.4.4.2 Payload Transmitted as Two Spatial Streams (Transmissions between G.hn MIMO Nodes)

A G.hn MIMO node transmitting to other G.hn MIMO nodes, and wanting to fully exploit the full spatial diversity of the MIMO PLC channel, will use transmission schemes in which the payload is created as two SSs. The G.hn specifications include three optional transmission schemes for this case:

1. Two variants of an SM MIMO with precoding, detailed in Section 12.4.4.2.1. This section also explains the difference between the two variants
2. SM MIMO without precoding (open-loop MIMO)

Selection between these schemes is at the hands of the receiver, indicating the requested 'MIMO mode' in the 'channel estimation' management messages it sends to the transmitter. The G.hn specifications include these options in order to allow a variety of implementation possibilities. Each of the schemes has its strengths and weaknesses. Typically a G.hn node vendor will implement one of the schemes taking into considerations practical considerations such as complexities in the receiver and transmitter, memory requirements and performance gains.

In OFDM systems, MIMO can be applied on a subcarrier level. One of the interesting and unique features of MIMO transmissions with two SSs, in G.hn, which cannot usually be found in wireless MIMO OFDM systems, is that this capability is really put into practice, by allowing the G.hn MIMO receiver to control the MIMO Tx port mapping, that is, deciding whether to transmit on one/two ports and with/without precoding, and conveying the precoding parameters (if used), on a per subcarrier basis, in addition to the ability to control the bit loading (modulation) of the transmitted signal on a per subcarrier basis.

12.4.4.2.1 Spatial Multiplexing MIMO with Precoding in G.hn MIMO

This section describes the SM MIMO scheme with precoding (or shortly, the 'precoding scheme') as implemented in the G.hn MIMO specifications. This scheme actually includes two operational modes as described hereafter. Some practical issues needs to be solved when devising this scheme, which was theoretically described in Section 12.3.4.1.

12.4.4.2.1.1 Feedback Format   There are several options for an explicit feedback of channel state information from the receiver to the transmitter:

- *Non-compressed feedback*: Feeding back either the full channel response matrix, H, or the full precoding matrix V.
- *Compressed feedback*: Feeding back only the precoding matrix, V. There are also several options for a compressed feedback:
  - *Angle-based (parametric) approach*: Due to the fact that V is a unitary matrix, it can be fully described with just two coefficients. There are different options for the definition of the two coefficients that build the matrix, for example, sending an amplitude and phase or sending two angles, as described later on.
  - *Codebook-based approach*: In this approach, the precoding matrix is quantised and a *look-up-table* (LUT) of quantised precoding matrices is prepared (this is the 'codebook'). The receiver selects the quantised precoding matrix that is closest (according to some error metric) to the calculated one and sends the index of this matrix in the LUT. The codebook is designed so that some error metric is minimised (usually the SNR degradation).

The approach selected for G.hn MIMO is the compressed feedback with an angle-based (parametric) approach. The major reasoning behind this selection is as follows:

- Using a non-compressed feedback usually reduces the efficiency of the MIMO scheme since the amount of information conveyed as feedback from the receiver to the transmitter is considerable; therefore, a compressed feedback is preferable.
- Comparing the compressed approaches (parametric vs. codebook-based approaches),
  - The difference between the two lies in the fact that in the parametric approach, the parameters are quantised directly, which may not give a uniform bound for the SNR degradation, while in the codebook approach, construction of the codebook tries to uniformly bound this error (which may result in codebook entries which are non-uniformly spaced). Potentially, the codebook approach requires less feedback bits for the same quality.
  - Practically, the codebook approach suffers from the following:
    - Creating a codebook is not a trivial task. It requires a large measurement database which will cover different geographies. It is not obvious that a single codebook can achieve good results in different installations, geographies, etc.
    - Complexity in the transmitter: The 'codebook-based' scheme requires storage of the codebook in the transmitter. Based on the size of the codebook, this can be a substantial amount of memory. The 'angle-based' scheme requires a negligible amount of memory at the transmitter (the precoding matrix can be calculated on-the-fly).
    - Complexity in the receiver: In the 'codebook-based' scheme, the receiver performs a complex, exhaustive search of the best fitting precoder among the codebook. In order to achieve reasonable performance, the codebook will need to be very large, which would make this search impractical.

- For the parametric approach, several parametric representations were checked, such as magnitude/phase and log-magnitude/phase. These turned out to exhibited inferior performance relative to the representation with two angles.

The precoding matrix (per subcarrier), that is, the matrix used for *Tx port mapping* in the G.hn MIMO transmitter, is identical to the matrix specified in IEEE Std802.11n, defined by the Givens rotation, for the special case of a 2 × 2 matrix. The precoding matrix, per a single subcarrier (usually for the cases where the bit loading for the two SSs is non-zero), is defined by the angles θ and φ:

$$TPM\#5 = \frac{1}{\sqrt{2}} \begin{bmatrix} e^{j\varphi}\cos\theta & -e^{j\varphi}\sin\theta \\ \sin\theta & \cos\theta \end{bmatrix}; \quad 0 \le \theta \le \frac{\pi}{2}; \quad 0 \le \varphi < 2\pi.$$

Practically, the receiver sends the transmitter, a vector containing the pair of angles (θ, φ) for each subcarrier in the OFDM symbol. The transmitter uses these angles, reconstructs the precoding matrix and precodes the transmitted data on each subcarrier using the appropriate matrix.

12.4.4.2.1.2 Quantisation of the Angles   The angles θ and φ are quantised to B_1 and B_2 bits, respectively. G.hn MIMO specifies two possible quantisation levels of either $B_1 = B_2 = 4$ bits or $B_1 = B_2 = 8$ bits. The quantisation level is selected by the receiver and applies to all of the precoding parameters (i.e. all of the subcarriers) in a single 'channel estimation' message used to deliver the precoding feedback for a set of subcarriers. The quantisation level is indicated in this message. Each message (feedback) can use a different quantitation level. The communication of the angle indices and quantisation level is described hereafter. Reconstructing the precoding matrix in the transmitter will be done in the following way: Given phase indices P_1 and P_2 for θ and φ, respectively, $0 \le P_1 \le 2^{B_1} - 1; 0 \le P_2 \le 2^{B_2} - 1$, for some subcarrier, the transmitter uses the mentioned precoding matrix, in which

$$\theta = \frac{\pi \cdot (2P_1 + 1)}{2^{B_1+2}}, \quad \varphi = \frac{\pi \cdot (2P_2 + 1)}{2^{B_2}}.$$

12.4.4.2.1.3 Precoding Grouping   Grouping of precoding parameters is very similar to the G.hn bit-loading (BAT) grouping. Precoding grouping decimates the amount of feedback (angles) sent in the reverse channel: instead of sending 2 angles per subcarrier, the feedback includes a single set of 2 angles per a group of subcarriers. This is in addition to the bit-loading grouping (i.e. the precoder group size *PG* may be different than the bit-loading group size *G*).

   Precoding grouping, as other precoding parameters, is determined by the receiver. Clearly, this factor should be considered along with the quantisation issue. For example, a precoding feedback scheme that uses quantisation of (8, 8) for the two precoder angles (θ, φ) and precoding grouping of *PG* = 2 will probably outperform a feedback scheme with quantisation of (4, 4) and no precoding grouping, though both consume the same feedback rate.

   One difficulty with precoding grouping is incurred by the cyclic shift applied to the 2nd transmit stream (Tx port). This cyclic shift imposes a rather large variation of the precoder parameters between adjacent subcarriers (the precoder parameters are not smooth), and consequently, a simple grouping scheme in which the same parameterisation of the precoder is used for a group of subcarriers is not feasible.

Let's assume that without applying the cyclic shift, the channel matrix response, H, at a given subcarrier, admits the SVD $H = UDV^H$, where the precoder matrix V is indexed on two angles:

$$V(\theta, \varphi) = \begin{bmatrix} e^{j\varphi}\cos\theta & -e^{j\varphi}\sin\theta \\ \sin\theta & \cos\theta \end{bmatrix}.$$

When a cyclic shift is used on the second Tx port, we get the equivalent channel:

$$\tilde{H} = H \begin{bmatrix} 1 & 0 \\ 0 & e^{jk\alpha} \end{bmatrix},$$

where $e^{jk\alpha}$ accounts for the cyclic shift (linear phase) on the kth subcarrier. (Practically $\alpha = 2*\pi*T_{CS}*F_{SC} = 0.098175$ radians, where T_{CS} is the cyclic shift in time units and F_{SC} is the subcarrier frequency spacing.) Now, using the decomposition of the original channel, we have

$$\tilde{H} = H \begin{bmatrix} 1 & 0 \\ 0 & e^{jk\alpha} \end{bmatrix} = UDV^H(\theta, \varphi) \begin{bmatrix} 1 & 0 \\ 0 & e^{jk\alpha} \end{bmatrix} = UD \begin{bmatrix} e^{j\varphi}\cos\theta & -e^{j\varphi}\sin\theta \\ e^{-jk\alpha}\sin\theta & e^{-jk\alpha}\cos\theta \end{bmatrix}^H,$$

$$= e^{jk\alpha}UD \begin{bmatrix} e^{j(\varphi+k\alpha)}\cos\theta & -e^{j(\varphi+k\alpha)}\sin\theta \\ \sin\theta & \cos\theta \end{bmatrix}^H = e^{jk\alpha}UDV^H(\theta, \varphi+k\alpha).$$

The previous derivation implies that given the cyclic-shift-free precoder parameterisation (θ, φ), when the cyclic shift (linear phase of $k\alpha$) is applied, the counterpart precoder parameters are given by $(\theta, \varphi + k\alpha)$. This new parameterisation actually counteracts the cyclic shift. Consequently, the G.hn MIMO specifications specify that if grouping is used, the parameterisation of the precoder, (θ, φ), that the receiver communicates to the transmitter refer to the first subcarrier in the group, and the transmitter should use the following precoder parameterisation (with compensation for the cyclic shift) for the group of PG subcarriers: (θ, φ), $(\theta, \varphi+\alpha)$, ..., $(\theta, \varphi+(PG-1)\alpha)$, where PG denotes the precoding group size.

12.4.4.2.1.4 Two Precoding Modes (Treating Specific Subcarriers Which Have Bit Loading on Only One SS)   As mentioned earlier, the G.hn MIMO specifications actually include two variants of the 'SM MIMO scheme with precoding'. The difference between the two variants (called in the G.hn MIMO specifications 'MIMO mode 1' and 'MIMO mode 2') is in the way they treat specific subcarriers for which the bit loading (BAT) tells the transmitter, for a specific subcarrier, to load bits only on one SS related to that subcarrier (i.e. on the other SS of that subcarrier, zero bits are to be loaded). The Tx port mapping operation is described by

$$\begin{bmatrix} S_{out,i}^{(1)} \\ S_{out,i}^{(2)} \end{bmatrix} = \begin{bmatrix} TPM_{11,i} & TPM_{12,i} \\ TPM_{21,i} & TPM_{22,i} \end{bmatrix} \cdot \begin{bmatrix} S_{in,i}^{(1)} \\ S_{in,i}^{(2)} \end{bmatrix}, \quad i = 0, \ldots, N-1.$$

We are dealing with cases in which for a specific subcarrier i, zero bits have to be loaded on a specific SS q, that is, $b_i^{(q)} = 0$. In this case, $S_{in,i}^{(q)} = 0$. The difference between the two precoding variants (modes) is as follows:

1. In the first approach ('MIMO mode 1'), the following matrices are used:

$$TPM\,\#6 = \begin{bmatrix} e^{j\varphi}\cos\theta & 0 \\ \sin\theta & 0 \end{bmatrix}, \quad TPM\,\#7 = \begin{bmatrix} 0 & -e^{j\varphi}\sin\theta \\ 0 & \cos\theta \end{bmatrix}; \quad 0 \le \theta \le \frac{\pi}{2}; \quad 0 \le \varphi < 2\pi.$$

2. In the second approach ('MIMO mode 2'), the following matrices are used:

$$TPM\,\#3 = \begin{bmatrix} 1 & 0 \\ 0 & 0 \end{bmatrix}, \quad TPM\,\#4 = \begin{bmatrix} 0 & 0 \\ 0 & 1 \end{bmatrix}.$$

In each case, the first matrix is used when $b_i^{(2)} = 0$ (i.e. loading on SS 2 is 0), and the second for the case when $b_i^{(1)} = 0$ (i.e. loading on SS 1 is 0).

When a specific transmission (PHY frame) is using a specific mode out of the two modes previously described, this mode applies to the entire set of subcarriers (and the entire frame). This means that all subcarriers having bit loading of 0 on one of their SSs will follow the previous rule, according to the selected mode.

12.4.4.2.1.5 On Power Allocation   The difference between the two precoding modes is that for the mentioned cases (in which, for a specific subcarrier, the bit-loading algorithm allocates zero bits to one SS and x bits to the other), in the first mode, all power is directed to a single SS, before applying the precoding matrix, while in the second mode, all power is directed to a single Tx port. The second mode is aimed to address potential concerns with the first mode: Although the first mode has the potential to achieve better performance than the second mode, the first mode might be violating radiation regulations. This is since in this mode, transmissions through the two Tx ports are correlated, and using full power on one of the SSs in this case might result (in some point in space) in a radiated power which is larger than that radiated from a single transmit port system by a factor that may reach 3 dB (this is in contrary to transmissions with the unitary precoding matrix and equal power of the SSs, which result in uncorrelated and equal-power transmit streams). Statistical probabilities of increased radiations (backed by field measurements) are discussed in Chapter 16. The second mode is more conservative by imposing the strict *equivalent isotropically radiated power* (EIRP) constraint (the sum of the absolute values of the transmitted signals through the two wire pairs should meet the total power, i.e. field strength, limitation). It can be shown that the optimum transmit strategy, for subcarriers with non-zero bit loading on a single SS, through multiple available transmit ports, under the EIRP constraint reduces to transmission through the Tx port with the largest transfer function for that subcarrier.

12.4.4.2.1.6 Virtual Grouping of the Tx Port Mappings   Some receivers may implement payload-based channel estimation (e.g. decision-directed loops), which may be needed for various tracking purposes. For this kind of channel estimation process, frequency–domain smoothing of the channel estimates will be needed. This kind of smoothing requires that groups of subcarriers will use the same Tx port mapping (i.e. the same type

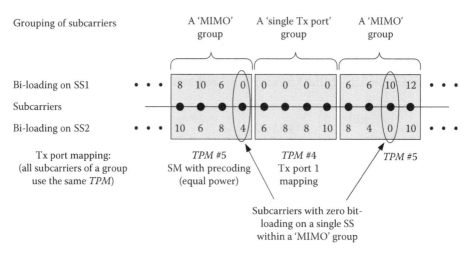

FIGURE 12.21
Tx port mapping example for MIMO mode 2. With virtual grouping of Tx ports mapping.

of *TPM* matrix). For this reason, the specification provides means for the receiver to ask (optionally) the transmitter, to use either the unitary precoding matrix (*TPM #5*) or the matrices unique to mode 1 (*TPM #6* and *TPM #7*) and mode 2 (*TPM #3* and *TPM #4*), on groups of subcarriers. If this is indeed asked by the receiver, a group using (only) the unitary precoding matrix (*TPM #5*) will use it even for those subcarriers which have bit loading on only one SS. In this case, for those subcarriers, 2 random bits will be loaded from an LFSR (the one described in Section 12.4.3.4) on the SS with zero bit loading. An illustration of this feature is given in Figure 12.21, with an example of the Tx port mapping for the case of MIMO mode 2. In this example, the receiver is using virtual Tx port mapping grouping of 4 subcarriers in a group (the grouping is done implicitly by the receiver asking the transmitter for the same *TPM* for consecutive groups of subcarriers). The example shows a snapshot of the subcarrier axis, where the receiver asked for a 'MIMO' group followed by a 'single Tx port' group and another 'MIMO' group. The 'MIMO' group is a group of subcarriers which are using *TPM* #5 (the precoding matrix with equal power on the two Tx ports), and the 'single Tx port' group is using *TPM* #4 (per each subcarrier the bits of SS2 are mapped to Tx port 2, where full power is allocated to Tx port 2, while no power is allocated to Tx port1). The example also shows specific subcarriers within a 'MIMO' group which have zero bit loading on one SS but are still using *TPM* #5.

There is of course a trade-off between the need to group subcarriers in order to allow frequency–domain smoothing (with the performance gains that this mechanism gives) and the loss of theoretical performance incurred by this grouping (since, performance-wise, the optimal approach is to use an independent Tx port mapping on each subcarrier). This is a choice left for the implementer to decide on.

12.4.4.2.1.7 Channel Estimation Message Conveying the MIMO Scheme Parameters to the Transmitter   The receiver, estimating the MIMO channel response, determines the MIMO configuration to be used by the transmitter (1 or 2 SSs, SM with/without precoding, and the precoding mode, if precoding is used), and its related parameters, and sends this information to the transmitter. This information is conveyed to the transmitter as part of the *channel estimation* protocol via special *channel estimation* management messages. This feedback, including the following information, in the case of SM MIMO with precoding, is as follows:

- A MIMO 'mode' indicator: Indicating MIMO mode 1 or 2 (SM with precoding) or MIMO mode 0 (SM without precoding)
- The BMAT which includes
  - The two BATs for the two SSs
  - The MAT, that is, matrices indices for all subcarriers (practically this is encoded as part of the two BATs)
- Bit-loading grouping (*G*)
- Precoding angles for all subcarriers
- *Precoding grouping (PG)*

12.4.4.2.2 Spatial Multiplexing MIMO without Precoding in G.hn MIMO

The SM MIMO scheme without precoding, specified in G.hn MIMO, is a vertical SM scheme (see Section 12.3.4.2). This mode is called 'MIMO mode 0' in the G.hn MIMO specifications. This scheme uses either equal-power MIMO transmission per subcarrier or transmission through a single port per subcarrier. In other words, the following matrices are assigned, per subcarrier, as *Tx port mappings*:

1. For subcarrier i, for which the bit loading for the two SSs is non-zero, that is, $b_i^{(q)} > 0$ ($q = 1, 2$), the following matrix is used:

$$TPM\ \#0 = \frac{1}{\sqrt{2}}\begin{bmatrix} 1 & 0 \\ 0 & 1 \end{bmatrix}.$$

   This matrix connects SS 1 to Tx port 1 and SS 2 to Tx port 2, with equal power (in each port the power is 3 dB less than the full 'single Tx port' power). **Note:** This mapping is also used for subcarriers for which the bit loading is zero for either of the SS, in case the receiver works with the *virtual Tx port mapping* feature (explained further on), and this subcarrier belongs to a 'MIMO' group of subcarriers.
2. For subcarrier i, with zero bit loading on one of the SSs (i.e. $b_i^{(q)} > 0$ for either $q = 1$ or 2), the following matrices are used (unless *virtual Tx port mapping* is used and these subcarriers are part of a 'MIMO' group):

$$TPM\ \#3 = \begin{bmatrix} 1 & 0 \\ 0 & 0 \end{bmatrix}, \quad TPM\ \#4 = \begin{bmatrix} 0 & 0 \\ 0 & 1 \end{bmatrix}.$$

   The first matrix is used when $b_i^{(2)} = 0$ (i.e. loading on SS 2 is 0), and the second for the case when $b_i^{(1)} = 0$ (i.e. loading on SS 1 is 0). These matrixes direct all of the power to a single Tx port.

12.4.4.2.2.1 On Power Allocation The described scheme mixes subcarriers with 'two Tx ports' transmissions (with equal power on each port and in each port the power is 3 dB less than the full 'single Tx port' power) and subcarriers with 'single Tx port' transmissions (where the full power is placed in a single Tx port). This mix is actually a simple and pragmatic mechanism to achieve an approximate 'waterfilling' power allocation (without a need for the receiver to communicate the power allocation to the transmitter per subcarrier) which approaches the capacity of the optimal MIMO system (provided that 'optimal' MIMO decoding is used).

12.4.4.2.2.2 Virtual Grouping of the Tx Port Mappings As explained for the precoding scheme, a G.hn MIMO receiver can ask the transmitter to use Tx port mappings which allocates the same type of TPM to groups of subcarriers. For the SM MIMO without precoding scheme, this means that the MAT will be composed of groups of subcarriers, and each group will use either matrix *TPM #0* or the single Tx port matrices (*TPM #3* and *TPM #4*).

12.4.4.2.2.3 Channel Estimation Message Conveying the MIMO Scheme Parameters to the Transmitter
In the case of SM MIMO without precoding, the following parameters are conveyed:

- The MIMO 'mode' indicator, indicating MIMO mode 0
- The BMAT which includes
  - The two BATs for the two SSs
  - The MAT, that is, matrices indices for all subcarriers (practically this is encoded as part of the two BATs)
- Bit-loading grouping (*G*)

12.5 Conclusions

The G.hn home networking technology was reviewed, with an emphasis on the G.hn MIMO enhancement. While a 'regular' G.hn transceiver transmits and receives over a single Tx and Rx port (usually a port is connected to a power line wire pair, e.g. the P-N pair), a G.hn MIMO transceiver allows transmission and reception on multiple Tx and Rx ports. This enhancement provides increased throughput and coverage, not only for G.hn MIMO transceivers communication with other G.hn MIMO transceivers but also when these transceivers communicate with non-MIMO G.hn transceivers.

G.hn MIMO transceivers can operate using several schemes. All of these schemes provide full interoperability with non-MIMO G.hn nodes, so that G.hn home networks may have a mix of MIMO and non-MIMO nodes. Following is a short summary of the possible transmission schemes:

For transmission to non-MIMO G.hn nodes, a G.hn MIMO node can use the following two schemes:

- A non-MIMO G.hn transmission (i.e. the same transmission format used by non-MIMO G.hn transceivers)
- A MIMO transmission in which the payload is created as a single SS which is duplicated and transmitted over the two Tx ports, with a cyclic shift applied to the second port

For transmissions between G.hn MIMO nodes, the G.hn MIMO transmitter will usually use a MIMO transmission scheme in which the payload is created as two SSs and transmitted over two Tx ports. In this case, the MIMO transmission will be done following one of three possible modes:

- MIMO mode 0: SM MIMO without precoding. This is an open-loop mode in which the transmitter independently transmits the two SSs over the two Tx ports (without any feedback from the receiver about the channel).
- MIMO modes 1 and 2: SM MIMO with precoding. These two modes are closed-loop modes in which the receiver conveys channel information to the transmitter. The transmitter uses this feedback information to 'precode' its transmitted data, so that the transmission is optimally adapted to the eigenmodes of the MIMO channel. The two modes differ in the way they handle specific subcarriers with bit loading on only one of the SSs, as explained in Section 12.4.4.2.1.

Table 12.7 summarizes the usage of the *TPMs*, used on a per subcarrier basis, for the different MIMO transmission mode.

In Table 12.7, the first line includes the cases in which the bit loading on a single subcarrier is greater than zero for both of the SSs (the bit loading on each SS in these cases can range between 1 to 12 bits). For these subcarriers, the payload bits (after the SS parser) are mapped to the two Tx ports using either the SM with precoding matrix (*TPM #5*) if MIMO modes 1 or 2 are used or the identity matrix (*TPM #0*) for SM without precoding if MIMO mode 0 is used.

The first line in Table 12.7 also includes cases in which the bit loading on a specific subcarrier, on at least one SS, is zero, and the receiver chooses to use *TPM #5* or *TPM #0* (depending on the MIMO mode) for this subcarrier. For this subcarrier, two random bits are loaded from an LFSR on the SSs with zero bit loading. This feature was referred to as *virtual grouping of the Tx port mappings* and is useful for cases where the receiver is performing frequency–domain smoothing of channel estimates on a group of subcarriers. In this case if most of the subcarrier in the group has bit loadings different than zero, the receiver will ask the transmitter that all of the subcarriers in this group (called a 'MIMO group') will use mappings of *TPM #5* or *TPM #0*, even for those single subcarriers within the group where the bit loading on one or two SSs is zero.

TABLE 12.7

Usage of *TPMs* for the Different MIMO Modes Used for MIMO Payload Transmissions (No. of SSs = 2)

Bit Loading for the Two Spatial Streams		MIMO Modes		
		SM w/o Precoding	SM with Precoding	
SS 1	SS 2	Mode 0	Mode 1	Mode 2
$0 \leq x_1 \leq 12$	$0 \leq x_2 \leq 12$	TPM #0	TPM #5	TPM #5
$x_1 = 0$	$0 \leq x_2 \leq 12$	TPM #4	TPM #7	TPM #4
$0 \leq x_1 \leq 12$	$x_1 = 0$	TPM #3	TPM #6	TPM #3

Source: Based on table 8-2 from ITU-T. 2011. G.9963, Unified high-speed wireline-based home networking transceivers – Multiple input/multiple output specification. With permission.

Lines 2 and 3 in Table 12.7 include alternative Tx port mapping options for the cases in which the bit loading on a specific subcarrier, on one SS, is zero (an alternative which optimises the performance per these subcarriers on the expanse of not being able to perform the mentioned frequency–domain smoothing). In these cases the payload bits allocated to the SS with non-zero bit loading are mapped either to a single Tx port using *TPM #3* or *TPM #4* (depending on which SS is with non-zero bit loading) if MIMO modes 0 or 2 are used or through a degenerated precoding matrix (a precoding matrix with only the column matching the SS with non-zero bit loading) to the two Tx ports using *TPM #6* or *TPM #7* (depending on which SS is with non-zero bit loading) if MIMO mode 1 is used.

The two MIMO modes with precoding differ in the way they treat the cases of zero bit loading on a single SS for a specific subcarrier (subcarrier not included in 'MIMO groups' of subcarriers, if the *virtual grouping of Tx port mappings* option is used): MIMO mode 1 directs all of the power to a single SS, before applying the precoding matrix, while in MIMO mode 2, all power is directed to a single Tx port. Although mode 1 has the potential to achieve better performance than the second mode, it might be violating radiation regulations, while mode 2 is more conservative by imposing the strict EIRP constraint.

The open-loop and two closed-loop (precoding) modes provide the designer of the G.hn transceiver with flexibilities to allow different receiver complexities versus performance trade-offs. In addition, the MIMO transmission modes allows the receiver to control both the transmitted MIMO Tx port mapping (according to the specific MIMO mode use) and the bit loading on a per subcarrier basis, in a way which allows the receiver to perform frequency–domain smoothing of channel estimates (channel estimates derived from the payload).

The MIMO transmission format is such that the G.hn preamble and G.hn PFH are duplicated and transmitted over the two Tx ports (i.e. the preamble and PFH transmitted on the second Tx port are copies of the preamble and header symbol transmitted on the first Tx port), with cyclic shift applied to the transmissions on the second Tx port. An ACE symbol is added following the PFH to allow the receiver to obtain MIMO channel estimations. Transmissions following these schemes are on the one hand transparent to non-MIMO G.hn transceivers and on the other hand provide the G.hn MIMO receiver with the ability to tune and train its AGC and other frame acquisition loops (timing, frequency, etc.) on a MIMO signal, on a frame-by-frame basis, in preparation for receiving the MIMO transmission of the payload.

References

1. ITU-T. 2010. G.9961. Unified high-speed wire-line based home networking transceivers – Data link layer specification.
2. ITU-T. 2012. G.9961 Amendment 1. Data link layer (DLL) for unified high-speed wire-line based home networking transceivers – Amendment 1.
3. ITU-T. 2011. G.9963. Unified high-speed wireline-based home networking transceivers – Multiple input/multiple output specification.
4. ITU-T. 2010. G.9964. Unified high-speed wireline-based home networking transceivers – Power spectral density specification.
5. Oksman V. and Galli S. October 2009. G.hn: The new ITU-T home networking standard. *IEEE Communications Magazine*, 47(10): 138–145.

6. ITU-T. 2010. G.9972. Coexistence mechanism for wireline home networking transceivers.

7. Schumacher L, Berger L.T, and Ramiro Moreno J. 2002. Recent advances in propagation characterisation and multiple antenna processing 430 in the 3GPP framework, in *XXVIth URSI General Assembly*, Maastricht, the Netherlands, August 2002, session C2. [Online] Available: http://www.ursi.org/Proceedings/ProcGA02/papers/p0563.pdf, accessed 29 September 2013.

8. Alamouti S.M. October 1998. A simple transmit diversity technique for wireless communications. *IEEE Journal on Selected Areas in Communications*, 16(8): 1451–1458.

9. Gallager R.G. 1968. *Information Theory and Reliable Communication*. John Wiley & Sons, New York.

10. Lozano A, Tulino A.M. and Verdu S. 2006. Optimum power allocation for parallel Gaussian channels with arbitrary input distributions. *IEEE Transactions on Information Theory*, 52(7): 3033–3051.

11. Stadelmeier L. et al. 2008. MIMO for in home power line communications, in *Seventh International ITG Conference on Source and Channel Coding (SCC 2008)*, Honolulu, HI.

12. ITU-T. 2011. G.9960. Unified high-speed wireline-based home networking transceivers – System architecture and physical layer specification.

13

IEEE 1901: Broadband over Power Line Networks*

Arun Nayagam, Purva R. Rajkotia, Manjunath Krishnam,
Markus Rindchen, Matthias Stephan and Deniz Rende

CONTENTS

13.1 Introduction

The rapid growth of Internet content and media streaming options, coupled with increasing bandwidth of broadband access (AC) technologies, has created an ecosystem of products and services that creates a demand for media in the home. Bottlenecks are the connection into the home and inside the home, wherein the data are forwarded to a gateway or router that has to distribute to various media sinks like televisions, smartphones, tablets, digital video recorders and cable set-top boxes. Wireless distribution is an obvious solution but faces many challenges like spotty whole-home coverage, high latency and jitter and difficult-to-maintain quality of service for multimedia streams. Coaxial cables are an alternative, but are not always conveniently located; there are usually only a few in a home, and running new cables is expensive and disruptive. Thus, *in-home* (IH) distribution of multimedia remains a challenge. An ubiquitous feature of all homes is the presence of electrical wiring. The electrical power line provides a large number of potential locations for media sinks. Thus, much effort has been invested in using power lines as a viable medium for IH data and multimedia distribution. Unlike a dedicated medium, for example, coaxial cables, the power line channel faces many unique challenges. Some of these challenges include the following:

- It is a pre-existing medium that was not designed for communication. Power line technology should be able to work in any configuration or topology of wiring that may exist in a house or on power/utility company lines.
- As network topology can change dynamically, the technology should be able to adapt. For example, turning on and off circuits/switches causes changes in loads and impedances which in turn changes the reflection properties, attenuation and noise characteristics of the channel (see Chapters 4 and 5).
- The network may lack full connectivity, creating hidden node issues.
- Power line communication signals are unprotected from other signals that may be using the medium.
- The power line is a shared medium, and this creates challenges when there are neighbouring networks in multi-dwelling units.
- The power line has asymmetric transmission properties and it is also time varying.
- The power line suffers from narrowband interference, coloured-noise and impulse noise.

Many vendors have attempted to solve the previous challenges in different ways, and this resulted in many noncompatible, disparate power line technologies. The non-interoperability of these products proved to be a big hurdle to the widespread adoptability of power line technologies.

To overcome these issues, the *Institute of Electrical and Electronics Engineers* (IEEE) Communications Society sponsored the IEEE 1901 programme to develop a global standard for high-speed communication over the power line. The programme was launched in 2005. In 2007, a consolidated proposal using two of the most popular power line networking technologies was selected. The IEEE 1901 standard was approved and published in December 2010. The two technologies that were chosen were the *fast Fourier transform* (FFT)-*orthogonal frequency division multiplexing* (OFDM)-based and *wavelet OFDM* (W-OFDM)-based technologies. These two technologies are both specified as optional. They are not interoperable and hence an *intersystem protocol* (ISP) (see Chapter 10) was developed to ensure *coexistence* (CX) of these two technologies. Having two non-interoperable technologies was determined to be a necessary compromise to buffer for market acceptance of one technology versus the other. This is not something unique to power line technology; it happened before in the original IEEE 802.11 standard that included both direct-sequence as well as frequency-hopping spread spectrum.

The 1901 standard is divided into three clusters:

1. IH: This cluster of requirements and functionalities deals with the distribution of content over low-voltage electric power lines in a dwelling.

2. This cluster of requirements and functionalities deals with the transmission of broadband signals over medium- and low-voltage lines in the power grid to deliver content to homes.

3. This cluster of requirements and functionalities focuses on the ability to make different PLC technologies coexist even if they are not based on IEEE 1901.

This chapter is organised as follows. Section 13.2 provides an overview of the IH cluster. Although the focus of this chapter will be on the more widely deployed FFT-based OFDM *physical layer* (PHY), an introduction will be provided for the wavelet-PHY and ISP that is used for CX. Section 13.3 discusses the FFT-PHY in detail. Despite some minor differences, IEEE 1901 IH and AC use the same FFT-PHY layer. A functional overview of the IH *medium access* (MAC) layer is given in Section 13.4. Sections 13.5 through 13.9 describe the IEEE 1901 AC part focusing mostly on the MAC layer and aspects of the higher layers.

13.2 IEEE 1901 In-Home Architecture

The IEEE 1901 standard specifies the PHY layer and MAC layer functionalities for high-speed communication over electrical power lines. The IEEE 1901 PHY and MAC layers correspond to the lowest two layers of the basic reference model for open systems interconnection [1] specified by the *International Organization for Standardization* (ISO). The bird's-eye view of the dual-PHY, single-MAC IEEE 1901 architecture is shown in Figure 13.1. For flexibility, IEEE 1901 supports two non-interoperable PHYs, the FFT-PHY and the

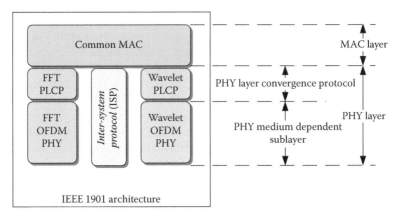

FIGURE 13.1
IEEE 1901 functional layers.

wavelet-PHY, that coexist via means of an ISP. The PHY layer is divided into two parts, a *PHY medium-dependent* (PMD) sub-layer and a *PHY layer convergence protocol* (PLCP). The PMD sub-layer generates the PHY signal that is transmitted on the power line medium using either FFT-OFDM or W-OFDM. The PMD functionality defines the method of transmitting and receiving data through a power line and the characteristics of the signal used to represent the data on the power line. For future reference, the terms PHY and PMD will be used interchangeably.

The PLCP defines methods for mapping *MAC protocol data units* (MPDUs) into a framing format that is suitable for the associated PMD system. This is an intermediate layer that helps translate data from the common MAC into two different PMD systems (FFT and wavelet).

The system architecture and the protocol entities of an IEEE 1901 FFT-OFDM station and how they relate to each other are described in Figure 13.2. The descriptions of the different protocol entities are as follows:

- The PHY layer is responsible for the PHY layer signal format such as the symbol generation, modulation and *forward error correction* (FEC).
- The MAC layer is responsible for the channel AC control and the data plane.
- The convergence layer provides support functions to the MAC such as bridging, packet classification, auto-connect and jitter control.
- The connection manager is used to set up connections to provide QoS for certain streams.
- The *basic service set (BSS) manager* (BM) is responsible for the set-up and maintenance of the network (a 1901 FFT-OFDM network is termed a BSS, see Section 13.4), managing the communication resource on the wire and coordinating with neighbouring BMs.

The *higher layer entity* (HLE) corresponds to all the layers responsible for generating traffic (both data and control messages) to be transmitted using the IEEE 1901 FFT-OFDM device. The HLE communicates the data and the control information (required to configure the station/BSS) using the *service access points* (SAPs).

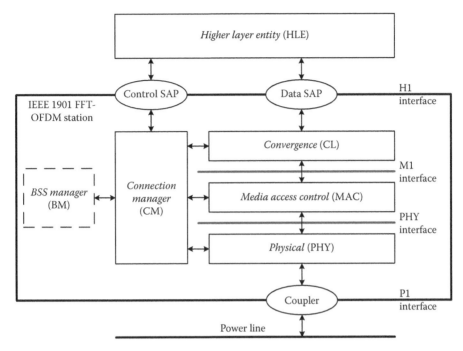

FIGURE 13.2
System architecture of a 1901 FFT-OFDM station.

13.2.1 FFT-OFDM Overview

In this section, a brief summary of the FFT-PHY is provided. The power line channel and noise have a few distinct characteristics (see Chapters 4 and 5):

- It is highly frequency selective with the average attenuation generally increasing with frequency.
- It suffers from narrowband interference due to ingress from existing radio services, wireless products and unintentional interference from devices like microwaves.
- There is substantial impulse noise on the power line generated by electrical loads from such items as halogen lamps and devices utilising motors (e.g. hair dryers).

OFDM [2,3] is a modulation technique that easily adapts to frequency-selective channels. Different carriers carry different amounts of information based on *signal-to-noise ratio* (SNR). OFDM is also inherently resilient to narrowband interference (only a few carriers are affected) and is robust in the presence of impulse noise. It also has a very efficient implementation using an *inverse fast Fourier transform* (IFFT).

The FFT-PHY uses 1974 carriers between 1.8 and 50 MHz with a carrier spacing of approximately 24.414 kHz. The use of frequencies greater than 30 MHz is optional. Certain frequency bands between 1.8 and 30 MHz are used for amateur (HAM) radio broadcasts. Various regulatory bodies – as described in Chapter 6 – in different parts of the world have imposed limits on intentional and unintentional transmit power in these bands to avoid interference to the radio services. Typically, the power, for example, in the HAM bands, should be 30 dB lower than the maximum transmit power allowed in the 1.8–50 MHz band. IEEE 1901 FFT-OFDM employs time domain pulse shaping of the OFDM symbols that enables achieving deep frequency notches without requiring explicit transmit notch filters. By just turning off carriers in the HAM bands and creating a 100 kHz guard band on either side of the HAM band, 30 dB notches are easily achieved. Turning off carriers to meet the notch requirements leaves 917 active carriers that can be used for data transmission in the 1.8–30 MHz band.

The carriers used for data transmission may be coherently modulated with a *binary phase-shift keying* (BPSK), a *quadrature phase-shift keying* (QPSK) and six different *quadrature amplitude modulation* (QAM) modes: 8-QAM, 16-QAM, 64-QAM, 256-QAM, 1024-QAM and 4096-QAM. Support for 4096-QAM is optional. The FFT-PHY uses a turbo code [4–7] for FEC. The turbo code provides good coding gain at low SNRs and hence optimises performance on poor power line channels. For additional robustness required for critical management messages and broadcast messages, IEEE 1901 FFT-OFDM also provides three additional signalling schemes called *robust OFDM* (ROBO), *mini ROBO* (MINI-ROBO) and *high-speed ROBO* (HS-ROBO). Additional details on modulation and coding are provided in Section 13.3.

13.2.2 Wavelet OFDM Overview

W-OFDM [8,9] has two main features that enable the achievement of a better spectral efficiency than the FFT-based OFDM system described in Section 13.2.1: lower side lobes and lack of *guard intervals* (GIs).

Lower side lobes can be attributed to the use of filters that span multiple symbol durations in W-OFDM when compared to the single symbol duration pulse-shaping filter in FFT-OFDM. Having low side lobes has two advantages: it does not require turning off additional carriers on either side of a HAM band to achieve deep spectral notches, and it also increases robustness to narrowband interference.

In FFT-OFDM, symbols do not overlap, and thus a GI is introduced to combat any symbol overlap or *inter-symbol interference* (ISI) introduced by the channel. In W-OFDM, consecutive symbols overlap in time by design (IEEE 1901 W-OFDM has an overlap factor of four symbols). Thus, a GI is not required to avoid symbol overlap. Although this does not eliminate ISI completely, it does lead to lower overhead in W-OFDM.

The IEEE 1901 W-OFDM PHY defines transmitting a W-OFDM frame in two different ways: (1) directly at baseband or (2) by modulating to a band-pass carrier. It is mandatory to support baseband transmission for IH and AC applications. Band-pass is an optional feature.

For baseband transmission, the IEEE 1901 W-OFDM PHY has 512 uniformly spaced carriers between DC (0 MHz) and 31.25 MHz. Only carriers between 1.8 and 28 MHz are used for data transmission. After accounting for HAM band notches, there are 312 active carriers remaining for data transmission. The carriers are modulated with five different *pulse-amplitude modulation* (PAM) modes: 2-PAM, 4-PAM, 8-PAM, 16-PAM and 32-PAM (optional). The mode where each tone carries independent data is called the high-speed mode. For more robust transmissions, IEEE 1901 W-OFDM also specifies a diversity mode where additional frequency domain copies are provided. Two types of FEC are specified: a mandatory concatenated code using Reed–Solomon and convolutional codes and an optional *low-density parity check* (LDPC) code.

The IEEE 1901 W-OFDM PHY defines 1024 evenly spaced carriers into any bands lying in the 1.8–50 MHz range for the optional band-pass transmission. Apart from allowing varying bands with a constant number of carriers, baseband transmission has the same coding, modulation and diversity definitions as in baseband transmission.

The IEEE 1901 W-OFDM PHY has a *PHY protocol data unit* (PPDU) that contains the following fields:

- W-OFDM PPDU preamble that contains 11–17 W-OFDM symbols
- *Tone map index* (TMI) field with one W-OFDM symbol that contains *tone map* (TM) information used in generating the frame body
- *Frame length* (FL) field with one W-OFDM symbol to indicate the frame body length
- *Frame control* (FC) field with more control information on the PPDU
- Frame body with variable data and tail bits to bring the encoding to a known state
- Pad bits

The TMI and FL fields are passed through a rate 1/2 convolutional code. The FC field is encoded with Reed–Solomon and convolutional concatenation. The frame body is first scrambled and then encoded with RS and convolutional concatenation or LDPC coding. After encoding, the TMI, FL, FC and frame body fields are passed through bit interleaving. Finally, interleaved bits for these fields are diversity copied and mapped to OFDM carriers. All the fields are then transmitted with base-band or band-pass W-OFDM transmission.

13.3 1901 FFT-PHY: A Functional Description

A block diagram of the different functional blocks that comprise an IEEE 1901 FFT-PHY transmitter is shown in Figure 13.3. Conceptually, the FFT-PHY transmitter can be thought of as three independent processing chains, although certain blocks may be shared between these chains to reduce implementation complexity:

- A preamble generator chain: Responsible for generating the IEEE 1901 FFT-OFDM preamble that provides packet synchronisation functionality.
- An IEEE 1901 FFT-OFDM FC encoder chain: Responsible for generating the IEEE 1901 FFT-OFDM FC symbol that is present in every PPDU transmission. The FC symbol carries addressing information and other information required to demodulate and decode the packet. Decoding of the FC is critical for successful decoding of the payload in a PPDU. If the FC decoding fails, then no information in the payload portion of the PPDU can be recovered.
- An IEEE 1901 FFT-OFDM payload encoder: Responsible for generating the payload portion of the PPDU.

The IEEE 1901 FFT-OFDM FC encoder and payload encoder receive MAC segments that are processed by the PLCP layer. The smallest unit that is processed by the FFT-PHY is referred to as a *PHY block* (PB). A *PB header* (PBH) is added to the MAC segment, and a *PB checksum* (PBCS) that is computed over the PBH and the body of the PB is appended to the end. Two different-length *cyclic redundancy checks* (CRCs), CRC-32 and CRC-24, are specified to compute the PBCS. The segmentation procedure is shown in Figure 13.11. The preamble generator does not need any input from the MAC; the only input it requires is

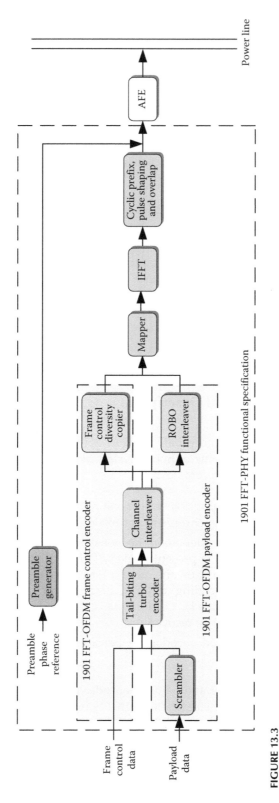

FIGURE 13.3

IEEE 1901 FFT transmitter. (Reprinted from IEEE Std 1901–2010: IEEE standard for broadband over power line networks: Medium access control and physical layer specifications, IEEE Standards Association, 2010. Available online standards.ieee.org/store. Copyright © 2010 IEEE. All rights reserved. With permission.)

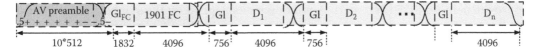

FIGURE 13.4
1901 FFT-OFDM PPDU. The numbers at the bottom indicate the number of samples for each portion of the PPDU (assumes a 100 MHz clock). (Reprinted from IEEE Std 1901–2010: IEEE standard for broadband over power line networks: Medium access control and physical layer specifications, IEEE Standards Association, 2010. Available online standards.ieee.org/store. Copyright © 2010 IEEE. All rights reserved. With permission.)

a reference phase table that is specified in the IEEE 1901 standard that is typically stored in *read-only memory* (ROM).

As shown in Figure 13.3, blocks like the turbo encoder and channel interleaver are shared between the 1901 FFT-OFDM FC encoder and payload encoder. The scrambler and ROBO interleaver blocks are only used by the payload encoder and are not used to generate the FC. The frequency domain FC data are obtained by passing the FC information from the MAC and PLCP through a turbo encoder and a FC diversity copier – which introduces some frequency diversity by repeating the FC bits on different carriers. The frequency domain payload data are obtained by passing payload PBs first through a scrambler, then a turbo encoder and finally a channel interleaver. For payloads transmitted using ROBO modes, the interleaver also introduces frequency and time domain redundancy. The frequency domain bits, irrespective of whether they are FC or data bits, pass through a common OFDM modulation chain consisting of a mapper, IFFT and a symbol shaping with overlap block. The mapper is responsible for converting the bitstream into modulation symbols. Each carrier can transport a different modulation. The vector specifying the mapping of modulation modes to carriers is referred to as a TM. The TM is estimated at the receiver and fed back to the transmitter. The mapper uses the TM to determine how many bits to put on each carrier. The IFFT is responsible for OFDM modulation and converts the frequency domain data into a time domain OFDM symbol. Then a cyclic prefix [2] is appended to the OFDM symbol before a pulse shaping is applied to the composite signal. The shaped signals are then overlapped with adjacent shaped symbols. These overlapped symbols are fed into the *analog front end* (AFE). The AFE module is not specified in the IEEE 1901 FFT-OFDM standard and is implementation dependent. However, the standard does specify the maximum level and fidelity of the signals at the output of the AFE.

The PHY entity comprising a concatenation of the preamble, FC and a sequence of payload symbols is referred to as the PPDU and is shown in Figure 13.4. Depending on the type of PPDU being transmitted, the payload portion may or may not be present. For instance, the *selective acknowledgment* (SACK) is conveyed as a PPDU with only a preamble and a FC.

13.3.1 IEEE 1901 FFT-OFDM Preamble Generator

The preamble provides the receiver with a reference symbol for packet synchronisation. The preamble signal is chosen to have good autocorrelation properties and is typically much more robust than the packet body (a preamble can be detected even under very low SNR conditions where payload communication is not possible). As shown in Figure 13.4, the preamble consists of a sequence of 'plus' and 'minus' short OFDM symbols that are repeated. The 'plus' symbol is called the SYNCP symbol and is generated by modulating as set of reference phases onto carriers between 1.8 and 30 MHz that are spaced approximately 195.3125 kHz apart. A windowing is applied on the symbol, and the symbols are overlapped during concatenation. The 'minus' or SYNCM symbols are obtained as a negation of the SYNCP symbol. The preamble consists of seven and a half SYNCP symbols (starting with the first half of the SYNCP followed by seven SYNCP symbols) and two and a half SYNCM symbols (starting with two SYNCM symbols and ending with the first half of a SYNCM symbol).

At the receiver, a correlation is performed of the received signal against a SYNCP signal. The presence of multiple SYNCP symbols can be used to improve detection performance by combining the outputs of consecutive correlation operations. When the correlator fires, there may still be an ambiguity as to which SYNCP symbol is detected. Then the receiver can search for the transition between the SYNCP and SYNCM symbols to find out exactly where on the packet the correlator fired and where the packet boundaries are.

13.3.2 Scrambler

The scrambler is only applicable to the payload portion of the PPDU. The FC bits are not scrambled. It is possible that the bits obtained from the MAC and PLCP are correlated. The scrambler helps to randomise the input sequence to the turbo encoder. The scrambler is implemented as a simple shift register module operating on the input bits. The generator polynomial for the scrambler is given by polynomial $S(x) = x^{10} + x^3 + 1$. The operation of the scrambler is shown in Figure 13.5.

13.3.3 IEEE 1901 FFT-OFDM Turbo Encoder

IEEE 1901 FFT-OFDM specifies a turbo code that is a parallel concatenation of two duo-binary tail-biting convolutional codes, as shown in Figure 13.6. The turbo code consists of two identical constituent *recursive systematic convolutional* (RSC) codes that are separated by an interleaver. The turbo interleaver is an algorithmic interleaver that is specified by a single equation and is parameterised by a seed table and two variables. Depending on the input block size to the decoder, a particular seed table and values for the other two parameters are specified and uniquely determine the interleaver table. Details of the interleaver can be found in the IEEE 1901 standard [10].

The constituent RSC code is shown in Figure 13.7. It is a rate 2/3 code producing one parity bit for every two information bits. Thus, the code rate of the turbo code is rate 1/2 before puncturing (for every two information bits, each RSC code produces one parity, and the

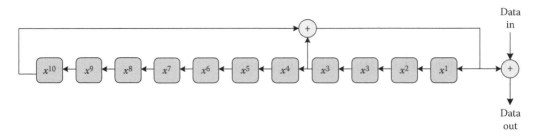

FIGURE 13.5
IEEE 1901 FFT-OFDM data scrambler. (Reprinted from IEEE Std 1901–2010: IEEE standard for broadband over power line networks: Medium access control and physical layer specifications, IEEE Standards Association, 2010. Available online standards.ieee.org/store. Copyright © 2010 IEEE. All rights reserved. With permission.)

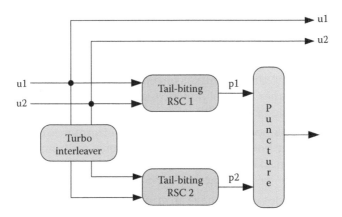

FIGURE 13.6
IEEE 1901 FFT-OFDM turbo code. (Reprinted from IEEE Std 1901–2010: IEEE standard for broadband over power line networks: Medium access control and physical layer specifications, IEEE Standards Association, 2010. Available online standards.ieee.org/store. Copyright © 2010 IEEE. All rights reserved. With permission.)

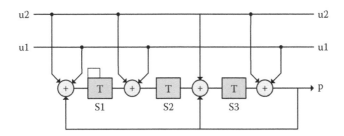

FIGURE 13.7
Constituent RSC code. (Reprinted from IEEE Std 1901–2010: IEEE standard for broadband over power line networks: Medium access control and physical layer specifications, IEEE Standards Association, 2010. Available online standards.ieee.org/store. Copyright © 2010 IEEE. All rights reserved. With permission.)

output of the turbo code is the two information bits + two parity bits leading to code rate of 2/4). Each constituent code is easily implemented using the eight-state linear feedback shift register structure that is shown in Figure 13.7.

For efficient decoding of turbo codes, the ending state of the RSC has to be known. This is achieved in conventional turbo codes by feeding in additional termination bits to force the end state of the shift register to a known value. The extra bits constitute a rate loss. This is avoided by using a tail-biting RSC. Tail-biting encoding (see [11–14] and references therein) is an encoding approach that ensures that the starting and ending states of the RSC are the same. Tail biting typically requires a two-pass encoding approach. The shift register is initialised to the all-zero state, and the information bits are run through the encoder. Depending on the ending state of the shift register, a *look-up table* (LUT) provides a new starting state. The shift register is then initialised to the new starting state, and then another round of encoding with the same information bits is performed. The ending state after the second round of encoding is the same as the new starting state specified by the LUT. The fundamentals behind the approach to generating the new starting state LUT are given in [12]. The IEEE 1901 FFT-OFDM standard [10] specifies the starting state LUT for the turbo code presented in this section.

Turbo codes with duo-binary RSC constituent codes have been well studied in the open literature [11–14] and have many other advantages in addition to not requiring termination bits. Some of them include the following:

- Less sensitive to puncturing: Unlike conventional turbo codes, tail-biting turbo codes can tolerate higher levels of puncturing. In fact, IEEE 1901 FFT-OFDM has a default operating code rate of 16/21.

- Larger minimum distance: This affects the error performance of the turbo code at low SNRs. The power line channel is a pretty adverse medium, and typically, the operating region for packet error rates lays around 10^{-2}. Having a larger minimum distance improves the performance on low-SNR, high-BLER region.

- Duo-binary turbo codes usually have an error floor that is multiple orders of magnitude lower than binary turbo codes [11]. The power line channel is also a very noisy medium, and in the typical operating range of frame error rates (10^{-1}–10^{-3}), the HPAV turbo code can be considered nearly error-floor free.

- Better convergence properties of iterative decoding.

- Lower latency: For each clock cycle, the encoder/decoder outputs two bits instead of one (such as in a conventional binary turbo code), leading to lower latency.

Code rates and block sizes that are supported in IEEE 1901 FFT-OFDM are summarised in Table 13.1.

13.3.4 Frame Control Diversity Copier

As seen in Table 13.1, the FC is a 16-byte (128-bit) PB that is encoded at a code rate of 1/2 to produce 256 output bits. These bits are then interleaved by the channel interleaver. Since the bits are already interleaved, the only purpose of the FC diversity copier is to maximally spread copies of the bits amongst all the available (not notched) carriers. FC information is conveyed using QPSK, and hence the encoded bits are repeated on both the *in-phase* (I) and *quadrature* (Q) components of certain carriers. Assuming that the interleaved bits indexed from 0 to 255, then interleaved bits are copied to I and Q components with an offset of 128 between the Q and I channels, as shown in Table 13.2.

13.3.5 Channel Interleaver

The channel interleaver randomises the location of the information and parity bits to protect against burst errors. The IEEE 1901 FFT-OFDM channel interleaver is essentially a rectangular interleaver. The bits are written sequentially in columns, and then the

TABLE 13.1

Block Size and Code Rate Support in the IEEE 1901 FFT-OFDM

PB Type	PB Size (Octets)	Code Rate
FC	16	1/2
Payload	136	16/21, 8/9 (optional)
	520	16/21, 8/9 (optional)

TABLE 13.2

Bit Addressing for the FC Diversity Copier

Used Carrier #	Encoded Bit Index Carrier on the I-Channel	Encoded Bit Index Carrier on the Q-Channel
0	0	128
1	1	129
...	...	...
i	i mod 256	(i+128)mod 256
...		

Source: Reprinted from IEEE Std 1901–2010: IEEE standard for broadband over power line networks: MAC control and PHY layer specifications, IEEE Standards Association, 2010. Available online standards.ieee.org/store. Copyright © 2010 IEEE. All rights reserved. With permission.

interleaved output is read out in rows. The information bits and parity bits are interleaved separately and then interlaced together. The information (or parity) bits are written sequentially into the columns of an $N \times 4$ interleaver table, where N depends on the block size and code rate being used. All the block size and code rate combinations produce encoded bits in multiples of four, and hence the interleaver table is always fully populated. The information bits are read out starting at Row 0 and then progress incrementally to the next row that is a specific step-size away, and so forth. When the last row is reached, the process is repeated starting at Row 1, and this continues until all the information bits are read out. The process for the parity bits is exactly the same, but instead of starting at Row 0, the process is initiated at a row that is at a specific amount offset from Row 0. The two sets of interleaved information bits and interleaved parity bits are interlaced as specified in the IEEE 1901 FFT-OFDM standard [10] to arrive at the interleaved output. The parameters for the rectangular interleaving of information and parity bits are summarised in Table 13.3.

In addition to the rectangular interleaving, the bits that are read out from each row are interleaved amongst each other to further randomise the output stream. For more details on this process, called sub-bank switching, the interested reader may refer to the actual standard document [10].

TABLE 13.3

Channel Interleaver Parameters

PB Size (Octets)	Code Rate	Step Size	Row Offset (for Parity Bits)
16	1/2	4	16
136	1/2	16	136
	16/21	8	40
	8/9	11	16
520	1/2	16	520
	16/21	16	170
	8/9	11	60

Source: Reprinted from IEEE Std 1901–2010: IEEE standard for broadband over power line networks: MAC control and PHY layer specifications, IEEE Standards Association, 2010. Available online standards.ieee.org/store. Copyright © 2010 IEEE. All rights reserved. With permission.

13.3.6 ROBO Interleaver

IEEE 1901 FFT-OFDM uses ROBO modes for various purposes such as session set-up, broadcast and multicast communication, beacon transmissions and exchange of management messages. Three different ROBO modes, namely, ROBO, HS-ROBO and MINI-ROBO, are specified.

All ROBO modes use rate 1/2 FEC and QPSK modulation. A single PB is first encoded and interleaved. Then, the ROBO interleaver operates on the output of the channel interleaver. The ROBO interleaver introduces frequency and time diversity by repeating the interleaver output on different carriers and different OFDM symbols, respectively. The number of copies depends on the type of ROBO mode. The ROBO interleaver creates the required redundancy by reading the output of the channel interleaver multiple times with different cyclic shifts. The different ROBO modulations, the corresponding PB sizes and redundancy factors are specified in Table 13.4. Note that the data rate depends on the number of used carriers in the 1.8–30 MHz band. For the default 917 used carriers that remain after all the HAM bands are notched, the data rates are 3.8, 4.9 and 9.8 Mbps for MINI-ROBO, ROBO and HS-ROBO, respectively. It is seen that MINI-ROBO has the lowest PHY rate and most robustness (number of copies). Therefore, this is typically used for transmissions that are critical to maintaining a network. For instance, beacons are transmitted using MINI-ROBO. Details of the ROBO interleaving algorithm can be found in the IEEE 1901 standard [10].

13.3.7 Mapper

The mapper is responsible for converting the input bitstream into constellation symbols for each carrier. The operation of the mapper can be summarised as follows:

- FC symbol: Pairs of bits are mapped using QPSK onto the used carriers.
- ROBO, HS-ROBO and MINI-ROBO: Pairs of bits are mapped using QPSK onto the used carriers.
- Payload symbols: The number of bits on each carrier depends on the TM. Based on the TM, an appropriate number of bits are mapped using BPSK, QPSK, 8-QAM, 16-QAM, 64-QAM, 256-QAM, 1024-QAM or 4096-QAM onto each of the used carriers.

If the TM indicates that a particular carrier has zero bits allocated to it, then the mapper uses a *pseudo-noise* (PN) generator with the polynomial $S(x) = x^{10} + x^3$ to generate a random

TABLE 13.4

IEEE 1901 FFT-OFDM ROBO Mode Parameters

ROBO Mode	PB Size (Octets)	Number of Copies
MINI-ROBO	136	5
ROBO	520	4
HS-ROBO	520	2

Source: Reprinted from IEEE Std 1901–2010: IEEE standard for broadband over power line networks: MAC control and PHY layer specifications, IEEE Standards Association, 2010. Available online standards.ieee.org/store. Copyright © 2010 IEEE. All rights reserved. With permission.

bit that is modulated using BPSK. It then assigns the bit to that carrier. This is to aid in estimating the SNR on that carrier in case the SNR eventually improves enough to support any of the supported modulation levels, which would allow the carrier to be used for transmitting payload data. The bits in the generated PN sequence are initialised to all 1s at the start of the first OFDM payload symbol. When a used (not notched) carrier is assigned 0 bits, the existing value in the first register (x^1) is used for BPSK modulation, and then the PN sequence generator is advanced.

In addition to modulating each carrier with a constellation symbol, the mapper also rotates the constellation on each carrier by a reference phase. The phase reference vector specified in IEEE 1901 FFT-OFDM was chosen to maintain a certain peak-to-average ratio. The mapper finally scales the constellation symbol to maintain unity average power for each modulation level.

13.3.8 Symbol Generation (Cyclic Prefix, Pulse Shaping and Overlap)

This block is responsible for generating the sequence of OFDM symbols in a PPDU for transmission over the power line. We will assume a sampling frequency of 100 MHz in the section. The functionality of this block is to generate a windowed OFDM symbol, as shown in Figure 14.12. The operation of this block can be broken into the following steps:

1. Add a cyclic prefix to the OFDM symbol obtained at the output of the IFFT. The duration of the OFDM symbol at the output of the IFFT is T = 40.96 μs (4096 samples). The duration of the cyclic prefix, t_{prefix}, is split into a *roll-off interval* (RI) that is used for windowing/pulse shaping and a GI that is used for combating ISI. The duration of the cyclic prefix depends on the GI used. The RI spans 4.96 μs. Three mandatory GIs of 5.56, 7.56 and 47.12 μs are specified. Apart from these, eight other optional GIs are also supported. The extended OFDM symbol with the cyclic prefix spans a duration of $T_E = T + t_{prefix}$.

2. Apply the pulse-shaping window specified in the IEEE 1901 FFT-OFDM standard [10] over the first RI and last RI symbol of the extended OFDM symbol obtained earlier. The pulse shaping helps to achieve the 30 dB notch depth required in the HAM bands.

3. Overlap the first RI/2 samples of the current windowed symbol with the last RI/2 samples of the previous windowed symbol. This results in an OFDM symbol duration of $T_S = T + GI = 40.96$ μs + GI.

The sequence of overlapped symbols is then sent to the AFE and transmitted on the power line.

13.4 IEEE 1901 In-Home MAC: Functional Overview

IEEE 1901 IH networks are designed to support audio/video streaming applications within the home. This section gives an overview of the 1901 MAC. Both the 1901 FFT-OFDM and the 1901 W-OFDM systems support similar MAC functionalities. However, the details of each of these functionalities are different. This section discusses the details of the different

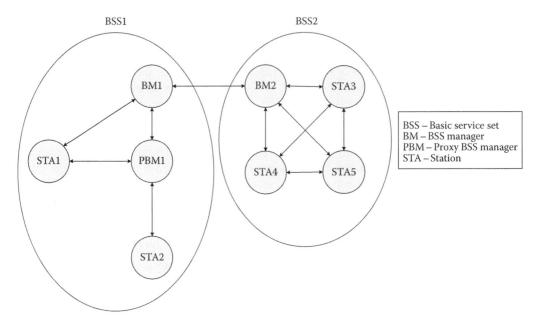

FIGURE 13.8
Network topology and components. (Reprinted from IEEE Std 1901–2010: IEEE standard for broadband over power line networks: Medium access control and physical layer specifications, IEEE Standards Association, 2010. Available online standards.ieee.org/store. Copyright © 2010 IEEE. All rights reserved. With permission.)

MAC functionalities as applicable to the 1901 FFT-OFDM systems. A brief description of an IEEE 1901 network is provided initially along with the protocol entities in a 1901 station. This is followed by details of the formation of the network, channel AC mechanisms, channel estimation to utilise the medium effectively, the MAC data plane and the different MAC packet transmission mechanisms.

A 1901 network (also termed a BSS) consists of a set of member stations that can communicate with one another exclusively. Each BSS is a centrally managed network, managed by a station called the BM. The BM is responsible for the creation, maintenance and operation of a BSS. A BM controls the channel AC in the BSS. The BM periodically transmits a beacon which contains essential information for the functioning of the BSS such as the list of stations in the network and channel AC schedule information.

Figure 13.8 describes an example of two neighbouring BSSs. BSS1 consists of a BM, a *proxy BM* (PBM) and two stations, while BSS2 consists of a BM and three stations. The two BSSs are neighbouring, and the BMs of the two BSSs can coordinate with each other with respect to sharing resources and channel AC. In the case of BSS1, station STA2 is hidden from its BM. In such circumstances, one of the stations can take on the role of a PBM to provide the BM functionalities to STA2.

13.4.1 Formation of BSS

A BSS is a set of stations with the same *network membership key* (NMK). A NMK is a 128-bit *advanced encryption standard* (AES) key. Stations with the same NMK automatically form a BSS. It is up to the user to provide the NMK to all of the stations intended to be a part of the BSS.

The provisioning of this NMK to each of the stations in the BSS is termed authorisation. There are multiple ways in which a NMK can be provided to various stations in a BSS:

- Direct entry: NMK provisioning using this technique requires an available user interface on each station. A user can enter the NMK using this interface on each station. An alternative, user-friendly technique is entering a *network password* (NPW) on each station. This NPW can be hashed to produce the NMK.

- Remote entry: This requires a user interface on a remote station. The user enters the *device password* (DPW) or a *device access key* (DAK) of a new station to this remote station. The remote station can send the NMK to the new station encrypting it using the DAK. The DPW is a user-friendly alternative to the DAK that can be derived by hashing the DPW.

- Simple connect: This is a user-friendly way of forming a BSS. This technique involves pushing buttons on the two devices within a very short time span either to form a network or to add a new station into the network. When the buttons are pushed, temporary encryption keys are exchanged between the two nodes using unicast control messages. These temporary encryption keys are used to exchange the NMK.

The direct entry and remote entry methods of authorisation provide the highest security but require a fairly sophisticated user interface. The simple connect method of authorisation is very user-friendly and requires a simple user interface. However, this technique is less secure.

Any of the stations in the BSS can become the BM. When a network is formed, one of the stations must become the BM. The selection methods for determining the BM station are as follows:

- User-appointed BM: The user can appoint a particular station to be made the BM. Care should be taken by the user to only appoint one station as the BM in a BSS.

- BM bandwidth management capability: Each station has different bandwidth management capabilities (*carrier sense multiple access* [CSMA] only, uncoordinated mode or coordinated mode) if functioning as a BM. A station that can operate a BSS in the coordinated mode has the highest preference of becoming a BM. Details of the different modes of operation are described in Section 13.2.

- BM based on topology: A station can become a BM depending on the topology of the BSS. A station with the best connection to every other station in the network is the best candidate for becoming a BM.

The BM is responsible for maintaining the function of a BSS. When another station becomes more capable than the existing BM in a BSS, then the BM functionality can be handed over to the other station. This can be done in a seamless fashion, without disturbing the operation of the BSS. Another station in the BSS will always be appointed as a backup BM to enable recovery in the event of a BM failure. The backup BM will maintain the information required to manage the BSS, so that recovery is seamless.

Once a BM is appointed in the BSS, the BM initiates the association process. To accomplish this, the BM provides an 8-bit *terminal equipment identifier* (TEI) for every station in the BSS including itself. Every transmission on the power line has a FC associated with it. This TEI is used for addressing in all FCs instead of the 6-octet MAC address associated with the station. The TEI has a finite amount of lease time. It is up to the station to renew the lease time.

An IEEE 1901 network is a secure network. The BM generates a random key called the *network encryption key* (NEK). This key is distributed to all the stations in the BSS using the NMK, thereby authenticating the stations in the BSS. All the transmissions in the network except for a select few control messages (e.g. messages required for joining/forming a new BSS or communications between nodes belonging to different BSSs) are encrypted using this NEK. To maintain a high level of security, the NEK is periodically updated and provided to all the stations using the NMK.

The BM maintains a list of valid TEIs in the BSS and updates every station with the latest list of valid TEIs regularly. A BM can remove a station from the BSS by providing a new NMK to every station in the network except for the station that needs to be removed.

13.4.2 Channel Access Mechanisms

IEEE 1901 FFT-OFDM supports both *time division multiple access* (TDMA) and CSMA. The BM determines the mode of channel AC and also the duration of time for which each AC method is valid. The time is divided into beacon cycles. Each beacon cycle is two AC line cycles long (33.33 ms where the AC line cycle frequency is 60 Hz and 40 ms where the AC line cycle frequency is 50 Hz). Each BSS can operate in any of the following three modes:

1. CSMA-only mode: In this mode of operation, the entire beacon cycle is available for CSMA channel AC only. All of the transmissions including the beacon transmissions use CSMA channel AC.

2. Uncoordinated mode: This mode of operation is present in the case of single BSS scenarios. In this mode of operation, the beacon cycle is divided into at least two regions: beacon region (reserved for transmissions of beacons) and persistent CSMA region (reserved for CSMA AC to be shared by all the nodes in the network). The remaining cycle is split into multiple TDMA regions and nonpersistent CSMA regions.

3. Coordinated mode: This mode of operation is present whenever multiple BSSs exist and TDMA channel AC is required. The beacon cycle in each of the BSSs has a beacon region, a shared CSMA region (which is common for all the networks), a stay-out region (wherein all the stations belonging to the BSS will restrain from transmissions) and a reserved region (which consists of all the TDMA allocations and nonpersistent CSMA allocations). The BMs of the neighbouring BSSs coordinate with each other to carve out their individual stay-out regions. A reserved region in one BSS may align with the stay-out regions of other BSSs. The beacon period will have multiple beacon slots with one slot reserved for the transmission of beacons of a particular coordinating BSS.

13.4.2.1 CSMA Channel Access

The CSMA channel AC mechanism used in IEEE 1901 FFT-OFDM networks is a prioritised channel AC. Each channel AC during CSMA consists of a priority resolution period, followed by the random back-off contention period, which is then followed by actual transmission.

Figure 13.9 describes a CSMA/CA channel AC followed by a packet transmission (this is described in detail in Section 13.4.5). When a station is ready to transmit, signalling is done in the two priority slots. Depending on the priority of the intended transmission, a station

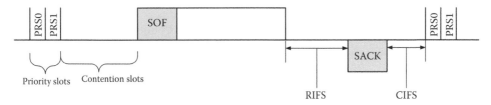

FIGURE 13.9
CSMA/CA channel AC. (Reprinted from IEEE Std 1901–2010: IEEE standard for broadband over power line networks: Medium access control and physical layer specifications, IEEE Standards Association, 2010. Available online standards.ieee.org/store. Copyright © 2010 IEEE. All rights reserved. With permission.)

TABLE 13.5

Priority Signalling for Different MAC Priorities

MAC Priority	PRS Signal in PRS 0	PRS Signal in PRS1
MA3	Transmit	Transmit
MA2	Transmit	Listen
MA1	Listen	Transmit
MA0	Listen	Listen

Source: Reprinted from IEEE Std 1901–2010: IEEE standard for broadband over power line networks: MAC control and PHY layer specifications, IEEE Standards Association, 2010. Available online standards.ieee.org/store. Copyright © 2010 IEEE. All rights reserved. With permission.

either transmits a *priority resolution signal* (PRS) or listens in a priority slot. A station loses its priority contention if it detects a PRS in one of the priority slots. Table 13.5 describes the different types of priority signalling (with MA3 the highest and MA0 the lowest priority) for different MAC priorities. *Start of frame* (SOF), SACK, *response interframe space* (RIFS) and CIFS are detailed when introducing Figure 13.12.

The priority with which a packet transmission occurs depends on the payload of the transmissions. A user can assign a priority to different types of traffic using the QoS field in the VLAN tag [15]. This can be mapped into one of the four MA priorities as described in subclause 7.7.3 of IEEE 802.1D [16]. Beacons, which are essential for the functioning of the BSS, can be transmitted using MA3.

IEEE 1901 FFT-OFDM stations use a modified exponential back-off procedure during CSMA. Each station maintains a *backoff procedure event counter* (BPC) for each priority level. When a station is ready to transmit, it backs off for a random number of contention slots, each slot being 35.84 μs long. The maximum contention window depends on the BPC. The BPC depends on priority of transmission, the number of successive collisions and the number of times the station had to defer to transmissions of the same priority. The dependence on the number of times transmissions had to defer is to further reduce the collision probability. Upon gaining priority AC to the channel, the station performs a random back-off. If no other transmissions are detected during the back-off, the station proceeds with the transmission. Otherwise, it increments its defer counter. If the transmission results in a collision or if the defer counter reaches a limit corresponding to the current BPC, its BPC counter is incremented. The defer counter is reset to zero each time the BPC is incremented or upon a transmission. The successive collision counter and the BPC are reset to zero upon a successful transmission. Table 13.6 provides the details of the different counters used for the priority back-off for MA0 and MA1.

TABLE 13.6

Max Contention Window as a Function of Defer Counter
and Successive Collisions

BPC	Successive Collisions	Defer Counter	Contention Window (# of Contention Slots)
0	0	0	7
1	1	1	15
2	2	3	31
>= 3	>= 3	15	63

13.4.2.2 TDMA Channel Access

During TDMA channel AC, the AC to the power line medium is reserved specifically for a source destination node pair. During this reserved period, the source controls the entire AC to the channel. The different kinds of MPDU transmissions as described in Section 13.4.5 are all permitted during a TDMA channel AC. It is important to note that all transmissions between the node pair will be restricted to the time allocated for this pair.

13.4.3 Channel Estimation

The power line has many appliances and devices connected on it. This results in impedance mismatches. The signals transmitted on this channel get reflected multiple times, resulting in multipath delay spread at the receiver. There is significant non-AWGN noise introduced to the channel by these appliances and devices. The noise introduced on this channel varies as a function of the AC line cycle. This phenomenon is also valid for lower frequencies and shown in detail in Chapter 2. Additionally, there is a long-term variation of the channel characteristics resulting from the various appliances powering on and off. The IEEE 1901 FFT-OFDM channel estimation is designed to handle these dynamic properties of the power line channel. The stations also operate in two different bands: 1.8–30 MHz and 1.8–50 MHz.

As a part of the channel estimation process, the transmitter and the receiver determine the common band of operation, a set of tonemaps and the different portions of AC line cycle during which each of the tonemaps can be used. A tonemap corresponds to a unique combination of modulation per carrier, FEC code rate to be used and the GI length to be used.

IEEE 1901 FFT-OFDM has a predefined set of tonemaps called the ROBO tonemaps. These tonemaps are designed to be robust enough to handle the harshest channel conditions and are therefore not efficient for data transmissions. They range in data rate from 5 to 10 Mbps. When a station has data to transmit and does not possess any user-defined tonemaps, it initiates the channel estimation process. As a part of this process, the station transmits a special MPDU (sound MPDU) to the receiver. When the receiver has collected a sufficient number of sound MPDUs, it generates a default tonemap along with AC line cycle region-specific tonemaps. The default tonemap can be used in any part of the AC line cycle. Once the transmitter receives the set of tonemaps, it will start transmitting data packets using these tonemaps. To handle dynamic channel changes, the receiver continues to monitor the channel and continuously updates the transmitter with new sets of tonemaps. As previously mentioned, the power line channel characteristics are different for different portions of the line cycle. Figure 13.10 describes an example with an AC line cycle divided into four tonemap regions and the different tonemaps (T1, T2, T3 and T4) associated

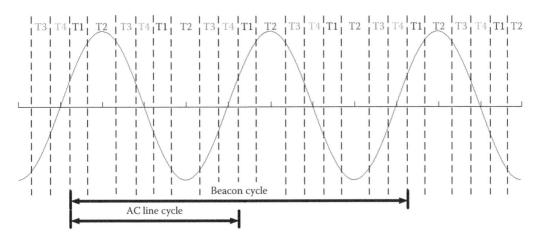

FIGURE 13.10
Tonemap regions w.r.t. AC line cycle and beacon cycle.

with each region. Transmissions using a particular tonemap have to be restricted to the region where the tonemap is valid. They cannot span multiple regions.

13.4.4 MAC Data Plane

The IEEE 1901 system supports the transmission of Ethernet packets. It limits the type of data that can be transmitted between stations that are not associated or authenticated to a subset of management messages. Data exchange between such stations is strictly prohibited. Any Ethernet packet received from the data SAP is called a *MAC service data unit* (MSDU), and the MAC sub-layer transports these MSDUs to the intended receiver.

Each transmitter maintains several queues as described here:

1. Unicast queues for each receiver
   a. Four priority queues for data
   b. One queue for management/control frames
2. Broadcast queues
   a. Four priority queues for data
   b. One queue for management/control frames
3. Specific queues for established connections

Any incoming MSDU belongs to an established connection or is independent. These MSDUs are classified upon arrival based on the destination address, the connection type and whether it is a management frame or a data frame. Connection-oriented MSDUs are passed on to the specific queue corresponding to that connection. The independent data MSDUs are passed on to one of the priority queues maintained for the specific receiver. The user-defined priority of the MSDU is mapped to a particular channel AC priority as specified by IEEE 802.1D [17].

The flow of data through an IEEE 1901 FFT-OFDM station is shown in Figure 13.11. Each arriving MSDU is prepended with a MAC frame header and appended with an *integrity check vector* (ICV) to form a MAC frame. The ICV is used to verify that the MAC frame is

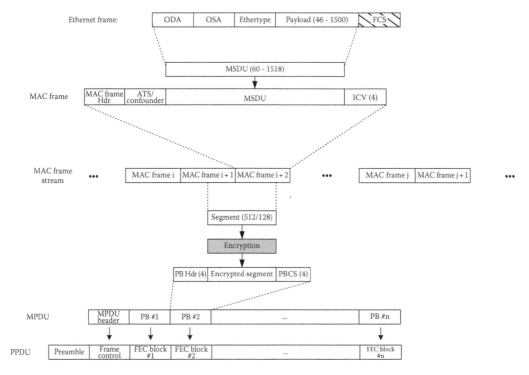

FIGURE 13.11
MAC framing, segmentation and MPDU generation. (Reprinted from IEEE Std 1901–2010: IEEE standard for broadband over power line networks: Medium access control and physical layer specifications, IEEE Standards Association, 2010. Available online standards.ieee.org/store. Copyright © 2010 IEEE. All rights reserved. With permission.)

received correctly. The arriving MSDUs have the *original source address* (OSA), the MAC address from where the MAC frame originated, the *original destination address* (ODA) (the MAC address where the particular MAC frame is eventually intended to be consumed) and an Ethertype. The MAC frame header is a two-octet field that carries information about the validity of the MAC frame, the presence of an *arrival time stamp* (ATS) and the length of the MAC frame. Management MSDUs have a four-octet confounder (which is a pseudorandom value) in their MAC frame. The ICV is four octets long and is a CRC-32 computed over the entire MAC frame. The MAC frames arriving at each of the queues are concatenated to form the *MAC frame stream* (MFS).

The MFSs are then segmented into 512-octet blocks. These segments are transported as a part of the *MAC protocol data unit* (MPDU) payload. Each 512-octet segment maps into a single FEC block of a PPDU at the PHY layer. Since PHY errors occur on a FEC block basis, only those FEC blocks in error need to be retransmitted. This MAC framing and segmentation makes the protocol very efficient. Each segment is mapped into a PB. The segment is encrypted using the NEK and is inserted into the *PHY block body* (PBB). A four-octet PHY block header is prepended and a four-octet *PHY block check sequence* (PBCS) is appended to the PBB to form a PB. The PBH contains information such as *segment sequence number* (SSN) and information about a MAC frame boundary, which can be used by the receiver to reassemble the segments. The SSN is a 16-bit field that is initialised to 0 and is incremented for every new segment that needs to be transmitted. The PBCS contains a 32-bit CRC computed over the PBH and the PB body. A PB in error is identified using this field.

A successful PB is sent to the appropriate reassembly buffer at the receiver, and a PB in error is indicated to the transmitter using a SACK MPDU. Figure 13.11 describes the MAC framing, MAC frame streaming, segmentation and MPDU generation.

An MPDU typically contains multiple PBs. Upon receipt, the PBCS is used to determine whether or not the PB was received successfully. In the case of broadcast transmissions, one of the stations is nominated as a proxy that responds with a SACK transmission. This increases the reliability of the broadcast transmissions.

13.4.5 MPDU Transmissions

The unit of transmission on power line is termed as a *MAC packet data unit* (MPDU). Each MPDU consists of a preamble and a FC, collectively called a delimiter. Furthermore, certain MPDUs also have payload transmissions following the delimiter. MPDUs with only the delimiters are termed short MPDUs, and those with a payload are termed long MPDUs. Data are transmitted using long MPDUs as shown in Figure 13.9. The data are transmitted as the payload of a long MPDU with a SOF delimiter. Every long packet transmission is followed by a RIFS gap and the transmission of a SACK MPDU by the receiving station. The SACK MPDU contains the reception status of the FEC blocks contained in the payload of the preceding SOF MPDU.

As described in Section 13.4.3, the payload of long MPDUs transmitted with user-defined tonemaps cannot cross tonemap region boundaries. If the data transmission is restricted to a single long MPDU, the MAC efficiency would be very low. To improve the MAC efficiency, MPDU bursting is employed. In a MPDU burst, for every channel AC, the transmitter sends multiple long MPDUs with very low interframe space (*burst interframe space* [BIFS]) between successive long MPDUs. A single SACK MPDU acknowledges the entire burst of MPDUs. MPDU bursting significantly improves the MAC efficiency during CSMA channel AC. IEEE 1901 FFT-OFDM allows up to four MPDUs to be transmitted in a single burst. Figure 13.12 depicts an example of MPDU bursting with two MPDUs in the burst.

Another MPDU bursting mechanism that may be supported is bidirectional bursting. A receiver can transmit data in the reverse direction along with the SACK. Figure 13.13 depicts a bidirectional bursting. In this example, the *reverse start of frame* (RSOF) FC of MPDU B and D acknowledges the payload in MPDU A and MPDU C, respectively. Similarly, the SOF of MPDU C acknowledges the payload of MPDU B. The number of bits available in a SOF delimiter to acknowledge a preceding RSOF transmission is very few. This limits the amount of data that can be transmitted using a RSOF in the case of a noisy channel. This bidirectional exchange of data is eventually terminated with the transmission of a SACK MPDU. Bidirectional bursting reduces the round-trip latency and overhead for TCP transmissions and other bidirectional applications (such as VoIP), thereby improving efficiency.

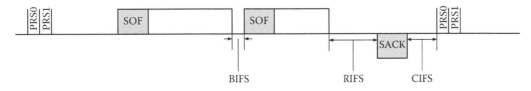

FIGURE 13.12
MPDU bursting. (Reprinted from IEEE Std 1901–2010: IEEE standard for broadband over power line networks: Medium access control and physical layer specifications, IEEE Standards Association, 2010. Available online standards.ieee.org/store. Copyright © 2010 IEEE. All rights reserved. With permission.)

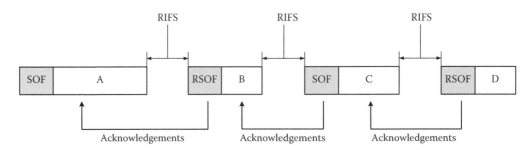

FIGURE 13.13
Bidirectional bursting. (IEEE 1901 1901-2010: IEEE Standard for Broadband over Powerline Networks; IEEE 2010.)

13.5 IEEE 1901 FFT-OFDM Access System

Writing about IEEE 1901 FFT AC systems needs a context about preceding technologies like *narrowband power line communication* (NB PLC), which uses the AC power line as a medium for communication. The differences between AC and IH systems need to be clearly outlined to understand the differing approaches and philosophies behind the actual technical implementation in these two domains.

This chapter gives an overview on the relevant features of the IEEE 1901 FFT AC system and points out the differences between NB PLC and IEEE 1901 FFT IH systems. Despite some minor differences, IEEE 1901 IH and AC use the same FFT-PHY layer. The essential difference between IH and AC is the number of *stations* (STA) that are connected to one cell.

13.5.1 Access BPL versus NB PLC

Over the past several years, various NB PLC systems have been developed and used. In Europe, they are operating in a frequency range from 9 to 150 kHz (ITU Region 1). The initial design for those PLC systems was heavily dependent on the application the whole system should provide. In most cases, NB PLC was used to create low-bandwidth connections to very few and selected remote endpoints. Low channel attenuation and high transmission power levels made point-to-point communication between concentrators and endpoints feasible. In this kind of application, the PLC-based communication system had some advantages due to lower channel attenuation and high transmission power levels. This made end-to-end communication between concentrators and endpoints feasible. Meanwhile, the noise level in the frequency band up to 150 kHz has increased due to man-made noise, and higher bandwidth and stricter latency requirements are inherent to actual and future Smart Grid requirements. NB PLC system standards are described in Chapter 11.

Broadband power line (BPL), however, utilises a larger frequency range (2–30 MHz). The noise level of higher frequencies typically tends to be lower and can be mitigated more easily.

The increasing number of endpoints poses another problem to the narrowband approach, since more devices will compete for the medium resource, which will limit the available bandwidth per device to a range below any usability. The wider frequency band of the BPL technology enables more freedom in the selection of the signal carriers and is, as a consequence, more suitable to a larger scale of an AC network.

13.5.2 Access Requirements

Compared to AC networks, the topology of IH networks is relatively simple since such networks are, by definition, restricted to small areas. Due to short link distances occurring in an IH environment, two stations can usually communicate with each other by the use of a direct power line link. In the case of high signal attenuation, a single repeater may be used to extend the range of a link. With that basic mechanism, it is possible to establish full coverage of the whole IH network area. Chapter 20 describes the efficiency of IH relaying protocols.

In AC networks, the situation is complex. In general, the link distances between stations are larger, and therefore, a single repetition is insufficient to obtain full coverage. As a consequence, multiple repetition is used, which leads to complex network structures with ring, tree-like or mesh topologies.

Another general challenge power line networks face is the dynamic link behaviour. Electric switches being opened or closed, nomadic stations like electric automobiles and general interference phenomena result in frequently varying link states, opposed to ordinary *local area networks* (LAN) where the link states are usually quite stable.

Both the complex network structure and the dynamic link behaviour result in a generic network topology that imposes unique requirements on an AC network system based on BPL. On top of the aforementioned PHY requirements that derive from the size of the power line network and the number of stations required to build a fully covering communication network, some application-oriented requirements need to be fulfilled by the AC network, as pointed out in the following.

Over the last few years, the usage of BPL AC networks has been strongly associated to Smart Grid applications such as smart metering, substation automation and remote sensor networks for power grid monitoring. The concept behind that approach is simple but effective, since most communication sources or sinks in a Smart Grid will be connected to the power networks anyway. In that case, the medium that transports the energy, and is subject to the Smart Grid, is also the communication network. The usage for Smart Grid applications implies additional requirements, such as high numbers of devices in an AC network, provided bandwidth and robustness of communication.

Large networks used for Smart Grid applications may consist of several hundred nodes, of which the majority might only be used for a few minutes per hour. To be able to take advantage of this circumstance, a sleep mode is required for such stations.

13.6 IEEE 1901 FFT-OFDM Access: System Description

The IEEE 1901 standard defines the system architecture of an AC BPL network that shares some common mechanisms with the architecture of an IH network. The FFT-OFDM PHY layer of the AC system is identical to the one of the IH system (see Section 13.3). This section covers the differences between IH and AC systems that apply to the MAC layer and partially to higher layers in terms of routing and bridging.

13.6.1 Attenuation Domains

Generally, the power line network is a shared medium: all stations connected to the network compete for bandwidth and potentially interfere with each other. Thus, the power

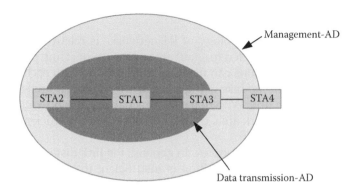

FIGURE 13.14
Example for a STA's AD. (IEEE 1901 1901-2010: IEEE Standard for Broadband over Powerline Networks; IEEE 2010.)

line network acts like a bus system. The only PHY restriction limiting this setting is the frequency- and length-dependent attenuation of the power lines.

The boundary created by the attenuation is unique for every station connected to the power line network and depends on the SNR required for exchanging information. For system information like beacons, the required SNR is lower, since robust modulation is used to send the beacons. For data exchange, the SNR requirements are higher, which results in a smaller *attenuation domain* (AD). Each station 'sees' its own ADs. Figure 13.14 shows an example of the ADs for a *single station 1* (STA1).

The *management attenuation domain* (M-AD) of STA1 is including STA2, STA3 and STA4. Those STAs receive management messages like an AC beacon sent by STA1. The *data transmission attenuation domain* (DT-AD) includes STA2 and STA3. This area provides sufficient S/N to allow satisfying data exchange rates.

13.6.2 Network Architecture

The operation of a BPL AC system needs a dedicated uplink to the backbone; therefore, the IEEE 1901 FFT AC system organises the stations in a cell structure. A cell is a group of stations managed by a single station, the *head end* (HE). The HE also provides and manages the connection to the backbone/backhaul network. The *network termination units* (NTUs) function as gateways between the AC BPL cell and the connected *customer premises equipment* (CPE). HE and NTU are the edge devices of a BPL AC network. Inside the BPL AC network, the connection between HE and NTUs is supported by *repeater stations* (RP).

Figure 13.15 shows a simple example of a meshed AC network topology. In reality, and especially in Smart Grid applications, the number of STAs deployed in a single cell is substantially bigger. This results in even more connections and possible paths for communication.

13.6.3 Access Subcell

Certain security requirements for AC networks providing smart metering functionality demand cryptographic isolation of a subset of stations within the cell. In this case, the concept of AC subcells can be applied. The core cell, including the HE, RPs and NTUs all sharing the same NEK, is extended by one or possibly more subcells. A subcell contains a set of AC CPEs that are connected to an NTU of the core cell. The MAC and PHY traffic classes from the NTU to the set of CPEs use a separate encryption, that is, the NEKs are

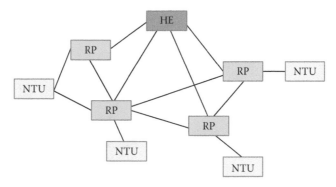

FIGURE 13.15
Example of an AC network.

different from the ones used in the core cell. Optionally, repeaters may be used in order to extend the range of the subcell. The cell management of a subcell is still performed on the basis of the HE within the core cell.

13.6.4 Addressing

In order to address stations in a network properly, every station needs a unique identifier. Because the use of standard 48-bit-long Ethernet MAC address would generate unnecessary frame overhead, the IEEE 1901 FFT AC system makes use of its local cell infrastructure and an own addressing method; every cell is given a *short network identification* (SNID). The six-bit-long SNID enables the differentiation of 63 neighbouring networks. Within a cell, a station is uniquely identified by the TEI. The 12-bit-long TEI supports 4078 stations within a single cell. The remaining address space is reserved for the indication of a new station, the HE and broadcast messages. The combination of SNID and TEI required to uniquely identify a station within an AC network is only 18 bits long. Compared to the 48-bit Ethernet MAC address, the use of the TEI/SNID combination leads to more efficient address spacing, especially when taking into account that the routing approach of IEEE1901 FFT AC systems needs to maintain and transmit the address information of up to four nodes.

13.6.5 Access Intercell Gateway

A station is usually associated with only one HE using a single TEI/SNID tuple. In order to improve the connectivity between two or more cells, an *access intercell gateway* (AIG) may be used. A station supporting AIG functionality performs association to all possible cells within range. As a consequence, the AIG obtains multiple identities, one per cell, and provides bridging functionality between these cells. With the use of AIGs, more flexible data transmission is possible by crossing different cells.

13.6.6 Cell Management

Management information, such as time synchronisation, is distributed within a cell using periodic AC beacons. In IH networks, the process of beacon transmission is relatively simple. In the event the cell manager cannot reach all stations, dedicated proxy stations are used to transmit beacon messages to unreached stations.

In AC networks, the beacon transmission is based on a multi-hop mechanism, and a dedicated AC beacon is used to supply the cell.

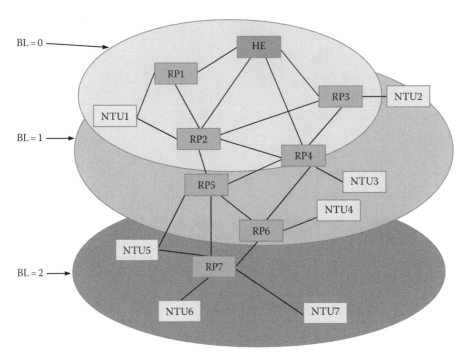

FIGURE 13.16
Example for BL AD.

The HE starts broadcasting an AC beacon message with the *beacon level* (BL) field set to zero. Every station that receives this beacon message processes the information contained in the beacon. A station that has already received a beacon message with a lower BL value will ignore the received AC beacon; all other stations will increase the BL by one and retransmit the AC beacon. This procedure is repeated until every station within the cell is reached. By using the BL technique, unnecessary beacon repetition is avoided. Figure 13.16 shows the resultant BLs for a specific network. As mentioned previously, the AD for AC beacons will include more STAs than the data connections in an IEEE 1901 FFT AC network.

In order to minimise the number of BLs, the beacon messages are transmitted using a robust modulation (see also Section 13.3.6) that has an increased transmission range and, respectively, a bigger AD. AC beacons are transmitted periodically to keep the network updated and synchronised and possibly to include new stations to the AC cell. Furthermore, the effective *bit-loading estimate* (BLE) (an estimate of link quality to the HE), TDMA schedule, state of the backup HE and certain station capabilities belong to the management information distributed by the AC beacon. Figure 13.17 shows the hierarchy resulting from the beacon distribution in Figure 13.16.

13.6.7 Cell Affiliation

The formation of new cells is based on the information distributed in the AC beacon. A new station scans for AC beacons for a period of time. It then uses the effective BLE contained in the AC beacon to derive the best connection to a HE. In order to obtain full functionality, the new station needs to associate and authenticate to a cell. During the association procedure, the HE grants permission to join the cell and provides the station with a TEI. The authentication procedure ensures that the station obtains the NEK that is

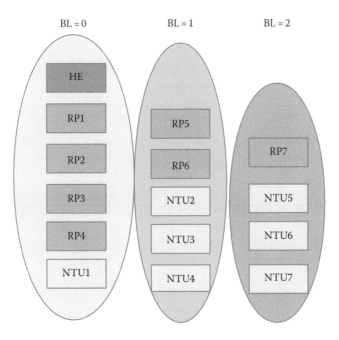

FIGURE 13.17
BL hierarchy.

only valid within a particular cell. The NEK used to encrypt intra-cell data transmission ensures data confidentiality (see also Section 13.4.1).

A station that has already successfully joined a cell keeps receiving AC beacons from neighbouring cells. It will check the information to find out if a neighbouring cell offers a better performance. In that case, it may improve its connectivity by hopping to the neighbouring cell.

13.6.8 Selection of Backup HE

In a meshed AC network, almost every device can be backed up using an alternative route. The dynamic construction of cells allows instant restructuring in case one of the devices fails. However, the HE providing the only connection point to the backbone may become a single point of failure. In order to cope with a HE failure, the IEEE1901 FFT AC standard suggests the installation of another HE in the same location working on standby mode. The two HEs appear as a single station, but only one of them is actually active. The other HE remains ready to take over the HE activity in case it detects that the active HE is not responding.

13.7 IEEE 1901 FFT-OFDM Access: Channel Access

13.7.1 Channel Access Management

IEEE 1901 FFT AC provides a mechanism to define time areas for different MAC types. This allows quasi-parallel operation of prioritised bandwidth allocation using carrier sense multiple AC with collision avoidance CSMA/CA and reserved bandwidth allocation using time divisional multiple AC TDMA.

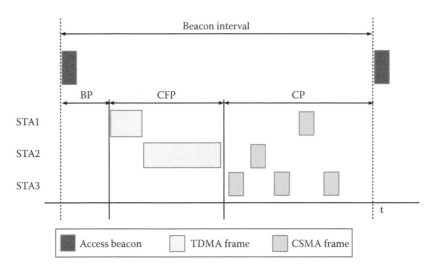

FIGURE 13.18
Channel AC managed by beacon interval.

Figure 13.18 shows the three different regions of a beacon interval, that is, the cell-wide synchronised time interval between two AC beacons is configured into three time slots. The first one, the beacon extent, is reserved to the duration of the beacon transmission.

The second time slot is the *contention-free period* (CFP), which is used for transmissions that require reserved bandwidth on the TDMA basis.

The third time slot, the *contention period* (CP), is open for contention channel AC using CSMA/CA.

The system is configurable in a way such that the reserved time for the CFP can be set to zero, and the whole channel AC is used with CSMA/CA.

Different service priorities demanded by HLE applications are mapped to the connection layer using prioritised traffic streams. Different traffic classes determine the level of priority and whether CSMA/CA or TDMA is used for bandwidth allocation.

Figure 13.19 shows a constellation where the CFP is not required, and the whole beacon interval is configured to be a contention period.

13.7.2 CSMA/CA

In the IEEE 1901 FFT AC standard, CSMA/CA is suggested as the primary protocol for channel AC management. However, the multi-hop structure of an AC network raises the problem of hidden nodes. In order to address this difficulty, a *request to send* (RTS)/*clear to send* (CTS) exchange is used to reduce the risk of data collision. Before a station starts a transmission, a RTS packet is sent to the receiving station. In case the required channel can be allocated, the receiving station acknowledges the transition with a CTS packet. Other stations in the same AD are now informed of the scheduled transmission and stop transmitting data into the used AD for a certain period of time.

13.7.3 TDMA

A major disadvantage of CSMA/CA is the fact that it is based on a statistical algorithm, which does not allow a deterministic channel AC or guarantee of a bandwidth allocation that may be required by certain HLE applications such as audio or video data streams.

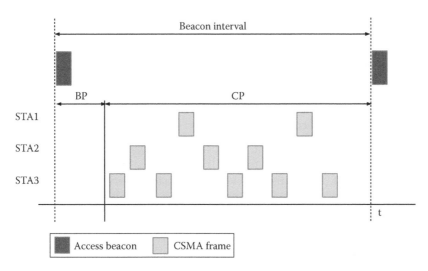

FIGURE 13.19
Pure CSMA mode.

The IEEE 1901 FFT AC standard supports the TDMA in order to meet these concerns and to provide highly consistent bandwidth with exclusive use of the medium.

A station that wants to use the TDMA service requests a channel based on the level of service required by its HLE. The HE then authorises the opening of a TDMA channel within its cell by initiating a remote and distributed procedure where the RPs allocate the required slots. Each RP is then responsible for its allocations according to its available time and bandwidth resources. In the case of a route change between two stations that communicate with each other on the basis of TDMA, a decentralised reallocation procedure ensures the adaption of the TDMA route to the altered situation.

13.8 IEEE 1901 FFT-OFDM Access: Routing

The routing algorithm is responsible for the automatic construction of optimised paths for data transmission based on the underlying network infrastructure. In a distributed network, every station maintains a routing table that maps the final destination of a data packet to its next hop destination. Since the link structure of a power line network is extremely dynamic, the routing algorithm has to adapt to changes of link quality as well as disappearance of stations. Since the IEEE 1901 standard focuses on the PHY and MAC layers, the standard does not specify, but only suggests two different routing methods compatible to the PHY/MAC architecture of the IEEE 1901 standard: the baseline approach and the distance vector approach. Both routing protocols are proactive, that is, the routes are constructed in advance and are updated dynamically.

13.8.1 Baseline Approach

The baseline approach focuses on building routes toward the HE. In analogy to the process of joining a cell, each station first listens to the AC beacons carrying the neighbours' connection level information such as the BLE of the link, the hop count toward the HE and the BL.

Based on the collected information, each station then selects the next hop station toward the HE. Each station continuously tries to optimise its route toward and from the HE by listening to AC beacons of neighbouring stations possibly carrying changed link-state information. As a result, the baseline routing algorithm dynamically reconstructs a hierarchical network tree optimised for data transmission.

13.8.2 Distance Vector Approach

If data transfer to the HE is dominant, the baseline approach is the best choice. The distance vector approach should be used in turn if communication between arbitrary stations within a cell is dominant. In the distance vector approach, every station maintains a routing table with entries for each station within the cell. The selection of the path is again derived on the basis of the information distributed by the AC beacon.

13.8.3 Loop Detection

An unwanted consequence of a varying link structure is the chance that frequent changes of the route may result in a loop. In that case, a data packet is continuously transmitted within a loop without reaching the intended destination. In order to deal with this issue, the header of the data packet contains a hop counter that is initially set to the maximum number of hops and decremented on each retransmission. If a station receives a packet with the hop counter set to zero, it detects the occurrence of the loop and reacts with a route repair procedure.

13.8.4 Bridging

The HE and the NTUs are located at the edges of the AC cell and provide connectivity to other (possibly non-BPL) networks. In order to support data transfer on the basis of MAC addresses, these stations should support a bridging functionality. To this end, the IEEE 1901 FFT AC frame format uses a dual-layer addressing mechanism. The MAC frame header determines the end-to-end communication by the use of the MAC addresses of the edge devices. The internal header contains the original *source TEI* (STEI) and the *original destination TEI* (ODTEI) of the edge stations as well as the STEI and *destination TEI* (DTEI) of the one-hop transmission. A station that receives a data frame where the ODTEI does not match the station's TEI transmits the data frame directly to the next hop station without informing the HLE. Bridging tables at the edge stations are used to map the MAC address to the ODTEI and vice versa.

13.9 IEEE 1901 FFT-OFDM Access: Power Management

BPL AC networks normally consist of a large number of stations. As a consequence, the costs of the overall energy supply generated by the whole AC system may become a major financial consideration for the network provider. To address this issue, the IEEE 1901 FFT AC standard defines a sleep mode feature. Due to the complexity of scheduling the sleeping periods over multi-hop distances, the power management in AC networks is more difficult to establish than the one in IH environments. A high number of stations set to sleep

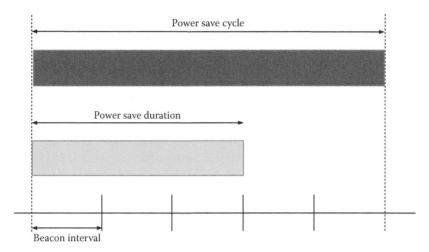

FIGURE 13.20
Timing for the centralised power management.

mode may increase the risk of network failures due to link changes. The standard suggests either a centralised or a distributed method of power management. In centralised power management, the HE controls the power-saving behaviour of every station within the cell. It determines a periodic power-save cycle as a multiple of the beacon period and assigns a part of this cycle to a station. During this time slot, the station may enter the power-save mode. In distributed power management, the station negotiates with the HE that it intends to be out of service for a certain time (Figure 13.20).

13.10 Conclusion

In this chapter, an overview of IEEE 1901 has been presented. The overall system architecture and the key features of the PHY and MAC layers have been described. IEEE 1901 specifies both high-speed BPL communications for IH and AC systems. IEEE 1901 supports two noncompatible PHY layers that can coexist using an ISP. Although this chapter focused on the widely deployed FFT-based OFDM PHY, a wavelet-PHY and ISP introduction was provided. IEEE 1901 IH and AC use the same FFT-PHY layer with only minor adaptations. The IEEE 1901 FFT-advanced IH PHY and MAC layer features support robust and high-speed transmission of up to 500 Mbit/s at the PHY layer. IEEE 1901 FFT AC defines advanced network architecture capable of fulfilling the present and future requirements for BPL AC systems. Its flexible concept including MAC layer encryption, power management capability and defect prevention predestines its usage in Smart Grid applications. BPL exceeds the narrowband power line technology in terms of robustness and scalability. With CSMA and TDMA, two different channel AC techniques are provided, and thus the support of sophisticated QoS requirements is realised. Since a major challenge of an AC network is the multi-hop environment resulting in a dynamic network topology, bridging and routing concepts are kept highly flexible.

IEEE 1901 FFT AC defines a system capable of fulfilling the present and future requirements of BPL AC systems and their usage in Smart Grid applications.

References

1. ISO/IEC 7498-1:1994: Information Technology – Open systems interconnection – Basic reference model: The basic model. ISO International Standards, 1994. Available online iso.org/iso/home/store/catalogue_ics
2. Z. Wang and G.B. Giannakis, Wireless multicarrier communications: Where Fourier meets Shannon, *IEEE Signal Processing Magazine*, 17(3): 29–48, May 2000.
3. R. Prasad, *OFDM for Wireless Communication Systems*, Artech House, Norwood, MA, 2004.
4. C. Berrou, A. Glavieux and P. Thitimajshima, Near Shannon limit error-correcting coding and decoding: Turbo-codes, *IEEE International Conference on Communications, (ICC'93)*, Geneva, Switzerland, pp. 1064–70, May 1993.
5. C. Berrou and A. Glavieux, Near optimum error correcting coding and decoding: Turbo-codes, *IEEE Transactions on Communication*, 44: 1261–1271, October 1996.
6. W.E. Ryan, Concatenated codes and iterative decoding, in *Wiley Encyclopedia of Telecommunications*, J.G. Proakis ed., Wiley & Sons, New York, pp. 556–570, 2003.
7. K. Gracie and M.H. Hamon, Turbo and turbo-like codes: Principles and applications in telecommunications, *Proceedings of the IEEE*, 95(5): 1228–1254, June 2007.
8. A.D. Rizos, J.G. Proakis and T.Q. Nguyen, Comparison of DFT and cosine-modulated filter banks in multicarrier modulation, in *Proceedings of the IEEE Global Communications Conference, (GLOBECOM'94)*, San Francisco, CA, pp. 687–691, 1994.
9. S. Galli, H. Koga and N. Kodama, Advanced signal processing for PLCs: Wavelet-OFDM, in *IEEE International Symposium on Power Line Communications and its Applications (ISPLC)*, Jeju Island, Korea, April 2–4, pp. 187–192, 2008.
10. IEEE Std 1901–2010: IEEE standard for broadband over power line networks: Medium access control and PHY layer specifications, IEEE Standards Association, 2010. Available online standards.ieee.org/store.
11. C. Douillard and C. Berrou, Turbo codes with rate-m/(m+1) constituent convolutional codes, *IEEE Transactions on Communications*, 53(10): 1630–1638, October 2005.
12. C. Weiss, C. Bettstetter, S. Riedel and D.J. Costello, Jr., Turbo decoding with tail-biting trellises, in *Proceedings of the International Symposium on Signals, Systems, and Electronics (ISSE-98)*, Pisa, Italy, pp. 343–348, 1998.
13. C. Douillard and C. Berrou, Turbo codes with rate-m/(m+1) constituent convolutional codes, *IEEE Transactions on Communications*, 53(10): 1630–1638, 2005.
14. T. Lestable, E. Zimmerman, M.-H. Hamon and S. Stiglmayr, BlockLDPC codes vs. Duo-binary turbo-codes for European next generation wireless systems, in *Proceedings of the 64th IEEE Vehicular Technology Conference (VTC'06)*, Montreal, Quebec, Canada, pp. 1–5, 2006.
15. IEEE Std 802.1Q-1998: IEEE standard for local and metropolitan area networks: Virtual bridged local area networks, IEEE Standards Association, 1998.
16. IEEE Std 802-2001: IEEE standard for local and metropolitan area networks: Overview and architecture, IEEE Standards Association, 2002.
17. IEEE Std. 802.1D-2004: IEEE standard for local and metropolitan area networks: Media access control (MAC) Bridges, IEEE Standards Association, 2004.

HomePlug AV2: Next-Generation Broadband over Power Line*

Larry Yonge, Jose Abad, Kaywan Afkhamie, Lorenzo Guerrieri,
Srinivas Katar, Hidayat Lioe, Pascal Pagani, Raffaele Riva,
Daniel M. Schneider and Andreas Schwager

CONTENTS

---

* This chapter is adapted and reprinted with permission from Yonge, L. et al., *Hindawi J. Electr. Comput. Eng.,*
2013, Article ID 892628, Copyright © 2013.

14.1 Introduction

The convergence of voice, video and data within a variety of multifunction devices, along with the evolution of *high-definition* (HD) and 3-dimensional (3D) video, is driving the demand for home connectivity solutions today. Home networks are required to support high-throughput connectivity, guaranteeing at the same time a high level of reliability and coverage (the percentage of links that are able to sustain a given throughput in two nodes or multi-node networks). Applications such as *HD television* (HDTV), *Internet protocol television* (IPTV), interactive gaming, whole-home audio, security monitoring and Smart Grid management have to be supported by these new home networks.

During the last decade, in-home *power line communication* (PLC) has received increasing attention from both the industry and research communities. The ability to reuse existing wires to deploy broadband services is the main attraction of in-home PLC technologies. Another major advantage of PLC is the ubiquity of the power lines that can be used to provide whole-home connectivity solutions. However, as the power line medium had not been originally designed for data communication, the frequency selectivity of the channel and different types of noise (background noise, impulsive noise and narrowband interferers) make power line a very challenging environment, requiring state-of-the-art design solutions.

In 2000, the HomePlug Alliance [1], an industry-led organisation, was formed with the scope of promoting power line networking through the adoption of HomePlug specifications. In 2001, the HomePlug Alliance released the HomePlug 1.0.1 specification and followed up in 2005 with a second release: HomePlug AV. (The letters 'AV' stand for 'audio, video'.) Following its release, HomePlug AV rapidly became the most widespread adopted solution for in-home PLC. To meet the future market needs, in January 2012, the HomePlug Alliance published the HomePlug AV2 specification [2]. HomePlug AV2 enables gigabit-class connection speeds by leveraging the existing power line wires while remaining fully interoperable with other technologies for in-home connectivity, such as HomePlug AV, *HomePlug Green PHY* [3] (see Chapter 10) and IEEE 1901 [3] (see Chapter 13). The alliance's *AV Technical Working Group* (AV TWG) defined new features at both the PHY and MAC layers. These were based on extensive field tests conducted in real-home scenarios across different countries and are included in the HomePlug AV2 specification. The field results obtained by the AV TWG validated HomePlug AV2's performance claims in terms of both achievable data rate and coverage.

In this chapter, we highlight the key differentiating HomePlug AV2 features as compared with HomePlug AV technology. At the *physical* (PHY) layer, HomePlug AV2 includes the following:

1. *Multiple-input multiple-output* (MIMO) signalling with beamforming to offer the benefit of improved coverage throughout the home, especially on highly attenuated channels. MIMO enables HomePlug AV2 devices to transmit on any two-wire pairs within three-wire configurations comprising *line* (L), *neutral* (N) and *protective earth* (PE) (the coupling is done in the MIMO *analogue front-end* [AFE] blocks as shown in Figure 14.1).

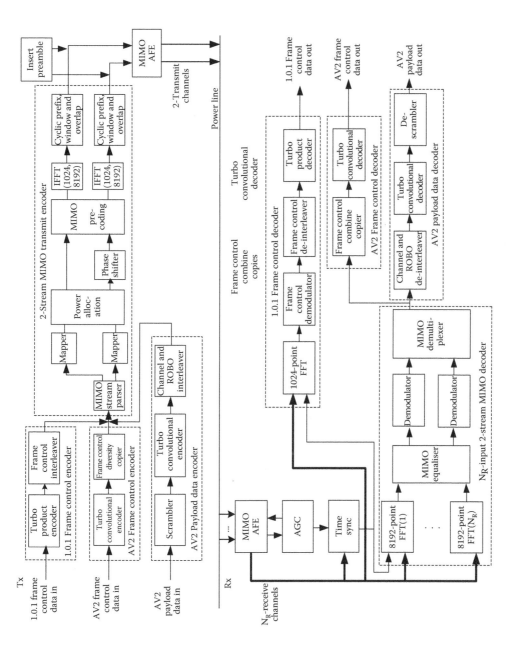

FIGURE 14.1
HomePlug AV2 Tx and Rx PHY layer. (From HomePlug Alliance, HomePlug AV Specification Version 2.0, HomePlug Alliance, Copyright © January 2012. With permission.)

2. Extended frequency band up to 86 MHz to increase the throughput especially at the low- to mid-coverage percentages (part of IFFT/FFT [1024; 8192] and mapper blocks as shown in Figure 14.1).

3. Efficient notching, which allows transmitters to create extremely sharp frequency notches. If *electromagnetic compatibility* (EMC) regulation will require a fragmented communication spectrum, the throughput loss by excluded frequencies can be minimised (part of *power allocation* [PA] and window and overlap blocks as shown in Figure 14.1).

4. Power back-off to increase the HomePlug AV2 data rate while reducing the electromagnetic emissions (part of PA blocks and the AFEs as shown in Figure 14.1).

5. EMC-friendly power boost to optimise the transmit power by monitoring the input port reflection coefficient (known as the S_{11} parameter) at the transmitting modem.

6. Additional PHY Improvements, comprising higher-order *quadrature amplitude modulation* (4096-QAM) (part of mapper and demodulator blocks as shown in Figure 14.1), higher code rates (8/9 code rate) (part of turbo convolutional decoder as shown in Figure 14.1) and smaller *guard intervals* (GIs) (part of cyclic prefix blocks as shown in Figure 14.1), to assist in better peak data rates.

At the *medium access control* (MAC) layer, HomePlug AV2 includes the following:

1. Power save mode to improve energy efficiency when the device is in standby

2. Short delimiter to reduce the overhead of the transmission shortening the preamble and *frame control* (FC) symbols

3. Delayed acknowledgements to increase the overall transmission efficiency by reducing the interframe spacings

4. Immediate repeating to expand the coverage by repeating the signal on paths with better *signal to noise ratio* (SNR) characteristics

All the features listed earlier improve the *quality of service* (QoS) of AV2 power line modems by improving coverage and robustness of communication links. Moreover, the AV TWG evaluated the performance of the HomePlug AV2 system. The adoption of all these features provides a significant advantage of using HomePlug AV2 versus HomePlug AV.

The following sections provide technical details for all the previously mentioned features. Section 14.2 begins with a brief overview of HomePlug AV technology and then presents the overall HomePlug AV2 system architecture. Sections 14.3 and 14.4 detail the new PHY and MAC layers. Section 14.5 summarises the performance gain of HomePlug AV2 over HomePlug AV. Finally, Section 14.6 concludes this introduction to HomePlug AV2 technology. The Inter System Protocol used by HomePlug AV2 to manage coexistence with other in-home technologies is described in details in Chapter 10.

14.2 System Architecture

14.2.1 Brief Overview of HomePlug AV

HomePlug AV employs PHY and MAC technologies that provide a 200 Mbps-class power line networking capability. The PHY operates in the frequency range of 2–28 MHz and uses windowed *orthogonal frequency division multiplexing* (OFDM) and a powerful *turbo convolutional code* (TCC) that provides robust performance within 0.5 dB of Shannon's limit. Windowed OFDM provides more than 30 dB spectrum notching. OFDM symbols with 917 usable carriers (tones) are used in conjunction with a flexible GI. Modulation densities from BPSK to 1024 QAM are adaptively applied to each carrier based on the channel characteristic between the *transmitter* (Tx) and the *receiver* (Rx).

On the MAC layer, HomePlug AV provides a QoS connection-oriented, contention-free service on a periodic *time division multiple access* (TDMA) allocation and a connectionless, prioritised contention-based service on a *carrier sense multiple access/collision avoidance* (CSMA/CA) allocation. The MAC receives *MAC service data units* (MSDUs) and encapsulates them with a header, optional *arrival time stamp* (ATS) and a checksum to create a stream of MAC frames. The stream is then divided into 512-octet segments, encrypted and encapsulated into serialised *PHY blocks* (PBs) and packed as *MAC protocol data units* (MPDUs) to the PHY unit, which then generates the final *PHY protocol data unit* (PPDU) to be transmitted onto the power line [5]. The HomePlug AV specification is included as one of the two physical layer protocols defined in the IEEE 1901 standard (see Chapter 13).

HomePlug AV2 significantly enhances the capability mentioned earlier to accommodate a new generation of multimedia applications with gigabit-class performance.

14.2.2 HomePlug AV2 Technology

The HomePlug system is specified for in-home communication via the power line channel. These in-home communications are established with the *time division duplexing* (TDD) mechanism to allow for symmetric communication between peers, as opposed to the classical access system in ADSL which uses two different downstream and upstream throughputs.

The PHY layer employs an OFDM modulation scheme for better efficiency and adaptability to the channel impairments (such as being frequency selective and suffering from narrowband interference and impulsive noise). The HomePlug AV2 OFDM parameters correspond to a system with 4096 carriers in 100 MHz, but only carriers from 1.8 to 86.13 MHz are supported for communication (3455 carriers). The subcarrier spacing of 24.414 kHz was chosen in the HomePlug AV system according to the power line coherence bandwidth characteristic and is maintained in HomePlug AV2 for interoperability.

More significantly, HomePlug AV2 incorporates MIMO capability (see details in the following sections) to improve throughput and coverage.

A block diagram of the PHY layer is shown in Figure 14.1. The HomePlug AV2 system is capable of supporting two network-operating modes: AV-only mode and hybrid mode (refer to 1.0.1 FC encoder in Figure 14.1). Hybrid mode is used for coexistence with HomePlug 1.0.1 stations. For that purpose, a 1.0.1 FC encoder is included for hybrid modes and an AV2 FC encoder is included for both hybrid and AV-only modes. The AV-only mode is used for communications in networks where only HomePlug AV and HomePlug AV2 stations are involved.

Two OFDM paths are also shown to demonstrate how to implement MIMO capabilities with two transmission ports.

Apart from the hybrid or AV-only FC symbols, the payload data can be sent using adaptive bit loading per carrier or *robust modes* (ROBO) with fixed *quadrature phase shift keying* (QPSK) constellation and several copies of data interleaved in both time and frequency.

Looking at the data path details in the block diagram (Figure 14.1), the following can be seen: at the Tx side, the PHY layer receives its inputs from the MAC layer. Three separate processing chains are shown because of the different encoding for HomePlug 1.0.1 FC data, HomePlug AV2 FC data and HomePlug AV2 payload data. AV2 FC data is processed by the AV2 FC encoder, which has a turbo convolutional encoder and FC diversity copier, while the HomePlug AV2 payload data stream passes through a scrambler, a turbo convolutional encoder and an interleaver. The HomePlug 1.0.1 FC data passes through a separate HomePlug 1.0.1 FC encoder.

The outputs of the FC encoders and payload encoder lead into a common MIMO OFDM modulation structure, consisting of the following:

- A *MIMO stream parser* (MSP) that provides up to two independent data streams to two transmit paths that include two mappers, a phase shifter that applies a 90° phase shift to one of the two streams (to reduce the coherent addition of the two signals)
- A MIMO precoder to apply Tx beamforming operations
- Two *inverse fast Fourier transform* (IFFT) processors
- Preamble and cyclic prefix insertion
- Symbol window and overlap blocks, which eventually feed the AFE module with one or two transmit ports that couple the signal to the power line medium

For potential MIMO coupler configuration, the interested reader may refer to Chapter 1.

At the receiver, an AFE with one, two, three or four (N_R) Rx ports operates with individual *automatic gain control* (AGC) modules and one or more time-synchronisation modules to feed separate FC and payload data recovery circuits. Rxs plugged into power outlets that are connected to the three wires (L, N and PE) might utilise up to three *differential-mode* Rx ports and one *common-mode* (CM) Rx port.

The FC data are recovered by processing the received signals through a 1024-point FFT (for HomePlug 1.0.1 delimiters) and multiple 8192-point FFTs and through separate FC decoders for the HomePlug AV2/HomePlug AV and HomePlug 1.0.1 modes. The payload portion of the sampled time domain waveform, which contains only HomePlug AV2 formatted symbols, is processed through the multiple 8192-point FFT (one for each receive port), a MIMO equaliser that receives N_R signals, performs receive beamforming and recovers the two transmit streams, two demodulators, a demultiplexer to combine the two MIMO streams and a channel de-interleaver followed by a turbo convolutional decoder and a de-scrambler to recover the AV2 payload data.

14.3 PHY Layer Improvements of HomePlug AV2

14.3.1 Multiple-Input Multiple-Output Capabilities with Beamforming

The HomePlug AV2 specification incorporates MIMO capabilities with beamforming, which offers the benefit of improved coverage throughout the home, particularly for hard-to-reach outlets. MIMO technology enables HomePlug AV2 devices to transmit and receive on any two-wire pairs within a three-wire configuration. Figure 14.2 shows a three-wire configuration with L, N and PE, as well as an exemplary schematic coupler design (see also Chapter 1). Whereas HomePlug AV always transmits and receives on line–neutral

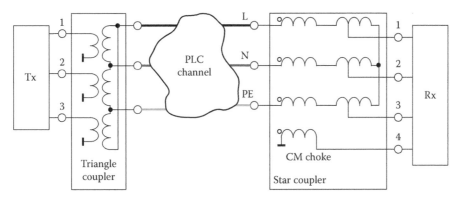

FIGURE 14.2
MIMO PLC channel: different feeding and receiving options.

pair (port 1 in Figure 14.2), HomePlug AV2 modems can transmit and receive signals on other pairs as well. Any two pairs formed by the L, N or PE wires (i.e. L–N, L–PE and N–PE) can be used at the Tx. At the Rx, up to four receive ports may be used (see Figure 14.2). The CM signal – indicated as port 4 in Figure 14.2 – is the voltage difference between the sum of the three wires and the ground.

The numbers of used transmit ports N_T and used receive ports N_R define the MIMO configuration, which is called $N_T \times N_R$ MIMO. For example, using L–N and L–PE to feed and receive signals results in a 2×2 MIMO configuration. HomePlug AV2 specification supports MIMO configurations with up to two Tx ports and up to N_R Rx ports.

Some regions and homes with older electrical installations do not have the third wire installed in private buildings. A survey on worldwide PE conductor availability may be found in Chapter 1. In this scenario, HomePlug AV2 automatically switches to *single-input single-output* (SISO) operating mode. HomePlug AV2 also incorporates selection diversity in SISO mode. The ports used for feeding and receiving might be different from the traditional L–N feeding. If, for example, the path from L–PE to L–N offers better channel characteristics than L–N to L–N, the Tx can choose to use the L–PE port for feeding.

A number of studies have been conducted to investigate and model the properties of MIMO transmission through power line channels [6–18]. More information about the MIMO PLC channel characteristics and channel modelling may be found in Chapters 1 through 5.

The MIMO PLC channel is described by a $N_R \times N_T$ channel matrix for each OFDM subcarrier c:

$$\mathbf{H}(c) = \begin{bmatrix} h_{11}(c) & \cdots & h_{1N_T}(c) \\ \vdots & \ddots & \vdots \\ h_{N_R 1}(c) & \cdots & h_{N_R N_T}(c) \end{bmatrix} \tag{14.1}$$

with c belonging to the set of used subcarriers (depending, e.g. on the frequency band and tone mask).

HomePlug AV2 supports one or two *streams* (see Section 14.3.1.1). As HomePlug AV2 supports MIMO configurations with up to two Tx ports and up to N_R Rx ports, the maximum number of supported *streams* in AV2 is two. The underlying *MIMO streams* are obtained by a *singular value decomposition* (SVD) of the channel matrix [19]:

$$\mathbf{H}(c) = \mathbf{U}(c)\mathbf{D}(c)\mathbf{V}(c)^H, \qquad (14.2)$$

where
   V and **U** are unitary matrices, that is, $\mathbf{V}^{-1} = \mathbf{V}^H$ and $\mathbf{U}^{-1} = \mathbf{U}^H$ (with ^{H} the Hermitian operator)
   D is a diagonal matrix containing the singular values of **H**

For details about the MIMO channel matrix and the underlying MIMO streams based on the SVD, refer to Chapter 8.

The decomposition into parallel and independent *streams* by means of the SVD illustrates the MIMO gain: instead of having one spatial *stream* in SISO, two independent spatial *streams* are available in a $2 \times N_R$ MIMO configuration with $N_R \geq 2$ on average doubling the SISO capacity [6,17,20]. An analysis of the MIMO PLC channel capacity and throughput investigations can be found in Chapter 9.

14.3.1.1 MIMO Stream Parser

Depending on how many *streams* are used in the transmission, the payload bits have to be split into different spatial *streams*. This task is performed by the MIMO stream parser (MSP). The MSP splits the incoming bits into one or two streams based on the MIMO mode of operation and the tone map information (see Figure 14.1). To transmit a single-stream payload with spotbeamforming (see Section 14.3.1.3) or for SISO transmissions, the MSP sends all the data at its input to the first mapper. In this case, the MSP operates as if it had only one output and it was connected to the first mapper (see Figure 14.1). To transmit a two-stream payload with eigenbeamforming (see Section 14.3.1.3), the MSP allocates the bits to the two streams.

14.3.1.2 Precoding

Precoded spatial multiplexing or beamforming was chosen as the MIMO scheme, as it offers the best performance by adapting the transmission in an optimal way to the underlying eigenmodes of the MIMO PLC channel. This performance is achieved in various channel conditions. On the one hand, the full spatial diversity gain is achieved in highly attenuated and correlated channels when each symbol is transmitted via each available MIMO path. On the other hand, a maximum bitrate gain is achieved for channels with low attenuation when all available spatial streams are utilised. Beamforming has also many advantages with respect to Rx design. The most simple detection or equaliser algorithm, which is the *zero-forcing* (ZF) detection (see Section 14.3.1.3), might be used in the case of beamforming. No performance gain is achieved for more sophisticated detection algorithms such as *minimum mean squared error* (MMSE) or *ordered successive interference cancellation* (OSIC) detection algorithms when optimum precoding is applied at the Tx [18]. Beamforming also offers flexibility with respect to the Rx configuration. Only one spatial stream may be activated by the Tx when only one receive port is available, that is, if the outlet is not equipped with the third wire or if a simplified Rx implementation is used that supports only one spatial stream. Since beamforming aims to maximise one MIMO stream, the performance loss of not utilising the second stream is relatively small compared to the spatial multiplexing schemes without precoding. This is especially true for highly attenuated and correlated channels, where the second MIMO stream carries only a small amount of information. These channels are most critical for PLC, and adequate MIMO schemes are important. A comparison and analysis of different MIMO schemes may be found in Refs. [18,20,21] and in Chapters 8 and 9.

Beamforming requires knowledge about the channel state information at the Tx to apply the optimum precoding. Usually, only the Rx has channel state information, that is, by channel estimation. Thus, the information about the precoding has to be fed back from the Rx to the Tx. The HomePlug AV2 specification supports adaptive modulation [22,23]. The application of adaptive modulation also requires feedback about the constellation of each subcarrier (tone map); that is, the feedback path is required. The information about the tone maps and the precoding are updated simultaneously upon a change in the PLC channel (realised by a channel estimation indication message). The information about the precoding is quantised very efficiently, and the amount required for the precoding information is in the same order of magnitude as the information about the tone maps. Thus, the overhead in terms of management messages and required memory can be kept low.

The optimal linear precoding matrix $\mathbf{F}$ for a precoded spatial multiplexing system can be factored into the two matrices $\mathbf{V}$ and $\mathbf{P}$ [24]:

$$\mathbf{F} = \mathbf{V}\mathbf{P}. \tag{14.3}$$

Note that the subcarrier index is omitted in Equation 14.3 and in the following section to simplify the notation and to allow better legibility. However, it has to be kept in mind that the following vector and matrix operations are applied to each subcarrier separately if not stated otherwise.

$\mathbf{P}$ is a diagonal matrix that describes the PA of the total transmit power to each of the transmit streams. PA is considered in more detail in Section 14.3.1.4. $\mathbf{V}$ is the right-hand unitary matrix of the SVD of the channel matrix (see Equation 14.2). Precoding by the unitary matrix $\mathbf{V}$ is often referred to as unitary precoding or eigenbeamforming.

Figure 14.3 shows the basic MIMO blocks of the Tx. The mappers of the two streams follow the MSP (see Figure 14.1). The two symbols of the two streams are

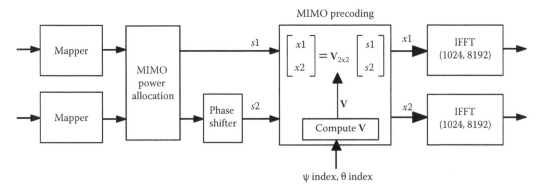

FIGURE 14.3
Precoding at the Tx. (From HomePlug Alliance, HomePlug AV Specification Version 2.0, HomePlug Alliance, Copyright © January 2012. With permission.)

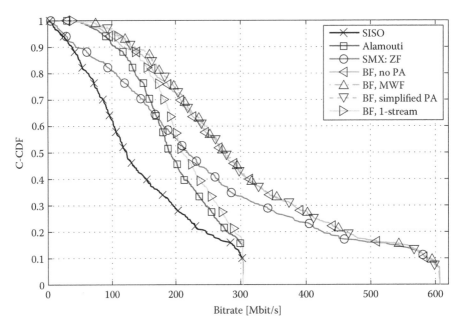

FIGURE 14.4
Performance comparison between SISO, MIMO 2 × 4 *spatial multiplexing without precoding* (SMX: ZF) and differ-ent modes of beamforming, transmit power to noise power of 65 dB, frequency range up to 30 MHz, adaptive modulation using constellations up to 4096-QAM at target BER of 10^{-3} (no error coding).

then weighted by the PA coefficients and multiplied by the matrix **V** (the computation of **V** is explained in Section 14.3.1.3 in more detail). Finally, each stream is OFDM-modulated separately.

Figure 14.4 shows the *complementary cumulative distribution function* (C-CDF) or cover-age of the bitrate for SISO and several MIMO modes of operation, over the (1–30 MHz) frequency band. The shown MIMO schemes include *spatial multiplexing without precod-ing* (SMX: ZF) [19], different modes of beamforming and, as an additional reference, the Alamouti scheme – a simple but effective space–time code for two transmit ports [Alamouti 1998]. The results of different PA schemes (labelled by *mercury water filling* [MWF] and simplified PA in the figure) will be discussed later. The MIMO PLC chan-nels used for this analysis (338 channels in total) were obtained during the ETSI STF410 measurement campaign [13–15]. At the high coverage point in Figure 14.4 (C-CDF values near 1), the spatial multiplexing without precoding and ZF detection shows no performance improvement compared to SISO. The best performance is achieved by beamforming and the Alamouti scheme. Both schemes exploit the full spatial diver-sity which is essential to obtain good performance in highly attenuated and correlated channels as shown at the high coverage point here. At a low coverage point (as C-CDF values approach 0) – which comprises good channels with low attenuation – the spa-tial multiplexing without precoding approaches the performance level of beamforming. The Alamouti scheme in this case approaches the SISO performance, since Alamouti transmits replicas of each symbol and therefore does not achieve a multiplexing gain by utilising the two streams. The spatial multiplexing scheme utilises two streams and thus achieves twice the bitrate performance as compared to SISO and the Alamouti. Overall, beamforming offers the best bitrate for all channel conditions.

14.3.1.3 Beamforming and Quantisation of the Beamforming Matrix

The precoding matrix $\mathbf{V}$ is derived by means of the SVD (see Equation 14.2) from the channel matrix $\mathbf{H}$.

There are two possible modes of operation. If only one spatial stream is utilised, single-stream beamforming (or spotbeamforming) is applied. In this case, the precoding is described by the first column vector of $\mathbf{V}$, that is, the precoding simplifies to a column vector multiplication. Note that despite the fact that only one spatial stream is used, both transmit ports are active as the precoding vector splits the signal to two transmit ports. If both spatial streams are used, two-stream beamforming (or eigenbeamforming) is used and the full precoding matrix $\mathbf{V}$ is applied in the MIMO precoding block (see Figure 14.3).

Since the information about the precoding matrix has to be fed back from the Rx to the Tx, an adequate quantisation is required. To achieve this goal, the special properties of $\mathbf{V}$ are utilised.

The unitary property of $\mathbf{V}$ consequences that the columns $\mathbf{v}_i$ (i = 1, 2) of $\mathbf{V}$ are orthonormal, that is, the column vectors are orthogonal and the norm of each column vector is equal to 1 [26]. There is more than one unique solution to the SVD: the column vectors of $\mathbf{V}$ are phase invariant, that is, multiplying each column vector of $\mathbf{V}$ by an arbitrary phase rotation results in another valid precoding matrix.

These properties allow us to represent the complex 2×2 matrix $\mathbf{V}$ by only the two angles θ and ψ:

$$\mathbf{V} = \begin{bmatrix} \mathbf{v}_1 & \mathbf{v}_2 \end{bmatrix} = \begin{bmatrix} v_{11} & v_{12} \\ v_{21} & v_{22} \end{bmatrix} = \begin{bmatrix} \cos\psi & \sin\psi \\ -e^{j\theta}\sin\psi & e^{j\theta}\cos\psi \end{bmatrix}, \tag{14.4}$$

where the range of θ and ψ to represent all possible beamforming matrices is $0 \le \psi \le \pi/2$ and $-\pi \le \theta \le \pi$.

According to the phase-invariance property, the first entry of each column (v_{11}, v_{12}) may be set to be real without loss of generality, as defined in Equation 14.4. It is easy to prove that the properties of the unitary precoding matrix are fulfilled. The norm of the column vectors is one: $|v_{11}|^2 + |v_{21}|^2 = |v_{12}|^2 + |v_{22}|^2 = \sin^2(\psi) + \cos^2(\psi) = 1$. Also, the two columns are orthogonal:

$$\mathbf{v}_1^H \mathbf{v}_2 = \sin(\psi)\cos(\psi) - \sin(\psi)\cos(\psi)e^{-j\theta}e^{j\theta} = 0. \tag{14.5}$$

In both modes, spotbeamforming and eigenbeamforming, the beamforming vector or beamforming matrix, respectively, is described by both angles θ and ψ. Thus, the signalling of θ and ψ is the same in both modes.

If the MIMO equaliser is based on ZF detection, the detection matrix

$$\mathbf{W} = \mathbf{H}^p = (\mathbf{H}^H\mathbf{H})^{-1}\mathbf{H}^H \tag{14.6}$$

is the pseudoinverse $\mathbf{H}^P$ of the channel matrix $\mathbf{H}$. In case of eigenbeamforming with the precoding matrix $\mathbf{V}$, $\mathbf{H}$ can be replaced by the equivalent channel $\mathbf{HV}$ in Equation 14.6, and the detection matrix can be expressed by

$$\mathbf{W} = \mathbf{V}^H\mathbf{H}^p = \mathbf{D}^{-1}\mathbf{U}^H. \tag{14.7}$$

The SNR of the MIMO streams after detection is calculated as

$$\text{SNR}_1 = \rho \frac{1}{\|\mathbf{w}_1\|^2} \quad \text{and} \quad \text{SNR}_2 = \rho \frac{1}{\|\mathbf{w}_2\|^2} \tag{14.8}$$

with ρ being the ratio of transmit power to noise power and $\|w_i\|$ the norm of the ith row of the detection matrix $\mathbf{W}$.

Figure 14.5a shows the SNR of the first stream of a MIMO PLC channel. The median attenuation of this link is 40 dB. The ratio of the transmit power to noise power is $\rho = 65$ dB. The dashed line represents the SNR of spatial multiplexing without precoding and the solid line represents the eigenbeamforming. The SNR of the second stream is displayed in Figure 14.5b.

The two markers X in Figure 14.5a mark a frequency (13 MHz) where good beamforming conditions are found. Different precoding matrices influence the SNR of the two MIMO streams according to Equations 14.6 and 14.8.

Figure 14.5c and d shows the level of the gain or signal elimination due to beamforming for the frequency marked by X in Figure 14.5a and b. The curves in Figure 14.5c and d indicate the SNR. Depending on the precoding matrix, the SNR varies between 6 and 35 dB. No beamforming ($\psi = 0$ and $\theta = 0$) would result in an SNR of 11 dB. As seen in Figure 14.5c, there is one SNR maximum in the area spanned by ψ and θ. As both streams are orthogonal, one shows an SNR minimum at the location where the other one shows its maximum. The SNR plot in Figure 14.5c and d is 2π periodic in θ, where the black horizontal lines indicate $\theta = \pm\pi$.

The beamforming matrix is quantised efficiently to reduce the amount of feedback required to signal $\mathbf{V}$. A total of 12 bit is used to describe the precoding matrix $\mathbf{V}$. A normal distribution of the real and imaginary part of the entries of the channel matrix $\mathbf{H}$ results in a uniform distribution of $\mathbf{V}$. This results in a uniform distribution of θ and a normal distribution of ψ. Optimal performance is achieved by quantifying θ with 7 bit and ψ with 5 bit. This quantisation is chosen such that the loss of the SNR after detection (refer also to Equations 14.6 and 14.8) is within 0.2 dB compared to the optimum beamforming without quantisation.

14.3.1.4 Power Allocation

MIMO PA is applied to two-stream MIMO transmissions. The PA adjusts the power of a carrier on one stream relative to the other stream. For SISO transmissions and MIMO spotbeamforming transmissions, MIMO PA can be bypassed, since there is only one transmit stream. In this case, the only available option is to allocate all the power to the single stream. The PA module is located between the mapper and the precoding block (see Figure 14.1) and performs on the two MIMO streams before eigenbeamforming. The PA in AV2 is designed in a way to not periodically feedback additional information from the Rx to the Tx. The PA evaluates the tone maps of the two streams to set their PA coefficients. It was shown that their performance is very close to the optimal PA (MWF) [27]. PA improves especially highly attenuated channels at the high coverage point.

According to the adaptive modulation, constellation mapping is applied individually for each carrier based on the tone map information for each stream. If the SNR of a carrier on one stream is low enough so that it does not carry any information, the power is allocated to the other stream. The level of power to be assigned is adjustable in 0.5 dB steps.

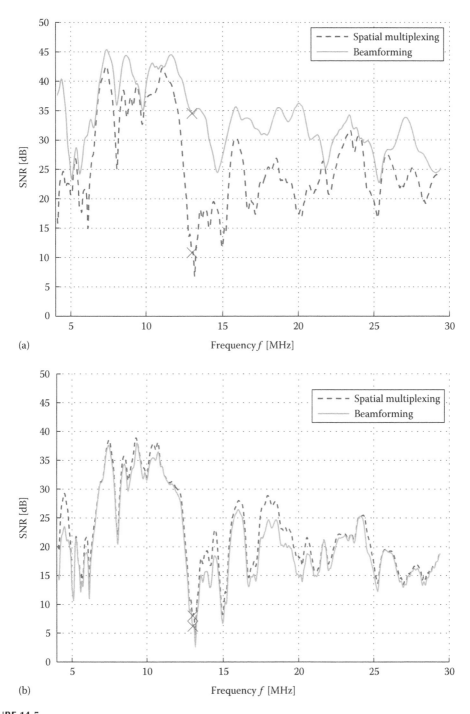

(a)

(b)

FIGURE 14.5
Influence of precoding on the SNR: SNR of the first (a) and second (b) stream with and without precoding, transmit power to noise power of $\rho = 65$ dB, 40 dB average channel attenuation. (From Yonge, L. et al., *Hindawi J. Electr. Comput. Eng.*, 2013, Article ID 892628, Copyright © 2013.)

(*continued*)

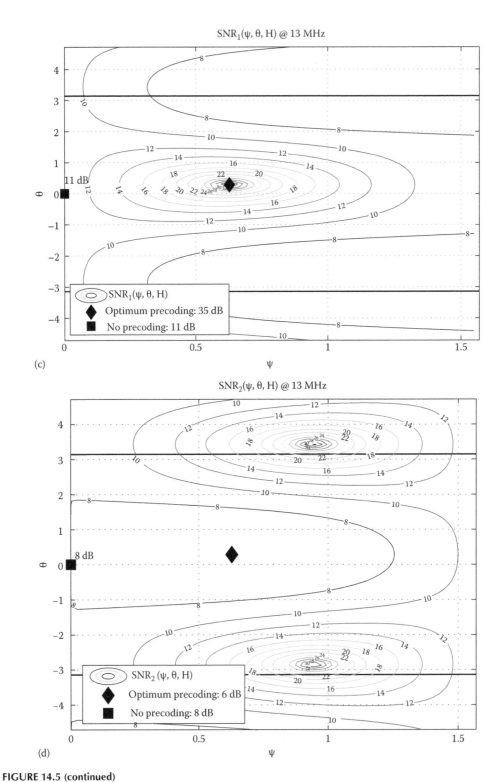

FIGURE 14.5 (continued)
Influence of precoding on the SNR: SNR depending on the precoding matrix at 13 MHz of the first (c) and second (d) stream. (From Yonge, L. et al., *Hindawi J. Electr. Comput. Eng.*, 2013, Article ID 892628, Copyright © 2013.)

Each AV2 Tx selects the level of power adjustment depending on regulatory requirements and its transmit implementation. This level is signalled once to the receive node to be considered at SNR estimation and the generation of the tone map.

14.3.1.5 Precoding Grouping

HomePlug AV2 devices support an expanded frequency spectrum (up to 86.13 MHz; see Section 14.3.2) and one additional stream (utilising MIMO) compared with the 30 MHz for HomePlug AV devices. Thus, HomePlug AV2 devices need to store two tone maps (one for each stream) and one *precoding matrix* (PCM) per carrier. This significantly increases the memory requirements for an AV2 modem. In order to save memory, HomePlug AV2 devices transmit and store the PCM only on a subset of carriers called *precoding pilot carriers* (PPCs). At the Tx, the PCMs for the carriers between two adjacent PPCs are obtained via interpolation. The spacing between the PPCs is selected out of a set of predefined values and may be adjusted depending on memory capabilities or channel conditions. The more memory embedded in the modems, the finer the granularity. The performance loss of precoding grouping compared to the quantisation of each subcarrier separately is marginal, due to the high correlation of the precoding matrices of neighboured subcarriers. An investigation about different precoding grouping algorithms for MIMO PLC may be found in [18].

Different classes of AV2 devices may have different memory implementations. Devices where a large memory is embedded may group the carriers with finer granularity. Devices with limited memory have to perform grouping at a coarser granularity. Since the Rx feeds back PCMs, it is important that the Rx is aware of the memory capabilities of the Tx. For instance, a Tx with limited memory will not be able to store and use the PCMs if they are transmitted at a finer granularity from a Rx with large memory. The PCM grouping supported by the Tx is indicated to the Rx via a management message sent by the Tx.

The Rx forwards the PCM for the pilot carriers to the Tx. The Tx reconstructs the pilot PCM information to obtain the precoding matrices for all carriers.

Figure 14.6 shows a reconstruction based upon interpolation in the angle domain (similar to Figure 14.5c and d). The markers labelled by 'x' represent the precoding matrix of 17 adjacent subcarriers of a MIMO PLC channel. The markers labelled by the filled circles show the quantised precoding matrices (quantisation with 12 bit per carrier). Some subcarriers use the same precoding matrix, and therefore, the number of circles is less than 17 in Figure 14.6. The linear interpolation with a grouping of eight subcarriers is shown by the unfilled circles in Figure 14.6. The precoding matrices of the interpolated subcarriers resemble the optimal precoding matrix here quite closely.

Figure 14.7 shows the behaviour of the interpolation near notched or masked subcarriers. The location of PPCs may occur in notched bands where the Rx does not have an estimate of the PCM. In such cases, the Rx extrapolates the PCMs on the carriers at the notch edges to get the precoding matrix for the PPC inside the notch. The Rx knows the interpolation applied at the Tx and uses this information to get an extrapolated precoding matrix on PPCs inside a notch.

14.3.2 Extended Frequency Band up to 86 MHz

During the specification development, the AV TWG conducted a measurement campaign, where power line channel and noise measurements were performed in 30 homes located

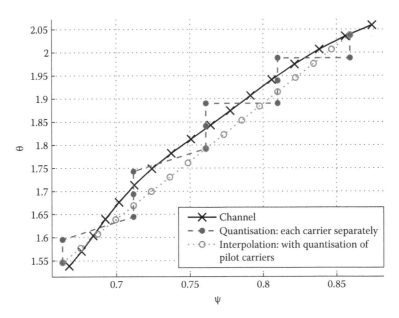

FIGURE 14.6
Linear interpolation of the precoding matrix in the angle domain.

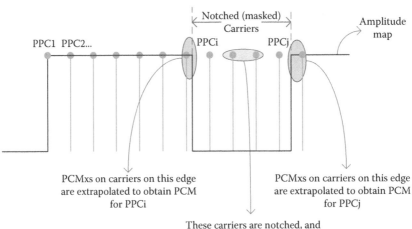

FIGURE 14.7
Interpolation of the precoding matrix at the edges of notches. (From HomePlug Alliance, HomePlug AV Specification Version 2.0, HomePlug Alliance, Copyright © January 2012. With permission.)

in different European countries as well as in the United States. Such variety was a key ingredient for obtaining insight into the power line at frequencies above those used in HomePlug AV (1.8–30 MHz). Specifically, in every home, all the possible links among at least five different nodes were measured.

Examples of channel attenuation and noise PSD are reported in Figures 14.8 and 14.9, respectively.

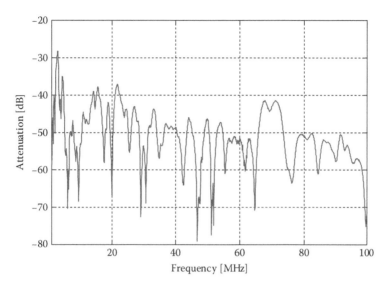

FIGURE 14.8
Example of measured channel transfer function.

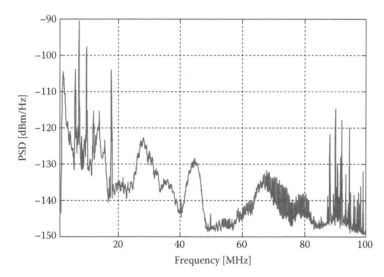

FIGURE 14.9
Example of measured noise PSD.

The final outcome of the analysis was the selection of the 30–86 MHz band. This was based upon the results of the measurement campaign, the present EMC regulations, coverage and complexity targets, as well as the following:

1. The *frequency modulation* (FM) band region 87.5–108 MHz must be avoided, as this frequency region presents higher attenuation and higher noise compared to the 30–86 MHz band (see also Figures 14.8 and 14.9). In addition, one can expect that lower transmit power levels will be required in this band in order to protect FM radio services. Therefore, very low operating SNRs could be obtained in this band with negligible coverage increase.

2. The 30–86 MHz frequency band appears to offer a throughput increase, especially at the low- to mid-coverage percentages; this is due to the fact that while the attenuation is greater compared to the 1.8–30 MHz band, noise is lower. In the past, EMC requirements for the 30–86 MHz frequency band were generally tighter than in the 1.8–30 MHz band. However, regulatory bodies such as CENELEC are currently working towards a stabilisation of the PLC regulation status in Europe (see details on standard FprEN 50561-1 in Chapter 6). Therefore, it can be expected that commonly agreed transmit power masks will be defined soon for the 30–86 MHz frequency range.

3. Within the HomePlug AV2 specification, it has been possible to provide flexibility in choosing the stop frequency in the 30–86 MHz interval. In particular, an AV2 device implementing a frequency band from 1.8 to X MHz (with $30 < X < 86$) will be interoperable with a device implementing a frequency band from 1.8 to Y MHz (with $30 < Y < 86$, $X \neq Y$).

4. The 30–86 MHz frequency band extension allows devices to be fully interoperable with the IEEE 1901 devices that use the 1.8–50 MHz frequency band. In fact, HomePlug AV2 devices that implement a frequency band extension shall support the IEEE 1901 bandwidth.

14.3.3 Efficient Notching

HomePlug AV2 increases throughput by allowing devices to minimise the overhead incurred due to EMC notching requirements. While in HomePlug AV the mechanism ('windowed OFDM') for creating the PSD notches is fixed and relatively conservative, HomePlug AV2 devices may gain up to 20% in efficiency if they implement additional techniques to accommodate sharper PSD notches. The 20% includes the gain of guard carriers that were excluded by HomePlug AV modems and the reduced *transition interval* (TI) in the time domain. Such devices gain additional carriers at the band edges and may utilise shorter cyclic extensions, which reduces the duration of the OFDM symbols.

14.3.3.1 Influence of Windowing on Spectrum and Notch Shape

The FFT process uses a rectangular window to cut data out of a continuous stream to convert them from time to frequency domain. The FFT of a rectangular function in the time domain is a $\sin(x)/x$ function in the frequency domain. The $\sin(x)/x$ becomes 0 at integer multiples of π. Some parts of the signal remain in between the zeros, which results in the unwanted side lobes of an FFT OFDM system. Figure 14.10 shows a $\sin(x)/x$ function marked with '+' symbols. Frequency is shown on the horizontal axis. The level of magnitude is shown on the vertical axis, in a logarithmic view.

The process of multiplying a window with an OFDM symbol (see Figure 14.11) in the time domain is aimed to suppress the sharp corners at the beginning and the end of the OFDM symbol in order to obtain smooth transitions. This affects the shape and distances of the side lobes in the frequency domain [29]. There are numerous types of window functions available for implementation such as Hamming, Barlett (triangular), Kaiser, Blackman and Raised Cosine (Hann). A comparison of various window waveforms with the achieved side lobe attenuation is shown in Figure 14.10, which also serves to illustrate the disadvantage of windowing. The side lobes are suppressed, but the main lobe stays wider.

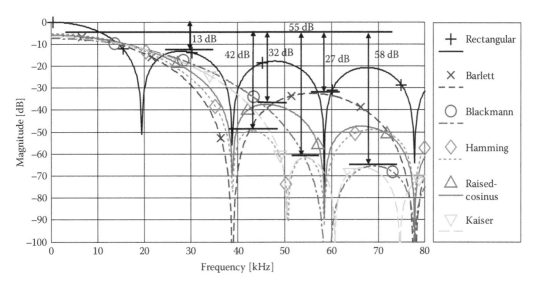

FIGURE 14.10
OFDM side lobes of a single carrier. Comparison of various window functions. (From Yonge, L. et al., *Hindawi J. Electr. Comput. Eng.*, 2013, Article ID 892628, Copyright © 2013.)

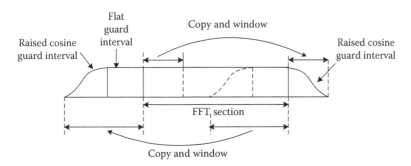

FIGURE 14.11
OFDM symbol with GI and window. (From Yonge, L. et al., *Hindawi J. Electr. Comput. Eng.*, 2013, Article ID 892628, Copyright © 2013.)

The first time the spectrum reaches zero, the distance for all windows is at least twice the frequency of the rectangular window used by the pure FFT. Windowing is considered to be state of the art in signal processing [30].

The process of how a window is applied is shown in Figure 14.11. The original OFDM symbol or the output of the IFFT at the Tx is marked with 'FFT section'. The GI is copied from the tail samples to the beginning of the symbol. In order to create a window, the symbol has to be expanded further at the beginning and at the end by copying the bits as done for the GI. This expansion is multiplied with the smoothly descending window. The more smoothly the signal approaches zero in the time domain, the lower the side lobes in the frequency domain. This expansion of a symbol is a waste of communication resources, as it does not carry useful information and has to be cut by the Rx.

The two descending slopes in the time domain could overlap, to save communication resources at consecutive OFDM symbols, as shown in Figure 14.12. The new symbol time

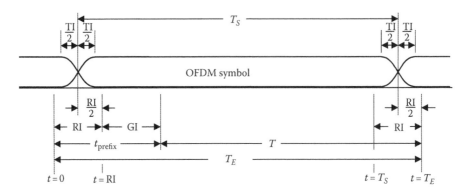

FIGURE 14.12
Consecutive OFDM symbols, GI, RI and windowing. (From HomePlug Alliance, HomePlug AV Specification Version 2.0, HomePlug Alliance, Copyright © January 2012. With permission.)

T_s is measured between the centres of the *roll-off intervals* (RIs) before and after the symbol. The overlap region is the RI (see Figure 14.12), which is related to the OFDM symbol duration T_s via parameter β

$$RI = \beta T_s \qquad (14.9)$$

and total symbol length T_E

$$T_E = (1 + \beta) \cdot T_s. \qquad (14.10)$$

The HomePlug AV specification allows the usage of several GIs [31,32]. If the shortest GI is selected, HomePlug AV uses a β of RI/(T + GI) = 4.96 μs/(40.96 + 5.56 μs) = 0.1066.

With HomePlug AV2, the TI is introduced. It is relevant for shaping the Tx implementation-dependent windowing, and it does not affect interoperability. The pulse-shaping window and GIs might even be reduced to zero to minimise overhead in the time domain and to efficiently notch. In order to guarantee backward compatibility with previous HomePlug versions, the definition and timings of the parameter RI as shown in Figure 14.12 must stay stable. In contrast to RI, the new parameter TI might be reduced to zero.

There is a balance between these intervals and obtaining a better side lobe attenuation.

Figure 14.13 shows the notch of any OFDM system spectrum that could be obtained by varying the roll-off factors (the β values) of the raised cosine window. Increasing the roll-off factor directly increases the amount of attenuation achieved inside the notches. However, this has the drawback of increased symbol length.

To create notches in the OFDM spectrum, a model is implemented using QAM modulation and notches of omitting a various number of carriers: 1–5 and 10. A max-hold function is implemented in order to create these figures. The 10-carrier notch shows the spectral benefit of windowing. The influence of windowing is hardly visible up to the point of the five-carrier notch. At the five-carrier notch, the difference between no window and the highest simulated β is around 5 dB. The trade-off for this β is an 11% longer symbol time. At the 10-carrier notch, the centre frequency is suppressed by around 15 dB without windowing and almost 30 dB if the HomePlug AV window is applied.

The additional overhead in the time domain is extremely large compared to the improved notch depth. Windowing with a small roll-off factor is sufficient in order to suppress the

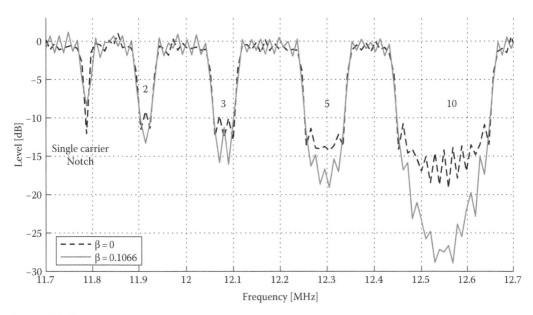

FIGURE 14.13
OFDM spectra with different notch widths and depths achieved with different roll-off.

side lobes outside the used spectrum, but is not recommended to increase the depth of a single carrier notch.

In the case of the HomePlug AV specification using a β of 0.1066, some guard carriers on the left and right side of a protected frequency range have to be omitted to guarantee the depth of the notch. The North American Carrier Mask requests 10 notches in between 1.7 and 30 MHz. The first carriers as well the last carriers of this spectrum are notched. These two notches have only one slope to the carriers allocated for communications. All other notches have 2 slopes, resulting in a total of 18 notch slopes. The spectrum loses almost 6% of communication resources in the frequency domain. Additionally, the β causes almost 11% of wasted resources in the time domain. Assuming an ideal implementation with maximal sharpness in time as well as frequency domain, it would be possible to regain these communication resources when applying the North American Carrier Mask. If the frequency spectrum becomes more fragmented because of additional notches like those requested by the newly upcoming European regulations [33], these losses will become even more obvious.

14.3.3.2 Digital, Adaptive Band-Stop Filters Improve the Notch's Depth and Slopes

Although an ideal implementation as described earlier is not possible, digital band-stop filters increase the sharpness of the notches as well as the implementation efforts in hardware. Shrinking the semiconductor manufacturing process to smaller structures, allowing integration of additional functions on the same die size, shifts the balance towards hardware implementation efforts.

The HomePlug AV2 specification gives maximum freedom to the chip implementer. The filter algorithm, order and structure are implementation dependent. An example is documented in [34]. The more efforts placed in the filtering processing, the better the sharpness of the notch slopes. This in turn leads to shorter GI lengths, and thus, more resources are made available for communication. A reduction of the GI down to zero (see Section 14.3.5.1) is possible at short PLC channels without multipath reflections or intersymbol interference.

14.3.4 HomePlug AV2 Power Optimisation Techniques

The HomePlug AV2 standard introduces two novel techniques that can be used to optimise the use of transmit power. The first one, 'transmit power back-off', is a technique that reduces the transmitted *power spectral density* (PSD) for a selected set of carriers when this can be done without adversely affecting performance. Conversely, the second, 'EMC-friendly power boost' is a technique that allows the Tx to increase the power on some carriers with the knowledge that this can be done without exceeding regulatory limits.

14.3.4.1 Power Back-Off

In PLCs, the transmit power limit is typically defined as a PSD mask applicable over the range of frequencies used in the standard. And since power line modems are directly connected to the electrical wiring, they are traditionally designed to transmit with the maximum allowed transmit PSD on each frequency (i.e. they do not need to be sensitive to a limited battery supply). In many cases, maximising transmit power leads to the best performance; however, certain definitions of PSD masks combined with certain channel conditions can produce cases where modems can benefit from transmitting at less than the maximum allowed power level.

We illustrate the benefits of transmit power control using the North American regulatory limits as an example. FCC regulations that are applicable to power line devices in North America are commonly interpreted to allow a transmit PSD of −50 dBm/Hz from 1.8 to 30 MHz and −80 dBm/Hz from 30 up to 86 MHz. More details about regulatory limits can be found in Chapter 6. This 30 dB drop in the PSD (at 30 MHz) causes the signal from the higher frequency carriers (above 30 MHz) to have much smaller amplitude than the signal for the low-band carriers (up to 30 MHz). Consequently, when the overall signal is represented in a quantised digital domain, the high-band signal has lower resolution than the signal in the lower band and will therefore also have a limited SNR. This will be evident at the Tx where the 30 dB drop will result in a reduction of 5 bit of resolution for the high-band signal. If the transmit power is backed off in the low band so that the PSD drop is reduced, then the high-band signal will be represented with an increased number of bits of resolution and enjoy increased SNRs out of the Tx.

Another limiting factor is the limited dynamic range of the *analogue to digital converter* (ADC). We illustrate its impact once again using the example of the North American PSD limits. For the sake of simplicity, we assume a flat power line channel and flat noise spectrum in Figure 14.14. For convenience, the power line channel (curves labelled Rx signal: the transmitted signal after channel attenuation) and noise (curves labelled Rx noise) contributions to the received signal are shown separately. Moreover, the different signals are shown before (dashed line curves) and after (continuous line curves) the analogue amplifier. On the left side of Figure 14.14, the scenario without power back-off is shown: the transmitted signal is 30 dB greater in the 1.8–30 MHz band compared to the 30–86 MHz band. The analogue amplifier brings the received signal to a Level A tailored to optimise the ADC conversion. Since the dominating noise is the ADC noise (black curve), after the ADC converter SNRs of 35 and 5 dB are found below and above 30 MHz, respectively. On the right side of Figure 14.14, the scenario with power back-off is shown: the transmitted signal is reduced by 10 dB in the 1.8–30 MHz frequency band so that it is only 20 dB greater compared to the 1.8–86 MHz band. Again, the analogue

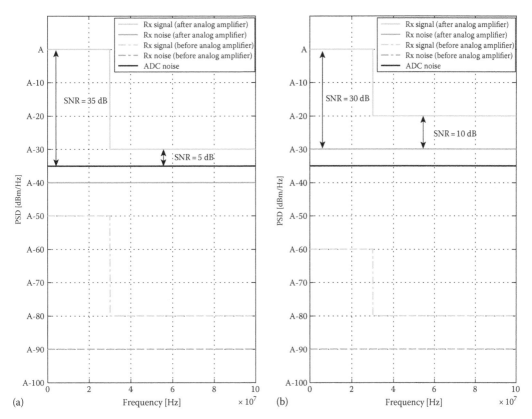

FIGURE 14.14
Benefits of power back-off. (a) No power back-off and (b) 10 dB power back-off. (From Yonge, L. et al., *Hindawi J. Electr. Comput. Eng.*, 2013, Article ID 892628, Copyright © 2013.)

amplifier brings the received signal to Level A. However, in this case, the dominating noise is no longer the ADC noise, and the obtained SNRs are 30 and 10 dB in the 1.8–30 and 30–86 MHz, respectively.

In this example, the power back-off technique results in a 5 dB SNR reduction for the carriers in the lower frequency band and a 5 dB SNR increase for the carriers in the upper frequency band. Given the larger bandwidth of the upper frequency band, there will be an overall throughput gain due to transmit power back-off.

Transmit power back-off is also an effective interference mitigation technique. For instance, in Europe, the ability of a PLC Tx to reduce the transmit power depending on the attenuation link is a possible requirement considered in [33], though the procedure described in CENELEC [33] does not consider QoS requirements of PLC modems.

14.3.4.2 EMC-Friendly Power Boost Using S_{11} Parameter

The EMC-friendly power boost is a mechanism introduced in the HomePlug AV2 specification to optimise the transmit power by monitoring the input port reflection coefficient at the transmitting modem. This coefficient is known as the S_{11} parameter. PSD limits as proposed in the HomePlug AV2 specification are based on representative statistics of the impedance match at the interface between the device port and the power line network. In practice, the input impedance of the power line network is frequency selective and varies for different network configurations. This leads to part of the transmit power being dissipated within the Tx. The input port reflection coefficient, or input return loss, is characterised by the S-parameter S_{11}, and the part of transmit power dissipated at the Tx is given by the amount $20 \cdot \log_{10}(|S_{11}|)$ in dB. In situations where the impedance mismatch is large (and thus the S_{11} parameter is large), only a small part of the input power is effectively transferred to the power lines. In those cases, the *electromagnetic interference* (EMI) induced by the PLC modem is reduced and can be much lower than the values recommended by EMC regulation limits.

In order to compensate for the frequency-selective impedance mismatch at the interface between the device port and the power line network, HomePlug AV2 modems adapt their transmission mask upon measurement of the S_{11} parameter at the Tx. The transmitted signal power is increased by an *impedance mismatch compensation* (IMC) factor, which leads to a more effective power transmission to the power line medium. While increasing the Tx power leads to an increase of the radiated EMI, the design of the IMC factor ensures that the resulting EMI continuously falls below the targeted EMC regulation limits.

A statistical analysis was conducted by the HomePlug Technical Working Group on the practical values of the S_{11} parameter and the effectiveness of the EMC-friendly power boost technique, based on a series of measurements performed by the ETSI Specialist Task Force 410 [13–17,35]. S_{11} parameter and EMI measurements were taken over the 1–100 MHz range in six countries: Germany, Switzerland, Belgium, the United Kingdom, France and Spain. The modem used for measurements is described in [13].

The S_{11} measurements considered in this study consist of three differential feeding possibilities (L–N, N–PE and PE–L) and one CM feeding. For the EMI measurements, we consider the measurements taken with different feeding possibilities, that is, differential feeding, with the other differential ports being unterminated or terminated with 50 Ω, and CM feeding.

The measurement set used in this analysis consists of 478 frequency sweeps that can be categorised as follows:

1. Six different locations in Germany and three different locations in France
2. 264 measurements outdoors at 10 m, 43 measurements outdoors at 3 m and 171 measurements indoors

Statistical analysis of this experimental data allowed us to design the practical implementation of the EMC-friendly power boost technique. In the following, we define the IMC factor as (in dB)

$$IMC(k) = \min\left(\max\left(10 \cdot \log_{10}\left(\frac{1}{1 - |S_{11}(k)|^2} \right) - M(k), 0 \right), IMC_{\max} \right), \tag{14.11}$$

where
$S_{11}(k)$ is the carrier-dependent estimate of the S_{11} parameter
M is a margin accounting for possible estimate uncertainties in the measurement of parameter S_{11}
$IMC_{\max}$ is the maximum allowed value of the IMC factor in dB

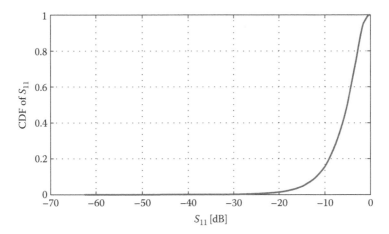

FIGURE 14.15
CDF of S_{11} parameter. (From Yonge, L. et al., *Hindawi J. Electr. Comput. Eng.*, 2013, Article ID 892628, Copyright © 2013.)

Figure 14.15 represents the statistical CDF of the S_{11} parameter as well as the corresponding IMC_0 factor, computed with a margin M of 0 dB. These statistics are based on the experimental data collected during the ETSI STF 410 measurement campaign.

In Figure 14.15, one can read that the S_{11} parameter is larger than −10 dB in 80% of the cases. This means that for 80% of the records, more than 10% of the energy is reflected back towards the Tx.

More interestingly, Figure 14.16 gives an idea of the potential power increase offered by the EMC-friendly power boost technique:

1. For 40% of the records, the Tx power could be increased by more than 2 dB to compensate for the impedance mismatch.
2. For 10% of the records, the Tx power could be increased by more than 4 dB to compensate for the impedance mismatch.

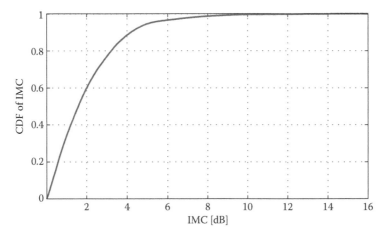

FIGURE 14.16
CDF of IMC parameter. (From Yonge, L. et al., *Hindawi J. Electr. Comput. Eng.*, 2013, Article ID 892628, Copyright © 2013.)

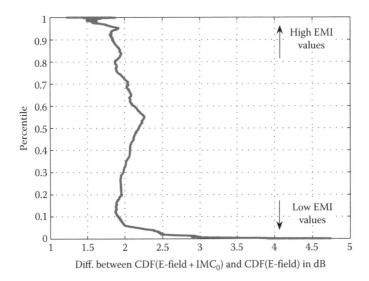

FIGURE 14.17
Difference in dB between the CDFs of the radiated EMI before and after applying the EMC-friendly power boost. (From Yonge, L. et al., *Hindawi J. Electr. Comput. Eng.*, 2013, Article ID 892628, Copyright © 2013.)

We then focused on the effect of the EMC-friendly power boost technique on the radiated EMI statistics. The recorded values allowed the computation of two statistics:

1. The statistical CDF of the recorded EMI in terms of E-field for all frequencies and feeding possibilities without applying any power boost
2. The statistical CDF of the recorded EMI in terms of E-field for all frequencies and feeding possibilities when applying the EMC-friendly power boost

Figure 14.17 presents, for each percentile of the CDF, the difference in dB between the E-field CDFs for the two methods of transmission.

Different conclusions can be drawn from this figure. First, the application of the EMC-friendly power boost leads to an increase of the radiated field CDF between 1.2 and 4.8 dB. Note that the extreme value of 4.8 dB arises for one of the lowest values of radiated field and hence is not relevant. Indeed, the lowest values of radiated fields are far from the limits imposed by regulation bodies. Therefore, even with an increase of a few dB, the radiated field will not exceed the limit. Secondly, in general, the application of the EMC-friendly power boost increases the radiated power CDF by about 2 dB. More importantly, the increase of the radiated power CDF is lower than 2 dB for the 25% most radiating cases. This practically means that in the worst-case scenarios where the modems produce the largest EMI, the application of the IMC factor does not increase the EMI by more than 2 dB. This value can be compared with the CDF of the IMC factor given in Figure 14.16. Although the IMC factor is larger than 2 dB in 40% of the cases and larger than 4 dB in 10% of the cases, the application of the EMC-friendly power boost technique does not increase the radiated power CDF extreme values by more than 2 dB. Of course, even an increase of the EMI by 2 dB is not acceptable. Therefore, a margin M of 2 dB is applied when increasing the Tx power using the IMC factor. In practice, this means that the Tx power will be boosted by the value indicated by the IMC factor, minus 2 dB. As a result, even in the most radiating cases, the application of the EMC power boost will not increase the perceived EMI.

Based on this study, we conclude that the application of the EMC-friendly power boost technique provides a significant gain in terms of transmit power increase for a large number of configurations, where the impedance mismatch causes the dissipation of the signal at the Tx. In addition, the statistical analysis shows that this technique will not lead to an increase of the undesirable radiated interference, in particular in the worst EMI scenarios, as long as a margin M of 2 dB is used in the computation of the IMC factor, as specified in Equation 14.11. Finally, a recommended limit for the maximum allowed value of the IMC factor is $IMC_{max} = 6$ dB.

14.3.5 Additional PHY Improvements

In addition to the MIMO technology, the frequency band extension, the efficient notching and power optimisation techniques such as the power back-off and the EMC-friendly power boost, other elements of the PHY layer were modified as presented in the following sections.

14.3.5.1 New Time Domain Parameters

In the HomePlug AV2 specification, a number of time domain parameters were refined. As the sampling frequency has increased from 75 to 200 MHz, the number of time samples for a given symbol duration is increased by a factor 8/3. The IFFT interval is 8192 samples in length, and the number of samples in the HomePlug AV GI has increased accordingly. In addition, new features have been added:

1. The TI defines the part of the RI dedicated to the transition window, allowing more flexibility in the choice of the window (see Section 14.3.3).
2. A new GI has been defined for the HomePlug AV2 short delimiter (see Section 14.4.2.1).
3. The payload symbol GI has been made variable and can be as short as 0 µs. It can also be increased up to 19.56 µs. This allows adaptation to a wide range of channel conditions and removes the overhead of the GI for channels that have either very low multipath dispersion or that are completely limited by the receive noise and not by ISI.

14.3.5.2 Additional Constellations

In HomePlug AV, the maximum constellation size is 1024-QAM, corresponding to 10 coded bits per carrier. HomePlug AV2 also provides support for 4096-QAM, which corresponds to 12 bit per carrier. The higher constellation size increases the peak PHY rates by 20%. Practically, the increased throughput will be available mostly on average to very good channels, but even some of the poorer channels sometimes have frequency bands in which high SNRs can be achieved, and thus, the increased constellation size can be used.

14.3.5.3 Forward Error Correction Coding

HomePlug AV2 uses the same duobinary turbo code as HomePlug AV. In addition to the code rates of 1/2 and 16/21, HomePlug AV2 also provides support for a 16/18 code rate. This allows more granularity in the compromise between robustness and throughput degradation. For this new code rate, a new puncturing structure and a new channel interleaver are defined. In addition, a new Physical Block size of 32 octets is defined, which includes specification of a new termination matrix for the *forward error correction* (FEC) as well as a new interleaver seed table. The 32 octet PBs are used in the PHY level acknowledgements and allow for the acknowledgement of much larger packet sizes that are supported with the increased PHY rates possible in HomePlug AV2.

14.3.5.4 Line Cycle Synchronisation

The HomePlug AV2 specification also describes the device operation in scenarios where there is no *alternating current* (AC) line cycle (e.g. a *direct current* [DC] power line) or when the AC line cycle is different from 50 or 60 Hz. In this case, the *central coordinator* (CCo) is preconfigured to select a beacon period matching either 50 Hz (i.e. beacon period is 40 ms) or 60 Hz (i.e. beacon period is 33.33 ms). One key use case where this feature is useful is the transfer of data towards a multimedia-equipped electrical vehicle during the electrical charging phase (using DC power).

14.4 MAC Layer Improvements of HomePlug AV2

14.4.1 Power Save Modes

HomePlug AV2 stations improve their energy efficiency in standby mode through the adoption of the specific power save mode already defined in the HomePlug Green PHY specification [3] (see Chapter 10 for details). In power save mode, stations reduce their average power consumption by periodically transitioning between awake and sleep states. Stations in the awake state can transmit and receive packets over the power line. In contrast, stations in sleep state temporarily suspend transmission and reception of packets over the power line.

We introduce some basic terms useful to describe the power save mode:

1. *Awake window*: period of time during which the station is capable of transmitting and receiving frames. The awake window has a range from a few milliseconds to several beacon periods (a beacon period is two times the AC line cycle: 40 ms for a 50 Hz AC line and 33.3 ms for a 60 Hz AC).

2. *Sleep window*: period of time during which the station is not capable of transmitting or receiving frames.

3. *Power save period (PSP)*: interval from the beginning of one awake window to the beginning of the next awake window. PSP is restricted to $2k$ multiples of beacon periods (i.e. one beacon period, two beacon periods, four beacon periods).

4. *Power save schedule (PSS)*: the combination of the values of the PSP and of the awake window duration. To communicate with a station in power save mode, other stations in the logical network (AVLN) need to know its PSS.

Potentially, the specification allows aggressive PSSs constituted by an awake window duration of 1.5 ms and a PSP of 1024 beacon period, which will result in over 99% energy savings compared to HomePlug AV. In practice, some in-home applications will require lower latency and response time, and a balance will take place reducing the mentioned gain. This is particularly appealing for applications that foresee a PLC utilisation that is variable during the day (e.g. large utilisation during daytime and small utilisation during nighttime). It is worth highlighting that the HomePlug AV2 specification is flexible in allowing each station in a network to have a different PSS. Given these remarks, in order to

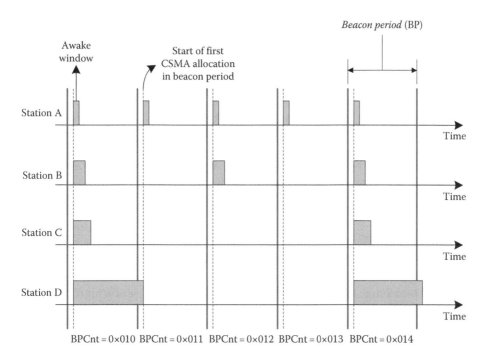

FIGURE 14.18
Example of power save operation in HomePlug AV2 and HPGP. (From HomePlug Alliance, HomePlug GreenPHY, The standard for in-home Smart Grid powerline communications, http://www.homeplug.org, Copyright © 2012.)

enable efficient power save operation without causing difficulties to regular communication, all the stations in a network need to know the PSSs of the other stations. The network CCo has a key role since it grants the requests of the different stations to enter and exit from power save mode operation. Moreover, it distributes the different PSSs to all the stations in the network. When needed, a CCo can

1. Optionally disable power save mode for all stations of the AVLN
2. Optionally wake up a station in power save mode

The shared knowledge of the PSS allows stations communicating during the common awake windows (the HomePlug AV2 and HPGP specifications have structured the protocols to ensure at least one superposition of all the awake windows occurs). This overlap interval can also be used for transmission of information that needs to be received by all stations within the AVLN.

Figure 14.18 shows an example of PSS of the four stations (A, B, C and D). All stations save more than 75% of energy compared to HomePlug AV. Note that in this example, stations A and B can communicate once every two beacon periods. Moreover, all the stations are always awake at the same time once every four beacon periods, thus preserving communication possibility.

14.4.2 Short Delimiter and Delayed Acknowledgement

The short delimiter and delayed acknowledgement features were added to HomePlug AV2 to improve efficiency by reducing the overhead associated with transmitting payloads over the power line channel. In HomePlug AV, this overhead results in relatively poor efficiency for *transmission control protocol* (TCP) payloads. One goal that was achieved with these new features was TCP efficiency that improved to be relatively close to that of UDP.

In order to send a packet-carrying payload data over a noisy channel, signalling is required for an Rx to detect the beginning of the packet and to estimate the channel so that the payload could be decoded, and additional signalling is needed to acknowledge the payload was received successfully. Interframe spaces are also required between the payload transmission and the acknowledgement for processing time reasons. Indeed, the interframe duration covers the time needed at the Rx to decode the payload, to check accurate reception and to encode the acknowledgement message. This overhead is even more significant for TCP payload, since the TCP acknowledgment payload must be transmitted in the reverse direction.

14.4.2.1 Short Delimiter

The delimiter specified in HomePlug AV contains the preamble and FC symbols and is used for the beginning of data PPDUs as well as for immediate acknowledgements. The length of the HomePlug AV delimiter is 110.5 µs and can represent a significant amount of overhead for each channel access. A new single OFDM symbol delimiter is specified in HomePlug AV2 to reduce the overhead associated with delimiters, by reducing the length to 55.5 µs. Figure 14.19 shows that every fourth carrier in the first OFDM symbol is assigned as a preamble carrier, and the remaining carriers encode the FC. The following OFDM symbols encode data the same as in HomePlug AV.

Figure 14.20 demonstrates the efficiency improvement when the HomePlug AV2 short delimiter is used for the acknowledgement of a CSMA long MPDU, compared to the HomePlug AV delimiter. Not only is the length of the delimiter reduced from 110.5 to 55.5 µs, the *response interframe space* (RIFS) and *contention interframe space* (CIFS) can also be reduced to 5 and 10 µs (which is around the fifth part of HomePlug AV's timing), respectively. Reduction of RIFS requires delayed acknowledgement, which is described in Section 14.4.2.2. Backward compatibility when contending with HomePlug AV devices is maintained by indicating the same length field for virtual carrier sense in both cases, so that the position of the *priority resolution symbols* (PRS) contention remains the same. A field in the FC of the long MPDU indicates the short delimiter format to a HomePlug AV2 device so that it can correctly determine the length of the payload.

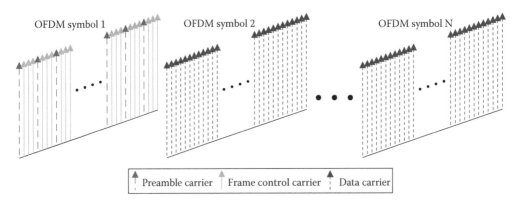

FIGURE 14.19
Short delimiter. (From Yonge, L. et al., *Hindawi J. Electr. Comput. Eng.*, 2013, Article ID 892628, Copyright © 2013.)

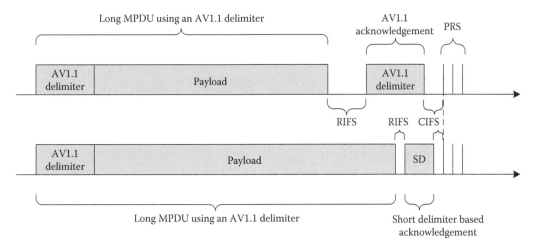

FIGURE 14.20
Short delimiter efficiency improvement. (From Yonge, L. et al., *Hindawi J. Electr. Comput. Eng.*, 2013, Article ID 892628, Copyright © 2013.)

14.4.2.2 Delayed Acknowledgement

The processing time to decode the last OFDM symbol and encode the acknowledgement can be quite high, thus requiring a rather large RIFS. In HomePlug AV, since the preamble is a fixed signal, the preamble portions of the acknowledgement can be transmitted while the Rx is still decoding the last OFDM symbol and encoding the payload for the acknowledgement. With the short delimiter, the preamble is encoded in the same OFDM symbol as the FC for the acknowledgement, so the RIFS would need to be larger than for HomePlug AV, eliminating much of the gain the short delimiter provides.

Delayed acknowledgement solves this problem by acknowledging the segments ending in the last OFDM symbol in the acknowledgement transmission of the next PPDU, as shown in Figure 14.21. This permits practical implementation with a very small RIFS, reducing the RIFS overhead close to zero. HomePlug AV2 also allows the option of delaying acknowledgement for segments ending in the second to the last OFDM symbol to provide flexibility for implementers.

14.4.3 Immediate Repeating

HomePlug AV2 supports repeating and routing of traffic not only to handle hidden nodes but also to improve coverage (i.e. performance on the worst channels).

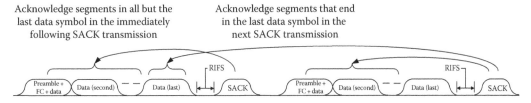

FIGURE 14.21
Delayed acknowledgement. (From Yonge, L. et al., *Hindawi J. Electr. Comput. Eng.*, 2013, Article ID 892628, Copyright © 2013.)

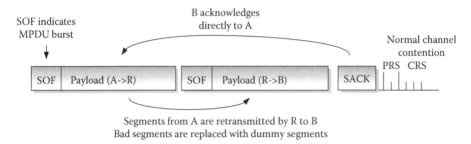

FIGURE 14.22
Immediate repeating channel access for CSMA. (From Yonge, L. et al., *Hindawi J. Electr. Comput. Eng.*, 2013, Article ID 892628, Copyright © 2013.)

With HomePlug AV2, hidden nodes are extremely rare. However, some links may not support the data rate required for some applications, such as a 3D HD video stream. In a network where there are multiple HomePlug AV2 devices, the connection through a repeater typically provides a higher data rate than the direct path for the poorest 5% of channels.

Immediate repeating is a new feature in HomePlug AV2 that enables highly efficient repeating. Immediate repeating provides a mechanism to use a repeater with a single channel access, and the acknowledgement does not involve the repeater. This is shown in Figure 14.22, where station A transmits to repeater R. In the same channel access, repeater R transmits all payload received from station A to station B. B sends an acknowledgement directly to A. With this approach, latency is actually reduced with repeating, assuming the resulting data rate is higher, the obvious criterion for using repeating in the first place. Also, resources required by the repeater are minimised since the repeater uses and immediately frees memory it would require for receiving payload destined for it. Also, there is no retransmission responsibility for failed segments.

14.5 Gain of HomePlug AV2 Compared to HomePlug AV

The AV TWG has evaluated the performance of the HomePlug AV2 specification: this activity was required in order to see if the produced specification would meet the requirements of all stakeholders. The following tables show the performance improvement as compared to HomePlug AV in terms of coverage. These preliminary results are based on a six home field test in Florida, with home sizes 170–300 m². Table 14.1 presents the results in a two-node network scenario: 95% of nodes experienced a throughput improvement greater than 136% compared to HomePlug AV (which means a performance enhancement by a factor of nearly 2.4!). Benefits are even higher when considering the most favourable connections (see the improvement at the 5% coverage value). Table 14.2 considers a four-node scenario where three streams carrying different data are transmitted from one source (e.g. a set-top box) to three different destinations (e.g. TVs). In this case, the greater than 131% improvement in the aggregate throughput is relevant for 99% of networks compared to HomePlug AV.

TABLE 14.1

Improvement of HomePlug AV2 in a 2-Node Network

Coverage Based on UDP Throughput (%)	Percentage of Throughput Improvement of HomePlug AV2 Compared to HomePlug AV (%)
95	>136
5	>220

Source: Yonge, L. et al., *Hindawi J. Electr. Comput. Eng.*, 2013, Article ID 892628, Copyright © 2013.

TABLE 14.2

Improvement of HomePlug AV2 in a 4-Node Network

Coverage Based on UDP Throughput (%)	Percentage of Throughput Improvement of HomePlug AV2 Compared to HomePlug AV (%)
99	>131
5	>173

Source: Yonge, L. et al., *Hindawi J. Electr. Comput. Eng.*, 2013, Article ID 892628, Copyright © 2013.

TABLE 14.3

Maximum PHY Rate Computation

System Configuration (North American Tone Mask)	Max PHY Rate (Mbps)
HomePlug AV (1.8–30 MHz)	197
(917 carriers, 10 bit per carrier, 5.56 μs GI)	
IEEE 1901 (1.8–50 MHz)	556
(1974 carriers, 12 bit per carrier, 1.6 μs GI)	
HomePlug AV2 SISO (1.8–86.13 MHz)	1012
(3455 carriers, 12 bit per carrier, 0.0 μs GI)	
HomePlug AV2 MIMO (1.8–86.13 MHz)	2024
(3455 carriers, 12 bit per carrier, 0.0 μs GI, 2 streams)	

Source: Yonge, L. et al., *Hindawi J. Electr. Comput. Eng.*, 2013, Article ID 892628, Copyright © 2013.

Note that the benefits of the HomePlug AV2 technology are expected to be greater than the ones shown in Tables 14.1 and 14.2 since, for instance, a 2 × 2 MIMO was tested in Florida. A 2 × 3 or 2 × 4 MIMO would likely provide better performance.

Another interesting figure is the theoretical maximum PHY throughput for the system for different options of the standard (Table 14.3). This number represents the throughput of transmitted bits on the PHY layer for optimum channel conditions and gives an idea of the benefits of different features. It can be seen that if the full frequency range is used, HomePlug AV2 provides a 1 Gbps throughput in SISO configuration and 2 Gbps in a MIMO configuration.

14.6 Conclusion

In this chapter, an overview of HomePlug AV2 has been presented. The overall system architecture and the key technical HomePlug AV2 improvements introduced at PHY and MAC layers have been described. It has also been shown that the related performance improvements were achieved by HomePlug AV2 while ensuring backward compatibility with HomePlug AV. In addition, the coexistence with other power line technologies is ensured through the use of the Inter System Protocol (see Chapter 10).

The HomePlug AV2 performance presented in this work has been assessed by AV TWG through simulations based on field measurements.

The results show the significant benefits introduced by the new set of HomePlug AV2 features, both in terms of achievable data rate and coverage.

References

1. HomePlug Alliance, http://www.homeplug.org, accessed 15 October 2013.
2. HomePlug Alliance, HomePlug AV Specification Version 2.0, HomePlug Alliance, January 2012.
3. HomePlug Alliance, HomePlug GreenPHY, The Standard for In-Home Smart Grid Powerline Communications, http://www.homeplug.org, 2010.
4. IEEE Standard 1901-2010, IEEE standard for broadband over power line networks: Medium access control and physical layer specifications, http://standards.ieee.org/findstds/standard/1901-2010.html (accessed 15 October 2013), 2010.
5. HomePlug Alliance, HomePlug AV White Paper, http://www.homeplug.org.
6. R. Hashmat, P. Pagani, A. Zeddam and T. Chonavel, MIMO communications for inhome PLC networks: Measurements and results up to 100 MHz, in *Proceedings of the IEEE International Symposium on Power Line Communications and Its Applications (ISPLC'10)*, Rio de Janeiro, Brazil, March 2010, pp. 120–124.
7. R. Hashmat, P. Pagani, A. Zeddam and T. Chonavel, A channel model for multiple input multiple output in-home power line networks, in *Proceedings of the IEEE International Symposium on Power Line Communications and Its Applications (ISPLC'11)*, Udine, Italy, April 2011, pp. 35–41.
8. R. Hashmat, P. Pagani, A. Zeddam and T. Chonavel, Analysis and modeling of background noise for in-home MIMO PLC channels, in *Proceedings of the IEEE International Symposium on Power Line Communications and Its Applications (ISPLC'12)*, Beijing, China, March 2012, pp. 120–124.
9. R. Hashmat, P. Pagani, T. Chonavel and A. Zeddam, A time domain model of background noise for in-home MIMO PLC networks, *IEEE Transactions on Power Delivery*, 27(4), 2082–2089, 2012.
10. D. Rende, A. Nayagam, K. Afkhamie et al., Noise correlation and its effect on capacity of inhome MIMO power line channels, in *Proceedings of the IEEE International Symposium on Power Line Communications and Its Applications (ISPLC'11)*, Udine, Italy, April 2011, pp. 60–65.
11. D. Veronesi, R. Riva, P. Bisaglia et al., Characterization of in-home MIMO power line channels, in *Proceedings of the IEEE International Symposium on Power Line Communications and Its Applications (ISPLC'11)*, Udine, Italy, April 2011, pp. 42–47.
12. F. Versolatto and A. M. Tonello, A MIMO PLC random channel generator and capacity analysis, in *Proceedings of the IEEE International Symposium on Power Line Communications and Its Applications (ISPLC'11)*, Udine, Italy, April 2011, pp. 66–77.

13. ETSI TR 101 562, PowerLine Telecommunications (PLT); MIMO PLT; Part 1: Measurement Methods of MIMO PLT, http://www.etsi.org/deliver/etsi_tr/101500_101599/10156201/01.0 3.01_60/tr_10156201v010301p.pdf (accessed 15 October 2013), 2012.

14. ETSI TR 101 562, PowerLine Telecommunications (PLT); MIMO PLT; Part 2: Setup and Statistical Results of MIMO PLT EMI Measurements, http://www.etsi.org/deliver/etsi_tr/101500_1015 99/10156202/01.02.01_60/tr_10156202v010201p.pdf (accessed 15 October 2013), 2012.

15. ETSI TR 101 562, PowerLine Telecommunications (PLT); MIMO PLT; Part 3: Setup and Statistical Results of MIMO PLT Channel and Noise Measurements, http://www.etsi.org/ deliver/etsi_tr/101500_101599/10156203/01.01.01_60/tr_10156203v010101p.pdf (accessed 15 October 2013), 2012.

16. P. Pagani, R. Hashmat, A. Schwager, D. Schneider and W. Bäschlin, European MIMO PLC field measurements: Noise analysis, in *Proceedings of the IEEE International Symposium on Power Line Communications and Its Applications (ISPLC'12)*, Beijing, China, March 2012.

17. D. Schneider, A. Schwager, W. Bäschlin and P. Pagani, European MIMO PLC field measurements: Channel analysis, in *Proceedings of the IEEE International Symposium on Power Line Communications and Its Applications (ISPLC'12)*, Beijing, China, March 2012.

18. Schneider, D., Inhome power line communications using multiple input multiple output principles, Doctoral thesis, University of Stuttgart, Stuttgart, Germany, 2012.

19. A. Paulraj, R. Nabar and D. Gore, *Introduction to Space-Time Wireless Communications*, Cambridge University Press, New York, 2003.

20. L. Stadelmeier, D. Schneider, D. Schill, A. Schwager and J. Speidel, MIMO for Inhome power line communications, in *Proceedings of the Seventh International ITG Conference on Source and Channel Coding (SCC'08)*, Ulm, Germany, January 2008.

21. A. Canova, N. Benvenuto and P. Bisaglia, Receivers for MIMO PLC channels: Throughput comparison, in *Proceedings of the IEEE International Symposium on Power Line Communications and Its Applications (ISPLC'10)*, Rio de Janeiro, Brazil, March 2010, pp. 114–119.

22. S. Katar, B. Mashburn, K. Afkhamie, H. Latchman and R. Newman, Channel adaptation based on cyclo-stationary noise characteristics in PLC systems, in *Proceedings of the IEEE International Symposium on Power Line Communications and Its Applications (ISPLC'06)*, Orlando, FL, March 2006, pp. 16–21.

23. A. M. Tonello, A. Cortés and S. D'Alessandro Optimal time slot design in an OFDM-TDMA system over power-line time-variant channels, in *Proceedings of the IEEE International Symposium on Power Line Communications and Its Applications (ISPLC'09)*, Dresden, Germany, April 2009, pp. 41–46.

24. A. Scaglione, P. Stoica, S. Barbarossa, G. B. Giannakis and H. Sampath, Optimal designs for space-time linear precoders and decoders, *IEEE Transactions on Signal Processing*, 50(5), 1051–1064, 2002.

25. Alamouti, S. M., A simple transmit diversity technique for wireless communications, *IEEE Journal on Selected Areas in Communications*, 16(8), 1451–1458, 1998.

26. R. A. Horn and C. R. Johnson, *Matrix Analysis*, Cambridge University Press, New York, 1985.

27. A. Lozano, A. M. Tulino and S. Verdu, Optimum power allocation for parallel Gaussian channels with arbitrary input distributions, *IEEE Transactions on Information Theory*, 52(7), 3033–3051, 2006.

28. L. Yonge, J. Abad, K. Afkhamie, L. Guerrieri, S. Katar, H. Lioe, P. Pagani, R. Riva, D. Schneider, and A. Schwager, An overview of the HomePlug AV2 technology, *Hindawi Journal of Electrical and Computer Engineering*, 2013, Article ID 892628, 2013.

29. S. D'Alessandro, A. M. Tonello and L. Lampe, Adaptive pulse-shaped OFDM with application to in-home power line communications, *Telecommunications Systems Journal*, 51(1), 3–13, 2011.

30. Harris, F. J., On the use of windows for harmonic analysis with the discrete Fourier transform, *Proceedings of the IEEE*, 66(1), 51–83, 1978.

31. K. H. Afkhamie, H. Latchman, L. Yonge, T. Davidson and R. Newman, Joint optimization of transmit pulse shaping, guard interval length, and receiver side narrow-band interference mitigation in the HomePlugAV OFDM system, in *Proceedings of the IEEE Sixth Workshop on Signal Processing Advances in Wireless Communications (SPAWC'05)*, New York, June 2005, pp. 996–1000.

32. A. M. Tonello, S. D'Alessandro and L. Lampe, Cyclic prefix design and allocation in bit-loaded OFDM over power line communication channels, *IEEE Transactions on Communications*, 58(11), 3265–3276, 2010.

33. CENELEC, Power line communication apparatus used in low voltage installations – Radio disturbance characteristics – Limits and methods of measurement – Part 1: Apparatus for in-home use, Final Draft European Standard FprEN 50561-1, June 2011.

34. Schwager, A., Powerline communications: Significant technologies to become ready for integration, Doctoral thesis, University of Duisburg-Essen, Essen, Germany, 2010.

35. A. Schwager, W. Bäschlin J. L. Gonzalez Moreno et al., European MIMO PLC field measurements: Overview of the ETSI STF410 campaign & EMI analysis, in *Proceedings of the IEEE International Symposium on Power Line Communications and Its Applications (ISPLC'12)*, Beijing, China, March 2012.

15

IEEE 1905.1: Convergent Digital Home Networking[*]

Etan G. Cohen, Duncan Ho, Bibhu P. Mohanty and Purva R. Rajkotia

CONTENTS

---

[*] IEEE 1905.1-2013. Copyright © 2013 IEEE. All rights reserved. Reprinted with permission from IEEE.

15.1 Introduction

With the increase of bandwidth-intensive home networking applications and consumers' endless appetite for accessing services anytime, anywhere, home networking technologies have become the latest frontier in the evolution of service delivery.

Both wired and wireless home networking technologies have significant market presence due to the value they create for end users, in terms of coverage and performance. Wireless networks offer mobility, and wired technologies offer extensive bandwidth and outlet ubiquity for transmissions. Wired and wireless technologies complement each other to provide full home coverage.

To address a wide variety of applications, regions, environments and topologies, multiple connectivity technologies must be used. However, no standard solution exists for creating networks that fully utilise all of the available technologies concurrently to provide higher bandwidth and robust transmissions. Over the last 10 years, more than one billion home networking devices have been deployed in the market, and any proposed solution must therefore interoperate with this deployment.

A typical hybrid home network is illustrated in Figure 15.1. As with any network deployment, many problems need to be addressed for the network to deliver performance to end users. In fact, these problems are more challenging in the context of home networks due to the fact that the end user may not be tech savvy; additionally, carriers and retailers are looking for solutions that do not require a home visit from qualified technical support

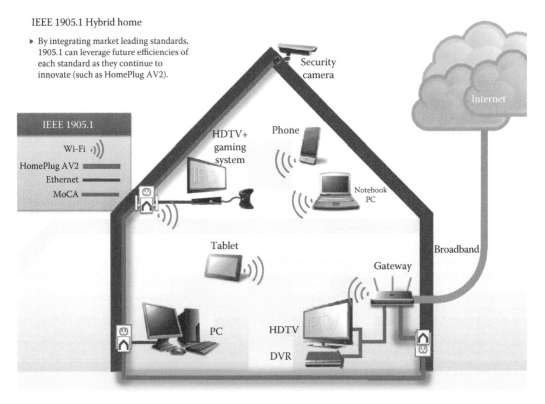

FIGURE 15.1
Illustration of a typical hybrid home network.

personnel and that minimise return of devices. These solutions thus must be simple and secure and provide required performance – all with minimal or no user intervention.

The problems to be addressed can be broadly categorised as setup/installation, configuration, performance management and maintenance/troubleshooting. Although core solutions and algorithms may be vendor specific for product differentiation, standardised inter-device interfaces are preferable, so networks consisting of hybrid devices from multiple vendors can implement effective solutions. As described in the following sections, IEEE 1905.1 specifies a set of interfaces which enable vendor-specific implementation of solutions.

Section 15.2 describes the IEEE 1905.1 architecture and basic message-handling procedures. Section 15.3 specifies the topology discovery protocol, which can be used by implementations for setup, operations and maintenance. Section 15.4 specifies the link metrics and how they are distributed for network algorithms to use for various purposes, such as achieving application performance. Section 15.5 describes the security procedures for easy configuration across multiple technologies without any changes in the underlying technology specifications. Section 15.6 provides data model specifications which can be useful for provisioning and maintenance. Finally, the coverage and capacity benefits of a power line and Wi-Fi hybrid network are illustrated in Section 15.7.

15.2 IEEE 1905.1

15.2.1 IEEE 1905.1 Abstraction Layer

IEEE 1905.1 defines an *abstraction layer* (AL) (1905.1 AL) for multiple home networking technologies that provides a common interface to the prevalent home networking technologies deployed in the field today: power line communications (IEEE 1901), Wi-Fi (IEEE 802.11), Ethernet (IEEE 802.3) and MoCA 1.1 (see Figure 15.2). IEEE 1901 specifies data communication over power line media, whereas MoCA 1.1 over coaxial cables. MoCA is especially

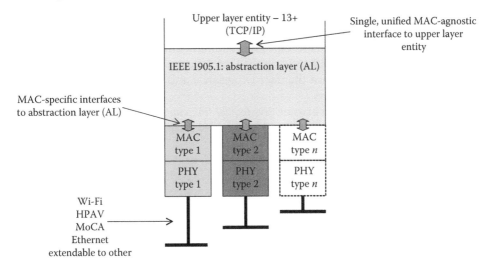

FIGURE 15.2
Overview of IEEE 1905.1 abstraction layer. (Reprinted with permission from IEEE 1905.1-2013, 1, IEEE standard for a convergent digital home network for heterogeneous technologies, 2013, P1905.1™/D08 draft. Copyright © 2013 IEEE.)

interesting for US homes, where most rooms are supplied with a coaxial cable. The design of IEEE 1905.1 is flexible and scalable to accommodate future home networking technologies.

The 1905.1 AL supports interface selection for the transmission of packets arriving from any interface or application.

The 1905.1 layer does not require modification of the underlying home networking technologies and hence does not change the behaviour or implementation of existing home networking technologies.

The 1905.1 layer benefits all the players in the service connectivity ecosystem – end users, service providers, chip suppliers and *original equipment manufacturers* (OEMs):

- For end users, 1905.1 specifies protocols and guidelines which provide a simplified user experience to add devices to the network, to set up security and to extend network coverage and application performance.

- For service providers, 1905.1 provides network management features to enable enhanced features for neighbour discovery, topology discovery, path selection, *quality of service* (QoS), network control and management.

- For chip suppliers and OEMs, the 1905.1 AL provides quick delivery of interoperable enhanced user experience products to the market and improved performance for the applications they support.

The IEEE 1905.1 AL addresses the following:

- Facilitates the installation and the operation/management of multi-connectivity technology devices in-home networks

- Improves the end-user experience by providing higher and more reliable bandwidth

- Provides a scalable and reusable specification for existing and future technologies

- Enhances the opportunity to use QoS-sensitive applications over a multi-technology home network

IEEE 1905.1 greatly enhances the networking experience for consumers by simplifying network installation, while at the same time providing seamless user experience. A broad base of industry-leading chipmakers, equipment manufacturers and service providers are collaborating to bring IEEE 1905.1 to fruition. This widespread support demonstrates the technology's potential to enhance the user experience and enable the next generation of connected services for consumers.

Technically, an AL is a layer between the *logical link control* (LLC) and one or multiple *media access control* (MAC) *service access points* (SAPs) of 1905.1 supported MAC/*physical* (PHY) standards.

The 1905.1 AL appears to the LLC as a single MAC, as it features a single MAC address and hides the heterogeneous nature of the home network. It, however, does export the underlying link metrics to allow high-level application to perform path selection.

The native bridging function of the underlying MAC is untouched by the inclusion of an AL.

The 1905.1 AL within an IEEE 1905.1 device uses an *extended unique identifier* (EUI)-48 value (an IEEE-defined EUI-48, 1905.1 AL MAC address) for identification. This 1905.1 AL MAC address may be used as a source or *destination address* (DA) for data and *control message data units* (CMDUs) originating or destined for the IEEE 1905.1 device, respectively.

Each 1905.1 AL shall locally administer its 1905.1 AL MAC address so that it does not conflict with any other MAC address or 1905.1 AL MAC address in the 1905.1 network to which it connects.

15.2.2 Architecture

The 1905.1 AL is an intermediate layer between the LLC L2 layer and underlying MAC layer(s) as illustrated in Figure 15.3.

The 1905.1 AL abstracts the heterogeneous MAC and PHY technologies of the converged home network by creating a single virtual MAC on top of the underlying MAC/PHY of the respective network technologies.

The 1905.1 AL provides SAPs to higher layers:

- For the data plane, a 1905.1 MAC SAP to the LLC
- For the management plane, a 1905.1 *abstraction layer management entity* (ALME) *Control* (CTRL) SAP to invoke the AL management functions

The 1905.1 AL is able to forward 802.3 *MAC protocol data units* (MPDUs) between

- The 1905.1 MAC SAP and the underlying 1905.1 interfaces
- The underlying 1905.1 interfaces

The forwarding entity, if present, shall be interoperable with IEEE 802.1 bridging among various networks inside or outside the home.

A 1905.1 network is formed by IEEE 1905.1 devices interconnected through 1905.1 links as illustrated in Figure 15.4.

CMDUs are exchanged between 1905.1 ALs. All CMDUs are received by neighbouring 1905.1 ALs. Some types of received CMDUs are relayed to other 1905.1 ALs.

CMDUs are used to carry 1905.1 protocol *type–length–value* (TLV) entries from a transmitting IEEE 1905.1 device to one or more receiving IEEE 1905.1 devices (depending on

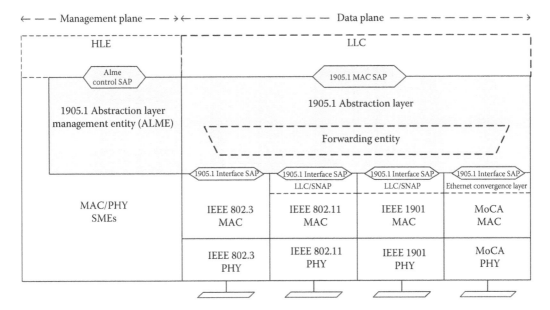

FIGURE 15.3
1905.1 abstraction layer model. (Reprinted with permission from IEEE 1905.1-2013, 1, IEEE standard for a convergent digital home network for heterogeneous technologies, 2013, P1905.1™/D08 draft. Copyright © 2013 IEEE.)

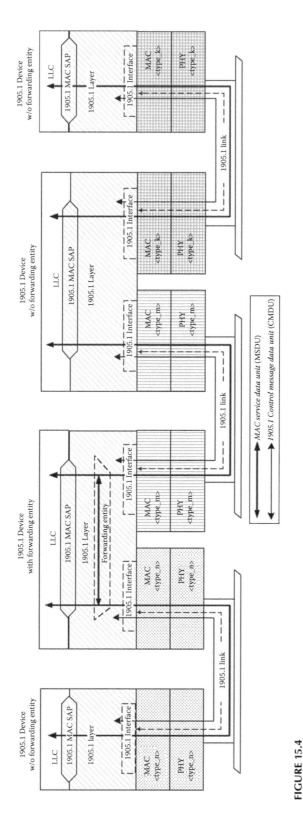

FIGURE 15.4
IEEE 1905.1 devices interconnected. (Reprinted with permission from IEEE 1905.1-2013, 1, IEEE standard for a convergent digital home network for heterogeneous technologies, 2013, P1905.1™/D08 draft. Copyright © 2013 IEEE.)

whether the DA is a unicast address or a group address). If a message is too large to fit within an Ethernet frame, multiple fragments can be created at the TLV boundaries to form multiple messages. More details on the contents of the CMDU fields are given in the IEEE 1905.1 specification [1].

Ethernet *LLC data units* (LLCDUs) are exchanged between 1905.1 ALs.

15.2.3 CMDU Transmission

The IEEE 1905.1 protocol makes use of CMDUs that are transmitted either with a unicast DA or with the 1905.1 multicast address, 01-80-C2-00-00-13. This multicast address is in a range defined by 802.1D/Q compliant bridges as forwarding – that is, such CMDUs are never blocked by 802.1D/Q bridges.

Additionally, the IEEE 1905.1 message header defines a relay indicator bit. There are three types of IEEE 1905.1 control CMDU propagation behaviours:

1. *Point to point*: Such CMDUs are addressed with a unicast DA.

2. *Advertisement to neighbours*: Such CMDUs are addressed with the 1905.1 multicast address as the DA and the relay indicator bit set to 0.

3. *Advertisement to the whole network*: Such CMDUs are addressed with the 1905.1 multicast address as the DA and the relay indicator bit set to 1.

When an IEEE 1905.1 device receives a CMDU, it checks to see if the CMDUs have not been previously processed using the information contained in the CMDU: message ID and the source of the message. If this is a new CMDU (to this device), then the device processes the CMDU and inspects the relay indicator bit. If the relay indicator bit is 1, then the device must retransmit this CMDU to the rest of the network – that is, on all of its interfaces – except the one on which it arrived.

15.3 Topology Discovery Protocol

The IEEE 1905.1 AL provides services to higher-level applications which enable them to manage networks of devices with multiple interfaces spanning different technology types. Such higher-level applications may be network management applications or applications optimising network traffic, which monitor available paths to destinations and their capacities, and route traffic optimally. One of the services required for such applications is topology discovery.

15.3.1 Topology Discovery

The IEEE 1905.1 AL provides the following information to applications:

- What are the IEEE 1905.1 devices in the network?

- What type of interfaces does each of the IEEE 1905.1 devices contain, and what networks are they connected to? Note: The network identification is medium specific, for example, *service set identifier* (SSID) for Wi-Fi networks.

- How are these IEEE 1905.1 devices connected to their neighbour IEEE 1905.1 devices, and are these connections direct or through 802.1D/Q compliant bridges?

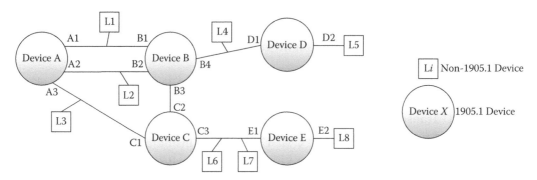

FIGURE 15.5
Sample topology.

The topology discovery process involves the following building blocks, which are described in detail in subsequent paragraphs:

- Devices must advertise themselves by multicasting a topology discovery message.
- Devices may query other devices using a topology query message.
- Devices must respond to queries using a topology response message.
- Devices must send notifications using a topology notification message if any of their local topology information changes.

These mandatory requirements enable other devices to discover a particular IEEE 1905.1 device. The optional elements of topology discovery enable an application on a device to discover as much or as little of the network topology as required for its function – with a corresponding difference in network overhead and processing effort. For example, a management application may be required to discover the full network topology, whereas a bandwidth management application may need to discover the topology of the immediate neighbours of this device. The details and examples of both are described next using the sample topology shown in Figure 15.5.

15.3.1.1 Topology Discovery Message

Each IEEE 1905.1 device must regularly (at least every 60 s) transmit a topology discovery message on each of its interfaces. The messages are transmitted as an advertisement to neighbours – that is, using the 1905.1 multicast address with the relay indicator bit set to 0. Each topology discovery message contains two TLVs:

- TLV identifying the transmitting device (using its AL MAC address)
- TLV identifying the interface on which the message is transmitted (using its MAC address)

For example, Device B of Figure 15.5 transmits a set of four topology discovery messages – one on each of its interfaces. Each message contains two TLVs: the first identifying Device B and the second identifying the interface on which that message is transmitted, that is, B1, B2, B3 or B4.

Devices receiving a topology discovery message can then deduce the following:

- They have a direct IEEE 1905.1 device as a neighbour: Topology discovery messages are not relayed by other IEEE 1905.1 devices.
- The connectivity to the IEEE 1905.1 device, by looking at which interface it was transmitted from and which local interface it was received on.

Using the previous sample topology, Device C, upon receiving a topology discovery message on its C2 interface transmitted from Device B on its B3 interface, can deduce that Device B is a direct neighbour and that its C2 interface is connected to Device B's B3 interface.

Whenever a device transmits a topology discovery message on any of its interfaces, it also transmits a standard 802.1AB *Link Layer Discovery Protocol* (LLDP) message. Standard compliant 802.1D or 802.1Q bridges will bridge IEEE 1905.1 topology discovery messages, but will not bridge 802.1AB messages. Thus, a device receiving a 1905.1 topology discovery message may deduce the presence or absence of an 802.1D/Q bridge between the neighbour IEEE 1905.1 device and itself on that particular interface.

Topology discovery messages are lightweight: they contain minimal information and are relatively small.

15.3.1.2 Topology Query/Response Messages

IEEE 1905.1 devices may use the topology query message to request more detailed information from any other IEEE 1905.1 device in the network (not just their neighbours). An IEEE 1905.1 device receiving such a topology query must respond with a topology response. Both topology query and topology response are unicast messages – that is, point-to-point messages.

The topology response message contains the following TLVs:

- TLV containing a list of all local interfaces, their types and the networks to which they are connected (specified as SSID for Wi-Fi networks and *network identifier* [NID] for IEEE 1901 networks).
- TLV specifying the set of IEEE 1905.1 neighbour devices connected to each of the interfaces of this device.
- TLV specifying the set of non-IEEE 1905.1 neighbour devices connected to each of the interfaces of this device. The mechanism to identify such neighbours may include address learning.
- TLV specifying the bridging behaviour of the device: The set of interfaces that acts as a broadcast domain – where a broadcast (or multicast) packet received on one interface is flooded to the others.

For example, Device B of Figure 15.5 may send a topology query message to Device D. Device D must respond with a topology response, which includes the following TLVs:

- *Local interface information*: A list of D1 and D2 with information about their type (e.g. Wi-Fi or IEEE 1901)
- *List of IEEE 1905.1 neighbours per interface*: (D1, {Device B}), (D2, {null}) – that is, from D1, it sees Device B as a direct IEEE 1905.1 neighbour, and from D2, it does not have IEEE 1905.1 neighbours.

- *List of non-IEEE 1905.1 neighbours per interface*: (D2, {L5}), (D1, {L1,L2,L3,L4,L6,L7,L8}). The actual list of non-IEEE 1905.1 neighbours seen by Device D on its D1 interface could be smaller depending on the overall bridging topology of the network – that is, if all packets from all these devices are bridged to it.
- *Bridging behaviour of device*: (D1, D2) – Device D (in most likelihood) is a bridge between its two interfaces, so any broadcast packet arriving on one is flooded to the other.

Note that devices do not have to send topology query messages, though they do have to respond to them. Devices may query none, some or all other IEEE 1905.1 devices in the network, depending on their requirements.

Topology query messages are small. Topology response messages are heavyweight – they may contain significant amounts of topology information.

15.3.1.3 Topology Notification Message

Finally, IEEE 1905.1 provides a mechanism for devices to inform other devices in the network of any changes in their topology information. The purpose of this message is to eliminate the need for polling for topology information. Whenever an IEEE 1905.1 device learns of any change to the information it has sent (or would have sent) in a topology response message, it must send a topology notification message. Topology notification messages are transmitted as an advertisement to the network – that is, using the 1905.1 multicast address with the relay indicator bit set to 1. This guarantees that every device in the network receives these messages. The topology notification message contains a single TLV identifying the transmitting device.

Devices receiving a topology notification message may decide to send a topology query message to get updated topology information if they are interested. For example, if Device E of Figure 15.5 learns of a new non-IEEE 1905.1 device connected to its E2 interface (e.g. by seeing a packet arriving on E2 with a new MAC address), it must send a topology notification message. Such a message will not contain any information about this new non-IEEE 1905.1 device – it will only contain an identifier for Device E. Devices that are interested in obtaining more information can send a topology query message to Device E directly.

Topology notification messages are also lightweight: they contain minimal information and are relatively small.

15.3.2 Topology Discovery Scenarios

The fine granularity of the basic building blocks of 1905.1 topology discovery enables a scalable approach to topology discovery: devices may decide to discover as much or as little of the network as needed for their application. Two examples are given in Sections 15.3.2.1 and 15.3.2.2.

15.3.2.1 Local Topology Discovery

Path selection applications may only need to perform next hop routing of packets. This requires discovery of the local topology – not the full network. Such a process may involve the procedure shown in Figure 15.6 (in addition to the base requirements of 1905.1).

As seen from this procedure, the device is relatively passive – it discovers its neighbours and queries them for information but goes no further. Using the sample topology of Figure 15.5, this implies that Device A queries Devices B and C but does not query Devices D and E.

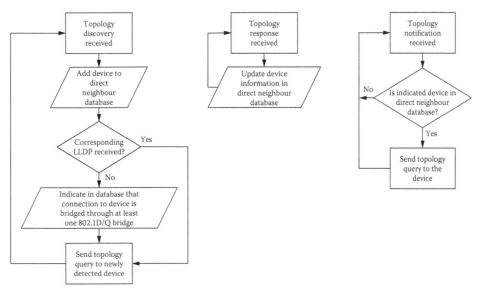

FIGURE 15.6
Local topology discovery procedure.

15.3.2.2 Full Topology Discovery

Network management applications may need information on the full network topology. In particular, this is required by applications that construct the data model defined in the IEEE 1905.1 standard, for example, TR-069 agents. Such an application may involve the procedure shown in Figure 15.7 (in addition to the base requirements of 1905.1)

As seen from the previous procedure, this is an iterative process whereby gradually every device in the network is discovered: as each device is discovered (starting from direct

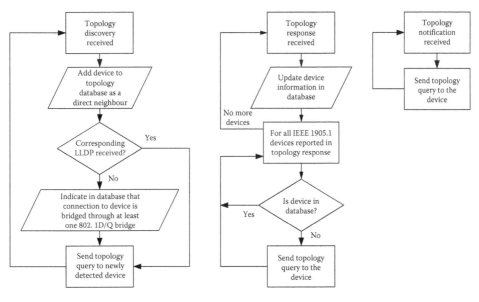

FIGURE 15.7
Global topology discovery procedure.

neighbours), we query them for their neighbours and their neighbours for their neighbours–continuing until no new devices are discovered. Using the sample topology of Figure 15.5, this means that Device A queries Device B, from which it learns of Device D, which it then queries as well (and similarly for Device C through which it learns of and queries Device E).

Once the initial discovery is done, the maintenance phase is based on the reception of topology notification messages. These cause the topology information to be queried as needed.

15.3.3 Trade-Offs in IEEE 1905.1 Topology Discovery Design

The messages provided as part of the 1905.1 topology discovery process are designed to reduce the overhead of the topology discovery and support multiple scenarios.

Mandatory topology discovery messages, including topology discovery and topology notification (and the 802.1AB LLDP message), are small messages by design. Optional topology query is small and topology response (the response is not optional but is only triggered by the optional query packet) is heavy-duty. These carry the main overhead of topology discovery. Different topology discovery requirements use them to a lower degree (local topology) or higher degree (full topology) – paying the cost as required.

15.4 Link Metrics and Information Dissemination Protocol

In order to facilitate load balancing and path selection to satisfy QoS requirements of a particular stream within the home network, 1905.1 supports a link metric information dissemination protocol. The protocol allows any IEEE 1905.1 device to query another IEEE 1905.1 device to get the link metric information from the latter.

The 1905.1 transmitter link metrics include the following:

- *Packet errors*: Estimated number of lost packets on the transmit side of the link during the measurement period.
- *Transmitted packets*: Estimated number of packets transmitted by the transmitter of the link on the same measurement period used to estimate packet errors.
- *MAC throughput capacity*: The maximum MAC throughput of the link *estimated* at the transmitter and expressed in Mb/s.*
- *Link availability*: The estimated average percentage of time that the link is available for data transmissions.
- *PHY rate*: For IEEE 802.3, IEEE 1901 or MoCA 1.1, this is the PHY rate estimated at the transmitter of the link expressed in Mb/s.

The 1905.1 receiver link metrics include the following:

- *Packet errors*: Estimated number of lost packets during the measurement period.
- *Packets received*: Number of packets received at the interface during the same measurement period used to count packet errors.
- *Received signal strength indicator (RSSI)*: For IEEE 802.11, this is the estimated RSSI in dB at the receive side of the link.

---

* The MAC throughput capacity is a function of the PHY rate and of the MAC overhead.

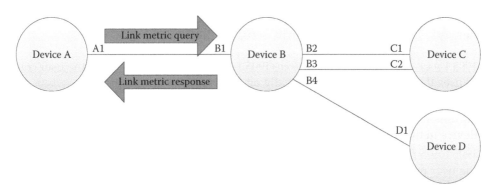

FIGURE 15.8
Link metric query examples.

Also note that some metrics are applicable only to some media: for example, PHY rate is only applicable to Ethernet and IEEE 1901 but not Wi-Fi, for which RSSI is more representative of the quality of the medium.

The link metric information dissemination protocol enables a 1905.1 management entity to obtain link metric information stored at another IEEE 1905.1 device using query and response messages. An IEEE 1905.1 device receiving a query provides information regarding link metrics related to all of its interfaces to a particular 1905.1 neighbour device or all of its 1905.1 neighbour devices.

The 1905.1 link metric information dissemination protocol consists of a procedure with link metric query message and link metric response message.

Figure 15.8 shows a Device A querying Device B for link metric information. Device A has one interface: A1. Device B has four interfaces: B1, B2, B3 and B4. Devices C and D also have their own interfaces, as shown. In this example, Device A may send a link metric query message to its neighbour, Device B, to ask for the link metric information on all the interfaces of Device B (i.e. links B1-A1, B2-C1, B3-C2 and B4-D1). Device A may also send a link metric query message to Device B to ask for the link metric info on all the interfaces between Device B and Device C (i.e. links B2-C1 and B3-C2).

Device A can also send a link metric query message to Device B to ask for the link metric information on all the interfaces between Device B and one of its neighbours, Device A (i.e. link B1-A1). Note that devices are not restricted to querying their direct neighbours – Device A can query Device C for its links to Device D (if such a link existed).

In the previous example, all the transmitter link metrics included in the link metric response message refer to the measurements for packet transmissions on Device B's interfaces. Similarly, all the receiver link metrics included in the link metric response message refer to the measurements for packet receptions on Device B's interfaces.

15.5 IEEE 1905.1 Security Setup

15.5.1 Joining a 1905.1 Device to the Network

The 1905.1 provides a unified security setup aiming to allow an IEEE 1905.1 device to join a given 1905.1 network through a single action for all the 1905.1 underlying network technologies supported by the device. The 1905.1 security reuses and relies on the security mechanisms of each underlying networking technology. To provide flexibility in configuring a network, 1905.1 offers three security setup methods:

- 1905.1 *user-configured passphrase/key* (1905.1 UCPK) method, in which the user configures the 1905.1 network key locally on each IEEE 1905.1 device
- 1905.1 *push button configuration* (1905.1 PBC) method, in which the user pushes a button on one IEEE 1905.1 device and then pushes a button on another IEEE 1905.1 device in quick succession to cause these two devices to join the same network
- 1905.1 *near field communication* (NFC) *network key* (1905.1 NFCNK) method, in which the user uses a NFC-enabled 'key-carrying device (KCD)' to touch and transfer the 1905.1 network key between the devices

The support of the 1905.1 PBC method is mandatory, and the support of 1905.1 UCPK and NFCNK is optional. Any given 1905.1 network may use 1905.1 PBC, 1905.1 UCPK or NFCNK to establish the security of the devices.

15.5.1.1 *1905.1 User-Configured Passphrase/Key Setup Method*

In a 1905.1 network, multiple devices are typically used to provide networking coverage for the whole house. Each device may use interfaces with different technologies, which have different security credentials (e.g. Wi-Fi password, IEEE 1901 password). It is cumbersome for the user to enter and manage these different passwords separately for each device. To solve this problem, 1905.1 supports the 1905.1 UCPK setup method.

In the 1905.1 UCPK setup method, the same single 256-bit 1905.1 network key is locally configured on each IEEE 1905.1 device. The user may choose to come up with 64 hex characters as the 1905.1 network key, which is very secured but may be inconvenient to input manually to the devices.

Alternatively, the user may choose to come up with a 1905.1 network passphrase (between 8 and 63 characters) and a 1905.1 network name (between 1 and 63 characters) that are easier to remember and input manually. While the network passphrase has to be kept secret, the network name may be made public without compromising security. Those two inputs are then converted to a 256-bit 1905.1 network key by a PBKDF2 method (see [1]).

After the 1905.1 network key is generated, the following network technology-dependent user keys (u-keys) will be generated and supplied to the underlying interfaces automatically:

- The WPA/WPA2 passphrase u-key
- The 1901 in-home shared key *device-based security network* (DSNA), *network master key* (NMK) and direct entry NMK-HS u-key
- The 1901 *pairwise security network* (PSNA) *pairwise key* (PWK) u-key
- The MoCA 1.1 privacy password u-key

15.5.1.2 *1905.1 Push Button Configuration Setup Method*

The UCPK setup method described previously greatly simplifies the security setup of hybrid networks by allowing the user to use one network key or network passphrase for all devices in the network. However, UCPK requires the user to enter that key or passphrase manually to each device. To further simplify the security setup, 1905.1 supports the 1905.1 push button configuration setup method.

The 1905.1 PBC method works between any in-network IEEE 1905.1 device and an out-of-network IEEE 1905.1 device. The button referred to hereafter can mean a physical button or a logical button (e.g. a user interface that shows a button on the screen that the user can click on).

When the user pushes the button on the in-network device, the in-network device's interfaces will start the underlying PBC sequence, and it will relay that event to other in-network devices using the *push button event notification* (PBEN) message. When the user pushes the button on the out-of-network device, the underlying network technology-specific push button configuration sequence on each interface of the device will be initiated.

When an in-network device receives such PBEN message, the device will start the PBC sequence of all of its interfaces except for 802.11 interfaces. For 802.11 interfaces, the interface will initiate the 802.11 PBC sequence if the following apply:

- It is an *access point* (AP) and is configured as the registrar and the PBEN message does not indicate the 802.11 PBC has already started. This is to avoid PBC failure due to session overlap when more than one AP is performing PBC.
- It is a *station* (STA) that is not associated with any AP.

Upon successful PBC completion, the device sends a 1905.1 push button join notification message to the whole network.

Note that the push button of each network technology will generate its own key that is not derived from a single key like in the UCPK method, so it is advisable not to use both UCPK and push button on the same 1905.1 network.

15.5.1.3 1905.1 Near Field Communication Network Key Setup Method

NFC can be used to transfer the 1905.1 network key between IEEE 1905.1 devices. Such a transfer can be achieved by using a KCD, which is an NFC device that can store, generate and display a 1905.1 network key to the user.

NFCNK can be used in situations where physical access to the device may be cumbersome. For example, a satellite disk on the roof or door camera installed high on a wall may be hard to access. NFCNK provides a more user-friendly way to integrate devices to a home network.

With NFCNK, the following use cases may apply:

Does NFC KCD Contain a 1905.1 Network Key?	Is 1905.1 Network Key Locally Configured on the IEEE 1905.1 Device?	Action
No	Yes	KCD copies the 1905.1 network key from the 1905.1 device.
Yes	No	The IEEE 1905.1 device copies the 1905.1 network key from the KCD.
Yes	Yes, but different than that of the KCD	The higher-layer entity is notified.
No	No	The higher-layer entity is notified.

Further, before copying the 1905.1 network key to an IEEE 1905.1 device, validation of the key may be necessary (e.g. NFC may provide a secured way to transfer or copy the IEEE 1905.1 device key and the higher-layer entities of KCD and the IEEE 1905.1 device to authenticate each other).

The higher-layer entity on the KCD may provide a selection possibility to the user if the network key to be used is newly generated, taken from the IEEE 1905.1 device or taken from the key storage. If the user wants to set up a new 1905.1 network, the initial network key has to be generated. To add additional devices to the 1905.1 network, this network key is stored and used when devices are added.

15.5.2 Autoconfiguring an AP

To provide whole home coverage, multiple APs may exist in a 1905.1 network. These APs should be automatically configured and kept synchronised without user intervention. The 1905.1 AP-autoconfiguration protocol provides this configuration and synchronisation functionality. The AP-autoconfiguration protocol allows an IEEE 1905.1 device that has an unconfigured AP interface (AP enrollee) to obtain 802.11 parameters from an entity called the registrar, which contains 802.11 credentials of the 1905.1 network. A 1905.1 network may have at most one registrar.

The AP-autoconfiguration process operates in two phases:

1. *Registrar discovery phase*: To get information about the registrar available on the 1905.1 network.
2. *802.11 parameter configuration phase*: To transfer configuration data (ConfigData) (as specified in Wi-Fi Simple Configuration) between the registrar and the AP enrollee. The 1905.1 AL provides a transparent transport protocol for *Wi-Fi Simple Config* (WSC) frames M1 and M2.

This process is defined per interface and is repeated for each of the unconfigured IEEE 802.11 AP interfaces.

During the registrar discovery phase, the AP enrollee sends out an AP-autoconfiguration search message to which the registrar responds with an AP-autoconfiguration response message. When the AP enrollee receives an AP-autoconfiguration response message from the registrar, it continues to the second phase – 802.11 parameter configuration phase.

During the 802.11 parameter configuration phase, the AP enrollee and registrar exchange two messages: AP-autoconfiguration WSC (M1) and AP-autoconfiguration WSC (M2). The 802.11 ConfigData are delivered to the AP enrollee in M2 (see [1] for details on the ConfigData fields).

After the registrar receives M1, it sends the ConfigData encrypted by the KeyWrapKey (as specified in the Wi-Fi Simple Configuration) in M2.

If the 802.11 parameter configuration phase is not completed successfully, the AP enrollee restarts the registrar discovery phase.

If and when the 802.11 parameters in the registrar change, the registrar informs the AP enrollees of this change by sending an AP-autoconfiguration renew message to all the devices in the network. Upon reception of this message, an AP enrollee will (re)start the 802.11 parameter configuration phase.

This AP-autoconfiguration process is defined per frequency band and is repeated multiple times if the registrar supports configuration for multiple bands.

15.6 1905.1 Data Models

IEEE 1905.1 facilitates deployment of a complex home network topology through interconnection of interoperable hybrid network devices. These devices may come from different vendors as long as they are compliant with IEEE 1905.1 and if the network can meet the consumer application/service requirements from an operational point of view. However, the network needs provisioning and diagnostics for management and maintenance. This is quite critical for the user as well as service provider and requires interoperability for successful deployment of multivendor devices. Ability to provide rich capability for diagnostics and troubleshooting reduces operator expense and increases customer satisfaction. To achieve this, IEEE 1905.1 adopts the TR-069 management architecture and protocols defined by the *Broadband Forum* (BBF) and specifies IEEE 1905.1 data models (see [2]). These models, when implemented using the BBF framework, enable configuration, collection of performance and operational statistics, diagnostics and troubleshooting of the network, typically through the use of an *auto configuration server* (ACS). In the following, we provide a description of these models and their capabilities.

This section describes the IEEE 1905.1 device data model for the *customer premises equipment* (CPE) *wide area network* (WAN) *management protocol* (CWMP). TR-069 defines the generic requirements of the management protocol methods that can be applied to any TR-069 CPE. The 1905.1 data model is in addition to any other data models supported by an IEEE 1905.1 device (e.g. TR-098, TR-181 i2 data models). The architecture follows the specification in [3].

The data models follow the notations and the data types defined in [4,5]. Here, we present the overall structure of the device model and the submodels/elements within a hierarchical order; the detailed specification can be found in [1]. The IEEE 1905.1 data model structure is illustrated in Figure 15.9.

- *Device object*: This is the root-level data model object in the IEEE 1905.1 data model hierarchy. It provides the specification version.
- *Abstraction layer object*: This object describes the AL of the device protocol stack. It is composed of an AL identifier, a status (Up/Down/Unknown/Dormant/NotPresent/LowerLayerDown/Error as per [6]), information on how long the device has been in this operational state, a list of underlying interfaces and the number of entries in the interface table.
- *Interface object*: This interface object contains information on its identifier, its status (as described previously for the abstraction layer object), the last time it changed status, a list of lower layers, the media type, the power state and the number of links on this interface.
- *Link object*: This object describes and identifies a 1905.1 link in the network. It consists of the following elements: an interface identifier to which the link belongs, a neighbour identifier to which the link connects and a media type.
- *Link metric object*: The link metrics are very useful for diagnosing possible trouble spots and also for ensuring the health of the network. For each of the links, the link metrics are contained in a link metric object, which contains the following metrics as per the IEEE 1905.1 specification: packet errors, transmitted packets, packets received, MAC throughput capacity, link availability, PHY rate and RSSI when applicable.
- *Forwarding table object*: This object can be very useful in diagnosing loops and other packet transport/routing-related issues. This object consists of a list of forwarding rules, each of which contains a list of interfaces (on which a packet is to be forwarded), a destination MAC address, the source MAC address, the Ethertype, the *virtual LAN* (VLAN) ID and the *priority code point* (PCP).

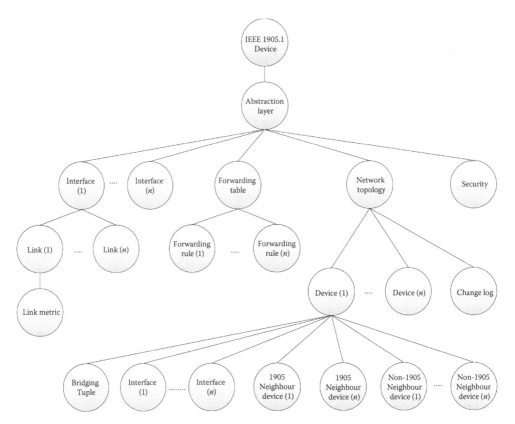

FIGURE 15.9
IEEE 1905.1 data model structure.

- *Topology objects*: Knowledge of the overall network topology and a log of the changes in topology are invaluable tools for operations as well as diagnostics. Overall network topology objects contain information on devices, interfaces, neighbours and bridging tuple for interfaces. Topology change logs contain information on the time stamp of a change event, the event type, the reporting device, the interface on which detection occurs and the neighbour on the interface that changed.

- *Security object*: This object describes the relevant security configuration and consists of a setup method (UCPK or PBC) and an IEEE 1905.1 common password for generating the security keys.

IEEE 1905.1 defines several data model profiles for ease of use as per [5]. These are collections of relevant objects/elements that represent certain attributes of the network/device. Examples of specified profiles are the device profile, the power profile, the interface selection profile, the link metric profile and the network topology profile.

15.7 Performance of Hybrid Networks

In this section, we address the following question: can a *hybrid* (Hy-Fi) network meet stream rate requirements of clients for some representative use cases currently prevalent?

The hybrid network consists of hybrid devices that support Wi-Fi and HomePlug AV technologies, which is compliant to the 1901 (see Chapter 13) specifications. An auxiliary question is: what are the advantages of a hybrid network compared to a Wi-Fi or *power line communication* (PLC)-alone network? We focus on some typical home networking scenarios. For each scenario, the network topology and the rate requirements of the individual streams are specified. We consider in our study multiple houses with given layouts (floor plans). For each house, we consider various possible placements of the *gateway* (GW) AP and the client devices. We call each of the possible placements a configuration. The placements of the devices will affect the data rates of the Wi-Fi links due to path loss and the data rates of the PLC links due to the house wiring. Then we define some useful quantitative performance metrics such as the coverage and residual capacity (which are defined later) to characterise the performance and show the advantage of hybrid networks over Wi-Fi or PLC alone networks.

Section 15.7.1 describes the key assumptions used in our analysis. We first consider the Wi-Fi coverage available in a single house in Section 15.7.2 followed by coverage for applications such as video streaming (*digital video recorders* [DVRs] and *set-top boxes* [STBs]) and data (portable devices) for a number of different houses in Section 15.7.3.

15.7.1 Simulation Parameters

HomePlug AV raw PHY data rates in Mbps between the outlets (or locations), obtained through measurements, are shown in Table 15.1. The Wi-Fi analysis was done using a 3D ray-tracing model/software [7], which models the indoor propagation more accurately than a simple propagation model.

The parameters related to the wireless path loss are shown in Table 15.2. The number of *transmitter* (Tx) antenna was set to two for the DVR and STB and three for the GW. The number of receiver antennas varied between 1 (laptop) and 2 (DVR, STB). The antenna Tx power was set to 27 dBm, and the antenna gains were set to 0 dBi at Rx and 3 dBi at Tx.

In the whole simulation, a noise figure of 5 dB and a thermal noise level of -92 dBm were assumed. The minimum Wi-Fi contention window was set to 16.

15.7.2 Wi-Fi Coverage in a Single House

In this section, we compare the wireless coverage in a house between a single 5 GHz wireless router and a hybrid network. A 202 m^2 (~2000 sq. ft.) house was chosen and configured with five hybrid devices placed in different locations as shown in Figure 15.10. The house is a typical Florida one-storey house that is constructed with wood, stucco and dry walls.

TABLE 15.1

HomePlug AV Data Rate in Mbps between Outlets

Outlet	1	2	3	4	5
1		219	18	32	74
2	178		53	88	238
3	39	74		48	101
4	58	104	88		232
5	78	206	37	208	

TABLE 15.2

Path Loss Parameters

Parameter	Value
Breakpoint distance	10 m
Shadowing loss	0 dB
Penetration loss, interior dry wall, 2.4 GHz	3.5 dB
Penetration loss, exterior dry wall, 2.4 GHz	6.5 dB
Penetration loss, interior dry wall, 5 GHz	7 dB
Penetration loss, exterior dry wall, 5 GHz	16 dB
Penetration loss, exterior concrete wall, 2.4 GHz (6 in. thick)	23 dB
Penetration loss, exterior concrete wall, 5 GHz (6 in. thick)	40 dB

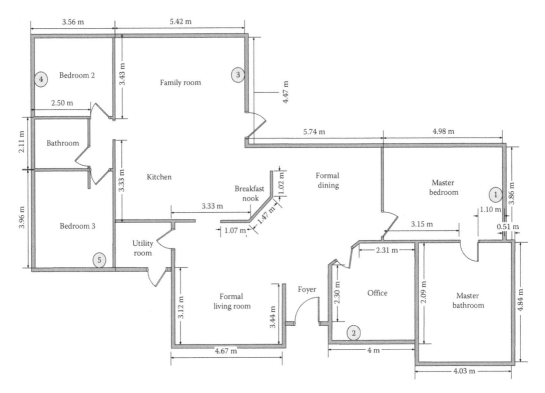

FIGURE 15.10
House layout with power outlets.

The wireless environment of the house is then simulated by WinProp® on the 5 GHz band. WinProp has its own path loss model. The path loss parameters are listed in Table 15.2. We compare the two scenarios:

1. A single 5 GHz wireless router at location 2 which is connected to the GW
2. A hybrid system that has a 5 GHz wireless router at location 2 and two HomePlug AV-to-wireless relays at locations 1 and 5

For scenario 1, the wireless coverage in terms of path loss from the strongest AP is shown in Figure 15.11, and the UDP throughput is shown in Figure 15.12. Similar for scenario 2,

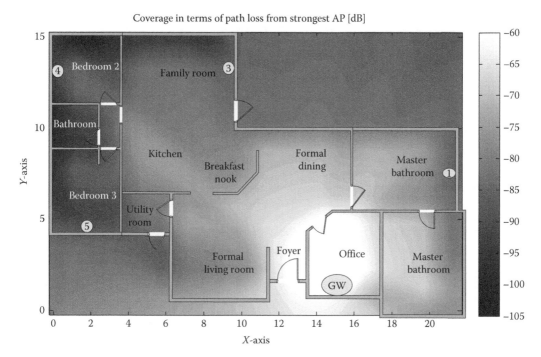

FIGURE 15.11
Path loss in dB from the strongest AP for scenario 1.

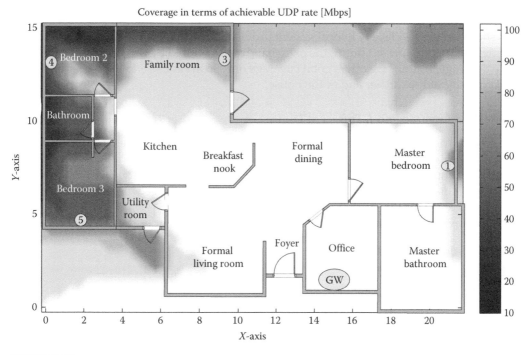

FIGURE 15.12
UDP data rate in Mbps for scenario 1.

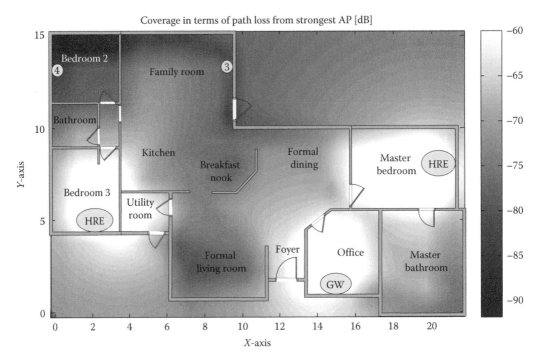

FIGURE 15.13
Path loss in dB from the strongest AP for scenario 2.

the corresponding plots are shown in Figures 15.13 and 15.14. Note all the units of the X-axis and Y-axis are in metres.

As shown in the results, the hybrid system provides a much more uniform high data rate coverage as can be expected due to the extra bandwidth offered by the *hybrid relays* (HREs), while the single router case offers good coverage only in areas close to the router but lacks good coverage at the farther end of the house.

15.7.3 Data Coverage on 6 Different Houses

This section tries to answer the question 'If a vendor sells these devices to a user and the user were to place these devices randomly throughout the house, how much data rate can the network support?' To simulate this case, we selected six different houses (representing six different users) and studied the performance of data coverage when devices are randomly placed in a house.

Then we evaluate how much data the network can support under these scenarios:

1. Wi-Fi alone (2.4 GHz)
2. HomePlug AV alone
3. Hybrid (Wi-Fi and HomePlug AV)

We considered a typical use case scenario as depicted in Figure 15.15. We loaded the system with two variable rate *high-definition* (HD) streams ($2x$ Mbps per stream) and two variable rate *standard-definition* (SD) streams (x Mbps per stream) between the GW and the other devices and studied whether a configuration can meet these stream requirements as the rate (x) increases. Since we do not know in which location the user will put the GW

Coverage in terms of achievable UDP rate [Mbps]

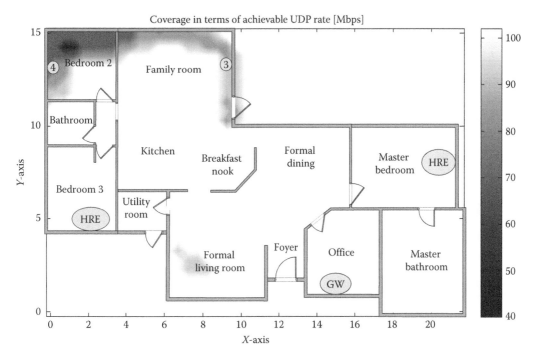

FIGURE 15.14
UDP data rate in Mbps for scenario 2.

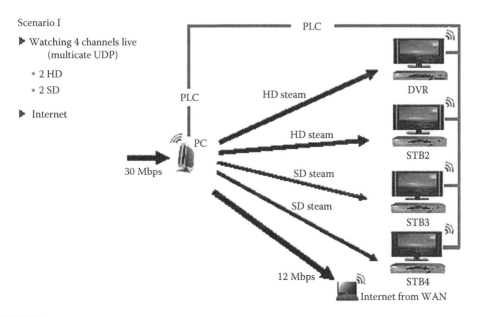

FIGURE 15.15
A scenario with example stream rate requirements of the clients.

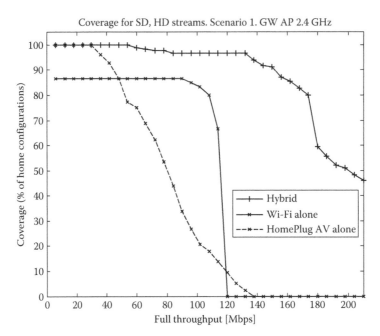

FIGURE 15.16
Coverage results for SD and HD streams (GW AP 2.4 GHz).

(depending on where the WAN access comes into the house), we went through all possible combinations exhaustively by placing the GW and the rest of the devices at all different locations of the house. Next, we check if the devices can support the load on the system. If it can, it means the configuration is successful.

Then we repeated this for all the six houses and plotted the percentage of successful configurations as a function of total data rate. Figure 15.16 shows the percentage of successful configuration vs. the total data rate. For example, at 95% successful configuration, it shows Wi-Fi alone can never satisfy such requirement, whereas HomePlug AV alone can support 40 Mbps total traffic and hybrid networks can support 140 Mbps total traffic.

Also note that when x = 3 Mbps video streams (standard rates for a single SD stream of 3 Mbps and two HD streams with 6 Mbps each), total traffic is 18 Mbps, and both HomePlug AV alone and hybrid can support such data rates 100% of the configuration, whereas Wi-Fi alone cannot.

Table 15.3 shows the improvements of hybrid over Wi-Fi alone or HomePlug AV alone.

We also show in Figure 15.17 the residual Wi-Fi data rate that is available for the laptop. It shows that the hybrid network offers a much higher throughput than the Wi-Fi only

TABLE 15.3

Improvements of Hybrid Network

Scenario	Can Support the Standard SD and HD Rates (x = 3 Mbps) 95% of the Configuration?	Total Traffic Supported	Improvement over HomePlug AV Alone
Wi-Fi only (2.4 GHz)	No	N/A	N/A
HomePlug AV alone	Yes	40 Mbps	N/A
Hybrid	Yes	140 Mbps	250%

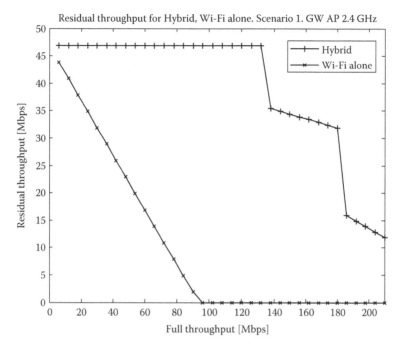

FIGURE 15.17
Residual throughput for hybrid and Wi-Fi alone (GW AP 2.4 GHz).

network as the total offered traffic increases. The reason is that for a Wi-Fi only network, the Wi-Fi bandwidth is completely consumed by the SD and HD streams (e.g. when the total throughput is beyond 96 Mbps), whereas in a hybrid network, the load can be shared among the HomePlug AV network and the Wi-Fi network, hence leaving more Wi-Fi bandwidth for the laptop to use.

15.8 Conclusion

In a typical home environment, one finds a large variety of end devices, such as STBs, smart TVs, desktop computers, laptops, peripherals, handheld devices (tablets, smartphones, etc.) and other consumer devices connected to the home network which may constitute of both wired and wireless technologies. In the future, the list of end devices is expected to include home appliances, home automation and security to provide a rich application/services environment. These devices running different applications will have varied requirements in terms of throughput and delay. Examples include streaming multimedia delivery to DVRs and STBs and delivery of data streams to laptops and other handheld devices which have very different performance requirements. A home networking solution must be capable of simultaneously meeting the throughput (rate) requirements of all the streams with desired latency. Some of these clients are portable, whereas others will be placed at fixed locations in the house.

In order to satisfy these requirements, a global solution is necessary where data can be seamlessly routed through the wired and wireless media available at home. The IEEE

1905.1 specification presented in this chapter provides an efficient and evolutionary AL for the hybridization of Wi-Fi (IEEE 802.11), power line communications (IEEE 1901), Ethernet (IEEE 802.3) and multimedia over coaxial cable (MoCA 1.1). The AL acts as an intermediate protocol between the MAC layers of each particular technology and the upper-layer entities. It allows a seamless merging of heterogeneous technologies without requiring any change in the underlying home networking systems. As a result, the end-user experience is enhanced and advanced QoS and network management features are facilitated.

In addition to throughput optimisation and coverage extension, IEEE 1905.1 provides new services to higher-level applications. The topology discovery protocol provides information on the IEEE 1905.1 devices in the network and identifies their medium-dependent connection features, thereby enabling optimised routing and bridging throughout the network. The link metric information dissemination protocol shares quantitative information on each available link, such as packet error and link capacity, which allows selecting the most appropriate paths and balancing the loads of the data flow. Efficient security protocols have also been defined to provide contained security within the heterogeneous network.

Simulations presented in this chapter illustrated how the IEEE 1905.1 specification significantly increases the Wi-Fi coverage in a typical home by using HomePlug AV power line technology as a backbone network. In addition, a statistical analysis performed on measurements from six different houses demonstrated that IEEE 1905.1-enabled hybrid networking enhances the user experience in typical scenarios, by providing increased throughput and coverage: at 95% coverage, the total supported traffic increased from 40 to 140 Mbps. Therefore, IEEE 1905.1 is seen as a key specification for home networking with unprecedented quality of experience.

Acronyms

AL	Abstraction layer
ALME	Abstraction layer management entity
AP	Access point
CMDU	Control message data unit
DSNA	Device-based security network
HLE	Higher-layer entity
KCD	Key-carrying device
LLC	Logical link control
LLCDU	LLC data unit
LLDP	Link Layer Discovery Protocol
LLDPDU	LLDP data unit
MPDU	MAC protocol data unit
MSDU	MAC service data unit
NDEF	NFC data exchange format
NFC	Near field communication
NFCNK	NFC network key
PDU	Protocol data unit
PSNA	Pairwise security network
SAP	Service access point
SME	Station management entity

TLV	Type–length–value
WPA	Wireless Protected Access
WPA2	Wireless Protected Access 2
WSC	Wi-Fi Simple Configuration

References

1. IEEE 1905.1-2013, 1. IEEE standard for a convergent digital home network for heterogeneous technologies, 2013, P1905.1™/D08 draft.
2. TR-069 Amendment 4, July 2011, CPE WAN Management Protocol, Broadband Forum. Available online http://www.broadband-forum.org/technical/download/TR-069_Amendment-4.pdf, accessed 28 September 2013.
3. TR-181 Issue: 2, May 2010, Device Data Model for TR-069, Broadband Forum Technical Report. www.broadband-forum.org/technical/download/TR-181_Issue-2.pdf, accessed 28 September 2013.
4. Simple Object Access Protocol *(SOAP) 1.1*.
5. TR-106 Amendment 6, July 2011, Data Model Template for TR-069-Enabled Devices, Broadband Forum. www.broadband-forum.org/technical/download/TR-106_Amendment-6.pdf, accessed 28 September 2013.
6. The Interfaces Group MIB; http://tools.ietf.org/html/rfc2863 (accessed 2013).
7 AWE Communications: WinProp Software Package. Free evaluation version and user manual of a rigorous 3D ray tracing tool for urban and indoor environments, http: www.awe-communications.com.

Part IV

Advanced PHY and MAC Layer Processing

16

Smart Beamforming: Improving PLC Electromagnetic Interference

Daniel M. Schneider and Andreas Schwager

CONTENTS

16.1 Introduction

Multiple-input multiple-output (MIMO) technology enhances the performance of PLC systems by utilising multiple ports for sending and receiving signals (see Chapters 8, 9, 12, 14 and 24). Beamforming or precoding is used in several current *power line communication* (PLC) systems (see Chapters 12 and 14) and significantly improves their performance as seen in Chapters 8 and 9. The influence of MIMO transmission on *electromagnetic interference* (EMI) is addressed in Chapter 7. In particular, the influence on EMI, when feeding signals on transmit ports other than the traditional differential feeding between L and N, is compared and analysed. However, Chapter 7 does not touch on the effects on EMI in detail, when using a particular MIMO scheme at the transmitter and simultaneous transmission on several ports. A comparison of MIMO schemes for PLC can be found in Chapter 8. This chapter discusses how precoding or beamforming influences EMI. The analysis, in this chapter, is based on the MIMO EMI measurements performed at the ETSI STF410 measurement campaign (see Chapters 5, 7 and [1–3]). Section 16.2 briefly recalls the EMI measurement setup. Section 16.3 provides a theoretical analysis of expected EMI for different beamforming modes based on the correlation of transmit signals. Next, a detailed statistical analysis of EMI for different beamforming modes is given in Section 16.4. A point of interest is whether appropriate precoding can mitigate EMI effects. This question opens the second part of this chapter where ideas and first results of an EMI-friendly beamforming are presented (see Section 16.5). Section 16.6 addresses the issue of how PLC modems can obtain information about EMI properties of the PLC network.

458 MIMO Power Line Communications

16.2 Measurement and System Setup

The EMI measurement setup is described in detail in Chapter 7, Figure 7.1. Figure 16.1 is a block diagram of the basic setup, focusing on the functional aspects of the transmitter. The left-hand side of the block diagram shows the eigenbeamforming transmitter with two spatial streams (Figure 16.1a), spotbeamforming with one spatial stream (Figure 16.1b) and *single input single output* (SISO) (Figure 16.1c), respectively. Figure 16.1a and b depict typical MIMO-*orthogonal frequency division multiplexing* (OFDM) transmitters as introduced in Chapter 8. The right-hand side of each block diagram depicts the antenna used to evaluate the EMI of the MIMO transmission. The k-factor for each MIMO feeding option was measured, in magnitude and phase, for frequencies up to 100 MHz in horizontal and vertical polarisation of the antenna, during the ETSI STF 410 measurement campaign (see Chapter 7). The k-factor is measured in dBµV/m-dBm. This factor can be used to calculate the electrical field strength in dBµV/m assuming a given transmit power level (in dBm). In case of MIMO transmission (see Figure 16.1a and b), the EMI of the two signals fed in the two transmit ports interfere with each other. Let h_1 be the k-factor of the first transmit port to the antenna location and h_2 the k-factor of the second transmit port of a given frequency. Assume further that s_1 and s_2 are signals of a given frequency, where the two signals are combined into the vector $\mathbf{s} = \begin{bmatrix} s_1 \\ s_2 \end{bmatrix}$ (e.g. the symbols of one OFDM subcarrier as shown in Figure 16.1a and b or one frequency measurement point of the *network analyser* [NWA]). Then, the resulting field at the antenna location of the given frequency is given by $h_1 s_1 + h_2 s_2$.

Figure 16.1a shows beamforming in the precoding matrix **F**. If $\mathbf{b} = \begin{bmatrix} b_1 \\ b_2 \end{bmatrix}$ is the vector of the two signals of a given frequency before precoding, then the vector of the two transmit

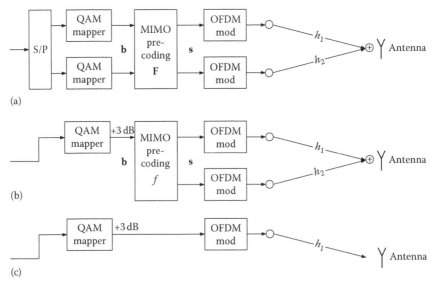

FIGURE 16.1
Setup: transmitter and EMI measurement. (a) Two-stream MIMO: Eigenbeamforming, (b) one-stream MIMO: Spotbeamforming, and (c) SISO.

signals is given by $\mathbf{s} = \mathbf{Fb}$ (refer also to Chapter 8). In the case of spotbeamforming (refer to Figure 16.1b), only one spatial stream is used, that is, $\mathbf{b} = b_1$, and the precoding matrix simplifies to a precoding vector $\mathbf{f}$. Figure 16.1c shows SISO transmission.

The total transmit signal energy is identical for all setups (Figure 16.1a through c). OFDM and the precoding blocks are also identical in the three setups. The *quadrature amplitude modulation* (QAM) mapper is the source of the transmit signals and defines the level of transmit energy. The output energy of spotbeamforming (Figure 16.1b) and SISO (Figure 16.1c) configurations has been boosted by 3 dB to make the comparison more clear, as each configuration only has one QAM mapper. MIMO precoding splits input energy into two output ports but does not amplify or attenuate the signals. The OFDM units also do not change the energy between input and output. This is how identical transmit signal energy is guaranteed for all three setups.

16.3 Difference between Eigen- and Spotbeamforming

The relation or correlation between the signals transmitted on the two transmit ports influences the electrical field levels at the antenna and determines if the signals interfere constructively or destructively. The following is a discussion of the spatial correlation of transmitted signals where beamforming has been applied. Note that there is a similar discussion already in Section 8.5.2, which is recalled here for convenience to describe the special case of two transmit ports.

In Figure 16.1a, both MIMO streams are modulated independently by the two QAM mappers. If each MIMO stream is modulated independently, the signals are spatially uncorrelated

$$E\{\mathbf{bb}^H\} = \mathbf{I}_2, \tag{16.1}$$

where
  $E\{\cdot\}$ represents expectation over time samples
  $\mathbf{I}_2$ is the 2×2 identity matrix

Thus, where no precoding is applied, $\mathbf{s} = \mathbf{b}$, the transmitted signals are also uncorrelated $E\{\mathbf{ss}^H\} = E\{\mathbf{bb}^H\} = \mathbf{I}_2$. If precoding is applied and the entries of $\mathbf{b}$ are uncorrelated, the correlation matrix results in

$$E\{\mathbf{ss}^H\} = E\left\{(\mathbf{Fb})(\mathbf{Fb})^H\right\} = \mathbf{F}E\{\mathbf{bb}^H\}\mathbf{F}^H = \mathbf{FF}^H. \tag{16.2}$$

For eigenbeamforming (two spatial streams) with $\mathbf{F} = \mathbf{V}$, the unitary property of $\mathbf{V}$ simplifies Equation 16.2 to $\mathbf{VV}^H = \mathbf{I}_2$ and the transmit signals remain uncorrelated. Thus, eigenbeamforming does not influence the electrical field levels. Spatial multiplexing without precoding and beamforming with an arbitrary unitary precoding matrix results in the same EMI.

If one-stream beamforming (spotbeamforming) is used with $\mathbf{F} = \mathbf{v}_1$ (where $\mathbf{v}_1$ is the first column vector of $\mathbf{V}$), the simplification $\mathbf{v}_1\mathbf{v}_1^H = \mathbf{I}_2$ is not valid. This causes a correlation between the transmit signals, which means that the precoding vector influences the electrical field levels (see next section).

16.4 Results: Radiation Depending on Beamforming

The unitary precoding matrix **V** can be represented by two angles ψ and ϕ:

$$
\mathbf{V} = [\mathbf{v}_1 \quad \mathbf{v}_2] = \begin{bmatrix} \cos(\psi) & \sin(\psi) \\ -e^{j\phi}\sin(\psi) & e^{j\phi}\cos(\psi) \end{bmatrix}, \tag{16.3}
$$

where the range of ψ and ϕ to represent all possible beamforming matrices is $0 \le \psi \le \pi/2$

and $-\pi \le \phi \le \pi$ (refer also to Chapter 14). The precoding vector for spotbeamforming is given by $\mathbf{v}_1 = \begin{bmatrix} \cos(\psi) \\ -e^{j\phi}\sin(\psi) \end{bmatrix}$. Figure 16.2 shows an example of one measurement at one

frequency. The *x*- and *y*-axes represent the angles ψ and ϕ in the range described earlier. The contour lines in the figure show the level of the electrical field relative to SISO (D1 port of the triangle style coupler when feeding at the same outlet, see Chapter 5) depending on the precoding vector represented by ψ and ϕ. The distance between the contour lines is 1 dB. If the signals of the two transmit ports interfere destructively, almost complete signal elimination may be achieved (see $\psi = 1.1$ and $\phi = -2.8$). On the other hand, the signals may interfere constructively for some precoding vectors. This is the case especially for $\psi < 1.1$ and $-1 < \phi < 1$. The bold contour line marks the median line, that is, 50% of the beamforming angles result in an electrical field level higher than the median value of -3.5 dB

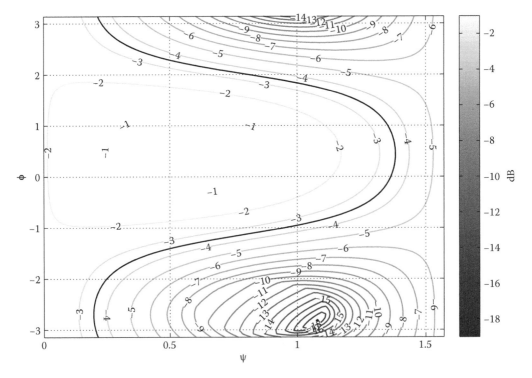

FIGURE 16.2
Electrical field levels relative to SISO depending on the precoding vector, one channel at one frequency.

(relative to SISO), while 50% of the beamforming angles result in lower radiation compared to the median value. The maximum (–0.5 dB) is 3 dB higher than the median value. In Figure 16.2, there is a small peak denoting that the electrical fields have been completely eliminated, while the flat area describes levels slightly higher than the median value. The phenomena of deep canyons and flat hills in signal level readings can be monitored with many applications with multipath fading channels (e.g. HF radio broadcast transmissions; see Chapter 22, Figure 22.2).

Figure 16.3 shows the *cumulative distribution frequency* (CDF) of the radiation compared to SISO with the same channel and frequency as in Figure 16.2. To obtain this CDF, all possible precoding vectors of one-stream beamforming were used. Figure 16.3 comprises the 20%, 80% and 50% (median) values. The median value shows a radiation of 3.5 dB less than the SISO case seen in Figure 16.1c for this channel and frequency. The maximum radiation is reached with low statistical probability and is 3 dB higher than the median value (see upper right corner in Figure 16.3). Here, the radiation level of the SISO transmission is never reached. On the other hand, signal elimination (–18 dB) may be achieved with a very low statistical probability (see lower left corner in Figure 16.3). The median value of spotbeamforming radiation is equal to the radiation levels of two-stream MIMO (eigenbeamforming or spatial multiplexing without precoding). This was validated for every measurement (see next section).

The calculation as shown for Figure 16.3 was performed for each measurement (i.e. each measurement site, each frequency and each antenna location). Overall, 100,863 measurement points are used in these statistics (1,601 frequency points between 1 and 100 MHz at 63 sweeps). Figure 16.4 shows the CDF of the radiation for all measurements using different MIMO setups where D1 and D3 (see Chapter 1, Figure 1.2) were used as feeding ports. The level of radiation is given relative to the radiation of SISO (D1 port) where the difference between MIMO and SISO is calculated for each pair of measurements before determining

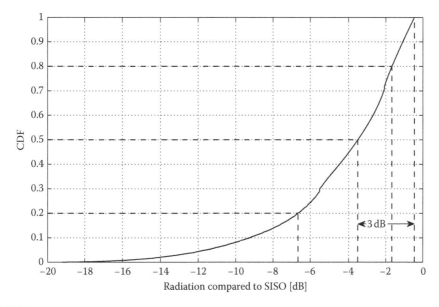

FIGURE 16.3
CDF of the electrical field levels of all possible one-stream beamforming vectors relative to SISO, one channel at one frequency, same channel as in Figure 16.2.

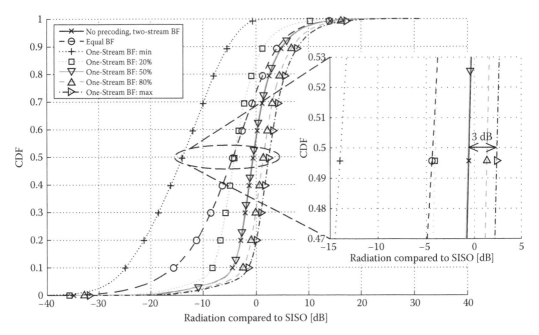

FIGURE 16.4
CDF of the radiation of all feeding outlets, all antenna positions and all frequencies, D1 and D3 feeding.

the CDF. As shown in Section 16.3, the radiation of two-stream MIMO is independent of the beamforming matrix and is the same as spatial multiplexing without precoding. Since these cases all yield the same levels of radiation, they are shown by only one single line (indicated by 'no precoding or two-stream BF' in Figure 16.4). The min, 20%, 50%, 80% and max lines of spotbeamforming (one-stream BF) are related to the corresponding CDF values of the different beamforming angles (refer also to Figures 16.2 and 16.3). The median value of spotbeamforming is also identical to two-stream MIMO or spatial multiplexing. The maximum radiation of spotbeamforming is shifted by +3 dB compared to two-stream MIMO but is only 2.4 dB higher compared to SISO. This is depicted in the zoomed version of the graph in Figure 16.4. The figure also comprises the special spotbeamforming $\mathbf{f} = \begin{bmatrix} 1/\sqrt{2} \\ 1/\sqrt{2} \end{bmatrix}$, where the signal is transmitted via both transmit ports with the same power and no phase offset (denoted by 'equal BF' in the figure). Identical signals are transmitted here via both transmit ports. Interestingly, the electrical field level is significantly reduced compared to two-stream MIMO and SISO.

Figure 16.5 shows a similar plot as Figure 16.4 where T1 and T2 are used as feeding ports. The figure basically shows the same properties as the D1 and D3 feeding in Figure 16.4. The zoom in Figure 16.5 confirms the worst-case 3 dB increase but an optimum case signal elimination of 12 dB relative to two-stream MIMO.

The CDFs of Figures 16.4 and 16.5 show records of some channels providing higher EMI in the MIMO cases, while the majority of measurements in the SISO configuration show higher radiation. Figure 16.6 shows an example of the k-factor of one measurement sweep over the frequency. The figure compares the k-factor of SISO (solid line, feeding port D1) and spotbeamforming (dashed line, one-stream BF, D1 and D3 feeding). At this channel, SISO

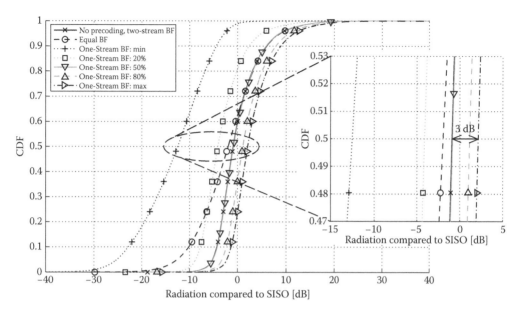

FIGURE 16.5
CDF of the radiation of all feeding outlets, all antenna positions and all frequencies, T1 and T2 feeding.

radiates higher at some frequencies, while at other frequencies MIMO provides higher EMI. This explains the CDF shape of Figures 16.4 and 16.5.

It is assumed that MIMO PLC modems operate in eigenbeamforming mode at almost all channels and frequencies. Spotbeamforming is used only at high attenuated channels or frequencies. Simulations were performed to investigate how many channels and frequencies

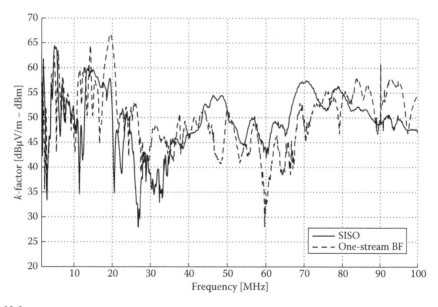

FIGURE 16.6
Example of the *k*-factor depending on the frequency of one measurement, comparison between SISO and spotbeamforming.

use spotbeamforming if the *signal to noise ratio* (SNR) conditions do not support eigenbeam-forming. The following assumptions are used for these simulations:

- The transmit to noise power ratio is set to 75 dB, corresponding to a transmit power level of −55 dBm/Hz and a flat noise power spectral density of −130 dBm/Hz (which is the median value of noise below 30 MHz; see Chapter 5).
- The system parameters are the same as described in Chapter 9.
- The channels of the ETSI measurement campaign were used (more than 340 channels; see Chapter 5).

As a result of these simulations, 5% of the subcarriers of all channels use spotbeamforming due to the insufficient SNR of the second stream. 95% of these subcarriers support eigenbeamforming.

16.5 EMI-Friendly Beamforming

As shown in the previous section, beamforming may be used to improve interference (EMI) from PLC by eliminating signals at some locations or reducing the electrical or magnetical field of the radiated signals in the air. If the level of interference is improved as described earlier, experimentally, the level of transmit power of PLC modems may be enhanced to have the equivalent EMI potential. The beamforming vector for minimising EMI will usually not be the optimum beamforming vector for a PLC receiver on the network. Thus, there is a loss of throughput rates compared to beamforming optimised for the communication link. On the other hand, the gain of increased feeding levels might compensate for this loss. Of course, given EMC regulatory requirements have to be considered before bringing such an experiment to maturation.

Figure 16.7 shows the electrical field in the air (solid contour lines, denoted by 'E-field depending on precoding' in the legend), radiated from the power lines. The field depends on the precoding vector described by the two angles ψ and ϕ (similar to the plot shown in Figure 16.2). The level of the E-field is shown relative to the E-field of SISO-PLC radiation at 30 MHz. The antenna was located indoors. The highest increase of radiation (less than +3 dB compared to SISO in this example) is obtained by the precoding vector marked with the black square ■. The maximal signal elimination (maximal reduction compared to SISO's transmission E-field, −14 dB) is obtained by the precoding vector marked with grey star ★ (see $\psi \approx 1$ and $\phi \approx −3$ in the figure). The communication links to other outlets in the building were also measured for the feeding outlet where the E-field measurements were performed. Two links are available in this example. The optimum (spotbeamforming) precoding vector for these two receiving outlets is marked by the two diamond symbols ◆ labelled 1 and 2. The dashed contour lines in Figure 16.7 show the available SNR for the first link depending on the precoding vector. Of course, the highest SNR of 30 dB (see the grey circle • in Figure 16.7) is achieved if the precoding vector is optimised for this link. The SNR decreases when 'moving away' from this optimum precoding. The precoding vector optimised for the communication link and the precoding vector optimised to reduce the E-field are usually not identical. The example shown in Figure 16.7 provides quite some distance between the two vectors: The subscript number (0.88) at the first

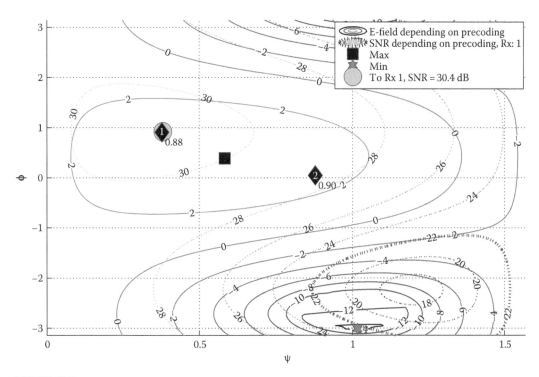

FIGURE 16.7
Electrical field levels relative to SISO (solid contour lines) and SNR (dashed contour lines) depending on different precoding vectors, 30 MHz.

diamond symbol indicates a measure of distance to the optimum precoding vector to minimise the E-field. This distance might vary between 0 (same precoding vector) and 1 (most distant precoding vector or orthogonal vector); details of the definition of this distance may be found in [4].

Assume that the precoding vector is optimised to achieve elimination of the E-field (star symbol in Figure 16.7). The radiated E-field is E_{reduced} = 2 dB − (−14 dB) = 16 dB below the radiation level of the case where the precoding vector is used, which has been optimised for this link. Meaning, the level of transmit power may be increased by this amount to obtain the same E-field level. On the other hand, the loss in SNR by not using the optimum precoding vector for this link is SNR_{loss} = 8 dB (30 dB − 22 dB), as observed by the intersection of the dashed contour lines with the star symbol. Overall, an SNR increase of $\text{SNR}_{\text{gain}} = E_{\text{reduced}}$ − SNR_{loss} = 8 dB may be obtained in this example which results in an increased bitrate.

Figure 16.8 shows similar plots as Figure 16.7, with the same transmitting outlet location, building and two different antenna locations 10 m from the exterior wall at the outside of the building. The E-field contour lines are quite similar to Figure 16.7 (the E-field maximum ■ is in the centre of the figure, while the minimum precoding vector ★ is also centralised but nearer the bottom line). However, the distance between the two optimum E-field vectors ★ is not large. This indicates that the optimal E-field precoding is independent, to an extent, of the antenna location.

In the United States, the assessment of the PLC product certification (see also Chapter 6) is done by checking the electrical field level on the outside of a building. If this level is lowered by, for example, 10 dB by smart beamforming, the transmit level of PLC modems may also be increased by 10 dB.

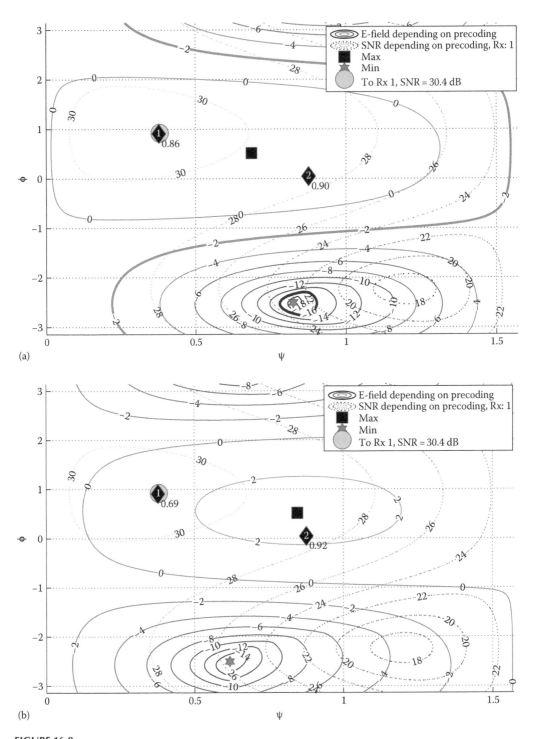

FIGURE 16.8
Same settings as in Figure 16.7 at two different outdoor antenna locations at 10 m distance (compared to the indoor antenna location in Figure 16.7). (a) First antenna location at 10 m distance and (b) second antenna location at 10 m distance.

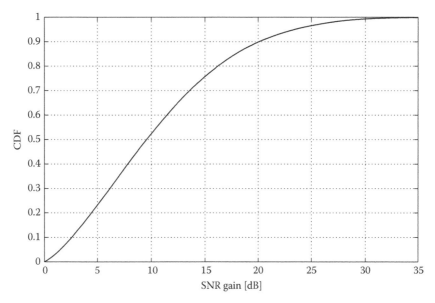

FIGURE 16.9
CDF of SNR gain due to EMI-friendly beamforming.

Per Figures 16.7 and 16.8, the EMI could be mitigated completely, as shown by the deep notches. Theoretically, as an extreme example, the electrical field level of the MIMO modems could be lower than the ambient noise level. As a result, the feeding level could be increased accordingly. Of course, in reality, several limiting factors have to be considered. The EMI information has to be quite accurate and the beamforming angles are usually quantised to a set of certain values. To investigate the potential of EMI-friendly beamforming and corresponding SNR gain, the algorithm earlier is applied to all available EMI and corresponding S21 measurements. The quantisation of the beamforming angles is set to 12 bits (5 bits for ψ and 7 bits for ϕ) which are the beamforming parameters of HomePlug AV2 (see Chapter 14). The SNR gain of each frequency and measurement site is influenced by two factors. First, the loss of SNR due to non-optimised beamforming of this link, and second, feeding level gain from EMI-friendly beamforming. If the SNR gain is less than 0 dB, classical spotbeamforming is applied in the following simulations. Figure 16.9 shows the CDF of the signal to noise level gain SNR$_{gain}$ for all measurements. In median, an SNR gain of almost 10 dB is observed. At the 90% point, that is, for 10% of the sweeps, the SNR gain is 20 dB and higher. The SNR gain is always larger than 0 dB as the scheme defaults to normal spotbeamforming if no SNR gain is achieved from EMI-friendly beamforming.

16.6 Obtaining EMI Properties without Measuring Field Levels

One important question has to be answered if beamforming is adjusted to mitigate the EMI: How does a modem know the EMI properties? Usually the modem cannot measure the radiation or the EMI properties of the actual power line grid. One interesting approach could be to use current probes to measure the currents on each wire. According to Biot–Savart's law, the CM current is responsible for radiation. The current probe might be located in the transmitter or somewhere else on the network, depending on how the EMI depends on the outlet properties or on the properties of the power line grid. In a training phase, the transmitting modem toggles all possible precoding vectors in a predefined way and the probes signal which precoding vector gives balanced currents on the three wires.

To investigate the potential of this idea, measurements were performed in an anechoic chamber. Figure 16.10 shows the setup. An artificial mains network including a fuse cabinet, wires with similar length as in a small flat, power strips and several appliances or impedances connected was laid out in the chamber. The electrical field probe [5] was located at a distance of approximately 3 m from the artificial mains. The MIMO current probe consists of three inductors, inserted into the *live* (L), the *neutral* (N) and the *protective earth* (PE) wire and connected at various locations in the network. The probe's transfer impedance is

$$Z_T(\mathrm{dB\Omega}) = V(\mathrm{dB\mu V}) - I(\mathrm{dB\mu A}) = 21.5 \ \mathrm{dB\Omega}. \tag{16.4}$$

A multiport NWA feeds signals using the PLC coupler described in Chapter 1 into the mains and simultaneously records the E-field, as well as the currents on the wires. The coupling factor (*k*-factor in dBμV/m – dBm) in the chamber is derived using Equation 7.1 in Chapter 7. The current I_{wire} is calculated as

$$I_{\mathrm{wire}} = P_{\mathrm{Tx}} + S_{21} - Z_T + \mathrm{Conv}_{\mathrm{dBm2dB\mu V}}, \tag{16.5}$$

where
$P_{\mathrm{Tx}} = 12$ dBm is the NWA's injected feeding power
S_{21} is the scatter factor recorded by the NWA
$\mathrm{Conv}_{\mathrm{dBm2dB\mu V}} = 107$ dBm – dBμV is the conversion factor from dBm to dBμV

Figure 16.11a shows the currents on L, N and PE of the current probe placed somewhere on the mains grid. Equal beamforming was applied in this example, that is, the same signal is transmitted on both transmit ports. In a next step, the beamforming is adjusted to balance the currents on N and PE by minimising the current on L. Minimising the current on L is selected, since D1 (L–N) and D3 (L–PE) are used for feeding. Figure 16.11b shows the influence of this

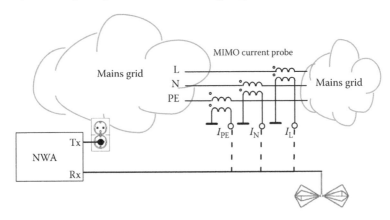

FIGURE 16.10
Setup of EMI current measurements.

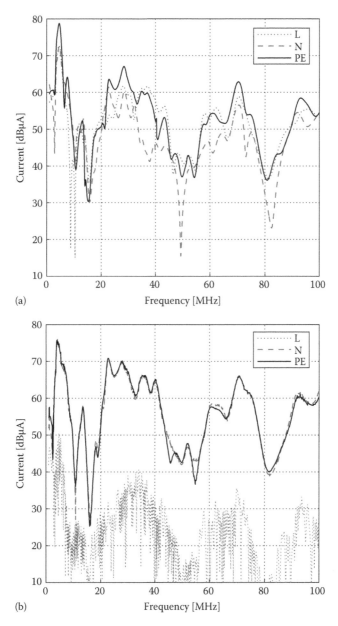

FIGURE 16.11
Magnitude of currents on L, N and PE for different precoding. (a) Equal precoding, (b) precoding to balance currents on N and PE by minimising current on L.

beamforming on the currents measured by the current probe. As expected, the current on L is minimised, while the currents on N and PE have the same magnitude.

Figure 16.12 shows the *k*-factor versus frequency of the two different beamforming vectors introduced before, that is, equal precoding (solid line) and precoding to balance currents on N and PE (dashed line). For most frequencies, the *k*-factor is decreased by the beamforming based on the balanced currents.

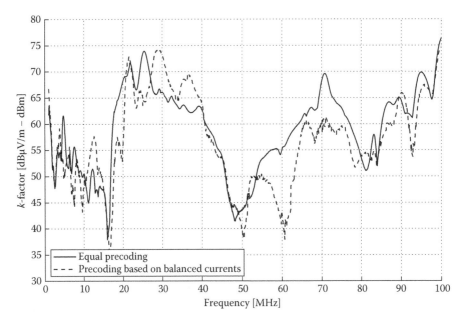

FIGURE 16.12
k-factor for different types of beamforming: equal precoding and precoding based on balanced currents.

Figure 16.13 shows the corresponding CDF, that is, the CDF of the *k*-factor of Figure 16.12. In median, an improvement of 4 dB is observed for beamforming based on balanced currents compared to equal beamforming.

Another idea is to use the S11 parameter at the transmitting modem to identify the optimum beamforming vector in order to mitigate the EMI. The S11 parameter measures the reflected energy. By toggling all possible beamforming vectors (or a subset of beamforming vectors) and monitoring the corresponding S11 value, a correlation to the expected EMI might be revealed in future research.

The transfer functions to other modems on the network might also give an indication of the beamforming to mitigate EMI. Beamforming vectors providing minimal signal level at one or more receiving PLC modems are assumed to cause low signal radiation. Beamforming vectors providing maximal signal level at one or more receiving PLC modems are assumed to cause higher signal radiation. There are devices or resistors on the network which absorb communication signal energy. If beamforming is directed to these devices, radiation is also reduced. The variation of signal level at the receiving modem might indicate how much energy is lost due to signal radiation. An antenna or level meter providing a feedback channel might check the level of interference for different beamforming vectors and indicate the optimum beamforming vector.

The location where the interference should be reduced might also be another PLC modem operating in a neighbour's flat. This PLC modem might signal the channel parameters to the interfering transmit modem. The transmitter might easily calculate the beamforming angles to eliminate interference at the neighbour's flat.

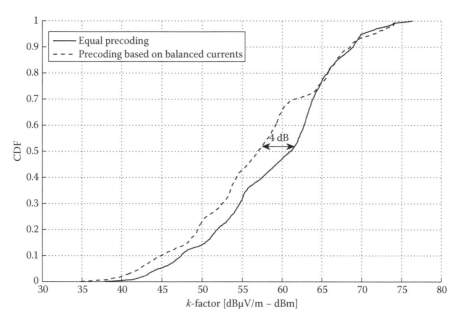

FIGURE 16.13
CDF of the k-factor for different types of beamforming: equal precoding and precoding based on balanced currents, measurement in anechoic chamber of an artificial mains network.

16.7 Conclusions

This chapter aimed to answer the question how MIMO and especially beamforming influence the levels of the electrical field compared to SISO. The results showed that in median MIMO radiates less compared to SISO. This result suggests that the transmit power level of each MIMO port may not have to be reduced by 3 dB compared to SISO but may be chosen to be less. It was shown that the levels of the electrical field are identical for

- Two-stream MIMO (independent of any arbitrary unitary precoding matrix, eigenbeamforming)
- Spatial multiplexing without precoding (special case of two-stream MIMO)
- Median of one-stream beamforming

Spatial multiplexing and eigenbeamforming yield the same probability of EMI. In the case of one-stream beamforming or spotbeamforming, the precoding vector influences the EMI. The beam may result in a destructive or constructive interference. The worst-case constructive interference results in a 3 dB higher radiation compared to two-stream MIMO. However, this special case is unlikely since only one special beamforming vector may give this result. It is also assumed that MIMO PLC modems operate in eigenbeamforming mode most of the time and for almost all channels. On the other hand, beamforming may be used to almost eliminate the radiation.

This conclusion motivated the second part of this chapter: Beamforming may be used in future PLC systems to mitigate the effects of EMI and unwanted radiation. If beamforming is optimised to minimise the radiation, the question arises how this would influence the modem's throughput performance. Usually, beamforming is used at the transmitter to maximise the performance (throughput) to the receiver. However, if beamforming is chosen to minimise EMI, the transmit power level might be increased which leads to an increase in performance. First results were presented which showed an SNR gain when this concept is applied. This gives another degree of freedom in the design of the PLC modem.

An interesting question for future research is the following: How does a modem obtain the properties of EMI, in order to correctly adjust beamforming to mitigate it? One interesting approach could be to use a current probe to measure the currents on each wire and to toggle the beamforming vectors until the currents on the wires are balanced. First measurements in the laboratory indicated that there is a relation between the currents on the wires and EMI which can be used to adapt the beamforming to reduce EMI.

Applying this concept to remove interferences at PLC modems in a neighbour's flat, optimal beamforming can be calculated using the channel parameter's feedback information from a neighbour's modem.

References

1. ETSI, TR 101 562-1 v1.3.1, PowerLine Telecommunications (PLT), MIMO PLT, Part 1: Measurement methods of MIMO PLT, Technical Report, 2012.
2. ETSI, TR 101 562-2 v1.2.1, PowerLine Telecommunications (PLT), MIMO PLT, Part 2: Setup and statistical results of MIMO PLT EMI measurements, Technical Report, 2012.
3. ETSI, TR 101 562-3 v1.1.1, PowerLine Telecommunications (PLT), MIMO PLT, Part 3: Setup and statistical results of MIMO PLT channel and noise measurements, Technical Report, 2012.
4. D. Schneider, Inhome power line communications using multiple input multiple output principles, Dr.-Ing. dissertation, Verlag Dr. Hut, Munich, Germany, January 2012.
5. SCHWARZBECK MESS-ELEKTRONIK; EFS 9218: Active electric field probe with biconical elements and built-in amplifier 9 kHz–300 MHz. http://www.schwarzbeck.com/Datenblatt/m9218.pdf, accessed 16 October 2013.

17

Radiation Mitigation for Power Line Communications Using Time Reversal*

Pascal Pagani, Amilcar Mescco, Michel Ney and Ahmed Zeddam

CONTENTS

17.1 Introduction

With the increasing demand for both high data rate applications and reliable links for command and control systems, *power line communication* (PLC) has emerged in recent years as an attractive communication technique [2]. The main advantage of this technology is its ability to benefit from the existing electrical network infrastructure for the transmission of electromagnetic signals. Hence, it becomes possible to build large communication networks without the need for installing new wires.

In the home or office environment, indoor PLC uses the *low-voltage* (LV) infrastructure. The presence of several electrical outlets in each room of the house allows ubiquitous coverage of the communication network. In addition, the relatively short distance between different outlets allows the system to operate within a limited attenuation. Current *broadband* (BB) in-home systems primarily operate in the frequency range from 2 to 30 MHz. However, recent

---

specifications, such as IEEE 1901 [3] or ITU-T G.9960 [4], allow signal transmission at higher frequencies up to 100 MHz. On the other hand, *narrowband* (NB) PLC systems are deployed at frequencies below 500 kHz, in both indoor and outdoor configurations, using LV or *medium-voltage* (MV) infrastructures [5]. These systems allow the transmission of command and control information over longer distances for Smart Grid applications. The ITU-T G.9955 standard is an example of such NB PLC systems [6]. More information on current NB and BB PLC systems is given in Chapter 10 and subsequent chapters.

LV or MV electrical wires were not initially designed to propagate communication signals at frequencies above 1 kHz. As a consequence, the communication channel between the *transmitter* (Tx) and the *receiver* (Rx) is a difficult channel, generating attenuation and multiple propagation paths. The channel capacity is, hence, limited, and signal processing needs to be optimised so as to maximise the offered data rate and *quality of service* (QoS).

This chapter focuses on one of the main limitations related to the PLC technology, namely, the generation of unintentional radiated signal. This phenomenon is mainly due to the unbalanced nature of the electrical network [7]. The variation of the impedances of the loads connected to the network as well as the unequal length of the live and neutral wires (due to single-phase switches) converts the differential PLC signal into common-mode current flowing through the network. Consequently, the copper wires used for transmitting the useful signal act as an antenna, and part of the transmitted power is radiated. This not only results in stronger signal attenuation at the Rx but also leads to *electromagnetic compatibility* (EMC) issues, as the radiated signal may interfere with other existing services, such as amateur radio (HAM) or *short-wave* (SW) broadcasting. The impact of PLC transmission on EMC has been studied, for example, within the ICT FP7 project OMEGA [8] and through the ETSI Specialist Task Force 410 [9]. Chapter 7 presents a detailed study of the EMC impact of PLC based on a series of field measurements.

In order to avoid interference between PLC systems and other users of the spectrum, regulation authorities impose strict emission masks for the transmission of electromagnetic signals on the power lines. In the United States, the *Federal Communications Commission* (FCC) Part 15 [10] specifies a maximum level of radiated field for carrier-current systems (including PLC), leading system specifications to define constrained power transmission masks. Figure 17.1 represents an example of *power spectral density* (PSD) limits provided in the IEEE 1901 standard for North America. The observed notches are defined to protect specific systems, such as HAM bands. In Europe, CENELEC is currently developing a draft regulation standard applying to in-home PLC systems [11]. Chapter 6 discusses the current regulation status for both NB and BB PLC systems.

Regardless of the regulation limits in place, the research presented in this chapter focuses on the mitigation of unintentional radiation due to PLC systems. Several attempts to solve this problem have been presented in the literature. Reference [12] presents a method to reduce radiated emissions by applying an auxiliary signal cancelling the electromagnetic field on a given point in space. Simulations demonstrated good performance, with the drawback that the *electromagnetic interference* (EMI) could only be mitigated at a single location. The authors in Ref. [13] used additional hardware connected at the wall outlets, in order to reduce asymmetries on power lines, at the cost of an increased complexity of the PLC network.

The presented approach tries to simultaneously reach two complementary goals by means of digital signal processing. The first intention is to focus the transmitted signal at the Rx location. The power gain linked to this energy focalisation allows in turn relaxing the required power level at the Tx, hence generating less EMI. A second target is to reduce the level of energy dissipated at any location except the intended Rx. In particular, it is

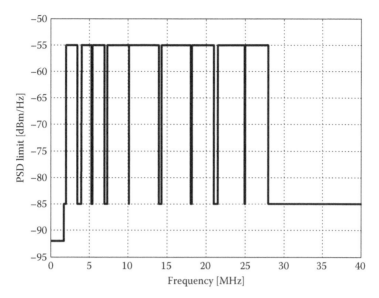

FIGURE 17.1

Example of PSD limit for North America. (From Mescco, A. et al., *Hindawi J. Electr. Comput. Eng.*, 2013, Article ID 402514, Copyright © 2013.)

desirable to minimise the level of radiated power from the electrical wires. These two benefits already appear as features of a known technique in the field of wireless transmission: *time reversal* (TR) [14]. Experimental investigations conducted using *ultra-wideband* (UWB) radio waves demonstrated both the focusing and interference mitigation properties of this technique [15,16].

This chapter presents an experimental analysis of TR as a means to mitigate radiation effects for wired signal transmission. Our investigation focuses on *high-frequency* (HF) BB PLC signals but could be extended to NB PLC and other wired transmission systems, such as *digital subscriber line* (DSL) access. The chapter is organised as follows. Section 17.2 presents the concepts of the TR technique and its application to wired systems. Section 17.3 details the experiment conducted to assess the merits of TR for BB PLC, and the results are statistically analysed in Section 17.4. Section 17.5 validates the concept on the basis of frequency domain measurements. Finally, conclusions are drawn in Section 17.6.

17.2 Time Reversal for Power Line Communications

17.2.1 Time Reversal for Wireless Transmission

The TR technique, also known as phase conjugation in the frequency domain, was first used in the fields of acoustics [17,18]. More recently, this concept has been successfully extended to electromagnetic waves, where the rich multipath channel provides excellent conditions for its application [14]. The basic concept of TR is simple. Let $\delta(\tau)$ be an ideal, Dirac impulse emitted by a Tx antenna (Figure 17.2a). By definition, at any Rx location r_0, the received signal is given by the *channel impulse response* (CIR) $h(\tau, r_0)$. The CIR is composed of multiple echoes reflecting the multiple propagation paths of the propagation channel.

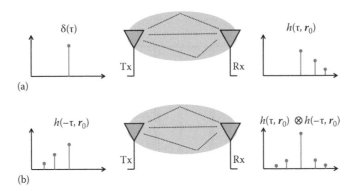

FIGURE 17.2
Transmission over an ideal multipath propagation channel: (a) of a Dirac impulse and (b) of the time reversed CIR. (From Mescco, A. et al., *Hindawi J. Electr. Comput. Eng.*, 2013, Article ID 402514, Copyright © 2013.)

TR uses this *channel state information* (CSI) at the Tx to prefilter the signal to be transmitted. More specifically, the CIR $h(\tau, r_0)$ is time reversed and normalised to serve as an input filter for the signal to be transmitted (Figure 17.2b). Physically, each delayed echo constituting the TR filter travels, among other multiple paths, through its original propagation path. As a result, the multiple echoes sum up coherently at the Rx, hence focusing the received energy in time.

Mathematically, applying TR leads for any Rx situated at an arbitrary location r to the equivalent perceived CIR $h_{TR}(\tau, r)$ (note that this formulation holds for a real valued CIR) [19]:

$$h_{TR}(\tau, r) = \frac{h(-\tau, r_0)}{\sqrt{\int |h(\tau, r_0)|^2 \, d\tau}} \otimes h(\tau, r), \tag{17.1}$$

where the symbol $\otimes$ denotes time domain convolution. Formulating Equation 17.1 in the frequency domain leads to

$$H_{TR}(f, r) = \frac{H^*(f, r_0)}{\sqrt{\int |H(f, r_0)|^2 \, df}} \times H(f, r), \tag{17.2}$$

where
$H(f, r)$ represents the complex valued *channel transfer function* (CTF)
the superscript * denotes the complex conjugate operation

For this reason, TR is sometimes called frequency domain phase conjugation.

Two conclusions can be drawn from Equations 17.1 and 17.2. First, at the intended Rx location r_0, the perceived CIR simplifies to

$$h_{TR}(\tau, r_0) = \frac{1}{\sqrt{\int |h(\tau, r_0)|^2 \, d\tau}} \times R_h(\tau, r_0), \tag{17.3}$$

where $R_h(\tau, r_0)$ denotes the time domain autocorrelation of the function $h(\tau, r_0)$. Similarly, the perceived CTF simplifies to

$$H_{TR}(f,r_0) = \frac{1}{\sqrt{\int |H(f,r_0)|^2 \, df}} \times |H(f,r_0)|^2 . \tag{17.4}$$

In the time domain, the effect of the TR filter is to transform the CIR into its autocorrelation. For a rich multipath environment, the autocorrelation of the CIR presents a large peak at $\tau = 0$, with reduced side echoes. Experimental studies demonstrated that the resulting channel is less spread in time [15,16], hence reducing the possible *intersymbol interference* (ISI). In the frequency domain, the perceived CTF is proportional to the square of the magnitude of the actual CTF. Besides the fact that TR provides a real valued CTF, which could be exploited at the Rx, this also leads to a significant gain in terms of Rx power, due to a better exploitation of the frequency selective nature of the channel. It was demonstrated in Ref. [19] that the application of TR in a flat channel (i.e. without frequency domain power decay) under Rayleigh fading leads to a gain of 3 dB in the total received power. This gain was increased to 5 dB when considering the frequency domain power decay observed in practical UWB radio channels.

The second conclusion drawn from Equations 17.1 and 17.2 is that for any other location r different from r_0, TR creates a mismatch between the Tx filter and the channel. This is particularly observable in the frequency domain representation of TR given in Equation 17.2. The perceived CTF corresponds to the product of two independent CTF, $H(f, r)$ and $H^*(f, r_0)$, with possibly very different frequency-fading structures. More precisely, minima of the first CTF can happen randomly at maxima of the second CTF. Hence, averaging over all frequencies, the total received power at untargeted locations is reduced. In wireless TR analysis, this effect is called spatial focusing and is generally assessed as the ratio between the maximum of $h_{TR}(\tau, r)$ and the maximum of $h_{TR}(\tau, r_0)$ for a given distance $\|r-r_0\|$. Spatial focusing factors of -10 dB have been reported in Refs. [19,20].

17.2.2 Extension of Time Reversal to Wired Transmission

As observed through the study of TR for wireless transmission, the TR scheme provides two main features, namely, an increase of the Rx power at the intended Rx location and a decrease of the Rx power at any other location. These features are highly desirable in the context of wired transmission, where the level of Tx power is constrained by the unintentional radiation from the wires, causing possible EMI to other systems.

Based on this observation, experimental studies were conducted to analyse the potential of TR to mitigate unwanted emissions for PLC systems. The main principles of the extension of TR to PLC transmission can be explained with the help of Figure 17.3.

An intended transmission is assumed between a Tx PLC modem and an Rx PLC modem, over a LV indoor electrical network. Different experimental investigations reported the PLC channel as a rich multipath propagation channel, due to the multiple branches present in a classical electrical network and to the impedance mismatch occurring at the network terminations (outlets) and nodes [21–24]. This similarity of the PLC channel with wireless channels suggests promising results when applying TR to PLC.

With reference to Figure 17.3, the Rx modem is situated at the intended location r_0, and the Tx modem is situated at the origin. By applying TR filtering at the Tx, the Rx power will be increased at location r_0; hence, the Rx modem will benefit from an increased

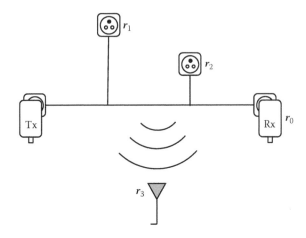

FIGURE 17.3
Principle of the extension of TR to wired transmission. (From Mescco, A. et al., *Hindawi J. Electr. Comput. Eng.*, 2013, Article ID 402514, Copyright © 2013.)

signal-to-noise ratio (SNR). This power increase can also in turn be applied as a reduction of the Tx power to achieve similar performance. At other outlets in the network, situated, for example, at locations r_1 or r_2, the Rx power will be reduced. This effect can be further exploited in the design of multi-user transmission schemes.

For our purpose of radiation mitigation, let us now consider a location r_3, situated at any point in space in the vicinity of the electrical network. The level of radiated field at this location can be evaluated, for instance, by means of an equivalent transfer function $H(f, r_3)$ between the Tx modem and an ideal antenna situated at location r_3. By the virtue of the TR scheme (Equation 17.2), the perceived transfer function at location r_3 after applying TR will be proportional to the product $H^*(f, r_0) \times H(f, r_3)$. As the functions $H^*(f, r_0)$ and $H(f, r_3)$ are not correlated, their frequency fading structures are different. In particular, the deep notches due to frequency-selective fading do not appear at the same frequencies. As a result, the product $H^*(f, r_0) \times H(f, r_3)$ will provide more average attenuation when compared to $H(f, r_3)$ alone, and therefore, the total power radiated at location r_3 will be reduced. A similar observation is made in previous studies dedicated to wireless channels [14,15,18]. Hence, TR appears as an efficient method to mitigate EMI for wired communications.

The theoretical basis for the application of TR to wired transmission being set, the experimental assessment of this method will be described in the next sections.

17.3 Experimental Setup

17.3.1 Equipment

In order to experimentally assess the use of TR as a method to mitigate EMI for wired communication, the experimental setup presented in Figure 17.4 was used. In this setup, a signal generator Tektronix AWG7082C was used as a generic Tx. A *digital sampling oscilloscope* (DSO) LeCroy WaveRunner 715Zi-A was used to sample the received signal at Rx. The signal generated at Tx is denoted $s(t)$, and the signal received at the DSO is denoted $y(t)$. Two universal PLC couplers were used as baluns to couple the Tx and Rx signals with

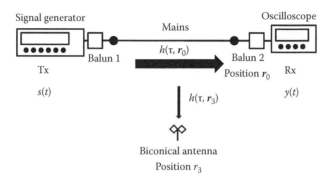

FIGURE 17.4
Equipment used in the experimental setup. (From Mescco, A. et al., *Hindawi J. Electr. Comput. Eng.*, 2013, Article ID 402514, Copyright © 2013.)

the power lines. The Tx coupler is situated at the origin, and the Rx coupler is connected to a plug at location r_0. These couplers were developed within the ETSI Specialist Task Force 410 [25]. In order to measure the power density of the radiated emission received at any arbitrary location r_3, a biconical antenna Schwarzbeck EFS921 was connected to a second port of the DSO.

17.3.2 Calibration

With reference to Figure 17.5, two measurement paths cans be distinguished. Path 1 serves for the measurement of the CTF and is composed of the following elements: the Tx, coaxial cable 1, a 30 dB amplifier IFIM50, coaxial cable 2, balun 1, mains, balun 2, coaxial cable 3, a 20 dB attenuator Radial R412720000 and the Rx. The baluns were considered as part of the channel. The calibration for measurement path 1 was made by directly connecting coaxial cable 2 and coaxial cable 3.

Path 2 is composed by the Tx, coaxial cable 1, a 30 dB amplifier IFIM50, coaxial cable 2, the wire to free space propagation channel (represented by the CTF $H(f, r_3)$), the biconical antenna, coaxial cable 4 and the DSO. The calibration for this second path was made by connecting coaxial cable 2 and coaxial cable 4. The gain of the antenna was removed from the measurements by postprocessing.

The signal generator and the DSO were synchronised by a direct connection of their 10 MHz reference clocks. The sampling rates of both devices were set to $f_s = 100$ MHz.

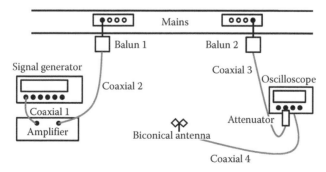

FIGURE 17.5
Experimental setup calibration. (From Mescco, A. et al., *Hindawi J. Electr. Comput. Eng.*, 2013, Article ID 402514, Copyright © 2013.)

17.3.3 Signal Processing

After initial system calibration, the measurements are performed in three steps:

1. The CTF $H(f, r_0)$ is evaluated using a specific Tx frame (see succeeding text). The considered frequency band extends from 2.8 to 37.5 MHz.
2. The TR filter is generated using the phase and magnitude of $H(f, r_0)$.
3. The CTFs $H(f, r_0)$ and $H(f, r_3)$ as well as the perceived CTFs $H_{TR}(f, r_0)$ and $H_{TR}(f, r_3)$ are measured using a single Tx frame (see succeeding text). Measurements at the location of the Rx outlet (r_0) and at arbitrary locations in space (r_3) are made simultaneously using two ports of the oscilloscope, one connected to the Rx balun and the other connected to the biconical antenna.

The Tx signal is generated according to the HomePlug standard [26]. The used frame is called *PHY protocol data unit* (PPDU), and it is composed by a preamble (including a frame control) and a number of payload symbols as shown in Figure 17.6. Each payload symbol consists of a 3072-sample *orthogonal frequency division multiplexing* (OFDM) symbol as defined in the HomePlug specification [26]. In order to estimate the CTFs $H(f, r)$ and $H_{TR}(f, r)$, the OFDM symbols were loaded with predefined constellations.

The frame used for the calibration and the initial measurement of the CTF before computation of the TR filter is represented in Figure 17.7.

The frame used for the measurement of the CTF and EMI after computation of the TR filter is represented in Figure 17.8. Note that in this frame, the TR filter is applied on the last three symbols only (represented by the signal $s'(t)$). From this particular frame scheme, the CTF and EMI can be evaluated with and without application of TR quasi-simultaneously. Recomputing the CTF at this stage also allows monitoring any possible temporal evolution of the channel between calibration and measurement.

In essence, the overall channel estimation process is similar to the channel sounding procedure defined in the HomePlug AV specification. In our experiment, the computation of the CTF from the received signal uses a classical *zero-forcing* (ZF) channel estimation procedure.

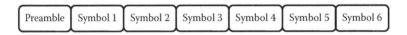

FIGURE 17.6
HomePlug frame. (From Mescco, A. et al., *Hindawi J. Electr. Comput. Eng.*, 2013, Article ID 402514, Copyright © 2013.)

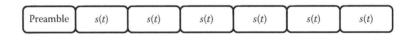

FIGURE 17.7
HomePlug frame for calibration and CTF measurement before computation of the TR filter. (From Mescco, A. et al., *Hindawi J. Electr. Comput. Eng.*, 2013, Article ID 402514, Copyright © 2013.)

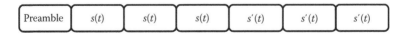

FIGURE 17.8
HomePlug frame for CTF and EMI measurement after application of TR. (From Mescco, A. et al., *Hindawi J. Electr. Comput. Eng.*, 2013, Article ID 402514, Copyright © 2013.)

This is suitable in our experiment involving high levels of SNR at the Rx. In practical systems operating at lower SNR, more sophisticated methods such as the minimum mean square error channel estimation would be more efficient against noise enhancement. Note that the TR filter was purposely implemented in the time domain at the Tx using a programmable waveform generator, in order to measure results as close as possible to a realistic implementation of the TR scheme in practical PLC modems.

In order to estimate channel attenuation continuously over the measured frequency band, no spectral notches were implemented for this study. In practical systems, a Tx PSD mask is defined, where the Tx power is notched at predefined frequencies, to protect existing services using the same spectrum. The HomePlug specification uses windowing in order to better exploit the power allocated within the PSD mask while protecting out-of-band services. The present study concentrates on reducing the EMI within the band effectively used by PLC systems. Therefore, the results also hold for practical systems including notches.

17.3.4 Measured Environment

The measurement campaign was conducted using 13 different topologies of 230 V mains networks within the premises of Orange Labs in Lannion. The campaign took place in different rooms of about 5×4 m². Figure 17.9 presents a picture of the experimental setup in an exemplar location.

The Tx and Rx modems were connected to two outlets in the same room, with distances varying between 2 and 8 m. In general, the rooms are equipped with several other electrical outlets (between 4 and 10). About half of the outlets were connected to classical office appliances (lamps, desktops, etc.). For each topology, one CTF was measured, first without applying TR and then after applying TR filtering. In addition, for each topology, between three and five locations were selected to measure the received electrical field with the help of the biconical antenna. In total, 13 CTF and 43 measurements of the electrical field were collected for statistical analysis.

FIGURE 17.9
Picture of experimentation. (From Mescco, A. et al., *Hindawi J. Electr. Comput. Eng.*, 2013, Article ID 402514, Copyright © 2013.)

17.4 Results and Statistical Analysis

17.4.1 Preliminary Results

The preliminary results for an exemplar network will be presented in two parts: first, the CTF at r_0 and, second, the electrical field and its associated power density at $r_3 \neq r_0$.

The attenuation characteristics of the CTF at r_0 are shown in Figure 17.10.

Let us first consider the measured CTF (solid line). The deep notches at some frequencies are due to reflections at the terminations of the network and reflect the multipath nature of the PLC network. The average attenuation before TR $\overline{H(r_0)}$ is defined in dB as follows:

$$\overline{H(r_0)} = 10\log_{10}\left(\frac{1}{f_{max} - f_{min}} \int_{f_{min}}^{f_{max}} \left| H(f, r_0) \right|^2 df \right), \tag{17.5}$$

where f_{min} and f_{max}, respectively, represent the minimum and maximum sounded frequencies. The average attenuation corresponds to the signal attenuation perceived by an Rx capable of exploiting all the power received in the frequency band from f_{min} to f_{max}. Typically, an OFDM system like the HomePlug AV specification is able to exploit the total received power over a wide frequency band. In our example, the average attenuation is about 11 dB.

Let us now focus on the perceived CTF after applying TR. Owing to the mathematical definition of the TR filter, TR allocates more power to frequencies showing minimal attenuation, while strongly attenuated frequencies are more power constrained. In particular, for all frequencies where the attenuation of the channel $H(f, r_0)$ is higher than the average attenuation $\overline{H(r_0)}$, the perceived channel is more attenuated after applying TR. This can be clearly seen in the frequency range from 26 to 37.5 MHz. For the frequencies where

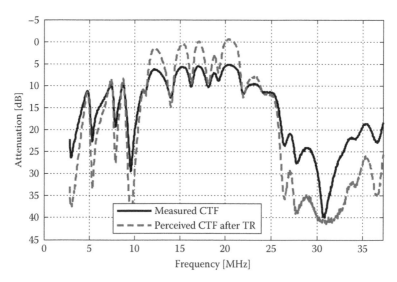

FIGURE 17.10

Channel attenuation before and after TR. (From Mescco, A. et al., *Hindawi J. Electr. Comput. Eng.*, 2013, Article ID 402514, Copyright © 2013.)

the attenuation of the channel is less than $\overline{H(r_0)}$, the response of the channel is improved using TR. This is clearly observable in the frequency range from 11 to 22 MHz. Defining the average attenuation after TR $\overline{H_{TR}(r_0)}$ in dB as

$$\overline{H_{TR}(r_0)} = 10\log_{10}\left(\frac{1}{f_{max} - f_{min}} \int_{f_{min}}^{f_{max}} \left|H_{TR}(f,r_0)\right|^2 df\right), \tag{17.6}$$

it can be observed that the average attenuation after applying TR is about 7.5 dB. Hence, the application of TR provided a gain in the total received power of 3.5 dB in this particular example.

Note that only the measured CTF (solid line) is physical. Therefore, the measured CTF will always degrade the transmission with some degree of channel attenuation. On the contrary, the CTF perceived at Rx after application of TR (dashed line) corresponds to a logical channel, where the effects of the measured CTF are combined with the effects of the TR filter at Tx. Hence, the perceived CTF may exhibit some gain over a limited frequency range.

Let us now consider the electrical field and its associated power density at $r_3 \neq r_0$. The value of the electrical field $E(f, r_3)$ in dBµV/m was computed from the CTF $H(f, r_3)$ measured between the Tx balun and the antenna connector assuming an injected PSD $P_{feed} = -55$ dBm/Hz, and using the following formula [25]:

$$E(f,r_3) = P_{feed} + 20\log_{10}\left(\left|H(f,r_3)\right|\right) + 107 + AF(f), \tag{17.7}$$

where
$AF(f)$ represents the antenna factor
107 represents the conversion from dBm to dBµV

In addition, the average radiated power density $\overline{S(r_3)}$ was computed in dB (W/m²) as follows:

$$\overline{S(r_3)} = 10\log_{10}\left(\frac{1}{f_{max} - f_{min}} \int_{f_{min}}^{f_{max}} \frac{1}{120\pi}\left|E(f,r_3)\right|^2 df\right), \tag{17.8}$$

where
f_{min} and f_{max}, respectively, represent the minimum and maximum sounded frequencies
$E(f, r_3)$ is expressed in V/m

In Equation 17.8, the term 120π provides the value of the impedance of free space in Ohm.

Note that both $E(f, r_3)$ and $\overline{S(r_3)}$ can also be computed after applying TR filtering, using the following equations:

$$E_{TR}(f,r_3) = P_{feed} + 20\log_{10}\left(\left|H_{TR}(f,r_3)\right|\right) + 107 + AF(f) \tag{17.9}$$

and

$$\overline{S_{TR}(r_3)} = 10\log_{10}\left(\frac{1}{f_{max} - f_{min}} \int_{f_{min}}^{f_{max}} \frac{1}{120\pi}\left|E_{TR}(f,r_3)\right|^2 df\right). \tag{17.10}$$

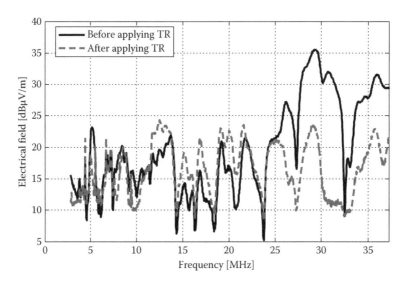

FIGURE 17.11
Electrical field before and after TR. (From Mescco, A. et al., *Hindawi J. Electr. Comput. Eng.*, 2013, Article ID 402514, Copyright © 2013.)

An example of radiated emission measurement is given in Figure 17.11. For this particular electrical network, the mitigation of radiated emissions is clear in the frequency band from 26 to 37.5 MHz, and the average radiated power density has reduced by about 7.5 dB. Thus, it can be observed with this example that the application of TR filtering can reduce significantly the level of undesired radiated power.

17.4.2 Statistical Analysis of the Time Domain Measurements

This section presents the statistical analysis of the measurement database collected within 13 rooms of the office building at Orange Labs in Lannion. The total measurement set is composed of 13 CTF and 43 measurements of the electrical field.

The channel gain G_{TR} observed on the perceived CTF $H_{TR}(f, r_0)$ after application of TR filtering is first computed. As OFDM systems can exploit the total power received over a given frequency band, this gain is computed in dB for the total received power as

$$G_{TR} = \overline{H(r_0)} - \overline{H_{TR}(r_0)} . \tag{17.11}$$

The *cumulative distribution function* (CDF) of G_{TR} is given in Figure 17.12. This parameter shows always a positive gain, between 1.4 and 6.6 dB in our experiment, which is in line with results reported in similar wireless experiments [19]. In about 60% of cases, the channel gain is higher than 3 dB. This means that at Rx, there is always a better reception using TR. This channel gain can in turn be used to reduce the injected PSD at Tx, hence reducing by the same factor the unwanted EMI.

The EMI reduction coefficient R_{TR} was then computed, corresponding to the reduction of the undesired radiated power due to the application of a TR filter. This figure of merit of the reduction of radiated signal is computed in dB as follows:

$$R_{TR} = \overline{S(r_3)} - \overline{S_{TR}(r_3)}. \tag{17.12}$$

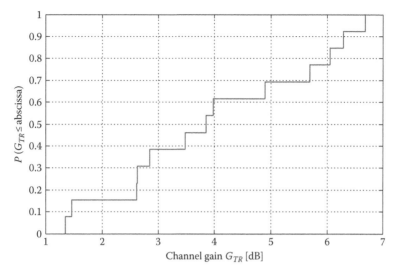

FIGURE 17.12
CDF of the channel gain G_{TR} for the time domain experiment. (From Mescco, A. et al., *Hindawi J. Electr. Comput. Eng.*, 2013, Article ID 402514, Copyright © 2013.)

The CDF of the EMI reduction coefficient R_{TR} is given in Figure 17.13. Results show that the simple application of TR reduces the EMI in more than 60% of cases. In the best case, the EMI reduced by more than 7 dB using TR. In the worst case, the EMI increased by 2 dB. The observations of particular cases indicates that the reduction of EMI is more effective when the CTFs $H(f, r_0)$ and $H(f, r_3)$ are highly decorrelated. This is more likely to happen in complex electrical network topologies, where the rich multipath environment results in different frequency fading structures for different Rx locations.

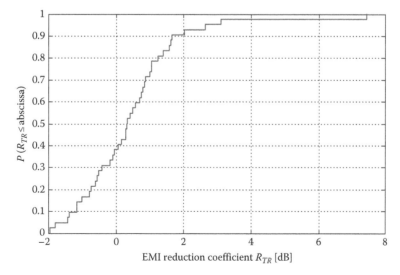

FIGURE 17.13
CDF of the EMI reduction coefficient R_{TR} for the time domain experiment. (From Mescco, A. et al., *Hindawi J. Electr. Comput. Eng.*, 2013, Article ID 402514, Copyright © 2013.)

Observing the statistics of the channel gain G_{TR} and of the EMI reduction coefficient R_{TR}, an optimal strategy can be proposed in order to minimise EMI for a PLC system. Indeed, Figure 17.12 shows that the application of TR provides better system performance due to the reduced channel attenuation. This gives us in turn the flexibility to reduce the level of the Tx power to further reduce EMI while keeping the system performance constant. More precisely, when TR is applied, a Tx power backoff of G_{TR} dB can be applied without modifying the total received power. Finally, following this power backoff strategy, the effective EMI mitigation factor M_{TR} can be computed as the sum of the power backoff and the EMI reduction coefficient:

$$M_{TR} = G_{TR} + R_{TR}. \qquad (17.13)$$

The CDF of the effective EMI mitigation factor M_{TR} is depicted in Figure 17.14. Several conclusions can be drawn from this statistical result:

1. First, the TR method is able to mitigate EMI generated by PLC transmission in 100% of our experimental observations. To this respect, one can conclude that the gain G_{TR} provided by the application of TR allows a Tx power backoff that largely compensates for the possible EMI increment observed in Figure 17.13.

2. Second, in 40% of the cases, the undesired radiated power is reduced by more than 3 dB.

3. Finally, in the most favourable configurations, a reduction of the EMI by more than 10 dB can be observed. Such configurations correspond to cases where the CTF between the Tx and Rx modem and the EMI spectrum are particularly decorrelated.

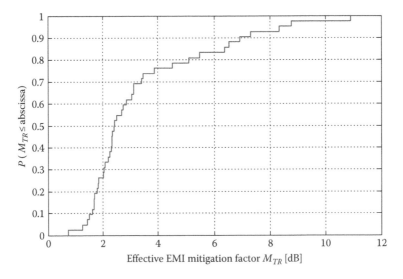

FIGURE 17.14
CDF of the effective EMI mitigation, M_{TR}, for the time domain experiment. (From Mescco, A. et al., *Hindawi J. Electr. Comput. Eng.*, 2013, Article ID 402514, Copyright © 2013.)

17.5 Concept Validation on Frequency Domain Measurements

17.5.1 Frequency Domain Measurements Database

In order to validate the gain of the TR technique in terms of mitigation of the radiation of PLC systems, a larger set of measurements was considered. This data set was collected in nine houses and flats located in France and Germany, in the framework of the ETSI STF 410 measurement campaign [9]. Measurements were recorded using a *vector network analyser* (VNA).

The PLC CTF $H(f, r_0)$ was first recorded between a Tx modem and an Rx modem, at random outlet locations in the house. In a second step, the CTF $H(f, r_3)$ between the Tx modem and an antenna situated at an arbitrary location r_3 was recorded. The small biconical antenna used for this purpose allowed an easy placement at different locations within the house or outside the building. Outdoor measurement locations were selected either at 3 m or at 10 m from the external wall. In total, 114 couples of CTFs $H(f, r_0)$ and $H(f, r_3)$ were available for processing the statistical analysis. For the purpose of comparison with the results obtained in the time domain experiment, the same frequency band extending from 2.8 to 37.5 MHz was considered in the computations.

More details on the ETSI STF 410 measurement campaign and on the resulting analysis of channel, noise characteristics can be found in Chapter 5. In addition, a study of the EMI properties of the PLC channel based on this measurement campaign is provided in Chapter 7.

17.5.2 Statistical Analysis of the Frequency Domain Measurements

The first analysis consisted of studying the channel gain G_{TR} observed on the perceived CTF $H_{TR}(f, r_0)$ after application of the TR filter, see Equation 17.11. Figure 17.15 presents the CDF of this parameter over the whole measurement campaign. On this larger data set, the channel gain G_{TR} of the TR technique is always greater than 0.9 dB and can be as

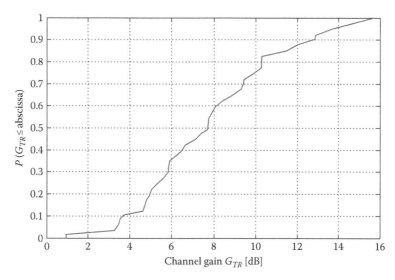

FIGURE 17.15
CDF of the channel gain G_{TR} for the frequency domain experiment.

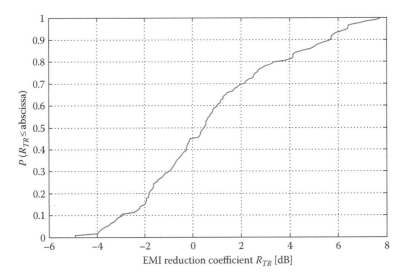

FIGURE 17.16
CDF of the EMI reduction coefficient R_{TR} for the frequency domain experiment.

high as 15.6 dB. About 95% of the channels present a gain larger than 3 dB and 70% of the channels present a gain larger than 6 dB.

The second characteristic related to the application of TR is the EMI reduction coefficient R_{TR}, as defined in Equation 17.12. The CDF of this parameter for the frequency domain campaign is presented in Figure 17.16. One can observe that in 55% of the cases, the total power density is reduced by the TR technique, up to a maximum of 8 dB. However, in 45% of the observations, the EMI is actually increased, with a maximal increase of 5 dB. Hence, a similar trend as for the time domain experiment is observed, with larger extreme values.

The statistics presented in Figure 17.16 tend to show that in a large number of cases, the EMI is actually increased by the application of TR. However, one has to note that in the same time, the total power received at the Rx modem is significantly increased, due to the TR channel gain. Again, the optimal strategy thus consists in reducing the Tx power by G_{TR} dB. This smart power backoff allows further reducing the EMI while preserving the power level observed at Rx, and hence keeping the same system performance. The final EMI reduction is then given by the EMI mitigation factor M_{TR} defined in Equation 17.13. The CDF of parameter M_{TR} for the frequency domain measurements is presented in Figure 17.17. In 98% of the cases, the resulting EMI is decreased, with a maximal reduction of 16.4 dB. Therefore, the good performance of the TR technique is confirmed on this larger data set collected in the frequency domain. One needs to note that in only 2% of the observations, the EMI is actually increased, up to a maximum of 1 dB. Further research will focus on the detection of such cases in order to avoid application of TR if it results in higher radiation.

17.6 Conclusion

This chapter proposed the application of TR in order to mitigate EMI generated by wired communication systems. TR was originally used in the field of wireless transmission as a mean to focus the transmitted signal in both time and space around the intended Rx.

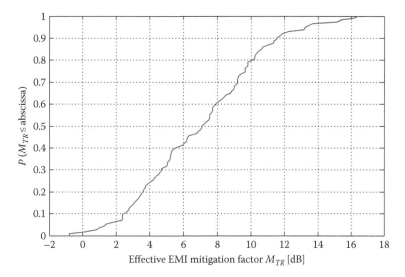

FIGURE 17.17
CDF of the effective EMI mitigation, M_{TR}, for the frequency domain experiment.

It is proposed to use the same property to focus the signal injected in a wired medium, such as the electrical network in the case of PLC, for instance. As a result, the energy lost through undesired radiation is expected to decrease significantly.

An experimental setup was presented in order to demonstrate this method. The experiment was conducted in the time domain using signal frames similar to the industrial specification HomePlug AV. In addition, the TR filter was actually implemented using an arbitrary wave generator, thus providing results encompassing the possible drawbacks of a practical implementation.

Results demonstrated that on the wired medium, TR could provide a transmission channel gain between 1 and 7 dB, which is similar to the gain observed in wireless transmission. In addition, the application of a TR filter alone could effectively reduce EMI for 60% of the observations, with a maximum EMI mitigation of 7 dB. Finally, by combining the channel gain and the EMI reduction features, it was experimentally demonstrated that TR was efficient in 100% of the observed cases to reduce EMI. In 40% of the cases, the EMI mitigation was larger than 3 dB, with maxima higher than 10 dB. These encouraging results were confirmed by a second statistical analysis performed on a larger set of measurements conducted in the frequency domain. These measurements were collected in France and Germany in the framework of the ETSI STF 410 campaign. This second statistical study confirmed the observed trends, with larger extreme values. First, TR channel gains between 1 and 16 dB were observed. Second, the optimal EMI mitigation strategy led to an effective EMI reduction in 98% of the observations, down to a maximum reduction of 16 dB. In 2% of the cases only, the EMI is actually increased, but by less than 1 dB. From these observations, TR is seen as a promising technique to help resolving EMC issues related to PLC and other wired media.

In the TR strategy presented in this chapter, the CTF gain provided by the TR technique was fully dedicated as a power backoff to minimise the EMI. Another strategy could target an increase of the offered capacity while maintaining a constant EMI. Further analyses will, therefore, be dedicated to analyse the trade-off between channel throughput increase and EMI reduction. In addition, future research will focus on the study of wired TR at higher frequencies and on other media, such as DSL cables. Finally, optimal protocols will be developed to practically implement TR in future standards. In particular, the application of TR to multicast or broadcast scenarios, involving one Tx modem and several Rx modems, could be further investigated.

References

1. A. Mescco, P. Pagani, M. Ney and A. Zeddam, Radiation mitigation for power line communications using time reversal, *Hindawi Journal of Electrical and Computer Engineering*, Article ID 402514, 2013.
2. H. C. Ferreira, L. Lampe, J. Newbury and T. G. Swart, eds., *Power Line Communications: Theory and Applications for Narrowband and Broadband Communications Over Power Lines*, Wiley, Chichester, U.K., 2010.
3. IEEE 1901-2010, IEEE standard for broadband over power line networks: Medium access control and physical layer specifications, December 2010. Copyright (c) 20101 IEEE. All rights reserved.
4. ITU-T G.9960, Unified high-speed wireline-based home networking transceivers – System architecture and physical layer specification, June 2010.
5. S. Galli, A. Scaglione and Z. Wang, For the grid and through the grid: The role of power line communications in the smart grid, *Proceedings of the IEEE*, 99(6), 998–1027, June 2011.
6. V. Oksman and J. Zhang, G.HNEM: The New ITU-T standard on narrowband PLC technology, *IEEE Communications Magazine*, 49(12), 36–44, December 2011.
7. M. Ishihara, D. Umehara and Y. Morihiro, The correlation between radiated emissions and power line network components on indoor power line communications, *IEEE International Symposium on Power Line Communications and Its Applications*, Orlando, FL, March 2006, pp. 314–318.
8. Seventh Framework Programme, Theme 3 ICT-213311 OMEGA, Deliverable D3.3, Report on electromagnetic compatibility of power line communications and its Applications, December 2009.
9. A. Schwager, W. Bäschlin et al., European MIMO PLC field measurements: Overview of the ETSI STF410 campaign & EMI analysis, *IEEE International Symposium on Power Line Communications and Its Applications (ISPLC)*, Beijing, China, March 2012, pp. 304–309.
10. Federal Communications Commission, Title 47 of the Code of Federal Regulations, part 15. 2007.
11. CENELEC, Final Draft European Standard FprEN 50561-1 Power line communication apparatus used in low voltage installations – Radio disturbance characteristics – Limits and methods of measurement – Part 1: Apparatus for in-home use, June 2011.
12. A. Vukicevic, M. Rubinstein, F. Rachidi and J.-L. Bermudez, On the impact of mitigating radiated emissions on the capacity of PLC systems, *IEEE International Symposium on Power Line Communications and Its Applications*, Pisa, Italy, March 2007, pp. 487–492.
13. P. Favre, C. Candolfi and P. Krahenbuehl, Radiation and disturbance mitigation in PLC networks, *20th International Zurich Symposium on Electromagnetic Compatibility*, Zürich, Switzerland, January 2009, pp. 5–8.
14. G. Lerosey, J. de Rosny, A. Tourin, A. Derode, G. Montaldo and M. Fink, Time reversal of electromagnetic waves, *Physical Review Letters*, 92, 193904, 1–3, May 2004.
15. A. E. Akogun, R. C. Qiu and N. Guo, Demonstrating time reversal in ultra-wideband communications using time domain measurements, *51st International Instrumentation Symposium*, Knoxville, TN, May 2005.
16. A. Khaleghi, G. El Zein and I. H. Naqvi, Demonstration of time-reversal in indoor ultra-wideband communication: Time domain measurement, *Fourth International Symposium on Wireless Communication Systems*, Trondheim, Norway, October 2007, pp. 465–468.

17. A. Derode, P. Roux and M. Fink, Acoustic time-reversal through high-order multiple scattering, *Proceedings of the IEEE Ultrasonics Symposium*, vol. 2, Seattle, WA, November 1995, pp. 1091–1094.
18. D. R. Jackson and D. R. Dowling, Phase conjugation in underwater acoustics, *Journal of the Acoustical Society of America*, 89, 171–181, January 1991.
19. P. Pajusco and P. Pagani, On the use of uniform circular arrays for characterizing UWB time reversal, *IEEE Transactions on Antennas and Propagation*, 57(1), 102–109, January 2009.
20. C. Zhou and R. C. Qiu, Spatial focusing of time-reversed UWB electromagnetic waves in a hallway environment, *Southeastern Symposium on System Theory*, Cookeville, TN, March 2006, pp. 318–322.
21. M. Tlich, A. Zeddam, F. Moulin and F. Gauthier, Indoor power line communications channel characterization up to 100 MHz – Part I: One-parameter deterministic model, *IEEE Transactions on Power Delivery*, 23(3), 1392–1401, July 2008.
22. M. Tlich, A. Zeddam, F. Moulin and F. Gauthier, Indoor power line communications channel characterization up to 100 MHz – Part II: Time-frequency analysis, *IEEE Transactions on Power Delivery*, 23(3), 1402–1409, July 2008.
23. A. M. Tonello and F. Versolatto, Bottom-up statistical PLC channel modeling – Part I: Random topology model and efficient transfer function computation, *IEEE Transactions on Power Delivery*, 26(2), 891–898, April 2011.
24. A. M. Tonello and F. Versolatto, Bottom-up statistical PLC channel modeling – Part II: Inferring the statistics, *IEEE Transactions on Power Delivery*, 25(4), 2356–2363, October 2010.
25. ETSI TR 101 562-1 V2.1.1 Technical Report, Powerline Telecommunications (PLT), MIMO PLT, Part 1: Measurements methods of MIMO PLT, Chapter 7.1, 2012.
26. HomePlug, HomePlug AV specification, version 1.1, May 21, 2007.

18

Linear Precoding for Multicarrier and Multicast PLC

Jean-Yves Baudais and Matthieu Crussière

CONTENTS

18.1 Introduction

Since their introduction during the 1960s as a solution to the issue of high-bitrate transmission over dispersive communication channels, multicarrier techniques have intensively been studied and proposed in many communication standards [1]. Well known under the acronym of OFDM, standing for orthogonal frequency division multiplexing, and first envisaged for wireless transmissions, multicarrier concepts have also been proposed for *digital transmission over copper wire subscriber loop* (DSL) systems under the *digital multitone* (DMT) terminology. For 10 years, multicarrier schemes have become the basis of the physical layer of all modern communication standards, from European *digital video broadcasting* (DVB) to American IEEE-802.11 *wireless local area network* (WLAN) and worldwide 4G cellular systems. In the *power line communication* (PLC) landscape, multicarrier schemes have also been adopted by the HomePlug Alliance and integrated as the baseline technology for the HomePlug AV specification.

The reasons for such a success story come from its robustness in the case of frequency-selective fading channels, its capability of portable and mobile reception and its flexibility. The multicarrier concept relies on the conversion of a serial high-rate stream into several low-rate streams distributed over closely spaced orthogonal subcarriers or tones. Using a sufficiently large number of subcarriers, any frequency-selective channel can hence be converted into non-interfering flat-fading sub-channels. As a result, single-tap frequency-based equalisation can be carried out to compensate for channel distortions instead of time-domain filter equalisation as needed in single carrier systems. Last but not least, multicarrier systems can be employed to realise a spectral efficiency benefit by adaptively sizing the data constellation used in each subcarrier with respect to its *signal-to-noise ratio* (SNR). This principle, referred to as adaptive loading, can be easily implemented in PLC systems since the SNR values can accurately be measured in the receiver and communicated to the transmitter over a feedback channel.

Apart from multicarrier schemes, spread spectrum is another communication strategy that takes on a significant role in cellular and personal communications [2]. The name spread spectrum stems from the fact that the transmitted signals occupy a much wider frequency band than is required to transmit the information. This confers many well-known advantages to spread spectrum signals such as immunity against multipath distortion, resistance to jamming, low power transmission possibility and signal hiding capability. Spread spectrum signals were originally designed by the military services to overcome either intentional or unintentional interference but are today used to provide reliable communications in a variety of commercial applications. For example, spread spectrum signals are used in the so-called unlicensed frequency ISM bands at 2.4 GHz for cordless telephones, wireless LAN's and Bluetooth. But, more interestingly, spread spectrum signals can also provide communications to several concurrent users through *code division multiple access* (CDMA) when using pseudorandom patterns. CDMA is employed in several wireless cellular standards such as IS-95, WCDMA and UMTS.

The advantages and success of multicarrier and spread spectrum techniques motivated many researchers to investigate the suitability at the combination of both techniques [3]. This combination known as *multicarrier spread spectrum* (MC-SS) benefits from the main advantages of both schemes: high spectral efficiency, high flexibility, multiple access capabilities, narrowband interference rejection, simple one-tap equalisation, etc. MC-SS schemes can actually be viewed as an extension of the multicarrier concept in case the information data stream to be transmitted is linearly precoded using spreading sequences.

Many variations around the MC-SS concept have been introduced during the 1990s trying to find the most efficient precoding function to use along with the multicarrier modulation, essentially depending on the way this function is carried out: time-wise or frequency-wise, before or after multicarrier modulation or even exploiting or not multiple access capabilities. Whereas this concept has been originally proposed for multicarrier access scheme, it can be extended to single-user multicarrier systems and it is referred as *linear precoded OFDM* (LP-OFDM) [4]. As far as wireline communications are concerned, MC-SS has been firstly proposed in [5,6] for xDSL transmission and later for PLC [7,8], hereby demonstrating the interest of the precoding concept in terms of bitrate maximisation.

In this chapter, we review the principles and the major results of the adaptive LP-OFDM communication system. Improvements encountered in adaptive multicarrier systems can in fact also be obtained in adaptive LP-OFDM systems. Therefore, the bitrate maximisation problem is described and analysed in the PLC context. Significant bitrate gains are reached integrating a simple precoding function in the transmission scheme and modifying conventional bit-loading algorithms in accordance with the waveform modification. To illustrate the benefit of the precoding concept, we propose to treat the problem of bitrate maximisation under the simple target *symbol error rate* (SER) or noise margin constraint at first and then under the more challenging constraint of target *bit error rate* (BER). Also, to take full advantage of the precoding function flexibility, we propose to optimise the LP-OFDM system using frequency, time and even 2D precoding. Point-to-point and multicast point-to-multipoint transmissions are simulated through in-home power line channels to quantify the bitrate gains.

The chapter is organised as follows. The multicarrier system and the linear precoding function are introduced and described in Section 18.2. In this section, we also propose a synthetic study on resource allocation considering the case of discrete bit allocation. Then, Section 18.3 analyses the problem of resource allocation for bitrate maximisation when discrete modulations are employed. A time-frequency application of the precoding is analysed in Section 18.4. The multicast context is investigated in Section 18.5 and numerical examples highlight the powerfulness of the linear precoding. Section 18.6 concludes the chapter.

18.2 Linear Precoding for Multicarrier Systems

18.2.1 System Model

As mentioned in Section 18.1, the reference model proposed herein is based on linearly precoded multicarrier modulation, referred to in the sequel as LP-OFDM. Let us first introduce the OFDM system and the related signal and then give the extension to LP-OFDM.

18.2.1.1 OFDM System

A simple representation of an OFDM system is given in Figure 18.1 where the precoding and deprecoding blocks have first to be ignored. In this model, basically, the binary information data stream enters a modulation and coding block which produces a series of n digitally modulated symbols, such as *quadrature amplitude modulations* (QAMs) mapped symbols. Those n symbols are then multiplexed in the frequency domain by the OFDM function which ensures their parallel transmission onto n orthogonal subcarriers. Assuming a convenient choice of the OFDM parameters with respect to the channel characteristics,

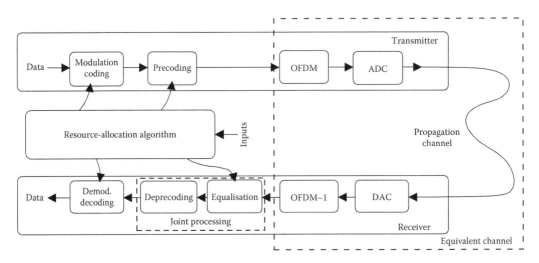

FIGURE 18.1
LP-OFDM communication system.

and perfect time and frequency synchronisation of the receiver, the transmission system can essentially be modelled as n independent and non-interfering flat-fading links or sub-channels in parallel associated to the n subcarriers of the OFDM modulation. Hence, the input–output relationship of the ith subcarrier can directly be written in the frequency domain as follows:

$$y_i = h_i s_i + v_i, \quad \text{for all } i = 1, \dots n \tag{18.1}$$

where
  $\{s_i\}_{i=1}^n$ and $\{y_i\}_{i=1}^n$ are the transmitted and received symbols in subcarrier i, respectively
  $\{v_i\}_{i=1}^n$ are related to white Gaussian noise samples with variance σ^2
  $\{h_i\}_{i=1}^n$ are the complex channel coefficients modelling the amplitude and phase rotations independently experienced by the symbols in each subcarrier i

The PLC channel varies in time with long-term and short-time variations [9]. This time-varying nature requires periodic estimation of the channel to perform equalisation or channel adaptation [10]. The channel can be considered quasi-static between two estimations and it is possible to integrate resource allocation functionalities at the transmitter side as depicted in the system model. This means that the OFDM transmitter will have the opportunity to adapt the order of the symbols of each subcarrier and to allocate an appropriate amount of power to these symbols according to the known channel gain. Following this idea, the transmitted symbols are written as

$$s_i = \sqrt{p_i} x_i, \tag{18.2}$$

where symbols x_i are assumed to be unit power whatever the order of the selected constellations, that is, $E[x_i x_i^*] = 1$, $\forall i$, and power scaling factors p_i are supposed to be obtained through a power allocation policy as it will be detailed in Section 18.3. Throughout the chapter, $(\cdot)^*$ will stand for conjugate transpose which reduces to conjugate for scalar and transpose for real matrix.

Using matrix representations finally leads to the following expression:

$$Y = HP^{1/2}X + N, \tag{18.3}$$

where
  Y, X and N are column vectors of length n with elements y_i, x_i and v_i, respectively
  H and P are diagonal matrices of size $n \times n$ filled up with the h_i and the p_i, respectively

Assuming that the *channel state information* (CSI) is made available at the receiver owing to a channel estimation function, channel gain h_i can easily be compensated for through a linear detector consisting in a one-tap equalisation per subcarrier. This leads to the following estimated symbols Z:

$$Z = WY = WHP^{1/2}X + WN, \tag{18.4}$$

where W is the equalisation matrix whose elements can be derived from various equalisation criteria as, for instance, *zero forcing* (ZF) or *mean square error* (MSE). In OFDM, all these criteria lead to equivalent system performance and W is usually chosen as

$$W = (H^* H)^{-1} H^*, \tag{18.5}$$

which is simply reduced to H^{-1}.

18.2.1.2 LP-OFDM System

The LP-OFDM system is simply obtained activating the precoding and deprecoding functions in the generic system model in Figure 18.1. Consequently, note that the resource allocation algorithm has to take into consideration the waveform structure modification. This aspect will be fully investigated in Section 18.3 in the case of system throughput maximisation.

As discussed in Section 18.1, many variations of the precoding function can be proposed, depending on which dimension (namely, time, frequency or both) it is applied to. Considering PLC applications, time precoding is proposed in [11] as in the well-known MC-DS-CDMA waveform, frequency precoding is studied in [12] leading to the so-called MC-CDMA waveform, and extension to 2D precoding is investigated in [13]. For the sake of clarity, we give a complete and comprehensive study in the case of frequency domain precoding in this part and further considerations about time and 2D precoding will be given in Section 18.4.

Considering frequency domain precoding, the mathematical expression of the OFDM signal has to be updated to yield that of the LP-OFDM signal. This is achieved by means of a precoding matrix denoted $C \in \mathbb{R}^{n \times n}$ which is applied to symbol vector S defined in Equation 18.2. The received signal then writes

$$Y = HCP^{1/2}X + N. \tag{18.6}$$

Matrix C contains the precoding sequences that realise the spreading of the set of symbols X over the n sub-channels defined by the multicarrier system structure. Many solutions can be envisaged to build or give a particular structure to the precoding matrix C as detailed in the following section. Note that simply choosing $C = I_n$, with I_n being the identity matrix of size $n \times n$, amounts, in fact, to the traditional OFDM system.

As in the OFDM case, linear detectors can be used at the receiver to deal with the channel distortions. However, the frequency selectivity of the propagation channel destroys the precoding structure embedded in the LP-OFDM signal, which translates into interference between the precoded symbols called *mutual code interference* (MCI). Hence, the detector performance strongly depends on its capability to handle the MCI while mitigating the Gaussian noise effect. For instance, an MRC detector yields poor performance in LP-OFDM [14], whereas ZF and *minimum mean square error* (MMSE) detectors are worth of interest. The ZF criteria lead to the following general expression of the detector [15]:

$$W_{ZF} = (P^{1/2}C^*H^*HCP^{1/2})^{-1}P^{1/2}C^*H^*, \tag{18.7}$$

where the equalisation and the precoding are jointly treated. Such a detector perfectly compensates for the MCI and is able to extract the information symbols x_i without any interference. It can then be viewed as a decorrelating detector but however leads to a significant noise term boost. On the other hand, the MMSE detector writes

$$W_{MMSE} = (P^{1/2}C^*H^*HCP^{1/2} + \sigma^2 I_n)^{-1}P^{1/2}C^*H^*. \tag{18.8}$$

The MMSE detector is the optimal linear detector since it trades off MCI reduction against noise effect mitigation. However, its complexity turns out to be much higher than that of the ZF detector. In the sequel, we will analyse in what extent the use of the ZF or the MMSE detector has impact on the resource allocation efficiency.

18.2.2 Optimum Linear Precoding

A first step in the LP-OFDM system design and resource allocation optimisation is to lead investigations about the kind of precoding sequences that should be used in the system. The optimal choice can be found by computing the constraint mutual information between the transmitted and received LP-OFDM signals as a function of matrix C. Assuming information vectors X and Y defined earlier as Gaussian vectors, and exploiting the multicarrier waveform structure of the LP-OFDM signal as n non-interfering Gaussian links, we have [12]

$$I(X \mid H, Y \mid H) = \log_2 \det\left(I_n + \frac{1}{\sigma^2}HCR_SC^*H^*\right), \tag{18.9}$$

where $R_S = E[SS^*]$ represents the data symbol covariance matrix of size $n \times n$. From the previous definition of S, R_S is then a diagonal matrix with diagonal elements $\{p_i\}_{i=1}^n$, which yields

$$I(X \mid H, Y \mid H) = \log_2 \det\left(I_n + \underbrace{(1/\sigma^2)HCPC^*H^*}_{A}\right). \tag{18.10}$$

Then, using the Hadamard inequality [16] to find an upper bound to the determinant of matrix A, it comes

$$|\det(A)| \le \prod_{i=1}^{n} \|A_i\|_2, \tag{18.11}$$

where
  A_i is the ith column of matrix A
  $\|\cdot\|_2$ the Euclidean norm

This determinant is maximised if and only if vectors A_i are mutually orthogonal, that is, $\forall i \ne j \; A_i^* A_j = 0$. Leading the calculation gives the following equation:

$$A_i^* A_j = \frac{2}{\sigma^2} \Re\left(h_i h_j^*\right) C_i^T PC_j + \frac{1}{\sigma^2} h_i h_j^* \sum_{l=1}^{n} C_l^* PC_i \, |h_i|^2 \, C_j^* PC_l, \tag{18.12}$$

from which it can be noticed that imposing P to be an identity matrix and C_i, C_j to be orthogonal sequences for all $i \ne j$ are sufficient conditions to have $A_i^* A_j = 0$. This is simply obtained choosing C as a Hadamard matrix while assigning the same amount of power to each code sequence C_i, that is, having $p_i = p$ for all $i \in [1,n]$. Note that there exist some other sequences, not necessarily orthogonal, that also lead to maximal constraint mutual information. Nevertheless, the proposed overview focuses on orthogonal sequences. To give more generality and flexibility to the precoding process, we however use the set C of orthogonal matrices which are composed of Hadamard matrices with the following definition.

Definition 18.1 *Let C be the set of optimal precoding matrices such that*

$$C \in \mathcal{C} \Leftrightarrow c_{i,j} \in \{-1,0,1\} \quad \text{and} \quad C_i^T C_j = 0, \forall i \ne j.$$

Following this definition, the precoding matrix can be sparse. The data symbols that are precoded by C are consequently not necessarily spread all over the multicarrier spectrum but can be rather limited to a subset of subcarrier. The corresponding system can then be viewed as a multiple block system as proposed in [17]. This idea can be exemplified by the following sparse precoding matrix:

$$C = \begin{pmatrix} 1 & 1 & 1 & 1 & 0 & 0 \\ 1 & -1 & 1 & -1 & 0 & 0 \\ 0 & 0 & 0 & 0 & 1 & 1 \\ 1 & 1 & -1 & -1 & 0 & 0 \\ 0 & 0 & 0 & 0 & 1 & -1 \\ 1 & -1 & -1 & 1 & 0 & 0 \end{pmatrix}. \tag{18.13}$$

This matrix is orthogonal and exhibits two precoding blocks: The first block, corresponding to the first four columns, spreads four data symbols across four subcarriers, and the second block, related to the last two columns, spreads two data symbols across two other subcarriers.

When the precoding matrix is not sparse, C is then reduced to a classical Hadamard matrix of order n, as stated by the following property.

Property 18.1 *If $\exists i$ such that $\|C_i\|_2^2 = n$, then $\|C_j\|_2^2 = n$ for all $j \in [1,n]$ and C is a Hadamard matrix with $n \in \{1, 2, 4s \mid s \in \mathbb{N}\}$ and $CC^T = nI_n$.*

The Hadamard conjecture proposes that Hadamard matrices exist for sizes n with $n = 1$, $n = 2$ or n being a multiple of 4. Hence, the well-known Sylvester method [18], which gives a simple solution for the construction of Hadamard matrices of size $n = 2^s$, is not sufficient to build all possible Hadamard matrices. The Paley, Turyn or Williamson methods [18] help to find other sizes of matrices. For example, for $n = 12$, the Hadamard matrix can be

$$
C = \begin{pmatrix}
+1 & -1 & -1 & -1 & -1 & -1 & -1 & -1 & -1 & -1 & -1 & -1 \\
+1 & +1 & -1 & +1 & -1 & -1 & -1 & +1 & +1 & +1 & -1 & +1 \\
+1 & +1 & +1 & -1 & +1 & -1 & -1 & -1 & +1 & +1 & +1 & -1 \\
+1 & -1 & +1 & +1 & -1 & +1 & -1 & -1 & -1 & +1 & +1 & +1 \\
+1 & +1 & -1 & +1 & +1 & -1 & +1 & -1 & -1 & -1 & +1 & +1 \\
+1 & +1 & +1 & -1 & +1 & +1 & -1 & +1 & -1 & -1 & -1 & +1 \\
+1 & +1 & +1 & +1 & -1 & +1 & +1 & -1 & +1 & -1 & -1 & -1 \\
+1 & -1 & +1 & +1 & +1 & -1 & +1 & +1 & -1 & +1 & -1 & -1 \\
+1 & -1 & -1 & +1 & +1 & +1 & -1 & +1 & +1 & -1 & +1 & -1 \\
+1 & -1 & -1 & -1 & +1 & +1 & +1 & -1 & +1 & +1 & -1 & +1 \\
+1 & +1 & -1 & -1 & -1 & +1 & +1 & +1 & -1 & +1 & +1 & -1 \\
+1 & -1 & +1 & -1 & -1 & -1 & +1 & +1 & +1 & -1 & +1 & +1
\end{pmatrix}. \tag{18.14}
$$

For $n > 12$, there exist multiple distinct Hadamard matrices. The first order of unknown Hadamard matrix is 668 and, up to now, only 13 matrices of size lower than 2000 are not known. In the sequel and for the theoretical analysis, we will admit that all sizes of Hadamard matrices are available. In practical systems, there are enough Hadamard matrices to design the precoded systems.

18.2.3 Power Line Constraints and Bitrate Maximisation Problem Formulation

In the perspective of resource allocation optimisation, let us remind the constraints encountered in the power line context [19].

18.2.3.1 Power Spectrum Mask

The first constraint is the power spectrum mask that limits the transmitted power for each subcarrier of the OFDM signal. Consequently, PLC transmissions have to fulfil peak power rather than sum-power conditions. With a power upper limit of p on each subcarrier, we can derive the peak-power constraint for an OFDM signal as follows:

$$
E\left[\left\| P^{1/2} X \right\|_\infty^2 \right] \le p. \tag{18.15}
$$

Using Equation 18.2, it yields

$$p_i \leq p \quad \forall i \in [1;n]. \tag{18.16}$$

Now considering the LP-OFDM system, the precoding matrix C is added in Equation 18.15 and the peak-power constraint writes

$$E\left[\left\|CP^{1/2}X\right\|_{\infty}^2\right] \leq p, \tag{18.17}$$

which leads to

$$\sum_{i=1}^{n} p_i \leq p \tag{18.18}$$

for non-sparse precoding matrices. With uniform power allocation, this constraint leads to $p_i \leq p/n$ for all i. Interestingly, from this equation, we can notice that the precoding function brings about the following property.

Property 18.2 *Orthogonal precoding converts the peak-power constraint into a sum-power constraint.*

The precoding, in fact, introduces some dependency between the data symbols at the transmitter side that transforms the peak constraint to sum constraint. Contrary to peak constraint, the sum constraint permits flexibility in allocation to fully exploit the power constraint p with multiple choices for the p_i.

Taking into account that many notches are specified in the PLC power mask, the effective peak-power constraint cannot be considered as uniform over the whole bandwidth. This can however be translated into a uniform peak-power constraint by integrating the power mask fluctuations into the channel transfer function, that is, defining new channel coefficients $h_i' = \alpha_i \cdot h_i$, where α_i correspond to weighting factors depending on the power mask. More importantly, for LP-OFDM, the set C in Definition 18.1 remains optimal even under non-uniform peak-power constraint. From these perspectives, the power mask will be considered as flat in the following without loss of generality.

18.2.3.2 Constellations of Discrete Orders

The second constraint is the discrete nature of the constellation size. The number r_i of bits per constellation symbol is the bitrate per 2D symbol and is defined in the set $\{0, \beta, 2\beta, 3\beta, \ldots, r_{max}\}$ with $\beta > 0$. The maximal constellation size is given by r_{max} and the granularity of the constellations is given by β which has the same unit as r_i. This has strong impact in practice when peak-power constraint is considered since the constellation size cannot be finely adjusted to the available amount of peak power p when the number of possible sizes is limited. In other words, the peak-power constraint imposes to use β as low as possible to reduce the bitrate loss due to the under-exploitation of the peak power. Considering the modulation and the channel coding as a couple, it is possible to define β lower than 1. At the contrary, without any channel coding, $\beta = 1$ for all sizes of square and rectangular QAM. In practice, the ultimate measure is the coded error rate of the communication system. However, the coded error rate is strongly related to the uncoded one and the channel coding and modulation designs are separable issues [20,21]. Therefore, we focus on the uncoded part of the system to design it in the following and we will assume β integer.

18.2.3.3 Target SNR Gap and BER

The last constraint to consider is the *quality of service* (QoS). We propose to consider best-effort applications, which means that the system has to operate at maximum bitrate. The QoS turns into a constraint of maximum target error rate from the physical layer point of view. In practice, a so-called SNR gap is usually introduced to give some confidence to the target error rate [22]. The SNR gap, or power gap, is a convenient mechanism for analysing systems that transmit below channel capacity, or constraint capacity, or without Gaussian input, and the bitrate becomes

$$r_i = \log_2\left(1 + \frac{\text{snr}_i}{\gamma_i}\right). \tag{18.19}$$

This gap γ_i depends on the constellation, coding format and error probability. The QAM approximation leads to $\gamma_i = \gamma$ independent of i under SER constraint [22]. With BER constraints, the QAM approximation is less accurate and using γ_i instead of γ [23] leads to higher bitrate. We consider both SER and BER constraints to design LP-OFDM systems. The results are mainly developed with SER constraint and the BER constraint is treated as an extension.

18.2.3.4 Problem Formulation

Eventually, the optimisation of the proposed PLC system consists in the maximisation of the bitrate taking into account the aforementioned constraints. This can be summarised in the form of the following problem.

Problem 18.1 *Maximise the bitrate under peak-power constraint p per carrier, under discrete and finite number of bits or bitrate r_i per 2D symbol, $r_i \in \{0, \beta, 2\beta, 3\beta, \ldots, r_{\max}\}$, under error rate or noise margin constraint, and with precoding matrices defined in C.*

18.2.4 Continuous Bitrate Considerations

Before solving the previously stated problem, it is interesting to lead analysis relaxing the constraint of discrete constellations. Hereby, we enable the use of continuous bitrate in the system, that is, $r_i \in \mathbb{R}_+$. It is then possible to make use of convex optimisation tools and analytically find the optimal set of power allocation $\{p_i\}_{i=1}^n$ to be used in the LP-OFDM system to maximise the total bitrate. The results are developed hereafter and will be used as benchmark for the constrained discrete case in the next section.

For the sake of clarity and without loss of generality, the analysis of the problem is firstly presented assuming non-sparse precoding matrix C, with C of Hadamard type. We focus on linear receivers as previously mentioned in Equations 18.7 and 18.8 where the matrix W of the detector jointly realises the equalisation and deprecoding process. Two receivers are analysed: the decorrelating detector and the optimal linear detector.

The predistortion is also considered in the case of ZF criterion. As the precoding is used at transmitter side to improve the bitrate, we may ask whether the channel can be also corrected at the transmitter side, that is, using predistortion. The predistortion is then compared to the equalisation with ZF criterion before analysing the optimal linear detector.

18.2.4.1 Decorrelating Detector

The general expression of the detector matrix is given in Equation 18.7. In our case, with invertible and square matrices, the matrix of the detector reduces to

$$W = \frac{1}{n} P^{-1/2} C^T H^{-1}, \tag{18.20}$$

which yields

$$Z = X + \frac{1}{n} P^{-1/2} C^T H^{-1} N. \tag{18.21}$$

Using this formulation, the equalisation and the deprecoding functions turn out to be separable as mentioned in Figure 18.1. Estimated symbols Z then writes as the transmitted ones, namely, X, corrupted by a coloured noise term. The total continuous bitrate achieved by the system can consequently be viewed as the sum of the bitrates $\{r_i\}_{i=1}^n$:

$$R = \sum_{i=1}^{n} \log_2 \left(1 + \frac{1}{\gamma} \frac{n^2}{\sum_{j=1}^{n} \left(1/|h_j|^2 \right)} \frac{p_i}{\sigma^2} \right). \tag{18.22}$$

Note that taking $\gamma = 1$ in this equation yields the system capacity, whereas R means the achievable bitrate at a given SER for $\gamma > 1$. It is then possible to state the bitrate maximisation problem in the case of continuous bitrates as follows.

Problem 18.2 *Maximise achievable bitrate of the system* $\max_{\{p_i, i \in [1,n]\}} R$ *under the power constraint* $\sum_{i=1}^{n} p_i \leq p$, *with R given in Equation 18.22.*

In this problem, the power constraint is stated as a sum constraint as explained earlier in the case of Hadamard precoding matrices (see Property 18.2). Since the term depending on the channel state in Equation 18.22, namely, $\sum_{j=1}^{n} |h_j|^{-2}$, does not depend on index i, it is straightforward to prove that the total achievable bitrate writes [7]

$$R \leq n \log_2 \left(1 + \frac{1}{\gamma} \frac{n}{\sum_{j=1}^{n} \left(1/|h_j|^2 \right)} \frac{p}{\sigma^2} \right). \tag{18.23}$$

This indicates that the optimal power allocation is obtained with $p_i = p/n$, $\forall i$, and the optimal bit allocation with R/n bits per precoding sequence. Interestingly, this result corroborates the previous analysis made in Section 18.2.2 concerning optimal precoding matrices.

From there, we can state the following proposition which will be used as a reference result throughout the chapter.

Proposition 18.1 *The maximal achievable continuous bitrate of an LP-OFDM system equipped with a ZF receiver and using Hadamard precoding matrices is reached with equal power and bit allocations under sum-power constraint.*

Let us now analyse the impact of the precoding factor n on the maximum bitrate and particularly compare two cases: OFDM and LP-OFDM. In the case of continuous bitrates, it can be seen that

$$
n \log_2 \left(1 + \frac{1}{\gamma} \frac{n}{\sum_{j=1}^{n} \left(1/|h_j|^2 \right)} \frac{p}{\sigma^2} \right) \le \sum_{i=1}^{n} \log_2 \left(1 + \frac{1}{\gamma} \frac{|h_i|^2}{\sigma^2} p \right), \tag{18.24}
$$

equality being obtained if and only if $|h_i|$ is constant for all i. This means that the continuous bitrate of precoded systems is lower than that of unprecoded ones as soon as frequency selectivity is experienced by the transmitted signal. In other words, the OFDM system provides the highest achievable throughput in that case. To go further, using the fact that the harmonic mean of an exponential distribution is zero and assuming $|h_i|$ are iid Rayleigh fading channel coefficients, it comes that the continuous bitrate of an LP-OFDM system is upper bounded by zero when the size of the precoding matrix tends to infinity. This very negative conclusion can be understood by the fact that the LP-OFDM system suffers from the noise boost induced by the ZF detector which in effect is all the more destructive when the channel frequency selectivity is high.

18.2.4.2 Predistortion versus Equalisation

Assuming CSI knowledge at the transmitter and under sum-power constraint, it is well known that optimal power allocation is obtained through the water-filling algorithm [24]. Water filling can in fact be viewed as a predistortion of the transmitted signal. With peak-power constraint, however, previous results show that full power allocation per subcarrier is the optimal allocation strategy for OFDM signalling. This eventually indicates that predistortion is not recommended in that case. Nevertheless, since we have shown that peak-power constraint is converted to a sum-power constraint when precoding is activated, it is interesting to raise the question whether it is useful to carry out predistortion with LP-OFDM. In the following, the question is addressed in the case of the ZF criterion.

Let W be the predistortion matrix instead of the equalisation matrix. With LP-OFDM, we then have

$$
Z = \frac{1}{n} P^{-1/2} C^T (HWCP^{1/2} X + N). \tag{18.25}
$$

Applying the ZF criterion at the transmitter side means that only the most attenuated subcarrier is assigned the maximum peak power p. The other subcarriers are attenuated such that the predistortion matrix is proportional to H^{-1}. This reads

$$
W = H^{-1} \times \| H \|_{-\infty}. \tag{18.26}
$$

Clearly, the disadvantage of this approach is that the available power is largely under-exploited at the transmitter. To mitigate this, the correction of the channel can be shared between the transmitter and the receiver applying the conventional square root function. In a more general approach, the correction is performed using W^α at the transmitter and $W^{1-\alpha}$ at the receiver guaranteeing a total correction of W, with $\alpha \in [0,1]$. The continuous bitrate is then

$$R = \sum_{i=1}^{n} \log_2 \left(1 + \frac{n^2 \max_{j \in [1,n]} |h_j|^{2\alpha}}{\sum_{j=1}^{n} \left(1/|h_j|^{2(1-\alpha)}\right) \sigma^2} \frac{p_i}{\sigma^2} \right). \tag{18.27}$$

Note that the full predistortion is obtained with $\alpha = 1$ and full equalisation with $\alpha = 0$, and uniform sharing is obtained with square root $\alpha = 1/2$. The study of the function $f(\alpha): \alpha \mapsto R$ actually shows that it is a decreasing function with maximal value achieved at $\alpha = 0$. Hence, the optimal strategy is to exploit all the available power and to treat the distortion introduced by the channel at the receiver side. This can be understood by the fact that predistortion always leads to transmit power reduction and, thus, bitrate limitation. We can then state the following proposition.

Proposition 18.2 *Considering a precoded multicarrier system and with respect to the ZF criterion, the continuous and discrete bitrates are higher with equalisation than with predistortion.*

Predistortion will then not be further investigated in the rest of the chapter.

18.2.4.3 Optimum Linear Detector

We may wonder whether conclusions drawn earlier in the ZF detector case would hold using the optimal MMSE linear detector. As such a detector has prohibitive complexity for practical implementations, simpler receivers have been proposed in the literature of wireless communications [14,25] applying the MSE criterion subcarrier per subcarrier. This yields

$$W = \frac{1}{n} P^{-1/2} C^T (HPH^* + \sigma^2 I_n)^{-1} H^*. \tag{18.28}$$

The resulting detected symbols Z consist of a useful term, an MCI term and a noise term. They all depend on channel coefficients h_i, power allocation p_i and noise variance σ^2. Assuming that the MCI is an additional noise component, the total continuous bitrate can be expressed after some mathematical manipulations as the sum of the bitrates over dimension i [26]:

$$R = \sum_{i=1}^{n} \log_2 \left(1 + \frac{1}{\gamma} \frac{\text{Tr}(HH^*(HH^* + \lambda I_n)^{-1})^2 p_i}{\sum_{\substack{j=1 \\ j \neq i}}^{n} \left(C_i^T HH^*(HH^* + \lambda I_n)^{-1} C_j\right)^2 p_j + \text{Tr}(HH^*(HH^* + \lambda I_n)^{-2})\sigma^2} \right), \tag{18.29}$$

with

$$\lambda = \frac{\sigma^2}{\sum_{i=1}^{n} p_i}. \tag{18.30}$$

It can first be shown in that case that the continuous bitrate obtained from Equation 18.29 is the same as that computed in the general case using Equation 18.8. However, whereas Equation 18.8 needs to merge deprecoding and channel equalisation functions, Equation 18.29 allows compatible architecture with conventional OFDM equalisation in which the receiver equalises the channel subcarrier per subcarrier and then deprecodes.

When the number of dimensions goes to infinity, that is, $n \to \infty$, and using the fact that half of the terms in $C_i^T C_j$ are equal to 1 and half equal to -1, the asymptotic limit of Equation 18.29 reduces to

$$\lim_{n\to\infty} \sum_{i=1}^{n} \log_2\left(1 + \frac{1}{\gamma \log_e 2} \frac{E\left[\frac{|h_i|^2}{|h_i|^2+\lambda}\right]^2 p_i}{\frac{n-1}{n}\left(E\left[\left(\frac{|h_i|^2}{|h_i|^2+\lambda}\right)^2\right] - E\left[\frac{|h_i|^2}{|h_i|^2+\lambda}\right]^2\right)(1-p_i) + E\left[\frac{|h_i|^2}{\left(|h_i|^2+\lambda\right)^2}\right]\sigma^2}\right). \tag{18.31}$$

The sum over i depends only on p_i and it is independent on other power p_j. The power allocation that maximises the bitrate is then the uniform one, as in the ZF case and as expected from Section 18.2.2. Hence, we can let $p_i = p/n$ for all i and

$$\max_{\{p_i\}_{i\in[1,n]}} \lim_{n\to\infty} R = \frac{1}{\gamma \log_e 2} \frac{E\left[\frac{|h_i|^2}{|h_i|^2+\lambda}\right]^2}{E\left[\left(\frac{|h_i|^2}{|h_i|^2+\lambda}\right)^2\right] - E\left[\frac{|h_i|^2}{|h_i|^2+\lambda}\right]^2 + \frac{\sigma^2}{p} E\left[\frac{|h_i|^2}{\left(|h_i|^2+\lambda\right)^2}\right]}. \tag{18.32}$$

Contrary to the ZF receiver, the continuous bitrate of MMSE receiver does not tend to zero but it remains lower than the OFDM continuous bitrate. In high SNR regime, it however tends to zero since the performance of the MMSE receiver tends to the performance of the ZF one. With a finite dimension n, the analysis is not tractable since the approximation no longer holds. Computing one simple case of two subcarriers shows that with high channel distortions, the bitrate is maximised when total power is gathered on one single dimension, that is, $p_1 = p$ and $p_2 = 0$, whereas with low channel distortions, the uniform power allocation strategy maximises the bitrate. In any case, the achievable continuous bitrate remains lower in LP-OFDM than in OFDM. Note that with $\gamma = 1$, the continuous OFDM bitrate is the channel capacity that cannot be exceeded by any linear and non-linear receiver. The same conclusion is obtained with $\gamma > 1$ for both OFDM and LP-OFDM systems.

If we stopped the analysis at this step, we would conclude that the precoding function is not interesting and introduces substantial throughput loss compared to unprecoded systems. We will prove that conclusions can be far different in the case of discrete modulations. We will however keep in mind the main properties developed in these sections to better understand the system behaviour in the following.

18.3 Resource Allocation for LP-OFDM

Let us now develop and give the solution to the resource allocation problem stated in Section 18.2.3.4 for LP-OFDM systems. In particular, we will show in what extent the precoding function provides additional flexibility to the bit and power allocation function and eventually leads to performance gain when adaptive modulations of finite order are employed.

Generally speaking, adaptive modulations allow to optimise the performance of a transmission scheme according to the link quality between the transmitter and the receiver. In multicarrier systems, for instance, it provides the ability to assign different numbers of bits to different subcarriers, hereby adapting the waveform parameters to the propagation conditions. By that means, different subcarriers experiencing different channel gains can be assigned an adequate number of bits and amount of power to finally achieve the same error rate. Many resource allocation algorithms designed for multicarrier systems can be found in the literature, with the aim of optimising either the bitrate or the robustness of the system. The related problems are commonly referred to as *rate maximisation* problem and *margin maximisation* optimisation problem [27], respectively. However, as soon as a precoding function is integrated to the multicarrier system, as proposed herein, new resource allocation algorithms have to be derived, taking into account the waveform and system particularities. Indeed, additional parameters have to be adaptively computed, such as the number or the length of the precoding sequences.

In this section, we propose to detail the key points of resource allocation mechanisms in LP-OFDM in comparison to more conventional OFDM systems. Perfect CSI will be assumed at the transmitter, which supposes perfect channel estimation at the receiver and perfect information feedback to the transmitter. In practice, however, perfect CSI is rarely achieved. The problem of imperfect CSI has already been discussed for OFDM systems and modified solutions for LP-OFDM systems can be found in [28,29], where the bitrate maximisation problem takes into account channel estimation inaccuracies.

18.3.1 Multicarrier System

To provide elements of comparison, let us first briefly derive the bitrate maximisation algorithm under peak-power constraint for the OFDM system. Using Equations 18.3 and 18.19, the highest achievable continuous bitrate of an OFDM system under maximum peak power p is

$$R = \sum_{i=1}^{n} \log_2\left(1 + \frac{|h_i|^2 p}{\gamma\sigma^2}\right). \tag{18.33}$$

The discrete version r of this continuous bitrate is then obtained as

$$r = \sum_{i=1}^{n} \lfloor r_i \rfloor_\beta = \sum_{i=1}^{n} \left\lfloor \log_2\left(1 + \frac{|h_i|^2 p}{\gamma\sigma^2}\right) \right\rfloor_\beta, \tag{18.34}$$

with $\lfloor x \rfloor_\beta = \beta \lfloor x / \beta \rfloor_1$, and where $\lfloor \cdot \rfloor_1$ is the conventional floor operation. In this equation, $\lfloor r_i \rfloor_\beta$ denotes the discrete bitrate achieved on subcarrier i based on continuous bitrate r_i. Accordingly, the effective power to assign to subcarrier i to guarantee $\lfloor r_i \rfloor_\beta$ can be computed as

$$p_i = \left(2^{\lfloor r \rfloor_{i\beta}} - 1\right) \frac{\gamma \sigma^2}{|h_i|^2} = p - \epsilon_i, \tag{18.35}$$

where ϵ_i corresponds to the residual power available on subcarrier i with respect to the peak-power limitation p. This amount of power is not sufficient to upgrade the constellation used on subcarrier i with an additional bit. Hence, it is concluded that the floor operation limits the system capability of exploiting all the available power. This limitation can be evaluated using a so-called power efficiency factor. Let us define this power efficiency factor q_i for subcarrier i as

$$q_i = 1 - \frac{\epsilon_i}{p} = \frac{2^{\lfloor r_i \rfloor_\beta} - 1}{2^{r_i} - 1}. \tag{18.36}$$

Under the assumption of a uniform distribution of $r_i - \lfloor r_i \rfloor_\beta$ in $[0,\beta)$, the mean value of q_i is

$$m_{q_i} = \left(2^{\lfloor r_i \rfloor_\beta} - 1\right)\left(\frac{1}{\beta}\log_2 \frac{2^{\lfloor r_i \rfloor_\beta + \beta} - 1}{2^{\lfloor r \rfloor_{i\beta}} - 1} - 1\right). \tag{18.37}$$

It can be computed that the mean power efficiency m_{q_i} remains lower than 72% for all modulation orders, that is, with $\beta = 1$, and lower than 54% with $\beta = 2$. These figures indicate that there is room for system improvement through better power exploitation. This can be obtained by choosing β as low as possible, which means using large varieties of modulation-coding couples. This however can lead to very complex receiver structures equipped with numerous channel decoders in parallel. Alternatively, we will show in the following section that linear precoding can provide better power efficiency factors with low complexity.

18.3.2 Linear Precoding and Multicarrier

It has been proven in Section 18.2.4 that the bitrate maximisation for continuous rates leads to equal power allocation and then equal bitrate per precoding sequence. As in the OFDM case, a simple allocation algorithm for discrete modulations would then consist in applying the floor operator to the continuous bitrates derived from either Equation 18.22 or Equation 18.29. In that case, the uniform distribution of bits and power would be preserved. With discrete bitrates, this is however shown as being suboptimal in the following.

18.3.2.1 Decorrelating Detector

From Proposition 18.1, the maximal continuous bitrate of a precoded system is

$$R = n\log_2 \left(1 + \frac{1}{\gamma} \frac{n}{\sum_{j=1}^{n} \left(1/|h_j|^2\right)} \frac{p}{\sigma^2}\right). \tag{18.38}$$

With discrete bitrates, it can be proven that the maximal achievable bitrate is computed according to the following proposition.

Proposition 18.3 *The maximal discrete bitrate of an LP-OFDM system is* [7]

$$r = n \left\lfloor \frac{R}{n} \right\rfloor_\beta + \beta \left\lfloor n \frac{2^{R/n - \lfloor R/n \rfloor_\beta} - 1}{2^\beta - 1} \right\rfloor_1,$$

with R given by Equation 18.39.

Comparing the results of this proposition to Equation 18.35, we notice that, contrary to OFDM, it is possible in LP-OFDM to allocate a number of bits higher than that obtained through a simple floor operation applied to continuous bitrates. In other terms, it becomes possible to improve the discrete bitrate with precoding. Let us have a numerical example to simply illustrate this idea. Let us consider a flat-fading channel and discrete modulations such that $\beta = 1$, and let the continuous bitrate per subcarrier be lower than 1 for all subcarriers, that is, $r_i = 1 - \varepsilon$ with $\varepsilon > 0$. The discrete bitrate that would be achieved by the OFDM system without precoding would be zero. Now, integrating a precoding function in the system, the continuous bitrate given by Equation 18.38 would become $R = n(1 - \varepsilon)$ and the discrete bitrate given by Proposition 18.3 would be $r = \left\lfloor n(2^{1-\varepsilon} - 1) \right\rfloor$. For every ε, if n is large enough, then the discrete bitrate r can exceed one, which enables the transmission of at least one bit. (Note that in practice, n is limited by the maximum number of subcarriers in the OFDM symbol.) Generalising this idea, it is understood that the precoding function has the capability to gather the pieces of power available on each subcarrier of the multicarrier spectrum and consequently find resource to allocate additional bits compared to the non-precoding case. In contrast to that, a conventional OFDM system exploits the power resource subcarrier by subcarrier without any gathering effect.

From Proposition 18.3, it is then possible to define the bitrate allocation policy that yields the maximum discrete bitrate for the LP-OFDM system. It is given by the following corollary.

Corollary 18.1 *The maximal discrete bitrate is obtained with* $\left\lfloor n(2^{R/n - \lfloor R/n \rfloor_\beta} - 1)/(2^\beta - 1) \right\rfloor_1$ *precoding sequences carrying* $\lfloor R/n \rfloor_\beta + \beta$ *bits and* $n - \left\lfloor n(2^{R/n - \lfloor R/n \rfloor_\beta} - 1)/(2^\beta - 1) \right\rfloor_1$ *precoding sequences carrying* $\lfloor R/n \rfloor_\beta$ *bits.*

Interestingly, the optimal allocation guarantees the lowest deviation of the bitrate per dimension or per precoding sequence.

As in the OFDM system, the power efficiency of the precoded system is now evaluated. The useful power is the necessary and sufficient power to transmit the bitrate given in Proposition 18.3. It then comes

$$\sum_{i=1}^{n} p_i = \left(2^{\lfloor R/n \rfloor_\beta} \left(n + (2^\beta - 1) \left\lfloor n \frac{2^{R/n - \lfloor R/n \rfloor_\beta} - 1}{2^\beta - 1} \right\rfloor_1 \right) - n \right) \sum_{i=1}^{n} \frac{\gamma \sigma^2}{n^2 |h_i|^2}, \qquad (18.39)$$

which leads to the following power efficiency factor:

$$q = \frac{2^{\lfloor R/n \rfloor_\beta}\left(1 + \frac{2^\beta - 1}{n}\left\lfloor n\frac{2^{R/n - \lfloor R/n \rfloor_\beta} - 1}{2^\beta - 1}\right\rfloor_1\right) - 1}{2^{R/n} - 1}, \tag{18.40}$$

and we note that

$$\lim_{n \to \infty} q = 1. \tag{18.41}$$

It finally appears that the precoding is all the more power efficient than the precoding matrix size increases. To illustrate this idea, Table 18.1 provides some numerical examples of power efficiency for $n \in \{1,2,16\}$. Note that the power efficiency factor q of an OFDM system is obtained with $n = 1$ in Equation 18.40. Two values of β and four modulation orders are considered. The power efficiency clearly increases with the modulation order or the precoding sequence length and decreases with β. Hence, the precoded system is unconditionally more efficient than the unprecoded one. The power efficiency even exceeds 90% with n higher than 16.

We then conclude that the precoding matrix has to be large enough to make the power efficiency factor tend to 100% or equivalently to make the discrete bitrate achieve the maximal continuous one. At the same time, however, as stated in Section 18.2.4, the continuous bitrate goes to zero as the precoding matrix size increases due to channel frequency selectivity effects. We then understand that a trade-off has to be found on the precoding matrix size between the power efficiency increase and the frequency selectivity minimisation. As this will be exemplified in the following, the trade-off must be sought regarding channel conditions.

18.3.2.2 Optimum Linear Detector

Contrary to the decorrelating ZF receiver, there is no analytical solution for discrete bitrate allocation with the MMSE receiver. The algorithmic approach is then investigated using an incremental allocation procedure. The basic idea is to iteratively load the precoding sequences with β bits, verifying at each iteration that the deviation of the discrete bitrate

TABLE 18.1

Mean Power Efficiency

$\lfloor R/n \rfloor$	$\beta = 1$			$\beta = 2$		
	$n = 1$	$n = 2$	$n = 16$	$n = 1$	$n = 2$	$n = 16$
1	0.58	0.75	0.96	0.40	0.60	0.93
2	0.67	0.81	0.97	0.48	0.66	0.94
5	0.72	0.84	0.98	0.53	0.70	0.95
10	0.72	0.84	0.98	0.54	0.70	0.95

allocation is minimised over the sequences. This is to obtain a bitrate allocation as close as possible from the continuous bitrate case. Obviously, the iterative allocation goes on as long as the power constraint is satisfied. To evaluate the power to be allocated to each additional β bit, let us define matrix $A = (a_{i,j})_{\{i,j\}\in[1,n]^2}$ such that

$$
\begin{cases}
a_{i,j} = (1 - 2^{r_i})\left(C_i^T H H^* (H H^* + \lambda I_n)^{-1} C_j\right)^2, & \forall i \neq j, \\
a_{i,i} = \dfrac{1}{\gamma} \mathrm{Tr}\left(H H^* (H H^* + \lambda I_n)^{-1}\right)^2
\end{cases}
\tag{18.42}
$$

and diagonal matrix $B = (b_{i,j})_{\{i,j\}\in[1,n]^2}$ as

$$
b_{i,i} = (2^{r_i} - 1)\sigma^2 \, \mathrm{Tr}(H H^* (H H^* + \lambda I_n)^{-2}),
\tag{18.43}
$$

where we recall that r_i is the continuous bitrate within the dimension i. Using Equation 18.29, it is obtained that

$$
a_{i,i} p_i + \sum_{\substack{j=1 \\ j \neq i}}^{n} a_{i,j} p_j = b_i, \quad \forall i \in [1,n].
\tag{18.44}
$$

Hence, for a given set of bitrate allocation $\{r_i\}_i \in [1,n]$, we can compactly write

$$
AP = B.
\tag{18.45}
$$

Finally, the convenient power allocation of an LP-OFDM system equipped with an MMSE detector has to respect the following proposition [26].

Proposition 18.4 *Assuming a bitrate allocation $\{r_i\}_i \in [1,n]$, the power allocation that minimises the sum power is the diagonal matrix P such that $P = A^{-1}B$.*

To reduce the complexity of the algorithm and the number of iterations, Corollary 18.1 can advantageously be used to initialise the procedure. In other words, the bit and power allocation is realised in a first step as if the ZF detector were used. In a second step, the iterative procedure is launched and β bits are added iteratively sequence by sequence, each time recomputing the power allocation according to Proposition 18.4. Following this strategy, the discrete bitrate achieved with the MMSE receiver is unconditionally higher or equal to than that obtained with the ZF receiver.

18.3.3 BER Constraint

As evident from the previous sections, resource allocation under SER constraint leads to simple optimisation strategies, since noise margin γ does not depend on the constellation order. In practical systems, however, the QoS is often expressed in terms of BER, or packet error rate derived from BER, instead of SER. BER constraint is usually meant as average BER and several definitions can be given [30]. We define the mean BER constraint as

$$\frac{\sum_{i=1}^{n} r_i \mathrm{ber}_i(r_i)}{\sum_{i=1}^{n} r_i} \leq \mathrm{ber}, \tag{18.46}$$

where

ber$_i(r_i)$ is the BER within dimension i associated to r_i bits
ber is the target BER

Recall that dimension i corresponds to the ith subcarrier of the OFDM system or the ith precoding sequence of the LP-OFDM one. One has to note that the BER resulting from streams of unequal bitrates over I is not the arithmetic mean of individual BER per dimension but rather expresses as a weighted mean BER.

Unfortunately, this mean BER constraint leads to high computation time for optimal solution which is obtained with greedy-type algorithms. Basically, these algorithms iteratively allocate β bits at a time to the dimension that minimizes the mean BER measure. Alternatively, suboptimal solutions can be sought considering peak BER rather than mean BER. In that case, the BER within each dimension is individually limited by the target BER. The constraint then reads

$$\mathrm{ber}_i(r_i) \leq \mathrm{ber}. \tag{18.47}$$

As introduced in Equation 18.19, the BER constraint can be taken into account through the SNR gap γ_i which depends on the constellation order, namely, bitrate r_i. Denoting $\gamma(r_i)$ this SNR gap, it is possible to update Proposition 18.3 to give the optimal bitrate allocation for LP-OFDM under target BER constraint. This is given by the following proposition.

Proposition 18.5 *The maximal discrete bitrate with precoding is* [31]

$$r = n \left\lfloor \frac{R}{n} \right\rfloor_\beta + \beta \left\lfloor n \frac{2^{R/n} - 2^{\lfloor R/n \rfloor_\beta}}{\left(2^{\lfloor R/n \rfloor_\beta + \beta} - 1\right) \dfrac{\gamma\left(\lfloor R/n \rfloor_\beta + \beta\right)}{\gamma\left(\lfloor R/n \rfloor_\beta\right)} - 2^{\lfloor R/n \rfloor_\beta} + 1} \right\rfloor_1 = n \left\lfloor \frac{R}{n} \right\rfloor_\beta + \beta n',$$

with R given by Equation 18.38 and calculated with $\gamma = \gamma(\lfloor R/n \rfloor_\beta)$.

Taking $\gamma\left(\lfloor R/n \rfloor_\beta\right) = \gamma\left(\lfloor R/n + \beta \rfloor_\beta\right) = \gamma$, that is, for constant SNR gap, Proposition 18.5 amounts to Proposition 18.3. Hence, Proposition 18.5 can be interpreted as an extension of Proposition 18.3 to the case where the SNR gap is not constant. Interestingly, one can notice that the optimal bit allocation once again respects the already mentioned general principle which encourages a minimal bitrate deviation over the dimension i. We can hence state the following corollary.

Corollary 18.2 *The maximal discrete bitrate is obtain with n' sequences carrying $\lfloor R/n \rfloor_\beta + \beta$ bits and $n - n'$ sequences carrying $\lfloor R/n \rfloor_\beta$ bits.*

At high SNR asymptotic regime with square modulations and high modulation orders, it can be shown that the SER and BER constraints lead to equivalent performance [32], essentially owing to the fact that SER to BER translation can easily be obtained. At non-asymptotic regime, it is difficult to convert SER to BER for the same operating point and the equivalence no more holds. In this case, performance comparison is irrelevant since the operating points differ.

18.3.4 Practical System Design

In the previous sections, allocation procedures that maximise continuous and discrete bitrates have been developed with non-sparse precoding matrices, that is, only considering Hadamard matrices of size n. To address the allocation problem, it remains yet to find the optimal matrix leading to a bitrate maximisation within the set $\mathcal{C}$ given in Definition 18.1. As already mentioned in Section 18.2.2, such sparse matrices lead to multiblock precoded systems, each block corresponding to a Hadamard matrix of size lower than n. Allocation algorithms have then to find the optimal size for each precoding block and how to interleave them over the n subcarriers. We already know that the optimal size of a non-sparse precoding matrix should answer the trade-off between the increase of the power efficiency and the degradation of the system continuous bitrate.

Unfortunately, there is no analytical solution to optimally design the sparse matrices. Indeed, the optimisation problem becomes a combinatorial optimisation problem that can be stated as follows: how to distribute n subcarriers within subsets of size 1, 2 and all multiples of 4. The optimal solution is obtained through an exhaustive search over the number of partitions of a set with n elements and with constrained sizes of subsets. The generating function of such a number of partitions is

$$\exp\left(x + \frac{x^2}{2!} + \sum_{n=1}^{\infty} \frac{x^{4n}}{4n!}\right) = 1 + x + x^2 + \frac{2}{3}x^3 + \frac{11}{4!}x^4 + \cdots = \sum_{n=1}^{\infty} q(n)\frac{x^n}{n!}, \tag{18.48}$$

where $q(n)$ is the number of partitions. Table 18.2 summarises some figures obtained with various values of n. It clearly appears that the problem becomes rapidly intractable for practical numbers of subcarriers.

To reduce the number of combinations, it is valuable to split the problem into two sub-problems. Let M be the matrix such that $C = \Pi M$ with Π a permutation matrix and

$$M = \begin{bmatrix} M_1 & & 0 \\ & \ddots & \\ 0 & & M_k \end{bmatrix}, \tag{18.49}$$

TABLE 18.2

Number of Partitions

N	1	2	5	10	20	50	100	1000
$q(n)$	1	2	31	28,696	$1.64 \cdot 10^{12}$	$7.18 \cdot 10^{42}$	$1.80 \cdot 10^{106}$	$7.87 \cdot 10^{1,835}$
$q'(n)$	1	2	4	14	71	1,780	73,486	$7.49 \cdot 10^{16}$
$q''(n)$	1	2	3	4	7	14	27	252

with $(M_l)_l \in [1,k]$ the Hadamard matrices of size $n_l \times n_l$ that verifies

$$\sum_{l=1}^{k} n_l = n. \qquad (18.50)$$

The problem then translates into the search of the k precoding matrices M_l on one hand and the search of the permutation matrix Π on the other hand. Analytical solution is obtained for the design of matrix Π as stated in the succeeding text [7].

Proposition 18.6 *For a given matrix M, the permutation matrix Π that maximises the continuous bitrate is such that if M_1 is distributed over the set of subcarriers $\mathcal{H}_l = \{h_{\pi(s_l+1)}, \ldots, h_{\pi(s_l+n_l)}\}$ and M_m over the set of subcarriers $\mathcal{H}_m = \{h_{\pi(s_m+1)}, \ldots, h_{\pi(s_m+n_m)}\}$, then for all $l \neq m$, $|h_l| \geq |h_m|$ with $h_l \in \mathcal{H}_l$ and $h_m \in \mathcal{H}_m$ and with $s_l = \sum_{i=1}^{l-1} n_i$.*

This result can simply be interpreted as follows: Π has to be built such that the first Hadamard matrix M_1 is distributed over n_l subcarriers corresponding to the highest amplitudes $|h_i|$, M_2 over the remaining highest n_l subcarriers and so on. Consider the following example to illustrate the proposition. Let

$$H = \begin{pmatrix} 1 & & & 0 \\ & 0.4 & & \\ & & 0.3 & \\ 0 & & & 0.6 \end{pmatrix} \qquad (18.51)$$

be the channel matrix and

$$M = \begin{pmatrix} 1 & 1 & & 0 \\ 1 & -1 & & \\ & & 1 & 1 \\ 0 & & 1 & -1 \end{pmatrix} \qquad (18.52)$$

be the M-precoding matrix, made of two block matrices M_l with $n_l = 2$. Proposition 18.6 indicates that the permutation matrix must be such that channel gains are gathered two by two in descending order. This reads

$$\Pi = \begin{pmatrix} 1 & 0 & 0 & 0 \\ 0 & 0 & 1 & 0 \\ 0 & 0 & 0 & 1 \\ 0 & 1 & 0 & 0 \end{pmatrix}. \qquad (18.53)$$

Finally, we obtain the following precoding matrix

$$C = \begin{pmatrix} 1 & 1 & 0 & 0 \\ 0 & 0 & 1 & 1 \\ 0 & 0 & 1 & -1 \\ 1 & -1 & 0 & 0 \end{pmatrix}. \tag{18.54}$$

Note that the same result would be obtained sorting channel gains in ascending order instead of descending order.

The first subproblem being solved, it remains now to find the optimal set of Hadamard matrices M_l, that is to say, find the number k of matrices and their size n_l to maximise the bitrate. This search is again a combinatorial optimisation problem. In this case, the number of solutions is given by the number of partitions of the integer n using as addends integers 1, 2 and all multiples of 4. The generating function of such a number of partitions is

$$\frac{1}{1-x} \frac{1}{1-x^2} \prod_{n=1}^{\infty} \frac{1}{1-x^{4n}} = 1 + x + 2x^2 + 2x^3 + 4x^4 + \cdots = \sum_{n=1}^{\infty} q'(n)x^n. \tag{18.55}$$

Even if $q'(n)$ is lower than $q(n)$, the search for optimal solution remains intractable in practical systems with high numbers of subcarriers, and Table 18.2 gives some values of $q'(n)$.

A heuristic solution is then proposed imposing $n_l = n_m$ for all $\{l,m\} \in [1,k]^2$. In that case, the number of solutions $q''(n)$ is the number of Hadamard matrices with size lower than n. This is the cardinal of the set $\mathcal{K}$ such that $\mathcal{K} = \{n_l \mid n_l \in \{1,2,4s\}, s \in \mathbb{N} \text{ and } n_l \leq n\}$. The search for the optimal solution within this subset of matrices becomes tractable as the number $q''(n)$ remains low enough. If n_l is well chosen, this heuristic approach, which is suboptimal, performs very close to the optimal solution obtained among the $q(n)$ possible ones [12,33]. Note that if n is not a multiple of n_l, the remaining subcarriers are also gathered within one or more matrices of different sizes to not waste these remaining subcarriers.

18.3.5 Bitrate Maximisation: Examples

In this section, we propose to lead comparative performance evaluation of the introduced adaptive multicarrier systems. In particular, we will assess their capability in terms of bitrate maximisation under peak-power constraint. For that purpose, we use the power line channel model developed in the context of home network applications [34]. This model defines nine classes of channel sorted by capacity and built up from measurement campaigns in various PLC environments. The useful channel bandwidth is set to 1.8–87.5 MHz and the power mask expressed in dBm/Hz is given in Figure 18.2 [35]. Concerning the multicarrier parameters, the subcarrier spacing and the guard time are compliant with the HPAV ones. An example of the received SNR per subcarrier is given in Figure 18.3 for the channel classes 2, 5 and 9. These SNR values reflect the channel transfer function but take also into account the power mask limitation as defined in Figure 18.2, the coloured background noise and the radio interference noise following the model defined in [36]. In addition, the analogue-to-digital conversion noise is also integrated in the SNR values as evident from the SNR saturation floor around 48 dB in the figure. Note that among the

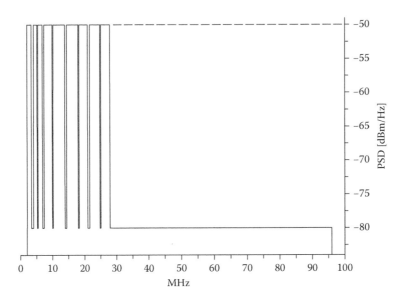

FIGURE 18.2
Power mask (see Ref. [35]).

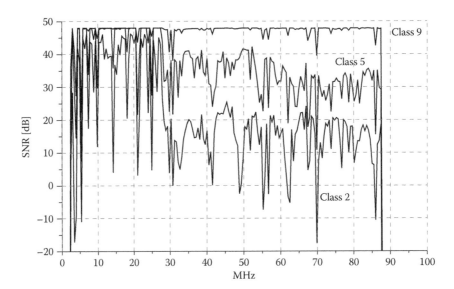

FIGURE 18.3
SNR per subcarrier through channel classes 2, 5 and 9.

three chosen classes, class 9 gives the highest SNR values, while class 2 corresponds to the worst channel environment. Concerning bit allocation, the maximal bitrate is set to $r = 10$ with granularity factor $\beta = 1$. In other terms, the utilised modulations are QAM from BPSK to 1024-QAM with all square and rectangular constellations. The SNR gap γ is chosen to 4 dB corresponding to a SER of 1% without channel coding. Note that the analogue-to-digital conversion noise clip at 48 dB is higher than the needed SNR for 1024-QAM at SER of 1%. Then, this noise will not limit the maximal achievable bitrate of the OFDM system.

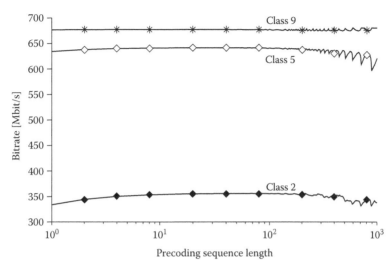

FIGURE 18.4
Bitrates versus precoding sequence lengths, $\gamma = 4$ dB.

Using these system and channel parameters, we first propose simulation results in terms of achieved bitrate under target SER constraint for LP-OFDM using various lengths of precoding sequences. Figure 18.4 gives the achieved bitrate versus the precoding sequence length for three examples of channel responses extracted from channel classes 2, 5 and 9. Note that the OFDM performance is obtained for a precoding sequence length equal to unity, while LP-OFDM performance corresponds to other lengths. The depicted curves clearly illustrate the trade-off that should be made about the precoding sequence length to maximise the available power exploitation while minimising the negative impact of the noise power boost due to ZF equalisation. The optimal precoding sequence length is between 12 and 100 which proves that OFDM is not, in fact, the optimal system when discrete modulations are used. This is essentially due to the power loss that occurs with finite-order modulations under peak-power constraint in OFDM. In contrast to that, the power-gathering capability of the precoding function can translate into additional bitrate transmission. However, the bitrate gain remains moderate and can even be almost null with very good channel conditions, namely, class 9. To go further in the analysis, Figure 18.5 shows the *cumulative distribution functions* (CDFs) of the precoding bitrate gain in Mbit/s and the relative bitrate gain in percent of the bitrate achieved with OFDM. These CDF are obtained fixing $n_l = 32$ and running 1000 trials of channel responses within each of classes 2, 5 and 9. The curves tend to show that the bitrate gain is more significant for poor channel classes than for good channels. This can be understood by the fact that power-gathering effects are all the more profitable than SNR values are low. The bitrate gains in percent remain lower than 10% but this corresponds in channel classes 2 or 5 to an additional bitrate of at least 20 Mbit/s compared to OFDM. Considering practical end-device applications, this latter figure is anything but negligible and can in fact be viewed as a new possibility for more high-definition TV streaming, for example. Note that the bitrates given in Figure 18.4 are around 350 Mbit/s for channel class 2, less than 650 Mbit/s for channel class 5 and more than 650 Mbit/s for channel class 9.

Similar simulation results are given in Figure 18.6 in the case of BER constraint of 1% instead of SER constraint, again without channel coding. The same precoding sequence length as in Figure 18.5 is used, namely, $n_l = 32$. As evident from the CDF curves

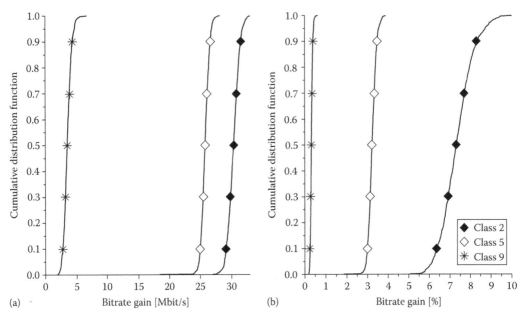

FIGURE 18.5
CDF of the precoding bitrate gain, $n_l = 32$, $\gamma = 4$ dB. Absolute bitrate gain (a) and relative bitrate gain (b).

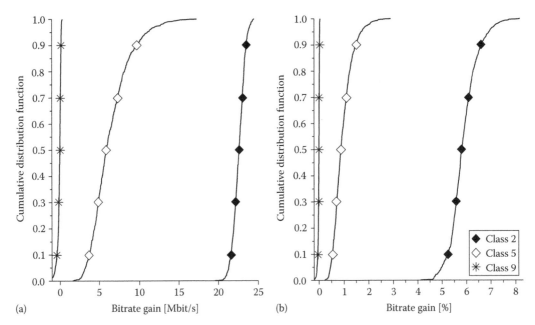

FIGURE 18.6
CDF of precoding bitrate gain, $n_l = 32$, BER constraint of 10^{-2}. Absolute bitrate gain (a) and relative bitrate gain (b).

corresponding to channel class 2, this size of precoding matrix is not a judicious choice since the precoding gain can become negative which means a bitrate reduction compared to OFDM. As shown in Figure 18.7, bitrate gains can be all the time positive if the precoding sequence length is adapted to the channel conditions. In this figure, we give the CDF of the optimal precoding sequence length exhaustively sought among all possible lengths

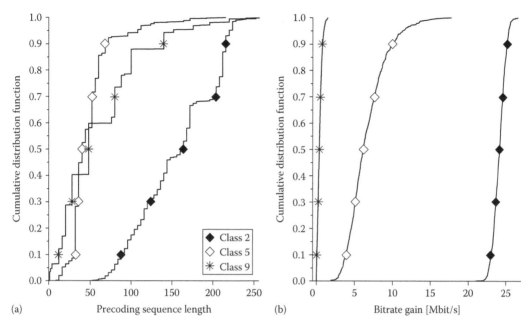

FIGURE 18.7
CDFs of optimal precoded sequence lengths (a) and of bitrate gains (b); BER constraint of 10^{-2}.

as discussed in Section 18.3.4. Accordingly, the CDF of the precoding bitrate gain is plotted considering that the optimal length is used. In average, one can notice that the precoding sizes are lower with good channel conditions than with poor channel conditions, which keeps in line with previous observations and comments. However, the optimal precoding sequence length is not unique within each channel class and should actually be adaptively searched to achieve maximal bitrate for each channel. As done for the bitrate and the power, the precoding should be adaptive to yield the highest gains. For complexity reason, however, fixed values can reasonably be envisaged. In the proposed example, 32 appears as a good trade-off except for channel class 9 that would rather need lower sizes. Last but not least, it is interesting to compare the bitrate gain obtained using the MMSE detector instead of the ZF one. Through simulations, it turns out that the MMSE-LP-OFDM system outperforms the ZF-LP-OFDM one, but in a range of less than 0.1% whatever the precoding sizes. This can be understood by the fact that the design of the permutation matrix Π, given in Proposition 18.6, leads to very low channel frequency selectivity. Consequently, the MMSE receiver cannot efficiently improve the ZF receiver performance and the impact on bitrates becomes negligible.

18.4 From Time Precoding to 2D-Precoding Extension

In the previous section, we have investigated the resource allocation issue in LP-OFDM systems when the precoding function is applied along the frequency axis as in MC-CDMA systems. It is now interesting to extend the study to the case when precoding is applied in the time domain [11] as in MC-DS-CDMA and even when applied in both time and frequency domains as firstly proposed for mobile communications in [37].

With precoding in the time domain, the data symbols are spread over n OFDM symbols, instead of n subcarriers with frequency-wise precoding. The discrete bitrate r_i on subcarrier i is straightforwardly derived from Proposition 18.3 and reads

$$r_i = n \left\lfloor \frac{R_i}{n} \right\rfloor_\beta + \beta \left\lfloor n \frac{2^{R_i/n - \lfloor R_i/n \rfloor_\beta} - 1}{2^\beta - 1} \right\rfloor_1, \tag{18.56}$$

with

$$R_i = n \log_2 \left(1 + \frac{|h_i|^2\, p}{\gamma \sigma^2} \right). \tag{18.57}$$

Note that Equation 18.57 takes the form of the achievable rate for one subcarrier of a classical OFDM system without precoding (see Equation 18.33), simply because the channel fading h_i in Equation 18.38 is supposed to remain constant over the precoding sequence length. Both systems, OFDM and LP-OFDM, have thus the same achievable rate considering continuous bitrates. With discrete modulations, the corresponding OFDM bitrate within subcarrier i is n times the bitrate given in Equation 18.34

$$r_i = n \left\lfloor \frac{R_i}{n} \right\rfloor_\beta. \tag{18.58}$$

Comparing Equations 18.56 and 18.58, we can conclude that LP-OFDM leads to higher throughput than OFDM if both systems are constrained to use discrete modulations. These results can also be derived with target BER constraint using γ_i instead of γ. The following corollary can then be established.

Corollary 18.3 *The discrete bitrate achieved by the time-precoded OFDM system is unconditionally higher that of the non-precoded OFDM system.*

Indeed, the OFDM performance is limited by the granularity of the constellation size, that is, β. The precoding artificially improves this granularity thanks to its power-gathering effect. It can hereby partially compensate for the power loss that occurs with OFDM. More precisely, the discrete bitrate approaches the continuous bitrate with increasing n, yet with a residual margin that depends on the continuous value. To evaluate this margin, let us use Equations 18.56 and 18.57 to write

$$\frac{R_i}{n} - \frac{\beta}{2^\beta - 1} + \frac{\log\left(\beta \log_e 2 / 2^\beta - 1\right) + 1}{\log_e 2} \leq \lim_{n \to \infty} \frac{r_i}{n} \leq \frac{R_i}{n}, \tag{18.59}$$

the right-hand equality being obtained when $R_i/(n\beta) \in \mathbb{N}$. Figure 18.8 shows the discrete bitrates r_i/n achieved on one subcarrier in OFDM (Equation 18.58), in LP-OFDM (Equation 18.56) with $n = 4$ and for upper bound LP-OFDM, that is, for $n = \infty$. Results are given in reference to the continuous bitrate case and highlight the evolution of the discrete bitrates

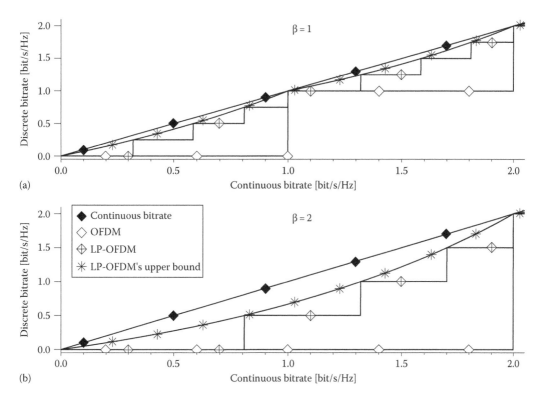

FIGURE 18.8
Discrete bitrate r_i/n versus continuous bitrate with $n \in \{1,4,\infty\}$ and $\beta = 1$ (a), $\beta = 2$ (b).

obtained for an increasing continuous bitrate. Two examples are proposed for $\beta = 1$ and $\beta = 2$. The staircase functions, which are typical of discrete modulation allocation, show that the precoding operation improves the bitrate granularity regardless of β, whereas the OFDM bitrate is directed by the value of β. The residual difference between the continuous bitrate and the upper-bound LP-OFDM bitrate given by Equation 18.59 is at most 0.0861 with $\beta = 1$ and 0.3378 with $\beta = 2$.

It is now interesting to evaluate the bitrate gain introduced by the precoding in time domain. Let q_i be this gain:

$$q_i = r_i(\text{LP-OFDM}) - r_i(\text{OFDM}). \tag{18.60}$$

The ratio q_i/n is upper bounded by

$$\lim_{n \to \infty} \frac{q_i}{n} = \frac{\beta}{2^\beta - 1} \left(2^{R/n - \lfloor R/n \rfloor_\beta} - 1 \right). \tag{18.61}$$

With the assumption of a uniform distribution of $R/n - \lfloor R/n \rfloor_\beta$ in $[0, \beta)$, the mean value of this upper bound reads

$$\mathrm{E}\left[\lim_{n \to \infty} \frac{q_i}{n} \right] = \frac{1}{\log_e 2} - \frac{\beta}{2^\beta - 1}, \tag{18.62}$$

which is equal to 0.44 with β = 1 and 0.77 with β = 2. In other words, this means that a gain of 0.44 bit per subcarrier can be expected from the use a precoding function if uncoded QAM modulations are considered in the transmission scheme.

The bitrate gain introduced by the precoding is upper bounded and, in average, increases with β. To reach the upper bound, large matrix sizes should be used. A strong drawback of this is that the data are decoded at the receiver once all the OFDM symbols linked by the precoding matrix are received. This can introduce very large delays for the signal decoding in the receiver. A possible solution is then to fold up a part of the precoding matrix in the frequency domain, thus leading to the so-called 2D precoding. Through this approach, the system can benefit from time-domain precoding advantages without suffering from too long decoding delay and, in the same way, can take advantage of the frequency domain precoding gain while limiting the effects of frequency selectivity. For that purpose, the 2D-precoding size in time and frequency has to be judiciously chosen through a trade-off between the following:

- Bitrate gain and delay, for time-domain precoding length
- Bitrate gain and frequency distortion, for frequency domain precoding length

The optimal choice is thus to use precoding sequences as long as possible in time and frequency while limiting the frequency distortion as much as possible and getting acceptable delays. In practice, multiple configurations exist that yield discrete bitrates higher than 99.5% of the OFDM continuous bitrate [13]. 2D precoding is thus an efficient solution to achieve almost optimal bitrates in multicarrier systems.

18.5 Multicast Scenarios

Multicasting is a technique that allows data, including packet form, to be simultaneously transmitted to a selected set of destinations [38]. In other terms, the same data are transmitted to several receivers at the same time. Multicasting is a more efficient method of supporting group communications than unicasting or broadcasting, as it allows transmission to multiple destinations using fewer network resource [39]. It is particularly useful for high data rate multimedia services due to its ability to save the network resource. Contrary to unicasting, the multicast service does not repeat multiple time the same transmission and, contrary to broadcasting, the multicast data are addressed to known receivers. With the knowledge of the receivers, the source, that is, the transmitter, can evaluate the channel conditions of each link. Based on the channel estimation and on the quality of service requirements, the source has the capability to adapt the multicast bitrate to the channel conditions.

18.5.1 Resource Allocation

As in Section 18.2.1.1, we will assume throughout this section that the multicarrier system is well adapted to the channel and perfectly synchronised in time and frequency. All CSI of the multicast environment are supposed to be known at both transmitter and receiver sides. We consider n subcarriers and u receivers. The signal received by receiver j then writes

$$Y_j = H_j P^{1/2} X + N, \tag{18.63}$$

where H_j conveys the channel coefficients for the communication link created between the transmitter and the jth receiver. The resulting OFDM multicast discrete bitrate then reads

$$r = u \sum_{i=1}^{n} \min_{j \in [1,u]} \left\lfloor \log_2 \left(1 + \frac{p}{\gamma \sigma^2} | h_{i,j} |^2 \right) \right\rfloor_\beta, \tag{18.64}$$

with $h_{i,j}$ the updated notation for channel gains h_i taking into account receiver index j. In the latter equation, it can be noticed that the multicast bitrate is governed by the worst channel conditions among the links defined by the multicast environment. Hence, the allocation can be computed using equivalent channel:

$$\left| \tilde{h}_i \right| = \min_{j \in [1,u]} | h_{i,j} |, \tag{18.65}$$

and the OFDM multicast discrete bitrate becomes

$$r = \sum_{i=1}^{n} u \lfloor r \rfloor_i = u \sum_{i=1}^{n} \left\lfloor \log_2 \left(1 + \frac{| \tilde{h}_i |^2 p}{\gamma \sigma^2} \right) \right\rfloor_\beta. \tag{18.66}$$

This bitrate is also given by Equation 18.34 using the multicast equivalent channel and without taking into account the multiplicative factor u. Through independent and identically distributed Rayleigh fading channels, and using order statistics [40], we have

$$\lim_{u \to \infty} u r_i = \frac{p}{\gamma \sigma^2 \log_e 2} \tag{18.67}$$

for all i. Consequently, it comes that

$$\lim_{u \to \infty} r = 0. \tag{18.68}$$

Hence, the drawback of such a multicast method based on the poorest channel conditions is that it can lead to dramatic degradation of the bitrate when the number of receivers increases. The so-called multirate multicast approach is a method that overcomes this drawback. The basic idea is to order the multicast data into layers so that each receiver can decode any combination of the layers depending on its link quality. This method is well adapted to OFDM transmission [41] and can be extended to LP-OFDM signal [26]. In the sequel, we will however demonstrate that the precoding function already brings bitrate improvement in the unirate multicast case.

18.5.1.1 Linear Precoding and Multicast

Let us first consider non-sparse precoding matrices applied in the frequency domain and ZF reception. The discrete multicast bitrate is then derived from Proposition 18.3 using $\tilde{H}$ in Equation 18.39 instead of H [33,42]. As mentioned before in the OFDM case, the discrete bitrate also converges to 0 when the number of receivers goes to infinity, regardless of the precoding sequence length. Note that with sparse precoding matrices, the method presented in Section 18.3.4 can be applied, however leading again to null bitrates for an infinite number of receivers. We may then ask whether we can expect better performance from the unirate multicast approach.

For that purpose, it is interesting to express the continuous multicast bitrate as the minimal continuous precoding bitrate. In this case, the computation of the bitrate takes into account the diversity of the receivers [43]. Using Equation 18.38 yields

$$R = u \min_{j \in [1,u]} n \log_2 \left(1 + \frac{1}{\gamma} \frac{n}{\sum_{i=1}^{n} \left(1/|h_{i,j}|^2 \right)} \frac{p}{\sigma^2} \right), \tag{18.69}$$

which can be lower bounded as

$$R \geq u n \log_2 \left(1 + \frac{1}{\gamma} \frac{n}{\sum_{i=1}^{n} \left(1/|\tilde{h}_i|^2 \right)} \frac{p}{\sigma^2} \right) \tag{18.70}$$

because $|\tilde{h}_i| \leq |h_{i,j}|$ for all $i \in [1,n]$ and all $j \in [1,u]$. It can be verified that this inequality remains valid considering discrete bitrate. The latter is computed using Equation 18.69 and Proposition 18.3 under SNR gap constraint or Proposition 18.5 under BER constraint. Interestingly, it appears that the multicast continuous bitrate can be improved employing precoding principles, whereas we have shown in Section 18.2.4 that precoding always reduces the unicast continuous bitrate. With a large number of receivers, we understand that the multicast bitrate will be degraded, yet keeping the advantage over the non-precoded system.

As in Section 18.3, it remains to find the optimal matrix C in the case of sparse matrices.

18.5.1.2 Practical Solution

The problem is similar to the one encountered in Section 18.3.4. It is then solved in the same way using the heuristic solution for which the equivalent channel is used for the design of the permutation matrix Π. We can then rapidly state the following result.

Corollary 18.4 *The multicast discrete bitrate is*

$$r = u \sum_{l=1}^{k} n_l \left\lfloor \frac{R_l}{u n_l} \right\rfloor_{\beta} + u \beta \left\lfloor n_l \frac{2^{R_l/u n_l - \lfloor R_l/u n_l \rfloor_{\beta}} - 1}{2^{\beta} - 1} \right\rfloor_1,$$

with

$$R_l = u \min_{j \in [1,u]} n_l \log_2 \left(1 + \frac{1}{\gamma} \frac{n_l}{\sum_{i=1}^{n_l} \left(1/ | h_{\pi(n_1 + \cdots + n_{j-1} + i),j} |^2 \right)} \frac{p}{\sigma^2} \right).$$

This result can also be extended in the case of BER constraint, and Proposition 18.5 should be used instead of Proposition 18.3.

18.5.2 Bitrate Maximisation: Examples

The multicast performance gains obtained with the precoding component are evaluated through the channel model presented in Section 18.3.5. The variations of the discrete multicast bitrate versus the precoding sequence lengths are similar to those encountered in the case of unicast shown in Figure 18.4. We consequently choose the spreading length $n_l = 64$ hereafter.

It is first interesting to observe the variation of the total multicast bitrate when the number of receivers increases as depicted in Figure 18.9. In this figure, we compare the bitrates obtained for three different classes of channels and carrying out OFDM, LP-OFDM with ZF and MMSE detectors. It can be viewed that the gain provided by the precoding function is much more visible for the most severe channels than for the lesser ones. As in the unicast scenario, the precoding does not improve significantly the bitrate in very good channels as evident from the results of class 9. For classes 2 and 5, one can note that the relative gain is quite independent of the number of receivers and that the absolute gain depends linearly on this number of receivers. This result remains true if the number of receivers is not too on high. Considering a very high number of receivers, the bitrate would saturate, and for even more receivers, it would decrease and tend to zero. The last important remark is that the

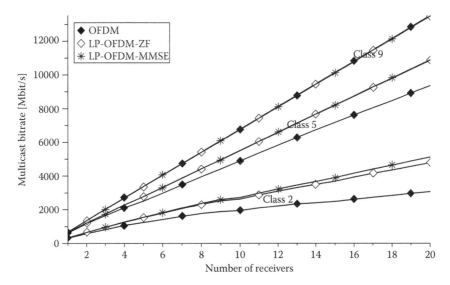

FIGURE 18.9
Multicast bitrates versus number of receivers, $n_l = 64$, $\gamma = 4$ dB.

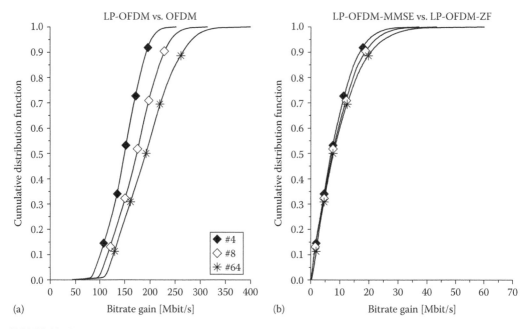

FIGURE 18.10
CDF of bitrate gains of LP-OFDM-ZF compared to OFDM (a) and LP-OFDM-MMSE compared to LP-OFDM-ZF (b), $u = 5$, $\gamma = 4$ dB.

additional gain provided by the MMSE detector over the ZF detector is very reduced. The ZF detector can thus be viewed as the best trade-off between performance and complexity.

In Figure 18.10, we give the CDF of the bitrate gains obtained with LP-OFDM through various channel realisations: each link is chosen within one of the nine channel classes following the channel percentage of classes [34]. Precoding lengths are set to 4, 8 and 64, and $u = 5$ receivers are used in the multicast network. On the left side of the figure, the gains are computed with reference to OFDM; on the right side, comparison is made between MMSE and ZF detectors. It clearly appears that LP-OFDM outperforms OFDM with bitrate gains higher than 100 Mbit/s in more than 90% of the cases. The highest gain is obtained with $n_l = 64$, that is, when the subcarrier gathering effect of the precoding component is already high. As far as the MMSE detector is concerned, one can notice that the gain over the ZF detector remains much more moderated. For example, MMSE brings less than 10 Mbit/s gain with a probability of around 50%. In addition, the comparative performance between MMSE and ZF has very weak dependence on the precoding length.

Finally, these results confirm that LP-OFDM is profitable in a multicast context compared to OFDM and that most of the expected gain is obtained using a simple ZF detector.

18.6 Conclusion

In this chapter, we have reviewed the principle and the major results of adaptive precoded OFDM communication system. Simple precoding added to conventional OFDM system leads to bitrate improvement in point-to-point communication as well as in multicast power line environment. These bitrate gains up to 30 Mbit/s, and up to 10% of the OFDM bitrates, allow transmission of more high-definition TV flows in home network. The simple solution for bitrate increase based on Hadamard matrices can benefit from fast Hadamard transform for efficient realisation and can be combined with enhanced OFDM systems.

Acknowledgements

The research leading to these results has partially received funding from the European Community's Seventh Framework Programme FP7/2007-2013 under grant agreement no. 213311 also referred to as OMEGA.

References

1. S. Weinstein, The history of orthogonal frequency-division multiplexing, *IEEE Communications Magazine*, 47(11), 26–35, November 2009.
2. M. Simon, J. Omura, R. Scholtz and B. Levitt, *Spread Spectrum Communications Handbook*. McGraw-Hill, New York, 2001.
3. S. Hara and R. Prasad, Overview of multicarrier CDMA, *IEEE Communications Magazine*, 35(12), 126–133, December 1997.
4. M. Debbah, W. Hachem, P. Loubaton and M. de Courville, MMSE analysis of certain large isometric random precoded systems, *IEEE Transactions on Information Theory*, 49(5), 1293–1311, May 2003.
5. S. Mallier, F. Nouvel, J.-Y. Baudais, D. Gardan and A. Zeddam, Multicarrier CDMA over lines – Comparison of performances with the ADSL system, in *IEEE International Workshop on Electronic Design, Test and Applications*, Christchurch, New Zealand, January 2002, pp. 450–452.
6. O. Isson and J.-M. Brossier et D. Mestdagh, Multi-carrier bit-rate improvement by carrier merging, *Electronics Letters*, 38(19), 1134–1135, September 2002.
7. M. Crussière, J.-Y. Baudais and J.-F. Hélard, Robust and high-bit rate communications over PLC channels: A bit-loading multi-carrier spread-spectrum solution, in *International Symposium on Power-Line Communications and Its Applications*, Vancouver, British Columbia, Canada, April 2005, pp. 37–41.
8. M. Crussière, J.-Y. Baudais and J.-F. Hélard, New loading algorithms for adaptive SS-MC-MA systems over power line channels: Comparisons with DMT, in *International Workshop on Multi-Carrier Spread Spectrum*, Oberpfaffenhofen, Germany, 14–16 September 2005, pp. 327–336.
9. M. Raug, T. Zheng, M. Tucci and S. Barmada, On the time invariance of PLC channels in complex power networks, in *IEEE International Symposium on Power Line Communications and Its Applications*, Copacabana Rio de Janeiro, Brazil, March 2010, pp. 56–61.
10. K.-H. Kim, H.-B. Lee, Y.-H. Kim and S.-C. Kim, Channel adaptation for time-varying powerline channel and noise synchronized with AC cycle, in *IEEE International Symposium on Power Line Communications and Its Applications*, Dresden, Germany, April 2009, pp. 250–254.
11. M. Crussière, J.-Y. Baudais and J.-F. Hélard, Improved throughput over wirelines with adaptive MC-DS-CDMA, in *International Symposium on Spread Spectrum Techniques and Applications*, Manaus-Amazon, Brazil, 28–31 August 2006, pp. 143–147.
12. M. Crussière, J.-Y. Baudais and J.-F. Hélard, Adaptive linear precoded DMT as an efficient resource allocation scheme for power-line communications, in *IEEE Global Communications Conference*, series 5, no. 1, San Francisco, CA, December 2006, pp. 1–5.

13. J.-Y. Baudais and M. Crussière, Resource allocation with adaptive spread spectrum OFDM using 2D spreading for power line communications, *EURASIP Journal on Advances in Signal Processing*, 2007, 1–13, 2007 (special issue on Advanced Signal Processing and Computational Intelligence Techniques for Power Line Communications).

14. S. Kaiser, OFDM code-division multiplexing in fading channels, *IEEE Transactions on Communications*, 50(8), 1266–1273, August 2002.

15. S. Verdú, *Multiuser Detection*. Cambridge University Press, New York, 1998.

16. I.S. Gradshteyn and I.M. Ryzhik, *Table of Integrals, Series, and Products*, 7th edn. Elsevier Academic Press publications, San Diego, CA, 2007.

17. M. Crussiére, J.-Y. Baudais and J.-F. Hélard, Adaptive spread spectrum multicarrier multiple access over wirelines, *IEEE Journal on Selected Areas in Communications*, 24(7), 1377–1388, July 2006 (special issue on Power Line Communications).

18. A.S. Hedayat, N.J.A. Sloane and J. Stufken, *Orthogonal Arrays: Theory and Applications*. Springer-Verlag, New York, 1999, Chapter 7.

19. TR 102 494, Powerline Telecommunications (PLT) technical requirements for in-house PLC modems, ETSI, June 2005.

20. A.J. Goldsmith and S.-G. Chua, Adaptive coded modulation for fading channels, *IEEE Transactions on Communications*, 45(5), 595–602, May 1998.

21. D.P. Palomar, J.M. Cioffi and M.A. Lagunas, Joint Tx-Rx beamforming design for multicarrier MIMO channels: A unified framework for convex optimization, *IEEE Transactions on Signal Processing*, 51(9), 2381–2401, September 2003.

22. J.M. Cioffi, A multicarrier primer, ANSI T1E1.4/91–157, Committee Contribution, Technical Report, November 1991.

23. A. Maiga, J.-Y. Baudais and J.-F. Hélard, An efficient channel condition aware proportional fairness resource allocation for powerline communications, in *International Conference on Telecommunications*, Marrakech, Morocco, 25–27 May 2009, pp. 286–291.

24. D.P. Palomar and J.R. Fonollosa, Practical algorithms for a family of waterfilling solutions, *IEEE Transactions on Signal Processing*, 53(2), 686–695, February 2005.

25. K. Fazel and S. Kaiser, *Multi-Carrier and Spread Spectrum Techniques*. John Wiley & Sons Ltd, Chichester, U.K., 2003.

26. A. Maiga, J.-Y. Baudais and J.-F. Hélard, Bit rate optimization with MMSE detector for multicast LP-OFDM systems, *Journal of Electrical and Computer Engineering*, 2012, 1–12, 2012.

27. N. Papandreou and T. Antonakopoulos, Bit and power allocation in constrained multicarrier systems: The single-user case, *EURASIP Journal on Applied Signal Processing*, 2008, 1–14, 2008.

28. F.S. Muhammad, J.-Y. Baudais and J.-F. Hélard, Rate maximization loading algorithm for LP-OFDM systems with imperfect CSI, in *IEEE Personal, Indoor and Mobile Radio Communications Symposium*, Tokyo, Japan, 13–16 September 2009, pp. 1–5.

29. F.S. Muhammad, J.-Y. Baudais and J.-F. Hélard, Bit rate maximization for LP-OFDM with noisy channel estimation, in *Third International Conference on Signal Processing and Communication Systems*, Omaha, NE, 28–30 September 2009, pp. 1–6.

30. S.T. Chung and A.J. Goldsmith, Degrees of freedom in adaptive modulation: A unified view, *IEEE Transactions on Communications*, 49(9), 1561–1571, September 2001.

31. A. Maiga, J.-Y. Baudais and J.-F. Hélard, Very high bit rate power line communications for home networks, in *IEEE International Symposium on Power Line Communications and Its Applications*, Dresden, Germany, March 2009, pp. 313–318.

32. J.-Y. Baudais, F.S. Muhammad and J.-F. Hélard, Robustness maximization of parallel multichannel systems, *Journal of Electrical and Computer Engineering*, 2012, 1–16, 2012.

33. A. Maiga, J.-Y. Baudais and J.-F. Hélard, Increase in multicast OFDM data rate in PLC network using adaptive LP-OFDM, in *Second International Conference on Adaptive Science and Technology (ICAST)*, Accra, Ghana, IEEE, New York, 14–16 December 2009, pp. 384–389.

34. M. Tlich, A. Zeddam, F. Moulin and F. Gauthier, Indoor power-line communications channel characterization up to 100 MHz–Part I: One-parameter deterministic model, *IEEE Transactions on Power Delivery*, 23(3), 1392–1401, July 2008.

35. P. Pagani, R. Razafferson, A. Zeddam, B. Praho, M. Tlich, J.-Y. Baudais, A. Maiga et al., Electromagnetic compatibility for power line communications. Regulatory issues and countermeasures, in *IEEE Personal, Indoor and Mobile Radio Communications Symposium*, Istanbul, Turkey, September 2010, pp. 1–6.

36. W.Y. Chen, *Home Network Basis: Transmission Environments and Wired/Wireless Protocols*. Prentice Hall PTR, Upper Saddle River, NJ, 2004.

37. A. Persson, T. Ottosson and E. Strom, Time-frequency localized CDMA for downlink multicarrier systems, in *International Symposium on Spread Spectrum Techniques and Applications*, Vol. 1, Sun City, South Africa, September 2002, pp. 118–122.

38. ATIS-0100523.2011, *ATIS Telecom Glossary*. Alliance for Telecommunications Industry Solutions, Washington, DC, 2011.

39. U. Varshney, Multicast over wireless networks, *Communications of the ACM*, 45(12), 31–37, December 2002.

40. H.A. David and H.N. Nagaraja, *Order Statistics*, 3rd edn. Probability and Statistics Series. Wiley-Interscience, Hoboken, NJ, 2003.

41. C. Suh and J. Mo, Resource allocation for multicast services in multicarrier wireless communications, in *25th IEEE International Conference on Computer Communications*, Barcelona, Catalunya, Spain, April 2006, pp. 1–12.

42. A. Maiga, J.-Y. Baudais and J.-F. Hélard, Bit rate maximization for multicast LP-OFDM systems in PLC context, in *Third Workshop on Power Line Communications*, Udine, Italy, October 2009, pp. 93–95.

43. A. Maiga, J.-Y. Baudais and J.-F. Hélard, Subcarrier, bit and time slot allocation for multicast precoded OFDM systems, in *IEEE International Conference on Communications*, Cape Town, South Africa, 23–27 May 2010, pp. 1–6.

19

Multi-user MIMO for Power Line Communications

Yago Sánchez Quintas, Daniel M. Schneider and Andreas Schwager

CONTENTS

19.1 Introduction

There are different channel access methods for shared medium networks. Different network users might be separated in the time domain (*time division duplex* [TDD]), in the frequency domain (*frequency division duplex* [FDD]) or in the code domain (*code division multiple access* [CDMA]). However, if the transmitters and receivers have multiple transmit and receive ports (*multiple-input multiple-output* [MIMO]), different users might also be separated by the spatial dimension. Using MIMO in the *multi-user* (MU) context is called MU-MIMO. MU-MIMO has already proven to be a successful way of enhancing the MIMO performance of wireless transmissions [1–4]. The use of MIMO algorithms allows the transmission of several simultaneous spatial streams to different users on the same frequency band and time slot. Potentially, the total throughput is increased.

MIMO has been successfully applied to *power line communications* (PLCs); see Chapters 8 and 9 for more details on MIMO signal processing strategies and the resulting capacity gains. Current implementations of MIMO PLC systems are presented in Chapters 12 and 14. However, so far, MIMO has only been applied to the *single-user* (SU) scenario, that is, for links between one transmitter and one receiver. This chapter aims to study the feasibility of MU-MIMO techniques for PLC.

There are several scenarios where MU-MIMO algorithms can be applied in a PLC network. For example, two different high-definition video streams could be transmitted simultaneously from a router to two different TVs placed in different rooms. This scenario is similar to the downlink from a *base station* (BS) of a cellular network to several users. Other scenarios might comprise several user pairs communicating in the same frequency band at the same time, where the multiplexing is achieved in the spatial domain.

One issue that one encounters when trying to adapt MU-MIMO strategies to the PLC scenario is the reduced number of transmit ports. As explained in Chapter 1, the number of transmit ports is limited to two for inhome PLC. This limited number of transmit ports reduces the possibilities of MU-MIMO coding strategies. On the other hand, having a relatively large number of up to four receive ports (see Chapter 1) offers several possibilities to cancel the *multi-user interference* (MUI).

Different solutions can be applied to deal with the spatial interference generated by several users transmitting on the same frequency band and time slot. On the one hand, MUI could be precancelled by the transmitter, who necessarily needs *channel state information* (CSI) to exploit the advantages of MU-MIMO strategies. In this way, no – or significantly reduced – interference is seen at the receiver side. The resulting MU-MIMO system will be decomposed into parallel uncoupled channels or streams, and users' data can be transmitted in disjoint spaces. On the other hand, due to the relatively large number of receive ports, the MUI could be cancelled at the receiver. This is similar to the uplink from several users to one BS in a cellular network, where usually the BS has more antennas than the mobile users. Both techniques will have a limited number of simultaneous spatial streams and therefore a limited number of simultaneous users due to the limited number of transmit and receive ports.

Information theory discussed in [5–7] shows that it is necessary to use Costa's *'dirty-paper' coding* (DPC) or Tomlinson–Harashima precoding to reach the sum capacity of an MU-MIMO downlink system where the sum capacity is the sum of the capacities of all independent links. However, these techniques require the use of a complex sphere decoder or an approximate closest-point solution, which makes them hard to implement in practice [8]. Thus, the MU-MIMO algorithms investigated in this chapter are limited to linear algorithms.

Figure 19.1 gives a top-level overview of the different channel access methods and shows where MIMO and MU-MIMO fit in. For TDD and FDD, the MIMO algorithms are applied to point-to-point or SU connections. As shown in Chapter 8, different MIMO schemes might be applied, for example, *space–time–frequency codes* (STFC) or *spatial multiplexing* (SMX) without and with precoding, that is, *beamforming* (BF). TDD with BF serves as reference for the performance evaluation of the MU-MIMO algorithms in this chapter. The MU-MIMO algorithms might be separated into algorithms which cancel MUI at the receiver and algorithms that apply a precoding at the transmitter. The precoding might be either nonlinear (e.g. DPC) or linear. One form of linear precoding is based on *block diagonalisation* (BD), and this chapter focuses in particular on *multi-user orthogonal space division multiplexing* (MOSDM) and an iterative computation of this algorithm which is NuSVD.

The outline of this chapter is as follows. Section 19.2 introduces two different PLC scenarios where MU-MIMO algorithms could be applied to PLC. An MU-MIMO algorithm based on BD, which is MOSDM, is introduced in Section 19.3 where the algorithm is applied to the two scenarios introduced in Section 19.2. The performance of the introduced algorithms is investigated and discussed in Section 19.4. Depending on the underlying MU-MIMO scenario, a performance gain of MU-MIMO compared to TDD with BF of

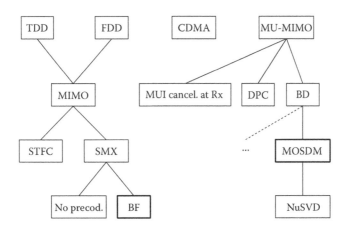

FIGURE 19.1
Channel access methods and MIMO and MU-MIMO algorithms.

up to 20% might be expected. The basis of these simulations is the MIMO PLC channels obtained in the *European Telecommunications Standards Institute* (ETSI) STF410 measurement campaign (refer to Chapter 5 [9–11]).

19.2 MU-MIMO Scenarios

In recent years, research and development of MU-MIMO algorithms focused mostly on wireless applications. In order to highlight the similarities and differences of MU-MIMO in the context of PLC compared to the wireless scenario, first, the typical wireless setup is recalled. Here, a BS typically offers service to a high number of users or *mobile stations* (MS). The BS might be the cell tower in a cellular network (e.g. a 4G or LTE network) or an access point on a wireless *local area network* (LAN). MU-MIMO algorithms at the BS allow for the transmission of spatial streams to the different users using the same frequency and time slot. Usually, the BS has a high number of antennas while the (mobile) users have a smaller and limited number of antennas. When applying MU-MIMO algorithms, the downlink (from the BS to the users) is considered to be the more challenging part, since the BS has to ensure that the spatial streams to one user do not cause interference to the other users [2,4]. For the uplink, the BS can use the larger number of antennas to cancel the interference from the different users.

In a PLC network, examples of the scenario described previously might be as follows: A home server transmits several, different video streams to different users in the home network or a router equipped with a PLC modem communicates to several users on the network. As explained in Chapters 1 and 5, up to 2 × 4 MIMO might be used for inhome PLC. The scenario of one transmitter and several receivers is illustrated in Figure 19.2a for two receivers where the transmitter comprises two transmit ports and the receivers have four receive ports. This scenario is referred to as MU in the following. As described in Chapter 8, the maximum number of spatial streams in SU-MIMO is limited by the minimum number of transmit and receive ports. In this example, two spatial streams are available where each spatial stream is assigned to one of the two users. The channels to users 1 and 2 are called $\mathbf{H}_1$ and $\mathbf{H}_2$ in Figure 19.2a. The transmission to user 1 causes interference to user 2 and vice versa. This interference needs to be handled either at the transmitter (T_x) or the receivers (R_{x1} and R_{x2}). In the uplink

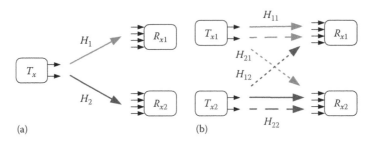

FIGURE 19.2
MU-MIMO scenarios. (a) MU system and (b) SDD system.

(from R_{x1} and R_{x2} to T_x in Figure 19.2a), basically more spatial streams are available since up to four receive ports may be used in receive mode.

In a meshed PLC network, the scenario described previously can be extended as shown in Figure 19.2b. Here, two independent communication links are illustrated where transmitter 1 (T_{x1}) communicates to receiver 1 (R_{x1}) and transmitter 2 (T_{x2}) sends data simultaneously to receiver 2 (R_{x2}). The transmission of T_{x1} causes interference to R_{x2} and the transmission of T_{x2} causes interference to R_{x1}. The idea in this scenario is to apply MU-MIMO algorithms to spatially multiplex the two links while transmitting at the same time and on the same frequency band. This scenario is called *spatial division duplex* (SDD) in the following.

The term MU-MIMO will be used throughout this chapter to describe MU-MIMO communication systems in a generic way, that is, simultaneous users receiving data in the same time slot and frequency by means of SMX. The term MU is used to describe the particular scenario where only one transmitter communicates with several receivers. The second scenario will always be referred to as SDD.

The MU and SDD scenarios yield differences in the implementation of the precoding algorithms. Also, the differences between the scenarios influence the obtained results with the same precoding techniques. Therefore, during the study of coding techniques in the next sections, a distinction will always be made between both cases. This difference will also be pointed out in the results section (Section 19.4).

Combinations between the described scenarios are also possible. However, due to the already large number of possibilities these two scenarios offer, combinations among them have not been included in the scope of this chapter. It has to be noticed that any other type of multiplexing, like TDD, could be applied together with MU-MIMO algorithms in case it becomes necessary to increase the flexibility of the system.

19.3 MU-MIMO Precoding

The adaptation of MU precoding techniques to the PLC environment was studied in [12]. Some investigated algorithms were discarded without the need of being implemented and simulated, mainly due to dimensionality constraints which did not fit the PLC channel. For instance, the algorithms suggested in [13–16], which have shown to yield successful results for the wireless environment are not feasible for PLC due to the limited number of ports at the transmitter. Among the investigated and simulated precoding algorithms in [12], two groups might be differentiated: interference cancellation algorithms at the transmitter, like BD, and interference cancellation at the receiver.

BD will be explained in detail in the course of this chapter (Sections 19.3.1 and 19.3.2), since the algorithms yielding the best simulation results for the PLC environment belong to this group. The interference decoding algorithms at the receiver are briefly discussed in Section 19.3.3.

As mentioned in the introduction, the sum capacity of the MIMO *broadcast* (BC) channel can be only achieved with DPC [6]. However, a practical scheme that approaches DPC is still unavailable, and worse, the encoding process to achieve the sum capacity is data dependent. This means that the cancellation needs to be done independently for every symbol. Several algorithms that approach the sum capacity exposed by DPC have been proposed in [17,18]; however, they are considered to be too complicated for cost-effective implementation. An alternative linear precoding technique to DPC (nonlinear), widely applied in the wireless environment, is BD. The main concept of BD consists in precoding each user's data with a linear matrix before transmission. This particular matrix lies in the null space of all other simultaneous user channel matrices. Hence, assuming the channel matrices of all simultaneous users are known at the transmitter, with perfect CSI, zero inter-user interference is achievable at every receiver. This enables the use of simple receiver structures. This group of algorithms has been described as a suboptimal solution in terms of total achievable throughput but also as a feasible solution in terms of complexity.

19.3.1 Block Diagonalisation

First, the BD system model is described for the MU scenario, that is, only one transmitter in the system. The differences to the SDD scenario will be shown in a second step. The system model presented here will be applied to the precoding algorithm described in this chapter. Several papers like [19,20] use this model for BD systems. Note that the matrix operations shown in the following are described for a single carrier system. However, it can be easily extended to an *orthogonal frequency division multiplexing* (OFDM) system where all the matrix operations have to be applied for each subcarrier separately.

19.3.1.1 MU Scenario

Consider a downlink MU-MIMO system with M users, where N_T indicates the number of transmit ports, N_R indicates the total number of receive ports among all users and $N_{R,j}$ denotes the number of receive ports at the jth user. The transmitted symbol vector of user j is denoted as a k_j-dimensional vector $\mathbf{s}_j$. Note that k_j indicates the number of spatial modes directed to the user j. $\mathbf{s}_j$ is precoded by a $N_T \times k_j$ precoding matrix $\mathbf{T}_j$ for each particular user. At the receiver j, a detection matrix $\mathbf{R}_j$ of size $N_{R,j} \times k_j$ is applied to the received signal in order to obtain the desired symbol. Thus, the post-detection symbol vector $\mathbf{y}_j$ for user j can be written as

$$\mathbf{y}_j = \mathbf{R}_j^H \left(\mathbf{H}_j \mathbf{T}_j \mathbf{s}_j + \sum_{m=1, m\neq j}^{M} \mathbf{H}_j \mathbf{T}_m \mathbf{s}_m + \mathbf{n}_j \right),$$

$$= \mathbf{R}_j^H \mathbf{H}_j \mathbf{T}_j \mathbf{s}_j + \mathbf{R}_j^H \sum_{m=1, m\neq j}^{M} \mathbf{H}_j \mathbf{T}_m \mathbf{s}_m + \mathbf{R}_j^H \mathbf{n}_j, \qquad (19.1)$$

where
$\mathbf{n}_j$ denotes the noise vector for user j
$(\cdot)^H$ indicates the Hermitian operator

The matrix $\mathbf{H}_j \in \mathbb{C}^{N_{R,j} \times N_T}$ indicates the channel matrix to the jth user. $\mathbf{T}_j$ and $\mathbf{R}_j$ are constructed to be unitary matrices (as explained in Section 19.3.2).

The overall number of spatial streams is limited by the number of transmit ports:

$$\sum_{j=1}^{M} k_j \leq N_T, \tag{19.2}$$

Details of the constraint in Equation 19.2 are discussed in Section 19.3.2.

Figure 19.3 shows the application to the MU scenario with $M = 2$ users. The transmitter has $N_T = 2$ transmit ports. According to Equation 19.2, the number of spatial streams for each user is $k_1 = k_2 = 1$, that is, one spatial mode is activated for each of the two users. The symbols $\mathbf{s}_1$ and $\mathbf{s}_2$ to the two users are weighted by the 2×1 precoding vectors $\mathbf{T}_1$ and $\mathbf{T}_2$, respectively, and the 2×1 symbol vector $\mathbf{x} = \mathbf{T}_1 \mathbf{s}_1 + \mathbf{T}_2 \mathbf{s}_2$ is transmitted to the channel. At the receivers, estimates of the transmitted symbols are obtained according to Equation 19.1.

Note that in Figure 19.3, $\hat{\mathbf{y}}_j$ represents the received vector before being decoded by $\mathbf{R}_j$.

The goal of BD is to find precoding matrices $\mathbf{T}_j$ for each user j such that no interference is generated to the other users. For the example shown in Figure 19.3 and according to Equation 19.1, an estimate of the symbol for user 1 is given by

$$\mathbf{y}_1 = \mathbf{R}_1^H \mathbf{H}_1 \mathbf{T}_1 \mathbf{s}_1 + \mathbf{R}_1^H \mathbf{H}_1 \mathbf{T}_2 \mathbf{s}_2 + \mathbf{R}_1^H \mathbf{n}_1 \tag{19.3}$$

and for user 2 by

$$\mathbf{y}_2 = \mathbf{R}_2^H \mathbf{H}_2 \mathbf{T}_2 \mathbf{s}_2 + \mathbf{R}_2^H \mathbf{H}_2 \mathbf{T}_1 \mathbf{s}_1 + \mathbf{R}_2^H \mathbf{n}_2, \tag{19.4}$$

In order to fully cancel the interference, the conditions $\mathbf{R}_1^H \mathbf{H}_1 \mathbf{T}_2 = \mathbf{0}$ and $\mathbf{R}_2^H \mathbf{H}_2 \mathbf{T}_1 = \mathbf{0}$ in Equations 19.3 and 19.4, respectively, have to be fulfilled. Note that the receivers in this example could comprise only one receive port since only one spatial mode is used per user. Of course, more receive ports would increase the performance due to the increased receive diversity.

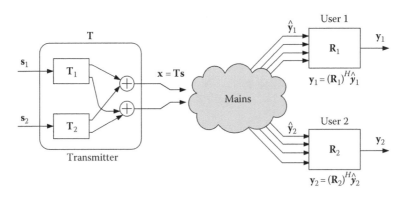

FIGURE 19.3
BD system model for PLC MU scenario.

Generally, the precoding matrices $\mathbf{T}_j$ of dimensions $N_T \times k_j$ and the decoding matrices $\mathbf{R}_j$ of dimensions $N_{R,j} \times k_j$ have to fulfil

$$\mathbf{R}_j^H \mathbf{H}_j \mathbf{T}_m = 0 \quad \text{for all} \quad j \neq m \quad \text{and} \quad 1 \leq j, m \leq M \tag{19.5}$$

Then, the post-detection symbol vector for user j is reduced to

$$\mathbf{y}_j = \mathbf{R}_j^H \mathbf{H}_j \mathbf{T}_j \mathbf{s}_j + \underbrace{\mathbf{R}_j^H \sum_{m=1, m \neq j}^{M} \mathbf{H}_j \mathbf{T}_m \mathbf{s}_m}_{=0} + \mathbf{R}_j^H \mathbf{n}_j,$$

$$= \mathbf{R}_j^H \mathbf{H}_j \mathbf{T}_j \mathbf{s}_j + \mathbf{R}_j^H \mathbf{n}_j, \tag{19.6}$$

In the noise-free case and for an appropriate design of $\mathbf{T}_j$ and $\mathbf{R}_j$ (see later), the entries of $\mathbf{y}_j$ are scaled versions of the corresponding entries of $\mathbf{s}_j$, that is, $\mathbf{y}_j$ needs to be equalised by a diagonal matrix to obtain estimates of $\mathbf{s}_j$.

As it can be observed in Equations 19.5 and 19.6, with the proper precoding matrices $\mathbf{T}_j$ and detection matrices $\mathbf{R}_j$, the inter-user interference can be cancelled. It is useful to define the total MU transmit weight matrix as

$$\mathbf{T} = \begin{bmatrix} \mathbf{T}_1 & \cdots & \mathbf{T}_M \end{bmatrix} \tag{19.7}$$

and the MU transmitted vector as

$$\mathbf{s} = \begin{bmatrix} \mathbf{s}_1 \\ \vdots \\ \mathbf{s}_M \end{bmatrix}. \tag{19.8}$$

For the example shown in Figure 19.3, the dimensions of $\mathbf{T}$ are 2×2 and the dimensions of $\mathbf{s}$ are 2×1.

If BD is applied successfully, an equivalent block diagonal model is obtained for the MU system. Figure 19.4 shows the result for the example introduced in Figure 19.3. λ_j indicates the equivalent channel gain for user j, and $\tilde{n}_j$ represents the equivalent noise sample after filtering.

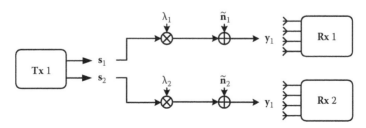

FIGURE 19.4
Equivalent block diagonal model for MU scenario.

19.3.1.2 SDD Scenario

Several differences should be considered between the adaptation of BD to the MU scenario and the adaptation to the SDD scenario. In the SDD scenario, several modems transmit on the same frequency band and time slot. In order to simplify the description of the model, a system with only two transmitters and two receivers is explained here. However, this model can easily be extended to the case where three or more transmitter–receiver pairs coexist together.

Figure 19.5 shows an example of $M = 2$ links, where transmitter 1 communicates to receiver 1 and transmitter 2 to receiver 2, respectively. In this example, each transmitter is equipped with two transmit ports and each receiver has four receive ports.

In the following, it is always assumed that transmitter 1 communicates to receiver 1, and transmitter 2 to receiver 2. In the SDD setup, the number of ports at the transmitter and the receiver remains the same as for the MU scenario, that is, up to $N_{T,j} = 2$ and $N_{R,j} = 4$ where j is the index of the link ($j = 1, \ldots, 2$). The channel between transmitter 1 and receiver 1 is represented by the channel matrix $\mathbf{H}_{11}$, and the channel between transmitter 2 and receiver 2 is represented by the channel matrix $\mathbf{H}_{22}$, respectively. Here, two interferences are possible. Transmitter 1, which is attempting to communicate with receiver 1, will generate interference to receiver 2, and transmitter 2 will generate interference to receiver 1. The interfering channels are denoted by $\mathbf{H}_{21}$ and $\mathbf{H}_{12}$ where the first index denotes the index of the receiver and the second index denotes the transmitter's index. As assumed for the MU scenario, each transmitter applies a precoding matrix $\mathbf{T}_j$ ($j = 1, \ldots, M$). According to this description, the precoding of transmitter 1 should be able to cancel any interference produced to receiver 2, and transmitter 2 should be able to cancel any interference generated to receiver 1.

In the SDD scenario, the transmitters need to be synchronised in order for the receivers to estimate the channels from the interfering modems. Also, the signals used for channel estimation have to be designed in a way that each receiver can estimate the channel from each transmitting modem. Similar to the channel estimation in MIMO systems, the training symbols need to be orthogonal to separate the different channels.

As a difference to the MU scenario, it should be noted that in this setup, the transmission of $k = 2$ spatial streams per user is physically possible. For example, two symbols could be sent simultaneously to user 1, at the same time and same frequency used for transmission

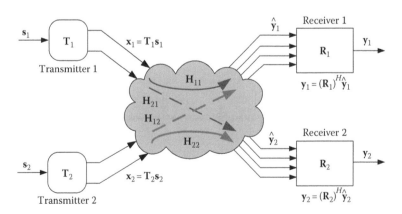

FIGURE 19.5
BD system model for PLC SDD scenario.

of two other symbols sent to user 2. However, as it will be discussed in the following sections, the number of spatial streams per link is limited due to constraints imposed by the interference cancellation.

According to the matrix operations illustrated in Figure 19.5 and similar to Equation 19.3, the equalised symbol vector of receivers 1 and 2 is given as

$$\mathbf{y}_1 = \mathbf{R}_1^H(\mathbf{H}_{11}\mathbf{T}_1\mathbf{s}_1 + \mathbf{H}_{12}\mathbf{T}_2\mathbf{s}_2 + \mathbf{n}_1),$$

$$= \mathbf{R}_1^H\mathbf{H}_{11}\mathbf{T}_1\mathbf{s}_1 + \mathbf{R}_1^H\mathbf{H}_{12}\mathbf{T}_2\mathbf{s}_2 + \mathbf{R}_1^H\mathbf{n}_1, \tag{19.9}$$

and

$$\mathbf{y}_2 = \mathbf{R}_2^H(\mathbf{H}_{21}\mathbf{T}_1\mathbf{s}_1 + \mathbf{H}_{22}\mathbf{T}_2\mathbf{s}_2 + \mathbf{n}_2),$$

$$= \mathbf{R}_2^H\mathbf{H}_{21}\mathbf{T}_1\mathbf{s}_1 + \mathbf{R}_2^H\mathbf{H}_{22}\mathbf{T}_2\mathbf{s}_2 + \mathbf{R}_2^H\mathbf{n}_2. \tag{19.10}$$

Generally, the equalised symbol vector of receiver n is given by

$$\mathbf{y}_n = \mathbf{R}_n^H\left(\sum_{m=1}^{M}(\mathbf{H}_{nm}\mathbf{T}_m\mathbf{s}_m) + \mathbf{n}_n\right), \tag{19.11}$$

$$= \mathbf{R}_n^H\mathbf{H}_{nn}\mathbf{T}_n\mathbf{s}_n + \mathbf{R}_n^H\sum_{m=1,m\neq n}^{M}(\mathbf{H}_{nm}\mathbf{T}_m\mathbf{s}_m) + \mathbf{R}_n^H\mathbf{n}_n, \tag{19.12}$$

where
$\mathbf{T}_m$ is the $N_{T,m} \times k_m$ precoding matrix
$\mathbf{s}_m$ is the $k_m \times 1$ transmit symbol vector of the mth transmitter which activates k_m spatial streams
$\mathbf{R}_n$ is the $N_{R,n} \times k_m$ receive matrix
$\mathbf{H}_{nm}$ is the channel matrix from transmitter m to receiver n

The aim of the precoding is to cancel the interference to other receivers. In the example shown in Figure 19.5, transmitter 1 should not cause interference to receiver 2 and transmitter 2 should not cause any interference to receiver 1. In terms of Equations 19.9 and 19.10, this requires $\mathbf{R}_1^H\mathbf{H}_{12}\mathbf{T}_2 = \mathbf{0}$ and $\mathbf{R}_2^H\mathbf{H}_{21}\mathbf{T}_1 = \mathbf{0}$, respectively.

Generally, the following equation has to be fulfilled for interference cancellation:

$$\mathbf{R}_n^H\mathbf{H}_{nm}\mathbf{T}_m = \mathbf{0} \quad \text{for } n, m = 1,\ldots,M, n \neq m. \tag{19.13}$$

If Equation 19.13 is fulfilled, Equation 19.11 reduces to

$$\mathbf{y}_n = \mathbf{R}_n^H(\mathbf{H}_{nn}\mathbf{T}_n\mathbf{s}_n + \mathbf{n}_n). \tag{19.14}$$

and the equalised symbol vector depends only on the desired transmit symbol vector.

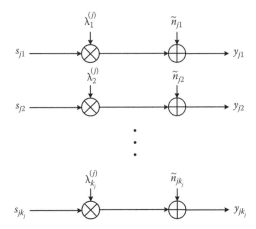

FIGURE 19.6
Equivalent block diagonal model of the jth user.

The equivalent block diagonal model for the SDD scenario is shown in Figure 19.6, where $\lambda_k^{(j)}$ indicates the equivalent gain for spatial stream k and link j. $\tilde{n}_{jk}$ is the equivalent noise sample after filtering of spatial stream k and link j.

The following section explains how the precoding and decoding matrices have to be designed in order to cancel the interference. The algorithm MOSDM based on BD will be presented.

19.3.2 Multi-user Orthogonal Space Division Multiplexing

As defined in [21], by jointly optimising the transmitter and receivers in the MU system, the MU signals can be projected onto orthogonal subspaces, so that the MUI experienced by each independent user is eliminated. The weight matrices in the transmitter and receivers are iteratively jointly optimised. This method assumes CSI at the transmitter side for all MU channels. This is necessary in order to have a mathematical representation of the subspaces where the different channel matrices lie and hence be able to project them onto equivalent orthogonal subspaces. The objective is to obtain an MU channel diagonalisation by finding the joint weight matrices $\mathbf{T}$, $\mathbf{R}_1$, $\mathbf{R}_2$, ..., $\mathbf{R}_M$. In the following, the algorithm is described for the MU scenario (i.e. with a single transmitter) according to [21] and is extended later to the SDD scenario.

19.3.2.1 MU Scenario

Recall Equation 19.5 which defined the necessary condition for interference cancellation. Equation 19.5 can be written in matrix notation as follows:

$$\mathbf{R}_j^H \mathbf{H}_j \mathbf{T} = \begin{bmatrix} \mathbf{0}_1 & \cdots & \mathbf{0}_{j-1} & \underbrace{\mathbf{\Lambda}_j}_{j\text{th subblock matrix}} & \mathbf{0}_{j+1} & \cdots & \mathbf{0}_M \end{bmatrix} \tag{19.15}$$

for the users $j = 1, ..., M$, where

$$\mathbf{\Lambda}_j = \mathrm{diag}\left(\lambda_1^{(j)} \quad \lambda_2^{(j)} \quad \cdots \quad \lambda_{k_j}^{(j)} \right) \tag{19.16}$$

is of dimension $k_j \times k_j$ and $\lambda_m^{(j)}$ represents the channel gain of the mth spatial stream of the jth user. Equations 19.15 and 19.16 not only show the interference cancellation but also show that the unitary precoding $\mathbf{T}_j$ and unitary decoding $\mathbf{R}_j^H$ decompose the link of user j into k_j parallel independent spatial streams. This is similar to eigenbeamforming (see e.g. Chapter 8).

Generally, if a solution to Equations 19.15 and 19.16 exists, there will be more than one solution. Between the possible solutions, the one that optimises the overall MIMO system performance is chosen [21]. This can be mathematically written as

$$(\mathbf{T}, \mathbf{R}_1, \mathbf{R}_2, \ldots \mathbf{R}_M)_{\text{opt}} = \arg \max_{\mathbf{T}, \mathbf{R}_1, \ldots \mathbf{R}_M} \sum_{j=1}^{M} \| \Lambda_j \|^2, \tag{19.17}$$

where $\|.\|$ denotes the Frobenius norm of a matrix.

Figure 19.6 illustrates the k_j parallel spatial streams of the jth user.

The calculation of the weighting matrices is approached as follows. First, the optimal transmit and receive weights are found for a particular user j, with the assumption that the receive weights for other users $\{\mathbf{R}_m\}_{m \neq j}$ are given. The weights should be found in a way that the equivalent channel gains are maximised and the *co-channel interference* (CCI) caused to other users is eliminated. Next, the weight matrices of other users are updated in a step-by-step approach until convergence.

First, the equivalent MU channel matrix $\mathbf{H}_e$ of dimensions $\left(\sum_{m=1}^{M} k_m \right) \times N_T$ is defined as

$$\mathbf{H}_e \triangleq \begin{bmatrix} \mathbf{R}_1^H \mathbf{H}_1 \\ \vdots \\ \mathbf{R}_M^H \mathbf{H}_M \end{bmatrix}. \tag{19.18}$$

Assuming that the receive matrices $\mathbf{R}_1, \ldots, \mathbf{R}_{j-1}, \ldots, \mathbf{R}_{j+1}, \ldots, \mathbf{R}_M$ ($m \neq j$) are known by the transmitter, it will try to find the best $\mathbf{T}_j$ and $\mathbf{R}_j$ that optimise the performance of the jth user link:

$$(\mathbf{T}_j, \mathbf{R}_j)_{\text{opt}} = \arg \max_{\mathbf{T}_j, \mathbf{R}_j} \| \Lambda_j \|^2, \tag{19.19}$$

while no CCI is caused to the other users according to

$$\mathbf{H}_e \mathbf{T}_j = \begin{bmatrix} \mathbf{0}_1^T \\ \vdots \\ \mathbf{0}_{j-1}^T \\ \Lambda_j \\ \mathbf{0}_{j+1}^T \\ \vdots \\ \mathbf{0}_M \end{bmatrix}. \tag{19.20}$$

Equation 19.20 is similar to Equation 19.15 and again shows the interference cancellation where each subblock represents the equivalent spatial streams associated with user j.

To satisfy Equations 19.19 and 19.20, the interference matrix $\tilde{\mathbf{H}}_e^{(j)}$ of dimensions $\left(\sum_{m=1,m\neq j}^{M} k_m\right) \times N_T$ is defined as

$$\tilde{\mathbf{H}}_e^{(j)} \triangleq \begin{bmatrix} \mathbf{R}_1^H \mathbf{H}_1 \\ \vdots \\ \mathbf{R}_{j-1}^H \mathbf{H}_{j-1} \\ \mathbf{R}_{j+1}^H \mathbf{H}_{j+1} \\ \vdots \\ \mathbf{R}_M^H \mathbf{H}_M \end{bmatrix}. \tag{19.21}$$

Once these definitions have been set, the condition to satisfy Equation 19.20 and therefore to eliminate the inter-user interference is

$$\mathbf{T}_j \in \text{null}\left\{\tilde{\mathbf{H}}_e^{(j)}\right\}, \tag{19.22}$$

where null{·} denotes the null space of a matrix.*

$\mathbf{T}_j$ according to Equation 19.22 only exists if the null space is not empty. This leads to the following constraint of the number of spatial streams. For a system with M users, each transmitting k_j spatial modes, the number of ports at the transmitter must be greater or equal to the total number of active spatial modes in the whole system [21]:

$$\sum_{j=1}^{M} k_j \leq N_T. \tag{19.23}$$

For each user taking part in the MU system, the number of ports on a particular receiver must be greater than or equal to the number of active spatial modes that this user is receiving [21]:

$$k_j \leq N_{R,j} \quad \forall j. \tag{19.24}$$

This is similar to the SMX case in SU-MIMO (see Chapter 8).

Consider the relevant case for inhome PLC of $N_T = 2$ transmit ports. According to Equation 19.23, two users are possible, each with one spatial stream. For one spatial stream, $\tilde{\mathbf{H}}_e^{(j)}$ is of dimensions 1×2 and the precoding vector according to Equation 19.22 is of dimensions 2×1.

Now, in order to optimise the system, we can write the basis of the null space as $\mathbf{Q}_j = \begin{bmatrix} \mathbf{q}_1^{(j)} & \mathbf{q}_2^{(j)} & \cdots \end{bmatrix}$, where $\mathbf{q}_m^{(j)}$ indicates the mth column vector of the matrix $\mathbf{Q}_j$. Next, we decompose the precoding matrix $\mathbf{T}_j$ into two submatrices as follows $\mathbf{T}_j = \mathbf{Q}_j \mathbf{B}_j$. In this decomposition, $\mathbf{B}_j$ indicates the coordinate transformation under the basis $\mathbf{Q}_j$. The matrix $\mathbf{Q}_j$ will be in charge of cancelling the interference in the MU system, and once the

---

* In linear algebra, the null space (or kernel) of a $m \times n$ matrix $\mathbf{A}$ is the set of all vectors $\mathbf{x}$ for which $\mathbf{A}\mathbf{x} = \mathbf{0}$: $\ker(\mathbf{A}) = \{\mathbf{x} \in \mathbb{C}^n : \mathbf{A}\mathbf{x} = \mathbf{0}\}$ where $\mathbf{0}$ is the $m \times 1$ zero vector. The kernel of a matrix is a linear subspace of the n-dimensional Euclidean space. The dimension of the null space of $\mathbf{A}$ is called the nullity of $\mathbf{A}$ [22].

interference is cancelled, the already independent channels can be optimised by applying the matrix $\mathbf{B}_j$. The next step is then to choose a matrix $\mathbf{B}_j$ such that

$$(\mathbf{B}_j, \mathbf{R}_j)_{\text{opt}} = \arg\max_{\mathbf{B}_j, \mathbf{R}_j} \|\Lambda_j\|^2, \tag{19.25}$$

and

$$\mathbf{R}_j^H \mathbf{H}_j \mathbf{Q}_j \mathbf{B}_j = \Lambda_j. \tag{19.26}$$

If the precoding matrix $\mathbf{T}_j$ for user j has been properly computed, we have ensured that there is no inter-user interference in the system. Due to the properties of the null space, precoding with the matrix $\mathbf{T}_j$ is actually projecting our channel matrix into a subspace which is orthogonal to all other users' subspaces.

Once the interference is cancelled (supposing the dimensionality constraints are satisfied), the rest of the problem is reduced to an optimisation task. Accordingly, this task is to find $(\mathbf{B}_j, \mathbf{R}_j)_{\text{opt}}$ of an SU-MIMO system supporting multiple spatial streams.

As shown in Chapter 8, the best precoding for SU-MIMO supporting several spatial modes is eigenbeamforming, and the precoding matrix is obtained by means of a *singular value decomposition* (SVD):

$$\mathbf{H}_j \mathbf{Q}_j = \mathbf{U}_j \Lambda_j \mathbf{V}_j^H. \tag{19.27}$$

Then, $\mathbf{R}_j$ and $\mathbf{B}_j$ are given by

$$(\mathbf{R}_j)_{\text{opt}} = \mathbf{U}_j \,|_{1 \leftrightarrow k_j} \tag{19.28}$$

and

$$(\mathbf{B}_j)_{\text{opt}} = \mathbf{V}_j \,|_{1 \leftrightarrow k_j}, \tag{19.29}$$

where the notation $|_{1 \leftrightarrow k_j}$ is used to indicate that only the k_j first column vectors corresponding to the k_j largest singular values are included in the matrices $\mathbf{R}_j$ and $\mathbf{B}_j$. Finally, the optimum precoding matrix is formulated as follows:

$$(\mathbf{T}_j)_{\text{opt}} = \mathbf{Q}_j (\mathbf{B}_j)_{\text{opt}}. \tag{19.30}$$

The final precoding matrix $\mathbf{T}$ obtained with this algorithm is not unitary. However, it presents normalised columns, which ensures equal transmit powers in all transmit ports.

19.3.2.2 NuSVD: Iterative Computation of MOSDM

An iterative way of approaching MOSDM is proposed in [21] and is called *iterative null space-directed SVD* (iterative NuSVD). Using the definitions and equations introduced before, the algorithm works iteratively as follows [21]:

1. The decoding matrices are initialised to the identity matrix, $\mathbf{R}_j = \mathbf{I}$.
2. The matrix $\tilde{\mathbf{H}}_e^{(j)}$ is formed for each user included in the MU system. The matrix $\mathbf{Q}_j$ is obtained as the null space basis of $\tilde{\mathbf{H}}_e^{(j)}$. Subsequently each link is optimised computing the SVD of $\mathbf{H}_j \mathbf{Q}_j$, which yields the optimum precoding matrices $\mathbf{B}_j$ according to Equation 19.29 and $\mathbf{R}_j$ using Equation 19.28, respectively.

3. The off-diagonal of the equivalent channel matrix $\mathbf{H}_e\mathbf{T}$ is computed. The off-diagonal norm indicates the level of interference in the system. As the iterations continue, the level of interference should decrease:

a. Compute $\varepsilon = \text{off}(\mathbf{H}_e\mathbf{T})$

b. where $\text{off}(\mathbf{A}) \triangleq \displaystyle\sum_{k,l,k \neq l} |a_{k,l}|$

in which $|\cdot|$ is the absolute value of the elements of the matrix $\mathbf{A}$. If ϵ is below a certain threshold $\epsilon \leq T_\epsilon$, go to step 4, otherwise go to step 2. The threshold was selected to $T_\epsilon = 10^{-12}$ since this value showed good results in our simulations.

4. For $\epsilon \leq T_\epsilon$, the convergence is said to be achieved. The columns of the precoding matrix $\mathbf{T}$ need to be then normalised in order to satisfy the power constraint.

It is important to note that this algorithm needs the CSI for all channels at the transmitter side in order to perform the described computations. This means that every receiver needs to feedback its corresponding channel matrix to the transmitter (if the channel is not reciprocal), before the whole process is started. Once convergence is reached and the decoding matrices are obtained, the transmitter must forward them to every receiver before the communication process starts.

19.3.2.3 SDD Scenario

The adaptation process of MOSDM and its iterative computation (NuSVD) to the SDD scenario share many similarities with the previous section for the MU scenario. The main difference is that two transmitters take part in SDD. Again, the derivations in the following are explained for the case of two transmitters and two receivers (refer to Figure 19.5).

Recall Equations 19.3 and 19.4 which show that the equalised symbol vector is disturbed by the interference from the second transmitter. Similar to Equation 19.18 in the previous subsection, equivalent channels can be defined as

$$\mathbf{H}_{e1} = \begin{bmatrix} \mathbf{R}_1^H\mathbf{H}_{11} & \mathbf{R}_1^H\mathbf{H}_{12} \end{bmatrix}, \tag{19.31}$$

$$\mathbf{H}_{e2} = \begin{bmatrix} \mathbf{R}_2^H\mathbf{H}_{21} & \mathbf{R}_2^H\mathbf{H}_{22} \end{bmatrix}. \tag{19.32}$$

In order to cancel the interference from the second transmitter, the precoding matrices are designed to fulfil the following conditions, analogue to Equation 19.20:

$$\mathbf{H}_{e1}\mathbf{T}_1 = [\Lambda_1 \quad \mathbf{0}],$$

$$\mathbf{H}_{e2}\mathbf{T}_2 = [\mathbf{0} \quad \Lambda_2]. \tag{19.33}$$

Analogue to Equation 19.21, the interference matrices are defined as

$$\tilde{\mathbf{H}}_{e1} = \mathbf{R}_2^H\mathbf{H}_{21},$$

$$\tilde{\mathbf{H}}_{e2} = \mathbf{R}_1^H\mathbf{H}_{12}. \tag{19.34}$$

Then, in order to cancel the interference at the transmitter, the precoding matrices have to lie in the null space of the corresponding equivalent inference matrix, analogue to Equation 19.22:

$$\mathbf{T}_1 \in \text{null}\{\tilde{\mathbf{H}}_{e1}\},$$

$$\mathbf{T}_2 \in \text{null}\{\tilde{\mathbf{H}}_{e2}\}. \tag{19.35}$$

Equation 19.35 shows how many spatial streams can be transmitted. It is assumed again that each transmitter has two transmit ports. If two spatial streams are transmitted by the two transmitters, $\tilde{\mathbf{H}}_{e1}$ and $\tilde{\mathbf{H}}_{e2}$, which are 2×2 in dimension according to Equation 19.34, then the null space according to Equation 19.35 is empty. Thus, the transmission of two spatial streams is not possible. Only one spatial stream can be utilised by each transmitter in this scenario. Then, the dimensions of $\tilde{\mathbf{H}}_{e1}$ and $\tilde{\mathbf{H}}_{e2}$ according to Equation 19.34 are 1×2, and the precoding vectors according to Equation 19.35 are of dimension 2×1, and one spatial stream is used.

The precoding matrices $\mathbf{T}_1$ and $\mathbf{T}_2$ will block diagonalise the matrix subspaces $\mathbf{H}_{e1}$ and $\mathbf{H}_{e2}$, obtaining the equivalent channels and cancelling the interference that transmitter 1 would generate to receiver 2 (and transmitter 2 receiver 1) according to Equation 19.33. Once the precoding matrices for interference cancellation have been acquired, the same channel optimisation process applied to the MU scenario can be applied here, that is, the steps described by Equations 19.25 through 19.30 and iterative computation according to NuSVD.

19.3.3 SU-Precoding: Eigenbeamforming

Assume that the receivers have more receive ports than the transmitter has transmit ports, as in the 2×4 MIMO setup. The additional receive ports can then be used to cancel the interference. Each link might be optimised by SU precoding, that is, by eigenbeamforming as introduced in Chapter 8. First, assume the MU scenario of one transmitter and two receivers. If the transmitter has two transmit ports, two spatial streams might be used, one for each of the users. For each link, the optimum precoding vector is calculated, for example, each receiver feedbacks the optimum precoding vector based on the CSI. The transmitter composes the final precoding matrix by combining the two precoding vectors into a matrix. However, the columns of this matrix will not be orthogonal to each other. This will generate more interference among the users. Although each receiver is able to detect the spatial stream directed at it, the detection matrix might enhance the noise and interference. Sanchez [12] investigated this algorithm for *zero-forcing* (ZF) detection. No performance gain could be achieved compared to SU-MIMO. More sophisticated detection algorithms compared to the simple ZF detection (see Chapter 8) might improve the performance.

The same concepts can be applied to the SDD scenario. Each of the links uses eigenbeamforming to maximise the performance. Again, the receivers use the higher number of receive ports to detect the spatial stream intended for the specific receiver and simultaneously cancel the interference. So, for the 2×4 MIMO setup, the 4 receive ports can be used to decode 4 spatial streams, 2 from the intended transmitter and 2 to remove the interference from other transmitters. The detection matrix is different compared to the SU-MIMO since the detection also has to consider the interference from other users. For ZF, the detection matrix for the two receivers is calculated as

$$\mathbf{W}_1 = \text{pinv}\left([\mathbf{H}_{11}\mathbf{T}_1, \mathbf{H}_{12}\mathbf{T}_2]\right),$$

$$\mathbf{W}_2 = \text{pinv}\left([\mathbf{H}_{22}\mathbf{T}_2, \mathbf{H}_{21}\mathbf{T}_1]\right). \tag{19.36}$$

where pinv $(\mathbf{A})$ is the pseudoinverse of the matrix $\mathbf{A}$, that is, pinv $(\mathbf{A}) = (\mathbf{A}^H \mathbf{A})^{-1} \mathbf{A}^H$.

Unfortunately, the signal to *interference plus noise ratio* (SINR) after detection is lower compared to the SINR for the SU-MIMO case, and the performance is not as good as for the BD algorithms [12].

19.4 Simulation Results

19.4.1 System Parameters

Simulations were performed in order to investigate the performance of the MU-MIMO algorithms introduced in Section 19.3. The MIMO-OFDM system introduced in Chapter 9 was used. The main system parameters are summarised in the following. The system deploys 1296 carriers in the frequency range between 4 and 30 MHz. The precoding and detection matrices are computed for each subcarrier. From there, the SINR after detection was calculated for each subcarrier according to the equivalent channels as introduced in Chapter 8. The adaptive modulation algorithm introduced in Chapter 9 was then used to derive the throughput rates for an uncoded *bit error ratio* (BER) of $P_b = 10^{-3}$. No *forward error correction* (FEC) is applied. Perfect channel knowledge is assumed. As a reference for the SU-MIMO case, eigenbeamforming (see Chapter 8) was used as this MIMO scheme showed the best performance for SU-MIMO (see Chapter 9). In order to compare the throughput results obtained for SU-MIMO eigenbeamforming with the bitrates obtained for MU-MIMO, the bitrates acquired for SU-MIMO have been divided by the number of users M with $M = 2$ in the inhome PLC scenario. This appears to be an acceptable estimation, considering that SU-MIMO needs to apply TDD (or FDD), using M time instants (or M times the bandwidth) to give the same service MU-MIMO would.

The MIMO PLC channels obtained in the ETSI measurement campaign (see Chapter 5) were used. Figure 19.7 shows an example of the measured channels in one home. In this example, the transfer functions (shown by the solid lines) between 5 outlets were recorded. An example of the MU scenario (refer also to Figure 19.3) is illustrated by the dashed lines, where the channels from outlet P4 to the outlets P2 and P3 are used. The SDD scenario (refer also to Figure 19.5) is illustrated by the dotted lines, where the channels from P1 to the P2 and from P3 to P5 are used. All possible combinations for each scenario (MU or SDD) for each home and all measurement sites were used for the simulations.

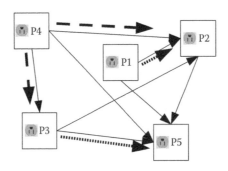

FIGURE 19.7
Depiction of channels measured in one house and possible MU (dashed lines) and SDD (dotted lines) setup.

This results in 357 different channel combinations for the MU scenario and 201 channel combinations for the SDD scenario. The noise is assumed to be *additive white Gaussian noise* (AWGN) and uncorrelated for the receive ports. The transmitters have $N_T = 2$ transmit ports and the receivers have $N_R = 4$ receive ports.

19.4.2 MU Scenario

Figure 19.8a shows the bitrate depending on the transmit to noise power level. The sum bitrate of both streams was averaged over all 357 channels. The number of users is $M = 2$, that is, one transmitter communicates to two receivers. Figure 19.8b shows

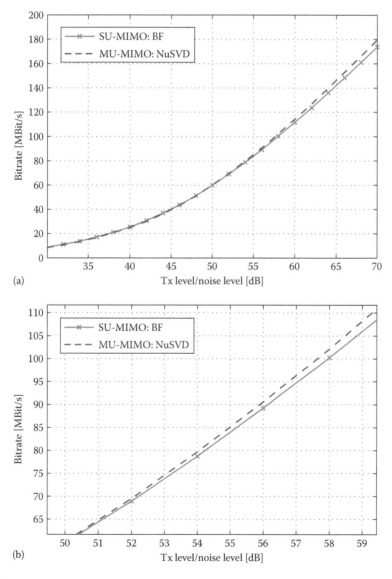

FIGURE 19.8
MU – throughput results averaged over 357 channels. (a) Averaged bitrate comparison for 357 channels. (b) Averaged bitrate comparison for 357 channels zoomed around 56 dB.

a zoomed plot of Figure 19.8a. There is only a marginal gain of MU-MIMO (with the NuSVD algorithm) compared to SU-MIMO with *eigenbeamforming* (BF). At a transmit to noise power level of 56 dB, the gain of MU-MIMO is 1.35 Mbit/s or 1.6% compared to SU-MIMO.

The main reason for the similar performance between MU-MIMO and SU-MIMO in the MU scenario is the high spatial correlation of the MIMO PLC channels (for details of the spatial correlation in MIMO PLC, refer to Chapters 4 and 5). Assume the following ideal case for MU-MIMO with two receivers. The MU-MIMO precoding vector of the first link is identical to the first column vector of the SU-MIMO eigenbeamforming matrix of this first link, that is, the precoding vector to cancel the interference to the other user is the same precoding vector which maximises the SINR of this spatial stream to the intended user. In addition, the MU-MIMO precoding vector of the second link is identical to the first column vector of the SU-MIMO eigenbeamforming matrix of the second link. If we further assume that the SINR conditions of the second spatial stream of the SU-MIMO eigenbeamforming allow no data transmission on the second spatial stream, then the total bitrate of MU-MIMO would be doubled compared to SU-MIMO eigenbeamforming. Expressed in simplified terms, the channels to the two receivers could be considered to be orthogonal and each link could be optimised independently. However, due to the high spatial correlation, this is not true, and the precoding to cancel out the interference to the other users is far from optimal for this link.

It was observed that the use of random channel matrices as MU-MIMO channels led to a much higher performance gain of MU-MIMO compared to SU-MIMO.

Figure 19.9 shows the *cumulative distribution function* (CDF) of the throughput for the 357 MU-MIMO channels. There is a small gain of 6.6 Mbit/s at the median point. For the high-throughput area (CDF values between 0.7 and 0.9), there is some gain of MU-MIMO compared to SU-MIMO which can be estimated in 9.6 and 12.1 Mbit/s, respectively, while SU-MIMO shows better results than MU-MIMO for low-throughput values (high coverage, low CDF values).

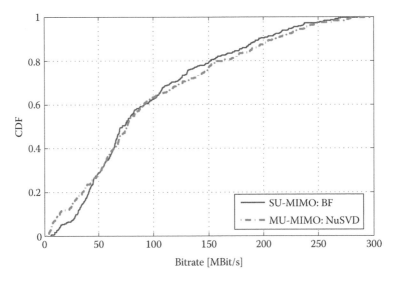

FIGURE 19.9
MU – CDF over 357 channels computed for a Tx/noise level of 56 dB.

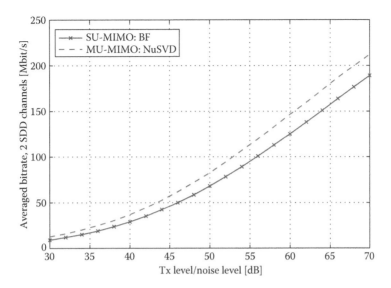

FIGURE 19.10
SDD – averaged bitrate comparison for 201 channels.

19.4.3 SDD Scenario

Following the model of the MU scenario, the throughput results obtained are averaged over 201 real measured channels. Figure 19.10 shows the comparison between SU-MIMO (with eigenbeamforming) and MU-MIMO (NuSVD). The results presented here belong to a scenario with two transmitter–receiver pairs.

Several differences can be observed in comparison with the results introduced for the MU scenario. Figure 19.10 shows a performance gain of MOSDM (NuSVD SDD) compared to SU-MIMO with eigenbeamforming. This BD method, which uses only one spatial stream between each transmitter–receiver pair, shows a throughput increase of approximately 20% in comparison to SU-MIMO, for a transmit to noise power level of 56 dB. This gain is also observed in Figure 19.11, which shows the CDF of the throughput.

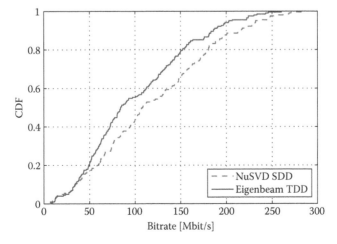

FIGURE 19.11
SDD – CDF over 201 channels computed for a Tx/noise level of 56 dB.

19.5 Conclusions

This chapter investigated the adaptation of MU-MIMO algorithms to MIMO PLC. The aim of this work was to analyse the suitability of MU-MIMO algorithms to MIMO PLC systems and study the benefits and improvements that they could bring to the PLC system. The adaptation of wireless MU-MIMO strategies to MIMO PLC faces some challenges, namely, the limited number of ports at the transmitter (only two) and the relatively high spatial correlation.

Two possible scenarios introduced in Section 19.2 have been analysed during this work. The MU scenario with one transmitter and several receivers presents a high degree of similarity to the wireless environment, where one BS offers service to several users located on one cell. On the other hand, the SDD scenario was introduced which involves at least two pairs of transmitters and receivers operating at the same time and on the same frequency band.

In particular, an algorithm called MOSDM [21] was applied to PLC. This algorithm turned out to be a good solution for MIMO PLC [12]. The algorithm applies precoding at the transmitter in order to cancel out inference to other users.

Simulations were performed to evaluate and compare MU-MIMO performance to SU-MIMO systems. The SU-MIMO system reference was based on eigenbeamforming. The channels obtained in the ETSI MIMO PLC measurement campaign (see Chapter 5) were used. A total set of 357 channels were used in the MU scenario, while 201 of them were used for the SDD scenario. The performance in terms of throughput of MU-MIMO was increased by 1.5% and 20% compared to SU-MIMO with eigenbeamforming, for the MU and SDD scenario, respectively. Clearly, only a small improvement is achieved for the MU scenario. However, 20% of throughput increase is reached when two transmitters are simultaneously working, that is, in the SDD scenario.

In the SDD scenario, a synchronisation between the transmitters is required in order for the receivers to estimate and cancel out the interference from the interfering transmitters. In the MU scenario, this synchronisation is not needed since only one transmitter is used.

Possible directions for future research in MU-MIMO for the PLC field could include the following:

- The investigation of the use of bands above 30 MHz. It could increase the efficiency of MU-MIMO since the spatial correlation above 30 MHz is reduced, compared to lower frequencies (see also Chapter 5).

- Finding a parameter or set of features that would detect, in advance, when MU-MIMO precoding techniques will yield better throughput results than SU-MIMO precoding techniques or vice versa. In this case, a hybrid algorithm could be used which toggles between MU-MIMO and SU-MIMO.

References

1. R. Heath, M. Airy and A. Paulraj, Multiuser diversity for MIMO wireless systems with linear receivers, in *Conference Record of the Thirty-Fifth Asilomar Conference on Signals, Systems and Computers*, Vol. 2, Pacific Grove, CA, pp. 1194–1199, 2001.
2. Q. Spencer, C. Peel, A. Swindlehurst and M. Haardt, An introduction to the multi-user MIMO downlink, *IEEE Communications Magazine*, 42(10): 60–67, 2004.

3. P. Fernandes, P. Kyritsi, L. T. Berger and J. Mártires, Effects of multi user MIMO scheduling freedom on cellular downlink system throughput, in *IEEE 60th Vehicular Technology Conference*, Vol. 2, Los Angeles, CA, pp. 1148–1152, September 2004.
4. D. Gesbert, M. Kountouris, R. Heath, C.-B. Chae and T. Salzer, Shifting the MIMO paradigm, *IEEE Signal Processing Magazine*, 24(5): 36–46, 2007.
5. S. Vishwanath, N. Jindal and A. Goldsmith, On the capacity of multiple input multiple output broadcast channels, in *IEEE International Conference on Communications (ICC)*, Vol. 3, New York, pp. 1444–1450, 2002.
6. M. Costa, Writing on dirty paper, *IEEE Transactions on Information Theory*, 29(3): 439–441, May 1983.
7. G. Caire and S. Shamai, On the achievable throughput of a multiantenna Gaussian broadcast channel, *IEEE Transactions on Information Theory*, 49(7): 1691–1706, 2003.
8. V. Stankovic and M. Haardt, Generalized design of multi-user MIMO precoding matrices, *IEEE Transactions on Wireless Communications*, 7(3): 953–961, 2008.
9. ETSI, TR 101 562-1 v1.3.1, PowerLine Telecommunications (PLT), MIMO PLT, Part 1: Measurement Methods of MIMO PLT, Technical Report, 2012.
10. ETSI, TR 101 562-2 v1.2.1, PowerLine Telecommunications (PLT), MIMO PLT, Part 2: Setup and Statistical Results of MIMO PLT EMI Measurements, Technical Report, 2012.
11. ETSI, TR 101 562-3 v1.1.1, PowerLine Telecommunications (PLT), MIMO PLT, Part 3: Setup and Statistical Results of MIMO PLT Channel and Noise Measurements, Technical Report, 2012.
12. Y. Sanchez, Multiuser MIMO for power line communications, Master's thesis, University of Stuttgart, Stuttgart, Germany, 2011.
13. M. Rim, Multi-user downlink beamforming with multiple transmit and receive antennas, *Electronics Letters*, 38(25): 1725–1726, 2002.
14. L.-U. Choi and R. Murch, A transmit preprocessing technique for multiuser MIMO systems using a decomposition approach, *IEEE Transactions on Wireless Communications*, 3(1): 20–24, 2004.
15. X. Chen, J. Liu, R. Xing and H. Xu, A suboptimal user selection algorithm for multiuser MIMO systems based on block diagonalization, in *Third International Conference on, Communications and Networking in China, ChinaCom 2008*, Hangzhou, China, pp. 877–881, 2008.
16. W. Liu, L. L. Yang, and L. Hanzo, SVD-assisted multiuser transmitter and multiuser detector design for MIMO systems, *IEEE Transactions on Vehicular Technology*, 58(2): 1016–1021, 2009.
17. R. Zamir, S. Shamai and U. Erez, Nested linear/lattice codes for structured multiterminal binning, *IEEE Transactions on Information Theory*, 48(6): 1250–1276, 2002.
18. M. Airy, A. Forenza, R. Heath and S. Shakkottai, Practical costa precoding for the multiple antenna broadcast channel, in *Global Telecommunications Conference, GLOBECOM'04, IEEE*, Vol. 6, Dallas, TX, pp. 3942–3946, 2004.
19. Z. Shen, R. Chen, J. Andrews, R. Heath and B. Evans, Sum capacity of multiuser MIMO broadcast channels with block diagonalization, *IEEE Transactions on Wireless Communications*, 6(6): 2040–2045, 2007.
20. S. Shim, J. S. Kwak, R. Heath and J. Andrews, Block diagonalization for multi-user MIMO with other-cell interference, *IEEE Transactions on Wireless Communications*, 7(7): 2671–2681, 2008.
21. Z. Pan, K.-K. Wong, and T.-S. Ng, Generalized multiuser orthogonal space-division multiplexing, *IEEE Transactions on Wireless Communications*, 3(6): 1969–1973, 2004.
22 S. J. Leon, *Linear Algebra with Applications*, 7th edn. Prentice Hall, Englewood Cliffs, NJ, 2006.

20

Relaying Protocols for In-Home PLC

Salvatore D'Alessandro and Andrea M. Tonello

CONTENTS

20.1 Introduction

Power saving is playing an important role in the development of advanced communication devices. For instance, the IEEE 802.3az Ethernet standard [1] and the HomePlug *Green physical* (PHY) (GP) *power line communication* (PLC) specifications (HomePlug GP [2]) have been developed to specifically address this problem. Not only power saving but also high transmission rate has to be granted, for instance, in multimedia applications such as *high-definition television* (HDTV) or 3D virtual video games. It therefore becomes essential to consider advanced communication techniques such as multi-carrier modulation with bit and power loading algorithms, cooperative communication algorithms and cross-layer optimisation.

In this chapter, we investigate the use of cooperative half-duplex time division relay protocols to possibly provide power savings, achievable rate improvements and coverage extension to the in-home PLC networks whose communication devices adopt multi-carrier modulation at the PHY layer, that is, *orthogonal frequency division multiplexing* (OFDM) [3].

20.1.1 Related Literature

In relay networks, the communication between the source and the destination nodes is helped by the use of one or more relays. More precisely, the relay receives the signal addressed to the destination node, processes it according to a given relay protocol and forwards it to the destination. The destination node combines the signals received from the source and from the relay. Many relay protocols have been proposed in the literature [4]: *amplify and forward* (AF), classic multi-hop, *compress and forward* (CF), *decode and forward* (DF) and multipath DF. In the following, we focus on AF and DF. In AF, the relay only amplifies and forwards the received signal to the destination, whereas in DF, the relays decode and re-encode the signal before forwarding it. AF and DF have been studied considering both half-duplex and full-duplex transmission, namely, a node can only transmit or receive at one time, or it can transmit and receive simultaneously. In the rest of this chapter, we consider the half-duplex modality, since full duplex is often not possible to be implemented due to high complexity [4].

The problem of resource allocation in relay networks has been thoroughly treated in the wireless literature. In the following, a number of relevant papers on the topic are reported.

The optimal power and time slot allocation for capacity maximisation over Rayleigh fading relay channels has been considered in Refs. [5–7]. The case of power allocation for capacity maximisation of single-hop parallel Gaussian relay channels (e.g. OFDM systems) under a total power constraint has been treated in Refs. [8–12]. In particular, in Ref. [8], the authors found a suboptimal power allocation considering half-duplex AF and a total power constraint at the source and relay nodes. The optimal solution to the previous problem has been found in Ref. [9]. In Ref. [11], the authors found the optimal power allocation for half-duplex AF and DF under a total power constraint (source plus relay) in each OFDM sub-channel. Both previous papers assume that the destination node is not directly reachable from the source. In Ref. [12], the power allocation for full-duplex DF [5] under a total power constraint (source and relay) is considered. The optimal power allocation for the hybrid use of AF, DF and direct link transmission is computed in Ref. [10] under a source plus relay power constraint in each sub-channel. In Ref. [13], the authors found the optimal power allocation for full- and half-duplex DF under a total power constraint at the source and destination nodes. Cooperation with multiple relays or relays with multiple antennas has been extensively treated in the wireless literature (cf., e.g. [14,15] and references therein).

In this chapter, we consider the specific and peculiar application of relaying in the in-home PLC scenario. In contrast to the wireless case, the use of cooperative communication schemes for PLC has not been deeply investigated yet. In the following, we report a list of relevant papers.

The use of repeaters over large-scale PLC networks, namely, networks where the source and destination nodes are far apart so that they cannot directly communicate, was proposed in Ref. [16] and extended in Ref. [17]. In particular, the use of the single-frequency network flooding approach was advocated to simplify the transmission of data by multiple relays. In Ref. [18], the authors considered the application of distributed space–time coding to the single-frequency network consisting of multiple relays to improve the network performances in terms of required transmit power and multi-hop delay. Different geographic routing schemes – namely, routing schemes that make use of the location of PLC devices – for low-data-rate Smart Grid applications over low- and medium-voltage distribution grids, were compared in Ref. [19] and optimised in Ref. [20]. In Ref. [21], the authors studied the diversity gains – namely, the asymptotic decay of the outage rate as a function of the *signal to noise ratio* (SNR) – over single-hop PLC relay networks. In particular, they found that no diversity gains are, in general, attainable in the examined context. This result is due to the peculiar electrical characteristics of the PLC network which behaves similarly to the keyhole channel observed in some MIMO wireless contexts. Despite the previous result, as it will be also shown in the following, in general, the use of relays in PLC networks leads to significant capacity gains w.r.t. the *direct transmission* (DT). Resource allocation algorithms were presented in Ref. [22], where the authors proposed practical sub-channel and power allocation algorithms for a two-hop DF relay scheme to improve the achievable rate of an *orthogonal frequency division multiple access* (OFDMA) PLC in-home network. The numerical results were obtained using a small number of measured channels and assuming a total power constraint at each network node. Achievable rate comparisons for AF and DF schemes over PLC channels were reported in Ref. [23], where it was assumed that the network nodes employed OFDM at the PHY layer and the power was equally distributed among the used OFDM sub-channels. Numerical results showed that the use of half-duplex single-relay schemes leads to marginal improvements of the achievable rate w.r.t. the DT. However, this result is only partially true, since the opportunistic use of the relay was not considered and the dependency on the relay position was not thoroughly investigated. Finally, in Ref. [24], the authors extended the work by considering the effect of channel estimation errors.

20.1.2 Chapter Contribution

In this chapter, we consider a network whose nodes have a PHY layer based on OFDM and where the communication between the source and the destination nodes follows an opportunistic protocol, namely, the relay is used whenever it allows (w.r.t. the DT): (1) for achievable rate improvements under a *power spectral density* (PSD) mask constraint or (2) for power saving under a PSD mask and a rate target constraint. *Opportunistic decode and forward* (ODF) and *opportunistic amplify and forward* (OAF) are considered. As it is typically required by state-of-the-art communication standards, for example, the wireless IEEE 802.11 standard, the power line IEEE 1901 standard and the twisted pair xDSL standard, we assume that the signal transmitted by the network nodes has to satisfy a PSD mask [25]. Under these assumptions and a Gaussian noise model, we find the optimal resource allocation, namely, the optimal power and time slot allocation, at the source and relay nodes that maximises the achievable rate or minimises the total transmitted power for both ODF and OAF.

Furthermore, for the specific and peculiar in-home PLC scenario, we consider the optimal relay positioning. In fact, differently from the wireless case, where the relay can be placed wherever between the source and the destination nodes, in in-home PLC networks, the relay can only be placed in accessible points of the network, that is, in the outlets or in the *main panel* (MP) or, in principle, in accessible derivation boxes (boxes where wires are connected to generate branches or extensions).

The remainder of the chapter is as follows. In Section 20.2, we describe the adopted PLC system model. Then, in Sections 20.3 and 20.4, we, respectively, consider the resource allocation problem of ODF and OAF. Section 20.5 discusses numerical results. Finally, a summary of the findings follows in Section 20.6.

20.2 PLC System Model

We consider an in-home PLC network where the communication between the *source* (S) and the *destination* (D) nodes exploits the use of a *relay* (R) (see Figure 20.1; the meaning of the variables in Figure 20.1 is explained in the following). In particular, we consider that the communication between the source and the destination nodes follows an opportunistic cooperative protocol, namely, the relay is used whenever it allows, according to the goal, for rate improvements or for power saving w.r.t. the DT. The multiplexing between the source and the relay nodes is accomplished via *time division multiple access* (TDMA). The time is divided in frames of duration T_f, and each frame is divided into two time slots whose durations are τ and $T_f - \tau$. When the relay is used, the source transmits its data to the relay and destination nodes during the first slot – although it is possible that the source cannot directly reach the destination – whereas, during the second slot, the source is silent and the relay transmits the received data to the destination according to the adopted opportunistic cooperative protocol, that is, ODF or OAF. When ODF is used, the relay decodes, re-encodes and forwards the received data using an independent codebook [6,26], whereas, in OAF, the relay only amplifies and forwards the data (see Figure 20.2).

At the PHY layer, we assume OFDM with M sub-channels. The channel frequency response between each pair of nodes is denoted as $H_{x,y}^{(k)}$, where the subscripts x and y denote the pairs $\{S,R\}$, $\{S,D\}$ or $\{R,D\}$ and k is the sub-channel index, that is, $k \in K_{on}$, where K_{on} is

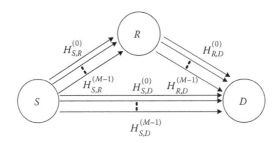

FIGURE 20.1
Two-hop relay network.

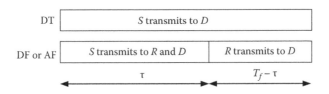

FIGURE 20.2
Time slot allocation for DT, DF and AF.

the subset of used (switched on) sub-channels that allows for satisfying a PSD mask with notches, as it is the case in broadband and narrowband PLC systems [27,28]. Therefore, the received signal in the kth sub-channel of the yth node reads

$$z_y^{(k)} = a_x^{(k)} H_{x,y}^{(k)} + w_y^{(k)}, \quad \{x,y\} \in \{\{S,D\};\{S,R\};\{R,D\}\}, \quad k \in K_{on}, \tag{20.1}$$

where
  $a_x^{(k)}$ is the symbol transmitted by node x in sub-channel k using DT, DF or AF modes
  $w_y^{(k)}$ is the background noise

We assume the noise and the transmitted symbols to be independent and identically distributed (i.i.d.) and drawn from normal Gaussian distributions with zero mean and power $P_{w,y}^{(k)}$ and $P_{x,mode}^{(k)}$, respectively.* In the remainder of this section, we assume the application of a PSD mask constraint for the signal transmitted by the network nodes. Furthermore, in order to simplify the notation, we assume the PSD to be constant over the sub-channels, that is, $P_{x,mode}^{(k)} \leq \bar{P} \; \forall \, k \in K_{on}$, $x \in \{S;R\}$ and $mode \in \{DT;DF;AF\}$. We highlight that all the power allocation algorithms that will be presented are also valid when a more general nonconstant PSD is considered.

20.2.1 In-Home Power Line Network Topology

As discussed in Section 20.1, we are interested to see whether achievable rate improvements, power savings and coverage extension are attainable through the use of ODF and OAF. To this end, in the following, we describe a typical PLC network topology, which allows for understanding where relays can be placed, and it highlights the differences with the wireless context. It is representative of the majority of Italian and EU residential wiring structures [29]. In particular, it is characterised by a wiring topology composed of two layers. As shown in Figure 20.3, the outlets are placed at the bottom layer and are grouped and fed by the same 'super node', which is referred to as derivation box. All the outlets fed by the same derivation box are nearby placed. Therefore, the location plan is divided into elements denoted as 'clusters' that contain a derivation box with the associated outlets. Each cluster represents a room or a small number of nearby rooms. Different clusters are usually interconnected through their derivation boxes with dedicated cables. This set of interconnections forms the second layer of the topology. We refer to the channels that connect a pair of outlets belonging to the same cluster as intracluster channels, whereas the channels associated to pairs of outlets that belong to different clusters are referred to as intercluster channels. An intercluster channel example is shown in Figure 20.3.

The MP plays a special role inasmuch as it connects the home network with the energy supplier network (not shown in Figure 20.3) through *circuit breakers* (CBs). We distinguish two cases. The first case, which we refer to as single-sub-topology networks, is when a single CB feeds all the derivation boxes of the home network. The second case, which we refer to as multi-sub-topology networks, is when many sub-topologies, each comprising a group of derivation boxes, have their own electrical circuit that is interconnected at

---

* The assumption on the transmitted symbols to be i.i.d. Gaussian distributed is meant to compute the capacity, that is, the maximum achievable rate (Cover and Thomas 2006).

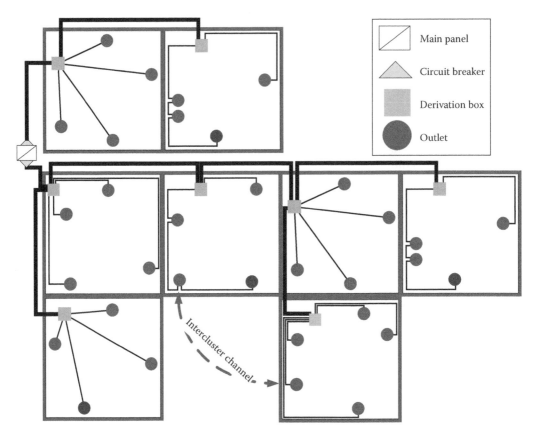

FIGURE 20.3
In-home network with two sub-topologies where each sub-topology is fed by a CB.

the MP through a CB. The latter case can be, for instance, representative of a multi-floor house, where each floor is a sub-topology. In Figure 20.3, we report an example of a two-sub-topology network.

Now, we consider the communication between source and destination nodes with the help of a relay. In particular, we consider source–destination channels defined between pairs of outlets that do not belong to the same cluster, that is, intercluster channels. As shown in Ref. [30] these channels experience higher attenuations than intracluster channels. Thus, they can benefit more from the presence of a relay. Clearly, these benefits are also dependent on the relay location. To this end, for single-sub-topology networks, the relay can be strategically placed in the following manners:

- *Outlet relay arrangement* (ORA). The relay is placed in a randomly selected network outlet.
- *Main panel single sub-topology* (MPS). The relay is placed immediately after the CB of the MP.
- *Random derivation box* (RDB). The relay is placed in a randomly selected network derivation box. In general, derivation boxes are accessible although they do not have an already installed outlet. Thus, a relay can be installed inside the box or nearby.

- *Backbone derivation box* (BDB). The relay is located in a randomly selected derivation box that belongs to the backbone between the source and the destination nodes. Note that, for intercluster channels, the source and destination nodes are at least divided by the source and *destination derivation boxes* (DDBs).
- *Source derivation box* (SDB). The relay is located in the derivation box that feeds the source node. Note that for intercluster channels, the path between source and destination includes at least the derivation box that feeds the source and the one that feeds the destination.
- DDB. The relay is located in the derivation box that feeds the destination node.

When we consider multi-sub-topology networks, we assume that the source and the destination node are located in two different sub-topologies. In such a case, we can consider the following strategical configurations for the relay:

- *Main panel multi-sub-topology* (MPM). The relay is located between the CBs that feed the sub-topologies.
- *Outlet relay arrangement source sub-topology* (ORAS). The relay is located in a randomly selected outlet belonging to the same sub-topology of the source node.
- *Outlet relay arrangement destination sub-topology* (ORAD). The relay is located in a randomly selected outlet belonging to the same sub-topology of the destination node.

20.3 Opportunistic Decode and Forward

In ODF, the source node sends data to the destination node according to two modes: DT or DF. Assuming a frame of normalised duration $T_f = 1$, we can compute the achievable rate of ODF as [6]

$$C_{ODF}(\tau) = \max\{C_{DT}, C_{DF}(\tau)\},\tag{20.2}$$

where C_{DT} and $C_{DF}(\tau)$, respectively, denote the capacity of DT and the achievable rate of DF [5]. These are given by

$$C_{DT} = C_{S,D},\tag{20.3}$$

$$C_{DF}(\tau) = \min\left\{\underbrace{\tau C_{S,R}}_{f_1(\tau,\mathrm{Ps,DF})}, \underbrace{\tau C_{S,D} + (1-\tau)C_{R,D}}_{f_2(\tau,\mathrm{Ps,DF},\mathrm{P_{R,DF}})}\right\}.\tag{20.4}$$

In Equations 20.3 and 20.4, $C_{S,D}$, $C_{S,R}$ and $C_{R,D}$ denote the capacities of the links $S–D$, $S–R$ and $R–D$, respectively; furthermore, the minimisation in Equation 20.4 is due to the fact that in DF mode, we require both the relay and the destination to decode the signal, and the unitary term is given by $T_f = 1$. Now, assuming the system model of Section 20.2, they are given by Ref. [26]

$$C_{x,y} = \frac{1}{MT} \sum_{k \in K_{on}} \log_2\left(1 + \text{SNR}_{x,y}^{(k)}\right), \quad \{x,y\} \in \{\{S,D\}; \{S,R\}; \{R,D\}\}, \qquad (20.5)$$

where

$$\text{SNR}_{x,y}^{(k)} = P_{x,mode}^{(k)} \frac{|H_{x,y}^{(k)}|^2}{P_{w,y}^{(k)}} = P_{x,mode}^{(k)} \eta_{x,y}^{(k)} \qquad (20.6)$$

is the SNR in sub-channel k for the link x–y, $k \in \{0, \ldots, M - 1\}$, T is the sampling period, and $\eta_{x,y}^{(k)}$ denotes the normalised SNR for the link x–y in sub-channel k. Furthermore, in Equation 20.4, $\mathbf{P}_{S,DF}$ and $\mathbf{P}_{R,DF}$ denote the vectors (with $|K_{on}|$ elements) of the sub-channel powers at the source and at the relay node, respectively. As it will be clear in the following, it is convenient to express the arguments of the minimisation in Equation 20.4 through the functions f_1 and f_2.

To simplify the notation, in Equation 20.5, we do not explicitly show the dependence of the capacity from the transmitted power distribution, which will be done if needed in the following. We notice that an SNR gap can be used in Equation 20.5 to take into account that practical coding and modulation schemes are used, for example, the HPAV broadband PLC system [31] employs turbo codes that allow having an SNR gap – namely, the amount of extra coding gain needed to achieve Shannon capacity [32] – of less than 3 dB vs. capacity. Furthermore, in Equation 20.5, we have implicitly assumed perfect channel state information; this is because we want to investigate theoretical performances. However, in this respect, we notice that PLC channels can be considered time invariant over the duration of several OFDM symbols, which allows accurate SNR estimation. It should also be noted that our analysis is in terms of achievable rate that corresponds to the definition of delay-limited capacity according to Refs. [6,33]. This capacity formulation is appropriate especially for delay-sensitive applications as voice and video where long delays cannot be tolerated. In this respect, sufficiently long codes can achieve the instantaneous capacity defined in Equation 20.5 since the PLC channel can be assumed constant for a long period of time. Channel variations are due to topology changes. In practice, moderate long codes, which introduce tolerable delay, should come close to the theoretical limit.

From Equations 20.2 through 20.4, it is interesting to note that a necessary condition to use the direct link is $C_{S,D} \geq \min \{C_{S,R}, C_{R,D}\}$. In the remaining cases, to see whether the communication follows the DT or the DF mode, we need to compute C_{DT}, $C_{DF}(\tau)$ and compare them as in Equation 20.2 to determine the largest. We also note that Equations 20.2 through 20.4 already take into account the case in which the destination cannot listen to the source.

In Sections 20.3.1 and 20.3.2, we will, respectively, deal with the power allocation for achievable rate improvements and power saving of ODF.

20.3.1 Rate Improvements with ODF

From Equation 20.2, we note that the achievable rate of ODF is a function of both the transmitted power distribution and the time slot allocation. In order to maximise it, when the DT is used, we only need to optimally allocate the power among the sub-channels of the source node. On the contrary, when the DF mode is used, we need to optimally allocate the power and the time slot of the source and relay.

Assuming that the network nodes have to satisfy a PSD mask constraint, it is known that the sub-channel power allocation that maximises the capacity for a point-to-point communication corresponds to the one given by the PSD constraint itself [34]. Therefore, for both ODF transmission modes, we set $P_{x,mode}^{(k)} = \bar{P}$, with $x \in \{S,R\}$, and $k \in K_{on}$. Now, to maximise the ODF achievable rate (Equation 20.2), we only need to compute the optimal time slot duration that can be found maximising (Equation 20.4), that is,

$$\tau_{mr}^* = argmax_{\tau \in [0,1]}\{C_{DF}(\tau)\}, \tag{20.7}$$

where we have used the subscripts mr to indicate that τ_{mr}^* is the time slot duration that maximises the achievable rate. To solve Equation 20.7, we observe that once the power transmitted by the source and the destination nodes is set, the arguments of the minimisation in Equation 20.4 are linear functions of τ. Assuming that $C_{S,R} \geq C_{S,D}$, the optimal time slot duration ($0 \leq \tau \leq 1$) is given by the intersection $f_1(\tau_{mr}^*, \mathbf{P_{S,DF}}) = f_2(\tau_{mr}^*, \mathbf{P_{S,DF}}, \mathbf{P_{R,DF}})$, with $P_{S,DF}^{(k)} = P_{R,DF}^{(k)} = \bar{P} \, \forall \, k \in K_{on}$.

20.3.2 Power Saving with ODF

We now consider the use of ODF for power saving and coverage extension. As discussed in the previous section, in ODF, the relay is used when the DF achievable rate is higher than that of DT. Now, let us suppose that the relay is used and we want to achieve a given target rate under a PSD constraint. Then, we can have three cases. The first case is when the target rate is reachable using either DT or DF. In such a case, since the DF achievable rate is higher than that of DT, the amount of power saved lowering the rate of DF to the target value will be higher than that saved lowering the rate of DT to the target value. The second case is when only the DT rate is lower than the target rate. In this case, the use of the relay can increase the network coverage. The third case is when the achievable rate of both modes is lower than the target rate so that the use of the relay increases the achieved rate possibly towards the target. We note that in the first considered case, the use of a relay can potentially lead to a transmitted power-saving w.r.t. DT. In our analysis, we focus on the transmitted power. Practical implementation issues and hardware/circuitry power consumption can also be studied [35]. For instance, the use of a sleep/awake protocol can lower the receiver power consumption over the listening periods. Now, to compute the power needed by ODF to achieve a target rate R when the communication is subject to a PSD constraint, we can solve the problem

$$P_{ODF} = \min\{P_{DT}, P_{DF}\}, \tag{20.8}$$

where P_{DT} and P_{DF}, respectively, denote the minimum power required by the DT and DF modes to achieve a rate R under a PSD constraint. Therefore, P_{DT} is the solution to the problem

$$P_{DT} = \min \sum_{k \in K_{on}} P_{S,DT}^{(k)},$$

$$\text{s.t.} \quad C_{S,D} = R,$$

$$0 \leq P_{S,DT}^{(k)} \leq \bar{P} \quad \forall \quad k \in K_{on}, \tag{20.9}$$

while P_{DF} is the solution to the problem

$$P_{DF} = \min \sum_{k \in K_{on}} \tau P_{S,DF}^{(k)} + (1-\tau)P_{R,DF}^{(k)},$$

s.t. $C_{DF}(\tau) = \min\{\tau C_{S,R}, \tau C_{S,D} + (1-\tau)C_{R,D}\} = R, \quad 0 \le \tau \le 1,$

$$0 \le P_{S,DF}^{(k)} \le \bar{P}, \quad 0 \le P_{R,DF}^{(k)} \le \bar{P}, \quad \forall \quad k \in K_{on}. \tag{20.10}$$

Starting from Equation 20.9, we note that its objective and its inequality constraint functions are convex, but its equality constraint is not an affine function. Therefore, it is not in general a convex problem [36] (pp. 136–137). Nevertheless, we note that the equivalent problem – see [36] (p. 67), for the definition of equivalent problems – obtained considering the change of variables $P_{S,DT}^{(k)} = (2^{b_{S,DT}^{(k)}} - 1)/\eta_{S,D}^{(k)}$, where $b_{S,DT}^{(k)} = \log_2\left(1 + P_{S,DT}^{(k)}\eta_{S,D}^{(k)}\right)$, is a convex optimisation problem. The solution to the equivalent problem (assuming that it exists) is well known and can be found imposing the *Karush–Kuhn–Tucker* (KKT) conditions (cf., e.g. [34,37]). Hence, the solution to the original problem can simply be found applying the inverse change of variables to the solution of the equivalent problem, and it is equal to

$$P_{S,DT}^{(k)} = P_{S,DT}^{(k)}(v) = \left[v - \frac{1}{\eta_{S,D}^{(k)}}\right]_0^{\bar{P}}, \tag{20.11}$$

where

$$[x]_a^b = \begin{cases} b, & x \ge b, \\ x, & a < x < b, \\ a, & x \le a, \end{cases} \tag{20.12}$$

and v is equal to the solution of the equality constraint of Equation 20.9, that is,

$$\frac{1}{MT} \sum_{k \in K_{on}} \log_2\left(1 + P_{S,DT}^{(k)}(v)\eta_{S,D}^{(k)}\right) = R. \tag{20.13}$$

It is interesting to note that when a non-uniform PSD mask has to be satisfied, the solution to problem 20.9 remains the same as in Equation 20.11, provided that the maximum allowable power in each sub-channel is set equal to the corresponding power constraint [34].

Problem 20.10 is more difficult to solve than problem 20.9 inasmuch its objective function is not in general convex. This can be proved observing that the Hessian associated to its objective function, for a given k, is neither semi-definite positive nor semi-definite negative; consequently, the Sylvester's criterion does not give any information regarding convexity [38].

The optimal solution to problem 20.10 can be found by splitting it into two convex subproblems [39]. The solution of each subproblem can be then found imposing the KKT conditions. However, it can be shown that the solution to the KKT conditions requires an iterative procedure. Consequently, its complexity is not less than that of conventional methods used for solving inequality-constrained minimisation problems, for example, the interior-point method [36] (Chapter 11).

To reduce the computational complexity, in the following, we summarise the simplified algorithm presented in Ref. [39]. It gives a suboptimal solution that gives results very close to the optimal ones.

20.3.3 Simplified Algorithm for DF Power Allocation

We assume the optimal time slot duration τ_{mp}^* is equal to the one computed in Equation 20.7, that is, $\tau_{mp}^* = \tau_{mr}^*$, where we have considered the achievable rate maximisation under a PSD constraint. Furthermore, we impose the constraint that for τ_{mp}^*, the arguments of the minimisation in the second line of Equation 20.10 are equal to R. Under these assumptions, Equation 20.10 can be divided into two subproblems where the first allows us to compute the power distribution of the source node independently from the power distribution of the relay node. It is obtained imposing in Equation 20.10 $C_{DF}(\tau_{mp}^*) = \tau_{mp}^* C_{S,R} = R$. Once we know the power distribution of the source, we can compute the power distribution of the relay solving the second subproblem, which is obtained imposing $C_{DF}(\tau_{mp}^*) = \tau_{mp}^* C_{S,D} + (1 - \tau_{mp}^*) C_{R,D} = R$ in Equation 20.10.

In particular, the power distribution at the source is given by

$$P_{S,DF}^{(k)} = P_{S,DF}^{(k)}(\nu) = \left[\nu - \frac{1}{\eta_{R,D}^{(k)}} \right]_0^{\bar{P}}, \tag{20.14}$$

where ν is given by the solution of

$$\sum_{k \in K_{on}} \log_2 \left(1 + P_{S,DF}^{(k)}(\nu) \eta_{S,R}^{(k)} \right) = \frac{MRT}{\tau_{mp}^*}. \tag{20.15}$$

The power distribution for the relay node is obtained solving the second subproblem, and it is given by

$$P_{R,DF}^{(k)} = P_{R,DF}^{(k)}(\nu) = \left[\nu - \frac{1}{\eta_{R,D}^{(k)}} \right]_0^{\bar{P}}, \tag{20.16}$$

where ν is given by the solution of

$$\sum_{k \in K_{on}} \log_2 \left(1 + P_{R,DF}^{(k)}(\nu) \eta_{R,D}^{(k)} \right) = MT \frac{R - \tau_{mp}^* C_{S,D}}{1 - \tau_{mp}^*}. \tag{20.17}$$

Eventually, the power needed by the DF mode to reach the rate R under the PSD constraint $\bar{P}$ is

$$P_{DF} = \tau_{mp}^* P_{S,DF} + \left(1 - \tau_{mp}^* \right) P_{R,DF}. \tag{20.18}$$

Therefore, we solve Equation 20.8 using Equations 20.11 and 20.18. It is worth noting that there could be cases where a solution to the power minimisation problem under a target rate and a PSD constraint does not exist. In particular, when only DT or DF admits a solution, the algorithm will choose the mode for which the solution exists. When the solution does not exist for both DT and DF, the algorithm will choose the mode that achieves the highest rate.

Finally, from Equations 20.11, 20.14 and 20.16, we note that the power allocation for the source node in both DT and DF modes, and for the relay node in DF mode, follows a typical water-filling shape, where the maximum allowable power in each sub-channel is limited by the power constraint.

20.4 Opportunistic Amplify and Forward

In order to compare the performance of ODF with a simpler relay scheme, we consider OAF.

For clarity, in the following, we describe the essence of the protocol.

Assuming the system model of Section 20.2, the achievable rate of OAF can be computed as

$$C_{OAF} = \max\{C_{DT}, C_{AF}\},\tag{20.19}$$

where C_{DT} is given by Equation 20.3. The achievable rate of AF can be computed as follows. We assume a frame-normalised duration $T_f = 1$, and further, we assume that the relay amplifies the signal received in sub-channel k by the quantity

$$g^{(k)} = \sqrt{\frac{P_{R,AF}^{(k)}}{P_{S,AF}^{(k)}\left|H_{S,R}^{(k)}\right|^2 + P_{w,R}^{(k)}}},\tag{20.20}$$

to assure that the relay transmits the power $P_{R,AF}^{(k)}$ in sub-channel k during the second half of the time frame. Finally, we assume the receiver adopts maximal ratio combining for the data received from the source and the relay in the two time slots. Therefore, the AF achievable rate can be written as [4,8,11]

$$C_{AF}(\mathbf{P_{S,AF}}, \mathbf{P_{R,AF}}) = \frac{1}{2MT}\sum_{k\in K_{on}}\log_2\left(1 + P_{S,AF}^{(k)}\eta_{S,D}^{(k)} + \frac{P_{S,AF}^{(k)}\eta_{S,R}^{(k)}P_{R,AF}^{(k)}\eta_{R,D}^{(k)}}{1 + P_{S,AF}^{(k)}\eta_{S,R}^{(k)} + P_{R,AF}^{(k)}\eta_{R,D}^{(k)}}\right).\tag{20.21}$$

From Equation 20.21, we note that the second and the third arguments of the log function, respectively, denote the SNR obtained with the direct link and the one obtained with the relay. Furthermore, the term 1/2 accounts for the slot duration. Clearly, the second term is null when the source cannot directly reach the destination.

20.4.1 Rate Improvements with OAF

To maximise the achievable rate of OAF, we need to optimally allocate the power for both DT and AF modes. As explained in Section 20.3.1, assuming that the network nodes have to satisfy a PSD mask constraint, the sub-channel power allocation that maximises the DT capacity corresponds to the one given by the same PSD constraint, namely, the DT capacity is maximised when $P_{S,DT}^{(k)} = \bar{P}, \forall k \in K_{on}$.

To maximise the achievable rate of AF subject to a PSD constraint, we notice that Equation 20.21 is the sum of monotonic increasing functions of both the power at the source and at the relay node. Therefore, since we have a constraint on the PSD, we can assert that the optimal power allocation is equal to that given by the same PSD, that is, $P_{x,AF}^{(k)} = \bar{P}$, with $x \in \{S,R\}$, and $k \in K_{on}$.

Eventually, to compute the OAF achievable rate when the system is subject to a PSD constraint, we can simply compute the DT and the AF achievable rates obtained setting the powers to $\bar{P}$, and then we can choose the mode that gives the highest achievable rate.

20.4.2 Power Saving with OAF

In Section 20.3.2, we have explained why the use of ODF can potentially bring power saving w.r.t. the DT. For the same reasons, OAF also can do that.

To compute the power used by OAF when the communication is subjected to a target rate R constraint and to a PSD constraint, we can solve the following problem:

$$P_{OAF} = \min\{P_{DT}, P_{AF}\}, \tag{20.22}$$

where P_{DT} and P_{AF}, respectively, denote the minimum power required by the DT and the AF modes to achieve a rate R under a PSD constraint. The optimal power allocation for DT can be found as in Equation 20.9, whereas P_{AF} is the solution to the following problem:

$$P_{AF} = \min \frac{1}{2} \sum_{k \in K_{on}} \left(P_{S,AF}^{(k)} + P_{R,AF}^{(k)} \right),$$

$$\text{s.t.} \quad C_{AF} = R,$$

$$0 \leq P_{S,AF}^{(k)} \leq \bar{P}, \quad 0 \leq P_{R,AF}^{(k)} \leq \bar{P}, \quad \forall k \in K_{on}. \tag{20.23}$$

Problem 20.23 is not convex because the equality constraint is not an affine function of the transmitted powers. We further report that we have not found a way to reduce the problem to an equivalent convex optimisation problem. Therefore, when showing numerical results, we solve Equation 20.23 using the interior-point method [36] (Chapter 11).

20.5 Numerical Results

In this section, we analyse the performance of ODF and OAF protocols for rate improvements, power saving and coverage extension over in-home PLC networks. To this end, we first describe the OFDM system parameters (Section 20.5.1). Then, in Section 20.5.2, we describe the statistical channel generator. Numerical results are finally presented and discussed in Sections 20.5.3 and 20.5.4.

20.5.1 Multi-Carrier System Parameters

We consider a multi-carrier scheme with parameters similar to that of HomePlug AV broadband PLC system [40]. In particular, we set $M = 1536$ sub-channels in the frequency band

0–37.5 MHz, unless otherwise stated. The set K_{on} of sub-channels that are switched on is defined so that the transmission band is 1–28 MHz. To respect the EMC norms [25], we consider a PSD mask constraint of –50 dBm/Hz, which is close to the one specified by Refs. [41]. Furthermore, we assume that the relay and the destination nodes experience white Gaussian noise both with PSD equal to –110 dBm/Hz (worst case) or –140 dBm/Hz (best case).

20.5.2 Statistical Channel Generator

According to experimental evidence and norms, a statistically representative in-home power line channel generator has been developed in Refs. [30,42,43]. It is representative of EU topologies and it uses a statistical topology model together with the computation of the channel responses through the application of transmission line theory. To be more precise, a location plan with a given area A_f contains $N_C = [A_f/A_C]$ clusters of area A_C (see Figure 20.3). The outlets are distributed only along the cluster perimeter. The number of outlets belonging to a given cluster is modelled as a Poisson variable with intensity $\Lambda_o A_C$. Furthermore, the outlets are uniformly distributed along the perimeter. The impact of the loads is also taken into account. In particular, a set of $N_L = 20$ measured loads for the in-home scenario, such as lamps or computer transformers, is considered. The load characterisation has been done as reported in Ref. [44], (Section 2.5.2). Furthermore, the impedance of the S, D and R nodes is set to 50 Ω.

To generate network topologies, we assume the home and the cluster areas equal to $A_f = 200$ m² and $A_C = 20$ m², for the single-sub-topology networks, whereas $A_f = 300$ m², for the two-sub-topology networks. We set the probability that no load is connected to a given outlet to 0.3. The intensity Λ_o is set to 0.33 (outlets/m²). Furthermore, for the two-sub-topology networks, we model each CB with a frequency attenuation, obtained from experimental measurements, that monotonically decreases from about –0.1 to –3.8 dB in the 1–28 MHz band.

More details regarding the network topology generator can be found in Ref. [42]. We report results for relatively large topology areas, since we have found that small gains are obtained with ODF and OAF in topologies of area smaller than 150 m². Finally, when showing results, we consider 100 network topologies. For each network topology, we consider 10 pairs of outlets (links S–D), and for each pair of randomly picked outlets, we place the relay according to the configurations presented in Section 20.2.1.

20.5.3 Achievable Rate Improvements with ODF and OAF

Table 20.1 lists the average achievable rate values considering all the strategic relay configurations presented in Section 20.2.1, for both noise levels and for ODF and OAF. The results are obtained computing the time slot τ according to Equation 20.7.

From Table 20.1, we notice that the average capacity values, for the link S–D, vary with the relay configuration. This is because the electrical properties of the network depend on the relay placement, which is different in each configuration.

Let us focus on ODF. From Table 20.1, we notice that in general there are two relay configurations for which we obtain high achievable rate gains with ODF and these are the SDB and the BDB. Now, in Figure 20.4, we show the *complementary cumulative distribution function* (CCDF) of the achievable rate for DT mode, when no relay is connected to the network, and for ODF according to the SDB and BDB configurations. The results obtained with the two best relay configurations for OAF are also reported. From Figure 20.4, we can see that with probability equal to 0.8, the SDB and BDB ODF relay configurations allow for

TABLE 20.1

Average Achievable Rate Values Using ODF and OAF for the Various Relay Configurations over the Generated Channels

Sub-Topology	Conf.	C_{DT} (Mbit/s)	C_{DF} (Mbit/s)	C_{ODF} (Mbit/s)	% Use of Relay	C_{AF} (Mbit/s)	C_{OAF} (Mbit/s)	% Use of Relay
				ODF			OAF	
Noise PSD = −110 (dBm/Hz)								
Single	SDB	182.8	220.2	220.2	99.9	128.9	183.8	12
	BDB	183.2	216.5	216.8	99.9	127	186.4	18.8
	RDB	190.3	104.3	207.6	52.8	113.9	191.9	11.1
	MPS	190.2	99.2	205.9	48.5	112.5	191.7	9.5
	ORA	193.6	70.6	202.7	29.8	107.7	194.2	6.6
	DDB	182.9	189	189.9	99.4	106.2	183	0.7
Two	MPM	116.4	131.5	148.6	91.3	87.6	120.2	28.6
	ORAS	121.2	73.2	128.2	33.8	71	122	7.5
	ORAD	121.3	51.8	127.9	31.5	70.7	121.9	6.1
Noise PSD = −140 (dBm/Hz)								
Single	SDB	421	454.8	454.9	99.9	255.7	421	0.3
	BDB	421.3	453.3	453.9	99.9	255.7	422	3.5
	RDB	429.4	238.8	447.6	53	240.4	430	2.3
	MPS	429.3	223.8	445	48.1	237.7	429.8	1.6
	ORA	433.8	181.3	444.7	32.5	233.7	434	1
	DDB	420.9	426.2	427.6	99.6	228.3	420.9	0
Two	MPM	341	339	376	91.7	215.6	342.3	5
	ORAS	348	232.8	357	38.4	191.9	348.3	1.5
	ORAD	348	161.7	356.4	40.1	191.2	348.2	1.2

an achievable rate improvement, w.r.t. DT, of about 50%. Furthermore, from Table 20.1, we notice that the ORA configuration gives small achievable rate improvements w.r.t. the DT for ODF. Although not shown, the same qualitative performance is obtained for the low noise scenario, where we found that the SDB gain, w.r.t. the DT, equals 20%.

Now, considering the two-sub-topology case, from Table 20.1, we notice that the MPM relay configuration gives the best performance. In particular, it gives an achievable rate improvement of 27% and 10%, respectively, for the high and for the low noise level. Another important observation regards the percentile usage of the relay for the various configurations. As explained in Section 20.3, the necessary condition not to use the direct link is $C_{S,D} < \min \{C_{S,R}, C_{R,D}\}$. This condition is satisfied when the relay lies on the backbone between the source and the destination nodes, which is always true for the SDB, BDB, DDB and MPM relay configurations. Therefore, these configurations are also the ones for which the relay is mostly used.

We now turn our attention to the achievable rate improvements provided by the OAF protocol. From Figure 20.4, we note that in general OAF does not bring appreciable achievable rate improvement w.r.t. DT. In particular, we notice that the best relay position, in terms of reliability, namely, minimum rate value, is the BDB. With probability equal to 0.8, it assures an achievable rate gain, w.r.t. the DT, equal to 5.6% and 0%, respectively, for the high and the low noise levels (although the low-noise-level results are not shown). This result agrees with what is reported in Refs. [24,45] where it is shown that in low-SNR scenarios, the AF protocol does not perform well. This is because the noise is also amplified

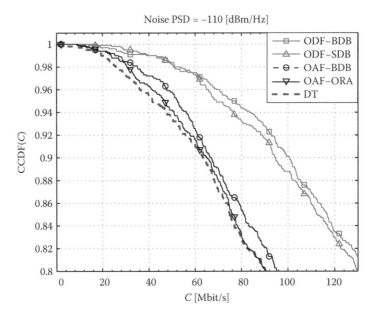

FIGURE 20.4
CCDF of achievable rate obtained using ODF and OAF with the relay located according to the best performers of the described configurations in single-sub-topology networks. The CCDF of capacity obtained assuming DT mode when no relay is connected to the network is also reported. $M = 1536$.

at the relay. It is interesting to note that the BDB relay configuration is not the one that gives the best performance in terms of average achievable rate. In fact, looking at Table 20.1, we can see that the relay configuration that yields the highest average achievable rate for OAF is the ORA.

From the previous results, we highlight that over the considered in-home power line network topologies, ODF gives good rate improvements w.r.t. DT over both single- and two-sub-topology networks, whereas OAF does not bring any substantial benefit. Furthermore, in ODF, the gains are more significant when the relay is placed in a backbone node.

20.5.4 Power Saving with ODF and OAF

In order to assess the performance of the described ODF power allocation algorithm, we set the target rate equal to the capacity of the DT link, that is, $R = C_{DT}$ when $P_{S,DT}^{(k)} = \bar{P} \ \forall \ k \in K_{on}$. Figure 20.5 shows the *cumulative distribution function* (CDF) of the total transmitted power for DT and for ODF when considering the various single-sub-topology relay configurations. The noise level is set to −110 dBm/Hz. From Figure 20.5, we can see that the best relay position is the SDB. With probability equal to 0.8, it allows for saving 2.6 dB. With the same probability, the BDB relay configuration allows for saving about 1.2 dB. We have found that similar results are obtained considering the low noise level. In particular, with probability equal to 0.8, the SDB and the BDB configurations lead to a power saving of 2 and 0.9 dB.

Although not shown, we highlight that the use of OAF yields to small power savings w.r.t. DT. We could have expected this result since the achievable rate of OAF is close to that of DT (see Figure 20.4).

We now turn our attention to the network coverage (number of links satisfying the target rate) improvements that can be obtained with the use of a relay. To this end, we consider the MPM, the ORAS and the ORAD relay configurations in a two-sub-topology network.

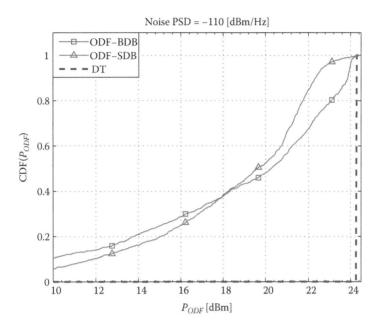

FIGURE 20.5

CDF of the total transmitted power using ODF with the relay located according to the various described configurations in single-sub-topology networks and the DT. $M = 1536$.

Since the $S–D$ links experience high attenuation given by the presence of CBs in the MP, we infer that the use of the relay should yield high coverage extension. To validate our conjecture, we consider the following scenario. We impose a target rate of 100 Mbit/s, for example, as required by a multimedia application to be delivered from the living room to the bedroom located in an upper floor. In Table 20.2, we report the percentage of links that satisfy the requirements for both measured and generated channels using DT and ODF with MPM, ORAS and ORAD relay configurations. We also report the average total transmitted power. Looking at the results, we note that the use of the relay does not substantially increase the coverage when we place it in a random outlet belonging to either the

TABLE 20.2

Percentage of Satisfied Links and Mean Transmitted Power for Two-Sub-Topology Networks Using DT and ODF

	DT		ODF	
Conf.	% of Satisfied Links	$E[P_{DT}]$ (dBm)	% of Satisfied Links	$E[P_{ODF}]$ (dBm)
Noise PSD = −110 (dBm/Hz)				
ORAS	54.9	21.8	59.3	21.5
ORAD	55.2	21.8	59.1	21.5
MPM	52.4	22	77	20.1
Noise PSD = −140 (dBm/Hz)				
ORAS	97.9	9.5	99	7.2
ORAD	97.9	9.5	98.7	7.6
MPM	97.8	9.8	99.9	2.6

Note: The rate target is 100 (Mbit/s).

source or the destination sub-topology (ORAS and ORAD configurations). Notably, when we place the relay in between the CBs of the MP (MPM configuration), for the high noise level, the coverage increases by 47% and the corresponding power saving is equal to 1.9 dB. When the noise level is low, we still have high power saving given by the use of a relay, but the gains associated to the coverage extension are reduced below 5%. This is simply explainable observing that for low noise levels, the imposed target rates are also achievable using the DT.

Regarding the OAF, we notice that it does not bring to any appreciable power saving and/or coverage increase.

20.6 Conclusion

The use of half-duplex time division ODF and OAF relay protocols can provide achievable rate improvements, power saving and coverage extension over in-home PLC networks.

An optimal power and time slot allocation algorithm can be used to maximise the ODF achievable rate with a PSD constraint. Furthermore, a simplified algorithm that is based on the solution of two convex subproblems can be adopted to address the power minimisation problem of ODF with a target rate constraint.

Numerical results obtained considering the specific and peculiar application of the algorithms to the in-home PLC scenario show that, in general, ODF performs better than OAF. Significant rate improvements and power savings can be obtained depending on the relay position and the size of the network. In the considered single-circuit (single-sub-topology) network, with high reliability, achievable rate gains (up to 50%), or power savings (up to 3 dB), are offered by ODF when the relay is placed in the derivation box that feeds the source node or in a derivation box that lies on the backbone link between the source and the destination nodes. In the considered two-circuit network connected at the MP, for example, a typical two-floor house network, the best relay location is in the MP. Also in this case, substantial achievable rate improvements (up to 27%), power savings (up to 1.9 dB) and coverage extension (up to 47%) have been found.

References

1. IEEE Std. 802.3az. 2010. Management Parameters for Energy Efficient Ethernet. ISBN 978-0-7381-6486-1.
2. HomePlug Alliance. 2012. HomePlug Gren PHY 1.1 – The Standard for In-Home Smart Grid Powerline Communications: An application and technology overview. Version 1.02, 3 October 2012.
3. Tonello, A.M., S. D'Alessandro and L. Lampe. 2010. Cyclic prefix design and allocation in bit-loaded OFDM over power line communication channels. *IEEE Trans. Commun.* 58(11) (November): 3265–3276.
4. Kramer, G., I. Maric, and R. Yates. 2006. *Cooperative Communications*. Hanover, MA: NOW Publishers Inc.; *Found. Trends Networking* 1(3): 271–425.
5. Host-Madsen, A. and J. Zhang. 2005. Capacity bounds and power allocation for the wireless relay channel. *IEEE Trans. Inf. Theory* 51(6) (June): 2020–2040.

6. Gündüz, D. and E. Erkip. 2007. Opportunistic cooperation by dynamic resource allocation. *IEEE Trans. Wireless Commun.* 6(4) (April): 1446–1454.
7. Xie, L. and X. Zhang. 2007. TDMA and FDMA based resource allocations for quality of service provisioning over wireless relay networks. In *Proceedings of the IEEE Wireless Communication and Networking Conference*, Hong-Kong, China, pp. 3153–3157.
8. Hammerström, I. and A. Wittneben. 2006. On the optimal power allocation for nonregenerative OFDM relay links. In *Proceedings of the IEEE International Conference on Communications*, Istanbul, Turkey, vol. 10, June 2006, pp. 4463–4468.
9. Zhang, W., U. Mitra and M. Chiang. 2011. Optimization of amplify-and-forward multicarrier two-hop transmission. *IEEE Trans. Commun.* 59(5): 1434–1445.
10. Li, X., J. Zhang and J. Huang. 2009a. Power allocation for OFDM based links in hybrid forward relay. In *Proceedings of the IEEE Vehicular Technology Conference*, Barcelona, Spain, April 2009.
11. Li, Y., W. Wang, J. Kong and M. Peng. 2009b. Subcarrier pairing for amplify-and-forward and decode-and-forward OFDM-relay links. *IEEE Commun. Lett.* 13(4): 209–211.
12. Ying, W., Q. Xin-Chun, W. Tong and L. Bao-Ling. 2007. Power allocation subcarrier pairing algorithm for regenerative OFDM relay system. In *Proceedings of the IEEE Vehicular Technology Conference*, Dublin, Ireland, April 2007, pp. 2727–2731.
13. Liang, Y., V. Veeravalli and V. Poor. 2007. Resource allocation for wireless fading relay channels: Max-min solution. *IEEE Trans. Inf. Theory* 53(10): 3432–3453.
14. Gesbert, D., S. Hanly, H. Huang, S. Shamai-Shitz and W. Yu. 2010. Guest editorial: Cooperative communications in MIMO cellular networks. *IEEE J. Sel. Areas Commun.* 28(9): 1377–1379.
15. Bakanoglu, K., S. Tomasin and E. Erkip. 2011. Resource allocation for the parallel relay channel with multiple relays. *IEEE Trans. Wireless Commun.* 10(3): 792–802.
16. Bumiller, G. 2002. Single frequency network technology for medium access and network management. In *Proceedings of the IEEE International Symposium on Power Line Communications and Its Applications*, Athens, Greece, March 2002.
17. Bumiller, G., L. Lampe and H. Hrasnica. 2010. Power line communication networks for large-scale control and automation systems. *IEEE Commun. Mag.* 48(4): 106–113.
18. Lampe, L., R. Shober and S. Yiu. 2006. Distributed space-time block coding for multihop transmission in power line communication networks. *IEEE J. Sel. Areas Commun.* 24(7): 1389–1400.
19. Biagi, M. and L. Lampe. 2010. Location assisted routing techniques for power line communications in smart grids. In *Proceedings of the IEEE International Conference on Smart Grid Communications*, Gaithersburg, MD, October 2010, pp. 274–278.
20. Biagi, M., S. Greco and L. Lampe. 2012. Neighborhood-knowledge based geo-routing in PLC. In *Proceedings of the IEEE International Symposium on Power Line Communications and Its Applications*, Beijing, China, April 2012, pp. 7–12.
21. Lampe, L. and A.J.H. Vinck. 2012. Cooperative multihop power line communications. In *Proceedings of the IEEE International Symposium on Power Line Communications and Its Applications*, Beijing, China, March 2012, pp. 1–6.
22. Zou, H., A. Chowdhery, S. Jagannathan, J.M. Cioffi and J.L. Masson. 2009. Multi-user joint sub-channel and power resource-allocation for powerline relay networks. In *Proceedings of the IEEE International Conference on Communications*, Dresden, Germany, June 2009.
23. Tan, B. and J. Thompson. 2011a. Relay transmission protocols for in-door powerline communications networks. In *Proceedings of the IEEE International Conference on Communications*, Kyoto, Japan, June 2011, pp. 1–5.
24. Tan, B. and J. Thompson. 2011b. Capacity evaluation with channel estimation error for the decode-and-forward relay PLC networks. In *Proceedings of the European Signal Processing Conference*, Barcelona, Spain, August 29–September 2, 2011, pp. 834–838.
25. Tlich, M., R. Razafferson, G. Avril and A. Zeddam. 2008. Outline about the EMC properties and throughputs of the PLC systems up to 100 MHz. In *Proceedings of the IEEE International Symposium on Power Line Communications and Its Applications*, Jeju Island, Korea, April 2008, pp. 259–262.

26. Cover, T.M. and J.A. Thomas. 2006. *Elements of Information Theory*. New York: Wiley & Sons.

27. Bumiller, G. 2012. Transmit signal design for NB-PLC. In *Proceedings of the IEEE International Symposium on Power Line Communications and Its Applications*, Beijing, China, March 2012, pp. 132–137.

28. D'Alessandro, S., A.M. Tonello and L. Lampe. 2011. Adaptive pulse-shaped OFDM with application to in-home power line communications. *Springer Telecommun. Syst. J.* 50: 1–11.

29. Comitato Elettrotecnico Italiano (CEI). 2007. *Norma CEI per impianti elettrici utilizzatori – CEI norm for electrical systems*. Milan, Italy: CEI, 2007.

30. Tonello, A.M. and F. Versolatto. 2010. Bottom-up statistical PLC channel modeling – Part II: Inferring the statistics. *IEEE Trans. Power Delivery* 25(4): 2356–2363.

31. Latchman, H. and R. Newman. 2007. HomePlug standards for worldwide multimedia in-home networking and broadband powerline access. In *International Symposium on Power Line Communications and Its Applications*, Speech II, Pisa, Italy. http://www.ieee-isplc.org/2007/docs/keynotes/latchman-newman.pdf (accessed September 23, 2012).

32. Campello, J. 1999. Practical bit loading for DMT, in *Proceedings of the IEEE International Conference on Communications (ICC '99)*, Vancouver, Canada, June 1999, vol. 2, pp. 801–805.

33. Hanly, S. and D. Tse. 1998. Multiaccess fading channels – Part II: Delay-limited capacities. *IEEE Trans. Inf. Theory.* 44(7): 2816–2831.

34. Papandreou, N. and T. Antonakopoulos. 2008. Bit and power allocation in constrained multi-carrier systems: The single-user case. *EURASIP J. Adv. Signal Process.* Article ID 643081: 2008: 1–15.

35. Cui, S., R. Madan, A. Goldsmith and S. Lall. 2005. Energy-delay tradeoffs for data collection in TDMA-based sensor networks. In *Proceedings of the IEEE International Conference on Communications*, Seoul, South Korea, pp. 3278–3284.

36. Boyd, S. and L. Vandenberghe. 2004. *Convex Optimization*. Cambridge, MA: Cambridge University Press.

37. Kuhn, H.W. and A.W. Tucker. 1951. Nonlinear programming. In *Proceedings of Second Berkeley Symposium on Mathematical Statistics and Probability, California*, Berkeley, CA, August 1951, pp. 481–492.

38. Weisstein, E.W. Sylvester's criterion. http://mathworld.wolfram.com/SylvestersCriterion.html (accessed September 23, 2012).

39. D'Alessandro, S. and A.M. Tonello. 2012. On rate improvements and power saving with opportunistic relaying in home power line networks. *EURASIP J. Adv. Signal Process.* 194: 1–16.

40. Afkhamie, K., S. Katar, L. Yonge and R. Newman. 2005. An overview of the upcoming HomePlug AV Standard. In *Proceedings of the IEEE International Symposium on Power Line Communications and Its Applications*, Vancouver, British Columbia, Canada, April 2005, pp. 400–404.

41. CENELEC. 2012. Final Draft of EN-50561-1 Standard. Power line communication apparatus used in low-voltage installations – Radio disturbance characteristics – Limits and methods of measurement – Part 1: Apparatus for in-home use. Brussels, Belgium: CENELEC, 2012.

42. Tonello, A.M. and F. Versolatto. 2010. Bottom-up statistical PLC channel modeling – Part I: Random topology model and efficient transfer function computation. *IEEE Trans. Power Delivery* 26(2): 891–898.

43. Versolatto, F. and A.M. Tonello. 2012. On the relation between geometrical distance and channel statistics in in-home PLC networks. In *Proceedings of the IEEE International Symposium on Power Line Communications and Its Applications*, Beijing, China, March 2012, pp. 280–285.

44. Ferreira, H.C., L. Lampe, J. Newbury and T.G. Swart. 2010. *Power Line Communications: Theory and Applications for Narrowband and Broadband Communications over Power Lines*. New York: Wiley & Sons.

45. Laneman, J., D. Tse and G. Wornell. 2004. Cooperative diversity in wireless networks: Efficient protocols and outage behavior. *IEEE Trans. Inf. Theory* 50(12): 3062–3080.

Part V

Implementations, Case Studies and Field Trials

21

Narrowband PLC Channel and Noise Emulation

Klaus Dostert, Martin Sigle and Wenqing Liu

CONTENTS

With the advent and popularity of Smart Grids and advanced meter reading, more and more NB-PLC systems are available on the market. It becomes more difficult for customers to choose the best solution for their application. At the same time, system designers and researchers are developing a new generation of PLC modems. For system verification and benchmark testing, engineers are no longer satisfied with test methods and equipment adopted from other communication industries. For example, a vector signal generator is very popular in mobile phone and wireless tests. You can build a transmitter in it, generate the test signal to investigate the receiver and add noise and attenuation, so that the transmit signal seems to be distorted by a real channel. This method is accurate and flexible. However, it needs exact details of the *device under test* (DUT). If you want to test several devices of different technologies, you have to build a transmitter for each one. You can also connect the transmitter and receiver with a switchable attenuator; noise can be generated by using a low-cost function generator. Spectrum analysers and oscilloscopes can be used to measure signals and noise at the receiver. If the attenuation can work for both

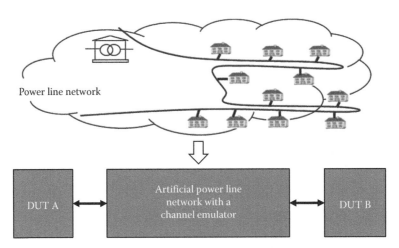

FIGURE 21.1
Test PLC systems using emulated channels.

directions, a bidirectional communication can be tested. However, usually, the attenuator is flat for wide frequency range, so the performance against frequency-selective attenuation cannot be evaluated. It has low flexibility and also high cost. Alternatively, you can connect the DUT to mains network directly. Although such test results are highly realistic, they suffer from the lack of reproducibility and low flexibility.

Traditional equipment and approaches make the test of NB-PLC systems quite challenging. The new idea is to develop a hardware device, with which the sophisticated behaviours of real power line networks can be reproduced in a laboratory any time, as shown in Figure 21.1. Since it simulates the real-life power line channel in hardware, it can be called a channel emulator. The channel emulator indeed opens a new path to the flexible, reliable and technique-independent performance evolution of PLC modems.

21.1 NB-PLC Channel Emulator

An NB-PLC channel emulator has been developed according to the channel model mentioned in Chapter 2. As shown in Figure 21.2, the hardware design is composed of four modules: the input module emulates the access impedance with digitally switchable passive LCR networks. The *analogue front end* (AFE) contains a 12-bit *analogue-to-digital converters* (ADC) and two 14-bit *digital-to-analogue converters* (DACs). The sampling rates of the ADC and the DACs are configurable and set to 2 MS/s. Two additional digital-controlled attenuators are utilised to achieve a wide range of the *signal-to-noise ratio* (SNR) value with high accuracy. Each of the attenuators has a dynamic range of more than 75 dB with a resolution of 0.37 dB. The *field-programmable gate array* (FPGA) is the key component for the signal-processing part of the channel emulation. It implements digital filters and stores the impulse response values for multiple different channels. The frequency resolution of the reproduced transfer function can reach 2 kHz. The filter can be reloaded with a new set of coefficients within 10.5 μs during the filtering operation. The noise unit reproduces the noise scenario. The time variance control unit manages the switching among the access impedances, the updating of the digital filter and the modification of the noise parameters.

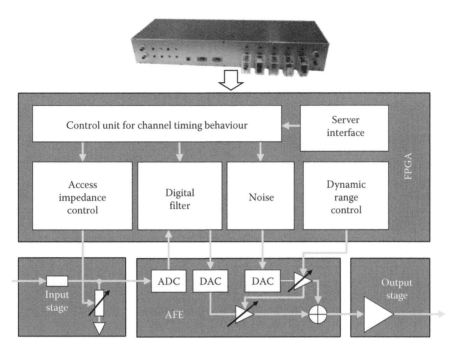

FIGURE 21.2
Block diagram of NB-PLC channel emulator.

The event table holds the time schedule of the switching events. The SNR unit determines the attenuation values for the signal and the noise paths. The output module scales the output to an appropriate voltage level. The emulator is connected to a computer via a standard serial interface, so that the emulation process can be controlled externally.

21.2 Emulating Narrowband Interferer

The emulation of narrowband noise covers noise types with sinusoidal waveform. According to [1], this kind of noise has a bandwidth of several kHz. The amplitude level is greater than that of background noise. A narrowband noise model features three parameters: the centre frequency f_m, the bandwidth Δf and the amplitude A_{NBN}. The centre frequency refers to the fundamental oscillation of the noise. The centre frequency may change permanently, and its variation range is indicated by the bandwidth. The amplitude is a measure of the noise voltage level. The generation of a sinusoidal waveform is based on the phase accumulation method. As shown in Figure 21.3, this method utilises a phase accumulator and a look-up table. The phase accumulator consists of an adder and a phase register. The current phase $P(n + 1)$ results from the modulo-N sum of the previous phase $P(n)$ with the phase increment I:

$$P(n+1) = Mod_N\left[P(n)+I\right]. \tag{21.1}$$

The look-up table is also called phase-to-amplitude converter. It contains one period of a sinusoidal waveform. The phase $P(n)$ is fed to the look-up table as an address input, under

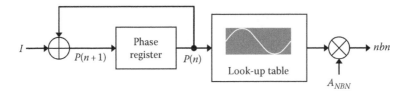

FIGURE 21.3
Phase accumulation unit.

which the amplitude value appears at the output. Multiplied with the factor A_{NBN}, the amplitude values finally form the narrowband noise *nbn* with its fundamental oscillation.

The centre frequency f_m depends on the clock frequency f_a, on the number of stored amplitude values N as well as on the phase increment I_m, that is,

$$f_m(I_m) = \frac{f_a}{N} \cdot I_m. \tag{21.2}$$

The variation of the centre frequency can be realised by incrementing or decrementing the phase increment I, that is,

$$I = \begin{cases} I_m + \Delta I, \\ I_m - \Delta I. \end{cases} \tag{21.3}$$

When I reaches its lower limit I_{max-}, it will be incremented by ΔI with each clock. When I reaches the upper limit I_{max+}, it is decremented by ΔI with each clock. The resulting instantaneous frequency $f(I)$ is

$$f(I) = \frac{f_a}{N} \cdot I. \tag{21.4}$$

As a result, the bandwidth can be calculated by

$$\Delta f = \frac{2f_a}{N} \cdot (I_{max+} - I_{max-}). \tag{21.5}$$

21.3 Emulating SFN Using Chirp Function

Due to the distinct spectral features of the SFN, it is reasonable to treat SFN in the frequency domain. The SFN can be modelled by a superposition of multiple chirp functions, such as

$$y_{chirp} = \sum_{n=1}^{N} m_n(t) \cdot \sin\left[2\pi \int_0^t f_n(\tau)d\tau \right], \tag{21.6}$$

where
    N is the number of fundamental chirp waveforms
    $m_n(t)$ and $f_n(\tau)$ are the envelope and the instantaneous frequency of the nth chirp, respectively

For the periodical noise with superimposed linear chirps and time-varying envelopes, the instantaneous frequency is obtained by

$$f_n(\tau) = f_0 + \frac{f_1 - f_0}{\Delta\tau} \cdot \tau, \tag{21.7}$$

where
f_0 and f_1 are the start and stop frequencies, respectively
$\Delta\tau$ is the duration within which the frequency changes from f_0 to f_1
$m_n(t)$ can be approximated by the scaled waveform of the mains voltage

The emulated noise together with coloured background noise is shown in Figure 21.4.

The chirp (see Chapter 2) has an instantaneous frequency which changes nonlinearly. By curve fitting the frequency trace, the instantaneous frequency can be approximated by

$$f_n(\tau) = f_1 \cdot \frac{(\tau - \tau_0)^{1.8}}{\tau_0^{1.8}} + f_0, \tag{21.8}$$

where
τ_0 is the time corresponding to the minimum frequency f_0
f_1 is the maximum frequency

The harmonic frequencies can also be modelled in Equation 21.8 conveniently. The envelope can be approximated by a Chebyshev window function whose Fourier transform side lobe magnitude is 120 dB below the main lobe magnitude. The synthesised noise waveform is shown in Figure 21.5.

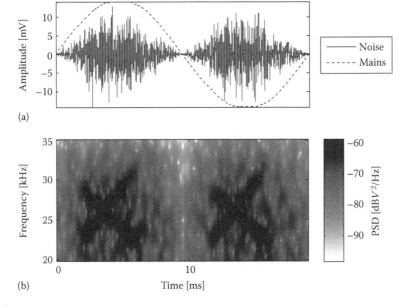

FIGURE 21.4
Synthesised periodic noise with rising and falling swept frequencies. (a) Waveform in time domain and (b) *short-term Fourier transform* (STFT).

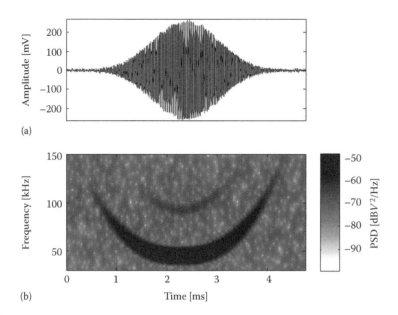

FIGURE 21.5

Synthesised periodic noise with nonlinearly swept frequencies. (a) Waveform in time domain and (b) STFT.

21.4 Emulating Time-Domain Waveform of Impulsive Noise

Since many impulsive noise types have similar magnitudes as the background noise, in the case where the time-domain noise waveform plays a less important role as the noise *power spectral density* (PSD), it can be assumed that the impulsive noise has a random waveform, and the appearance of such an impulse only raises the spectral level of the background noise shortly. This kind of impulsive noise can be generated by switching the background noise on and off according to a control pulse sequence. As shown in Figure 21.6, coloured background noise serves as noise source. The scaling unit determines the maximum amplitude of the noise. A switch is inserted between the noise source and the scaling unit. The switching sequence generator launches a sequence of '1' or '0'. '1' switches the noise source on to the scaling unit. The width of the pulse is determined by the duration of a '1' level. If the '1' level repeats within constant intervals, periodic impulsive noise is produced. Otherwise we get aperiodic impulsive noise. The former can be defined as synchronous with the mains frequency, if it repeats, for example, every 10 or 20 ms.

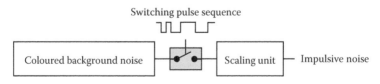

FIGURE 21.6

Generation of impulsive noise with stochastic waveform.

All other rates correspond to periodic impulsive noise asynchronous with the mains frequency. The emulation of the time behaviour will be introduced later in the next section.

When an oscillation with exponential decay is expected, the impulsive waveform can be controlled by using three parameters: the pulse width, the maximal amplitude and the frequency of the oscillation. The noise can be modelled by modulating a sinusoidal waveform

$$y_0 = \sin(\omega_0 \cdot t) \tag{21.9}$$

with an exponentially attenuated envelope

$$x_0 = A_0 \cdot e^{-a_0 \cdot t}, \tag{21.10}$$

where
$\omega_0 = 2\pi f_0$ is the circular frequency of the sinusoidal signal
A_0 is the initial amplitude
a_0 is the time constant of the exponential decay

The modulated signal can be expressed as

$$z_0 = x_0 \cdot y_0 = A_0 \cdot e^{-a_0 \cdot t} \cdot \sin(\omega_0 \cdot t). \tag{21.11}$$

The implementation of the noise is shown in Figure 21.7. A segment of an exponential function with 1000 samples as well as a period of a sinusoidal signal with the same number of samples is created in MATLAB® first and stored as a look-up table within the FPGA hardware. The step width of the generated addresses can be configured with the parameters Δm and Δn. x_1 and y_1 are outputs of both look-up tables. They are multiplied with each other, and the product sequence forms the modulated exponentially decaying oscillation z.

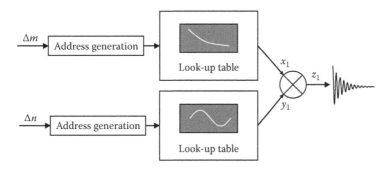

FIGURE 21.7
Emulation of impulsive noise with exponentially decayed waveform.

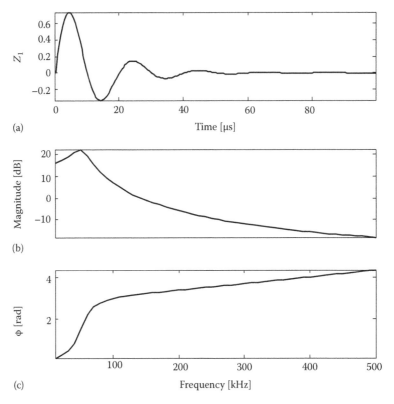

FIGURE 21.8
Emulated impulsive noise (a) waveform, (b) magnitude of the FFT and (c) phase vector of the FFT.

Combining both into the expressions of both signals, we get

$$x_1 = A_0 \cdot e^{-a_0 \cdot (\Delta m \cdot T_s)} = A_0 \cdot e^{-a_1 \cdot T_s} \tag{21.12}$$

and

$$y_1 = B_1 \cdot \sin\left(\omega_0 \cdot \Delta n \cdot T_s\right) = B_1 \cdot \sin\left(\omega_1 \cdot T_s\right), \tag{21.13}$$

where T_s is the period of the clock, with which samples are read out of the look-up tables. Changing the step widths of both addresses provides a variation of the sampling period. This can also be interpreted as an alteration of the time constant of the exponential decay and the frequency of the oscillation, respectively. In this way, both signals can be configured conveniently. An example of the emulated impulsive noise is shown in Figure 21.8.

21.5 Emulating Time-Varying Background Noise

Figure 21.9a shows the simulated PSDs; plot (b) shows the time–frequency representation of the simulated PSDs. It can be seen that the fundamental features of the time variance are reconstructed.

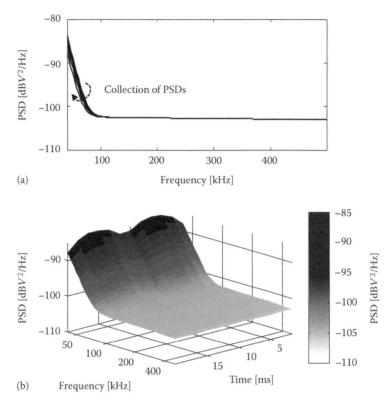

FIGURE 21.9
Emulated background noise. (a) PSD overlapped over time and (b) time–frequency representation of PSD.

21.6 Emulation of Channel Transfer Function

This part deals with the emulation of *channel transfer function* (CTF). From the signal processing's point of view, the CTF influences the transmit signal in the form of a convolution of the transmit signal with the channel impulse response. To emulate the CTF means to implement the convolution algorithm. A starting point is the measured CTF in the form of magnitude, phase response and group delay denoted by $|H_{dB}(i)|$, $\varphi(i)$ and $\tau(i)$, respectively.

21.6.1 Direct Convolution and FFT Convolution

There are two possibilities to implement the convolution. One is the conventional convolution in the time domain using a digital filter. The other is the block convolution in the frequency domain. The first method obtains the channel impulse response samples and saves them as coefficients of the FIR filter. The transmit signal is correlated with the impulse response samples. The block convolution is based on the principle that a convolution in the time domain corresponds to a multiplication in the frequency domain. It converts the transmit signal in the frequency domain, multiplies the signal spectrum by the CTF and converts the products back in the time domain via IDFT. It takes advantages of the highly efficient *fast Fourier transform* (FFT), thus it is also called FFT convolution. By using the FFT algorithm to calculate the DFT, this kind of convolution can be faster than directly convolving the time-domain signals. Common methods to implement the FFT convolution are the overlap-add, the weighted overlap-add and the overlap-save methods. Details of these methods can be found in Ref. [2]. By comparing the number of major operations, it has been realised that the overlap-add method can be faster and more efficient than the FIR-based time-domain convolution if the filter order exceeds 20 or 40 for complex- and real-valued filters, respectively. Besides the implementation efficiency, the implementation must be able to reproduce time-varying convolution because typical NB-PLC channels exhibit time-varying responses. For this purpose, the FFT convolution could be quite simple and straightforward. Multiple CTFs can be stored in advance. A switch from one CTF to the other refers to a change of channel response. The switch rate depends on the length of each FFT block. In principle, the CTF shall not change during a FFT block. Therefore the time resolution can be as high as one FFT block for the emulation of time-varying CTF [3].

One disadvantage of the FFT convolution is an inherent delay within the signal-processing chain. To investigate this delay in detail, we use an overlap-add-based system as an example. Figure 21.10 illustrates a simplified block diagram. The system uses a N_{FFT} point – FFT convolution to emulate the filtering effect of power line channels. The clock cycles that the FPGA spends in these modules and the corresponding redundancies are denoted by N_1 through N_4 and τ_1 through τ_4, respectively. The channel is supposed to have an impulse response of N_{ir} points. It is padded with $(N_{FFT} - N_{ir})$ zeros, converted to frequency response H and stored in the system. The transmit signal is converted to digital samples by the ADC. The 'overlap-add input' module collects $(N_{FFT} - N_{ir})$ samples, adds N_{ir} zeros to them to build a FFT block and delivers this block to the 'FFT' module. As a result, the first sample $s_i(1)$ must wait for at least N_1 sampling periods until it can be processed by the FFT module, and

$$N_1 = N_{FFT} - N_{ir}. \tag{21.14}$$

The 'FFT' module performs a N_{FFT} point FFT on the block. The FPGA needs N_2 clock cycles to finish the FFT. The frequency of this clock can be different from the signal sampling rate. The FFT results are delivered to the next module one by one. At the same time, they are multiplied by H. In this case, the multiplication lasts for at least N_{FFT} clock cycles. The product vector is converted to waveform in the time domain by an IFFT of the same length as the FFT. The IFFT needs the same clock cycles as the FFT module. The 'overlap-add output' module adds the current IFFT block to the previous one with an appropriate superimposition. Finally, the output value corresponding to $s_i(1)$ is obtained. The delays caused by each module are listed in Table 21.1.

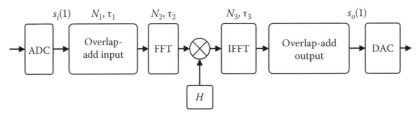

FIGURE 21.10
Implementation of overlap-add method.

TABLE 21.1

Delays Caused by Modules

N_{FFT}	N_{ir}	N_1	$\tau_1(\mu s)$	N_2	$\tau_2(\mu s)$	N_3	$\tau_3(\mu s)$	N_4	$\tau_4(\mu s)$
4,096	1,536	2,560	25.6 (100 M)	12,445	31.51 (395 M)	4,096	10.37 (395 M)	12,445	31.51 (395 M)
			1,280 (2 M)						

The represented system uses 4096 point FFT and IFFT. The signal is sampled at 100 MS/s. The channel impulse response is supposed to be 15.36 µs; therefore, the length of the impulse response is 1536 samples. Each signal block contains 2560 samples, leading to a latency of 2560 sample periods in the input module. The FPGA needs 12,445 clock cycles for calculating FFT or IFFT [4]. Suppose the FFT, IFFT and the multiplications are performed at the maximum clock rate, for example, at 395 MHz for a Virtex 5, as a result, the latencies of the FFT and IFFT are both 31.51 µs, and the multiplication lasts for 10.37 µs. The minimum total latency reaches 99 µs, more than six times the channel impulse response. Applying the same conditions for convolution with a sample rate of 2 MS/s, the total latency is about 1.35 ms. Note that this large latency is not caused by the data transmission channel, but by the convolution algorithm itself. It may not impair the data transmission which deploys a preamble for the frame synchronisation, because both the preamble signal and the transmit signal are delayed identically. However, there are systems that use the zero crossings of the mains voltage for the purpose of bit and symbol synchronisation. Systems based on this kind of synchronisation are most S-FSK and other single-carrier systems as well as some *orthogonal frequency division multiplexing* (OFDM) systems [5]. The latency in the convolution can introduce an 'artificial' synchronisation error which degrades the data transmission in an unwanted manner. Therefore, it has serious consequences for the investigation of communication systems.

In contrast to the FFT convolution, the direct convolution does not have the inherent delay. The latency can be reduced by shifting the filter kernel properly. Nowadays, the development of silicon makes the implementation of complicated signal-processing algorithms possible for real-time applications. As a result of the comparison of both convolutions, we have selected direct convolution in the time domain for emulating the CTF.

21.6.2 Digital Filters

Most efforts are made in the design of the filter kernel for direct convolution applications. The simplest and most straightforward way is the windowed frequency sampling method. The filter kernel can be obtained by performing an IFFT on the desired frequency response.

21.6.3 Frequency Sampling Method

The filter length is denoted by N_{fir}; therefore, the filter has an order of $N_{fir} - 1$. The frequency samples are expressed in linear scale

$$H(i) = 10^{\frac{|H_{dB}(i)|}{20} \cdot \exp[j\varphi(i)]}. \tag{21.15}$$

The sampled values are given for the frequency of interest. The vector for the IFFT should be filled carefully so that the output of the IFFT is a real vector. For this purpose, the real and the imaginary parts should have an even and odd symmetry around the sample 0, respectively [6]. These symmetries are illustrated in Figure 21.11 where the zero-frequency bin is shifted to the centre. The real and imaginary parts are both normalised, and the plot has been blown up so that the significant values of both parts can be seen clearly.

The direct result of the IFFT cannot be used as a filter kernel. It may cause aliasing in the time domain during a convolution. The dashed curve in Figure 21.12 presents the IFFT result of the CTF that is shown in Figure 21.11. The IFFT result needs to be shifted and windowed to obtain a valid filter kernel [6]. Obviously, the direct IFFT has significant side lobes. The shifted IFFT does not go towards zero on both sides, leading to aliasing in the time domain, side lobes in the frequency domain and a loss of dynamic range of the filter response.

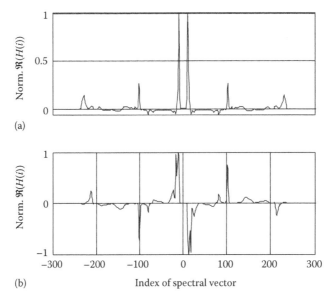

FIGURE 21.11
Frequency domain vector: (a) even symmetry for real part and (b) odd symmetry for imaginary part. The vector has 1000 samples, and the plot is enlarged, so that the frequency interval with significant energy, that is, between −240 and 240, is clearly visible.

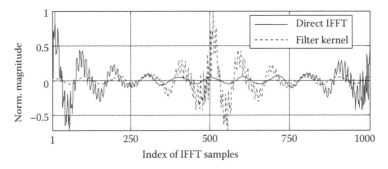

FIGURE 21.12
Direct output of IFFT and corresponding filter kernel.

To reduce the unwanted effects, the shifted IFFT vector can be multiplied by a window function. A window function has a bell-like waveform with its maximum in the middle, and both ends decline to zero. In this work, the Kaiser window function is used. The Kaiser window approximates the prolate-spheroidal wave functions in terms of the zero-order modified Bessel function. It tries to maximise the restricted energy in the frequency band of interest [7]. The Kaiser window is defined by

$$w(i) = \begin{cases} \dfrac{I_0\left[\pi\alpha\sqrt{1-\left(i/\left(N_{fir}/2\right)\right)^2}\right]}{I_0[\pi\alpha]}, & 0 \le |i| \le \dfrac{N_{fir}}{2}; \\ 0, & \text{otherwise,} \end{cases} \qquad (21.16)$$

$I_0(x)$ is the zero-order modified Bessel function of the first kind

$$I_0(x) = \sum_{k=0}^{\infty}\left[\frac{(x/2)^k}{k!}\right]^2, \qquad (21.17)$$

where α is the parameter which defines the width of the window. In the frequency domain, it determines the attenuation of the side lobes. Figure 21.13 shows the Kaiser window with $\alpha = 32$ and the windowed filter kernel. The side lobes are reduced largely, while the most significant part between index 250 and 750 is well preserved.

The original filter kernel itself causes a delay of $N_{fir}/2$. The delay can be reduced by padding the right side with zeros and shifting the filter kernel to the left. The delay is constant for all frequencies. It is actually the latency of the FIR filter. Suppose this latency is denoted by t_L. It must be considered in the filter design by adding it as a constant to the desired group delay. The delayed version of the desired complex CTF is obtained by

$$H_{shift}(i) = H(i)\cdot\exp\left(j2\pi\cdot i\cdot\left(1-\frac{N_{shift}}{N_{fir}}\right)\right), \qquad (21.18)$$

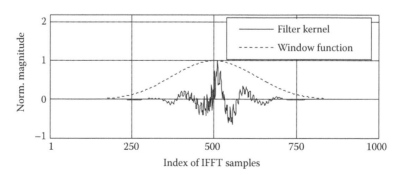

FIGURE 21.13
Kaiser window function with $\alpha = 32$ and windowed filter kernel.

where N_{shift} denotes the number of samples by which the sampled CTF vector should be shifted. Let f_s be the sampling rate, N_{shift} can be obtained by

$$N_{shift} = \left\lfloor \frac{t_L}{f_s} \right\rfloor. \tag{21.19}$$

The filter kernel should also be shifted to the left by N_{shift} samples in the time domain. And the resulting filter kernel is

$$h_{shift}(i) = \begin{cases} h(i + N_{shift}), & 1 \le i \le N_{fir} - N_{shift}, \\ 0, & N_{fir} - N_{shift} < i \le N_{fir}. \end{cases} \tag{21.20}$$

The operation (Equation 21.20) discards the part $h(1)$ through $h(N_{shift})$ and leads to the loss of signal energy. If the shifting is driven too much to the left or to the right, a large part of the filter coefficients with significant energy may get lost. The deviation from the desired CTF becomes unacceptable. The phase error reaches 35° at 199 kHz. The maximum relative error of the group delay exceeds 18%. Another problem caused by windowing is the badly controlled error on either side of a discontinuity of the desired magnitude response. The frequency response between two adjacent samples is not constrained; therefore, the resulting frequency can deviate from the desired response between the samples. The obtained filter kernel is supposed to be suboptimal.

21.6.4 Extension of the Frequency Sampling Method

Sharp corners of the magnitude response can be smoothed by the aforementioned processing steps. Except that, most parts of the magnitude response can be preserved. It could be less critical for real PLC channels, because those sharp corners as shown in the first plot of Figure 21.14 are quite rare to see in the frequency range of interest. Therefore, this work focuses on reducing the error of the phase and the group delay responses. An iterative algorithm is presented to reduce the phase error below a defined boundary $e_b(f)$:

$$\left| \Delta \varphi(f) \right| \le e_b(f). \tag{21.21}$$

The boundary is defined by

$$e_b(f) = \frac{e_1}{|H(f)|^{e_2}} + e_3, \tag{21.22}$$

where
e_1, e_2 and e_3 are tunable parameters
$|H(f)|$ is the desired magnitude response in linear scale

Obviously, the boundary is frequency dependent if e_1 is a nonzero value. The phase errors at frequencies with high magnitude attenuation are less critical than those at other frequencies. Therefore, they are constrained with fewer efforts. In principle, the lower the

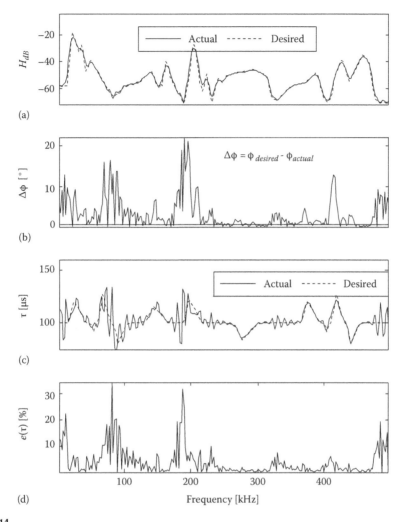

(a)

(b)

(c)

(d)

FIGURE 21.14
Deviation from desired CTF due to shifting, truncating and zero padding, as described by Equation 21.20. (a) Magnitude response, (b) error of phase response and threshold, (c) group delay and (d) relative error of group delay.

magnitude attenuation, the lower the boundary of the phase error. The parameter e_3 is constant over all sampled frequencies and determines the offset component of the phase error boundary. e_2 adjusts the boundary difference between frequencies with high and low attenuation levels. A larger e_2 leads to a larger boundary difference. e_1 determines the ratio of the frequency-selective boundary component to the offset component. The steps listed in Table 21.2 are to be considered for the error reduction.

Figure 21.15 shows an example of the result. The boundary is frequency dependent. The peak values of the boundary have good correlation to high magnitude attenuation. Compared with the plots shown in Figure 21.14, the phase and group delay errors are largely reduced at the cost of errors on sharp turns of the magnitude response. Large errors of phase and group delay occur at frequencies around 16, 84, 188, 230, 328, 404 and 490 kHz, at which the magnitude suffers from high attenuation and meanwhile features sharp corners. The group delays corresponding to moderate and low magnitude attenuation are

TABLE 21.2

Steps to Reduce CTF Error

Step Index	Operation
1	Perform shift and truncate on the filter kernel using Equation 21.20 and obtain h_{shift1}.
2	Perform FFT on h_{shift1}, obtain H_{shift1} and calculate phase error $\Delta \varphi_1$.
3	Multiply H_{shift1} by $\exp(j \cdot \Delta \varphi_1)$, perform IFFT and obtain h_{shift2}.
4	Repeat steps 1 and 2 once and calculate H_{shift2}, $\Delta \varphi_2$ and $|H/H_{shift2}|$.
5	Multiply H_{shift2} by $|H/H_{shift2}|$, perform IFFT and obtain h_{shift3}.
6	Repeat steps 1 and 2 and calculate H_{shift3} and $\Delta\varphi_3$.
7	Check inequality of Equation 21.21, and find frequency bins f_x at which the inequality is not fulfilled.
8	Multiply $H_{shift3}(f_x)$ by $\exp[j \cdot \Delta\varphi_3(f_x)]$, perform IFFT and obtain h_{shift4}.
9	Repeat steps 6, 7 and 8 until the phase errors at all frequencies fulfil Equation 21.21.

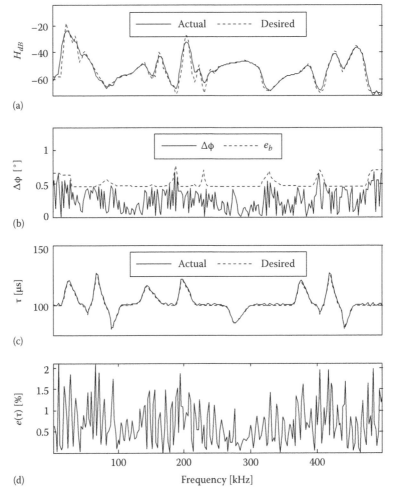

FIGURE 21.15
Improved results by applying iterative error reduction, $e_1 = 10^{-9}$, $e_2 = 1.9$, $e_3 = 7.9 \times 10^{-3}$, $E\{e(\tau)\} = 1.71\%$. (a) Magnitude response, (b) error of phase response and threshold, (c) group delay and (d) relative error of group delay.

even lower than 1%. The expected value of the relative error is as low as 1.71%. This coefficient set has a relatively large e_2 in comparison with e_3. The boundary differences at different frequencies dominate the boundary curve.

The iterative extension to the frequency sampling method provides a flexible and straightforward way to design FIR filters with custom frequency responses. In addition, the redundancy of the implemented filter can also be reduced to fulfil requirements in real-time applications. It is suitable for both short and long FIR filters with respect to the filter length. Nevertheless, unlike in conventional optimisation algorithms, the obtained filter coefficients using this method are still suboptimal. The user has to tune parameters e_1 through e_3 to obtain satisfied coefficients. The quality also depends on the specification of the error boundary. The iteration could have convergence problem if the boundary is not defined appropriately.

21.6.5 Verification

The emulated CTF is verified with a *vector network analyser* (VNA). Figure 21.16 compares the desired CTF (desired), the designed CTF (SIM), using the extended frequency sampling approach and the measured CTF (MEA) of the hardware implementation. For the magnitude response, the overall quality of the emulation is high. There are deviations of the measured magnitude from the design in the frequency range 400–500 kHz. This is caused by the antialiasing filter which has a 6 dB drop around 800 kHz. The measured group delay exhibits good agreement.

21.6.6 Realisation of Short-Term Time Variance

It has been mentioned that the channel response features both the long- and the short-term variances. This part focuses on the emulation of the latter situation. Reference [3] models the time variance with eight fundamental states per mains cycle. To improve the

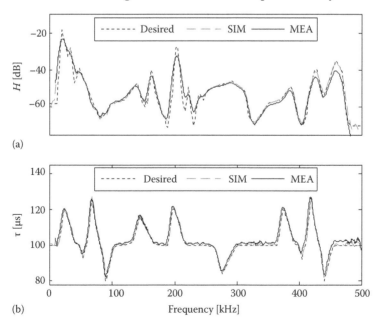

FIGURE 21.16
Comparison of desired, simulated (SIM) and measured (MEA) CTF. (a) Magnitude response and (b) group delay.

time-frame resolution, a linear interpolation is performed on the fundamental states. As a result, more frequency responses are created – one for each frame – and the resolution reaches one response per time frame. With the help of the linear interpolation, the change of frequency response reaches a smooth transition. The FFT-based fast convolution provides a convenient way for the linear interpolation in the frequency domain. A new frequency response can be calculated by simply adding an increment to the previous one at each frequency. Consider the linear interpolation

$$H_n(k) = H_{n-1}(k) + \Delta H_m(k), \qquad (21.23)$$

where n, m and k are indexes for the interpolated state, the fundamental states and the frequency bin, respectively. The increment $\Delta H_m(k)$ is obtained by

$$\Delta H_m(k) = \frac{H_{m+1}(k) - H_m(k)}{T_m}, \qquad (21.24)$$

where T_m is the duration of the mth fundamental state. Obviously, $\Delta H_m(k)$ remains constant within a fundamental state.

For the FIR filter-based direct convolution which is implemented in this work, it is desirable to realise the spectral interpolation with the help of operations in the time domain. In the time domain, the impulse response $h_n(i)$ of $H_n(k)$ is obtained by

$$h_n(i) = \frac{1}{N} \cdot \sum_{k=0}^{N-1} H_n(k) \cdot e^{j \cdot \frac{2\pi \cdot k \cdot i}{N}}. \qquad (21.25)$$

Applying Equation 21.23 in Equation 21.25, a new impulse response can be expressed by the impulse response of the previous state and an increment

$$h_n(i) = \frac{1}{N} \cdot \sum_{k=0}^{N-1} \left[H_{n-1}(k) + \Delta H_m(k) \right] \cdot e^{j \cdot \frac{2\pi \cdot k \cdot i}{N}} = h_{n-1}(i) + \Delta h_m(i), \qquad (21.26)$$

where $\Delta h_m(i)$ is the inverse DFT of $\Delta H_m(k)$. Since $\Delta H_m(k)$ does not change within the mth state, $\Delta h_m(i)$ also remains unchanged. Figure 21.17 shows the frequency responses of six fundamental states, denoted by S1 through S6. These frequency responses are not from real-life measurements; instead, they are created artificially so that the state transitions can be observed clearly. S1, S2, S3 and S4 have band-pass characteristics. The attenuation level in S5 decreases linearly with frequency. S6 has a random frequency response. These fundamental states are interpolated linearly with 49 additional states between two consecutive fundamental states. Figure 21.18 shows the spectra of the fundamental and the interpolated states. The transmissions are smoothed and the boundaries are blurred.

The direct convolution delivers the output for each new input at the sampling rate. Theoretically, the resolution can be as high as the sample period if the new filter coefficients are also available within each sample period. Therefore, the resolution is only determined by the speed at which a complete coefficient set can be delivered to the FIR filter.

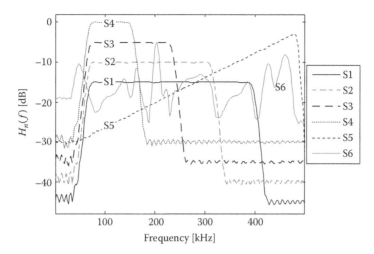

FIGURE 21.17
Frequency responses of the fundamental states S1 through S6.

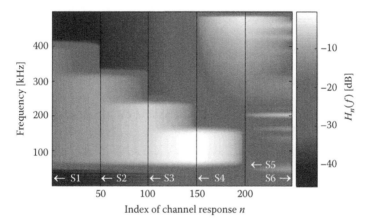

FIGURE 21.18
Interpolated states of frequency response.

21.7 Emulator-Based Test Platform

Figure 21.19 shows an emulator-based test environment. It consists of a PC, a channel emulator, two *coupling circuits* (CCs), two *line impedance stabilisation networks* (LISNs) and an *uninterruptible power supply* (UPS). DUTs are connected to the platform by plugging the transmitter and the receiver in the reserved Tx/Rx sockets, respectively, corresponding to path P_2 and P_7. At the same time, interfaces of the DUTs for digital values are connected to the test server via P_1 and P_8. The PC is a test server. It generates diverse channels, configures the DUTs, controls the emulator and manages the whole test process. The UPS provides mains voltage as a power supply and a synchronisation source for zero-crossing detection-based DUTs. Meanwhile, it isolates the test environment from real mains networks. Each of the LISN provides well-defined stable impedance to the test environment and prevents the transmitted signal from reaching the receiver via the mains voltage path P_9–P_{10}–P_{11}.

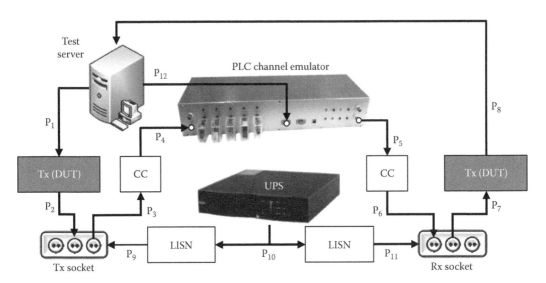

FIGURE 21.19
Channel emulator-based test platform.

High-frequency noise coming from the UPS can also be filtered by the LISNs. The CCs are necessary to separate the emulator from the high-voltage side and to exchange the 'pure' communication signals between the emulator and the DUTs. From the signal/data flow point of view, there is a closed loop: digital values, also called test patterns, are generated by the test server and transferred to the transmitter via P_1. The analogue transmit signal is generated, and it reaches the emulator via P_3 and P_4.

21.8 Performance Evaluation of an OFDM System

Due to the capability of reproducing real-world channel characteristics, NB-PLC systems can be evaluated in laboratorial conditions using the proposed test platform. In order to investigate how well the laboratorial performance matches the system performance in the real channels, a case study has been performed. This case study consists of three steps. First of all, relevant channel characteristics such as the attenuation and noise scenarios are measured. Data transmissions are made using two OFDM-based PLC systems in each measurement pause. The *bit error rate* (BER) in each data transmission is recorded for an assessment of the link quality. Subsequently, the channel properties are characterised and reproduced using a PLC channel emulator. The same PLC systems are connected to the emulator. The same data transmissions are made, and the BER values are measured as well. Finally, the BER values obtained at the platform are compared with those BER values measured in the real PLC channels. The detailed procedure and the comparison result will be given in this part.

21.8.1 Device under Test

As mentioned in Chapter 2, the FPGA used for the measurement platform also hosts a flexible and easily adaptable OFDM modem core. Different multi-carrier systems can be configured conveniently. A simple OFDM realisation is configured on the FPGA for the case study. The system is using 48 carriers, ranging from 79 to 95 kHz. Each carrier is modulated with *differential binary phase shift keying* (DBPSK) for a robust data transmission. Each platform is equipped with a zero-crossing detector for mains voltage. The OFDM symbols are synchronised to the falling edges of the mains zero crossings. The total symbol duration – including a guard interval – is set to 1/6 of the mains period, that is, 3.3 ms in a 50 Hz environment. The frame length is nine OFDM symbols. 200 frames of random values are generated for each BER test. This corresponds to 86,400 bits binary data and results in accuracies of 1% and 3.4% for measuring BER values of 0.1 and 0.001, respectively [8].

21.8.2 Channel Transfer Function

It has been shown in Chapter 2 that the attenuation in NB-PLC channels exhibits small variation and weak time-varying feature. The difference between the highest and the lowest attenuation values is smaller than 10 dB. The variation of the attenuation at the same frequency does not exceed 6 dB, and it becomes even smaller towards higher frequencies. Therefore, the measured attenuation profiles are averaged first, and then only the averaged attenuation is emulated for each data link. Furthermore, the measured attenuation profiles are interpolated so that the frequency resolution of the measurement can be matched to that of the emulation. Figure 21.20 shows both the measured and the reproduced attenuation profiles. The asterisks correspond to the measured attenuation values, and the solid lines are the emulated CTFs. The reproduced channels have almost the same frequency-selective attenuation as their real-world counterparts. There is an error of 1.5 dB at around 30 kHz for the link from S1 to S3. However, its influence is neglectable since it is relatively small compared to the attenuation value (more than 50 dB) at this frequency, and it lies outside of the transmission band. The coupling loss is relatively small since the transmitter of our platform has very low output impedance. Therefore, its influence is neglected, and the access impedance is not emulated for this work.

21.8.3 Emulating Noise Scenario

For the noise emulation, we take the scenario measured at S2 as an example. Since the data transmission is made between 60 and 100 kHz, the noise is analysed and emulated for this frequency range. Due to the time-varying and frequency-selective nature, the analysis is performed in time and frequency domains using STFT. Figure 21.21 shows the waveform and the PSD in (a) and (b), respectively. Narrowband interferers with periodical fading can be observed at around 65.5, 67.5, 71, 76.5 and 88 kHz in (b). All the fadings seem to be synchronous with mains voltage and appear every 10 ms. A total of 12 narrow spectral peaks are found within 40 ms. They are the spectral components of broadband impulsive noises. By applying the analysis approach introduced in Chapter 2, these impulses can be divided into three groups. The group index and the arrival times of the individual impulses are shown in Table 21.3. It is obvious that the impulse in each group has a repetition rate of 100 Hz. Last but not least, the coloured background noise has weak time-varying feature in this frequency range. It can be modelled by a time-invariant exponential function of the frequency. By applying the emulation approaches for each noise type, the overall noise scenario is reproduced for S2. As shown in Figure 21.22, the emulated noise scenario retains the essential characteristics.

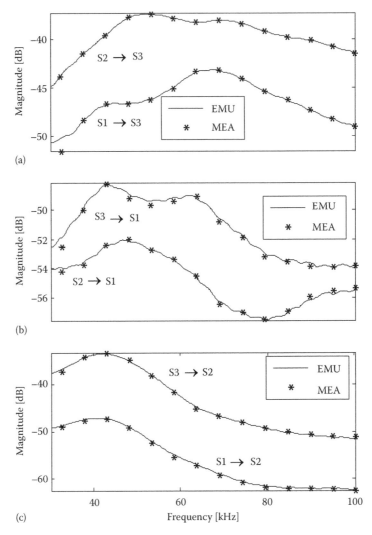

FIGURE 21.20
Comparison of measured and emulated transfer functions. The asterisks correspond to the measured attenuation values, and the solid lines are the emulated CTFs. Transfer functions for links (a) S2 → S3 and S1 → S3, (b) S3 → S1 and S2 → S1 and (c) S3 → S2 and S1 → S2.

21.8.4 Test Results

Figure 21.23 illustrates the BER results measured during the channel measurement ('MEA' bars) and in the emulator-based test platform ('EMU' bars). The link from S3 to S2 has the best communication quality, while the two links between S1 and S2 as well as the link from S2 to S1 have very high BER values. The BER results obtained by using the test platform are very close to those of the field tests. Therefore, the laboratorial test platform can also be used to predict the system performance in the real-life channel. We can even produce worst-case channel conditions obtained from measurements and apply them to test different PLC systems. Systems that pass through these worst cases could have reliable and robust performance later in the real-life application. In this way, the complexity and period of selecting the best PLC solution can be largely reduced.

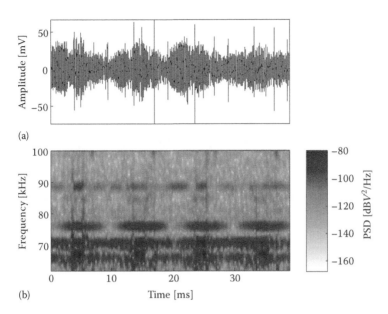

(a)

(b)

FIGURE 21.21
Measured noise scenario for S2. (a) Noise waveform in the time domain and (b) STFT of the noise waveform.

TABLE 21.3

Arrival Time and Group Index

Group Index	Arrival Time (ms)
1	3.3, 13.3, 23.3, 33.3
2	5.5, 15.5, 25.5, 35.5
3	6.8, 16.8, 26.8, 36.8

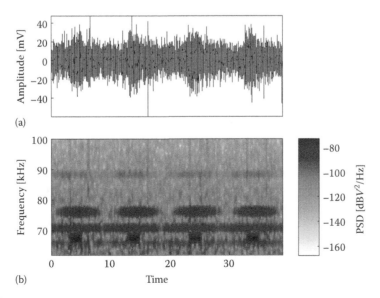

(a)

(b)

FIGURE 21.22
Emulated noise scenario for S2. (a) Amplitude over time and (b) PSD over frequency and time.

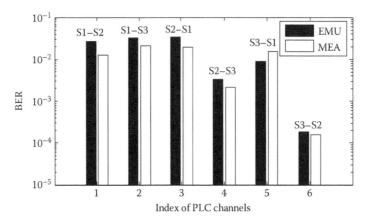

FIGURE 21.23
Comparison of BER results.

21.9 Conclusion

This chapter has introduced a channel emulator for NB-PLC channels in LV mains networks. The channel emulation is based on a three-stage equivalent electrical model. A test bed has been built by using this emulator for the evaluation of data transmission on the *physical* (PHY) layer. Digital FIR filters are utilised to emulate the CTF. An extended frequency sampling algorithm provides a flexible and straightforward way to design such FIR filters for emulating any customised complex-valued CTF. In addition, recommendations are given to reduce the redundancy of the implemented filters, so that requirements for real-time applications can be met.

Narrowband interference features significant noise levels in the frequency domain in comparison with background noise. Envelopes of a number of narrowband interferers exhibit strong dynamic and time-varying features. A band-pass filtering-based approach is proposed for the estimation of the time-varying envelopes. This approach can detect broadband impulsive noise and eliminate its influences on the estimation. The typical sinusoidal waveform of narrowband interference can be generated by using the phase accumulation method. The estimated envelopes can be modelled by unsymmetrical triangular functions. In addition to the typical narrowband interferer, a class of interferers with time-varying frequencies, called swept-frequency noise in this thesis, has been observed both in indoor and access domain. Such interference is usually caused by active *power factor correction* (PFC) circuits in power supply units of many end-user appliances, such as fluorescent lamps and PCs. Emulation is possible by using chirp functions.

For impulsive noise, detection can be made by comparing the power of segmented noise waveforms with a constant threshold. In cases where the exact course of a noise waveform is unimportant, impulsive noise can be generated by switching the background noise on and off according to a control pulse sequence. Otherwise, it can be emulated as an exponentially attenuated oscillation. For coloured background noise, an accurate estimation can only be made after all the other noise classes have been removed. Its power level depends on the number and type of connected and active electrical devices. Coloured background noise also exhibits cyclo-stationary characteristics synchronous with the mains frequency, and its smoothed spectrum can be approximated by the sum of two exponential functions.

A case study made in a small low-voltage grid on a university campus has shown that the test results obtained by using the test platform are very close to those of the field tests. Therefore, the laboratorial test platform can also be used to predict the system performance in the real-life channel. The channel emulator indeed opens a new path to the flexible, reliable and technology-independent performance evaluation of PLC modems.

References

1. J. Bausch, T. Kistner, M. Babic and K. Dostert, Characteristics of indoor power line channels in the frequency range 50–500 kHz, in *IEEE International Symposium on Power Line Communications and Its Applications*, Orlando, FL, 2006, pp. 86–91.
2. L.R. Rabiner and B. Gold, *Theory and Application of Digital Signal Processing*, Prentice-Hall, Englewood Cliffs, NJ, 1975.
3. F.J. Cañete, L. Díez, J.A. Cortés, J.J. Sánchez-Martínez and L.M. Torres, Time-varying channel emulator for indoor power line communications, in *IEEE Global Communications Conference*, New Orleans, LA, 2008, pp. 2896–2900.
4. XILINX, LogiCORE IP fast Fourier transform v7.1, product specification, DS260, March 2011.
5. T. Kistner, M. Bauer, A. Hetzer and K. Dostert, Analysis of zero crossing synchronization for OFDM-based AMR systems, in *Proceedings of the IEEE International Symposium on Power Line Communications and Its Applications*, Jeju Island, Korea, April 2008, pp. 204–208.
6. S.W. Smith, *The Scientist and Engineer's Guide to Digital Signal Processing*, 2nd edn., California Technical Publishing, San Diego, CA, 1999.
7. F.J. Harris, On the use of windows for harmonic analysis with the discrete Fourier transform, *Proc. IEEE*, 66(1), 51–83, January 1978.
8. M.G. Bulmer, *Principles of Statistics*, Dover Publications, Inc., New York, 1979.

22

Cognitive Frequency Exclusion in EN 50561-1:2012

Andreas Schwager

CONTENTS

This chapter describes the 'cognitive frequency exclusion' concept in detail, specified in *European Norm* (EN) 50561-1. After an overview of the standardisation history, the *high-frequency* (HF) radio transmissions, the influence on the *power line communication* (PLC) spectrum, are shown. Thereafter, an excursion into the sensitivity of *shortwave* (SW) radio receivers follows. When the sensitivity of these devices is known, the threshold can be defined for receivable radio broadcast signals. If the frequencies to be excluded are known, requirements for a notch are defined. The process of adaptive modulation is described before calculating the impact of notching on the throughput of PLC. This allows the realisation of the notching concept with a minimal loss of throughput and implementation effort. The concept is verified by a hardware demonstrator system after discussing the theoretical aspects.

22.1 Introduction

The EN 50561-1 [1] was approved in autumn 2012 by European national members. As described in Chapter 6, it includes new requirements for conducted disturbances and communications signals at PLC ports, which were never included in a harmonised *electromagnetic compatibility* (EMC) standard before.

In the past, EMC standards specified fixed and permanent limits which usually were described using a straight line for the given frequency range in a linear or logarithm graph (Figure 6.1). Manufacturers implementing such a norm guaranteed compliance by shielding their device at the time of production. The straight line must not be crossed at any time.

The new requirements in EN 50561-1 are adaptive, flexible and cognitive and have to function only when there is a risk of interference. This type of intelligent concept is also called 'dynamic spectrum management', 'cognitive PLC' or 'smart notching' and runs autonomous on the PLC modems without any user interaction. Compliance is guaranteed at the time and location, when and where, and the PLC modem is operated by the behaviour of the communication system. The principal philosophy is 'a service which is not present needs not to be protected'. Therefore, EN 50561-1 requests flexible and/or permanent frequency exclusions in the *power spectral density* (PSD) mask of PLC modems. Compared to state-of-the-art PLC modem design, enhanced protection of radio services is provided in buildings where PLC is operated.

The general requirements of EN 50561-1 request the exclusion of sensitive frequency areas which are allocated by ITU-R Radio Regulations [2] to aeronautical, amateur radio and broadcasting services. The frequency exclusions for aeronautical and amateur spectra are permanent, where the radio broadcast frequencies could be notched permanently or dynamically. Permanent notching results in a significant loss of PLC communication resources versus dynamic notches. They only have to be applied at the few broadcast frequencies when the service is receivable at the location of operation.

22.2 Standardisation of 'Cognitive Frequency Exclusions'

Reports about interference from PLC to HF radio broadcasts were published quite early by Stott and Salter [3]. The new high-bitrate PLC modems using multiple communication carriers at various frequencies with adaptive constellations enable the realisation of the cognitive frequency exclusion concept. The idea was discussed lively at ETSI PLT [4]. In order to prove that the new technology works reliably, prototype implementations demonstrated their functions under the umbrella of the ETSI special task force (STF 332) performing a plugtest with PLC modems and SW radio receivers. During the ETSI plugtest in October and November 2007, various tests and measurements were performed to stress the concept. A description of the demonstrator system and the plugtest highlights are given in the succeeding text.

After the plugtest and publishing the results, the *European Broadcasting Union* (EBU) issued a public statement that they are content that any system fully meeting the specification will offer adequate protection to HF broadcast transmissions. This led to the unanimous approval of ETSI TS 102 578 [5] specifying the 'smart notching' concept.

The *Special International Committee on Radio Interference* (CISPR) announced that they will work on adaptive dynamic notching in any future *committee draft* (CD) or *committee draft for voting* (CDV) related to PLC equipment in CISPR/I/257/CD [6]. Finally, they adopted the adaptive *electromagnetic interference* (EMI) mitigation techniques as a normative annex to their latest CIS/I/301/CD [7] and CIS/I/302/DC [8] documents. Accepting adaptivities is almost 'revolutionary' in the world of EMI standardisation.

The IEEE included the concept as 'stand-alone dynamic notching' to the IEEE 1901 standard [9].

22.3 Overview of Shortwave Radio Broadcasting

The frequency range of conventional PLC modems (2–30 MHz) overlaps with HF radio broadcast frequencies defined by ITU-R [2]. Power line wires in private homes are not shielded, and due to branches, distribution boxes, etc., the power line network is structured with a certain amount of asymmetry. As discussed in Chapters 1 and 5, the asymmetries of the power line network convert the differentially fed signals into common-mode signals (see Figure 1.3), which tend to interfere with radio devices. If an SW radio receiver (*amplitude modulation* [AM] or *digital radio mondiale* [DRM] [10]) is operated indoors where a PLC is active, the radio reception quality might suffer. When the radio device is connected to the mains power supply and the radio has an insufficient decoupling at its mains port, the conducted path is dominant in terms of interference. By design, SW radio receivers have insufficient decoupling at the mains port, because they use the mains grid – in the same way that PLC does – for something which it was not designed for: they use it as an antenna. Usually an SW radio receiver is equipped with a whip or monopole antenna. When the receiver is connected to the mains, the counterpoise of the mains grid allows it to generate a dipole antenna improving the reception quality of radio services.

Of course, some might argue that HF radio broadcast is becoming less important. *Frequency modulation* (FM) radio provides a significantly better signal quality, but it is not a worldwide service. Satellite radio or the upcoming web-radio services compete with today's AM transmissions. This might be true for the industrial countries or in the developed countries but the demand for information exists globally. HF frequencies have the unique property that a transmission can pass halfway around the globe. However, if the transmission goes halfway around the globe, so does the interference.

HF radio broadcast is also relatively cheap: the costs of constructing an HF radio transmission station, as well as the annual operational costs, are unrivalled by modern technologies. If a single HF transmission station is located in Thailand, the broadcast could reach 60% of the world's population. Anybody who would like to receive this broadcast needs a receiver costing around 10€. HF radio broadcast is especially important in developing countries where there is little or no infrastructure in rural areas and HF radio is often the only option. Today, HF radio broadcast is used for the following reasons:

- In newly industrialising and developing countries where the transmission distances are very large and the installation of FM transmission infrastructure is too expensive.
- Tourists who like to receive their home services in a habitual manner. However, satellite or web services are available in most hotels of industrialised countries.
- Amateur radio listeners or hams. The PLC interference situation is different for them. Usually they do not use a kitchen radio equipped with a whip antenna. Furthermore, they have the knowledge of how to protect their equipment from interference by using additional filters, for example. However, due to their desire to receive extremely weak signals and the sensitivity of their equipment, interferences from PLC are relevant.
- Military services permanently use the HF frequency range but they also use professional antennas. Their operational area is often far abroad where no alternative to HF transmissions exists. Of course, militaries train the operation in their home countries. This should be done as closely as possible to real conditions. PLC is an interference source to an airplane flying over Europe. Tanks or marines usually do not operate in the vicinity of PLC modems.
- New upcoming digital services such as DRM [10]. Sales of DRM receivers are not as high as initially expected. However, it will be interesting to see how these services develop. If the frequency resources are polluted, once, a future installation is no longer possible.
- It is a fundamental freedom in democracies to receive information from everyone and everywhere. Broadcasters such as Deutsche Welle, BBC World Service, Voice of America and others transmit information in multiple languages using HF bands. The most frequently listened SW radio service today in Europe is Radio China. Unlike the Internet, it is difficult to censor HF radio broadcast; only jamming is possible. But more importantly, radio listeners cannot be monitored by government authorities.
- In the event of a crisis, disaster or earthquake, satellite dishes may no longer be aligned. In this case, HF radio broadcast is the most robust and proven technology and expected to be the first source of information to be reconstructed.

Countries without a well-developed wired telecommunication infrastructure expect to support people with Internet services over PLC. Such countries are also the main target areas of HF radio transmissions. Therefore, the coexistence between PLC and HF radio reception is very important.

EN 50561-1 specifies the solution to solve interference problems from PLC to SW radio receivers.

A permanent notching of all HF radio bands would result in the loss of 21% of the communication spectrum of a conventional PLC modem using the frequency range from 2 to 30 MHz.

Whether or not the HF frequencies can be used by radio transmissions depends on weather conditions and the reflection quality of the ionosphere. The structure of the ionisation layers in the ionosphere vary according to the time of day and seasonal changes. An 11-year sunspot cycle also affects radio reception (Annex A of ETSI TR 102 616 [11]). Radio broadcasters permanently operate monitoring stations in their target area to measure reception quality and schedule their services accordingly.

A service description channel is included in the specification for the new digital radio service DRM [12] for HF bands, which informs the receiver on which frequency the transmission will continue before a change in the transmission schedule is performed.

Usually, an HF transmission band is either fully allocated with radio services or relatively empty. This is why a permanent default notching of all HF bands is unnecessary and would result in too much throughput loss for PLC modems.

The cognitive frequency exclusion specified in EN 50561-1 provides optimum reduction of interference between PLC and HF radio broadcast and minimum impact on data throughput and *quality of service* (QoS) requirements of PLC.

22.4 Concept of 'Cognitive Frequency Exclusion'

'Cognitive frequency exclusion' is an adaptive process which automatically excludes all frequencies – from PLC – being used by receivable radio services, without any user or network operator interaction.

The presence of broadcasting signals can be detected by PLC modems by sensing the 'noise' (including radio broadcasts picked up on the mains cabling) in an electrical socket. Frequencies where HF radio broadcasting signals are identified can then be omitted from the transmitted signal by inserting notches into the transmitting PLC spectrum.

Radio broadcast signals transmitted with a high power from the antenna of a radio station will be received by any wire acting as an antenna, for example, an electrical power grid (see Figure 22.1). The ingress of a broadcast signal can be detected within the reception range of the radio broadcast signals. The power line wires are passive and therefore reciprocity is valid. The transfer function, or the antenna gain, is identical for radiation as well as for signal reception. At frequencies with a high potential for interference, signal ingress is excellent and there is likely no ingress at non-radiating frequencies.

PLC modems are connected to the mains and are equipped with a very sensitive analogue front-end. To achieve adaptive modulations (see Figure 22.10), PLC modems monitor the noise signal on the mains. This noise information could be reinterpreted by the modems in order to identify a receivable radio station. If done, the frequencies of all receivable radio stations could be excluded from PLC transmission.

22.4.1 Radio Signal Spectrum on Power Lines

Of initial interest is a comparison of the signals measured when connected to the mains and the field measured with an antenna in the air. Comparing the signal strength of an HF radio broadcast station at two locations at once is very difficult. Due to the strong dynamic fading effects of radio transmissions in the time domain, a received HF signal never has the same level it had a moment ago. This time-variant effect was further investigated by monitoring the level of HF radio stations over a period of time.

FIGURE 22.1
Radio signals ingressing a house's mains wiring.

22.4.1.1 Fading Effects of Radio Stations

Figure 22.2 shows the levels of two radio stations (6918 and 7106 kHz) which were monitored for a long time. The horizontal axis represents the time from 0 to 1000 s. The vertical axis is a relative level in dB. Both transmissions show strong fading effects. The transmission at 7106 kHz (bottom, dotted curve) is particularly interesting. 700 s after the recording started, the signal level dropped by 40 dB. Obviously, there was a change in the

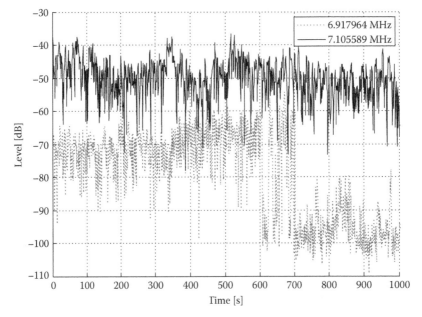

FIGURE 22.2
Fading of HF radio broadcast services in time domain.

transmission schedule of the broadcaster. Changes such as these can be monitored quite frequently in HF broadcasting. An overview of frequency scheduling in HF broadcasting is given in ETSI TR 102 616 [11], Annex A.

The service at 7106 kHz (top line) shows relatively less fading compared to the 6918 kHz station with typical characteristics. These are flat hills and deep canyons in the shape of the curve. Fading effects are caused by multipath propagation between the transmitter and the receiver. The waves from individual paths may overlap constructively or destructively, depending on the phase difference when they arrive at the receiver. The phase of a path may vary with the dynamics in the ionosphere. Depending on weather conditions, an ionisation layer can move with more than 100 km/h. This is why Doppler effects impair HF reception and an *automatic frequency control* (AFC) in SW radio receivers is beneficial, even if the transmitter and receiver do not move. The signal level of a station changes by more than 30 dB within a few seconds. The AGC of an SW radio receiver must be very dynamic in order to follow such level fluctuations. If HF radio stations are to be detected on the mains, PLC modems have to consider such dynamics of the signal levels.

22.4.1.2 Setup to Record a Signal Spectrum

Figure 22.3 shows the equipment required to compare the levels of the electromagnetic field in the air and the ingress signals on the mains. Such measurements were performed in private flats, hotels and office buildings. The antenna was located in the centre of the room with a vertical alignment of the biconal probe. Photographs of these measurements can be found in ETSI TR 102 616 [11]. The Schwarzbeck electrical field probe EFS 9218 [13] allows calibrated field measurements. It provides a constant antenna factor for the frequencies of interest. A spectrum analyser is connected, using a probe, to the mains or alternatively using the antenna. For the measurements in Figures 22.4 and 22.5, the maximum signal level is recorded using the max-hold function of the spectrum analyser for about 1 min.

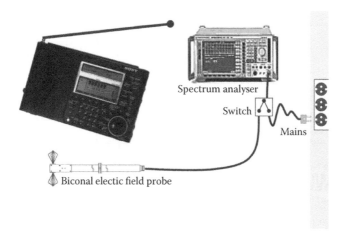

FIGURE 22.3
Detection of radio services. Setup for measurements in buildings.

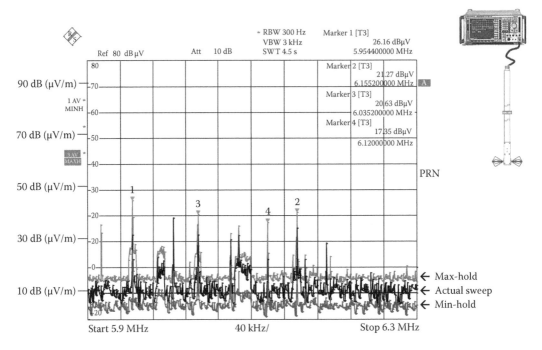

FIGURE 22.4
Indoor electrical field snapshot of the 49 m band, any location.

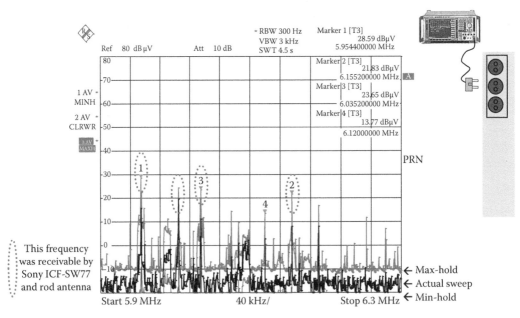

FIGURE 22.5
Signal ingress in the 49 m band, measured connected at a power outlet.

22.4.1.3 Recorded Spectrums

Figure 22.4 shows a snapshot of the electrical field in one building of the 49 m band. The HF range is allocated by ITU-R [2] into multiple bands. Each individual band has its pros and cons in transmission properties. Some of them have better daytime reception (e.g. 19, 16 m), others perform better at night (41, 31 m). Some bands are used in summer, others in tropical regions. The *x*-axis of Figure 22.4 represents the frequency range from 5.9 to 6.3 MHz. There are 40 kHz per division. The *y*-axis in Figure 22.4 is interpreted by the spectrum analyser in dB µV, but considering the antenna factor of 18 dB(/m) [13], the *y*-axis is converted to the E-field from −2 dB (µV/m) to 98 dB (µV/m) using a scale of 10 dB per division. Some AM services can easily be identified: at 5954, 6035, 6155, 6120 kHz and more. At 5950 kHz, for instance, a DRM transmission is visible.

All sweeps in Figures 22.4 and 22.5 were recorded with max hold (top line) and min hold (bottom line) to get an impression of the fading of the individual services. In Figure 22.4, the AM service at 5920 kHz was switched off during the recording period. It is neither visible at the min-hold line nor in Figure 22.5. The service at 5954 kHz shows an average fading (around 20 dB), and the service at 6005 kHz shows almost no fading behaviour. The black line represents the latest frequency scan before the snapshot was recorded.

Taking the fading into account and that some stations might have been switched on or off during the measurements, Figures 22.4 and 22.5 look virtually identical.

Figure 22.5 recorded the conducted signals on the mains. The vertical axis represents the voltage in dB µV.

22.4.1.3.1 Characteristics of a DRM Spectrum

DRM transmissions use *orthogonal frequency division multiplex* (OFDM) – a modulation scheme with 88–228 carriers in a 10 kHz spectrum. The number of carriers used by DRM depends on the transmission mode corresponding to typical propagation conditions. An HF-DRM spectrum appears as a 10 kHz wide rectangle spectrum. The DRM specification [12] also allows bandwidth allocations of 5 and 20 kHz. In the HF band, 10 kHz spectra are usually used. Some DRM transmissions may also have a very high single carrier in the centre of their spectrum. This is necessary to ensure that legacy transmitting amplifiers keep their linearity. DRM was especially designed in order to use portions of older AM Tx facilities, such as antennas and amplifiers, avoiding major new investment. Simulcast transmissions use an AM-modulated centre carrier with the surrounding DRM carriers (e.g. at 5954 kHz in Figure 22.4). DRM has a more variable *peak-to-average ratio* (PAR) compared to AM services. This used to cause problems at some old transmitting stations. It explains why the carrier in the centre of the spectrum (known from an AM 30% modulation depth) has been kept. In this case, some OFDM carriers in the middle of such a DRM channel are not used and the central carrier has to be filtered away by the receiver.

22.4.1.3.2 DRM and AM

If a PLC modem has to detect these signals, it does not differentiate between AM or DRM. Today's PLC modems have an OFDM carrier spacing f_{CS} of around 20 kHz. The carrier spacing is provided by the system's Nyquist frequency $f_{Nyquist}$ (half of ADC/DAC sampling clock frequency) divided by the *FFT_size* implemented by the PLC modem. The parameters of the system described in [14] are as follows:

$$f_{CS} = \frac{f_{Nyquist}}{FFT\_size} = \frac{40\,\text{MHz}}{2048} = 19.531\,\text{kHz}. \tag{22.1}$$

The carrier spacing f_{CS} of a transmitting OFDM system is identical to the resolution bandwidth of the receiving system using the same *FFT_size* and sample frequency. The resolution bandwidth of the PLC system is also relevant for noise measurements. Reference [14] presents methods to enhance the resolution bandwidth of noise measurements. If the bandwidth of the signal to be detected (AM or DRM) is smaller than the resolution bandwidth of the measurement system, its shape does not matter. Due to the fact that DRM is designed to reuse the transmission facilities (amplifiers, antennas) of AM equipment, it has the same signal power as an AM carrier.

The stations marked with a dotted ring in Figure 22.5 were receivable using the automatic frequency scan function of the Sony ICF-SW77 SW receiver. In total, at a single time instant during the day and at this location, 22 radio stations were receivable with the Sony ICF-SW77 with field strengths between 29 and 68 dB (µV/m). Test results in the same order of magnitude were measured during the ETSI plugtest and published in ETSI TR 102 616 [11].

J. Stott's BBC R&D White Paper 114 [15] posts similar measurement results.

All these measurements help to answer the question: Which air-based HF carriers can be detected on the mains network? All radio signals in the air which are stronger than 20 dB (µV/m) (measured with a *RBW* of 9 kHz) significantly increase the noise floor on a quiet power line.

Due to the strong variations in the amplitude of HF radio broadcast signals over time, PLC modems should periodically sense the ingress level of the radio signals. The level of these signals also depends on the modem's location and the structure of the wiring in the electricity grid.

22.4.1.3.3 *Noise on Power Lines*

As described and measured in Chapter 5, many other noise signals may appear on the mains spectrum. Further noise sources such as switching power supplies frequently enhance the ingress level of radio broadcast services. Later in this chapter, both an absolute and a relative threshold are defined, when such signal ingress is receivable by a radio device.

22.4.1.3.4 *Additional PLC Transmission*

For further studies of the interference potential of PLC, the measurement setup from Figure 22.3 was used to check the level of radiation due to PLC in a building. Figure 22.6 – where a PLC transmission was also set up in parallel – shows an identical spectrum to the one measured in Figure 22.4. The radiated noise level covers most of the radio services.

22.4.2 Sensitivity of SW Radio Receivers

Reference [14] evaluated the sensitivity of HF radio receivers. A couple of AM receivers were stressed in an anechoic isolation chamber to derive the *signal-to-noise ratio* (SNR) in the demodulated audio signal. Additional tests were performed by checking at which level the automatic station scan stops. Furthermore, theoretical studies were

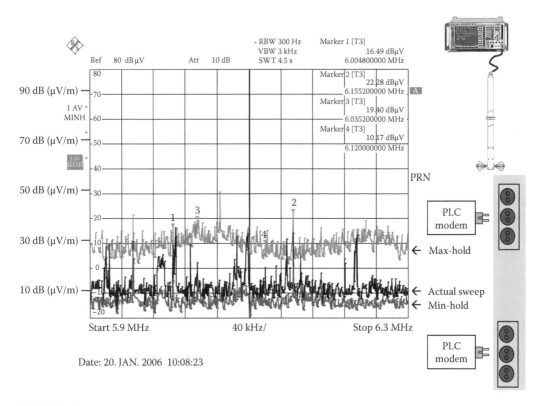

FIGURE 22.6
Radiation from PLC.

performed based on [16,17] and the DRM radio planning parameters. All evaluations showed the minimum sensitivity of HF radio receivers to be at 22 dB (µV/m).

22.4.3 Threshold to Detect Radio Services

Propagation characteristics of HF radio transmissions are not stable. As already seen in Figure 22.2, the fading in the time domain generates heavy variations in the signal level at the receiver side. ITU-R specifies the probability of a transmission being receivable by the target device in [18]. If PLC modems monitor the ingress of HF radio broadcast signals using a max-hold detector, the short-term signal variations might be considered. Figure 22.7 presents a sketch of these calculations.

The x-axis of Figure 22.7 represents the time in arbitrary units, the y-axis the electrical field in the air. The bottom line '<' is the intrinsic noise level that might be expected in a high-quality receiver. The 7 dB (µV/m) line '>' takes the added man-made noise into consideration, and the E_{min} = 22 dB (µV/m) line 'o' includes the theoretical DRM [16] or measured AM receiver's minimum sensitivity [14]. The fading line '+' shows the expected signal at the receiver's location with the statistics from [18]. Such signals are inevitably subject to a larger or smaller degree of fading. The maxima of the top fades are $D_u S_h$ = 5 dB higher than the minimum receiver's sensitivity [18]. The interleaver of DRM is designed to make receivers immune to such fading. If PLC modems need a margin $M_{to\_detect\_threshold}$ of 1 dB for detection, the fading statistics specify the threshold to be exceeded with a probability of 30% in any interval longer than 10 s. PLC modems have to detect the fading line '+' raising

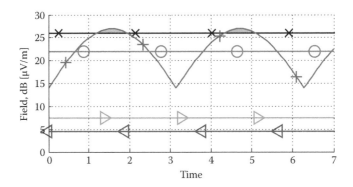

FIGURE 22.7
Threshold to detect HF radio broadcast ingress:
'x' 26 dB(μV/m), max hold with 1 dB headroom;
'+' 14–27 dB(μV/m), fading of HF broadcast;
'o' 22 dB(μV/m), minimum receiver's sensitivity;
'>' 7 dB(μV/m), intrinsic noise plus man-made noise;
'<' bottom line at 4 dB(μV/m), receiver's intrinsic noise level (identical to man-made noise).

the 26 dB (μV/m) line 'x'. In order to detect the top level of fading, the detection threshold of field strength $E_{field\_to\_detect}$ in the air is calculated as follows:

$$E_{field\,to\,detect} = E_{min} + D_u S_h - M_{to\_detect\_threshold},$$

$$= 22\ dB(\mu V/m) + 5\ dB - 1\ dB = 26\ dB(\mu V/m). \qquad (22.2)$$

The threshold of 26 dB (μV/m) is given by the top line 'x' in Figure 22.7. If PLC modems use an average detector instead of the max-hold one, the threshold has to be lowered by 5 dB. This is implementation dependent for the PLC modem manufacturer.

The reception factor (defined in the plugtest report ETSI TR 102 616 [11]) describes the relationship between the electrical field strength of a radio broadcast station in the air and the received power to be measured at outlets. The setup to measure the reception factor is visualised in Figure 22.8. An SW radio receiver is used to scan the spectrum. If a station is receivable, its E-field and its signal ingress level at the mains are verified.

Measurements deliver a cumulative statistical probability of the reception factor shown in Figure 22.9.

The lower the value in Figure 22.9, the better the antenna gain from the mains wiring. The median value of the *reception factor (ReFa)* is found to be 114 dB (μV/m) – dBm. The 80% worst case value is $ReFa_{80\%} = 121$ dB (μV/m) – dBm. A reception factor covering 80% of the cases, with an 80% confidence level, can be derived from the distribution function shown in Figure 22.9. With this value, the threshold of the signal level connected to the mains $P_{detect\_on\_mains}$ can be derived with

$$P_{detect\,on\,mains} = E_{field\,to\,detect} - ReFa_{80\%},$$

$$= 26\ dB(\mu V/m) - 121\ dB(\mu V/m) - dBm = -95\ dBm. \qquad (22.3)$$

This level can be verified with a spectrum analyser using a resolution bandwidth of 9 kHz and an average detector.

Besides the threshold level, a criterion has to be developed for the separation of SW radio stations from disturbance sources operated at mains. As shown in Figure 22.5, the ingress

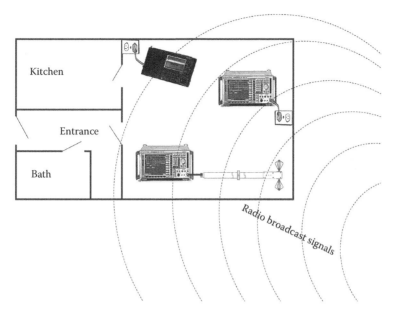

FIGURE 22.8
Definition of reception factor, measurement setup in a flat.

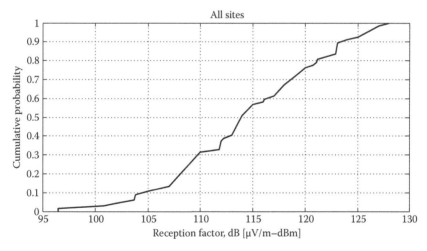

FIGURE 22.9
Cumulative probability of reception factor.

of a broadcast radio station appears as a needle in the noise measurement of a PLC modem. Today's PLC modems are not able to demodulate the AM or DRM signals. However, a needle and even a very stable one in the frequency domain is something unique within all noise sources at power lines (see Chapter 5) and can be detected by modems. Two criteria must be fulfilled for the level of the needle to be identified as a receivable service and worth being protected by PLC modems. The first criterion is a relative threshold because the usable signal must have a minimum SNR. The second is an absolute threshold when the signal passes the minimum sensitivity level of radio receivers.

The PLC modem is not able to apply weighting windows on the demodulated AM signal and to measure SNR as specified in [19]. A simpler approach is required. A PLC modem might measure the noise as well as the peak level of signals. As described in ETSI TR 102 616 [11], the minimum SNR of an HF service is 14.6 dB for the most robust DRM transmissions. This is the distance between the noise and the peak signal level.

Taking the typical noise sources into consideration, there is not only one noise level. The noise is time and frequency dependent. To check the SNR of a broadcast service, the noise next to the service shall be used. As long as the transmission conditions are good, an HF transmission band is densely allocated with radio stations. Measuring the noise inside the band would result in a value accumulating all radio services, not in the surrounding noise. This is why EN 50561-1 specifies the noise floor to be measured at adjacent frequencies lower and higher than the HF radio bands. The adjacent frequency blocks must be completely monitored by the PLC modems without any gaps in order to avoid cherry picking of noise values by a PLC modem. Finally, the noise floor is the median value of all measured values. The median value is not affected by individual peaks of, for example, a strong 'out of band' radio station.

In EN 50561-1, two criteria or thresholds are given, when a broadcast radio station is defined to be receivable:

- Criterion (1): 14 dB above the noise floor. It is 3 dB lower than the desired SNR of an AM receiver to understand voice and around 11 dB lower than that required by a normal DRM transmission (out of [16]).
- Criterion (2): the absolute threshold of −95 dBm which is derived from Equation 22.3.

For activation, a processing time of 15 s is conceded to PLC modems. If the two criteria were met once, the notch will be kept for at least 3 min. This timing hysteresis is a trade-off between consumer acceptance listening to a fading broadcast service and the processing capabilities of PLC modems creating a notch. The timings were found by the STF332 during plugtests where the concept was verified in ETSI TR 102 616 [11].

22.4.4 Requirements of a Notch

When signal ingress is identified as a receivable broadcast service, its frequency should be excluded from the PLC communication. The process of frequency exclusion in OFDM communication systems is called notching. EN 50561-1 characterises a notch by the bottom level and its width or slopes. In the case of a MIMO PLC modem, the notches have to be applied to both transmission channels.

22.4.4.1 Bottom Level of the Notch

The bottom level of the notch was derived from two different approaches: first, checking the levels of EMC standards [20] and secondly, feeding signals with this level into power outlets and checking whether it affects radio reception.

22.4.4.1.1 Checking Approach

CISPR 22 [20] specifies the mains class B (5–30 MHz) level to be $U_{AMN} = 50$ dB μV (RBW: 9 kHz, average detector).

An *artificial mains network* (AMN) (specified in CISPR 16 [21]) is used to verify mains port limits. It measures half of the differentially fed voltage at a measurement output. It follows that at the outlet U_{outlet} twice the differential voltage is allowed where the PLC modem is connected:

$$U_{outlet} = U_{AMN} \cdot 2 = 50 \text{ dB } \mu V + 6 \text{ dB} = 56 \text{ dB } \mu V. \tag{22.4}$$

When dB μV is converted to dBm using a characteristic impedance $Z = 100 \ \Omega$, the power allowed at a given outlet is

$$P_{outlet} = 56 \text{ dB } \mu V - 110 \text{ dB (mW}/\mu V) = -54 \text{ dBm}. \tag{22.5}$$

The PSD_{outlet} of PLC modem at the bottom level of the notch using Equation 22.1 is as follows:

$$PSD_{outlet} = -54 \text{dBm} - 10 \cdot \log_{10}(9 \text{kHz}),$$

$$= -54 \text{dBm} - 39.5 \text{dB(Hz)} = -93.5 \text{dBm/Hz}. \tag{22.6}$$

22.4.4.1.2 Subjective Evaluation

The noise is fed to the mains in the vicinity of the outlet where the radio receiver is connected. The level of noise is varied to check when interference is noticeable on the SW receiver. Human ears monitor reception to verify if the additional noise influences SW radio reception. The signal level is recorded, when the reception quality is deemed to be impaired (assessing the quality of *signal, interference, noise, propagation and overall* (SINPO) [22]). This is performed at many outlets in several buildings with the radio receiver tuned to various frequencies. ETSI TR 102 616 [11] documents some of these assessments. Noise levels up to those in Equation 22.4 (U_{outlet} = 56 dB μV) were not noticeable by the receiver. Noise was hardly noticeable when signals were fed into the mains at exactly this level. If the noise was increased by another 10 dB, human ears were able to audibly detect the interference. However, if the radio was disconnected from the mains – and therefore battery powered – the interference was gone. Often, connecting the radio receiver to another outlet also solved the interference problem.

The value from Equation 22.4 was found to be a good choice for the bottom level of the notch, where EN 50561-1 specifies a verification setup for PLC modems to confirm this level. A resolution bandwidth of 300 Hz is selected to make the bottom level visible with the spectrum analyser performing a sweep. Care must be taken when comparing the absolute values of the ingress signal level, the values given in CISPR 22 and the bottom level of the notch. Individual resolution bandwidths are used.

PLC modems with a bottom level of a notch as specified in EN 50561-1 no longer cause interference to an SW radio receiver.

22.4.4.2 Width or Slopes of the Notch

To avoid interfering with the bandwidth of an identified radio broadcast service, the minimum width of a notch should be at least 10 kHz (±5 kHz around the carrier frequency of the radio broadcast). Usually, the channels of radio broadcast services are allocated with a minimum spacing of 5 kHz. The centre frequency is a multiple of 5 kHz. If several neighbouring radio broadcast services are identified by the PLC system, the width of one notch may be scaled to integer multiples of 5 kHz.

In radio broadcasting, *intercarrier interference* (ICI) from other radio stations allocating adjacent channels is a serious problem. Signal amplitudes of adjacent carriers often differ by more than 30 dB. This is why slopes of potential ICI are precisely specified. References can be found in [10,16,23,24]. EN 50561-1 defines a notch where the slopes are approximated to the requirements of SW receivers' (AM, DRM) protection ratios.

The resolution bandwidth of PLC modems' noise measurements is usually identical to the width of an OFDM carrier. To protect one broadcast station with a single carrier notch, a PLC modem has to enhance the resolution bandwidth of the receiving *fast Fourier transformation* (FFT), in order to precisely locate the frequency position of the HF carrier.

22.4.5 Adaptive OFDM, Channel and Noise Estimation

Adaptive communications with feedback information was first presented in 1968 in [25]. Today's PLC modems provide good starting conditions for implementing notching with minimal effort. Carrier adaptive OFDM [26] is used in wired and wireless communications to match the bit loading of a carrier (*quadrature amplitude modulation* [QAM] constellation) to the conditions on the channel attenuation and noise. The process of channel adaptation is dynamic. When the channel changes (e.g. a light switch is shifted), the adaptation process has to be retriggered. Some communication systems can do this within milliseconds. Adaptive modulation systems require the knowledge of the *receiver's* (Rx) SNR at the *transmitter* (Tx). Its transfer function is measured at the Rx and fed back to the Tx. Adaptive modulation systems improve the rate of transmission and bit error rates by exploiting channel information that is present at the Tx. They exhibit great performance enhancements compared to other systems, especially for fading channels. As channel and noise are already estimated to realise an adaptive OFDM, noise information might be reinterpreted to identify receivable radio broadcast stations. In an adaptive OFDM system, every carrier loads an individual amount of information. It is no additional burden for the system if individual carriers do not carry information because they are notched. The concept is adaptive. The method of using a carrier, notch it later and reuse it again, does not cause any additional work.

Figure 22.10 shows an example of an SNR estimation of a PLC link. The horizontal axis represents the frequency where the vertical axis represents the SNR in dB estimated by the system. Frequencies with excellent SNR utilise 4096-QAM and are marked with '12 bit/carrier'. Frequencies with lower SNR are allocated with lower constellations from 1024-QAM down to *binary phase shift keying* (BPSK). Frequencies with less SNR than what the most robust implemented constellation requires can no longer be used for communication and can be notched or suppressed. This is the case for carriers around 10 MHz in Figure 22.10.

There are two cases for omitting a carrier from communication: (1) the desired SNR for the minimum constellation is not available at this frequency or (2) a radio service has to be protected at this frequency. OFDM only provides low side-lobe suppression (see Chapter 14, Section 14.3.3), where the shape of the notch might be improved by additional filters. These filters will only be applied if the notch was initiated to protect a radio service.

Both the channel transfer function and the noise have to be measured in order to estimate the SNR. This is done by re-encoding the Rx's demodulated data and comparing them

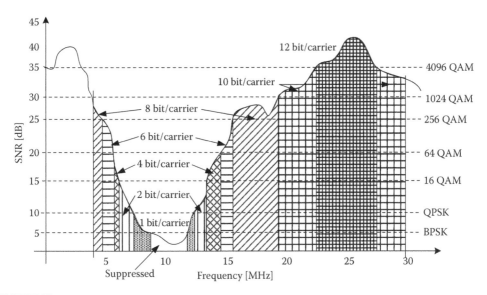

FIGURE 22.10
Adaptive channel modulation. Simulation of a frequency scan using the channel data of a measured in-house PLC channel.

with the received signal. The noise measurement during communication even works in notched carriers. If the noise is measured by calculating the variance of the received OFDM symbols, it does not matter if the carrier is allocated or notched. Figure 22.11 shows a result of a channel and noise measurement performed with the use of four training symbols per data burst. Figure 22.11 is a snapshot from a PLC modem prototype implementation [14]. The horizontal axis shows the index of the OFDM carriers. It represents the frequency range from 0 Hz up to 40 MHz with a carrier spacing of 19.53 kHz. The vertical

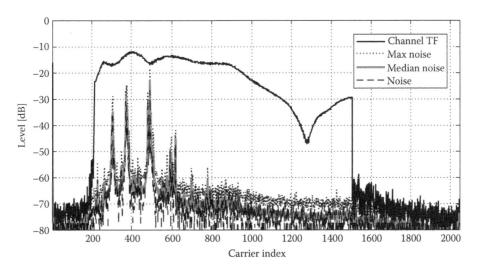

FIGURE 22.11
Channel and noise measurement of the PLC demonstrator performed during communication.

axis shows a relative level in dB. The carrier with the index of 211 is the first carrier used for communication; the one with index 1506 is the last. The topmost curve (solid black) is the channel transfer function derived by averaging four training symbols. It shows a relatively flat channel with only one fading of 30 dB. The three other curves are the noise measured by calculating the variance values of the four received training symbols. The tiny, dashed curve shows the noise signal last measured before the screenshot was taken. It shows a high variation (>10 dB) of the individual values. The fat line represents the median with the latest 20 noise shots and the third noise curve (dotted, black) is the max hold of these 20 shots. The average and max-hold values are better suited for detecting thresholds. The 'needles' of the ingress from HF radio broadcast are clearly visible.

When multiple OFDM symbols are sequentially chained, there is the disadvantage that noise measurements performed during communication do not detect frequencies at integer multiples exactly matching the training symbol's repetition frequency. Such noise does not enhance the variance between the received training symbols. To overcome this, variable guard intervals have to be inserted before preamble, frame control and data symbols. The length of each guard interval should be unique to avoid that a constant sine wave interfere with a frequency of a multiple of the training sequence repetition frequency influencing each training symbol identically. The HomePlug [27] or IEEE 1901 specifications support multiple guard intervals within a data burst. The IEEE 1901 (see Chapter 13, Figure 13.4 – *PHY protocol data unit* [PPDU]) data burst starts with the preamble (10 repetitions of a 5.12 μs sequence), followed by the frame control symbol with a 18.32 μs guard interval, followed by two data symbols with 7.56 μs of protection and further data symbols with a guard interval that can be selected out of 13 alternatives depending on channel characteristics.

If the noise is measured during communication, a further disadvantage is that it is recorded at the receiving modem. The transmitting modem is responsible for the detection of radio broadcast ingress and notching. However, in a time-division duplexing communication – which is usually the case with PLC – the return path of the noise measurement can be used. In order to capture the communication signal as well as the noise signal within the dynamic range of the Rx ADC, the implementation of dynamic power control as specified in this chapter is a prerequisite for measuring the noise during communication.

Alternatively, the noise could also be measured during any period of no PLC. The minimum length of such a quiet period is given by the basic OFDM system parameters such as carrier spacing and symbol duration. Equation 8 calculates the carrier spacing of an OFDM system. The OFDM symbol duration is reciprocal to the carrier spacing. For example, a carrier spacing of f_{CS} = 24.41 kHz like in IEEE 1901 (see Chapter 13) results in a symbol duration T_{Symbol} of

$$T_{Symbol} = \frac{1}{f_{CS}} = \frac{1}{24.41\,\text{kHz}} = 40.96\,\text{μs}, \tag{22.7}$$

The FFT of the system has to be filled once, in order to capture a noise shot. The time needed here is equal to the system's symbol duration T_{Symbol}. In a multi-node communication system, the MAC layer organises by device when the resources are allocated. For example, in a *carrier sense multiple access* (CSMA) *medium access layer* (MAC) layer, the contention-free interframe spacing is often longer than the symbol duration. All such gaps could be used to measure the noise.

To conclude, to activate a notch, the noise can be measured during communication or within a transmission break. For a PLC system to reuse the notched frequency, the noise has to be measured in a quiet period or within a notch during communication, but no signal may be transmitted on the notched frequency measuring the noise.

22.5 Implementation in a Demonstrator System

A feasibility study is implemented in order to prove the concept of 'cognitive frequency exclusions'. The PLC system is described in detail in [14]. Its main focus is desired applications and rapid development. The system is a proprietary PLC technology and does not follow any PLC standards such as HomePlug [27], HD-PLC [28] or ITU-G.Hn [29]. As an application, the system transports a high-definition video stream from Tx to Rx and measures maximum payload data. Its maximum throughput on PHY layer is 212 Mbps.

22.5.1 PLC Modem System

The PLC demonstrator was implemented using a similar platform as described in Chapter 24 but without MIMO features. Here, statistical evaluations were implemented from various noise recording techniques: max hold, median and latest shot (as shown in Figure 22.11). The main unit distinguishing this feasibility study from a conventional PLC modem today is the additional notch filtering function.

22.5.2 Notch Filter Environment

The implementation of the notch filtering function requires a noise measuring unit, a function to detect the presence of radio signals and the notch filter.

22.5.3 Detection of the Presence of Radio Services

A trade-off has to be found in the number of noise shots to be recorded and the detection speed to activate a notch. Timings in EN 50561-1 are 15 s to activate a notch. As discussed in connection with Figure 22.7, it is recommended to implement a max-hold detector to limit the huge number of noise records and to ensure that the top 1 dB margin of the fading signals is captured. These values have to be compared if the threshold is exceeded. The notch has to be activated at frequencies where the threshold is passed.

22.5.4 Notching

There are various techniques for creating notches. One simple technique for creating notches is using wavelet transformation [30], where omitting a carrier from the communication is sufficient and the spectrum is notched with a depth depending on the side lobes of the applied wavelet. The waveform of the wavelet is responsible for the shape of the notch. The notching feasibility study described here uses an FFT transformation process. The output of the FFT is very sharp in the time domain but provides weak slopes in the frequency domain compared to wavelet transmissions. Windowing can be applied for spectral shaping of FFT systems. Chapter 14 discusses the influence of windowing on the shape of a notch. Another alternative would be to design notches with additional filter stages in the Tx spectrum of the data. This solution is presented in the following.

22.5.4.1 Influence of Windowing on Spectrum and Notch Shape

The technique of efficient notching was considered when drafting the HomePlug AV2 specification. It is described in Chapter 14 (Section 14.3.3).

22.5.4.2 Adaptive Band-Stop Filters

Another alternative method of creating notches is to filter the unintended frequencies using tunable band-stop filters. In the feasibility study described here, a cascaded structure of second-order *infinite impulse response* (IIR) filters is implemented. A filter block consists of five multipliers, four delay lines and an adder. Depending on the frequency, up to three filter blocks are needed to create a notch as specified in EN 50561-1. The algorithm to calculate the filter coefficients can be explained using the filters unity circle. The zeros are on the unity circle where the angle specifies the frequency. The poles are close to the zeros at identical angles, inside the unity circle to guarantee stability. The distance between zero and pole defines the attenuation of the notch. Reference [14] described the algorithm in detail.

22.6 Verification of 'Cognitive Frequency Exclusions' in Buildings

The prototype system should be tested in buildings under real and noisy conditions in order to verify the implementation of 'cognitive frequency exclusions'.

As shown earlier, HF radio broadcast transmissions change their transmission frequency from time to time. Frequency hopping by an HF radio station requires a 'cognitive frequency exclusion' implementation in order to comply with the EN 50561-1 under all criteria. Such a scenario was very interesting to monitor under live broadcasting conditions in a private building. During the ETSI plugtest, a radio broadcast station in Skelton (United Kingdom) was available to schedule the transmission of any radio service according to the plugtest demands. To verify the dynamic behaviour of the PLC system, the Skelton transmission toggled from 7225 to 7320 kHz. Two SW radio receivers were used to monitor this event at the test location in Stuttgart (Germany). Each of the radios is tuned to one of the frequencies. There also was a PLC transmission running in parallel inside the building from the 'cognitive frequency exclusion' demonstrator system.

Figure 22.12 shows a timetable overview of the actions occurring at the test site. The horizontal axis represents the time in seconds. Before the station hop took place, the first radio receiver which was tuned to 7225 kHz received a good-quality AM signal. This frequency was notched by the PLC system. There was no interference from the PLC system to this radio station. The PLC signal was clearly noticeable on the second radio receiver, which was tuned to 7320 kHz.

The trigger for the time axis in Figure 22.12 was set to 0 s when the radio broadcast signal went silent on the first radio receiver. The Skelton transmitter had stopped its broadcast. Thirteen seconds later, the start of the transmission at 7320 kHz was noticed

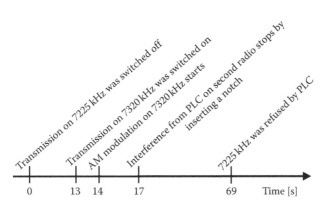

FIGURE 22.12
Frequency hopping of a notch recorded at verification of the demonstrator.

on the second radio receiver. A further second later, the AM started at 7320 kHz. The service could be heard on the second radio receiver, but it was still interfered with by the PLC transmission. The PLC device detected the presence of this radio station another 3 s later and inserted a notch to protect 7320 kHz. This was noticed when the interference stopped on the second radio device. Around 1 min later, the frequency at 7225 kHz was reused by the PLC system. The PLC signal was now noticed on the first radio receiver which was still tuned to this frequency. EN 50561-1 today specifies a time period of 3 min until a frequency might be reused by the PC system. In the early days when the draft specification of ETSI TS 102 578 [5] was still undergoing modifications, a time hysteresis of 1 min was sufficient.

To assess the quality of an AM radio station, the SINPO assumption was standardised in [22] with the properties of signal strength, interference, noise and propagation individually estimated. The signal strength and propagation can be measured using a spectrum analyser. Noise and interference levels are estimated by human ears. This way, it is difficult to identify the source of the noise. The listener's impression is noted without taking any further action. PLC was often the dominant interferer during the ETSI plugtest, when the 'cognitive frequency exclusions' were not activated. Finally, the overall estimation is given as an average of individual properties. Radio signal quality assessment was conducted according to SINPO [22] 168 times during the plugtest. Each of them is noted in the plugtest report ETSI TR 102 616 [11] including the signal levels at the mains as well as the E-field. Figure 22.13 shows a histogram of the occurrence of the overall SINPO estimation. The horizontal axis represents the SINPO level:

- Level 1: Unusable. No listener will stay tuned to a service with such bad quality.
- Level 2: Poor. A human voice might be understood.
- Level 3: Fair. Music might be enjoyed, with limited quality.
- Level 4: Good. AM audio quality.
- Level 5: Excellent. Usually, an AM service will never reach level 5. Only DRM supports this level in the HF band.

The SINPO assumption was done three times for each radio station received at the site of the plugtest: initially with PLC switched off (right [bright] column at each SINPO level in Figure 22.13) and later with a PLC transmission running with the 'cognitive frequency exclusion'

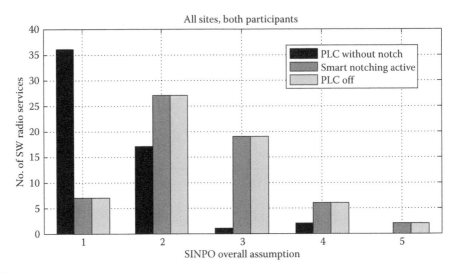

FIGURE 22.13
Histogram of subjective assessment of sound quality.

concept activated (middle column) and finally, when PLC transmission was running, but no notches were in place (left [black] column). To ensure that every station was captured, a station scan was executed in all HF transmission bands using the Sony ICF-SW77. The SINPO assumptions were performed at every station where the station scan stopped.

Figure 22.13 represents a histogram of how frequent an overall signal quality was monitored.

Figure 22.13 shows that there was no difference in the reception quality of HF radio services when the PLC system was transmitting data with 'smart notching' activated and when the PLC was off. The right and middle columns have an identical height. When the PLC system was transmitting data without inserting notches, many radio stations degraded to an unusable quality.

22.7 Outlook for Future EMC Coordination

Cognitive radio or dynamic spectrum allocations is a huge area in research today. There are many ideas at DVB, digital dividend, OFDM overlay systems, frequency management with mobile satellite services, dynamic spectrum management for DSL, etc., where this concept is relevant. Frequency resources are an extremely rare good. Motivation is high to allocate these resources as efficiently as possible. If intelligent devices are able to adapt to the local situation without causing interferences to other applications, the classical concept of EMI is overthrown. In the past, there were always constant limits for EMI emissions as well as immunity. Cognitive systems change this paradigm. Thanks to adaptive OFDM transmission with a high number of carriers, 'cognitive frequency exclusions' or 'smart notching' only causes a minor decrease in transmission bitrate as only low SNR carriers are lost. Continuous analysis allows the system to minimise interference and to optimise throughput depending on the current conditions.

EN 50561-1 is the first EMC standard that embeds dynamic cognitive interference mitigation technologies.

Some videos illustrate the implementation of the PLC 'smart notching' system in a flat:

- PLC with Sony ICF SW 77 [31]
- PLC with Sangean ATS 909 [32]
- PLC with Roberts DRM receiver [33]

References

1. EN 50561-1, Power line communication apparatus used in low-voltage installations – Radio disturbance characteristics – Limits and methods of measurement – Part 1: Apparatus for in-home use.
2. ITU-R Radio Regulations, edition of 2004.
3. Stott, J. and Salter, J., BBC R&D White Paper WHP067, The effects of powerline telecommunications on broadcast reception: Brief trial in Crieff. http://downloads.bbc.co.uk/rd/pubs/whp/whp-pdf-files/WHP067.pdf, accessed May 2013.
4. European Telecommunication Standardization Institute, Technical Committee on Power Line Transmissions (ETSI TC PLT). http://www.etsi.org/WebSite/Technologies/Powerline.aspx, accessed May 2013.
5. ETSI TS 102 578 V1.2.1 (2008-08), PowerLine Telecommunications (PLT); Coexistence between PLT modems and short wave radio broadcasting services.
6. IEC, CISPR/I/257/CD, CISPR22 – Limits and method of measurement of broadband telecommunication equipment over power lines, February 2008.
7. IEC, CIS/I/301/CD, Amendment 1 to CISPR 22 Ed. 6.0: Addition of limits and methods of measurement for conformance testing of power line telecommunication ports intended for the connection to the mains, July 2009.
8. IEC, CIS/I/302/DC, Comparison of the RF disturbance potential between Type 1 and Type 2 PLT devices compliant with the provisions of CISPR/I/301/CD and EUTs compliant with the limits in CISPR 22 Ed. 6.0, July 2009.
9. IEEE Std 1901-2010, IEEE Standard for Broadband over Power Line Networks: Medium Access Control and Physical Layer Specifications. http://grouper.ieee.org/groups/1901/, accessed May 2013.
10. Minimum Receiver requirements for DRM, Draft version 1.5.
11. ETSI TR 102 616 V1.1.1 (2008-03) PowerLine Telecommunications (PLT); Report from Plugtests™ 2007 on coexistence between PLT and short wave radio broadcast; Test cases and results. http://www.etsi.org/plugtests/plt/plt1.htm, accessed May 2013.
12. ETSI ES 201 980 (V2.2.1). Digital Radio Mondiale (DRM); System Specification. http://www.drm.org/, accessed May 2013.
13. Schwarzbeck EFS 9218, Active electric field probe with biconical elements EFS 9218 and built-in amplifier. http://www.schwarzbeck.com/Datenblatt/m9218.pdf, accessed May 2013.
14. Schwager, A., Powerline communications: Significant technologies to become ready for integration. Doctoral thesis, University of Duisburg-Essen, Essen, Germany, 2010. http://duepublico.uni-duisburg-essen.de/servlets/DerivateServlet/Derivate-24381/Schwager_Andreas_Diss.pdf, accessed May 2013.
15. Stott, J. H., BBC R&D White Paper WHP114, Co-existence of PLT and radio services – A possibility? June 2005. http://downloads.bbc.co.uk/rd/pubs/whp/whp-pdf-files/WHP114.pdf, accessed May 2013.

16. ITU-R Rec. BS.1615, Planning parameters for digital sound broadcasting at frequencies below 30 MHz.

17. ITU-R Rec. P.372-8, Radio Noise.

18. ITU-R Rec. P.842-2, Compotation of reliability and compatibility of HF radio systems.

19. EN 60315-3:2000, Methods of measurement on radio receivers for various classes of emission. Receivers for amplitude-modulated sound-broadcasting emissions.

20. CISPR 22:1997, Information technology equipment – Radio disturbance characteristics – Limits and methods of measurement.

21. CISPR 16-1-1, Specification for radio disturbance and immunity measuring apparatus and methods – Part 1-1: Radio disturbance and immunity measuring apparatus – Measuring apparatus.

22. ITU-R Rec. BS.1284, General methods for the subjective assessment of sound quality. See http://stason.org/TULARC/radio/shortwave/08-What-is-SINPOSIO-Shortwave-radio.html, accessed October 2007.

23. ITU-R Rec. BS.703, Characteristics of AM sound broadcasting reference receivers for planning purposes.

24. ITU-R Rec. 560-3 1, Radio-frequency protection ratios in LF, MF and HF broadcasting.

25. Hayes, J. F., Adaptive feedback communications, *IEEE Transactions on Communication Technology*, COM-16, 29–34, February 1968.

26. Lee, J.-J., Cha, J.-S., Shin, M.-C. and Kim, H.-M., Adaptive modulation based power line communication system, *Advances in Intelligent Computing* (Lecture Notes in Computer Science), Springer, Berlin, Germany, 2005, Vol. 3645, pp. 704–712.

27. Homeplug. http://www.homeplug.org/, accessed May 2013.

28. HD-PLC Alliance. http://www.hd-plc.org/, accessed May 2013.

29. ITU-T. 2011. G.9960, Unified high-speed wireline-based home networking transceivers – System architecture and physical layer specification.

30. Sandberg, S. D. and Tzannes, M. A., Overlapped discrete multitone modulation for high speed copper wire communications, *Journal on Selected Areas in Communications*, 13(9), 1571–1585, December 1995.

31. Video 'Smart Notching' demonstrator and AM receiver Sony ICF-SW77. http://plc.ets.uni-duisburg-essen.de/sony/SmartNotching_ICF-SW77.wmv, accessed May 2013.

32. Video 'Smart Notching' demonstrator and AM receiver Sangean ATS 909. http://plc.ets.uni-duisburg-essen.de/sony/SmartNotching_Sangean.wmv, accessed May 2013.

33. Video 'Smart Notching' demonstrator and DRM receiver Roberts MP-40. http://plc.ets.uni-duisburg-essen.de/sony/SmartNotching_DRM.wmv, accessed May 2013.

23

Mitigating PLC Interference to Broadcast Radio

Yang Lu and Weilin Liu

CONTENTS

23.1 Introduction*

In recent years, Smart Grid has received widespread attention from both academia and industrial communities. As the basis for its implementation, multiple communication technologies will be employed. However, *power line communication* (PLC) can present a more extensive and pervasive solution [2]. Up to now, *narrow-band* (NB) (3–500 kHz) and *broad-band* (BB) (2–100 MHz) PLC have been developed progressively [3,4]. Especially in the last half decade, multi-carrier NB PLC with a relatively high data rate ranging from 10 to 500 kbps has emerged and aroused immense industry interest as it is considered to be suitable for a part of the Smart Grid applications.

Although multi-carrier NB PLC is sufficient for automatic meter reading (AMR), future Smart Grid services that have rigorous real-time demand may not be supported if the communication bandwidth is restricted below 500 kHz. Moreover, due to high-noise power, low access impedance and marked time variation, a power line channel at low frequencies is characterised as a rather hostile medium for data transmission. As is well known, high-frequency power line channels have much lower noise level; on the other hand, their attenuation gradually increases with frequency (particularly for underground cables). Even so, both low- and high-frequency channels are still indispensable for PLC links within the access domain. Beyond that, note that the *medium-wave* (MW) band (500 kHz–1.6 MHz) may not only achieve an attractive trade-off between noise and attenuation, but also expand the channel capacity of NB PLC and enhance the coverage of PLC access networks. Therefore, it may be deemed as another potential opportunity for PLC transmission.

---

* Chapter in parts based on [1], Copyright © 2013 IEEE, with permission.

As a preliminary example, to show the potential of the MW frequencies for PLC access systems, a simple comparison between the MW band and low-frequency channels below 500 kHz has been made by investigating noise and attenuation characteristics simultaneously, where the corresponding channel measurements have been performed in typical *low-voltage* (LV) power line access networks in China. For convenience of comparison, the concept of *link quality index* (LQI), LQI(f,t), has also been introduced and defined as described in Chapter 3:

$$LQI(f,t) = Noise(f,t) + Loss(f,t),\tag{23.1}$$

where both Noise(f,t) and Loss(f,t) are obtained at the PLC receiver. Thus, LQI(f,t) can be interpreted as the transmitted signal level required to achieve an equivalent 0 dB *signal-to-noise ratio* (SNR) at the receiver for a specified resolution bandwidth, at a certain frequency f and time t. Evidently, the smaller the LQI(f,t), the lower the transmitted signal level needed to obtain a required SNR at the receiver, which may indicate that the channel is better. Based on the measurement results, Figures 23.1 and 23.2 show a group of LQI curves with a 10 kHz resolution bandwidth for the 30–500 kHz and 500 kHz–1.6 MHz band, respectively, where the units for noise and attenuation are unified as dBμV and dB in this case, and noise is measured by root mean square detection using a spectrum analyser. It can be seen from the statistical results that the MW band has much lower LQI values, which shows the potential of this new frequency range for PLC transmission.

However, note that when the 500 kHz–1.6 MHz band is employed, possible interferences to MW broadcast radio stations may be caused. This is because the unshielded LV power lines are not designed for data transmission at such high frequencies, and their electromagnetic radiation may disturb the surrounding radio services. Table 23.1 shows the radio frequency division regulation of China for the 505 kHz–1.6065 MHz band [5], which confirms broadcast as the main application of this frequency range.

In fact, many existing literature have paid close attention to the radiation effects of PLC. For example, a thorough investigation of the electromagnetic compatibility issues

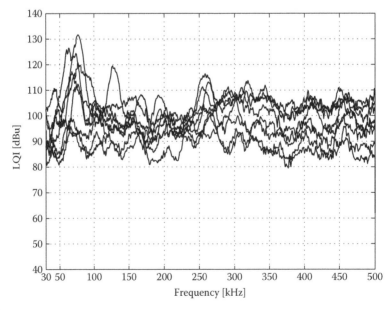

FIGURE 23.1
A group of LQI curves with 10 kHz resolution bandwidth for the 30–500 kHz band.

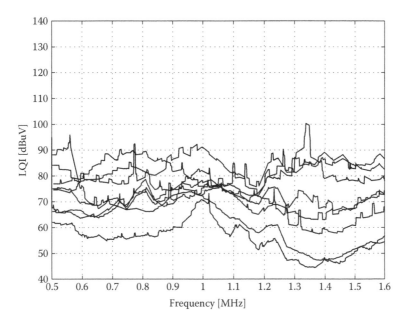

FIGURE 23.2
A group of LQI curves with 10 kHz resolution bandwidth for the 500 kHz–1.6 MHz band.

TABLE 23.1

Radio Frequency Division Regulation of China
for the 505 kHz–1.6065 MHz Band

Frequency Range	Application
505–526.5 kHz	Maritime mobile service, aeronautical radio navigation
526.5–535 kHz	Broadcast, aeronautical radio navigation
535 kHz–1.6065 MHz	Broadcast

associated with a PLC system was presented based on laboratory experiments [6]. In China, the State Grid has specified the PLC radiation limit in the AMR specifications [7].

With the aim of tackling the coexistence problem described earlier, cognitive PLC has been adopted as a potential detection and mitigation solution, which introduces the notching schemes to protect the valid radio services by not emitting at their frequencies. This idea was first proposed by Schwager [8] and similar research could also be found in references Weling [9] and Praho et al. [10]. Especially spectrum analyser- and SNR-based detection were presented for cognitive PLC as dynamic notching approaches [11,12]. Furthermore, ETSI TS 102 578 and the recently approved CENELEC FprEN50561-1 (see Chapters 6 and 22) have also specified the detection and notching criteria to achieve a harmonious coexistence between in-home PLC and *short-wave* (SW) radio broadcast [13,14].

Note that the scope of most of the research detailed earlier is limited to in-home PLC rather than outdoor power line access networks for Smart Grid applications, where the latter indicates a different coexistence scenario. On the one hand, the degree of impact caused by the outdoor PLC access system on indoor radio listeners has not been thoroughly investigated yet. On the other hand, cognitive PLC can be also utilised to protect the PLC access system from external radio interferences. Specifically, in this chapter, cognitive PLC is employed to explore the potential of the MW band for PLC transmission, which has not been studied in detail in the past.

The remainder of this chapter is organised as follows: Section 23.2 describes the test scenario and the measurement setup in detail. Section 23.3 details the analysis of the measurement results achieved in two typical LV power line access networks in China. Then the *adaptive detection* (AD) method as a cognitive PLC approach is proposed in Section 23.4. Finally, Section 23.5 concludes this chapter.

23.2 Test Scenario and Measurement Setup

Measurements were performed in two typical LV power line access networks in China.

1. Test site 1 is a village located in a suburb in the northeast part of Yiwu city (Zhejiang province) and represents the rural and overhead line environment.

   Part of the LV power line access network at test site 1 is shown in Figure 23.3. According to the network topology, two representative *measurement points* (MPs) were chosen to investigate the ingress caused by the MW broadcast radio stations on power lines, where MP1 represents the LV outlet end of the transformer in the distribution room and MP2 is a single-phase meter panel, which is installed on the outer wall of a two-storey house.

2. Test site 2 is a residential area located in the eastern part of Handan city (Hebei province) and represents the urban and underground cable environment.

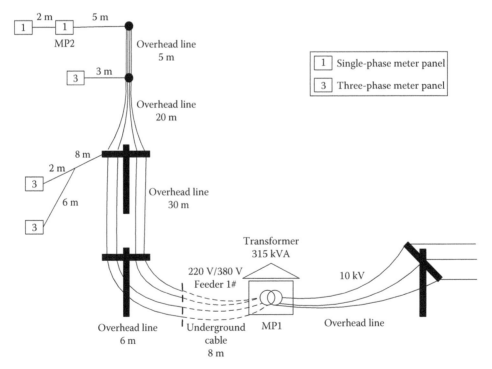

FIGURE 23.3
Part of the LV power line access network at test site 1.

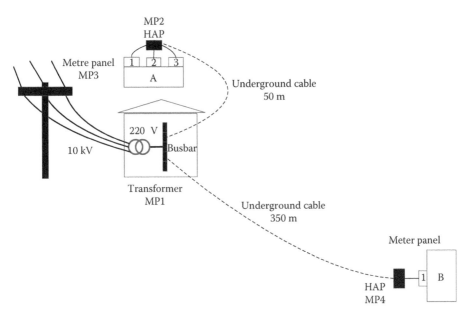

FIGURE 23.4
Part of the LV power line access network at test site 2.

Figure 23.4 shows the corresponding LV power grid topology. It can be seen that the underground cables link the transformer with the *house access point* (HAP) of building A and building B. Cable lengths are approximately 50 and 350 m, respectively. Building A has three units where each has six floors and a meter panel including 12 single-phase meters, whereas building B has only one unit with eight floors and a meter panel that has 16 single-phase meters. The typical distance between the meter panel and the HAP is approximately 5–20 m. Under this test scenario, four MPs are selected for measurement, where MP1 is the LV outlet end of the transformer in the distribution room, MP2 is the HAP of building A, MP3 represents the meter panel of the second unit of building A and MP4 represents the HAP of building B.

The measurement equipment include a passive coupler, an external attenuator, a spectrum analyser and a laptop. In order to verify the measurement results achieved from power lines, a MW radio receiver and a loop antenna are also required. Figure 23.5 shows the measurement setup and dedicated software have been installed on the laptop to collect and process data. Since the spectrum analyser and the laptop are powered by their own batteries, possible interferences introduced by the measurement equipment can be avoided.

The insertion loss of the passive coupler is shown in Figure 23.6, which indicates that the MW band signals ranging from 500 kHz to 1.6 MHz can be passed without significant loss. The unified parameters set for the spectrum analyser during the test are summarised in Table 23.2.

At both test sites, the following procedures were performed sequentially at each MP:

1. The MW radio receiver was used to record broadcast frequencies with audible sound from 500 kHz to 1.6 MHz.

2. For confirmation, the MW radio signals were measured based on the loop antenna.

3. The noises between phase and neutral on the LV power lines were captured by the spectrum analyser, and the results were compared with those obtained in steps (1) and (2) to verify whether the MW radio stations caused any narrow-band interferences on the power lines.

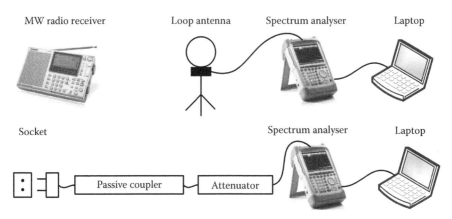

FIGURE 23.5
Spectrum analyser-based measurement setup.

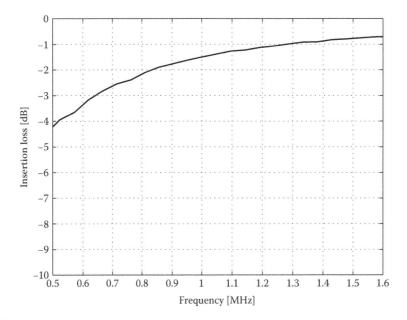

FIGURE 23.6
Insertion loss of the passive coupler for the MW band.

TABLE 23.2

Parameter Setup for the Spectrum Analyser

Resolution Bandwidth [kHz]	Video Bandwidth [kHz]	Trace Mode	Detector
10	10	Average 10 times	Max peak

23.3 Measurement Results and Analysis

This section includes two parts: In the first part, the measurement results obtained at each test site are shown and analysed. To investigate the MW radio interference to power lines more thoroughly, detailed information about the MW stations around these two test sites have also been provided for reference. Moreover, in the second part, the *electromagnetic interference* (EMI) issue of PLC is assessed by a simple field trial at MP2 of test site 1.

23.3.1 Spectrum Analyser-Based Measurements on Power Lines

The measurement results obtained at MP1 of test site 1 are shown in Figures 23.7 and 23.8, respectively, where the former corresponds to the results achieved according to test procedures (1) and (2), and the latter focuses on the noise measurement on the power lines. During the test, the MW radio receiver was utilised, and the circles labelled in Figure 23.7 indicate that some broadcast radio stations with audible sound could be detected at those frequencies. It can be seen that 10 MW radio stations were found by the receiver, the frequencies of which coincide very well with those of the signal peaks measured by the loop antenna. In this section, the unit of the vertical axis for the loop antenna measurement results is unified as dBm by using the free space impedance 377 Ω, the antenna factor and the input impedance of the measurement receiver (50 Ω). Since the orientation of the loop antenna also significantly influences the measurement results, the antenna has been rotated in the three-dimensional space to determine the arrival direction of radio signals beforehand. Note that the first two MW radio stations below 600 kHz present the stronger power level above −60 dBm, whereas the signals of the others are relatively weak. In fact, by referring to the local frequency regulation, there are no MW broadcasting stations in Yiwu city. Detailed information about the MW radio stations near Yiwu city are

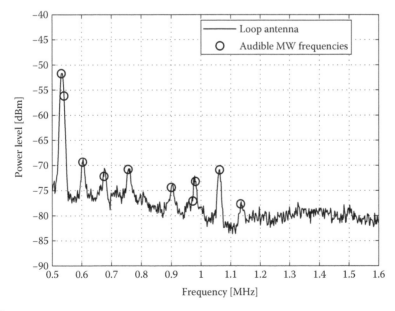

FIGURE 23.7
Loop antenna measurement results with 10 kHz resolution bandwidth for the MW band at MP1 of test site 1.

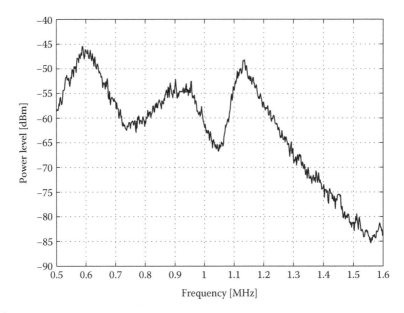

FIGURE 23.8

Power line measurement results with 10 kHz resolution bandwidth for the MW band at MP1 of test site 1.

TABLE 23.3

Detailed Information about the MW Radio Stations near Yiwu City

Frequency [kHz]	Province	City	Power [kW]	Radio Station	Time	Station Address	Distance to Yiwu [km]
531	Zhejiang	Jinhua	10	Voice of Zhejiang	24 h	Hutou village	45
540	Zhejiang	Ningbo	10	Voice of China	04:00–13:35	Wangjiang village	153
603	Zhejiang	Jinhua	1	Voice of Economic	24 h	Hutou village	45
675	Zhejiang	Jinhua	1	Jinhua News	05:30–00:00	Hutou village	45
756	Zhejiang	Jinhua	1	Voice of China	05:00–00:00	Hutou village	45
1134	Zhejiang	Lishui	—	Voice of Zhejiang	24 h	Daji village	91

summarised in Table 23.3. It shows part of the 10 audible frequencies determined in the measurement process, whereas the others may correspond to the radio stations located in the neighbouring province, whose detailed information has been omitted here.

However, as shown in Figure 23.8, the MW radio stations do not cause any clear ingress on the noise floor of the power lines. Possible explanation is that the power line noise level at MP1 is quite high for the MW band, where the maximum recorded noise power reaches −45 dBm (−85 dBm/Hz). Therefore, it submerges the weak radio signals. Another potential reason is that the electrical grid possibly does not provide loop characteristics according to the access network topology. Due to this antenna effect, the weak MW radio signals may be difficult to be detected on the power lines.

Figures 23.9 and 23.10 further show investigation of the measurements performed at MP2 of test site 1, which represents a single-phase meter panel of a residential house. As mentioned in Section 23.2, MP2 is closer to the power consumer than MP1 and the outdoor overhead line directly links with the meter panel. Note that the difference of the

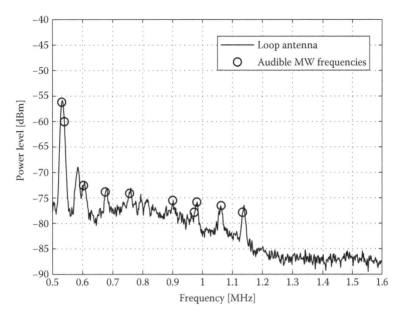

FIGURE 23.9
Loop antenna measurement results with 10 kHz resolution bandwidth for the MW band at MP2 of test site 1.

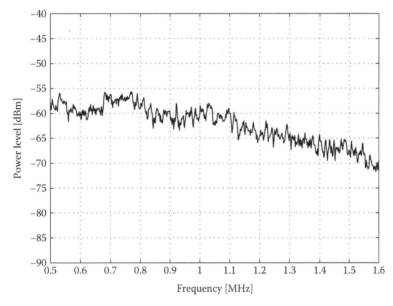

FIGURE 23.10
Power line measurement results with 10 kHz resolution bandwidth for the MW band at MP2 of test site 1.

power line noise floor between these two MPs may lead to different radio station detection results. However, similar to the results shown in Figure 23.8, the MW radio stations are still undetectable on the power lines.

In order to study the MW radio interferences to the power lines more carefully, another group of measurement results obtained at each MP of test site 2 is shown. The crosses labelled in Figure 23.11 represent the MW radio frequencies according to the local

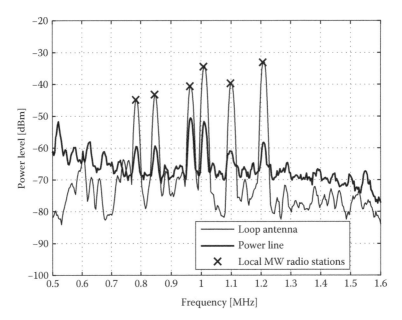

FIGURE 23.11
Measurement results with 10 kHz resolution bandwidth for the MW band at MP1 of test site 2.

TABLE 23.4

Detailed Information about the MW Radio Stations in Handan City

Frequency [kHz]	Province	City	Power [kW]	Radio Station	Time	Station Address	Distance to Test Site 2 [km]
783	Hebei	Handan	—	Hebei News	05:00–15:00	Nanbao village	6.5
846	Hebei	Handan	—	Handan Opera	05:00–22:00	Nanbao village	6.5
963	Hebei	Handan	10	Handan News	05:00–00:00	Nanbao village	6.5
1008	Hebei	Handan	10	Handan Traffic	05:00–00:00	Nanbao village	6.5
1098	Hebei	Handan	—	Handan Traffic	05:00–00:00	Nanbao village	6.5
1206	Hebei	Handan	10	Handan Economic	05:00–00:00	Nanbao village	6.5

frequency regulation, and detailed information about the MW radio stations in Handan city is presented in Table 23.4. Compared with the power level measured by the loop antenna in Figures 23.7 and 23.9, it can be seen that the signals corresponding to the local MW broadcasts are much stronger in this test site. This is because the radio stations are not located far from the test site. Therefore, the power line narrow-band noises caused by the MW radio interferences become more visible as well. On that basis, it shows that MW radio signals may be detected on power lines in some cases. However, there are also some peaks detected on the power lines, but they do not represent actual MW stations, which may result in false detection.

Figures 23.12 and 23.13 show additional measurement results obtained at MP2 and MP3 of test site 2, respectively. Taking MP2 as an example, due to variation of the radio signal strength, the power level measured by the loop antenna, which corresponds to the 846 kHz

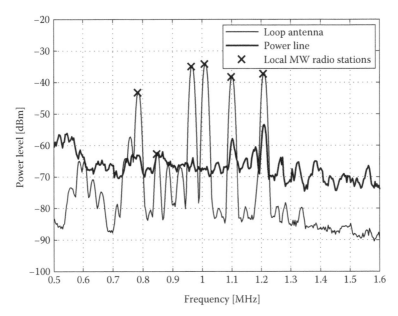

FIGURE 23.12
Measurement results with 10 kHz resolution bandwidth for the MW band at MP2 of test site 2.

radio station, is nearly 20 dB lower compared with the corresponding curves in Figure 23.11. This trend is more obvious at MP3 shown in Figure 23.13. Since the meter panels are installed inside the second unit of building A, the outer wall of the building will prevent the radio signals from being detected by the antenna, which results in even lower power levels. When considering noise measurement on the power lines, part of the local MW radio frequencies can be identified. However, in some cases, certain radio stations may

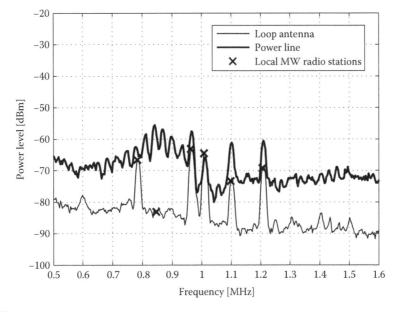

FIGURE 23.13
Measurement results with 10 kHz resolution bandwidth for the MW band at MP3 of test site 2.

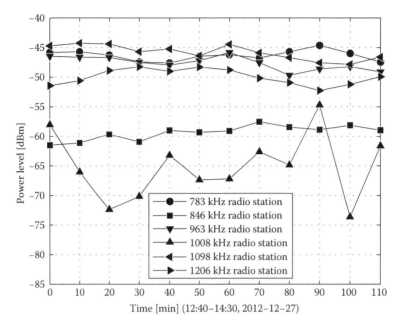

FIGURE 23.14
A series of power values measured by the loop antenna with 10 kHz resolution bandwidth for the local MW radio stations at MP4 of test site 2.

remain undetected on the power lines. Note that the radio frequencies not detected at MP2 and MP3 can be observed at MP1 and vice versa. Therefore, it may become an incentive for implementing a more reliable detection by merging the detection results obtained at different points of the access network.

As is well known, SW broadcast radio stations may have a very dynamic signal level due to reflections in the ionosphere. Since MW radio signals mainly propagate through the ground wave, the fading of their signal strengths in the time domain may be less obvious compared with SW stations. To verify this characteristic, a series of measurement results obtained at MP4 of test site 2 is shown in Figures 23.14 and 23.15. The measurements were performed every 10 min during the time interval 12:40–14:30 on 27 December 2012, and 12 sets of data were obtained for the loop antenna and the LV power line, respectively. It can be seen in Figure 23.14 that the curves corresponding to the power levels of the six local MW radio stations listed in Table 23.4 are almost flat, except the 1008 kHz frequencies. These results show that the power levels of the MW stations measured by the loop antenna are relatively stable. However, according to the power line measurement results in Figure 23.15, part of the local MW radio stations can be clearly seen across the 2 h, whereas the others may only appear for a certain period of time. It visually illustrates that the time-varying power line noise floor of one measurement location may lead to different detection results for the MW stations.

23.3.2 Field Trial for EMI Issues of PLC

As a preliminary example, this section evaluates the EMI issues of PLC through a simple field trial at MP2 of test site 1. The measurement setup is shown in Figure 23.16, where a signal generator is utilised to inject the swept-frequency signals ranging from 500 kHz to 1.6 MHz into the LV power line and the loop antenna is employed to capture the

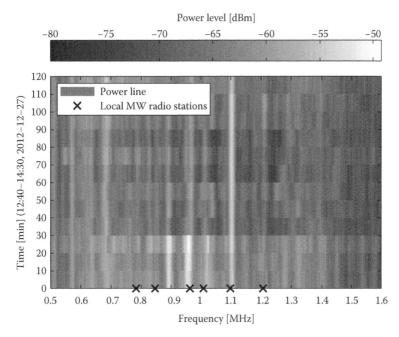

FIGURE 23.15
A series of power line measurement results with 10 kHz resolution bandwidth for the MW band at MP4 of test site 2.

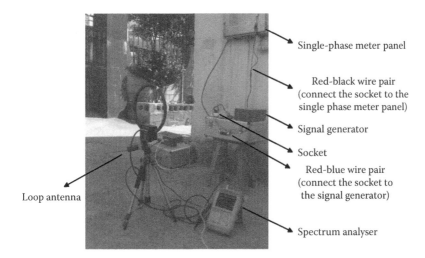

FIGURE 23.16
Measurement setup for PLC EMI issue assessment.

electromagnetic radiations. During the test, the resolution bandwidth of the spectrum analyser was set at 30 kHz in the Max Hold trace mode and using the Max Peak detector. Through placing the antenna at different distances in a vertical direction from the single-phase meter panel, a group of measurement results corresponding to distinct distances could be recorded.

The measurement results are shown in Figure 23.17. As the radiation source, the top curve shows the power level of the injected signal measured on the power line. In this

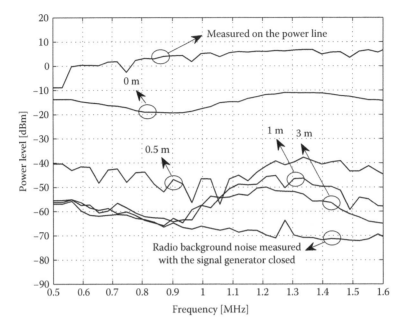

FIGURE 23.17
Measurement results with 30 kHz resolution bandwidth for electromagnetic radiation of PLC at MP2 of test site 1.

example, it can be seen that the level of the injected signal, which is between 0 and 7 dBm/30 kHz, is rather strong. The rest curves in Figure 23.17 show the power levels measured with antenna as a function of its distance to the power line. It suggests that the shorter the distance from the antenna to the power line is, the stronger the radiation is. Note that when the distance is 3 m, the radiated signal strength quickly decreases to below −50 dBm according to the measurement result; however, it still submerges most parts of the radio background noise measured by the antenna when the signal generator is closed. Although both electric wire pairs in Figure 23.16, the red–black one as well as the red–blue one, do not look very symmetrical, where high radiation may be expected, this field trial still indicates that potential EMI issues may occur near the power line, which renders certain detection and mitigation solutions indispensable to avoid harmful interferences.

23.4 Adaptive Detection for Cognitive PLC

From the measurement results, it shows that MW radio stations can be detected in the power line access networks in some situations. Although the power levels of some detectable MW stations are quite weak, cognitive capability may still be essential for such PLC access systems to prevent themselves from suffering external radio interferences. Note that ETSI TS 102 578 has specified two thresholds with the 9 kHz resolution bandwidth for SW radio detection on in-home power lines, namely, an absolute (−95 dBm) and a relative (14 dB) threshold above the noise floor, where the noise floor of each SW radio band is defined as the median power value of the adjacent frequency blocks [13]. However, measurement results in this chapter show its inapplicability for the MW band in some cases because of two reasons:

1. The power line noise level of the MW band is much higher than that of the SW band. In many situations, the noise powers across the whole MW band are all above −95 dBm with the resolution bandwidth as required, which makes the absolute threshold specified in ETSI TS 102 578 inapplicable.

2. The measurement results show that the narrow-band interferences due to MW stations may not reach 14 dB above the noise floor for most of the time, which will lead to high miss detection rate if this detection criterion is utilised. To be more intuitive, the power line noises recorded at both test sites are utilised for a statistical analysis of the MW radio interference level above the noise floor. For consistency, the noise floor is determined similarly to ETSI TS 102 578 by employing the whole MW band for median calculation. Figure 23.18 shows the corresponding statistical histogram, where the indistinguishable weak narrow-band noises possibly caused by the MW stations have not been included in the statistics. It can be seen that for nearly 80% cases the observable radio interference levels are less than 14 dB above the noise floor. This trend can be seen more clearly in Figure 23.19 by utilising the cumulative distribution function.

Based on the previous analysis, a different detection method for identifying MW stations on power lines should be devised. Note that the method to determine the noise floor by employing the median value is quite simple, however, it may not reflect the real noise floor of the power lines in some cases because the median of noise powers is closely related to the bandwidth used for calculation and the number of radio stations appearing

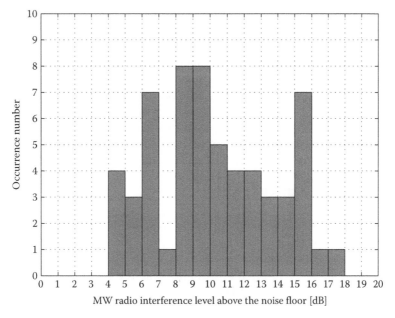

FIGURE 23.18
Histogram of the MW radio interference level above the noise floor.

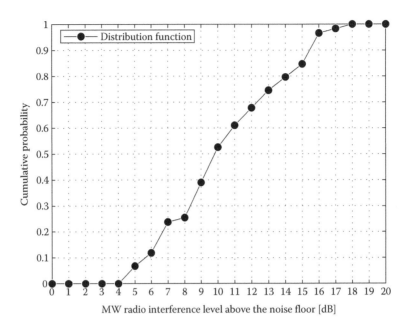

FIGURE 23.19
Distribution function for the MW radio interference level above the noise floor.

within this sub-band. If the narrow-band interferences caused by the radio signals occupy only less than half of the concerned sub-band and not the other part as power line background noise, the median value can be deemed as the proper noise floor of this sub-band. Otherwise, median value is not a good choice.

In this section, AD has been proposed as a detection approach for cognitive PLC. The main idea of this method is to first determine the appropriate noise floor of power lines by employing the iterative average, and then adding an empirical relative threshold to it to yield the final detection threshold. Moreover, the detected radio frequencies should be notched from the PLC transmission spectrum.

Let $\mathbf{F}$ be regarded as a string of numbers that correspond to the power values of different frequencies. Since the power line noise floor over the whole MW band may fluctuate with the frequency, the set $\mathbf{F}$ is first divided into N subgroups according to the sub-band division, denoted by $\mathbf{F}_1, \mathbf{F}_2, \ldots, \mathbf{F}_N$, with N is chosen such that the variation of the noise floor in each subgroup is small. For example, in the case of an orthogonal frequency division multiplexing system, which is employed by most BB PLC solutions, this sub-band division can be naturally realised by dividing the signals according to their sub-carrier frequencies. For the next procedure, the proposed AD algorithm performs the following iterative average steps for each sub-band:

1. Computes the initial threshold ε_n ($n = 1, 2, \ldots, N$) as the mean power within sub-band $\mathbf{F}_n$ ($n = 1, 2, \ldots, N$).

2. The power value of each sample in sub-band $\mathbf{F}_n$ ($n = 1, 2, \ldots, N$) is compared with the threshold ε_n ($n = 1, 2, \ldots, N$). If it is smaller than the threshold, the index of this sample should be included in set $\mathbf{H}_0$.

3. Computes λ_n ($n = 1, 2, \ldots, N$) as the mean power of the samples, which corresponds to set $\mathbf{H}_0$.

4. If the following termination condition is met, the algorithm stops and λ_n ($n = 1, 2, ..., N$) represents the final noise floor of this sub-band. Otherwise, it updates ε_n ($n = 1, 2, ..., N$) with λ_n ($n = 1, 2, ..., N$) and returns to step (2) for iteration:

$$\left| \varepsilon_n - \lambda_n \right| \leq \tau, \quad n = 1, 2, ..., N. \tag{23.2}$$

In the previously given equation, τ is a proper stop criterion. Note that in the ideal case, if the noise floor is strictly flat, $\left| \varepsilon_n - \lambda_n \right|$ ($n = 1, 2, ..., N$) will tend to be zero with the iterations obviously, namely $\tau \to 0$ can be deemed as a proper stop indicator. However, for real power line background noises, τ can be chosen as an empirical value determined by the measurement results. The earlier iterative average process is shown by the flow chart in Figure 23.20.

After determining the noise floor, a relative threshold has to be added to it. Based on the measurement results, 7 dB is chosen as the proper relative threshold for the MW band. Therefore, the final detection threshold can be achieved.

To validate the effectiveness of the proposed algorithm, the measurement results at MP1, MP2 and MP3 of test site 2 are utilised for assessment and the power line noises ranging from 500 kHz to 1.6 MHz are investigated to detect the radio stations. For simplicity, this whole MW band can be divided into four sub-bands by employing the division criteria as mentioned, namely, 0.5–0.75 MHz, 0.75–1 MHz, 1–1.25 MHz and 1.25–1.6 MHz. For each sub-band, the noise floor can be determined individually based on the iterative average process detailed earlier and the 7 dB relative threshold is then added to it.

The detection results for MP1 of test site 2 are shown in Figure 23.21 as an example, where the detection actually equals to performing a binary hypothesis test based on the thresholds determined by the presented algorithm. It can be seen that the local MW radio stations are all successfully identified except one that corresponds to 1098 kHz. There are also three falsely detected stations, which do not represent real MW radio stations. In summary, the detection results of the proposed AD algorithm for each MP of test site 2 are shown in Table 23.5. Note that ETSI TS 102 578 has further specified certain time requirements to minimise misdetection rate, namely, the criteria of −95 dBm and 14 dB above the noise floor should both be met for more than 30% of the time in any 10 s interval [13]. Since the research work

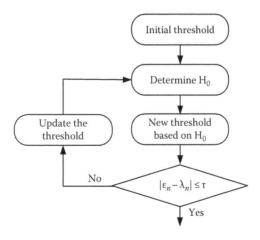

FIGURE 23.20
Flow chart of the iterative average process in the proposed AD algorithm.

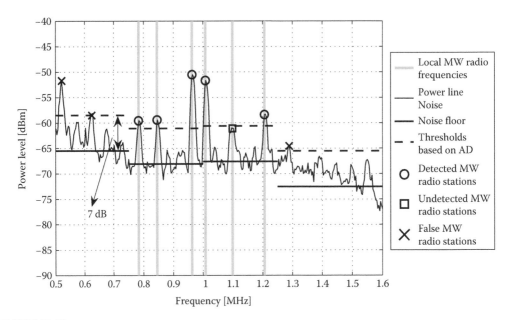

FIGURE 23.21
Detection results for the MW radio stations at MP1 of test site 2 (10 kHz resolution bandwidth).

TABLE 23.5

Detection Results of the Proposed AD
Algorithm

	Detected MW Radio Stations	Falsely Detected MW Radio Stations
MP1	5	3
MP2	2	2
MP3	5	3

has just begun, similar procedures have not been included in the proposed AD algorithm, which is of great interest and essentiality to be investigated in the future.

Furthermore, as discussed in Section 23.3, detection of MW radio stations at one place is not reliable enough. Therefore, cooperative detection in cognitive radio can be introduced by merging the detection results at transformer (MP1), HAP (MP2) and meter panels (MP3) to deliver better performance where the AND, MAJORITY or OR rule may be applied:

1. AND rule: An MW radio station is considered to be present if and only if each MP detects it.

2. MAJORITY rule: An MW radio station is considered to be present if over half of the MPs detect it.

3. OR rule: An MW radio station is considered to be present as long as one MP detects it.

In Table 23.6, the cooperative detection results of AD based on different fusion rules are shown. In this case, it can be seen that although the number of falsely detected ones reaches 7, the OR rule still achieves best performance because all local MW radio stations are correctly

TABLE 23.6

Cooperative Detection Results of AD Based on Different
Fusion Rules

	Detected MW Radio Stations	Falsely Detected MW Radio Stations
AND rule	1	0
MAJORITY rule	5	1
OR rule	6	7

detected. For these falsely detected frequencies, they may also be notched by PLC access systems to make communication more reliable. From this point of view, the OR rule may be considered as the preferred candidate for the fusion method of cooperative detection.

Finally, as far as design aspects are concerned, since the logical topology of the power line access network often shows master–slave characteristics, the proposed radio detection capability for cognitive PLC may be implemented in the concentrator of the network. Therefore, dynamic notching for protecting valid radio services and keeping PLC away from narrow-band interferences can be controlled by the central node more efficiently.

23.5 Conclusions

In this chapter, the coexistence issue between MW broadcast radio stations and PLC access systems that use the frequency band covering 500 kHz–1.6 MHz has been studied. First, spectrum analyser-based measurement results for the MW band are shown, where both overhead line and underground cable environment as typical power line access networks in China have been covered. It shows that the presence of MW broadcast radios in the LV access power line network cannot be easily detected as the SW broadcast radios. One reason can be that the radiation effect of power line network as a potential antenna in the MW band is not as strong as that in the SW band. The high noise floor in the MW band is another factor. Nevertheless, this needs further in-depth investigation. Statistical analysis shows that the MW radio interference level above the power line noise floor is usually less than 14 dB. Therefore, AD for identifying MW stations is presented as a detection approach for cognitive PLC, where an iterative average process is employed to determine the noise floor and a relative threshold above the noise floor smaller (e.g. 7 dB for the examples in this chapter) than 14 dB as specified in ETSI TS 102 578 for the SW radios is proposed to get the absolute detection threshold. Moreover, cooperative detection has also been discussed by merging the detection results to get performance promotion. Since the research work is ongoing, power line noise measurement and analysis on the MW band need further investigation and the interference level between PLC access systems and indoor radio listeners also calls for assessment.

Acknowledgement

This work was funded by the research project 'New Generation Smart PLC Key Technology Research' of the State Grid Corporation of China (SGCC).

References

1. Lu Y. and Liu W. 2013. Spectrum analyzer based measurement and detection of MW/SW broadcast radios on power lines for cognitive PLC. *Proceedings of IEEE Seventeenth International Symposium on Power Line Communications and Its Applications (ISPLC), 2013*, pp. 103–108, 24–27 March 2013.
2. Galli S., Scaglione A. and Wang Z. F. 2011. For the grid and through the grid: The role of power line communications in the smart grid. *Proceedings of the IEEE* 99(6): 998–1027.
3. Latchman H. and Yonge L. 2003. Power line local area networking (Guest Editorial). *IEEE Communications Magazine* 41(4): 32–33.
4. Pavlidou N., Vinck A. H., Yazdani J. et al. 2003. Power line communications: State of the art and future trends. *IEEE Communications Magazine* 41(4): 34–40.
5. Ministry of Industry and Information Technology of China. 2010. Radio frequency division regulation of China.
6. Pagani P., Razafferson R., Zeddam A. et al. 2010. Electromagnetic compatibility for power line communications. *Proceedings of IEEE 21st International Symposium on Personal Indoor and Mobile Radio Communications (PIMRC)*, Istanbul, Turkey, pp. 2799–2804.
7. Q/GDW 374.3. 2009. Power user electric energy data acquire system technic specification, part 3: Communication unit. State Grid Corporation of China.
8. Schwager A. 2010. Powerline communications: Significant technologies to become ready for integration. PhD dissertation, University of Duisburg-Essen, Essen, Germany.
9. Weling N. 2011. Expedient permanent PSD reduction table as mitigation method to protect radio services. *Proceedings of IEEE 15th International Symposium on Power Line Communications and Its Applications (ISPLC)*, Udine, Italy, pp. 305–310.
10. Praho B., Tlich M., Pagani P. et al. 2010. Cognitive detection method of radio frequencies on power line networks. *Proceedings of IEEE 14th International Symposium on Power Line Communications and Its Applications (ISPLC)*, Rio de Janeiro, Brazil, pp. 225–230.
11. Weling N. 2011. Feasibility study on detecting short wave radio stations on the powerlines for dynamic PSD reduction as method for cognitive PLC. *Proceedings of IEEE 15th International Symposium on Power Line Communications and Its Applications (ISPLC)*, Udine, Italy, pp. 311–316.
12. Weling N. 2012. SNR-based detection of broadcast radio stations on powerlines as mitigation method toward a cognitive PLC solution. *Proceedings of IEEE 16th International Symposium on Power Line Communications and Its Applications (ISPLC)*, Beijing, China, pp. 52–59.
13. ETSI TS 102 578 v1.2.1. 2008. Powerline telecommunications (PLT); coexistence between PLT modems and short wave radio broadcasting services.
14. CENELEC FprEN 50561-1. 2012. Power line communication apparatus used in low-voltage installations – Radio disturbance characteristics – Limits and methods of measurement – Part 1: Apparatus for in-home use.

24

MIMO PLC Hardware Feasibility Study

Daniel M. Schneider and Andreas Schwager

CONTENTS

24.1 Introduction

In order to prove the concept of *multiple-input multiple-output* (MIMO) for *power line communications* (PLCs), a feasibility study was being implemented. The MIMO PLC system was designed at Sony EuTEC laboratories and its basic parameters are based on an existing *single-input single-output* (SISO) PLC system implementation [1]. The design focuses on the desired application of comparing SISO and MIMO and rapid development. Hence, the design is not based on any PLC standard. It is a proprietary system that serves to understand and investigate some implementation-specific issues related to current broadband PLC systems as introduced in Chapters 12 through 14. The implemented demonstrator system consists of two modems, a transmitter and a receiver. It does not include a full multi-node *media access control* (MAC) layer. As an application, the system allows *high-definition* (HD) video streaming and monitoring several system parameters such as measured maximum bitrate and *bit error ratio* (BER). First, an overview of the implemented system is provided (see Section 24.2). A detailed description of the implemented MIMO blocks is given in Sections 24.3 through 24.6. Section 24.7 discusses possible changes of the implemented system architecture. Finally, the verification of the system (Section 24.8) concludes the chapter.

24.2 System Architecture

Figure 24.1 shows the setup of the demonstrator system. The left PC serves as transmitter and the PC shown on the right acts as the receiver. Four analog front-ends are mounted on each of the two PCs, allowing each PC to act as a transmitter or receiver. Only two of the front-ends are used in transmit mode. Two coaxial cables connect the front-ends of the transmitter to the Tx MIMO coupler (delta-style coupler). The transmit coupler is connected to an artificial MIMO channel (*MIMO artificial mains* [MAM] network; for details see Section 24.8). The receiver coupler is also connected to the MAM (bottom right-hand side in the figure) and couples the signals of the four receive ports to the four analog front-ends of the receiver. Each of the transmitter and receiver modems consists of a standard *personal computer* (PC) running Linux as an operating system. Each PC includes a *peripheral component interconnect* (PCI) board containing two Xilinx Virtex 5 *field programmable gate arrays* (FPGAs). The FPGAs realise the real-time functions of the demonstrator system, while the PC implements the control functions in software. The functions implemented in software are described in [1]; the key parameters are briefly summarised here. MATLAB® applications running on a third remote PC (not depicted in Figure 24.1) allow the control and monitoring of the PLC modems via *hypertext transfer protocol* (HTTP). Each embedded MIMO PLC modem computer includes a web server running on the Linux system, which communicates with the PLC drivers. Several applications like video streaming sources generate *user datagram protocol* (UDP) payload data to be transmitted and received via power line. The PLC driver is an extension of a standard Linux Ethernet driver implementing additional *application programming interfaces* (APIs) and hardware access functions.

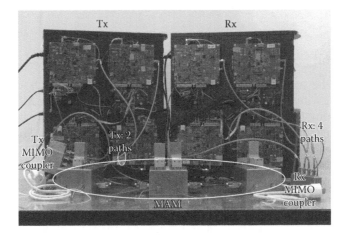

FIGURE 24.1
Hardware setup.

The PLC driver implements both a network driver as an interface to the host PC's *Internet protocol* (IP) stack and a device driver as an interface to the FPGA hardware. The remote PC might render the video stream.

The system architecture shown in Figure 24.2 focuses on the MIMO-specific blocks. The top row illustrates the blocks of the transmitter, while the bottom row illustrates the receiver (from right to left). Both, transmitter and receiver chain, are implemented on the FPGAs. The MIMO PLC channel connects the two transmit ports of the transmitter to the four receive ports of the receiver. As described earlier, a MATLAB application controls and monitors the PLC driver which, in turn, controls the FPGA of the transmitter and receiver. Note that although the PLC driver and interface are depicted only once in the figure, they are physically running on each of the PCs, whereas the MATLAB *graphical user interface* (GUI) is located on the third, remote PC. The input bit stream is first processed by the *forward error correction* (FEC) block which basically inserts redundancy bits (details see later). From a functional point of view, at this location, the bit stream would be split into the two streams of the two (logical) MIMO streams. However, the two MIMO streams are multiplexed into one signal path at a higher clock rate. The number of MIMO streams is indicated at the signal paths in Figure 24.2. By choosing this design, only one *quadrature amplitude modulation* (QAM) mapper and only one *orthogonal frequency division multiplex* (OFDM) modulation block are required. The adaptive QAM assigns the input bits to complex QAM symbols according to the constellation of each subcarrier. The PLC driver registers the constellation to be used by each subcarrier. Depending on the QAM constellation, the block decides how many input bits are modulated for each subcarrier. The QAM mapping is realised by a *look-up table* (LUT) which assigns to each input bit combination a complex QAM symbol. The next unit is the *power allocation* (PA). Here, three different PA coefficients $0, 1, \sqrt{2}$ are used (refer to the simplified PA algorithm in Chapter 8). The PA coefficient is derived from the constellations. If one stream is not bit loaded, the other stream assigns twice the power. The complex symbols of each stream are assigned to a vector which is multiplied by the precoding matrix $\mathbf{V}$ depending on the codebook index of each subcarrier (see Section 24.6). The codebook comprises a predefined set of different precoding matrices, where the index defines which precoding matrix is selected out of this set of predefined precoding matrices. The codebook indices are programmed by the PLC driver. The next block inserts training symbols before entering the OFDM modulation. Each burst consists of 4 training OFDM symbols and up to 20 data OFDM symbols. The OFDM modulation applies a 2048 *inverse fast Fourier transform* (IFFT) and inserts an OFDM guard interval of length 1/16. The guard interval consists of a cyclic prefix, which is a copy of the tail of the OFDM symbol. The guard interval is needed to prevent *intersymbol interference* (ISI) that would otherwise be caused by the multipath channel. The block *transmitter* (Tx) front-end summarises several blocks: A digital quarter period mixer shifts the complex baseband signal to 0–40 MHz to obtain the real-valued signal to be transmitted. A preamble is inserted before each burst. The preamble consists of a *constant amplitude with zero auto correlation* (CAZAC) sequence and is needed for time synchronisation in the receiver. Finally, the analog front-ends feed the signals to the channel. They consist of two *digital-to-analogue* (D/A) converters, low-pass filters, line driver amplifiers and couplers.

Similar blocks follow in reverse order in the receiver. The *receiver* (Rx) front-end includes four analog front-ends (couplers, band-pass filters, *automatic gain control* [AGCs] and *analog-to-digital* [A/D] converters) and the quarter period mixers. The band-pass filter limits the input signal to the interesting frequencies between 4 and 30 MHz. The AGC includes a *programmable gain amplifier* (PGA) to control the signal level at the A/D. A quarter period mixer converts the received signal to the complex baseband. A correlation

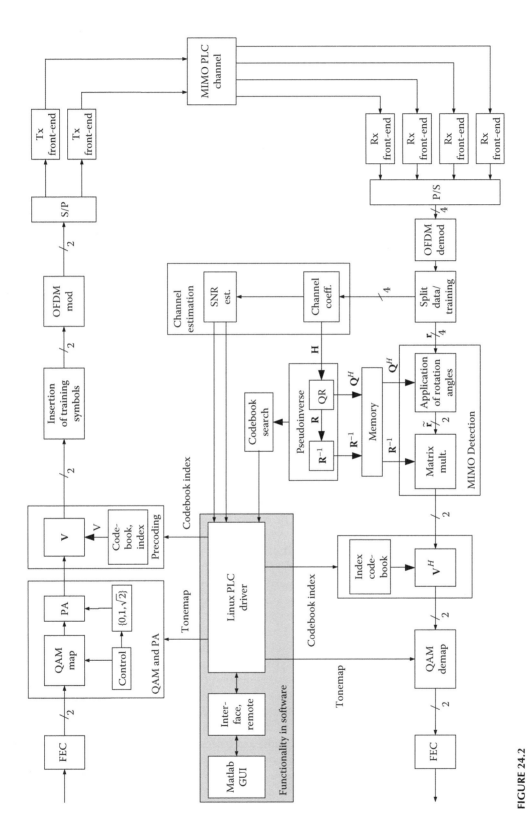

FIGURE 24.2

System: Functional overview.

function individually for each reception path detects the CAZAC sequence of the preamble, which is needed for time synchronisation and to define the individual gain level for the AGC of each Rx path. The OFDM demodulation cuts the cyclic prefix and transforms the signal back to the frequency domain by applying the fast *Fourier transform* (FFT). A splitter separates the training and data symbols. Beside channel estimation, the training symbols are used to estimate the sample clock offset (not shown in the figure). The clock offset arises from different oscillator frequencies in the transmitter and receiver. The clock offset estimation controls the *voltage-controlled crystal oscillator* (VCXO) of the A/D and ensures that the clock frequencies of the transmitter and receiver are synchronised. Instead of using an expensive VCXO, digital clock offset compensation might be realised by phase rotating the QAM symbols, depending on the subcarrier index and OFDM symbol index. Based on the received training symbols, the channel and *signal-to-noise ratio* (SNR) estimation derive the channel matrix of each subcarrier and SNR values of the two logical MIMO streams, respectively (see Section 24.5). The channel matrices serve as input to the pseudoinverse block which calculates the detection matrices. The detection (or equaliser) matrices are applied on the received data symbols by the MIMO detection unit (see Section 24.3). The training symbols are not precoded. Hence, MIMO detection is unable to take the precoding into account and separate compensation of the precoding is required. Changes in the design if the training symbols are also precoded are discussed in Section 24.7. The QAM demodulation recovers the bit sequence obtained from the received QAM symbols, depending on the constellations programmed by the PLC driver. The FEC corrects bit errors and removes the redundancy bits. Based on the detection matrix, the codebook index of each subcarrier is derived in the codebook search (see Section 24.6). The codebook indices and SNR values are passed to the PLC driver. The PLC driver determines the constellations, according to the SNR values, and programmes the constellations and codebook indices into the transmitter and receiver. It is also possible to monitor several internal signals, for example, the received training symbols or channel estimation results.

Figure 24.3 shows details of the FEC chain in the transmitter. Basically, the reverse order is used in the receiver. First, the incoming bits to be transmitted are added with redundancy by a *Reed–Solomon* (RS) code [2]. The RS operation is a block code, which corrects a maximum number of errors depending on the length of the code. The bit errors can be located anywhere inside the block. The RS code serves as an outer code to correct remaining errors which could not be corrected by the inner code. The time interleaver rearranges the order of the bits to spread burst errors into separated and isolated bit errors, which are then corrected by the RS code. Burst errors may occur in the case of impulsive noise. A convolution encoder is used as an inner code where redundancy is added to the input bits. A Viterbi decoder [2] is applied in the receiver. The frequency interleaver changes the order of the bits to ensure that adjacent bits are not mapped to adjacent subcarriers. Frequency-selective noise or deep fades of the transfer function may only affect a few subcarriers, resulting in burst errors if no frequency interleaver is used. More advanced

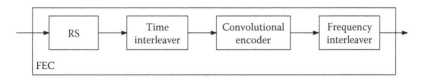

FIGURE 24.3
FEC in the transmitter.

TABLE 24.1

Basic Physical Layer Parameters

MIMO parameters	
MIMO setup	Up to 2 Tx and up to 4 Rx ports
MIMO modes	Two-stream or one-stream beamforming (based on 7 bit codebook for matrix quantisation), spatial multiplexing (beamforming switched off), SISO
OFDM parameters	
Sampling frequency (MHz)	80
Frequency band (MHz)	0–40, active subcarriers: 4–30
FFT points	2048
Number of active subcarriers (4–30 MHz)	1296
Carrier spacing (kHz)	19.53
Symbol length (µs)	51.2
Guard interval (µs)	3.2 (1/16)
QAM and FEC parameters	
Adaptive modulation (per subcarrier)	QPSK, 16-, 64-, 256-, 1024-QAM
Forward error correction	RS (204, 188), convolutional coding (Viterbi, code rate 1/2 or 3/4)
Maximum transmission speed, gross *physical* (PHY) layer bitrate (Mbit/s)	506

codes like turbo codes or *low-density parity check* (LDPC) codes may be used to improve the performance further. However, as already discussed, this demonstrator implementation focuses on the MIMO-specific blocks.

The basic system parameters are summarised in Table 24.1. Note that the OFDM parameters are the same as used in Chapter 9.

24.3 Detection and Pseudoinverse Calculation

24.3.1 Introduction

Figure 24.2 shows the *zero-forcing* (ZF)-detection process at the receiver. First, the pseudoinverse of the estimated channel matrix of each subcarrier is calculated. These detection matrices are applied to the received data vectors. According to Equation 8.18 in Chapter 8, the pseudoinverse $[\cdot]^\dagger$ is calculated as

$$
\begin{aligned}
\mathbf{H}^\dagger &= (\mathbf{H}^H\mathbf{H})^{-1} \quad \mathbf{H}^H, \\
(2 \times 4) &= ((2 \times 4)(4 \times 2))^{-1} \quad (2 \times 4).
\end{aligned}
\tag{24.1}
$$

The calculation involves a matrix inversion of the 2 × 2 matrix:

$$
\mathbf{H}^H\mathbf{H} = \begin{bmatrix} \mathbf{h}_1^H \\ \mathbf{h}_2^H \end{bmatrix} \begin{bmatrix} \mathbf{h}_1 & \mathbf{h}_2 \end{bmatrix} = \begin{bmatrix} \mathbf{h}_1^H\mathbf{h}_1 & \mathbf{h}_1^H\mathbf{h}_2 \\ \mathbf{h}_2^H\mathbf{h}_1 & \mathbf{h}_2^H\mathbf{h}_2 \end{bmatrix} = \begin{bmatrix} \mathbf{h}_1^H\mathbf{h}_1 & \mathbf{h}_1^H\mathbf{h}_2 \\ \left(\mathbf{h}_1^H\mathbf{h}_2\right)^H & \mathbf{h}_2^H\mathbf{h}_2 \end{bmatrix},
$$

$$
\doteq \mathbf{A} = \begin{bmatrix} a_{11} & a_{12} \\ a_{12}^* & a_{22} \end{bmatrix}.
\tag{24.2}
$$

NOTE: a_{11} and a_{22} are real. There is a closed-form expression of the inverse of **A**:

$$\mathbf{A}^{-1} = \frac{1}{a_{11}a_{22} - |a_{12}|^2}\begin{bmatrix} a_{22} & -a_{12} \\ -a_{12}^* & a_{11} \end{bmatrix}. \tag{24.3}$$

Equations 24.1 through 24.3 describe a possible implementation. However, a fixed-point implementation of this approach faces numerical problems. In particular, the calculation of the matrix inverse can be numerically unstable. If the two products $a_{11}a_{22}$ and $|a_{12}|^2$ are in the same order of magnitude, the difference becomes very small and $1/(a_{11}a_{22} - |a_{12}|^2)$ becomes large. Calculations of 'square products' of the form $\mathbf{H}^H\mathbf{H}$ should be also avoided for numerical reasons: The word width of the multiplication output is doubled compared to the input word width for full precision. Also, the calculation requires many multiplications which consume many hardware resources. These reasons motivate the use of another, more efficient algorithm for detection, based on a QR decomposition. The implementation of this algorithm uses numerically stable unitary matrix operations and is well suited for a parallel and efficient realisation [3,4]. The QR decomposition decomposes the channel matrix **H** into an upper triangular matrix **R** and a unitary matrix **Q**:

$$\begin{matrix} \mathbf{H} & = & \mathbf{Q} & \mathbf{R} & = \mathbf{Q}\begin{bmatrix} r_{11} & r_{12} \\ 0 & r_{22} \end{bmatrix}. \\ (4\times 2) & = & (4\times 2) & (2\times 2) & \end{matrix} \tag{24.4}$$

Replacing Equation 24.1 with Equation 24.4 results in the pseudoinverse

$$\mathbf{H}^\dagger = \mathbf{R}^{-1}\mathbf{Q}^H. \tag{24.5}$$

The calculation of **Q** involves only unitary matrix operations or rotations. Although the inversion of the triangular matrix **R** can be computed efficiently, the fixed-point parameters have to be chosen carefully to achieve good numerical results. More sophisticated algorithms avoid the calculation of $\mathbf{R}^{-1}$ [5,6]. These algorithms also support more complex detection algorithms as *minimum mean squared error* (MMSE) or *ordered successive interference cancellation* (OSIC). However, the increased computational complexity requires more hardware resources.

The detection algorithm can be expressed using the results earlier (the noise is neglected):

$$\check{\mathbf{s}} = \mathbf{H}^\dagger\mathbf{r} = \mathbf{R}^{-1}\underbrace{\mathbf{Q}^H\mathbf{r}}_{\tilde{\mathbf{r}}} = \mathbf{R}^{-1}\tilde{\mathbf{r}}\left(= \mathbf{R}^{-1}\mathbf{Q}^H\underbrace{\mathbf{H}}_{\mathrm{QR}}\mathbf{s} = \mathbf{s}\right). \tag{24.6}$$

The intermediate result $\tilde{\mathbf{r}} = \mathbf{Q}^H\mathbf{r}$ is introduced in Equation 24.6 since **Q** is not calculated explicitly. Therefore, $\mathbf{H}^\dagger$ is also not calculated explicitly. **Q** is characterised by several rotation angles, and the multiplication of **r** by $\mathbf{Q}^H$ to obtain $\tilde{\mathbf{r}} = \mathbf{Q}^H\mathbf{r}$ is actually a rotation of **r** by these rotation angles describing **Q** and $\mathbf{Q}^H$, respectively (as described in Section 24.3.2). Figure 24.2 shows the block diagram of the described operations; see the blocks *pseudoinverse* and *MIMO detection*. First, the QR decomposition is applied on the estimated channel matrix **H** resulting in **R** and the rotation angles describing **Q**. Next, $\mathbf{R}^{-1}$ is calculated. $\mathbf{Q}^H$ and $\mathbf{R}^{-1}$ are stored into a memory. For detection, the stored results are applied to the received data vectors.

24.3.2 Pseudoinverse Calculation Based on QR Decomposition

Starting from a mathematical description of the idea of the QR decomposition, the hardware architecture is derived in the following.

The goal of the QR decomposition is to transform $\mathbf{H}$ into an upper triangular matrix $\mathbf{R}$:

$$\mathbf{H} = \begin{bmatrix} h_{11} & h_{12} \\ h_{21} & h_{22} \\ h_{31} & h_{32} \\ h_{41} & h_{42} \end{bmatrix} \rightarrow \mathbf{R} = \begin{bmatrix} r_{11} & r_{12} \\ 0 & r_{22} \\ 0 & 0 \\ 0 & 0 \end{bmatrix}. \tag{24.7}$$

The algorithm works iteratively, introducing zeros in each step by applying unitary rotation matrices (so-called Givens rotations).

In a first step, the entries of the first column of $\mathbf{H}$ are turned into real values by multiplying by the matrix

$$\begin{bmatrix} e^{j\alpha_1^{(1)}} & 0 & 0 & 0 \\ 0 & e^{j\alpha_2^{(1)}} & 0 & 0 \\ 0 & 0 & e^{j\alpha_3^{(1)}} & 0 \\ 0 & 0 & 0 & e^{j\alpha_4^{(1)}} \end{bmatrix} \begin{bmatrix} h_{11} & h_{12} \\ h_{21} & h_{22} \\ h_{31} & h_{32} \\ h_{41} & h_{42} \end{bmatrix} = \begin{bmatrix} \bar{h}_{11} & h'_{12} \\ \bar{h}_{21} & h'_{22} \\ \bar{h}_{31} & h'_{32} \\ \bar{h}_{41} & h'_{42} \end{bmatrix}, \tag{24.8}$$

where the bar $(\bar{x})$ indicates real numbers and the prime (x') indicates which matrix entries are also affected by the operations. The rotation angles are calculated as

$$\alpha_1^{(1)} = -\arctan \frac{\Im\{h_{11}\}}{\Re\{h_{11}\}},$$

$$\vdots \tag{24.9}$$

$$\alpha_4^{(1)} = -\arctan \frac{\Im\{h_{41}\}}{\Re\{h_{41}\}}.$$

The superscript (1) in Equation 24.9 indicates that the angles correspond to the first column of $\mathbf{H}$. Based on the results obtained by the right-hand side of Equation 24.8, the first 0 is introduced in the next step:

$$\begin{bmatrix} \cos\beta_1^{(1)} & -\sin\beta_1^{(1)} & 0 & 0 \\ \sin\beta_1^{(1)} & \cos\beta_1^{(1)} & 0 & 0 \\ 0 & 0 & 1 & 0 \\ 0 & 0 & 0 & 1 \end{bmatrix} \begin{bmatrix} \bar{h}_{11} & h'_{12} \\ \bar{h}_{21} & h'_{22} \\ \bar{h}_{31} & h'_{32} \\ \bar{h}_{41} & h'_{42} \end{bmatrix} = \begin{bmatrix} \bar{h}'_{11} & h''_{12} \\ 0 & h''_{22} \\ \bar{h}_{31} & h'_{32} \\ \bar{h}_{41} & h'_{42} \end{bmatrix}, \tag{24.10}$$

$$\beta_1^{(1)} = -\arctan\frac{\overline{h}_{21}}{h_{11}}. \tag{24.11}$$

The next two zeros are introduced similarly by

$$\begin{bmatrix} \cos\beta_2^{(1)} & 0 & -\sin\beta_2^{(1)} & 0 \\ 0 & 1 & 0 & 0 \\ \sin\beta_2^{(1)} & 0 & \cos\beta_2^{(1)} & 0 \\ 0 & 0 & 0 & 1 \end{bmatrix} \begin{bmatrix} \overline{h}_{11}' & h_{12}'' \\ 0 & h_{22}'' \\ \overline{h}_{31} & h_{32}' \\ \overline{h}_{41} & h_{42}' \end{bmatrix} = \begin{bmatrix} \overline{h}_{11}'' & h_{12}''' \\ 0 & h_{22}'' \\ 0 & h_{32}'' \\ \overline{h}_{41} & h_{42}' \end{bmatrix}, \tag{24.12}$$

$$\beta_2^{(1)} = -\arctan\frac{\overline{h}_{31}}{\overline{h}_{11}'} \tag{24.13}$$

and

$$\begin{bmatrix} \cos\beta_3^{(1)} & 0 & 0 & -\sin\beta_3^{(1)} \\ 0 & 1 & 0 & 0 \\ 0 & 0 & 1 & 0 \\ \sin\beta_3^{(1)} & 0 & 0 & \cos\beta_3^{(1)} \end{bmatrix} \begin{bmatrix} \overline{h}_{11}'' & h_{12}''' \\ 0 & h_{22}'' \\ 0 & h_{32}'' \\ \overline{h}_{41} & h_{42}' \end{bmatrix} = \begin{bmatrix} \overline{h}_{11}''' & h_{12}'''' \\ 0 & h_{22}'' \\ 0 & h_{32}'' \\ 0 & h_{42}'' \end{bmatrix}, \tag{24.14}$$

$$\beta_3^{(1)} = -\arctan\frac{\overline{h}_{41}}{h_{11}''}. \tag{24.15}$$

After these steps, the manipulation of the first-column vector is finished and $r_{11} = \overline{h}_{11}'''$ and $r_{12} = h_{12}''''$ are determined. In the equations earlier, the number of the primes indicates the number of iterations which affects this variable. This iterative nature of the algorithm is reflected later by feedbacks in the hardware design.

Next, zeros have to be introduced into the second column. The operations described for the first column have to be repeated for the second column. The dimension is reduced by one. Again, first h_{22}'', h_{32}'' and h_{42}'' have to be made real delivering the rotation angles $\alpha_1^{(2)}$, $\alpha_2^{(2)}$ and $\alpha_3^{(2)}$. Introducing the zeros provides the rotation angles $\beta_1^{(2)}$ and $\beta_2^{(2)}$ and the last element of $\mathbf{R}$: r_{22}. The product of all rotation matrices is equal to $\mathbf{Q}^H$ since the rotation matrices are applied from the left-hand side:

$$\mathbf{Q}^H\mathbf{H} = \mathbf{R},$$
$$\Leftrightarrow \mathbf{H} = \mathbf{QR}. \tag{24.16}$$

The calculated rotation angles describe $\mathbf{Q}^H$. Note that r_{11} and r_{22} are real, while r_{12} is in general complex.

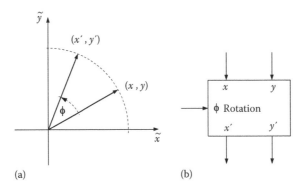

FIGURE 24.4
Rotation: Rotation of input vector (x, y) by ϕ (a) and corresponding block diagram (b).

Basically, the algorithm of the QR decomposition as described earlier includes two operations: on the one hand, the application of rotations by the given rotation angles and, on the other hand, the calculation of these rotation angles.

24.3.2.1 Rotation

The rotation of an input vector (x, y) by the angle ϕ is described by

$$x' = x\cos\phi - y\sin\phi,$$
$$y' = x\sin\phi + y\cos\phi. \tag{24.17}$$

Note that the application of the rotation matrices in Equations 24.8 and 24.10 through 24.14 affects two elements in each column. Figure 24.4a and b illustrates the rotation and the corresponding block diagram, respectively.

24.3.2.2 Vectoring

To calculate the rotation angles, one output of the rotation of (x, y) is set to 0. In Equations 24.8 and 24.9, the imaginary part is forced to be 0, while in Equations 24.10 through 24.15, zeros are introduced to the matrix. Forcing $y' = 0$ in Equation 24.17 results in

$$x' = \sqrt{x^2 + y^2} = x\cos\phi - y\sin\phi,$$
$$y' = 0 = x\sin\phi + y\cos\phi \tag{24.18}$$

with

$$\phi = -\arctan\frac{y}{x}. \tag{24.19}$$

The rotation shown in Figure 24.5a is performed in such a way that the output vector lies on the $\tilde{x}$-axis. Figure 24.5b shows another interpretation of the vectoring unit. Assuming that $\tilde{x}$ and $\tilde{y}$ represent the real and imaginary part in the complex plane, the equations earlier calculate the phase angle and absolute value of the complex input $x + jy$.

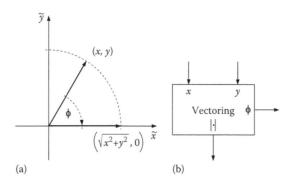

FIGURE 24.5
Vectoring: Rotation of the input vector (x, y) to the $\tilde{x}$ axis (a) and corresponding block diagram (b).

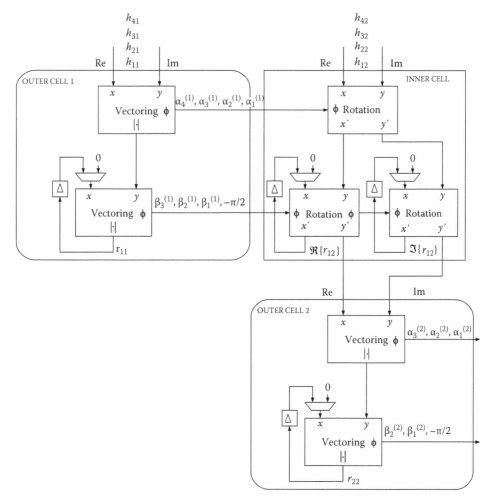

FIGURE 24.6
System architecture of the QR decomposition.

Figure 24.6 shows the architecture of the QR decomposition using the rotation and vectoring blocks of Figures 24.4 and 24.5, respectively [3,4]. The architecture is a so-called systolic array implementation, in which different processing units are placed in parallel. There are several data paths between the different processing units in contrast to a linear data path. The design includes two blocks labelled OUTER CELL 1 and OUTER CELL 2. An outer cell block consists of two vectoring blocks as defined in Figure 24.5b. Furthermore, the INNER CELL block consists of three rotation blocks as introduced in Figure 24.4b. The first column of the channel matrix **H** serves as input to OUTER CELL 1, while the second column serves as input to the INNER CELL. The feedback paths contain delay elements (labelled by Δ) which are needed to align the signals.

First, the entries of the first column of **H** are phase rotated to become real in the upper vectoring unit of OUTER CELL 1 according to Equation 24.8. The vectoring unit provides the corresponding rotation angles and the real entries of the first-column vector. Equation 24.8 shows that the second column of **H** has to be phase rotated by the same angles. This rotation is applied in the first rotation block of the INNER CELL. The output of the first vectoring unit serves as input to the second vectoring unit (input y) of OUTER CELL 1. For the processing of the first element $\bar{h}_{11}$, the x input of the vectoring unit is set to 0 via the multiplexer. Since this input is 0, $\bar{h}_{11}$ is guided through the vectoring unit without change resulting in an output angle of $-\pi/2$. When the next element $\bar{h}_{21}$ arrives at the y input, $\bar{h}_{11}$ is available at the x input via the feedback and the first 0 is introduced according to Equations 24.10 and 24.11. The result of this operation is $\bar{h}'_{11}$, which will be combined with the next input $\bar{h}_{31}$ to introduce the second 0 according to Equations 24.12 and 24.13. After the next step, the third 0 is introduced resulting in r_{11}. The obtained rotation angles are used to manipulate the second column of **H** according to Equations 24.10, 24.12 and 24.14 in the two lower rotation units of the INNER CELL. Since the entries of the second column are complex, two rotation units are needed, each for real and imaginary parts. The output of the inner cell provides the manipulated second-column vector which includes r_{12}. The output is connected to the second outer cell which introduces the remaining zeros and calculates the corresponding rotation angles.

The vectoring and rotation algorithms can be efficiently implemented by the *coordinate rotation digital computer* (CORDIC) algorithm [7]. The CORDIC algorithm iteratively solves trigonometric equations.* The CORDIC algorithm applies iteratively predefined rotation angles, which become smaller in each iteration by only using shift and add operations. In a pipelined implementation, cascaded CORDIC stages realise the iterations. The higher the number of stages, the higher the precision. To satisfy an m-bit precision CORDIC operation, $m + 1$ iterations are needed and the data path word length has to be $m + 2 + \log_2(m)$ [9]. The number of CORDIC stages mainly determines the latency. The vectoring units use the CORDIC algorithm. The latency L of the vectoring unit has to be considered in the design of the pseudoinverse calculation of Figure 24.6. The feedback signal of the second vectoring (input x) and the second input signal (input y) have to be aligned. Because of the feedback loop and the latency of the vectoring unit ($L > 1$), the input signals $\bar{h}_{11}$, $\bar{h}_{21}$, $\bar{h}_{31}$ and $\bar{h}_{41}$ of the input y have to be spread by the latency of the vectoring unit. This spreading would decrease the throughput by a factor of the latency. However, the calculations have to be performed for each subcarrier and changing the order of the input avoids any idling cycles of the vectoring unit. First, $\bar{h}_{11}$ of the first L subcarriers serve as input, followed by $\bar{h}_{21}$ of the first L subcarriers and so on. By changing the order in that way, the feedback signal of the vectoring unit is aligned and the design is fully pipelined without any idling cycles. The rotation unit could also be implemented with

---

* The CORDIC algorithm can also be used to solve a broad range of equations, including hyperbolic and square root equations [8].

the help of CORDIC. However, here, the rotation according to Equation 24.17 is realised by multipliers and LUTs to implement the sine and cosine functionality. The FPGA provides *digital signal processing* (DSP) slices, which are used to implement the multipliers. The combination of multipliers and sine/cosine LUTs has the advantage of smaller latency compared to the CORDIC algorithm and the available resources of DSP slices of the FPGA are fully utilised. The implementation makes use of the Xilinx Cores available for the CORDIC algorithm [10], the sine/cosine LUTs [11] and multipliers [12].

24.3.2.3 Calculation of R^{-1}

So far, the matrix $\mathbf{R}$ has been calculated. Now, the inverse of $\mathbf{R}$ has to be found. The inverse of the upper triangular 2×2 matrix $\mathbf{R} = \begin{bmatrix} \tilde{r}_{11} & \tilde{r}_{12} \\ 0 & \tilde{r}_{22} \end{bmatrix}$ is calculated by

$$\mathbf{R}^{-1} = \begin{bmatrix} \tilde{r}_{11} & \tilde{r}_{12} \\ 0 & \tilde{r}_{22} \end{bmatrix} = \begin{bmatrix} \dfrac{1}{r_{11}} & -r_{12} \cdot \dfrac{1}{r_{11}} \cdot \dfrac{1}{r_{22}} \\ 0 & \dfrac{1}{r_{22}} \end{bmatrix}. \tag{24.20}$$

The implemented pipeline design uses one divider [13] and three multipliers to realise the calculations according to Equation 24.20. Details can be found in [14].

24.3.3 ZF Detection

Figure 24.2 shows the ZF detection in hardware. First, $\mathbf{Q}^H$ is applied to the received data vector $\mathbf{r}$. Figure 24.7 shows the implementation of the application of the rotation angles; r_m denotes the received symbol of receive port m ($m = 1,\ldots,4$). Two inner cells are cascaded. The inner cells have the same structure as the inner cells introduced in Figure 24.6.

The multiplication by $\mathbf{R}^{-1}$ requires eight real multiplications. The eight multiplications are realised by two hardware multipliers utilising the four-times-multiplexed format of the signals. This serialisation at a higher clock rate is similar to the multiplication by the precoding matrix (see later in Section 24.4). Details can be found in [14].

24.3.4 Hardware Parameters

The pipeline structure of the different processing units described in the previous sections require some further data processing to make sure that the signals have the

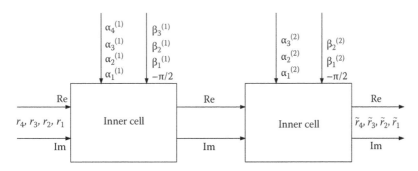

FIGURE 24.7
Detection: Rotation by $\mathbf{Q}^H$.

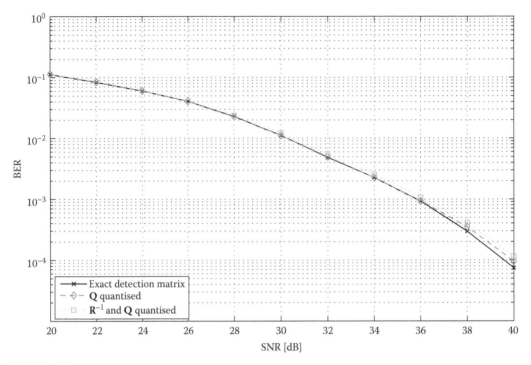

FIGURE 24.8
BER performance of the fixed-point pseudoinverse implementation.

suitable format when passing to the several processing units. Also, the latency of the blocks has to be considered when aligning the signals at the different points of the design. Details of the implementation, including the data formats and the control logic, can be found in [14].

To verify the fixed-point implementation, the model under test is embedded into the double-precision environment. Figure 24.8 shows the BER, depending on the SNR, for the fixed-point implementation of the pseudoinverse. 1024-QAM is chosen as modulation scheme since it is the most sensitive one to errors. No FEC and no precoding is applied. The figure shows the average performance for randomly generated channel matrices. The channels were normalised in such a way that the norm of the rows is equal to 1, that is, the SNR value corresponds to the SNR per receiving port. Figure 24.8 compares the double-precision (64 bit) detection matrix (exact detection matrix) to the detection matrix where only the rotation angles are obtained from the fixed-point model ($\mathbf{Q}$ quantised) and to the complete fixed-point detection matrix ($\mathbf{R}^{-1}$ and $\mathbf{Q}$ quantised). The performance loss is very small especially for operation points of the BER between 10^{-3} and 10^{-2}.

24.4 Precoding: Multiplication by V

Figure 24.9 shows the codebook-based precoding at the transmitter (for details about the codebook, see Section 24.6). The codebook indices of all subcarriers are stored in the INDEX MEMORY (block 'index memory' in Figure 24.9). This memory is written by the PLC driver every time when new codebook indices are available. The codebook index

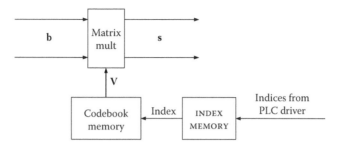

FIGURE 24.9
Codebook-based precoding at the transmitter.

determines the precoding matrix **V** out of the codebook. In the matrix multiplication block, the precoding matrix **V** is applied to the transmit symbol vectors. The compensation of the precoding at the receiver works similarly.

The matrix multiplication is described by

$$\mathbf{s} = \mathbf{Vb} = \begin{bmatrix} v_1 & -v_2^* \\ v_2 & v_1 \end{bmatrix} \begin{bmatrix} b_1 \\ b_2 \end{bmatrix} = \begin{bmatrix} v_1 b_1 - v_2^* b_2 \\ v_2 b_1 + v_1 b_2 \end{bmatrix}, \tag{24.21}$$

where the precoding matrix is constructed from the codebook entries v_1 and v_2 (for details, see [14]).

In order to derive the design of the implemented matrix multiplication, Equation 24.21 is rewritten with real and imaginary part notation:

$$\mathbf{s} = \mathbf{Vb} = \begin{bmatrix} \Re\{s_1\} + j\Im\{s_1\} \\ \Re\{s_2\} + j\Im\{s_2\} \end{bmatrix},$$

$$= \begin{bmatrix} \Re\{v_1\} + j\Im\{v_1\} & -\Re\{v_2\} + j\Im\{v_2\} \\ \Re\{v_2\} + j\Im\{v_2\} & \Re\{v_1\} + j\Im\{v_1\} \end{bmatrix} \begin{bmatrix} \Re\{b_1\} + j\Im\{b_1\} \\ \Re\{b_2\} + j\Im\{b_2\} \end{bmatrix},$$

$$= \begin{bmatrix} \Re\{v_1\}\Re\{b_1\} - \Im\{v_1\}\Im\{b_1\} - \Re\{v_2\}\Re\{b_2\} - \Im\{v_2\}\Im\{b_2\} \\ \Re\{v_2\}\Re\{b_1\} - \Im\{v_2\}\Im\{b_1\} + \Re\{v_1\}\Re\{b_2\} - \Im\{v_1\}\Im\{b_2\} \end{bmatrix},$$

$$+j\begin{bmatrix} \Im\{v_1\}\Re\{b_1\} + \Re\{v_1\}\Im\{b_1\} + \Im\{v_2\}\Re\{b_2\} - \Re\{v_2\}\Im\{b_2\} \\ \Im\{v_2\}\Re\{b_1\} + \Re\{v_2\}\Im\{b_1\} + \Im\{v_1\}\Re\{b_2\} + \Re\{v_1\}\Im\{b_2\} \end{bmatrix}. \tag{24.22}$$

The matrix multiplication requires 16 real multiplications. Figure 24.10 shows the design using four multiply and accumulate units. Due to the faster clock rate domain, four clock cycles are used to describe one subcarrier. The four values describing the transmit symbol vector **b** are multiplexed and serve as input to all of the four multipliers. The four values $\Re\{v_1\}$, $\Im\{v_1\}$, $\Re\{v_2\}$ and $\Im\{v_2\}$ describing the precoding matrix have to be repeated and reordered to realise the matrix multiplication according to Equation 24.22. The reordering is implemented by an appropriate memory access of the codebook memory. The control logic provides the sign of each of the multiply and accumulate units, depending on the clock cycle, to realise the plus/minus signs in Equation 24.22.

Table 24.2 defines special codebook entries to support different MIMO modes without precoding. The corresponding codebook index allows to select several MIMO modes without

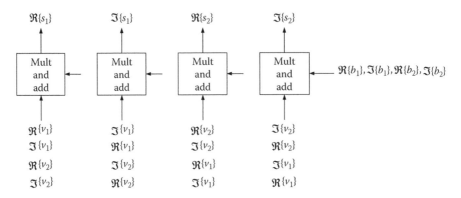

FIGURE 24.10
Block diagram of precoding: Multiplication by **V**.

TABLE 24.2

Special Codebook Entries

Codebook Index	1	2
Codebook entry $\begin{bmatrix} v_1 \\ v_2 \end{bmatrix}$	$\begin{bmatrix} 1 \\ 0 \end{bmatrix}$	$\begin{bmatrix} \dfrac{1}{\sqrt{2}} \\ \dfrac{1}{\sqrt{2}} \end{bmatrix}$
Description	No precoding	No precoding for MISO (second input must be 0)

additional implementation efforts. If the first entry is used, the precoding matrix is equal to the identity matrix and no precoding is applied. The second entry is interesting for the MISO mode where the second MIMO stream carries no information and the second transmit symbol is set to 0. In this case, this codebook entry applies no precoding for the MISO mode. By selecting the corresponding codebook index of Table 24.2 for all subcarriers, one of the two described MIMO modes without precoding is selected in the demonstrator system.

24.5 Channel and SNR Estimation

24.5.1 Estimation of the Channel Matrix

Channel estimation is based on four OFDM training symbols. These four training symbols are included at the beginning of each burst and are known at the receiver. MIMO requires a special format of the training symbols to estimate all MIMO paths simultaneously. In the following, only one receive port is considered. The calculation of the other receive ports proceeds likewise. Assume s_t to be the training symbol of one subcarrier. The following orthogonal sequence is transmitted:

$$\text{Tx port 1}: +s_t \quad +s_t \quad -s_t \quad -s_t,$$

$$\text{Tx port 2}: +s_t \quad -s_t \quad +s_t \quad -s_t.$$

The columns represent the time, that is, two positive training symbols are transmitted in the first two time instances, followed by two negative training symbols.

Basically, only two training symbols are needed to separate the two MIMO paths from each transmit port to one receive port. Averaging over four training symbols improves the performance of the estimation. $r^{(1)}$ and $r^{(2)}$ are the two received training symbols (the superscript denotes the time slot) which consecutively arrive at receive port one:

$$r^{(1)} = h_{11}s_t + h_{12}s_t,$$
$$r^{(2)} = h_{11}s_t - h_{12}s_t.$$
(24.23)

h_{11} and h_{12} are the channel coefficients from transmit port 1 and 2 to receive port 1, respectively. The noise is neglected. It is assumed that the channel coefficients do not change during the four time instances. This assumption holds for the quasi-static PLC channel. Combining the two consecutively received symbols results in

$$r^{(1)} + r^{(2)} = 2h_{11}s_t \quad \Rightarrow \quad h_{11} = \frac{r^{(1)} + r^{(2)}}{2s_t},$$

$$r^{(1)} - r^{(2)} = 2h_{12}s_t \quad \Rightarrow \quad h_{12} = \frac{r^{(1)} - r^{(2)}}{2s_t}.$$
(24.24)

The division by the training symbol simplifies to a multiplication of the conjugate complex training symbol if the absolute value of the training symbol is equal to 1. The division is further simplified if the training symbols are *binary phase-shift keying* (BPSK) modulated, that is, $s_t \in \{+1, -1\}$. In this case, only the sign has to be changed. The channel estimation is implemented according to Equation 24.24 with $|s_t| = 1$. Training symbols are allocated to each subcarrier. The channel estimation may be improved by more advanced algorithms. In particular, the duration of the training phase may be reduced by utilising the correlation of neighboured subcarriers. Only a few subcarriers, which are called pilot subcarriers, are allocated by training symbols. The spacing of the pilot subcarriers should be linked to the coherence bandwidth of the channel. The channel coefficients between the pilot subcarriers have to be interpolated.

24.5.2 SNR Estimation

Adaptive QAM modulation requires knowledge about the SNR per subcarrier. Figure 24.11 shows the SNR estimation based on the received training symbols.

First, the received training symbols

$$\mathbf{r} = \mathbf{H}\mathbf{s}_t + \mathbf{n}$$
(24.25)

are equalised by the MIMO detection block, which is the same as used for the data symbols (see Section 24.3). In the next step, the training symbols are subtracted. After ZF detection and subtraction of the training symbols, only the effect of the detection on the noise remains:

$$\mathbf{H}^\dagger \mathbf{r} - \mathbf{s}_t = \mathbf{H}^\dagger \mathbf{H} \mathbf{s}_t + \mathbf{H}^\dagger \mathbf{n} - \mathbf{s}_t = \mathbf{H}^\dagger \mathbf{n}.$$
(24.26)

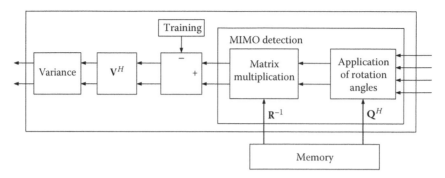

FIGURE 24.11
SNR estimation.

The training symbols are not precoded. Therefore, the effect of precoding on the SNR has to be considered by the multiplication of $\mathbf{V}^H$. This results in the detection matrix $\mathbf{W} = \mathbf{V}^H\mathbf{H}^\dagger$, taking into account the effect of precoding. The variance calculation (implemented by squaring the absolute values) results in the SNR of the two logical MIMO streams. The SNR values are averaged over several bursts to get more accurate results.

As shown in Figure 24.11, SNR estimation requires MIMO detection of the received training symbols. According to Figure 24.2, the system design comprises two instances of MIMO detection, one for the received data symbols and one for the received training symbols (within SNR estimation). The second detection unit for the training path may be unnecessary if the training symbols are equalised by the detection unit for the data symbols. The detection unit for the data symbols does not work during the processing of the training symbols and may be reused for the detection of the training symbols. This alternative design would require some reorganisation of the data flow of the design.

24.6 Codebook Search

As introduced in Section 24.2, the unitary precoding (beamforming) is based on a predefined set of precoding matrices (codebook). A matrix quantisation based on a codebook is the optimum quantisation of the precoding matrix since it takes into account the statistical distribution of the precoding matrices in the precoding space. Thus, the feedback overhead to represent the precoding matrix can be reduced to a minimum. A direct rounding scalar quantisation of the parameters describing the precoding matrix is also possible (see Chapter 14). In the implementation presented here, 7 bits per matrix are used to quantise the precoding matrix. For this small number of bits, the codebook-based quantisation achieves notable better performance compared to the direct quantisation. If the precoding matrix should be quantised with a higher resolution (e.g. 12 bit per matrix), the performance gain of the codebook-based approach gets smaller compared to the direct quantisation. The design of the codebook which is used here is described in [14].

Depending on the channel conditions of each subcarrier, the optimum precoding matrix out of the codebook (defined by the index within the codebook) has to be determined for each subcarrier separately. By exploiting the correlation between neighboured subcarriers, it may not be necessary to determine the precoding matrix for each subcarrier separately. The precoding matrix may be defined only for a subset of subcarriers (pilot subcarriers) and the precoding matrix of the other subcarriers may be obtained via interpolation. Alternatively, one precoding matrix may be defined for a group of neighboured subcarriers. This saves feedback information about the precoding matrices and saves memory to store the precoding matrices. A detailed analysis of these approaches can be found in [14]. This section explains how the best precoding matrix is found for each subcarrier based on the channel matrix of each subcarrier. Assume $\tilde{\mathbf{V}}$ is a precoding matrix out of the codebook and $\mathbf{H}$ is the channel matrix of one subcarrier. Multiplying the channel matrix $\mathbf{H}$ by $\tilde{\mathbf{V}}$ leads to the equivalent channel $\tilde{\mathbf{H}} = \mathbf{H}\tilde{\mathbf{V}}$. ZF detection applies the pseudoinverse of the equivalent channel matrix to obtain the detection matrix (refer also to Chapter 8):

$$\mathbf{W} = \tilde{\mathbf{H}}^{\dagger} = \left(\tilde{\mathbf{H}}^{H}\tilde{\mathbf{H}}\right)^{-1}\tilde{\mathbf{H}}^{H} = \left(\left(\mathbf{H}\tilde{\mathbf{V}}\right)^{H}\mathbf{H}\tilde{\mathbf{V}}\right)^{-1}\left(\mathbf{H}\tilde{\mathbf{V}}\right)^{H},$$

$$= \left(\tilde{\mathbf{V}}^{H}\mathbf{H}^{H}\mathbf{H}\tilde{\mathbf{V}}\right)^{-1}\tilde{\mathbf{V}}^{H}\mathbf{H}^{H}. \tag{24.27}$$

If $\tilde{\mathbf{V}}$ is a unitary matrix, the inverse matrix $\tilde{\mathbf{V}}^{-1} = \tilde{\mathbf{V}}^{H}$ exists. Then, Equation 24.27 is equivalent to

$$\mathbf{W} = \tilde{\mathbf{H}}^{\dagger} = \tilde{\mathbf{V}}^{H}(\mathbf{H}^{H}\mathbf{H})^{-1}\tilde{\mathbf{V}}\tilde{\mathbf{V}}^{H}\mathbf{H}^{H} = \tilde{\mathbf{V}}^{H}\mathbf{H}^{\dagger}, \tag{24.28}$$

that is, the precoding matrix can be separated from the pseudoinverse of the channel matrix.

If i ($1 \le i \le 2^{q}$) is the index within the codebook (with q the number of bits to quantise the precoding matrix), each matrix $\mathbf{T}_i$ out of the codebook will result in a different detection matrix $\mathbf{W}_i$ according to Equation 24.28. The SNR of the two MIMO streams depends on the norm of the rows $\|\mathbf{w}_{1,i}\|^2$ and $\|\mathbf{w}_{2,i}\|^2$ of the detection matrix $\mathbf{W}_i$. The lower $\|\mathbf{w}_{1,i}\|^2$ and $\|\mathbf{w}_{2,i}\|^2$, the higher SNR (see also Chapter 8). The optimum precoding matrix out of the codebook is then found by

$$c' = \arg\min_{\mathbf{T}_i}\left\{\|\mathbf{w}_{1,i}\|^2, \|\mathbf{w}_{2,i}\|^2\right\}, \quad 1 \le i \le 2^{q}. \tag{24.29}$$

The unitary precoding matrix $\tilde{\mathbf{V}}$ can be represented by two real parameters v_1 ($0 \le v_1 \le 1$) and ϕ_2 ($-\pi < \phi_2 \le \pi$). With v_1 and $v_2 = \sqrt{1 - v_1^2}e^{j\phi_2}$, $\tilde{\mathbf{V}}$ may be given by [14]

$$\tilde{\mathbf{V}} = \begin{bmatrix} \tilde{\mathbf{v}}_1 & \tilde{\mathbf{v}}_2 \end{bmatrix} = \begin{bmatrix} v_1 & -v_2^{*} \\ v_2 & v_1 \end{bmatrix}. \tag{24.30}$$

Using Equation 24.30 in Equation 24.28 results in

$$\mathbf{W}\mathbf{W}^{H} = \tilde{\mathbf{V}}^{H}\mathbf{H}^{\dagger}\left(\tilde{\mathbf{V}}^{H}\mathbf{H}^{\dagger}\right)^{H} \underset{\mathbf{H}^{\dagger}=\begin{bmatrix}\mathbf{p}_1\\\mathbf{p}_2\end{bmatrix}}{=} \begin{bmatrix}\tilde{\mathbf{v}}_1^{H}\\\tilde{\mathbf{v}}_2^{H}\end{bmatrix}\begin{bmatrix}\mathbf{p}_1\mathbf{p}_1^{H} & \mathbf{p}_1\mathbf{p}_2^{H}\\\mathbf{p}_2\mathbf{p}_1^{H} & \mathbf{p}_2\mathbf{p}_2^{H}\end{bmatrix}\begin{bmatrix}\tilde{\mathbf{v}}_1 & \tilde{\mathbf{v}}_2\end{bmatrix}, \tag{24.31}$$

and

$$\| \mathbf{w}_1 \|^2 = \left[\mathbf{W}\mathbf{W}^H \right]_{11} = \mathbf{p}_1\mathbf{p}_1^H v_1{}^2 + \mathbf{p}_1\mathbf{p}_2^H v_1 v_2 + \left(\mathbf{p}_1\mathbf{p}_2^H v_1 v_2 \right)^H + \mathbf{p}_2\mathbf{p}_2^H \left| v_2 \right|^2,$$

$$= \| \mathbf{p}_1 \|^2 v_1{}^2 + 2\Re\left\{ \mathbf{p}_1\mathbf{p}_2^H v_1 v_2 \right\} + \| \mathbf{p}_2 \|^2 \left| v_2 \right|^2,$$

(24.32)

$$\| \mathbf{w}_2 \|^2 = [\mathbf{W}\mathbf{W}^H]_{22} = \| \mathbf{p}_1 \|^2 \left| v_2 \right|^2 - 2\Re\left\{ \mathbf{p}_1\mathbf{p}_2^H v_1 v_2 \right\} + \| \mathbf{p}_2 \|^2 v_1{}^2,$$

(24.33)

where
   $\mathbf{p}_1$ and $\mathbf{p}_2$ are the two row vectors of $\mathbf{H}^\dagger$
   v_1 is real

In order to find the optimum precoding matrix (which maximises the SNR), the minimum of $\|\mathbf{w}_1\|^2$ and $\|\mathbf{w}_2\|^2$ has to be found for the different codebook entries.

Figure 24.12 visualises $\|\mathbf{w}_1\|^2$ (contour lines) depending on the precoding matrix, as described by the two parameters v_1 and ϕ_2. Note that ϕ_2 is periodic with 2π. The contour lines $\|\mathbf{w}_1\|^2$ differ for different channel matrices $\mathbf{H}$. There is one minimum and one maximum (for $-\pi < \phi_2 \le \pi$). The minimum (diamond symbol ◆) is achieved by the optimum precoding vector (or precoding matrix). The maximum of $\|\mathbf{w}_1\|^2$ (star symbol ★) is the minimum of $\|\mathbf{w}_2\|^2$ and represents the orthogonal vector (or optimum precoding matrix with changed columns). The squares ▫ represent the codebook entries. Here, the number of feedback bits is equal to $q = 7$, that is, 128 codebook entries. The codebook search delivers the codebook entry closest to the optimum precoding matrix and is marked by the filled black square ■. Evaluating $\|\mathbf{w}_2\|^2$ results in the codebook entry closest to the orthogonal vector (marked by

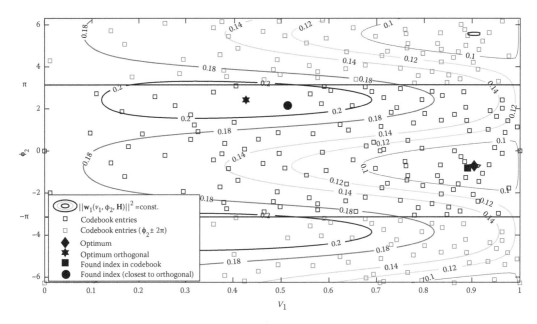

FIGURE 24.12
Codebook search: Influence of the precoding matrix on $\|\mathbf{w}_1\|^2$, number of codebook entries: $2^q = 2^7 = 128$.

the circle •). Here, the overall minimum is obtained by the first index (filled black square ■). Note that $\|\mathbf{w}_1\|^2$ gives an indication of the SNR at the receiver. If a ZF equaliser is used, the SNR is given as $\Lambda_1 = (1/(\|\mathbf{w}_1\|^2))(\rho/N_T)$ with ρ the transmit power to noise power ratio (refer also to Chapter 8). The lower $\|\mathbf{w}_1\|^2$, the higher is the SNR.

A search of all possible codebook entries would be computationally complex and time-consuming. A faster and more efficient codebook search was developed where first a subset of codebook entries is determined and the search to find the minimum is limited to this subset. Details of the algorithm may be found in [14].

24.7 Alternative Approach: Precoded Training Symbols

The training symbols are not precoded in the design described earlier. If the training symbols are also precoded, the $\mathbf{V}$ matrix block in the receiver is not needed anymore, since the channel estimation works on the equivalent channel matrix and the detection algorithm processes this equivalent channel. There is another advantage: The pseudoinverse calculation of the equivalent channel improves the numerical accuracy since the condition number of the equivalent channel matrix is increased compared to the actual channel matrix without precoding. This becomes especially important for correlated channels. The precoding of the training symbols or the preamble improves the coverage: The probability to reach a certain receiver is increased. It has to be ensured that no hidden nodes are created by optimising the preamble for only one receiver since the preamble has to be received by all modems in the network. The beamforming of the preamble might be optimised to the receiver with the worst channel conditions to improve the coverage and avoid potential hidden nodes. It has to be noted that an arbitrary precoding (no precoding) might also cause hidden nodes.

The codebook search has to be adapted since the precoding needs to be optimised to the actual channel and not to the equivalent channel.

Assume that $\mathbf{H}_1$ is the current channel matrix with the optimum precoding matrix $\mathbf{V}_1$. If the channel changes to $\mathbf{H}_2$, the channel estimation observes the equivalent channel matrix $\mathbf{H}_2\mathbf{V}_1$. The optimum precoding matrix $\mathbf{V}_2$ of $\mathbf{H}_2$ has to be found. Performing a *singular value decomposition* (SVD) of the new equivalent channel results in

$$\mathbf{H}_2\mathbf{V}_1 = \mathbf{UDV}^H. \tag{24.34}$$

Replacing $\mathbf{H}_2$ by the SVD yields

$$\mathbf{H}_2\mathbf{V}_1 = \mathbf{U}_2\mathbf{D}_2\mathbf{V}_2^H\mathbf{V}_1 = \mathbf{UDV}^H. \tag{24.35}$$

Note that $\mathbf{H}_2$ is not known in the receiver and is used only for the purpose of derivation. From Equation 24.35 follows

$$\mathbf{V}_2^H\mathbf{V}_1 = \mathbf{V}^H,$$
$$\Leftrightarrow \mathbf{V}_2 = \mathbf{V}_1\mathbf{V}. \tag{24.36}$$

The new precoding matrix $\mathbf{V}_2$ is calculated via the SVD of the new equivalent channel and $\mathbf{V}_1$. The receiver knows which precoding $\mathbf{V}_1$ is applied in the transmitter since the precoding information was sent back from the receiver to the transmitter.

The algorithm is easily extended to the codebook-based precoding. Assume *ind*1 is the codebook index corresponding to $\mathbf{V}_1$. The codebook search on the equivalent channel $\mathbf{H}_2\mathbf{V}_1$ derives *ind* corresponding to $\mathbf{V}$ according to Equation 24.35. The task is to derive *ind*2 corresponding to $\mathbf{V}_2$, according to Equation 24.36 if *ind*1 and *ind* are known. Because of the quantisation of the codebook, there is a finite set of possible combinations. For each possible combination of *ind*1 and *ind*, the best index *ind*2 is precalculated and stored in a 2D LUT. The receiver knows which precoding matrix is used in the transmitter ($\mathbf{V}_1$ or *ind*1) and the codebook search delivers *ind* of the equivalent channel. The LUT provides *ind*2 for the input combination *ind*1 and *ind*.

24.8 Verification of the MIMO PLC Demonstrator

This section describes the verification of the implemented prototype system. A configurable MIMO PLC channel (see Section 24.8.1) is used to test the system in the laboratory. Results of the demonstrator in the laboratory and in buildings under real conditions are presented in Section 24.8.2.

24.8.1 MIMO Artificial Mains Network

The MAM is an artificial and configurable MIMO PLC channel [15]. The basic schematic of the MAM is shown in Figure 24.13. The *common-mode* (CM) and *differential-mode* (DM) channels of each MIMO path can be configured by several filters (low-pass, high-pass or band-pass filters) or attenuators to model typical PLC transfer functions. The use of the MAM allows simple testing of the system in the laboratory. The MAM consist of three units, the first and third units are responsible to generate asymmetries to the mains network.

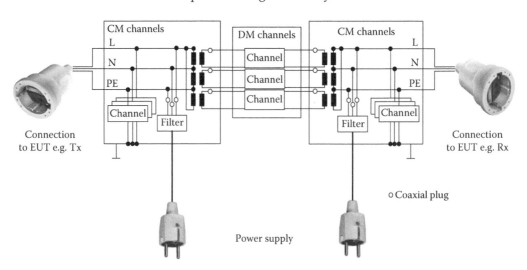

FIGURE 24.13
MAM network.

This causes CM currents (see Figure 1.3 in Chapter 1) flowing toward the ground. The other CM channel unit will pick up these currents from ground forwarding them to the second MIMO PLC modem. The DM channel unit provides the symmetrical channels from transmitter to receiver. The MIMO channel can be adjusted by pluggable filter units. As filter units, low-pass, high-pass, band-pass or band-gaps filters with one or many poles are available. Attenuators are also needed to model a typical PLC transfer function. Furthermore, the MAM includes several coaxial plugs where coax cables with individual lengths might be connected. These cables cause signal reflections analog to the stubs found at the mains grid in private homes. Each of the CM channel unit embeds a filtered connection to the power supply allowing to drive MIMO PLC modems connected to the MAM.

24.8.2 Results of the Demonstrator System

Figure 24.1 shows the demonstrator system connected to the MAM. In the following, results of the system are presented, in which the MAM was used as MIMO PLC channel. Figure 24.14 shows the absolute value of the channel estimation output which is proportional to the absolute value of the channel transfer function (magnitude response) from each transmit port (two columns) to each receive port (four rows). The scale of the y-axis is a result of the 16 bit signed fixed-point data format used here and has to be scaled by a factor depending on the AGC setting to obtain the magnitude responses. The magnitude responses show the typical frequency-selective shape of power line channels. A high variety between individual paths is visible.

Figure 24.15 shows some of the signals of the detection algorithm. The norm of the rows of the detection matrix is illustrated for the two MIMO streams. Figure 24.15a shows $\|\mathbf{w}_1\|^2$ and Figure 24.15b shows $\|\mathbf{w}_2\|^2$, each without and with precoding. The calculation is based on $\mathbf{R}^{-1}$ (see Section 24.3). $\mathbf{R}^{-1}$ is the result obtained from the pseudoinverse calculation. As explained in Section 8.5.1 of Chapter 8, $\|\mathbf{w}_p\|^2$, $p = 1, 2$, describes the two logical MIMO streams and is inversely proportional to the SNR. The solid lines show the results without precoding. The training symbols are not precoded and the results of the ZF detection do not include the precoding. The dashed lines show the results if precoding is included. The codebook indices obtained from the codebook search of the demonstrator system are used for this calculation. The difference between precoding and no precoding shows the gain of precoding. The first stream is increased significantly (up to 20 dB in some frequency ranges), while the second stream is decreased only slightly. This results in a gain of equivalent number of dB in SNR.

The demonstrator system supports also SISO transmission where only one transmit port and one receive port are used. The data throughput of SISO transmission is compared to the throughput of a MIMO transmission with two transmit and four receive ports. The adaptive modulation is adjusted for error-free transmission with maximum throughput. An HD video is streamed and no bit errors are monitored in the rendered video at the receiver. Also, the RS decoder provides an output which indicates if all bit errors are corrected. This flag is monitored continuously to ensure that the BER after the FEC is equal to 0.

Figure 24.16 shows a screenshot of the MATLAB application during SISO transmission. The application monitors the following key parameters of the transmission (four subplots in Figure 24.16 from top to bottom):

- SNR values and constellations of the OFDM subcarriers
- BER of the transmission
- Data throughput
- AGC settings

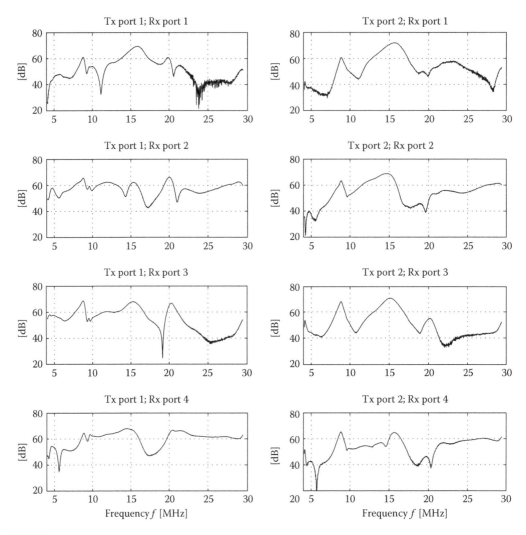

FIGURE 24.14
Absolute value of the channel estimation results of the demonstrator system which are proportional to the absolute value of the channel transfer functions from each transmit port (columns) to each receive port (rows).

The thin line of the top figure shows the SNR in dB of each subcarrier in the frequency range from 4 to 30 MHz. The SNR information is derived out of the training symbols. The selected constellations are shown by the bold line (for QAM steps, see right-hand side of the figure). The constellations of this transmission range from QPSK to 1024-QAM. If the SNR of a subcarrier is too low, the subcarrier is not used (e.g. some subcarriers around 18 MHz). The following three subplots show the past 60 s on the *x*-axis. The second figure shows several parameters of the BER. The BER is monitored at two different points: Firstly, before the Viterbi decoder (BER Viterbi) and, secondly, before the RS decoder (BER RS). The value BER RS indicates how many bits are corrected for one RS block. If the RS decoder is not able to correct an RS block, the corresponding

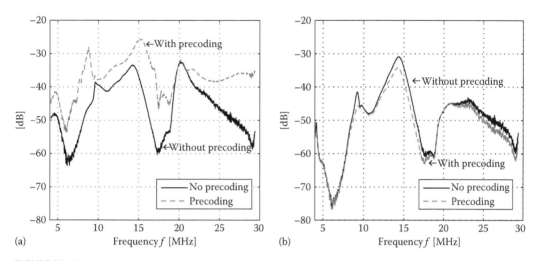

FIGURE 24.15

$\|\mathbf{w}_p\|^2$, $p = 1, 2$ of the two MIMO streams (a) stream 1 and (b) stream 2, depending on the frequency: without and with precoding.

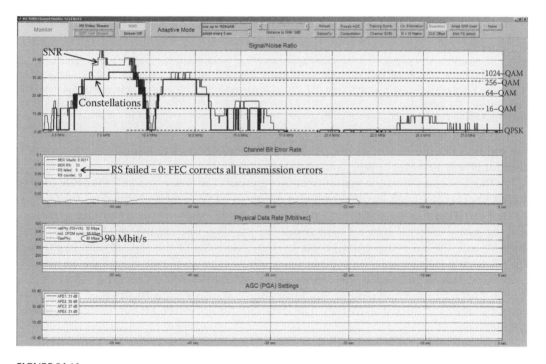

FIGURE 24.16

SISO transmission, screenshot of the monitor application: *top figure*: measured SNR (thin line) and selected constellations (bold line); *second figure*: actual BER of the past 60 s, BER before Viterbi decoder (BER Viterbi) and BER before RS decoder (BER RS), RS = 0 indicates that all remaining errors are corrected by the RS decoder; *third figure*: actual data rate of the past 60 s, raw PHY data rate, including OFDM preamble and including FEC; *bottom figure*: AGC settings of the past 60 s of all four receivers.

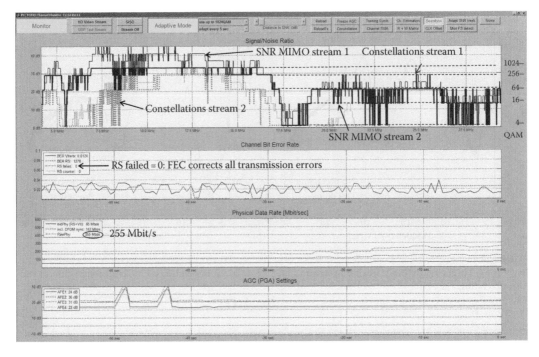

FIGURE 24.17
MIMO transmission, screenshot of the monitor application: *top figure*: measured SNR (thin line) and selected constellations (bold line) of the two MIMO streams (black and grey); *second figure*: actual BER of the past 60 s, BER before Viterbi decoder (BER Viterbi) and BER before RS decoder (BER RS), RS = 0 indicates that all remaining errors are corrected by the RS decoder; *third figure*: actual data rate of the past 60 s, raw PHY data rate, including OFDM preamble and including FEC; *bottom figure*: AGC settings of the past 60 s of all four receivers.

value shows the number of errors (RS failed). RS failed equal to 0 indicates that the RS decoder is able to correct every transmission error.*

The third figure shows the actual data throughput. The raw *physical* (PHY) rate (dash-dotted line) is calculated as the number of bits per OFDM symbol divided by the OFDM symbol duration. Taking the OFDM preamble (CAZAC sequence and training symbols) and the guard interval into account results in the data rate shown by the dashed line (legend: incl. OFDM sync). Multiplying by the code rates of the FEC results in the net PHY data rate (solid line). The bottom figure shows the AGC settings of the four front ends. Note that all front-ends are connected to the coupler and receive the signals of all receive ports. However, in case of SISO transmission, the signals of three of the receive ports are not used during detection. The data rate for error-free HD video transmission is 90 Mbit/s.

Figure 24.17 shows the screenshot of the MIMO transmission for the same channel directly after SISO transmission. No channel change was observed. Basically, the screenshot shows the same parameters as described before. Of course, MIMO transmission uses two (logical) MIMO streams. The SNR of these two streams is shown in the top figure (thin lines). The black line represents the SNR of the first MIMO stream, while the grey line

---

* RS failed is not detected if the RS decoder finds valid code words, although many bits are corrupted in a RS block. This case is extremely unlikely for large block sizes. In practice, this will happen only if there is a very erroneous transmission with high BER. Here, a high BER of the Viterbi decoder will indicate this case.

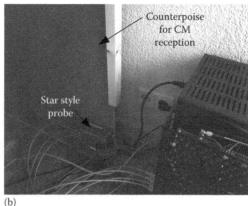

(a)　　　　　　　　　　　　　　　　　　　(b)

FIGURE 24.18
Setup of the field test: (a) transmitter and (b) receiver.

represents the second MIMO stream. The bold lines (black and grey) show the selected constellations of the two MIMO streams. The data rate is 255 Mbit/s.

The comparison between SISO and MIMO is not completely fair since the transmit power of each of the two transmit ports for MIMO transmission is the same as for SISO transmission at this demonstrator system. The total transmit power of MIMO is thus 3 dB higher, compared to SISO. To keep the total transmit power the same, the transmit power of SISO transmission has to be increased by 3 dB. As discussed in Chapter 7, a back-off of 3 dB is an expected upper limit. It was observed that increasing the transmit power by 3 dB enhances the data rate by approximately 15 Mbit/s.* Comparing the corrected SISO bitrate ($\approx$105 Mbit/s) to the bitrate of MIMO shows an improvement of the bit rate by more than a factor of 2. However, the different feeding power of the SISO and MIMO test in Figures 24.16 and 24.17 are balanced by the receivers AGC settings. The lowest graph in each figure shows these AGC settings. In the MIMO case, all AGC amplifiers are set to a 6 dB reduced (voltage) amplitude. So far the AGC is in operating range; it is ok to compare SISO and MIMO throughput rates.

The demonstrator system was also verified in private homes. Here, the apartment covers two levels with an area of 120 m². Figure 24.18 shows the transmitter (Figure 24.18a) and Figure 24.18b depicts the receiver located in the building. The transmitter is connected via a delta-style coupler to the mains network on the upper level, while the receiver is connected via a star-style coupler on the lower level of the apartment. Two transmit ports are used. The receiver uses all four receive ports (the three star-style ports and the CM port). To receive the CM signals, the coupler is mounted on a wooden board which is covered by a fleece made of copper. This construction can be easily transported. A metal plate could be used as well. A counterpoise with the size of around 1 m² was found to ensure a proper CM reception.

---

* The BER performance depending on the SNR of uncoded QAM differs by approximately 3 dB if the QAM is increased by 1 bit. Hence, the bit load of each subcarrier is increased by 1 bit if the transmit power is increased by $\approx$3 dB. This results in an increase of the bitrate of 1296/51.2 μs $\approx$ 25 Mbit/s. The use of only even QAM constellations, and the fact that not all subcarriers exceed the corresponding SNR thresholds, results in the usually observed enhancement of approximately 15 Mbit/s.

Schneider et al. [16] shows the results of the verification in this building. For SISO transmission, a bitrate of 143 Mbit/s and, for MIMO transmission, a bitrate of 315 Mbit/s were documented.

24.9 Conclusions

The implementation of a MIMO PLC feasibility study in hardware was described in this chapter. The demonstrator system allows up to 2×4 MIMO with beamforming. The systems allow monitoring several system parameters including the bit rate and channel estimation results. The system also supports a SISO mode. Comparing SISO and MIMO transmissions, the gain of MIMO is proven. The gain of precoding is shown by this demonstrator system by activating beamforming on the fly. Also, further aspects of MIMO PLC transmission, such as the influence of the number of receive ports and the influence of the noise, were investigated. The verification of the demonstrator system in buildings under real conditions shows the gain of MIMO versus SISO to be more than factor two. The demonstrator system proves the theoretical investigations of MIMO PLC and supported the standardisation work at HomePlug AV2.

References

1. A. Schwager, Powerline communications: Significant technologies to become ready for integration, Dr.-Ing. dissertation, Universität Duisburg-Essen, Essen, Germany, May 2010.
2. D. J. C. MacKay, *Information Theory, Inference and Learning Algorithms*. Cambridge University Press, New York, 2003.
3. C. Rader, VLSI systolic arrays for adaptive nulling, *IEEE Signal Processing Magazine*, 13(4), 29–49, 1996.
4. R. Walke, R. Smith and G. Lightbody, Architectures for adaptive weight calculation on ASIC and FPGA, in *Asilomar Conference on Signals, Systems, and Computers*, vol. 2, Pacific Grove, CA, 1999, pp. 1375–1380.
5. B. Hassibi, An efficient square-root algorithm for BLAST, in *IEEE International Conference on Acoustics, Speech, and Signal Processing*, vol. 2, Istanbul, Turkey, 2000, pp. 737–740.
6. Z. Guo and P. Nilsson, A VLSI architecture of the square root algorithm for V-BLAST detection, *The Journal of VLSI Signal Processing*, 44(3), 219–230, September 2006.
7. J. E. Volder, The CORDIC trigonometric computing technique, *IRE Transactions on Electronic Computers*, EC-8(3), 330–334, 1959.
8. J. S. Walther, A unified algorithm for elementary functions, in *Spring Joint Computer Conference*, Atlantic City, NJ, 1971, pp. 379–385.
9. J. Valls, M. Kuhlmann and K. K. Parhi, Evaluation of CORDIC algorithms for FPGA design, *Journal of VLSI Signal Processing Systems*, 32(3), 207–222, 2002.
10. Xilinx, *CORDIC v3.0*, Xilinx, 2005.
11. Xilinx, *Sine/Cosine Look-Up Table v5.0*, Xilinx, 2004.

12. Xilinx, *Multiplier v10.1*, Xilinx, 2008.

13. Xilinx, *Divider v2.0*, Xilinx, 2008.

14. D. Schneider, Inhome power line communications using multiple input multiple output principles, Dr.-Ing. dissertation, Verlag Dr. Hut, Munich, Germany, January 2012.

15. A. Schwager, D. Schneider, W. Bschlin, A. Dilly and J. Speidel, MIMO PLC: Theory, measurements and system setup, in *International Symposium on Power Line Communications and Its Applications*, Udine, Italy, 2011.

16. D. Schneider, A. Schwager, J. Speidel and A. Dilly, Implementation and results of a MIMO PLC feasibility study, in *International Symposium on Power Line Communications and Its Applications*, Udine, Italy, 2011.

Index